THE
MOBILE COMMUNICATIONS
HANDBOOK

Second Edition

The Electrical Engineering Handbook Series

Series Editor
Richard C. Dorf
University of California, Davis

Titles Included in the Series

The Avionics Handbook, Cary R. Spitzer

The Biomedical Engineering Handbook, Joseph D. Bronzino

The Circuits and Filters Handbook, Wai-Kai Chen

The Communications Handbook, Jerry D. Gibson

The Control Handbook, William S. Levine

The Digital Signal Processing Handbook, Vijay K. Madisetti & Douglas B. Williams

The Electrical Engineering Handbook, Richard C. Dorf

The Electric Power Engineering Handbook, L.L. Grigsby

The Electronics Handbook, Jerry C. Whitaker

The Engineering Handbook, Richard C. Dorf

The Handbook of Formulas and Tables for Signal Processing, Alexander D. Poularikas

The Industrial Electronics Handbook, J. David Irwin

The Measurement, Instrumentation, and Sensors Handbook, John G. Webster

The Mechanical Systems Design Handbook, Osita D.I. Nwokah

The Microwave Engineering Handbook, J. Michael Golio

The Mobile Communications Handbook, Jerry D. Gibson

The Ocean Engineering Handbook, Ferial El-Hawary

The Technology Management Handbook, Richard C. Dorf

The Transforms and Applications Handbook, Alexander D. Poularikas

The VLSI Handbook, Wai-Kai Chen

THE
MOBILE COMMUNICATIONS
HANDBOOK

Second Edition

Editor-in-Chief
JERRY D. GIBSON

Southern Methodist University
Dallas, Texas

Managing Editor
ELAINE M. GIBSON

 CRC PRESS

 IEEE PRESS

A CRC Handbook Published in Cooperation with IEEE Press

Acquiring Editor:	Nora Konopka
Project Editor:	Carol Whitehead
Production Manager:	Suzanne Lassandro
Marketing Manager:	Barbara Glunn, Jane Lewis, Arline Massey, Jane Stark
Cover design:	Jonathan Pennell
PrePress:	Gary Bennett
Manufacturing:	Carol Slatter

Library of Congress Cataloging-in-Publication Data

The mobile communications handbook / editor-in-chief, Jerry D. Gibson.
   -- 2nd ed.
          p.   cm.
   Includes bibliographical references and index.
   ISBN 0-8493-8597-0 (alk. paper)
   1. Mobile communication systems. I. Gibson, Jerry D.
   TK6570.M6M5934 1999
   621.3845--dc21   98-46558
                                                                                        CIP

Preface

To keep pace with the rapid evolution of the mobile and wireless communications field, this second edition of *The Mobile Communications Handbook* contains 12 revised chapters and 10 entirely new chapters to complement another 14 chapters carried forward from the first edition. We use the term mobile communications to include technologies ranging from cordless telephones, digital cellular mobile radio, and evolving personal communications systems to wireless data and wireless networks. To cover this range of topics, we present 36 concise chapters by experts from industry and academia. The chapters are written to provide a succinct overview of each topic, quickly bringing the reader up to date; additionally, the chapters contain sufficient detail and references to encourage further study. The chapters are more than a "just the facts" presentation, with many of the authors using their experience to provide insights into forthcoming developments in the various fields.

The Handbook is divided into two parts. The first part covers the basic principles of analog and digital communications that pertain to mobile communications, and consists of 14 tutorial/review chapters that lay a solid groundwork for the wide-ranging aspects of mobile communications technologies. The second part of the Handbook contains 22 chapters covering such topics as cellular mobile radio, Rayleigh fading channels, space-time processing, power control, personal communications systems, user location and addressing, wireless data, wireless local area networks (LANs), wireless ATM, and third generation wireless standards. The basic principles chapters readily allow the reader to jump right to the mobile communications topic of interest, with the option of efficiently filling in any gaps in his/her background by referring back to the Basic Principles section as needed, without searching through a textbook or shuffling off to the library.

Although there is an ordering to the articles, the sequence of articles does not have to be read from the beginning to end. Each article was written to be an independent contribution but with intentional overlap between some articles to prevent the reader from having to flip back and forth to extract the desired information. Interestingly, as the reader will discover, this overlap often admits alternative views of difficult issues.

As in the first edition, the lead-off article for the wireless section is written by an acknowledged leader and innovator in wireless communications, Dr. Don Cox of Stanford University. This chapter is unique in that Dr. Cox inserts comments and provides updates directly in the text of his chapter from the first edition. Besides being informative and easy to read, the article implicitly challenges those active in wireless communications to continually question the common knowledge in the field.

It has been a great pleasure to work with the authors to publish this second edition. The authors come from all parts of the world and constitute a literal who's who of workers in digital and mobile communications. Certainly, each article is an extraordinary contribution to the communications field, and the collection, I believe, is the most comprehensive treatment of the mobile communications field available in one volume today. I sincerely acknowledge the exceptional efforts and talents of my wife, Elaine M. Gibson, who served as Managing Editor for the project, and was the principal contact point for the authors and the publisher. Her energy and organizational skills in coordinating each contribution and in organizing the Handbook were essential to its timely delivery and to its quality. Elaine and I both appreciate the patience, support, and guidance of the staff at CRC Press during all stages of developing and producing this Handbook.

Jerry D. Gibson
Dallas, Texas

About the Editor

Jerry D. Gibson currently serves as Chairman of the Department of Electrical Engineering at Southern Methodist University in Dallas, Texas. He has held positions at General Dynamics-Fort Worth (1969–1972), the University of Notre Dame (1973–1974), and the University of Nebraska-Lincoln (1974–1976), and during the Fall of 1991, Dr. Gibson was on sabbatical with the Information Systems Laboratory and the Telecommunications Program in the Department of Electrical Engineering at Stanford University. From 1987–1997, he held the J. W. Runyon, Jr. Professorship in the Department of Electrical Engineering at Texas A&M University.

Dr. Gibson is coauthor of the book *Introduction to Nonparametric Detection with Applications* (Academic Press, 1975 and IEEE Press, 1995), the author of the textbook, *Principles of Digital and Analog Communications* (Prentice-Hall, second edition, 1993), and coauthor of the book *Digital Compression for Multimedia* (Morgan Kaufmann, 1998). He was Associate Editor for Speech Processing for the *IEEE Transactions on Communications* from 1981 to 1985 and Associate Editor for Communications for the *IEEE Transactions on Information Theory* from 1988–1991. He has served as a member of the Speech Technical Committee of the IEEE Signal Processing Society (1992–1995), on the Editorial Board for the *Proceedings of the IEEE* (1991–1997), and as a member of the IEEE Information Theory Society Board of Governors (1990-1998). He was President of the IEEE Information Theory Society in 1996. Dr. Gibson is Editor-in-Chief of *The Mobile Communications Handbook* (CRC Press, 1996) and Editor-in-Chief of *The Communications Handbook* (CRC Press, 1997).

In 1990, Dr. Gibson received The Fredrick Emmons Terman Award from The American Society for Engineering Education, and in 1992, was elected Fellow of the IEEE "for contributions to the theory and practice of adaptive prediction and speech waveform coding." He was corecipient of the 1993 IEEE Signal Processing Society Senior Paper Award for the Speech Processing area. His research interests include data, speech, image, and video compression, multimedia over networks, wireless communications, information theory, and digital signal processing.

Contributors

Melbourne Barton
Bellcore
Red Bank, New Jersey

V. K. Bhargava
University of Victoria
Department of Electrical and
    Computer Engineering
Victoria, Canada

Madhukar Budagavi
Texas Instruments
DSP Solutions R and D Center
Dallas, Texas

James J. Caffery, Jr.
Georgia Institute of Technology
School of Electrical and Computer
    Engineering
Atlanta, Georgia

Wai-Yip Chan
Electrical and Computer
    Engineering Department
Illinois Institute of Technology
Chicago, Illinois

Li Fung Chang
Bellcore
Morristown, New Jersey

Matthew Cheng
Bellcore
Morristown, New Jersey

Giovanni Cherubini
IBM Zurich Research Laboratory
Ruschlikon, Switzerland

Leon W. Couch, II
University of Florida
Electrical and Computer
    Engineering
Gainesville, Florida

Donald C. Cox
Stanford University
Stanford, California

Marc Delprat
Alcatel Telcom/MCD
Colombes, France

Spiros Dimolitsas
Lawrence Livermore National
    Laboratory
Livermore, California

I. J. Fair
Department of Electrical and
    Computer Engineering
Technical University of Nova Scotia
Halifax, Nova Scotia
Canada

Ira Gerson
Motorola Semiconductor Products
Schaumburg, Illinois

Steven D. Gray
Nokia Research Center
Irving, Texas

Lajos Hanzo
Department of Electrical and
    Computer Science
University of Southampton
Highfield, Southampton
United Kingdom

Tor Helleseth
Department of Informatics
University of Bergen
Bergen, Norway

Michael L. Honig
Northwestern University
Department of EECS
Evanston, Illinois

Hwei P. Hsu
Fairleigh Dickinson University
Teaneck, New Jersey

Bijan Jabbari
George Mason University
Department of Electrical and
    Computer Engineering
Fairfax, Virginia

Ravi Jain
Bell Communications Research
Red Bank, New Jersey

P. Vijay Kumar
University of Southern California
Los Angeles, California

Vinod Kumar
Alcatel Telecom
Mobile Communication Division
Colombes, France

Allen H. Levesque
GTE Laboratories
Waltham, Massachusetts

Yi-Bing Lin
National Chaio Tung University
Hsinchu, Taiwan
Republic of China

Joseph L. LoCicero
Illinois Institute of Technology
Armour College of Engineering
Department of Electrical and
    Computer Engineering
Chicago, Illinois

Paul Mermelstein
INRS—Telecommunications
Ile des Soeurs
Verdun, Quebec
Canada

Toshio Miki
Mobile Communication Network,
    Inc.
Yokosuka, Kanagawa
Japan

Laurence B. Milstein
University of California–San Diego
Department of Electrical and
    Computer Engineering
La Jolla, California

Seshadri Mohan
Bell Communications Research
Morristown, New Jersey

Tero Ojanperä
Nokia Research Center
Irving, Texas

Michael Onufry
COMSAT Laboratories
Clarksburg, Maryland

Geoffrey C. Orsak
Southern Methodist University
Electrical Engineering Department
Dallas, Texas

Kaveh Pahlavan
Worcester Polytechnic Institute
Worcester, Massachusetts

Bernd-Peter Paris
George Mason University
Department of Electrical and
    Computer Engineering
Fairfax, Virginia

Bhasker P. Patel
Illinois Institute of Technology
Chicago, Illinois

Arogyaswami J. Paulraj
Stanford University
Department of Electrical
    Engineering
Stanford, California

Roman Pichna
University of Oulu
Center for Wireless Communication
Oulu, Finland

John G. Proakis
Northeastern University
Electrical Engineering
Boston, Massachusetts

Bala Rajagopalan
NEC C&C Research Laboratories
Princeton, New Jersey

Daniel Reininger
NEC USA, Inc.
Princeton, New Jersey

Marvin K. Simon
Jet Propulsion Laboratory
Pasadena, California

Suresh Singh
Oregon State University
Electrical and Computer
    Engineering Department
Corvallis, Oregon

Bernard Sklar
Communications Engineering
    Services
Tarzana, California

Raymond Steele
MAC, Ltd.
Southampton
United Kingdom

Gordon L. Stüber
Georgia Institute of Technology
School of Electrical and Computer
    Engineering
Atlanta, Georgia

Raj Talluri
Texas Instruments
DSP Solutions R and D Center
Dallas, Texas

Qiang Wang
University of Victoria
Victoria, British Columbia
Canada

Michel Daoud Yacoub
The University of Campinas
San Paulo, Brazil

Contents

I

Basic Principles

1

Complex Envelope Representations for Modulated Signals[1]

Leon W. Couch, II
University of Florida

1.1 Introduction

What is a general representation for bandpass digital and analog signals? How do we represent a **modulated signal**? How do we evaluate the spectrum and the power of these signals? These are some of the questions that are answered in this chapter.

A *baseband* waveform has a spectral magnitude that is nonzero for frequencies in the vicinity of the origin (i.e., $f = 0$) and negligible elsewhere. A *bandpass* waveform has a spectral magnitude that is nonzero for frequencies in some band concentrated about a frequency $f = \pm f_c$ (where $f_c \gg 0$), and the spectral magnitude is negligible elsewhere. f_c is called the *carrier frequency*. The value of f_c may be arbitrarily assigned for mathematical convenience in some problems. In others, namely, **modulation** problems, f_c is the frequency of an oscillatory signal in the transmitter circuit and is the assigned frequency of the transmitter, such as 850 kHz for an AM broadcasting station.

In communication problems, the information source signal is usually a baseband signal— for example, a transistor-transistor logic (TTL) waveform from a digital circuit or an audio (analog) signal from a microphone. The communication engineer has the job of building a system that will transfer the information from this source signal to the desired destination.

---

[1]*Source:* Couch, Leon W., II. 1997. *Digital and Analog Communication Systems,* 5th ed., Prentice Hall, Upper Saddle River, NJ.

As shown in Fig. 1.1, this usually requires the use of a bandpass signal, $s(t)$, which has a bandpass spectrum that is concentrated at $\pm f_c$ where f_c is selected so that $s(t)$ will propagate across the communication channel (either a wire or a wireless channel).

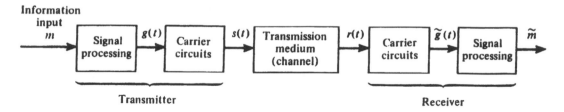

FIGURE 1.1   Bandpass communication system. *Source:* Couch, L.W., II. 1997. *Digital and Analog Communication Systems,* 5th ed., Prentice Hall, Upper Saddle River, NJ, p. 227. With permission.

Modulation is the process of imparting the source information onto a bandpass signal with a carrier frequency f_c by the introduction of amplitude and/or phase perturbations. This bandpass signal is called the *modulated* signal $s(t)$, and the baseband source signal is called the *modulating* signal $m(t)$. Examples of exactly how modulation is accomplished are given later in this chapter. This definition indicates that modulation may be visualized as a mapping operation that maps the source information onto the bandpass signal $s(t)$ that will be transmitted over the channel.

As the modulated signal passes through the channel, noise corrupts it. The result is a bandpass signal-plus-noise waveform that is available at the receiver input, $r(t)$, as illustrated in Fig. 1.1. The receiver has the job of trying to recover the information that was sent from the source; $\tilde{m}$ denotes the corrupted version of m.

1.2   Complex Envelope Representation

All bandpass waveforms, whether they arise from a modulated signal, interfering signals, or noise, may be represented in a convenient form given by the following theorem. $v(t)$ will be used to denote the **bandpass waveform** canonically. That is, $v(t)$ can represent the signal when $s(t) \equiv v(t)$, the noise when $n(t) \equiv v(t)$, the filtered signal plus noise at the channel output when $r(t) \equiv v(t)$, or any other type of bandpass waveform[2].

THEOREM 1.1   *Any physical bandpass waveform can be represented by*

$$v(t) = Re\left\{ g(t)e^{j\omega_c t} \right\} \tag{1.1a}$$

$Re\{\cdot\}$ denotes the real part of $\{\cdot\}$. $g(t)$ is called the complex envelope of $v(t)$, and f_c is the associated carrier frequency (hertz) where $\omega_c = 2\pi f_c$. Furthermore, two other equivalent representations are

---

[2]The symbol $\equiv$ denotes an equivalence and the symbol $\stackrel{\triangle}{=}$ denotes a definition.

$$v(t) = R(t) \cos \left[\omega_c t + \theta(t) \right] \tag{1.1b}$$

and

$$v(t) = x(t) \cos \omega_c t - y(t) \sin \omega_c t \tag{1.1c}$$

where

$$g(t) = x(t) + jy(t) = |g(t)|e^{j\angle g(t)} \equiv R(t)e^{j\theta(t)} \tag{1.2}$$

$$x(t) = Re\{g(t)\} \equiv R(t) \cos \theta(t) \tag{1.3a}$$

$$y(x) = Im\{g(t)\} \equiv R(t) \sin \theta(t) \tag{1.3b}$$

$$R(t) \triangleq |g(t)| \equiv \sqrt{x^2(t) + y^2(t)} \tag{1.4a}$$

$$\theta(t) \triangleq \angle g(t) = \tan^{-1}\left(\frac{y(t)}{x(t)}\right) \tag{1.4b}$$

The waveforms $g(t), x(t), y(t), R(t)$, and $\theta(t)$ are all **baseband waveforms**, and, except for $g(t)$, they are all real waveforms. $R(t)$ is a nonnegative real waveform. Equation (1.1) is a low-pass-to-bandpass transformation. The $e^{j\omega_c t}$ factor in (1.1a) shifts (i.e., translates) the spectrum of the baseband signal $g(t)$ from baseband up to the carrier frequency f_c. In communications terminology the frequencies in the baseband signal $g(t)$ are said to be *heterodyned* up to f_c. The **complex envelope**, $g(t)$, is usually a complex function of time and it is the generalization of the phasor concept. That is, if $g(t)$ happens to be a complex constant, then $v(t)$ is a pure sine wave of frequency f_c and this complex constant is the phasor representing the sine wave. If $g(t)$ is not a constant, then $v(t)$ is not a pure sine wave because the amplitude and phase of $v(t)$ varies with time, caused by the variations of $g(t)$.

Representing the complex envelope in terms of two real functions in Cartesian coordinates, we have

$$g(x) \equiv x(t) + jy(t) \tag{1.5}$$

where $x(t) = Re\{g(t)\}$ and $y(t) = Im\{g(t)\}$. $x(t)$ is said to be the *in-phase modulation* associated with $v(t)$, and $y(t)$ is said to be the *quadrature modulation* associated with $v(t)$. Alternatively, the polar form of $g(t)$, represented by $R(t)$ and $\theta(t)$, is given by (1.2), where the identities between Cartesian and polar coordinates are given by (1.3) and (1.4). $R(t)$ and $\theta(t)$ are real waveforms and, in addition, $R(t)$ is always nonnegative. $R(t)$ is said to be the *amplitude modulation* (AM) on $v(t)$, and $\theta(t)$ is said to be the *phase modulation* (PM) on $v(t)$.

The usefulness of the complex envelope representation for bandpass waveforms cannot be overemphasized. In modern communication systems, the bandpass signal is often partitioned into two channels, one for $x(t)$ called the I (in-phase) channel and one for $y(t)$ called the Q (quadrature-phase) channel. In digital computer simulations of bandpass signals, the sampling rate used in the simulation can be minimized by working with the complex envelope, $g(t)$, instead of with the bandpass signal, $v(t)$, because $g(t)$ is the baseband equivalent of the bandpass signal [1].

1.3   Representation of Modulated Signals

Modulation is the process of encoding the source information $m(t)$ (modulating signal) into a bandpass signal $s(t)$ (modulated signal). Consequently, the modulated signal is just a special application of the bandpass representation. The *modulated signal* is given by

$$s(t) = \text{Re}\left\{g(t)e^{j\omega_c t}\right\} \tag{1.6}$$

where $\omega_c = 2\pi f_c$. f_c is the carrier frequency. The complex envelope $g(t)$ is a function of the modulating signal $m(t)$. That is,

$$g(t) = g[m(t)] \tag{1.7}$$

Thus $g[\cdot]$ performs a mapping operation on $m(t)$. This was shown in Fig. 1.1.

Table 1.1 gives an overview of the *big picture* for the modulation problem. Examples of the mapping function $g[m]$ are given for amplitude modulation (AM), double-sideband suppressed carrier (DSB-SC), phase modulation (PM), frequency modulation (FM), single-sideband AM suppressed carrier (SSB-AM-SC), single-sideband PM (SSB-PM), single-sideband FM (SSB-FM), single-sideband envelope detectable (SSB-EV), single-sideband square-law detectable (SSB-SQ), and quadrature modulation (QM). For each $g[m]$, Table 1.1 also shows the corresponding $x(t)$ and $y(t)$ quadrature modulation components, and the corresponding $R(t)$ and $\theta(t)$ amplitude and phase modulation components. Digitally modulated bandpass signals are obtained when $m(t)$ is a digital baseband signal—for example, the output of a transistor transistor logic (TTL) circuit.

Obviously, it is possible to use other $g[m]$ functions that are not listed in Table 1.1. The question is: Are they useful? $g[m]$ functions are desired that are easy to implement and that will give desirable spectral properties. Furthermore, in the receiver the inverse function $m[g]$ is required. The inverse should be single valued over the range used and should be easily implemented. The inverse mapping should suppress as much noise as possible so that $m(t)$ can be recovered with little corruption.

1.4   Generalized Transmitters and Receivers

A more detailed description of transmitters and receivers as first shown in Fig. 1.1 will now be illustrated.

There are two canonical forms for the generalized transmitter, as indicated by (1.1b) and (1.1c). Equation (1.1b) describes an AM-PM type circuit as shown in Fig. 1.2. The baseband signal processing circuit generates $R(t)$ and $\theta(t)$ from $m(t)$. The R and θ are functions of the modulating signal $m(t)$, as given in Table 1.1, for the particular modulation type desired. The signal processing may be implemented either by using nonlinear analog circuits or a digital computer that incorporates the R and θ algorithms under software program control. In the implementation using a digital computer, one analog-to-digital converter (ADC) will be needed at the input of the baseband signal processor and two digital-to-analog converters (DACs) will be needed at the output. The remainder of the AM-PM canonical form requires radio frequency (RF) circuits, as indicated in the figure.

Figure 1.3 illustrates the second canonical form for the generalized transmitter. This uses in-phase and quadrature-phase (IQ) processing. Similarly, the formulas relating $x(t)$ and $y(t)$ to $m(t)$ are shown in Table 1.1, and the baseband signal processing may be implemented by using either analog hardware or digital hardware with software. The remainder of the canonical form uses RF circuits as indicated.

TABLE 1.1 Complex Envelope Functions for Various Types of Modulation[a]

Type of Modulation	Mapping Functions $g(m)$	Corresponding Quadrature Modulation	
		$x(t)$	$y(t)$
AM	$A_c[1 + m(t)]$	$A_c[1 + m(t)]$	0
DSB-SC	$A_c m(t)$	$A_c m(t)$	0
PM	$A_c e^{j D_p m(t)}$	$A_c \cos[D_p m(t)]$	$A_c \sin[D_p m(t)]$
FM	$A_c e^{j D_f \int_{-\infty}^t m(\sigma)d\sigma}$	$A_c \cos\left[D_f \int_{-\infty}^t m(\sigma)d\sigma\right]$	$A_c \sin\left[D_f \int_{-\infty}^t m(\sigma)d\sigma\right]$
SSB-AM-SC[b]	$A_c[m(t) \pm j\hat{m}(t)]$	$A_c m(t)$	$\pm A_c \hat{m}(t)$
SSB-PM[b]	$A_c e^{j D_p[m(t)\pm j\hat{m}(t)]}$	$A_c e^{\mp D_p \hat{m}(t)} \cos[D_p m(t)]$	$A_c e^{\mp D_p \hat{m}(t)} \sin[D_p m(t)]$
SSB-FM[b]	$A_c e^{j D_f \int_{-\infty}^t [m(\sigma)\pm j\hat{m}(\sigma)]d\sigma}$	$A_c e^{\mp D_f \int_{-\infty}^t \hat{m}(\sigma)d\sigma} \cos\left[D_f \int_{-\infty}^t m(\sigma)d\sigma\right]$	$A_c e^{\mp D_f \int_{-\infty}^t \hat{m}(\sigma)d\sigma} \sin\left[D_f \int_{-\infty}^t m(\sigma)d\sigma\right]$
SSB-EV[b]	$A_c e^{\{\ln[1+m(t)]\pm j\widehat{\ln}[1+m(t)]\}}$	$A_c[1 + m(t)] \cos\{\widehat{\ln}[1 + m(t)]\}$	$\pm A_c[1 + m(t)] \sin\{\widehat{\ln}[1 + m(t)]\}$
SSB-SQ[b]	$A_c e^{(1/2)\{\ln[1+m(t)]\pm j\widehat{\ln}[1+m(t)]\}}$	$A_c \sqrt{1 + m(t)} \cos\{\tfrac{1}{2}\widehat{\ln}[1 + m(t)]\}$	$\pm A_c \sqrt{1 + m(t)} \sin\{\tfrac{1}{2}\widehat{\ln}[1 + m(t)]\}$
QM	$A_c[m_1(t) + j m_2(t)]$	$A_c m_1(t)$	$A_c m_2(t)$

TABLE 1.1   Complex Envelope Functions for Various Types of Modulation[a] *(Continued)*

Type of Modulation	Corresponding Amplitude and Phase Modulation		Linearity	Remarks
	$R(t)$	$\theta(t)$		
AM	$A_c\lvert 1 + m(t)\rvert$	$\begin{cases} 0, & m(t) > -1 \\ 180^\circ, & m(t) < -1 \end{cases}$	L[c]	$m(t) > -1$ required for envelope detection
DSB-SC	$A_c\lvert m(t)\rvert$	$\begin{cases} 0, & m(t) > 0 \\ 180^\circ, & m(t) < 0 \end{cases}$	L	Coherent detection required
PM	A_c	$D_p m(t)$	NL	D_p is the phase deviation constant (rad/volt)
FM	A_c	$D_f \int_{-\infty}^{t} m(\sigma)d\sigma$	NL	D_f is the frequency deviation constant (rad/volt-sec)
SSB-AM-SC[b]	$A_c \sqrt{[m(t)]^2 + [\hat{m}(t)]^2}$	$\tan^{-1}[\pm \hat{m}(t)/m(t)]$	L	Coherent detection required
SSB-PM[b]	$A_c e^{\pm D_p \hat{m}(t)}$	$D_p m(t)$	NL	
SSB-FM[b]	$A_c e^{\pm D_f \int_{-\infty}^{t} \hat{m}(\sigma)d\sigma}$	$D_f \int_{-\infty}^{t} m(\sigma)d\sigma$	NL	
SSB-EV[b]	$A_c\lvert 1 + m(t)\rvert$	$\pm \hat{\ln}[1 + m(t)]$	NL	$m(t) > -1$ is required so that the $\ln(\cdot)$ will have a real value
SSB-SQ[b]	$A_c \sqrt{1 + m(t)}$	$\pm \frac{1}{2}\hat{\ln}[1 + m(t)]$	NL	$m(t) > -1$ is required so that the $\ln(\cdot)$ will have a real value
QM	$A_c \sqrt{m_1^2(t) + m_2^2(t)}$	$\tan^{-1}[m_2(t)/m_1(t)]$	L	Used in NTSC color television; requires coherent detection

Source: Couch, L.W., II, 1997, *Digital and Analog Communication Systems,* 5th ed., Prentice Hall, Upper Saddle River, NJ, pp. 231-232. With permission.

[a] $A_c > 0$ is a constant that sets the power level of the signal as evaluated by use of (1.11); L, linear; NL, nonlinear; and $[\hat{\cdot}]$ is the Hilbert transform (a -90° phase-shifted version of $[\cdot]$). For example,
$\hat{m}(t) = m(t) * \frac{1}{\pi t} = \frac{1}{\pi}\int_{-\infty}^{\infty} \frac{m(\lambda)}{t - \lambda}\, d\lambda.$

[b] Use upper signs for upper sideband signals and lower signals for lower sideband signals.

[c] In the strict sense, AM signals are not linear because the carrier term does not satisfy the linearity (superposition) condition.

Analogous to the transmitter realizations, there are two canonical forms of receiver. Each one consists of RF carrier circuits followed by baseband signal processing as illustrated in Fig. 1.1. Typically the carrier circuits are of the superheterodyne-receiver type which consist of an RF amplifier, a down converter (mixer plus local oscillator) to some intermediate frequency (IF), an IF amplifier and then detector circuits [1]. In the first canonical form of the receiver, the carrier circuits have amplitude and phase detectors that output $\tilde{R}(t)$ and $\tilde{\theta}(t)$, respectively. This pair, $\tilde{R}(t)$ and $\tilde{\theta}(t)$, describe the polar form of the received complex envelope, $\tilde{g}(t)$. $\tilde{R}(t)$ and $\tilde{\theta}(t)$ are then fed into the signal processor which uses the inverse functions of Table 1.1 to generate the recovered modulation, $\tilde{m}(t)$. The second canonical form of the receiver uses quadrature product detectors in the carrier circuits to produce the Cartesian form of the received complex envelope, $\tilde{x}(t)$ and $\tilde{y}(t)$. $\tilde{x}(t)$ and $\tilde{y}(t)$ are then inputted to the signal processor which generates $\tilde{m}(t)$ at its output.

Once again, it is stressed that any type of signal modulation (see Table 1.1) may be generated (transmitted) or detected (received) by using either of these two canonical forms. Both of these forms conveniently separate baseband processing from RF processing. Digital

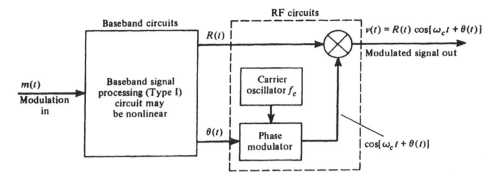

FIGURE 1.2 Generalized transmitter using the AM-PM generation technique. *Source:* Couch, L.W., II. 1997. *Digital and Analog Communication Systems,* 5th ed., Prentice Hall, Upper Saddle River, NJ, p. 278. With permission.

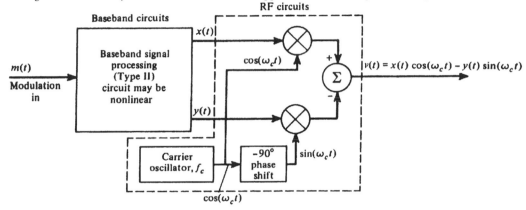

FIGURE 1.3 Generalized transmitter using the quadrature generation technique. *Source:* Couch, L.W., II. 1997. *Digital and Analog Communication Systems,* 5th ed., Prentice Hall, Upper Saddle River, NJ, p. 278. With permission.

techniques are especially useful to realize the baseband processing portion. Furthermore, if digital computing circuits are used, any desired modulation type can be realized by selecting the appropriate software algorithm.

1.5 Spectrum and Power of Bandpass Signals

The spectrum of the bandpass signal is the translation of the spectrum of its complex envelope. Taking the **Fourier transform** of (1.1a), the spectrum of the bandpass waveform is [1]

$$V(f) = \frac{1}{2} \left[G\left(f - f_c\right) + G^*\left(-f - f_c\right) \right] \tag{1.8}$$

where $G(f)$ is the Fourier transform of $g(t)$,

$$G(f) = \int_{-\infty}^{\infty} g(t)e^{-j2\pi ft} dt \; ,$$

and the asterisk superscript denotes the complex conjugate operation. The *power spectra density* (PSD) of the bandpass waveform is [1]

$$\mathcal{P}_v(f) = \frac{1}{4} \left[\mathcal{P}_g\left(f - f_c\right) + \mathcal{P}_g\left(-f - f_c\right) \right] \tag{1.9}$$

where $\mathcal{P}_g(f)$ is the PSD of $g(t)$.

The average power dissipated in a resistive load is V_{rms}^2/R_L or $I_{\text{rms}}^2 R_L$ where V_{rms} is the rms value of the voltage waveform across the load and I_{rms} is the rms value of the current through the load. For bandpass waveforms, Eq. (1.1) may represent either the voltage or the current. Furthermore, the rms values of $v(t)$ and $g(t)$ are related by [1]

$$v_{\text{rms}}^2 = \langle v^2(t) \rangle = \frac{1}{2} \langle |g(t)|^2 \rangle = \frac{1}{2} g_{\text{rms}}^2 \tag{1.10}$$

where $\langle \cdot \rangle$ denotes the time average and is given by

$$\left\langle \left[\quad \right] \right\rangle = \lim_{t \to \infty} \frac{1}{T} \int_{-T/2}^{T/2} \left[\quad \right] dt$$

Thus, if $v(t)$ of (1.1) represents the bandpass voltage waveform across a resistive load, the average power dissipated in the load is

$$P_L = \frac{v_{\text{rms}}^2}{R_L} = \frac{\langle v^2(t) \rangle}{R_L} = \frac{\langle |g(t)|^2 \rangle}{2R_L} = \frac{g_{\text{rms}}^2}{2R_L} \tag{1.11}$$

where g_{rms} is the rms value of the complex envelope and R_L is the resistance of the load.

1.6   Amplitude Modulation

Amplitude modulation (AM) will now be examined in more detail. From Table 1.1 the complex envelope of an AM signal is

$$g(t) = A_c[1 + m(t)] \tag{1.12}$$

so that the spectrum of the complex envelope is

$$G(f) = A_c \delta(f) + A_c M(f) \tag{1.13}$$

Using (1.6), we obtain the AM signal waveform

$$s(t) = A_c[1 + m(t)] \cos \omega_c t \tag{1.14}$$

and, using (1.8), the AM spectrum

$$S(f) = \frac{1}{2} A_c \left[\delta\left(f - f_c\right) + M\left(f - f_c\right) + \delta\left(f + f_c\right) + M\left(f + f_c\right) \right] \tag{1.15}$$

where $\delta(f) = \delta(-f)$ and, because $m(t)$ is real, $M^*(f) = M(-f)$. Suppose that the magnitude spectrum of the modulation happens to be a triangular function, as shown in Fig. 1.4(a). This spectrum might arise from an analog audio source where the bass frequencies are emphasized. The resulting AM spectrum, using (1.15), is shown in Fig. 1.4(b). Note that because $G(f - f_c)$ and $G^*(-f - f_c)$ do not overlap, the magnitude spectrum is

$$|S(f)| = \begin{cases} \frac{1}{2} A_c \delta\left(f - f_c\right) + \frac{1}{2} A_c \left|M\left(f - f_c\right)\right|, & f > 0 \\[2mm] \frac{1}{2} A_c \delta\left(f + f\right)_c + \frac{1}{2} A_c \left|M\left(-f - f\right)_c\right|, & f < 0 \end{cases} \tag{1.16}$$

The 1 in

$$g(t) = A_c[1 + m(t)]$$

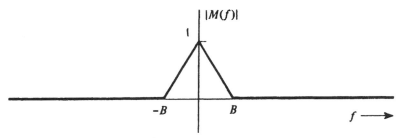

(a) Magnitude Spectrum of Modulation

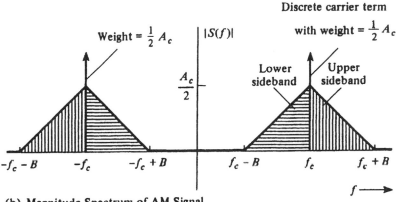

(b) Magnitude Spectrum of AM Signal

FIGURE 1.4   Spectrum of an AM signal. *Source: Couch, L.W., II. 1997. Digital and Analog Communication Systems,* 5th ed., Prentice Hall, Upper Saddle River, NJ, p. 235. With permission.

causes delta functions to occur in the spectrum at $f = \pm f_c$, where f_c is the assigned carrier frequency. Also, from Fig. 1.4 and (1.16), it is realized that the bandwidth of the AM signal is 2B. That is, the bandwidth of the AM signal is twice the bandwidth of the baseband modulating signal.

The average power dissipated into a resistive load is found by using (1.11).

$$P_L = \frac{A_c^2}{2R_L} \left\langle |1 + m(t)|^2 \right\rangle = \frac{A_c^2}{2R_L} \left[1 + 2\langle m(t) \rangle + \langle m^2(t) \rangle \right]$$

If we assume that the *dc* value of the modulation is zero, $\langle m(t) \rangle = 0$, then the average power dissipated into the load is

$$P_L = \frac{A_c^2}{2R_L} \left\langle 1 + m_{\text{rms}}^2 \right\rangle \tag{1.17}$$

where m_{rms} is the rms value of the modulation, $m(t)$. Thus, the average power of an AM signal changes if the rms value of the modulating signal changes. For example, if $m(t)$ is a sine wave test tone with a peak value of 1.0 for 100% modulation,

$$m_{\text{rms}} = 1/\sqrt{2} \ .$$

Assume that $A_c = 1000$ volts and $R_L = 50$ ohms, which are typical values used in AM broadcasting. Then the averagepower dissipated into the 50 Ω load for this AM signal is

$$P_L = \frac{(1000)^2}{2(50)} \left[1 + \frac{1}{2} \right] = 15,000 \text{ watts} \tag{1.18}$$

The Federal Communications Commission (FCC) rated carrier power is obtained when $m(t) = 0$. In this case, (1.17) becomes $P_L = (1000)^2/100 = 10,000$ watts and the FCC would rate this as a 10,000 watt AM station. The sideband power for 100% sine wave modulation is 5,000 watts.

Now let the modulation on the AM signal be a binary digital signal such that $m(t) = \pm 1$ where $+1$ is used for a binary one and -1 is used for a binary 0. Referring to (1.14), this AM signal becomes an *on-off keyed* (OOK) digital signal where the signal is on when a binary one is transmitted and off when a binary zero is transmitted. For $A_c = 1000$ and $R_L = 50\ \Omega$, the average power dissipated would be 20,000 watts since $m_{\text{rms}} = 1$ for $m(t) = \pm 1$.

1.7 Phase and Frequency Modulation

Phase modulation (PM) and *frequency modulation* (FM) are special cases of angle-modulated signalling. In angle-modulated signalling the complex envelope is

$$g(t) = A_c e^{j\theta(t)} \tag{1.19}$$

Using (1.6), the resulting *angle-modulated* signal is

$$s(t) = A_c \cos\left[\omega_c + \theta(t)\right] \tag{1.20}$$

For PM the phase is directly proportional to the modulating signal:

$$\theta(t) = D_p m(t) \tag{1.21}$$

where the proportionality constant D_p is the phase sensitivity of the phase modulator, having units of radians per volt [assuming that $m(t)$ is a voltage waveform]. For FM the phase is proportional to the integral of $m(t)$:

$$\theta(t) = D_f \int_{-\infty}^{t} m(\sigma)d\sigma \tag{1.22}$$

where the frequency deviation constant D_f has units of radians/volt-second. These concepts are summarized by the PM and FM entries in Table 1.1.

By comparing the last two equations, it is seen that if we have a PM signal modulated by $m_p(t)$, there is *also* FM on the signal corresponding to a *different* modulating waveshape that is given by

$$m_f(t) = \frac{D_p}{D_f}\left[\frac{dm_p(t)}{dt}\right] \tag{1.23}$$

where the subscripts f and p denote frequency and phase, respectively. Similarly, if we have an FM signal modulated by $m_f(t)$, the corresponding phase modulation on this signal is

$$m_p(t) = \frac{D_f}{D_p} \int_{-\infty}^{t} m_f(\sigma)d\sigma \tag{1.24}$$

By using (1.24), a PM circuit may be used to synthesize an FM circuit by inserting an integrator in cascade with the phase modulator input.

Other properties of PM and FM are that the **real envelope**, $R(t) = |g(t)| = A_c$, is a constant, as seen from (1.19). Also, $g(t)$ is a *nonlinear* function of the modulation. However, from (1.21) and (1.22), $\theta(t)$ is a linear function of the modulation, $m(t)$. Using (1.11), the average power dissipated by a PM or FM signal is the constant

$$P_L = \frac{A_c^2}{2R_L} \qquad (1.25)$$

That is, the average power of a PM or FM signal does not depend on the modulating waveform, $m(t)$.

The *instantaneous frequency deviation* for an FM signal from its carrier frequency is given by the derivative of its phase $\theta(t)$. Taking the derivative of (1.22), the *peak frequency deviation* is

$$\Delta F = \frac{1}{2\pi} D_f M_p \text{ Hz} \qquad (1.26)$$

where $M_p = \max[m(t)]$ is the peak value of the modulation waveform and the derivative has been divided by 2π to convert from radians/sec to Hz units.

For FM and PM signals, Carson's rule estimates the transmission bandwidth containing approximately 98% of the total power. This FM or PM signal bandwidth is

$$B_T = 2(\beta + 1)B \qquad (1.27)$$

where B is bandwidth (highest frequency) of the modulation. The modulation index β, is $\beta = \Delta F / B$ for FM and $\beta = \max[D_p m(t)] = D_p M_p$ for PM.

The AMPS (Advanced Mobile Phone System) analog cellular phones use FM signalling. A peak deviation of 12 kHz is specified with a modulation bandwidth of 3 kHz. From (1.27), this gives a bandwidth of 30 kHz for the AMPS signal and allows a channel spacing of 30 kHz to be used. To accommodate more users, narrow-band AMPS (NAMPS) with a 5 kHz peak deviation is used in some areas. This allows 10 kHz channel spacing if the carrier frequencies are carefully selected to minimize interference to used adjacent channels. A maximum FM signal power of 3 watts is allowed for the AMPS phones. However, hand-held AMPS phones usually produce no more than 600 mW which is equivalent to 5.5 volts rms across the 50 Ω antenna terminals.

The GSM (Group Special Mobile) digital cellular phones use FM with *minimum frequency-shift-keying* (MSK) where the peak frequency deviation is selected to produce orthogonal waveforms for binary one and binary zero data. (Digital phones use a speech codec to convert the analog voice source to a digital data source for transmission over the system.) Orthogonality occurs when $\Delta F = 1/4R$ where R is the bit rate (bits/sec) [1]. Actually, GSM uses Gaussian shaped MSK (GMSK). That is, the digital data waveform (with rectangular binary one and binary zero pulses) is first filtered by a low-pass filter having a Gaussian shaped frequency response (to attenuate the higher frequencies). This Gaussian filtered data waveform is then fed into the frequency modulator to generate the GMSK signal. This produces a digitally modulated FM signal with a relatively small bandwidth.

Other digital cellular standards use QPSK signalling as discussed in the next section.

1.8 QPSK Signalling

Quadrature phase-shift-keying (QPSK) is a special case of quadrature modulation as shown in Table 1.1 where $m_1(t) = \pm 1$ and $m_2(t) = \pm 1$ are two binary bit streams. The complex envelope for QPSK is

$$g(t) = x(t) + jy(t) = A_c \left[m_1(t) + j m_2(t) \right]$$

where $x(t) = \pm A_c$ and $y(t) = \pm A_c$. The permitted values for the complex envelope are illustrated by the QPSK **signal constellation** shown in Fig. 1.5a. The *signal constellation* is a plot of the permitted values for the complex envelope, $g(t)$. QPSK may be generated by using the quadrature generation technique of Fig. 1.3 where the baseband signal processor is a serial-to-parallel converter that reads in two bits of data at a time from the serial binary input stream, $m(t)$ and outputs the first of the two bits to $x(t)$ and the second bit to $y(t)$. If the two input bits are both binary ones, (11), then $m_1(t) = +A_c$ and $m_2(t) = +A_c$. This is represented by the top right-hand dot for $g(t)$ in the signal constellation for QPSK signalling in Fig. 1.5a. Likewise, the three other possible two-bit words, (10), (01), and (00), are also shown. The QPSK signal is also equivalent to a four-phase phase-shift-keyed signal (4PSK) since all the points in the signal constellation fall on a circle where the permitted phases are $\theta(t) = 45°$, $135°$, $225°$, and $315°$. There is no amplitude modulation on the QPSK signal since the distances from the origin to all the signal points on the signal constellation are equal.

For QPSK, the spectrum of $g(t)$ is of the $\sin x/x$ type since $x(t)$ and $y(t)$ consists of rectangular data pulses of value $\pm A_c$. Moreover, it can be shown that for equally likely independent binary one and binary zero data, the power spectral density of $g(t)$ for digitally modulated signals with M point signal constellations is [1]

$$\mathcal{P}_g(f) = K \left(\frac{\sin \pi f \ell T_b}{\pi f \ell T_b} \right)^2 \tag{1.28}$$

where K is a constant, $R = 1/T_b$ is the data rate (bits/sec) of $m(t)$ and $M = 2^\ell$. M is the number of points in the signal constellation. For QPSK, $M = 4$ and $\ell = 2$. This PSD for the complex envelope, $\mathcal{P}_g(f)$, is plotted in Fig. 1.6. The PSD for the QPSK signal ($\ell = 2$) is given by translating $\mathcal{P}_g(f)$ up to the carrier frequency as indicated by (1.9).

Referring to Fig. 1.6 or using (1.28), the first-null bandwidth of $g(t)$ is R/ℓ Hz. Consequently, the null-to-null bandwidth of the modulated RF signal is

$$B_{\text{null}} = \frac{2R}{\ell} \text{ Hz} \tag{1.29}$$

For example, if the data rate of the baseband information source is 9600 bits/sec, then the null-to-null bandwidth of the QPSK signal would be 9.6 Hz since $\ell = 2$.

Referring to Fig. 1.6, it is seen that the sidelobes of the spectrum are relatively large so, in practice, the sidelobes of the spectrum are filtered off to prevent interference to the adjacent channels. This filtering rounds off the edges of the rectangular data pulses and this causes some amplitude modulation on the QPSK signal. That is, the points in the signal constellation for the filtered QPSK signal would be fuzzy since the transition from one constellation point to another point is not instantaneous because the filtered data pulses are not rectangular. QPSK is the modulation used for digital cellular phones with the IS-95 Code Division Multiple Access (CDMA) standard.

Equation (1.28) and Fig. 1.6 also represent the spectrum for *quadrature modulation amplitude modulation* (QAM) signalling. QAM signalling allows more than two values for $x(t)$ and $y(t)$. For example QAM where $M = 16$ has 16 points in the signal constellation with 4 values for $x(t)$ and 4 values for $y(t)$ such as, for example, $x(t) = +A_c, -A_c, +3A_c, -3A_c$ and $y(t) = +A_c, -A_c, +3A_c, -3A_c$. This is shown in Fig. 1.5b. Each point in the $M = 16$ QAM signal constellation would represent a unique four-bit data word, as compared with the $M = 4$ QPSK signal constellation shown in Fig. 1.5a where each point represents a unique two-bit data word. For a $R = 9600$ bits/sec information source data rate, a $M = 16$ QAM signal would have a null-to-null bandwidth of 4.8 kHz since $\ell = 4$.

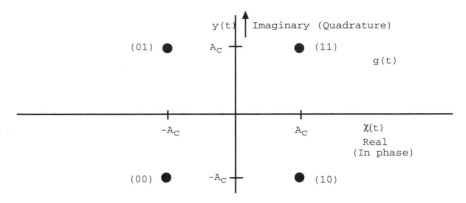

(a) QPSK Signal Constellation

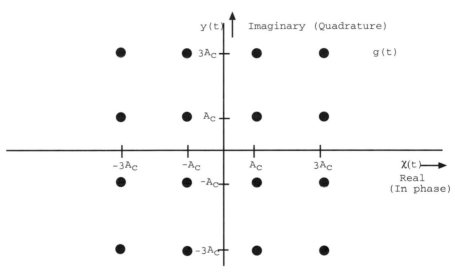

(b) 16 QAM Signal Constellation

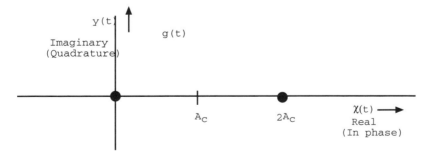

(c) OOK Signal Constellation

FIGURE 1.5 Signal constellations (permitted values of the complex envelope).

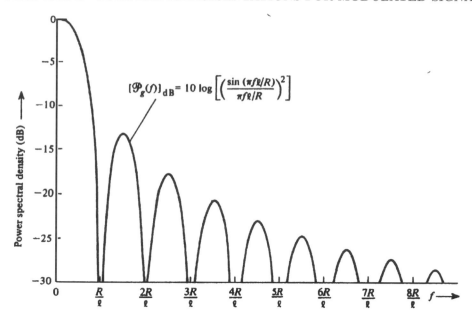

FIGURE 1.6   PSD for the complex envelope of MPSK and QAM where $M = 2^\ell$ and R is bit rate (positive frequencies shown). *Source:* Couch, L.W., II. 1997. *Digital and Analog Communication Systems,* 5th ed., Prentice Hall, Upper Saddle River, NJ, p. 350. With permission.

For OOK signalling as described at the end of Section 1.6, the signal constellation would consist of $M = 2$ points along the x axis where $x = 0, 2A_c$ and $y = 0$. This is illustrated in Fig. 1.5c. For a $R = 9600$ bit/sec information source data rate, an OOK signal would have a null-to-null bandwidth of 19.2 kHz since $\ell = 1$.

Defining Terms

Bandpass waveform: The spectrum of the waveform is nonzero for frequencies in some band concentrated about a frequency $f_c \gg 0$; f_c is called the carrier frequency.

Baseband waveform: The spectrum of the waveform is nonzero for frequencies near $f = 0$.

Complex envelope: The function $g(t)$ of a bandpass waveform $v(t)$ where the bandpass waveform is described by

$$v(t) = \text{Re}\left\{g(t)e^{j\omega_c t}\right\}$$

Fourier transform: If $w(t)$ is a waveform, then the Fourier transform of $w(t)$ is

$$W(f) = \Im[w(t)] = \int_{-\infty}^{\infty} w(t)e^{-j2\pi ft}dt$$

where f has units of hertz.

Modulated signal: The bandpass signal

$$s(t) = \text{Re}\left\{g(t)e^{j\omega_c t}\right\}$$

where fluctuations of $g(t)$ are caused by the information source such as audio, video, or data.

Modulation: The information source, $m(t)$, that causes fluctuations in a bandpass signal.

Real envelope: The function $R(t) = |g(t)|$ of a bandpass waveform $v(t)$ where the bandpass waveform is described by

$$v(t) = Re\left\{g(t)e^{j\omega_c t}\right\}$$

Signal constellation: The permitted values of the complex envelope for a digital modulating source.

References

[1] Couch, L.W., II, *Digital and Analog Communication Systems,* 5th ed., Prentice Hall, Upper Saddle River, NJ, 1997.

Further Information

[1] Bedrosian, E., The analytic signal representation of modulated waveforms. *Proc. IRE,* vol. 50, October, 2071–2076, 1962.

[2] Couch, L.W., II, *Modern Communication Systems: Principles and Applications,* Macmillan Publishing, New York, (now Prentice Hall, Upper Saddle River, NJ), 1995.

[3] Dugundji, J., Envelopes and pre-envelopes of real waveforms. *IRE Trans. Information Theory,* vol. IT-4, March, 53–57, 1958.

[4] Voelcker, H.B., Toward the unified theory of modulation—Part I: Phase-envelope relationships. *Proc. IRE,* vol.54, March, 340–353, 1966.

[5] Voelcker, H.B., Toward the unified theory of modulation—Part II: Zero manipulation. *Proc. IRE,* vol. 54, May, 735–755, 1966.

[6] Ziemer, R.E. and Tranter, W.H., *Principles of Communications,* 4th ed., John Wiley and Sons, New York, 1995.

2

Sampling

Hwei P. Hsu
Fairleigh Dickinson University

2.1 Introduction

To transmit analog message signals, such as speech signals or video signals, by digital means, the signal has to be converted into digital form. This process is known as analog-to-digital conversion. The sampling process is the first process performed in this conversion, and it converts a continuous-time signal into a discrete-time signal or a sequence of numbers. Digital transmission of analog signals is possible by virtue of the sampling theorem, and the sampling operation is performed in accordance with the sampling theorem.

In this chapter, using the Fourier transform technique, we present this remarkable sampling theorem and discuss the operation of sampling and practical aspects of sampling.

2.2 Instantaneous Sampling

Suppose we sample an arbitrary analog signal $m(t)$ shown in Fig. 2.1(a) instantaneously at a uniform rate, once every T_s seconds. As a result of this sampling process, we obtain an infinite sequence of samples $\{m(nT_s)\}$, where n takes on all possible integers. This form of sampling is called *instantaneous sampling*. We refer to T_s as the **sampling interval,** and its reciprocal $1/T_s = f_s$ as the **sampling rate.** Sampling rate (samples per second) is often cited in terms of sampling frequency expressed in hertz.

0-8493-8597-0/99/$0.00+$.50

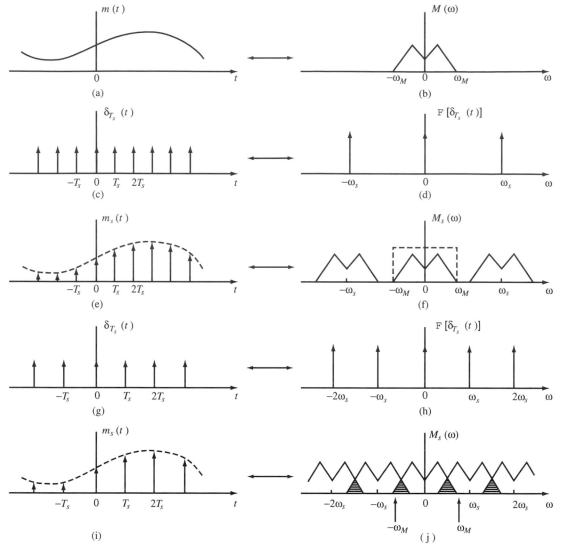

FIGURE 2.1   Illustration of instantaneous sampling and sampling theorem.

2.2.1   Ideal Sampled Signal

Let $m_s(t)$ be obtained by multiplication of $m(t)$ by the unit impulse train $\delta_T(t)$ with period T_s [Fig. 2.1(c)], that is,

$$
\begin{aligned}
m_s(t) &= m(t)\delta_{T_s}(t) = m(t) \sum_{n=-\infty}^{\infty} \delta(t - nT_s) \\
&= \sum_{n=-\infty}^{\infty} m(t)\delta(t - nT_s) = \sum_{n=-\infty}^{\infty} m(nT_s)\delta(t - nT_s)
\end{aligned}
\tag{2.1}
$$

where we used the property of the δ function, $m(t)\delta(t - t_0) = m(t_0)\delta(t - t_0)$. The signal $m_s(t)$ [Fig. 2.1(e)] is referred to as the **ideal sampled signal.**

2.2.2   Band-Limited Signals

A real-valued signal $m(t)$ is called a **band-limited signal** if its Fourier transform $M(\omega)$ satisfies the condition

$$M(\omega) = 0 \qquad \text{for } |\omega| > \omega_M \tag{2.2}$$

where $\omega_M = 2\pi f_M$ [Fig. 2.1(b)]. A band-limited signal specified by Eq. (2.2) is often referred to as a *low-pass signal*.

2.3   Sampling Theorem

The sampling theorem states that a band-limited signal $m(t)$ specified by Eq. (2.2) can be uniquely determined from its values $m(nT_s)$ sampled at uniform interval T_s if $T_s \leq \pi/\omega_M = 1/(2f_M)$. In fact, when $T_s = \pi/\omega_M, m(t)$ is given by

$$m(t) = \sum_{n=-\infty}^{\infty} m\left(nT_s\right) \frac{\sin \omega_M\left(t - nT_s\right)}{\omega_M\left(t - nT_s\right)} \tag{2.3}$$

which is known as the **Nyquist–Shannon interpolation formula** and it is also sometimes called the *cardinal series*. The sampling interval $T_s = 1/(2f_M)$ is called the *Nyquist interval* and the minimum rate $f_s = 1/T_s = 2f_M$ is known as the **Nyquist rate.**

Illustration of the instantaneous sampling process and the sampling theorem is shown in Fig. 2.1. The Fourier transform of the unit impulse train is given by [Fig. 2.1(d)]

$$\mathcal{F}\left\{\delta_{T_s}(t)\right\} = \omega_s \sum_{n=-\infty}^{\infty} \delta\left(\omega - n\omega_s\right) \qquad \omega_s = 2\pi/T_s \tag{2.4}$$

Then, by the convolution property of the Fourier transform, the Fourier transform $M_s(\omega)$ of the ideal sampled signal $m_s(t)$ is given by

$$
\begin{aligned}
M_s(\omega) &= \frac{1}{2\pi}\left[M(\omega) * \omega_s \sum_{n=-\infty}^{\infty} \delta\left(\omega - n\omega_s\right)\right] \\
&= \frac{1}{T_s} \sum_{n=-\infty}^{\infty} M\left(\omega - n\omega_s\right)
\end{aligned}
\tag{2.5}
$$

where $*$ denotes convolution and we used the convolution property of the δ-function $M(\omega) * \delta(\omega - \omega_0) = M(\omega - \omega_0)$. Thus, the sampling has produced images of $M(\omega)$ along the frequency axis. Note that $M_s(\omega)$ will repeat periodically without overlap as long as $\omega_s \geq 2\omega_M$ or $f_s \geq 2f_M$ [Fig. 2.1(f)]. It is clear from Fig. 2.1(f) that we can recover $M(\omega)$ and, hence, $m(t)$ by passing the sampled signal $m_s(t)$ through an ideal low-pass filter having frequency response

$$H(\omega) = \begin{cases} T_s, & |\omega| \leq \omega_M \\ 0, & \text{otherwise} \end{cases} \tag{2.6}$$

where $\omega_M = \pi/T_s$. Then

$$M(\omega) = M_s(\omega)H(\omega) \tag{2.7}$$

Taking the inverse Fourier transform of Eq. (2.6), we obtain the impulse response $h(t)$ of the ideal low-pass filter as

$$h(t) = \frac{\sin \omega_M t}{\omega_M t} \tag{2.8}$$

Taking the inverse Fourier transform of Eq. (2.7), we obtain

$$
\begin{aligned}
m(t) &= m_s(t) * h(t) \\
&= \sum_{n=-\infty}^{\infty} m\,(nT_s)\,\delta\,(t - nT_s) * \frac{\sin \omega_M t}{\omega_M t} \\
&= \sum_{n=-\infty}^{\infty} m\,(nT_s) \frac{\sin \omega_M\,(t - nT_s)}{\omega_M\,(t - nT_s)} \tag{2.9}
\end{aligned}
$$

which is Eq. (2.3).

The situation shown in Fig. 2.1(j) corresponds to the case where $f_s < 2f_M$. In this case there is an overlap between $M(\omega)$ and $M(\omega - \omega_M)$. This overlap of the spectra is known as *aliasing* or *foldover*. When this aliasing occurs, the signal is distorted and it is impossible to recover the original signal $m(t)$ from the sampled signal. To avoid aliasing, in practice, the signal is sampled at a rate slightly higher than the Nyquist rate. If $f_s > 2f_M$, then as shown in Fig. 2.1(f), there is a gap between the upper limit ω_M of $M(\omega)$ and the lower limit $\omega_s - \omega_M$ of $M(\omega - \omega_s)$. This range from ω_M to $\omega_s - \omega_M$ is called a *guard band*. As an example, speech transmitted via telephone is generally limited to $f_M = 3.3$ kHz (by passing the sampled signal through a low-pass filter). The Nyquist rate is, thus, 6.6 kHz. For digital transmission, the speech is normally sampled at the rate $f_s = 8$ kHz. The guard band is then $f_s - 2f_M = 1.4$ kHz. The use of a sampling rate higher than the Nyquist rate also has the desirable effect of making it somewhat easier to design the low-pass reconstruction filter so as to recover the original signal from the sampled signal.

2.4   Sampling of Sinusoidal Signals

A special case is the sampling of a sinusoidal signal having the frequency f_M. In this case we require that $f_s > 2f_M$ rather that $f_s \geq 2f_M$. To see that this condition is necessary, let $f_s = 2f_M$. Now, if an initial sample is taken at the instant the sinusoidal signal is zero, then all successive samples will also be zero. This situation is avoided by requiring $f_s > 2f_M$.

2.5   Sampling of Bandpass Signals

A real-valued signal $m(t)$ is called a **bandpass signal** if its Fourier transform $M(\omega)$ satisfies the condition

$$
M(\omega) = 0 \quad \text{except for} \quad \left\{ \begin{array}{l} \omega_1 < \omega < \omega_2 \\ -\omega_2 < \omega < -\omega_1 \end{array} \right. \tag{2.10}
$$

where $\omega_1 = 2\pi f_1$ and $\omega_2 = 2\pi f_2$ [Fig. 2.2(a)].

The sampling theorem for a band-limited signal has shown that a sampling rate of $2f_2$ or greater is adequate for a low-pass signal having the highest frequency f_2. Therefore, treating $m(t)$ specified by Eq. (2.10) as a special case of such a low-pass signal, we conclude that a sampling rate of $2f_2$ is adequate for the sampling of the bandpass signal $m(t)$. But it is not necessary to sample this fast. The minimum allowable sampling rate depends on f_1, f_2, and the bandwidth $f_B = f_2 - f_1$.

Let us consider the direct sampling of the bandpass signal specified by Eq. (2.10). The spectrum of the sampled signal is periodic with the period $\omega_s = 2\pi f_s$, where f_s is the sampling frequency, as in Eq. (2.4). Shown in Fig. 2.2(b) are the two right shifted spectra of the negative side spectrum $M_-(\omega)$. If the recovering of the bandpass signal is achieved by

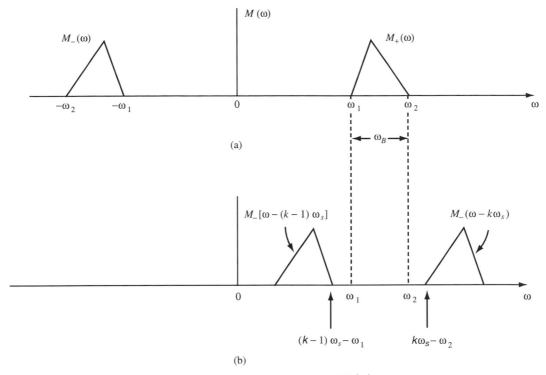

FIGURE 2.2 (a) Spectrum of a bandpass signal; (b) Shifted spectra of $M_-(\omega)$.

passing the sampled signal through an ideal bandpass filter covering the frequency bands $(-\omega_2, -\omega_1)$ and (ω_1, ω_2), it is necessary that there be no aliasing problem. From Fig. 2.2(b), it is clear that to avoid overlap it is necessary that

$$\omega_s \geq 2\left(\omega_2 - \omega_1\right) \tag{2.11}$$

$$(k-1)\omega_s - \omega_1 \leq \omega_1 \tag{2.12}$$

and

$$k\omega_s - \omega_2 \geq \omega_2 \tag{2.13}$$

where $\omega_1 = 2\pi f_1$, $\omega_2 = 2\pi f_2$, and k is an integer $(k = 1, 2, \ldots)$. Since $f_1 = f_2 - f_B$, these constraints can be expressed as

$$1 \leq k \leq \frac{f_2}{f_B} \leq \frac{k}{2}\frac{f_s}{f_B} \tag{2.14}$$

and

$$\frac{k-1}{2}\frac{f_s}{f_B} \leq \frac{f_2}{f_B} - 1 \tag{2.15}$$

A graphical description of Eqs. (2.14) and (2.15) is illustrated in Fig. 2.3. The unshaded regions represent where the constraints are satisfied, whereas the shaded regions represent the regions where the constraints are not satisfied and overlap will occur. The solid line in Fig. 2.3 shows the locus of the minimum sampling rate. The minimum sampling rate is given by

$$\min\{f_s\} = \frac{2f_2}{m} \tag{2.16}$$

where m is the largest integer not exceeding f_2/f_B. Note that if the ratio f_2/f_B is an integer, then the minimum sampling rate is $2f_B$. As an example, consider a bandpass signal with $f_1 = 1.5$ kHz and $f_2 = 2.5$ kHz. Here $f_B = f_2 - f_1 = 1$ kHz, and $f_2/f_B = 2.5$. Then from Eq. (2.16) and Fig. 2.3 we see that the minimum sampling rate is $2f_2/2 = f_2 = 2.5$ kHz, and allowable ranges of sampling rate are 2.5 kHz $\leq f_s \leq 3$ kHz and $f_s \geq 5$ kHz ($= 2f_2$).

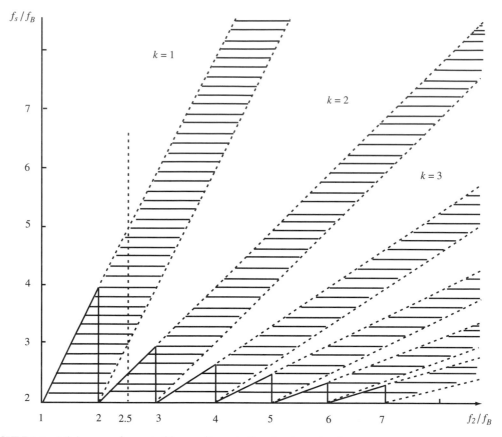

FIGURE 2.3   Minimum and permissible sampling rates for a bandpass signal.

2.6   Practical Sampling

In practice, the sampling of an analog signal is performed by means of high-speed switching circuits, and the sampling process takes the form of *natural sampling* or **flat-top sampling.**

2.6.1   Natural Sampling

Natural sampling of a band-limited signal $m(t)$ is shown in Fig. 2.4. The sampled signal $m_{\rm ns}(t)$ can be expressed as

$$m_{\rm ns}(t) = m(t)x_p(t) \tag{2.17}$$

where $x_p(t)$ is the periodic train of rectangular pulses with fundamental period T_s, and each rectangular pulse in $x_p(t)$ has duration d and unit amplitude [Fig. 2.4(b)]. Observe that the sampled signal $m_{ns}(t)$ consists of a sequence of pulses of varying amplitude whose tops follow the waveform of the signal $m(t)$ [Fig. 2.4(c)].

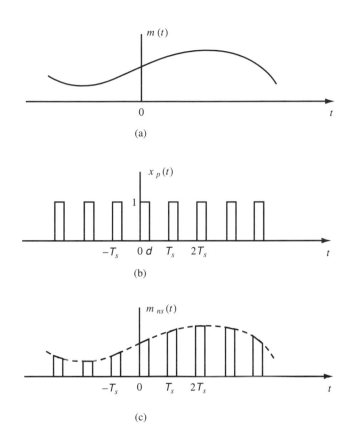

FIGURE 2.4 Natural sampling.

The Fourier transform of $x_p(t)$ is

$$X_p(\omega) = \sum_{n=-\infty}^{\infty} c_n \delta (\omega - n\omega_s) \qquad \omega_s = 2\pi/T_s \tag{2.18}$$

where

$$c_n = \frac{d}{T_s} \frac{\sin (n\omega_s d/2)}{n\omega_s d/2} e^{-jn\omega_s d/2} \tag{2.19}$$

Then the Fourier transform of $m_{ns}(t)$ is given by

$$M_{ns}(\omega) = M(\omega) * X_p(\omega) = \sum_{n=-\infty}^{\infty} c_n M (\omega - n\omega_s) \tag{2.20}$$

from which we see that the effect of the natural sampling is to multiply the nth shifted spectrum $M(\omega - n\omega_s)$ by a constant c_n. Thus, the original signal $m(t)$ can be reconstructed from $m_{ns}(t)$ with no distortion by passing $m_{ns}(t)$ through an ideal low-pass filter if the sampling rate f_s is equal to or greater than the Nyquist rate $2f_M$.

2.6.2   Flat-Top Sampling

The sampled waveform, produced by practical sampling devices that are the sample and hold types, has the form [Fig. 2.5(c)]

$$m_{\text{fs}}(t) = \sum_{n=-\infty}^{\infty} m\left(nT_s\right) p\left(t - nT_s\right) \tag{2.21}$$

where $p(t)$ is a rectangular pulse of duration d with unit amplitude [Fig. 2.5(a)]. This type of sampling is known as **flat-top sampling.** Using the ideal sampled signal $m_s(t)$ of Eq. (2.1), $m_{\text{fs}}(t)$ can be expressed as

$$m_{\text{fs}}(t) = p(t) * \left[\sum_{n=-\infty}^{\infty} m\left(nT_s\right) \delta\left(t - nT_s\right)\right] = p(t) * m_s(t) \tag{2.22}$$

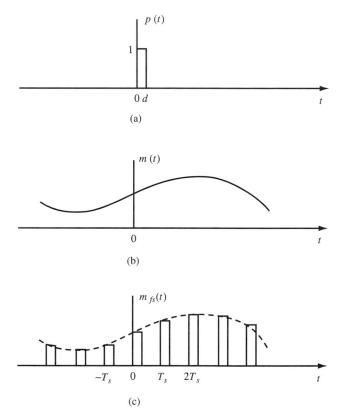

FIGURE 2.5   Flat-top sampling.

Using the convolution property of the Fourier transform and Eq. (2.4), the Fourier transform of $m_{\text{fs}}(t)$ is given by

$$M_{\text{fs}}(\omega) = P(\omega)M_s(\omega) = \frac{1}{T_s} \sum_{n=-\infty}^{\infty} P(\omega)M\left(\omega - n\omega_s\right) \tag{2.23}$$

where

$$P(\omega) = d\frac{\sin(\omega d/2)}{\omega d/2}e^{-j\omega d/2} \tag{2.24}$$

From Eq.(2.23) we see that by using flat-top sampling we have introduced amplitude distortion and time delay, and the primary effect is an attenuation of high-frequency components. This effect is known as the *aperture effect*. The aperture effect can be compensated by an equalizing filter with a frequency response $H_{\text{eq}}(\omega) = 1/P(\omega)$. If the pulse duration d is chosen such that $d \ll T_s$, however, then $P(\omega)$ is essentially constant over the baseband and no equalization may be needed.

2.7   Sampling Theorem in the Frequency Domain

The sampling theorem expressed in Eq. (2.4) is the time-domain sampling theorem. There is a dual to this time-domain sampling theorem, i.e., the sampling theorem in the frequency domain.

Time-limited signals: A continuous-time signal $m(t)$ is called **time limited** if

$$m(t) = 0 \qquad \text{for} \quad |t| > |T_0| \tag{2.25}$$

Frequency-domain sampling theorem: The frequency-domain sampling theorem states that the Fourier transform $M(\omega)$ of a time-limited signal $m(t)$ specified by Eq. (2.25) can be uniquely determined from its values $M(n\omega_s)$ sampled at a uniform rate ω_s if $\omega_s \leq \pi/T_0$. In fact, when $\omega_s = \pi/T_0$, then $M(\omega)$ is given by

$$M(\omega) = \sum_{n=-\infty}^{\infty} M(n\omega_s)\frac{\sin T_0(\omega - n\omega_s)}{T_0(\omega - n\omega_s)} \tag{2.26}$$

2.8   Summary and Discussion

The sampling theorem is the fundamental principle of digital communications. We state the sampling theorem in two parts.

THEOREM 2.1   *If the signal contains no frequency higher than f_M Hz, it is completely described by specifying its samples taken at instants of time spaced $1/2f_M$ s.*

THEOREM 2.2   *The signal can be completely recovered from its samples taken at the rate of $2f_M$ samples per second or higher.*

The preceding sampling theorem assumes that the signal is strictly band limited. It is known that if a signal is band limited it cannot be time limited and vice versa. In many practical applications, the signal to be sampled is time limited and, consequently, it cannot be strictly band limited. Nevertheless, we know that the frequency components of physically occurring signals attenuate rapidly beyond some defined bandwidth, and for practical purposes we consider these signals are band limited. This approximation of real signals by band limited ones introduces no significant error in the application of the sampling theorem. When such a signal is sampled, we band limit the signal by filtering before sampling and sample at a rate slightly higher than the nominal Nyquist rate.

Defining Terms

Band-limited signal: A signal whose frequency content (Fourier transform) is equal to zero above some specified frequency.

Bandpass signal: A signal whose frequency content (Fourier transform) is nonzero only in a band of frequencies not including the origin.

Flat-top sampling: Sampling with finite width pulses that maintain a constant value for a time period less than or equal to the sampling interval. The constant value is the amplitude of the signal at the desired sampling instant.

Ideal sampled signal: A signal sampled using an ideal impulse train.

Nyquist rate: The minimum allowable sampling rate of $2f_M$ samples per second, to reconstruct a signal band limited to f_M hertz.

Nyquist-Shannon interpolation formula: The infinite series representing a time domain waveform in terms of its ideal samples taken at uniform intervals.

Sampling interval: The time between samples in uniform sampling.

Sampling rate: The number of samples taken per second (expressed in Hertz and equal to the reciprocal of the sampling interval).

Time-limited: A signal that is zero outside of some specified time interval.

References

[1] Brown, J.L. Jr., First order sampling of bandpass signals—A new approach. *IEEE Trans. Information Theory,* IT-26(5), 613–615, 1980.

[2] Byrne, C.L. and Fitzgerald, R.M., Time-limited sampling theorem for band-limited signals, *IEEE Trans. Information Theory,* IT-28(5), 807–809, 1982.

[3] Hsu, H.P., *Applied Fourier Analysis,* Harcourt Brace Jovanovich, San Diego, CA, 1984.

[4] Hsu, H.P., *Analog and Digital Communications,* McGraw-Hill, New York, 1993.

[5] Hulthén, R., Restoring causal signals by analytical continuation: A generalized sampling theorem for causal signals. *IEEE Trans. Acoustics, Speech, and Signal Processing,* ASSP-31(5), 1294–1298, 1983.

[6] Jerri, A.J., The Shannon sampling theorem—Its various extensions and applications: A tutorial review, *Proc. IEEE.* 65(11), 1565–1596, 1977.

Further Information

For a tutorial review of the sampling theorem, historical notes, and earlier references see Jerri [6].

3

Pulse Code Modulation[1]

Leon W. Couch, II
University of Florida

3.1 Introduction

Pulse code modulation (PCM) is analog-to-digital conversion of a special type where the information contained in the instantaneous samples of an analog signal is represented by digital words in a serial bit stream.

If we assume that each of the digital words has n binary digits, there are $M = 2^n$ unique code words that are possible, each code word corresponding to a certain amplitude level. Each sample value from the analog signal, however, can be any one of an infinite number of levels, so that the digital word that represents the amplitude closest to the actual sampled value is used. This is called **quantizing**. That is, instead of using the exact sample value of the analog waveform, the sample is replaced by the closest allowed value, where there are M allowed values, and each allowed value corresponds to one of the code words.

PCM is very popular because of the many advantages it offers. Some of these advantages are as follows.

- Relatively inexpensive digital circuitry may be used extensively in the system.
- PCM signals derived from all types of analog sources (audio, video, etc.) may be time-division multiplexed with data signals (e.g., from digital computers) and transmitted over a common high-speed digital communication system.

---

[1]*Source:* Leon W. Couch, II. 1997. *Digital and Analog Communication Systems,* 5th ed., Prentice Hall, Upper Saddle River, NJ. With permission.

0-8493-8597-0/99/$0.00+$.50

- In long-distance digital telephone systems requiring repeaters, a *clean* PCM waveform can be regenerated at the output of each repeater, where the input consists of a noisy PCM waveform. The noise at the input, however, may cause bit errors in the regenerated PCM output signal.

- The noise performance of a digital system can be superior to that of an analog system. In addition, the probability of error for the system output can be reduced even further by the use of appropriate coding techniques.

These advantages usually outweigh the main disadvantage of PCM: a much wider bandwidth than that of the corresponding analog signal.

3.2 Generation of PCM

The PCM signal is generated by carrying out three basic operations: sampling, quantizing, and encoding (see Fig. 3.1). The sampling operation generates an instantaneously-sampled flat-top **pulse-amplitude modulated** (PAM) signal.

The quantizing operation is illustrated in Fig. 3.2 for the $M = 8$ level case. This quantizer is said to be *uniform* since all of the steps are of equal size. Since we are approximating the analog sample values by using a finite number of levels ($M = 8$ in this illustration), *error* is introduced into the recovered output analog signal because of the quantizing effect. The error waveform is illustrated in Fig. 3.2c. The quantizing error consists of the difference between the analog signal at the sampler input and the output of the quantizer. Note that the peak value of the error (± 1) is one-half of the quantizer step size (2). If we sample at the Nyquist rate ($2B$, where B is the absolute bandwidth, in hertz, of the input analog signal) or faster and there is negligible channel noise, there will still be noise, called *quantizing noise,* on the recovered analog waveform due to this error. The quantizing noise can also be thought of as a round-off error. The quantizer output is a *quantized* (i.e., only M possible amplitude values) PAM signal.

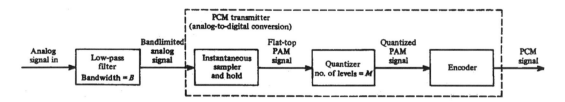

FIGURE 3.1   A PCM transmitter. *Source:* Couch, L.W. II 1997. *Digital and Analog Communication Systems,* 5th ed., Prentice Hall, Upper Saddle River, NJ, p. 138. With permission.

The PCM signal is obtained from the quantized PAM signal by encoding each quantized sample value into a digital word. It is up to the system designer to specify the exact code word that will represent a particular quantized level. If a Gray code of Table 3.1 is used, the resulting PCM signal is shown in Fig. 3.2d where the PCM word for each quantized sample is strobed out of the encoder by the next clock pulse. The Gray code was chosen because it has only 1-b change for each step change in the quantized level. Consequently, single errors in the received PCM code word will cause minimum errors in the recovered analog level, provided that the sign bit is not in error.

Here we have described PCM systems that represent the quantized analog sample values by *binary* code words. Of course, it is possible to represent the quantized analog samples by

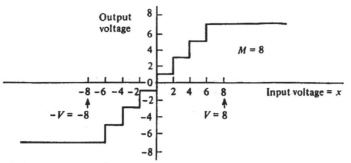

(a) Quantizer Output-Input Characteristics

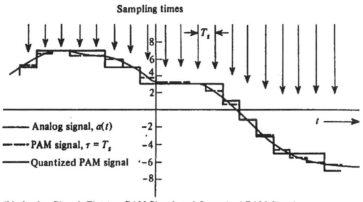

(b) Analog Signal, Flat-top PAM Signal, and Quantized PAM Signal

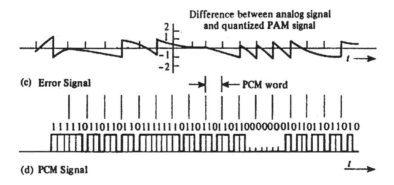

(c) Error Signal

(d) PCM Signal

FIGURE 3.2 Illustration of waveforms in a PCM system. *Source:* Couch, L.W. II 1997. *Digital and Analog Communication Systems,* 5th ed., Prentice Hall, Upper Saddle River, NJ, p. 139. With permission.

digital words using other than base 2. That is, for base q, the number of quantized levels allowed is $M = q^n$, where n is the number of q base digits in the code word. We will not pursue this topic since binary ($q = 2$) digital circuits are most commonly used.

3.3 Percent Quantizing Noise

The quantizer at the PCM encoder produces an error signal at the PCM decoder output as illustrated in Fig. 3.2c. The peak value of this error signal may be expressed as a percentage

TABLE 3.1   3-b Gray Code for $M = 8$ Levels

Quantized Sample Voltage	Gray Code Word (PCM Output)	
+7	110	
+5	111	
+3	101	
+1	100	
		Mirror image except for sign bit
−1	000	
−3	001	
−5	011	
−7	010	

Source: Couch, L.W., II. 1997. *Digital and Analog Communication Systems,* 5th ed., Prentice Hall, Upper Saddle River, NJ, p. 140. With permission.

of the maximum possible analog signal amplitude. Referring to Fig. 3.2c, a peak error of 1 V occurs for a maximum analog signal amplitude of $M = 8$ V as shown Fig. 3.1c. Thus, in general,

$$\frac{2P}{100} = \frac{1}{M} = \frac{1}{2^n}$$

or

$$2^n = \frac{50}{P} \tag{3.1}$$

where P is the peak percentage error for a PCM system that uses n bit code words. The design value of n needed in order to have less than P percent error is obtained by taking the base 2 logarithm of both sides of Eq. (3.1), where it is realized that $\log_2(x) = [\log_{10}(x)]/\log_{10}(2) = 3.32\log_{10}(x)$. That is,

$$n \geq 3.32\log_{10}\left(\frac{50}{P}\right) \tag{3.2}$$

where n is the number of bits needed in the PCM word in order to obtain less than P percent error in the recovered analog signal (i.e., decoded PCM signal).

3.4   Practical PCM Circuits

Three techniques are used to implement the analog-to-digital converter (ADC) encoding operation. These are the *counting* or *ramp, serial* or *successive approximation*, and *parallel* or *flash* encoders.

In the counting encoder, at the same time that the sample is taken, a ramp generator is energized and a binary counter is started. The output of the ramp generator is continuously compared to the sample value; when the value of the ramp becomes equal to the sample value, the binary value of the counter is read. This count is taken to be the PCM word. The binary counter and the ramp generator are then reset to zero and are ready to be reenergized at the next sampling time. This technique requires only a few components, but the speed of this type of ADC is usually limited by the speed of the counter. The Maxim ICL7126 CMOS ADC integrated circuit uses this technique.

The serial encoder compares the value of the sample with trial quantized values. Successive trials depend on whether the past comparator outputs are positive or negative. The trial values are chosen first in large steps and then in small steps so that the process will converge rapidly. The trial voltages are generated by a series of voltage dividers that are configured by (on-off) switches. These switches are controlled by digital logic. After the process converges, the value of the switch settings is read out as the PCM word. This technique requires more precision components (for the voltage dividers) than the ramp technique. The speed of the feedback ADC technique is determined by the speed of the switches. The National Semiconductor ADC0804 8-b ADC uses this technique.

The parallel encoder uses a set of parallel comparators with reference levels that are the permitted quantized values. The sample value is fed into all of the parallel comparators simultaneously. The high or low level of the comparator outputs determines the binary PCM word with the aid of some digital logic. This is a fast ADC technique but requires more hardware than the other two methods. The Harris CA3318 8-b ADC integrated circuit is an example of the technique.

All of the integrated circuits listed as examples have parallel digital outputs that correspond to the digital word that represents the analog sample value. For generation of PCM, the parallel output (digital word) needs to be converted to serial form for transmission over a two-wire channel. This is accomplished by using a parallel-to-serial converter integrated circuit, which is also known as a **serial-input-output** (SIO) chip. The SIO chip includes a shift register that is set to contain the parallel data (usually, from 8 or 16 input lines). Then the data are shifted out of the last stage of the shift register bit by bit onto a single output line to produce the serial format. Furthermore, the SIO chips are usually full duplex; that is, they have two sets of shift registers, one that functions for data flowing in each direction. One shift register converts parallel input data to serial output data for transmission over the channel, and, simultaneously, the other shift register converts received serial data from another input to parallel data that are available at another output. Three types of SIO chips are available: the *universal asynchronous receiver/transmitter* (UART), the *universal synchronous receiver/transmitter* (USRT), and the *universal synchronous/asynchronous receiver transmitter* (USART). The UART transmits and receives asynchronous serial data, the USRT transmits and receives synchronous serial data, and the USART combines both a UART and a USRT on one chip.

At the receiving end the PCM signal is decoded back into an analog signal by using a digital-to-analog converter (DAC) chip. If the DAC chip has a parallel data input, the received serial PCM data are first converted to a parallel form using a SIO chip as described in the preceding paragraph. The parallel data are then converted to an approximation of the analog sample value by the DAC chip. This conversion is usually accomplished by using the parallel digital word to set the configuration of electronic switches on a resistive current (or voltage) divider network so that the analog output is produced. This is called a *multiplying* DAC since the analog output voltage is directly proportional to the divider reference voltage multiplied by the value of the digital word. The Motorola MC1408 and the National Semiconductor DAC0808 8-b DAC chips are examples of this technique. The DAC chip outputs samples of the quantized analog signal that approximates the analog sample values. This may be smoothed by a low-pass reconstruction filter to produce the analog output.

The Communications Handbook [6, pp 107–117] and The Electrical Engineering Handbook [5, pp. 771–782] give more details on ADC, DAC, and PCM circuits.

3.5   Bandwidth of PCM

A good question to ask is: What is the spectrum of a PCM signal? For the case of PAM signalling, the spectrum of the PAM signal could be obtained as a function of the spectrum of the input analog signal because the PAM signal is a linear function of the analog signal. This is not the case for PCM. As shown in Figs. 3.1 and 3.2, the PCM signal is a nonlinear function of the input signal. Consequently, the spectrum of the PCM signal is not directly related to the spectrum of the input analog signal. It can be shown that the spectrum of the PCM signal depends on the bit rate, the correlation of the PCM data, and on the PCM waveform pulse shape (usually rectangular) used to describe the bits [2, 3]. From Fig. 3.2, the bit rate is

$$R = nf_s \tag{3.3}$$

where n is the number of bits in the PCM word ($M = 2^n$) and f_s is the sampling rate. For no aliasing we require $f_s \geq 2B$ where B is the bandwidth of the analog signal (that is to be converted to the PCM signal). The dimensionality theorem [2, 3] shows that the bandwidth of the PCM waveform is bounded by

$$B_{\text{PCM}} \geq \frac{1}{2}R = \frac{1}{2}nf_s \tag{3.4}$$

where equality is obtained if a $(\sin x)/x$ type of pulse shape is used to generate the PCM waveform. The exact spectrum for the PCM waveform will depend on the pulse shape that is used as well as on the type of line encoding. For example, if one uses a rectangular pulse shape with polar nonreturn to zero (NRZ) line coding, the first null bandwidth is simply

$$B_{\text{PCM}} = R = nf_s \ \text{ Hz} \tag{3.5}$$

Table 3.2 presents a tabulation of this result for the case of the minimum sampling rate, $f_s = 2B$. Note that Eq. (3.4) demonstrates that the bandwidth of the PCM signal has a lower bound given by

$$B_{\text{PCM}} \geq nB \tag{3.6}$$

where $f_s > 2B$ and B is the bandwidth of the corresponding analog signal. Thus, for reasonable values of n, the bandwidth of the PCM signal will be significantly larger than the bandwidth of the corresponding analog signal that it represents. For the example shown in Fig. 3.2 where $n = 3$, the PCM signal bandwidth will be at least three times wider than that of the corresponding analog signal. Furthermore, if the bandwidth of the PCM signal is reduced by improper filtering or by passing the PCM signal through a system that has a poor frequency response, the filtered pulses will be elongated (stretched in width) so that pulses corresponding to any one bit will smear into adjacent bit slots. If this condition becomes too serious, it will cause errors in the detected bits. This pulse smearing effect is called **intersymbol interference** (ISI).

3.6   Effects of Noise

The analog signal that is recovered at the PCM system output is corrupted by noise. Two main effects produce this noise or distortion: 1) quantizing noise that is caused by the M-step quantizer at the PCM transmitter and 2) bit errors in the recovered PCM signal. The bit errors are caused by *channel noise* as well as improper channel filtering, which causes ISI. In addition, if the input analog signal is not strictly band limited, there will be some

TABLE 3.2 Performance of a PCM System with Uniform
Quantizing and No Channel Noise

Number of Quantizer Levels Used, M	Length of the PCM Word, n (bits)	Bandwidth of PCM Signal (First Null Bandwidth)[a]	Recovered Analog Signal Power-to-Quantizing Noise Power Ratios (dB) $(S/N)_{out}$
2	1	2B	6.0
4	2	4B	12.0
8	3	6B	18.1
16	4	8B	24.1
32	5	10B	30.1
64	6	12B	36.1
128	7	14B	42.1
256	8	16B	48.2
512	9	18B	54.2
1,024	10	20B	60.2
2,048	11	22B	66.2
4,096	12	24B	72.2
8,192	13	26B	78.3
16,384	14	28B	84.3
32,768	15	30B	90.3
65,536	16	32B	96.3

[a]B is the absolute bandwidth of the input analog signal. *Source:*
Couch, L.W. II 1997. *Digital and Analog Communication Systems,* 5th ed., Prentice Hall, Upper Saddle River, NJ, p. 142. With
permission.

aliasing noise on the recovered analog signal [12]. Under certain assumptions, it can be
shown that the recovered analog *average* signal power to the average noise power [2] is

$$\left(\frac{S}{N}\right)_{out} = \frac{M^2}{1 + 4\left(M^2 - 1\right)P_e} \tag{3.7}$$

where M is the number of uniformly spaced quantizer levels used in the PCM transmitter
and P_e is the probability of bit error in the recovered binary PCM signal at the receiver
DAC before it is converted back into an analog signal. Most practical systems are designed
so that P_e is negligible. Consequently, if we assume that there are no bit errors due to
channel noise (i.e., $P_e = 0$), the S/N due only to quantizing errors is

$$\left(\frac{S}{N}\right)_{out} = M^2 \tag{3.8}$$

Numerical values for these S/N ratios are given in Table 3.2.

To realize these S/N ratios, one critical assumption is that the peak-to-peak level of the
analog waveform at the input to the PCM encoder is set to the design level of the quantizer.
For example, referring to Fig. 3.2, this corresponds to the input traversing the range $-V$ to
$+V$ volts where $V = 8$ V is the design level of the quantizer. Equation (3.7) was derived for
waveforms with equally likely values, such as a triangle waveshape, that have a peak-to-peak
value of $2V$ and an rms value of $V/\sqrt{3}$, where V is the design peak level of the quantizer.

From a practical viewpoint, the quantizing noise at the output of the PCM decoder can
be categorized into four types depending on the operating conditions. The four types are
overload noise, random noise, granular noise, and hunting noise. As discussed earlier, the
level of the analog waveform at the input of the PCM encoder needs to be set so that
its peak level does not exceed the design peak of V volts. If the peak input does exceed
V, the recovered analog waveform at the output of the PCM system will have flat tops
near the peak values. This produces *overload noise*. The flat tops are easily seen on
an oscilloscope, and the recovered analog waveform sounds distorted since the flat topping

produces unwanted harmonic components. For example, this type of distortion can be heard on PCM telephone systems when there are high levels such as dial tones, busy signals, or off-hook warning signals.

The second type of noise, *random noise,* is produced by the random quantization errors in the PCM system under normal operating conditions when the input level is properly set. This type of condition is assumed in Eq. (3.8). Random noise has a white hissing sound. If the input level is not sufficiently large, the S/N will deteriorate from that given by Eq. (3.8); the quantizing noise will still remain more or less random.

If the input level is reduced further to a relatively small value with respect to the design level, the error values are not equally likely from sample to sample, and the noise has a harsh sound resembling gravel being poured into a barrel. This is called *granular noise.* This type of noise can be randomized (noise power decreased) by increasing the number of quantization levels and, consequently, increasing the PCM bit rate. Alternatively, granular noise can be reduced by using a nonuniform quantizer, such as the μ-law or A-law quantizers that are described in Section 3.7.

The fourth type of quantizing noise that may occur at the output of a PCM system is *hunting noise.* It can occur when the input analog waveform is nearly constant, including when there is no signal (i.e., zero level). For these conditions the sample values at the quantizer output (see Fig. 3.2) can oscillate between two adjacent quantization levels, causing an undesired sinusoidal type tone of frequency $1/2f_s$ at the output of the PCM system. Hunting noise can be reduced by filtering out the tone or by designing the quantizer so that there is no vertical step at the constant value of the inputs, such as at 0-V input for the no signal case. For the no signal case, the hunting noise is also called *idle channel noise.* Idle channel noise can be reduced by using a horizontal step at the origin of the quantizer output–input characteristic instead of a vertical step as shown in Fig. 3.2.

Recalling that $M = 2^n$, we may express Eq. (3.8) in decibels by taking $10 \log_{10}(\cdot)$ of both sides of the equation,

$$\left(\frac{S}{N}\right)_{\mathrm{dB}} = 6.02n + \alpha \tag{3.9}$$

where n is the number of bits in the PCM word and $\alpha = 0$. This equation—called the 6-dB rule—points out the significant performance characteristic for PCM: an additional 6-dB improvement in S/N is obtained for each bit added to the PCM word. This is illustrated in Table 3.2. Equation (3.9) is valid for a wide variety of assumptions (such as various types of input waveshapes and quantification characteristics), although the value of α will depend on these assumptions [7]. Of course, it is assumed that there are no bit errors and that the input signal level is large enough to range over a significant number of quantizing levels.

One may use Table 3.2 to examine the design requirements in a proposed PCM system. For example, high fidelity enthusiasts are turning to digital audio recording techniques. Here PCM signals are recorded instead of the analog audio signal to produce superb sound reproduction. For a dynamic range of 90 dB, it is seen that at least 15-b PCM words would be required. Furthermore, if the analog signal had a bandwidth of 20 kHz, the first null bandwidth for rectangular bit-shape PCM would be 2×20 kHz $\times 15 = 600$ kHz. Consequently, video-type tape recorders are needed to record and reproduce high-quality digital audio signals. Although this type of recording technique might seem ridiculous at first, it is realized that expensive high-quality analog recording devices are hard pressed to reproduce a dynamic range of 70 dB. Thus, digital audio is one way to achieve improved performance. This is being proven in the marketplace with the popularity of the digital compact disk (CD). The CD uses a 16-b PCM word and a sampling rate of 44.1 kHz on each stereo channel [9, 10]. Reed–Solomon coding with interleaving is used to correct burst errors that occur as a result of scratches and fingerprints on the compact disk.

3.7 Nonuniform Quantizing: μ-Law and A-Law Companding

Voice analog signals are more likely to have amplitude values near zero than at the extreme peak values allowed. For example, when digitizing voice signals, if the peak value allowed is 1 V, weak passages may have voltage levels on the order of 0.1 V (20 dB down). For signals such as these with nonuniform amplitude distribution, the granular quantizing noise will be a serious problem if the step size is not reduced for amplitude values near zero and increased for extremely large values. This is called nonuniform quantizing since a variable step size is used. An example of a nonuniform quantizing characteristic is shown in Fig. 3.3.

The effect of nonuniform quantizing can be obtained by first passing the analog signal through a compression (nonlinear) amplifier and then into the PCM circuit that uses a uniform quantizer. In the U.S., a μ-law type of compression characteristic is used. It is defined [11] by

$$|w_2(t)| = \frac{\ln\left(1 + \mu\,|w_1(t)|\right)}{\ln(1 + \mu)} \tag{3.10}$$

where the allowed peak values of $w_1(t)$ are ± 1 (i.e., $|w_1(t)| \leq 1$), μ is a positive constant that is a parameter. This compression characteristic is shown in Fig. 3.3(b) for several values of μ, and it is noted that $\mu \to 0$ corresponds to linear amplification (uniform quantization overall). In the United States, Canada, and Japan, the telephone companies use a $\mu = 255$ compression characteristic in their PCM systems [4].

Another compression law, used mainly in Europe, is the A-law characteristic. It is defined [1] by

$$|w_2(t)| = \begin{cases} \dfrac{A\,|w_1(t)|}{1 + \ln A}, & 0 \leq |w_1(t)| \leq \dfrac{1}{A} \\[2ex] \dfrac{1 + \ln\left(A\,|w_1(t)|\right)}{1 + \ln A}, & \dfrac{1}{A} \leq |w_1(t)| \leq 1 \end{cases} \tag{3.11}$$

where $|w_1(t)| < 1$ and A is a positive constant. The A-law compression characteristic is shown in Fig. 3.3(c). The typical value for A is 87.6.

When compression is used at the transmitter, *expansion* (i.e., decompression) must be used at the receiver output to restore signal levels to their correct relative values. The *expandor* characteristic is the inverse of the compression characteristic, and the combination of a compressor and an expandor is called a *compandor*.

Once again, it can be shown that the output S/N follows the 6-dB law [2]

$$\left(\frac{S}{N}\right)_{\mathrm{dB}} = 6.02 + \alpha \tag{3.12}$$

where for uniform quantizing

$$\alpha = 4.77 - 20\log\left(V/x_{\mathrm{rms}}\right) \tag{3.13}$$

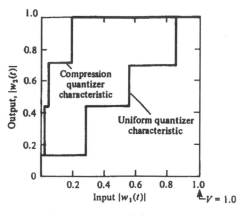

(a) $M = 8$ Quantizer Characteristic

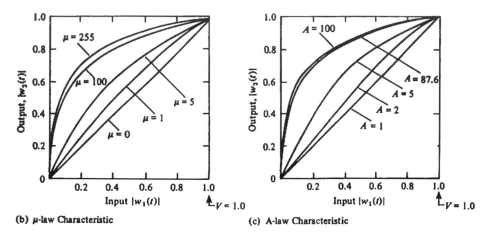

(b) μ-law Characteristic   (c) A-law Characteristic

FIGURE 3.3   Compression characteristics (first quadrant shown). *Source:* Couch, L.W. II 1997. *Digital and Analog Communication Systems,* 5th ed., Prentice Hall, Upper Saddle River, NJ, p. 147. With permission.

and for sufficiently large input levels[2] for μ-law companding

$$\alpha \approx 4.77 - 20\log[\ln(1+\mu)] \tag{3.14}$$

and for A-law companding [7]

$$\alpha \approx 4.77 - 20\log[1+\ln A] \tag{3.15}$$

n is the number of bits used in the PCM word, V is the peak design level of the quantizer, and x_{rms} is the rms value of the input analog signal. Notice that the output S/N is a function of the input level for the uniform quantizing (no companding) case but is relatively insensitive to input level for μ-law and A-law companding, as shown in Fig. 3.4. The ratio V/x_{rms} is called the *loading factor.* The input level is often set for a loading factor of 4 (12 dB) to

---

[2]See Lathi, 1998 for a more complicated expression that is valid for any input level.

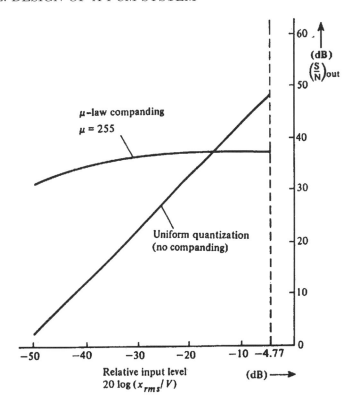

FIGURE 3.4 Output S/N of 8-b PCM systems with and without companding. *Source:* Couch, L.W. II 1997. *Digital and Analog Communication Systems,* 5th ed., Prentice Hall, Upper Saddle River, NJ, p. 149. With permission.

ensure that the overload quantizing noise will be negligible. In practice this gives $\alpha = -7.3$ for the case of uniform encoding as compared to $\alpha = 0$, which was obtained for the ideal conditions associated with Eq. (3.8).

3.8 Example: Design of a PCM System

Assume that an analog voice-frequency signal, which occupies a band from 300 to 3400 Hz, is to be transmitted over a binary PCM system. The minimum sampling frequency would be $2 \times 3.4 = 6.8$ kHz. In practice the signal is oversampled, and in the U.S. a sampling frequency of 8 kHz is the standard used for voice-frequency signals in telephone communication systems. Assume that each sample value is represented by 8 b; then the bit rate of the PCM signal is

$$
\begin{aligned}
R &= (f_s \text{ samples/s})(n \text{ b/s}) \\
&= (8 \text{ } k \text{ samples/s})(8 \text{ b/s}) = 64 \text{ kb/s} \qquad (3.16)
\end{aligned}
$$

Referring to the dimensionality theorem [Eq. (3.4)], we realize that the theoretically minimum absolute bandwidth of the PCM signal is

$$
B_{\min} = \frac{1}{2}D = 32 \text{ kHz} \qquad (3.17)
$$

and this is realized if the PCM waveform consists of $(\sin x)/x$ pulse shapes. If rectangular pulse shaping is used, the absolute bandwidth is infinity, and the first null bandwidth

[Eq. (3.5)] is

$$B_{\text{null}} = R = \frac{1}{T_b} = 64 \ \text{kHz} \tag{3.18}$$

That is, we require a bandwidth of 64 kHz to transmit this digital voice PCM signal where the bandwidth of the original analog voice signal was, at most, 4 kHz. Using $n = 8$ in Eq. (3.1), the error on the recovered analog signal is $\pm 0.2\%$. Using Eqs. (3.12) and (3.13) for the case of uniform quantizing with a loading factor, V/x_{rms}, of 10 (20 dB), we get for uniform quantizing

$$\left(\frac{S}{N}\right)_{\text{dB}} = 32.9 \ \text{dB} \tag{3.19}$$

Using Eqs. (3.12) and (3.14) for the case of $\mu = 255$ companding, we get

$$\left(\frac{S}{N}\right) = 38.05 \ \text{dB} \tag{3.20}$$

These results are illustrated in Fig. 3.4.

Defining Terms

Intersymbol interference: Filtering of a digital waveform so that a pulse corresponding to 1 b will smear (stretch in width) into adjacent bit slots.

Pulse amplitude modulation: An analog signal is represented by a train of pulses where the pulse amplitudes are proportional to the analog signal amplitude.

Pulse code modulation: A serial bit stream that consists of binary words which represent quantized sample values of an analog signal.

Quantizing: Replacing a sample value with the closest allowed value.

References

[1] Cattermole, K.W., *Principles of Pulse-code Modulation*, American Elsevier, New York, NY, 1969.

[2] Couch, L.W., *Digital and Analog Communication Systems*, 5th ed., Prentice Hall, Upper Saddle River, NJ, 1997.

[3] Couch, L.W., *Modern Communication Systems: Principles and Applications*, Macmillan Publishing, New York, NY, 1995.

[4] Dammann, C.L., McDaniel, L.D., and Maddox, C.L., D2 Channel Bank—Multiplexing and Coding. *B. S. T. J.*, 12(10), 1675–1700, 1972.

[5] Dorf, R.C., *The Electrical Engineering Handbook*, CRC Press, Inc., Boca Raton, FL, 1993.

[6] Gibson, J.D., *The Communications Handbook*, CRC Press, Inc., Boca Raton, FL, 1997.

[7] Jayant, N.S. and Noll, P., *Digital Coding of Waveforms*, Prentice Hall, Englewood Cliffs, NJ, 1984.

[8] Lathi, B.P., *Modern Digital and Analog Communication Systems*, 3rd ed., Oxford University Press, New York, NY, 1998.

[9] Miyaoka, S., Digital Audio is Compact and Rugged. *IEEE Spectrum*, 21(3), 35–39, 1984.

[10] Peek, J.B.H., Communication Aspects of the Compact Disk Digital Audio System. *IEEE Comm. Mag.*, 23(2), 7–15, 1985.

[11] Smith, B., Instantaneous Companding of Quantized Signals. *B. S. T. J.*, 36(5), 653–709, 1957.

[12] Spilker, J.J., *Digital Communications by Satellite*, Prentice Hall, Englewood Cliffs, NJ, 1977.

Further Information

Many practical design situations and applications of PCM transmission via twisted-pair T-1 telephone lines, fiber optic cable, microwave relay, and satellite systems are given in [2] and [3].

Baseband Signalling and Pulse Shaping

Michael L. Honig
Northwestern University

Melbourne Barton
Bellcore

Many physical communications channels, such as radio channels, accept a continuous-time waveform as input. Consequently, a sequence of source bits, representing data or a digitized analog signal, must be converted to a continuous-time waveform at the transmitter. In general, each successive group of bits taken from this sequence is mapped to a particular continuous-time pulse. In this chapter we discuss the basic principles involved in selecting such a pulse for channels that can be characterized as linear and time invariant with finite bandwidth.

4.1   Communications System Model

Figure 4.1a shows a simple block diagram of a communications system. The sequence of source bits $\{b_i\}$ are grouped into sequential blocks (vectors) of m bits $\{\boldsymbol{b}_i\}$, and each binary vector $\boldsymbol{b}_i$ is mapped to one of 2^m pulses, $p(\boldsymbol{b}_i; t)$, which is transmitted over the channel.

0-8493-8597-0/99/$0.00+$.50
© 1999 by CRC Press LLC

The transmitted signal as a function of time can be written as

$$s(t) = \sum_i p(\boldsymbol{b}_i; t - iT) \tag{4.1}$$

where $1/T$ is the rate at which each group of m bits, or pulses, is introduced to the channel. The information (bit) rate is therefore m/T.

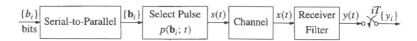

Figure 4.1a   Communication system model. The source bits are grouped into binary vectors, which are mapped to a sequence of pulse shapes.

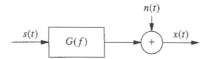

Figure 4.1b   Channel model consisting of a linear, time-invariant system (transfer function) followed by additive noise.

The channel in Fig. 4.1a can be a radio link, which may distort the input signal $s(t)$ in a variety of ways. For example, it may introduce pulse dispersion (due to finite bandwidth) and multipath, as well as additive background noise. The output of the channel is denoted as $x(t)$, which is processed by the receiver to determine estimates of the source bits. The receiver can be quite complicated; however, for the purpose of this discussion, it is sufficient to assume only that it contains a front-end filter and a sampler, as shown in Fig. 4.1a. This assumption is valid for a wide variety of detection strategies. The purpose of the receiver filter is to remove noise outside of the transmitted frequency band and to compensate for the channel frequency response.

A commonly used channel model is shown in Fig. 4.1b and consists of a linear, time-invariant filter, denoted as $G(f)$, followed by additive noise $n(t)$. The channel output is, therefore,

$$x(t) = [g(t) * s(t)] + n(t) \tag{4.2}$$

where $g(t)$ is the channel impulse response associated with $G(f)$, and the asterisk denotes convolution,

$$g(t) * s(t) = \int_{-\infty}^{\infty} g(t - \tau)s(\tau)\,d\tau$$

This channel model accounts for all linear, time-invariant channel impairments, such as finite bandwidth and time-invariant multipath. It does not account for time-varying impairments, such as rapid fading due to time-varying multipath. Nevertheless, this model can be considered valid over short time periods during which the multipath parameters remain constant.

In Figs. 4.1a, and 4.1b, it is assumed that all signals are **baseband signals**, which means that the frequency content is centered around $f = 0$ (DC). The channel passband, therefore, partially coincides with the transmitted spectrum. In general, this condition requires that the transmitted signal be modulated by an appropriate carrier frequency and demodulated at the receiver. In that case, the model in Figs. 4.1a, and 4.1b still applies; however, *baseband-equivalent* signals must be derived from their modulated (passband) counterparts.

Baseband signalling and *pulse shaping* refers to the way in which a group of source bits is mapped to a baseband transmitted pulse.

As a simple example of baseband signalling, we can take $m = 1$ (map each source bit to a pulse), assign a 0 bit to a pulse $p(t)$, and a 1 bit to the pulse $-p(t)$. Perhaps the simplest example of a baseband pulse is the *rectangular* pulse given by $p(t) = 1$, $0 < t \leq T$, and $p(t) = 0$ elsewhere. In this case, we can write the transmitted signal as

$$s(t) = \sum_i A_i p(t - iT) \tag{4.3}$$

where each symbol A_i takes on a value of $+1$ or -1, depending on the value of the ith bit, and $1/T$ is the *symbol rate*, namely, the rate at which the symbols A_i are introduced to the channel.

The preceding example is called *binary* **pulse amplitude modulation (PAM)**, since the data symbols A_i are binary valued, and they amplitude modulate the transmitted pulse $p(t)$. The information rate (bits per second) in this case is the same as the symbol rate $1/T$. As a simple extension of this signalling technique, we can increase m and choose A_i from one of $M = 2^m$ values to transmit at bit rate m/T. This is known as M-ary PAM. For example, letting $m = 2$, each pair of bits can be mapped to a pulse in the set $\{p(t), -p(t), 3p(t), -3p(t)\}$.

In general, the transmitted symbols $\{A_i\}$, the baseband pulse $p(t)$, and channel impulse response $g(t)$ can be *complex valued*. For example, each successive pair of bits might select a symbol from the set $\{1, -1, j, -j\}$, where $j = \sqrt{-1}$. This is a consequence of considering the baseband equivalent of passband modulation. (That is, generating a transmitted spectrum which is centered around a carrier frequency f_c.) Here we are not concerned with the relation between the passband and baseband equivalent models and simply point out that the discussion and results in this chapter apply to complex-valued symbols and pulse shapes.

As an example of a signalling technique which is not PAM, let $m = 1$ and

$$
\begin{aligned}
p(0; t) &= \begin{cases} \sqrt{2}\sin(2\pi f_1 t) & 0 < t < T \\ 0 & \text{elsewhere} \end{cases} \\
p(1; t) &= \begin{cases} \sqrt{2}\sin(2\pi f_2 t) & 0 < t < T \\ 0 & \text{elsewhere} \end{cases}
\end{aligned} \tag{4.4}
$$

where f_1 and $f_2 \neq f_1$ are fixed frequencies selected so that $f_1 T$ and $f_2 T$ (number of cycles for each bit) are multiples of $1/2$. These pulses are *orthogonal*, namely,

$$\int_0^T p(1; t) p(0; t)\, \mathrm{d}t = 0$$

This choice of pulse shapes is called binary **frequency-shift keying (FSK)**.

Another example of a set of orthogonal pulse shapes for $m = 2$ bits/T is shown in Fig. 4.2. Because these pulses may have as many as three transitions within a symbol period, the transmitted spectrum occupies roughly four times the transmitted spectrum of binary PAM with a rectangular pulse shape. The spectrum is, therefore, spread across a much larger band than the smallest required for reliable transmission, assuming a data rate of $2/T$. This type of signalling is referred to as **spread-spectrum**. Spread-spectrum signals are more

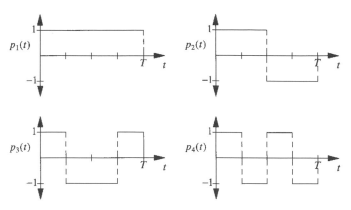

FIGURE 4.2   Four orthogonal spread-spectrum pulse shapes.

robust with respect to interference from other transmitted signals than are narrowband signals.[1]

4.2   Intersymbol Interference and the Nyquist Criterion

Consider the transmission of a PAM signal illustrated in Fig. 4.3. The source bits $\{b_i\}$ are mapped to a sequence of levels $\{A_i\}$, which modulate the transmitter pulse $p(t)$. The channel input is, therefore, given by Eq. (4.3) where $p(t)$ is the impulse response of the transmitter *pulse-shaping filter* $P(f)$ shown in Fig. 4.3. The input to the transmitter filter $P(f)$ is the modulated sequence of delta functions $\sum_i A_i \delta(t - iT)$. The channel is represented by the transfer function $G(f)$ (plus noise), which has impulse response $g(t)$, and the receiver filter has transfer function $R(f)$ with associated impulse response $r(t)$.

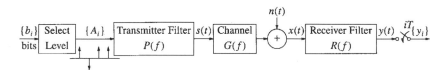

FIGURE 4.3   Baseband model of a pulse amplitude modulation system.

Let $h(t)$ be the overall impulse response of the combined transmitter, channel, and receiver, which has transfer function $H(f) = P(f)G(f)R(f)$. We can write $h(t) = p(t) * g(t) * r(t)$. The output of the receiver filter is then

$$y(t) = \sum_i A_i h(t - iT) + \tilde{n}(t) \tag{4.5}$$

where $\tilde{n}(t) = r(t) * n(t)$ is the output of the filter $R(f)$ with input $n(t)$. Assuming that samples are collected at the output of the filter $R(f)$ at the symbol rate $1/T$, we can write

---

[1] This example can also be viewed as coded binary PAM. Namely, each pair of two source bits are mapped to 4 coded bits, which are transmitted via binary PAM with a rectangular pulse. The current IS-95 air interface uses an extension of this signalling method in which groups of 6 b are mapped to 64 orthogonal pulse shapes with as many as 63 transitions during a symbol.

the kth sample of $y(t)$ as

$$
\begin{aligned}
y(kT) &= \sum_i A_i h(kT - iT) + \tilde{n}(kT) \\
&= A_k h(0) + \sum_{i \neq k} A_i h(kT - iT) + \tilde{n}(kT) \quad (4.6)
\end{aligned}
$$

The first term on the right-hand side of Eq. (4.6) is the kth transmitted symbol scaled by the system impulse response at $t = 0$. If this were the only term on the right side of Eq. (4.6), we could obtain the source bits without error by scaling the received samples by $1/h(0)$. The second term on the right-hand side of Eq. (4.6) is called **intersymbol interference**, which reflects the view that neighboring symbols interfere with the detection of each desired symbol.

One possible criterion for choosing the transmitter and receiver filters is to minimize intersymbol interference. Specifically, if we choose $p(t)$ and $r(t)$ so that

$$
h(kT) = \begin{cases} 1 & k = 0 \\ 0 & k \neq 0 \end{cases} \quad (4.7)
$$

then the kth received sample is

$$
y(kT) = A_k + \tilde{n}(kT) \quad (4.8)
$$

In this case, the intersymbol interference has been eliminated. This choice of $p(t)$ and $r(t)$ is called a **zero-forcing** solution, since it forces the intersymbol interference to zero. Depending on the type of detection scheme used, a zero-forcing solution may not be desirable. This is because the probability of error also depends on the noise intensity, which generally increases when intersymbol interference is suppressed. It is instructive, however, to examine the properties of the zero-forcing solution.

We now view Eq. (4.7) in the frequency domain. Since $h(t)$ has Fourier transform

$$
H(f) = P(f)G(f)R(f) \quad (4.9)
$$

where $P(f)$ is the Fourier transform of $p(t)$, the bandwidth of $H(f)$ is limited by the bandwidth of the channel $G(f)$. We will assume that $G(f) = 0$, $|f| > W$. The sampled impulse response $h(kT)$ can, therefore, be written as the inverse Fourier transform

$$
h(kT) = \int_{-W}^{W} H(f) e^{j2\pi f kT} \, df
$$

Through a series of manipulations, this integral can be rewritten as an inverse discrete Fourier transform,

$$
h(kT) = T \int_{-1/(2T)}^{1/(2T)} H_{eq}\left(e^{j2\pi fT}\right) e^{j2\pi f kT} \, df \quad (4.10a)
$$

where

$$
\begin{aligned}
H_{eq}(e^{j2\pi fT}) &= \frac{1}{T} \sum_k H\left(f + \frac{k}{T}\right) \\
&= \frac{1}{T} \sum_k P\left(f + \frac{k}{T}\right) G\left(f + \frac{k}{T}\right) R\left(f + \frac{k}{T}\right) \quad (4.10b)
\end{aligned}
$$

This relation states that $H_{\text{eq}}(z)$, $z = e^{j2\pi fT}$, is the discrete Fourier transform of the sequence $\{h_k\}$, where $h_k = h(kT)$. Sampling the impulse response $h(t)$ therefore changes the transfer function $H(f)$ to the *aliased* frequency response $H_{\text{eq}}(e^{j2\pi fT})$. From Eqs. (4.10), and (4.6) we conclude that $H_{\text{eq}}(z)$ is the transfer function that relates the sequence of input data symbols $\{A_i\}$ to the sequence of received samples $\{y_i\}$, where $y_i = y(iT)$, in the absence of noise. This is illustrated in Fig. 4.4. For this reason, $H_{\text{eq}}(z)$ is called the **equivalent discrete-time transfer function** for the overall system transfer function $H(f)$.

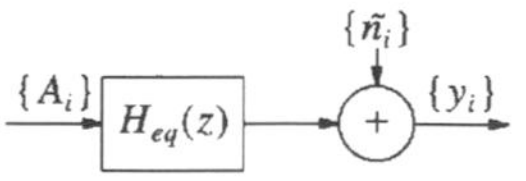

FIGURE 4.4   Equivalent discrete-time channel for the PAM system shown in Fig. 4.3 $[y_i = y(iT), \tilde{n}_i = \tilde{n}(iT)]$

Since $H_{\text{eq}}(e^{j2\pi fT})$ is the discrete Fourier transform of the sequence $\{h_k\}$, the time-domain, or sequence condition (4.7) is equivalent to the frequency-domain condition

$$H_{\text{eq}}\left(e^{j2\pi fT}\right) = 1 \tag{4.11}$$

This relation is called the **Nyquist criterion**. From Eqs. (4.10b) and (4.11) we make the following observations.

1. To satisfy the Nyquist criterion, the channel bandwidth W must be at least $1/(2T)$. Otherwise, $G(f + n/T) = 0$ for f in some interval of positive length for all n, which implies that $H_{\text{eq}}(e^{j2\pi fT}) = 0$ for f in the same interval.

2. For the minimum bandwidth $W = 1/(2T)$, Eqs. (4.10b) and (4.11) imply that $H(f) = T$ for $|f| < 1/(2T)$ and $H(f) = 0$ elsewhere. This implies that the system impulse response is given by

$$h(t) = \frac{\sin(\pi t/T)}{\pi t/T} \tag{4.12}$$

(Since $\int_{-\infty}^{\infty} h^2(t)\,dt = T$, the transmitted signal $s(t) = \sum_i A_i h(t - iT)$ has power equal to the symbol variance $E[|A_i|^2]$.) The impulse response in Eq. (4.12) is called a *minimum bandwidth* or Nyquist pulse. The frequency band $[-1/(2T), 1/(2T)]$ [i.e., the passband of $H(f)$] is called the **Nyquist band**.

3. Suppose that the channel is bandlimited to twice the Nyquist bandwidth. That is, $G(f) = 0$ for $|f| > 1/T$. The condition (4.11) then becomes

$$H(f) + H\left(f - \frac{1}{T}\right) + H\left(f + \frac{1}{T}\right) = T \tag{4.13}$$

Assume for the moment that $H(f)$ and $h(t)$ are both real valued, so that $H(f)$ is an even function of $f[H(f) = H(-f)]$. This is the case when the receiver filter is the matched filter (see Section 4.3). We can then rewrite Eq. (4.13) as

$$H(f) + H\left(\frac{1}{T} - f\right) = T, \quad 0 < f < \frac{1}{2T} \tag{4.14}$$

which states that $H(f)$ must have odd symmetry about $f = 1/(2T)$. This is illustrated in Fig. 4.5, which shows two different transfer functions $H(f)$ that satisfy the Nyquist criterion.

4. The pulse shape $p(t)$ enters into Eq. (4.11) only through the product $P(f)R(f)$. Consequently, either $P(f)$ or $R(f)$ can be fixed, and the other filter can be adjusted or adapted to the particular channel. Typically, the pulse shape $p(t)$ is fixed, and the receiver filter is adapted to the (possibly time-varying) channel.

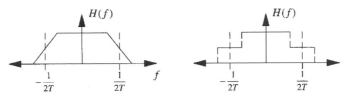

FIGURE 4.5   Two examples of frequency responses that satisfy the Nyquist criterion.

4.2.1   Raised Cosine Pulse

Suppose that the channel is ideal with transfer function

$$G(f) = \begin{cases} 1, & |f| < W \\ 0, & |f| > W \end{cases} \qquad (4.15)$$

To maximize bandwidth efficiency, Nyquist pulses given by Eq. (4.12) should be used where $W = 1/(2T)$. This type of signalling, however, has two major drawbacks. First, Nyquist pulses are noncausal and of infinite duration. They can be approximated in practice by introducing an appropriate delay, and truncating the pulse. The pulse, however, decays very slowly, namely, as $1/t$, so that the truncation window must be wide. This is equivalent to observing that the ideal bandlimited frequency response given by Eq. (4.15) is difficult to approximate closely. The second drawback, which is more important, is the fact that this type of signalling is not robust with respect to sampling jitter. Namely, a small sampling offset ε produces the output sample

$$y(kT + \varepsilon) = \sum_i A_i \frac{\sin[\pi(k - i + \varepsilon/T)]}{\pi(k - i + \varepsilon/T)} \qquad (4.16)$$

Since the Nyquist pulse decays as $1/t$, this sum is not guaranteed to converge. A particular choice of symbols $\{A_i\}$ can, therefore, lead to very large intersymbol interference, no matter how small the offset. Minimum bandwidth signalling is therefore impractical.

The preceding problem is generally solved in one of two ways in practice:

1. The pulse bandwidth is increased to provide a faster pulse decay than $1/t$.

2. A *controlled* amount of intersymbol interference is introduced at the transmitter, which can be subtracted out at the receiver.

The former approach sacrifices bandwidth efficiency, whereas the latter approach sacrifices power efficiency. We will examine the latter approach in Section 4.5. The most common example of a pulse, which illustrates the first technique, is the **raised cosine pulse**, given by

$$h(t) = \left[\frac{\sin(\pi t/T)}{\pi t/T} \right] \left[\frac{\cos(\alpha \pi t/T)}{1 - (2\alpha t/T)^2} \right] \qquad (4.17)$$

which has Fourier transform

$$H(f) = \begin{cases} T & 0 \leq |f| \leq \dfrac{1-\alpha}{2T} \\[2ex] \dfrac{T}{2}\left\{1 + \cos\left[\dfrac{\pi T}{\alpha}\left(|f| - \dfrac{1-\alpha}{2T}\right)\right]\right\} & \dfrac{1-\alpha}{2T} \leq |f| \leq \dfrac{1+\alpha}{2T} \\[2ex] 0 & |f| > \dfrac{1+\alpha}{2T} \end{cases} \qquad (4.18)$$

where $0 \leq \alpha \leq 1$.

Plots of $p(t)$ and $P(f)$ are shown in Figs. 4.6a, and 4.6b for different values of α. It is easily verified that $h(t)$ satisfies the Nyquist criterion (4.7) and, consequently, $H(f)$ satisfies Eq. (4.11). When $\alpha = 0$, $H(f)$ is the Nyquist pulse with minimum bandwidth $1/(2T)$, and when $\alpha > 0, H(f)$ has bandwidth $(1+\alpha)/(2T)$ with a raised cosine rolloff. The parameter α, therefore, represents the additional, or **excess bandwidth** as a fraction of the minimum bandwidth $1/(2T)$. For example, when $\alpha = 1$, we say that the pulse is a raised cosine pulse with 100% excess bandwidth. This is because the pulse bandwidth $1/T$ is twice the minimum bandwidth. Because the raised cosine pulse decays as $1/t^3$, performance is robust with respect to sampling offsets.

The raised cosine frequency response (4.18) applies to the combination of transmitter, channel, and receiver. If the transmitted pulse shape $p(t)$ is a raised cosine pulse, then $h(t)$ is a raised cosine pulse only if the combined receiver and channel frequency response is constant. Even with an ideal (transparent) channel, however, the optimum (matched) receiver filter response is generally not constant in the presence of additive Gaussian noise. An alternative is to transmit the *square-root raised cosine* pulse shape, which has frequency response $P(f)$ given by the square-root of the raised cosine frequency response in Eq. (4.18). Assuming an ideal channel, setting the receiver frequency response $R(f) = P(f)$ then results in an overall raised cosine system response $H(f)$.

4.3   Nyquist Criterion with Matched Filtering

Consider the transmission of an isolated pulse $A_0\delta(t)$. In this case the input to the receiver in Fig. 4.3 is

$$x(t) = A_0\tilde{g}(t) + n(t) \qquad (4.19)$$

where $\tilde{g}(t)$ is the inverse Fourier transform of the combined transmitter-channel transfer function $\tilde{G}(f) = P(f)G(f)$. We will assume that the noise $n(t)$ is white with spectrum $N_0/2$. The output of the receiver filter is then

$$y(t) = r(t) * x(t) = A_0[r(t) * \tilde{g}(t)] + [r(t) * n(t)] \qquad (4.20)$$

The first term on the right-hand side is the desired signal, and the second term is noise. Assuming that $y(t)$ is sampled at $t = 0$, the ratio of signal energy to noise energy, or signal-to-noise ratio (SNR) at the sampling instant, is

$$SNR = \frac{E\left[|A_0|^2\right]\left|\displaystyle\int_{-\infty}^{\infty} r(-t)\tilde{g}(t)\,\mathrm{d}t\right|^2}{\dfrac{N_0}{2}\displaystyle\int_{-\infty}^{\infty}|r(t)|^2\,\mathrm{d}t} \qquad (4.21)$$

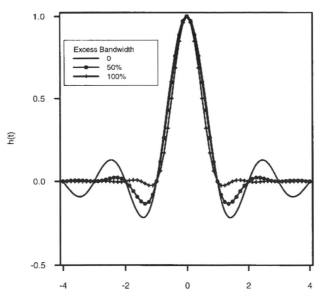

Figure 4.6a   Raised cosine pulse.

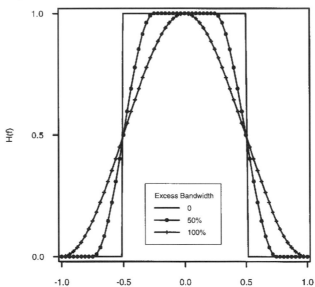

Figure 4.6b   Raised cosine spectrum.

The receiver impulse response that maximizes this expression is $r(t) = \tilde{g}^*(-t)$ [complex conjugate of $\tilde{g}(-t)$], which is known as the **matched filter** impulse response. The associated transfer function is $R(f) = \tilde{G}^*(f)$.

Choosing the receiver filter to be the matched filter is optimal in more general situations, such as when detecting a sequence of channel symbols with intersymbol interference (assuming the additive noise is Gaussian). We, therefore, reconsider the Nyquist criterion when the receiver filter is the matched filter. In this case, the baseband model is shown in Fig. 4.7, and the output of the receiver filter is given by

$$y(t) = \sum_i A_i h(t - iT) + \tilde{n}(t) \tag{4.22}$$

where the baseband pulse $h(t)$ is now the impulse response of the filter with transfer function $|\tilde{G}(f)|^2 = |P(f)G(f)|^2$. This impulse response is the *autocorrelation* of the impulse response of the combined transmitter-channel filter $\tilde{G}(f)$,

$$h(t) = \int_{-\infty}^{\infty} \tilde{g}^*(s)\tilde{g}(s+t)\,\mathrm{d}s \tag{4.23}$$

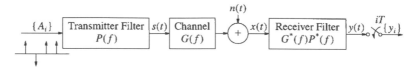

FIGURE 4.7   Baseband PAM model with a matched filter at the receiver.

With a matched filter at the receiver, the equivalent discrete-time transfer function is

$$
\begin{aligned}
H_{\mathrm{eq}}(e^{j2\pi fT}) &= \frac{1}{T}\sum_k \left|\tilde{G}\left(f-\frac{k}{T}\right)\right|^2 \\
&= \frac{1}{T}\sum_k \left|P\left(f-\frac{k}{T}\right)G\left(f-\frac{k}{T}\right)\right|^2
\end{aligned} \tag{4.24}
$$

which relates the sequence of transmitted symbols $\{A_k\}$ to the sequence of received samples $\{y_k\}$ in the absence of noise. Note that $H_{\mathrm{eq}}(e^{j2\pi fT})$ is positive, real valued, and an even function of f. If the channel is bandlimited to twice the Nyquist bandwidth, then $H(f) = 0$ for $|f| > 1/T$, and the Nyquist condition is given by Eq. (4.14) where $H(f) = |G(f)P(f)|^2$. The aliasing sum in Eq. (4.10b) can therefore be described as a folding operation in which the channel response $|H(f)|^2$ is folded around the Nyquist frequency $1/(2T)$. For this reason, $H_{\mathrm{eq}}(e^{j2\pi fT})$ with a matched receiver filter is often referred to as the folded channel spectrum.

4.4   Eye Diagrams

One way to assess the severity of distortion due to intersymbol interference in a digital communications system is to examine the **eye diagram**. The eye diagram is illustrated in Figs. 4.8a and 4.8b, for a raised cosine pulse shape with 25% excess bandwidth and an ideal bandlimited channel. Figure 4.8a shows the data signal at the receiver

$$y(t) = \sum_i A_i h(t - iT) + \tilde{n}(t) \tag{4.25}$$

where $h(t)$ is given by Eq. (4.17), $\alpha = 1/4$, each symbol A_i is independently chosen from the set $\{\pm 1, \pm 3\}$, where each symbol is equally likely, and $\tilde{n}(t)$ is bandlimited white Gaussian noise. (The received *SNR* is 30 dB.) The eye diagram is constructed from the time-domain data signal $y(t)$ as follows (assuming nominal sampling times at $kT, k = 0, 1, 2, \ldots$):

1. Partition the waveform $y(t)$ into successive segments of length T starting from $t = T/2$.
2. Translate each of these waveform segments $[y(t), (k+1/2)T \le t \le (k+3/2)T, k = 0, 1, 2, \ldots]$ to the interval $[-T/2, T/2]$, and superimpose.

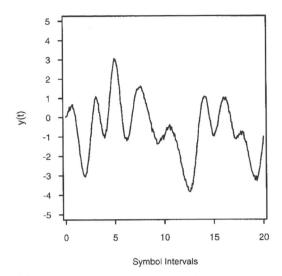

Figure 4.8a   Received signal $y(t)$.

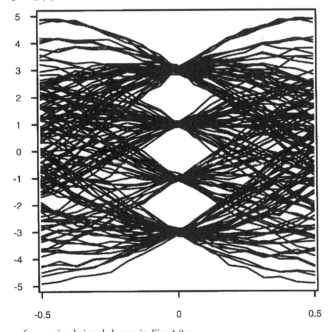

Figure 4.8b   Eye diagram for received signal shown in Fig. 4.8a.

The resulting picture is shown in Fig. 4.8b for the $y(t)$ shown in Fig. 4.8a. (Partitioning $y(t)$ into successive segments of length iT, $i > 1$, is also possible. This would result in i successive eye diagrams.) The number of eye openings is one less than the number of transmitted signal levels. In practice, the eye diagram is easily viewed on an oscilloscope by applying the received waveform $y(t)$ to the vertical deflection plates of the oscilloscope and applying a sawtooth waveform at the symbol rate $1/T$ to the horizontal deflection plates. This causes successive symbol intervals to be translated into one interval on the oscilloscope display.

Each waveform segment $y(t)$, $(k + 1/2)T \leq t \leq (k + 3/2)T$, depends on the particular sequence of channel symbols surrounding A_k. The number of channel symbols that affects a particular waveform segment depends on the extent of the intersymbol interference, shown in Eq. (4.6). This, in turn, depends on the duration of the impulse response $h(t)$. For example, if $h(t)$ has most of its energy in the interval $0 < t < mT$, then each waveform segment depends on approximately m symbols. Assuming binary transmission, this implies that there are a total of 2^m waveform segments that can be superimposed in the eye diagram. (It is possible that only one sequence of channel symbols causes significant intersymbol interference, and this sequence occurs with very low probability.) In current digital wireless applications the impulse response typically spans only a few symbols.

The eye diagram has the following important features which measure the performance of a digital communications system.

4.4.1   Vertical Eye Opening

The vertical openings at any time t_0, $-T/2 \leq t_0 \leq T/2$, represent the separation between signal levels with worst-case intersymbol interference, assuming that $y(t)$ is sampled at times $t = kT + t_0$, $k = 0, 1, 2, \ldots$. It is possible for the intersymbol interference to be large enough so that this vertical opening between some, or all, signal levels disappears altogether. In that case, the eye is said to be closed. Otherwise, the eye is said to be open. A closed eye implies that if the estimated bits are obtained by thresholding the samples $y(kT)$, then the decisions will depend primarily on the intersymbol interference rather than on the desired symbol. The probability of error will, therefore, be close to $1/2$. Conversely, wide vertical spacings between signal levels imply a large degree of immunity to additive noise. In general, $y(t)$ should be sampled at the times $kT + t_0$, $k = 0, 1, 2, \ldots$, where t_0 is chosen to maximize the vertical eye opening.

4.4.2   Horizontal Eye Opening

The width of each opening indicates the sensitivity to timing offset. Specifically, a very narrow eye opening indicates that a small timing offset will result in sampling where the eye is closed. Conversely, a wide horizontal opening indicates that a large timing offset can be tolerated, although the error probability will depend on the vertical opening.

4.4.3   Slope of the Inner Eye

The slope of the inner eye indicates sensitivity to timing jitter or variance in the timing offset. Specifically, a very steep slope means that the eye closes rapidly as the timing offset increases. In this case, a significant amount of jitter in the sampling times significantly increases the probability of error.

The shape of the eye diagram is determined by the pulse shape. In general, the faster the baseband pulse decays, the wider the eye opening. For example, a rectangular pulse produces a box-shaped eye diagram (assuming binary signalling). The minimum bandwidth pulse shape Eq. (4.12) produces an eye diagram which is closed for all t except for $t = 0$. This is because, as shown earlier, an arbitrarily small timing offset can lead to an intersymbol interference term that is arbitrarily large, depending on the data sequence.

4.5   Partial-Response Signalling

To avoid the problems associated with Nyquist signalling over an ideal bandlimited channel, bandwidth and/or power efficiency must be compromised. Raised cosine pulses compromise bandwidth efficiency to gain robustness with respect to timing errors. Another possibility is to introduce a controlled amount of intersymbol interference at the transmitter, which can be removed at the receiver. This approach is called **partial-response (PR) signalling**. The terminology reflects the fact that the sampled system impulse response does not have the full response given by the Nyquist condition Eq. (4.7).

To illustrate PR signalling, suppose that the Nyquist condition Eq. (4.7) is replaced by the condition

$$h_k = \begin{cases} 1 & k = 0, 1 \\ 0 & \text{all other } k \end{cases} \tag{4.26}$$

The kth received sample is then

$$y_k = A_k + A_{k-1} + \tilde{n}_k \tag{4.27}$$

so that there is intersymbol interference from one neighboring transmitted symbol. For now we focus on the spectral characteristics of PR signalling and defer discussion of how to detect the transmitted sequence $\{A_k\}$ in the presence of intersymbol interference. The equivalent discrete-time transfer function in this case is the discrete Fourier transform of the sequence in Eq. (4.26),

$$\begin{aligned} H_{eq}(e^{j2\pi fT}) &= \frac{1}{T} \sum_k H\left(f + \frac{k}{T}\right) \\ &= 1 + e^{-j2\pi fT} = 2e^{-j\pi fT} \cos(\pi fT) \end{aligned} \tag{4.28}$$

As in the full-response case, for Eq. (4.28) to be satisfied, the *minimum* bandwidth of the channel $G(f)$ and transmitter filter $P(f)$ is $W = 1/(2T)$. Assuming $P(f)$ has this minimum bandwidth implies

$$H(f) = \begin{cases} 2Te^{-j\pi fT} \cos(\pi fT) & |f| < 1/(2T) \\ 0 & |f| > 1/(2T) \end{cases} \tag{4.29a}$$

and

$$h(t) = T\{ \ \text{sinc} \ (t/T) + \ \text{sinc} \ [(t-T)/T]\} \tag{4.29b}$$

where sinc $x = (\sin \pi x)/(\pi x)$. This pulse is called a *duobinary* pulse and is shown along with the associated $H(f)$ in Fig. 4.9. [Notice that $h(t)$ satisfies Eq. (4.26).] Unlike the ideal bandlimited frequency response, the transfer function $H(f)$ in Eq. (4.29a) is continuous and is, therefore, easily approximated by a physically realizable filter. Duobinary PR was first proposed by Lender, [7], and later generalized by Kretzmer, [6].

The main advantage of the duobinary pulse Eq. (4.29b), relative to the minimum bandwidth pulse Eq. (4.12), is that signalling at the Nyquist symbol rate is feasible with zero excess bandwidth. Because the pulse decays much more rapidly than a Nyquist pulse, it is robust with respect to timing errors. Selecting the transmitter and receiver filters so that the overall system response is duobinary is appropriate in situations where the channel frequency response $G(f)$ is near zero or has a rapid rolloff at the Nyquist band edge $f = 1/(2T)$.

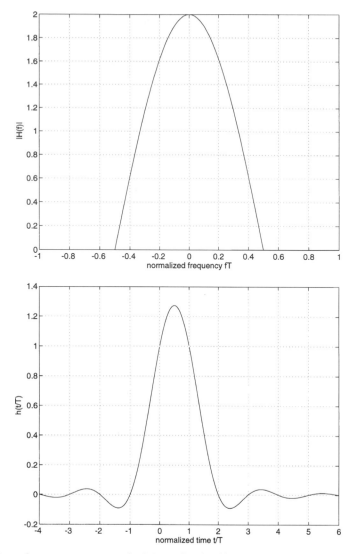

FIGURE 4.9   Duobinary frequency response and minimum bandwidth pulse.

As another example of PR signalling, consider the *modified* duobinary partial response

$$h_k = \begin{cases} 1 & k = -1 \\ -1 & k = 1 \\ 0 & \text{all other } k \end{cases} \tag{4.30}$$

which has equivalent discrete-time transfer function

$$\begin{aligned} H_{\text{eq}}\left(e^{j2\pi fT}\right) &= e^{j2\pi fT} - e^{-j2\pi fT} \\ &= j2\sin(2\pi fT) \end{aligned} \tag{4.31}$$

With zero excess bandwidth, the overall system response is

$$H(f) = \begin{cases} j2T\sin(2\pi fT) & |f| < 1/(2T) \\ 0 & |f| > 1/(2T) \end{cases} \tag{4.32a}$$

and

$$h(t) = T\{\text{sinc } [(t+T)/T] - \text{sinc } [(t-T)/T]\} \tag{4.32b}$$

These functions are plotted in Fig. 4.10. This pulse shape is appropriate when the channel response $G(f)$ is near zero at both DC ($f = 0$) and at the Nyquist band edge. This is often the case for wire (twisted-pair) channels where the transmitted signal is coupled to the channel through a transformer. Like duobinary PR, modified duobinary allows minimum bandwidth signalling at the Nyquist rate.

A particular partial response is often identified by the polynomial

$$\sum_{k=0}^{K} h_k D^k$$

where D (for delay) takes the place of the usual z^{-1} in the z transform of the sequence $\{h_k\}$. For example, duobinary is also referred to as $1 + D$ partial response.

In general, more complicated system responses than those shown in Figs. 4.9 and 4.10 can be generated by choosing more nonzero coefficients in the sequence $\{h_k\}$. This complicates detection, however, because of the additional intersymbol interference that is generated.

Rather than modulating a PR pulse $h(t)$, a PR signal can also be generated by filtering the sequence of transmitted levels $\{A_i\}$. This is shown in Fig. 4.11. Namely, the transmitted levels are first passed through a discrete-time (digital) filter with transfer function $P_d(e^{j2\pi fT})$ (where the subscript d indicates discrete). [Note that $P_d(e^{j2\pi fT})$ can be selected to be $H_{\text{eq}}(e^{j2\pi fT})$.] The outputs of this filter form the PAM signal, where the pulse shaping filter $P(f) = 1$, $|f| < 1/(2T)$ and is zero elsewhere. If the transmitted levels $\{A_k\}$ are selected independently and are identically distributed, then the transmitted spectrum is $\sigma_A^2 |P_d(e^{j2\pi fT})|^2$ for $|f| < 1/(2T)$ and is zero for $|f| > 1/(2T)$, where $\sigma_A^2 = E[|A_k|^2]$.

Shaping the transmitted spectrum to have nulls coincident with nulls in the channel response potentially offers significant performance advantages. By introducing intersymbol interference, however, PR signalling increases the number of received signal levels, which increases the complexity of the detector and may reduce immunity to noise. For example, the set of received signal levels for duobinary signalling is $\{0, \pm 2\}$ from which the transmitted levels $\{\pm 1\}$ must be estimated. The performance of a particular PR scheme depends on the channel characteristics, as well as the type of detector used at the receiver. We now describe a simple suboptimal detection strategy.

4.5.1 Precoding

Consider the received signal sample Eq. (4.27) with duobinary signalling. If the receiver has correctly decoded the symbol A_{k-1}, then in the absence of noise A_k can be decoded by subtracting A_{k-1} from the received sample y_k. If an error occurs, however, then subtracting the preceding symbol estimate from the received sample will cause the error to propagate to successive detected symbols. To avoid this problem, the transmitted levels can be **precoded** in such a way as to compensate for the intersymbol interference introduced by the overall partial response.

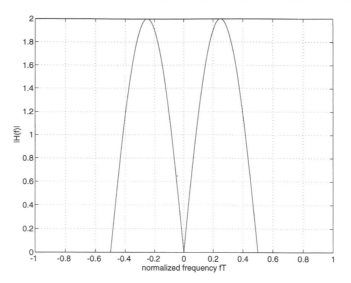

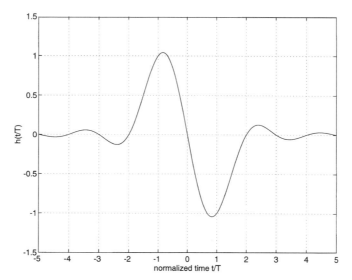

FIGURE 4.10   Modified duobinary frequency response and minimum bandwidth pulse.

FIGURE 4.11   Generation of PR signal.

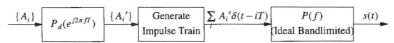

FIGURE 4.12   Precoding for a PR channel.

We first illustrate precoding for duobinary PR. The sequence of operations is illustrated in Fig. 4.12. Let $\{b_k\}$ denote the sequence of source bits where $b_k \in \{0, 1\}$. This sequence is transformed to the sequence $\{b'_k\}$ by the operation

$$b'_k = b_k \oplus b'_{k-1} \tag{4.33}$$

TABLE 4.1   Example of Precoding for Duobinary PR.

$\{b_i\}$:		1	0	0	1	1	1	0	0	1	0
$\{b_i'\}$:	0	1	1	1	0	1	0	0	0	1	1
$\{A_i\}$:	-1	1	1	1	-1	1	-1	-1	-1	1	1
$\{y_i\}$:		0	2	2	0	0	0	-2	-2	0	2

where $\oplus$ denotes modulo 2 addition (exclusive OR). The sequence $\{b_k'\}$ is mapped to the sequence of binary transmitted signal levels $\{A_k\}$ according to

$$A_k = 2b_k' - 1 \tag{4.34}$$

That is, $b_k' = 0$ ($b_k' = 1$) is mapped to the transmitted level $A_k = -1$ ($A_k = 1$). In the absence of noise, the received symbol is then

$$y_k = A_k + A_{k-1} = 2\left(b_k' + b_{k-1}' - 1\right) \tag{4.35}$$

and combining Eqs. (4.33) and (4.35) gives

$$b_k = \left(\frac{1}{2}y_k + 1\right) \bmod 2 \tag{4.36}$$

That is, if $y_k = \pm 2$, then $b_k = 0$, and if $y_k = 0$, then $b_k = 1$. Precoding, therefore, enables the detector to make *symbol-by-symbol* decisions that do not depend on previous decisions. Table 4.1 shows a sequence of transmitted bits $\{b_i\}$, precoded bits $\{b_i'\}$, transmitted signal levels $\{A_i\}$, and received samples $\{y_i\}$.

The preceding precoding technique can be extended to multilevel PAM and to other PR channels. Suppose that the PR is specified by

$$H_{\text{eq}}(D) = \sum_{k-0}^{K} h_k D^k$$

where the coefficients are integers and that the source symbols $\{b_k\}$ are selected from the set $\{0, 1, \ldots, M-1\}$. These symbols are transformed to the sequence $\{b_k'\}$ via the precoding operation

$$b_k' = \left(b_k - \sum_{i=1}^{K} h_i b_{k-i}'\right) \bmod M \tag{4.37}$$

Because of the modulo operation, each symbol b_k' is also in the set $\{0, 1, \ldots, M-1\}$. The kth transmitted signal level is given by

$$A_k = 2b_k' - (M-1) \tag{4.38}$$

so that the set of transmitted levels is $\{-(M-1), \ldots, (M-1)\}$ (i.e., a shifted version of the set of values assumed by b_k). In the absence of noise the received sample is

$$y_k = \sum_{i=0}^{K} h_i A_{k-i} \tag{4.39}$$

and it can be shown that the kth source symbol is given by

$$b_k = \frac{1}{2}\left(y_k + (M-1) \cdot H_{\text{eq}}(1)\right) \bmod M \tag{4.40}$$

Precoding the symbols $\{b_k\}$ in this manner, therefore, enables symbol-by-symbol decisions at the receiver. In the presence of noise, more sophisticated detection schemes (e.g., maximum likelihood) can be used with PR signalling to obtain improvements in performance.

4.6 Additional Considerations

In many applications, bandwidth and intersymbol interference are not the only important considerations for selecting baseband pulses. Here we give a brief discussion of additional practical constraints that may influence this selection.

4.6.1 Average Transmitted Power and Spectral Constraints

The constraint on average transmitted power varies according to the application. For example, low-average power is highly desirable for mobile wireless applications that use battery-powered transmitters. In many applications (e.g., digital subscriber loops, as well as digital radio), constraints are imposed to limit the amount of interference, or crosstalk, radiated into neighboring receivers and communications systems. Because this type of interference is frequency dependent, the constraint may take the form of a spectral mask that specifies the maximum allowable transmitted power as a function of frequency. For example, crosstalk in wireline channels is generally caused by capacitive coupling and increases as a function of frequency. Consequently, to reduce the amount of crosstalk generated at a particular transmitter, the pulse shaping filter generally attenuates high frequencies more than low frequencies.

In radio applications where signals are assigned different frequency bands, constraints on the transmitted spectrum are imposed to limit *adjacent-channel interference*. This interference is generated by transmitters assigned to adjacent frequency bands. Therefore, a constraint is needed to limit the amount of *out-of-band power* generated by each transmitter, in addition to an overall average power constraint. To meet this constraint, the transmitter filter in Fig. 4.3 must have a sufficiently steep rolloff at the edges of the assigned frequency band. (Conversely, if the transmitted signals are time multiplexed, then the duration of the system impulse response must be contained within the assigned time slot.)

4.6.2 Peak-to-Average Power

In addition to a constraint on average transmitted power, a *peak-power* constraint is often imposed as well. This constraint is important in practice for the following reasons:

1. The dynamic range of the transmitter is limited. In particular, saturation of the output amplifier will "clip" the transmitted waveform.
2. Rapid fades can severely distort signals with high peak-to-average power.
3. The transmitted signal may be subjected to nonlinearities. Saturation of the output amplifier is one example. Another example that pertains to wireline applications is the companding process in the voice telephone network [5]. Namely, the compander used to reduce quantization noise for pulse-code modulated voice signals introduces amplitude-dependent distortion in data signals.

The preceding impairments or constraints indicate that the transmitted waveform should have a low **peak-to-average power ratio (PAR)**. For a transmitted waveform $x(t)$, the PAR is defined as

$$\text{PAR} = \frac{\max |x(t)|^2}{E\left\{|x(t)|^2\right\}}$$

where $E(\cdot)$ denotes expectation. Using binary signalling with rectangular pulse shapes minimizes the PAR. However, this compromises bandwidth efficiency. In applications where PAR should be low, binary signalling with rounded pulses are often used. Operating RF power amplifiers with power back-off can also reduce PAR, but leads to inefficient amplification.

For an **orthogonal frequency division multiplexing (OFDM)** system, it is well known that the transmitted signal can exhibit a very high PAR compared to an equivalent single-carrier system. Hence more sophisticated approaches to PAR reduction are required for OFDM. Some proposed approaches are described in [8] and references therein. These include altering the set of transmitted symbols and setting aside certain OFDM tones specifically to minimize PAR.

4.6.3 Channel and Receiver Characteristics

The type of channel impairments encountered and the type of detection scheme used at the receiver can also influence the choice of a transmitted pulse shape. For example, a constant amplitude pulse is appropriate for a fast fading environment with noncoherent detection. The ability to track channel characteristics, such as phase, may allow more bandwidth efficient pulse shapes in addition to multilevel signalling.

High-speed data communications over time-varying channels requires that the transmitter and/or receiver adapt to the changing channel characteristics. Adapting the transmitter to compensate for a time-varying channel requires a feedback channel through which the receiver can notify the transmitter of changes in channel characteristics. Because of this extra complication, adapting the receiver is often preferred to adapting the transmitter pulse shape. However, the following examples are notable exceptions.

1. The current IS-95 air interface for direct-sequence code-division multiple access adapts the transmitter power to control the amount of interference generated and to compensate for channel fades. This can be viewed as a simple form of adaptive transmitter pulse shaping in which a single parameter associated with the pulse shape is varied.

2. Multitone modulation divides the channel bandwidth into small subbands, and the transmitted power and source bits are distributed among these subbands to maximize the information rate. The received signal-to-noise ratio for each subband must be transmitted back to the transmitter to guide the allocation of transmitted bits and power [1].

In addition to multitone modulation, *adaptive precoding* (also known as Tomlinson–Harashima precoding [4, 11]) is another way in which the transmitter can adapt to the channel frequency response. Adaptive precoding is an extension of the technique described earlier for partial-response channels. Namely, the equivalent discrete-time channel impulse response is measured at the receiver and sent back to the transmitter, where it is used in a precoder. The precoder compensates for the intersymbol interference introduced by the channel, allowing the receiver to detect the data by a simple threshhold operation. Both multitone modulation and precoding have been used with wireline channels (voiceband modems and digital subscriber loops).

4.6.4 Complexity

Generation of a bandwidth-efficient signal requires a filter with a sharp cutoff. In addition, bandwidth-efficient pulse shapes can complicate other system functions, such as timing and carrier recovery. If sufficient bandwidth is available, the cost can be reduced by using a rectangular pulse shape with a simple detection strategy (low-pass filter and threshold).

4.6.5   Tolerance to Interference

Interference is one of the primary channel impairments associated with digital radio. In addition to adjacent-channel interference described earlier, *cochannel interference* may be generated by other transmitters assigned to the same frequency band as the desired signal. Cochannel interference can be controlled through frequency (and perhaps time slot) assignments and by pulse shaping. For example, assuming fixed average power, increasing the bandwidth occupied by the signal lowers the power spectral density and decreases the amount of interference into a narrowband system that occupies part of the available bandwidth. Sufficient bandwidth spreading, therefore, enables wideband signals to be overlaid on top of narrowband signals without disrupting either service.

4.6.6   Probability of Intercept and Detection

The broadcast nature of wireless channels generally makes eavesdropping easier than for wired channels. A requirement for most commercial, as well as military applications, is to guarantee the privacy of user conversations (low probability of intercept). An additional requirement, in some applications, is that determining whether or not communications is taking place must be difficult (low probability of detection). Spread-spectrum waveforms are attractive in these applications since spreading the pulse energy over a wide frequency band decreases the power spectral density and, hence, makes the signal less visible. Power-efficient modulation combined with coding enables a further reduction in transmitted power for a target error rate.

4.7   Examples

We conclude this chapter with a brief description of baseband pulse shapes used in existing and emerging standards for digital mobile cellular and Personal Communications Services (PCS).

4.7.1   Global System for Mobile Communications (GSM)

The European GSM standard for digital mobile cellular communications operates in the 900-MHz frequency band, and is based on time-division multiple access (TDMA) [9]. The U.S. version operates at 1900 MHz, and is called PCS-1900. A special variant of binary FSK is used called *Gaussian minimum-shift keying (GMSK)*. The GMSK modulator is illustrated in Fig. 4.13. The input to the modulator is a binary PAM signal $s(t)$, given by Eq. (4.3), where the pulse $p(t)$ is a Gaussian function and $|s(t)| < 1$. This waveform frequency modulates the carrier f_c, so that the (passband) transmitted signal is

$$ w(t) = K \cos \left[2\pi f_c t + 2\pi f_d \int_{-\infty}^{t} s(\tau) \, \mathrm{d}\tau \right] $$

The maximum frequency deviation from the carrier is $f_d = 1/(2T)$, which characterizes minimum-shift keying. This technique can be used with a noncoherent receiver that is easy to implement. Because the transmitted signal has a constant envelope, the data can be reliably detected in the presence of rapid fades that are characteristic of mobile radio channels.

FIGURE 4.13   Generation of GMSK signal; LPF is low-pass filter.

4.7.2   U.S. Digital Cellular (IS-136)

The IS-136 air interface (formerly IS-54) operates in the 800 MHz band and is based on TDMA [3]. There is also a 1900 MHz version of IS-136. The baseband signal is given by Eq. (4.3) where the symbols are complex-valued, corresponding to quadrature phase modulation. The pulse has a square-root raised cosine spectrum with 35% excess bandwidth.

4.7.3   Interim Standard-95

The IS-95 air interface for digital mobile cellular uses spread-spectrum signalling (CDMA) in the 800-MHz band [10]. There is also a 1900 MHz version of IS-95. The baseband transmitted pulse shapes are analogous to those shown in Fig. 4.2, where the number of square pulses (chips) per bit is 128. To improve spectral efficiency the (wideband) transmitted signal is filtered by an approximation to an ideal low-pass response with a small amount of excess bandwidth. This shapes the chips so that they resemble minimum bandwidth pulses.

4.7.4   Personal Access Communications System (PACS)

Both PACS and the Japanese personal handy phone (PHP) system are TDMA systems which have been proposed for personal communications systems (PCS), and operate near 2 GHz [2]. The baseband signal is given by Eq. (4.3) with four complex symbols representing four-phase quadrature modulation. The baseband pulse has a square-root raised cosine spectrum with 50% excess bandwidth.

Defining Terms

Baseband signal: A signal with frequency content centered around DC.

Equivalent discrete-time transfer function: A discrete-time transfer function (z transform) that relates the transmitted amplitudes to received samples in the absence of noise.

Excess bandwidth: That part of the baseband transmitted spectrum which is not contained within the Nyquist band.

Eye diagram: Superposition of segments of a received PAM signal that indicates the amount of intersymbol interference present.

Frequency-shift keying: A digital modulation technique in which the transmitted pulse is sinusoidal, where the frequency is determined by the source bits.

Intersymbol interference: The additive contribution (interference) to a received sample from transmitted symbols other than the symbol to be detected.

Matched filter: The receiver filter with impulse response equal to the time-reversed, complex conjugate impulse response of the combined transmitter filter-channel impulse response.

Nyquist band: The narrowest frequency band that can support a PAM signal without intersymbol interference (the interval $[-1/(2T), 1/(2T)]$ where $1/T$ is the symbol rate).

Nyquist criterion: A condition on the overall frequency response of a PAM system that ensures the absence of intersymbol interference.

Orthogonal frequency division multiplexing (OFDM): Modulation technique in which the transmitted signal is the sum of low-bit-rate narrowband digital signals modulated on orthogonal carriers.

Partial-response signalling: A signalling technique in which a controlled amount of intersymbol interference is introduced at the transmitter in order to shape the transmitted spectrum.

Precoding: A transformation of source symbols at the transmitter that compensates for intersymbol interference introduced by the channel.

Pulse amplitude modulation (PAM): A digital modulation technique in which the source bits are mapped to a sequence of amplitudes that modulate a transmitted pulse.

Raised cosine pulse: A pulse shape with Fourier transform that decays to zero according to a raised cosine; see Eq. (4.18). The amount of excess bandwidth is conveniently determined by a single parameter (α).

Spread spectrum: A signalling technique in which the pulse bandwidth is many times wider than the Nyquist bandwidth.

Zero-forcing criterion: A design constraint which specifies that intersymbol interference be eliminated.

References

[1] Bingham, J.A.C., Multicarrier modulation for data transmission: an idea whose time has come. *IEEE Commun. Mag.*, 28(May), 5–14, 1990.

[2] Cox, D.C., Wireless personal communications: what is it? *IEEE Personal Comm.*, 2(2), 20–35, 1995.

[3] Electronic Industries Association/Telecommunications Industry Association. Recommended minimum performance standards for 800 MHz dual-mode mobile stations. Incorp. EIA/TIA 19B, EIA/TIA Project No. 2216, Mar.,1991

[4] Harashima, H. and Miyakawa, H., Matched-transmission technique for channels with intersymbol interference. *IEEE Trans. on Commun.*, COM-20(Aug.), 774–780, 1972.

[5] Kalet, I. and Saltzberg, B.R., QAM transmission through a companding channel—signal constellations and detection. *IEEE Trans. on Comm.*, 42(2–4), 417–429, 1994.

[6] Kretzmer, E.R., Generalization of a technique for binary data communication. *IEEE Trans. Comm. Tech.*, COM-14 (Feb.), 67, 68, 1966.

[7] Lender, A., The duobinary technique for high-speed data Transmission. *AIEE Trans. on Comm. Electronics,* 82 (March), 214–218, 1963.

[8] Muller, S.H. and Huber, J.B., A comparison of peak power reduction schemes for OFDM. *Proc. GLOBECOM '97*, (Mon.), 1–5, 1997.

[9] Rahnema, M., Overview of the GSM system and protocol architecture. *IEEE Commun. Mag.*, (April), 92–100, 1993.

[10] Telecommunication Industry Association. Mobile station-base station compatibility standard for dual-mode wideband spread spectrum cellular system. TIA/EIA/IS-95-A. May, 1995.

[11] Tomlinson, M., New automatic equalizer employing modulo arithmetic. *Electron. Lett.*, 7 (March), 138, 139, 1971.

Further Information

Baseband signalling and pulse shaping is fundamental to the design of any digital communications system and is, therefore, covered in numerous texts on digital communications. For more advanced treatments see E.A. Lee and D.G. Messerschmitt, *Digital Communication*, Kluwer 1994, and J.G. Proakis, *Digital Communications*, McGraw-Hill 1995.

5

Channel Equalization

John G. Proakis
Northeastern University

5.1 Characterization of Channel Distortion

Many communication channels, including telephone channels, and some radio channels, may be generally characterized as band-limited linear filters. Consequently, such channels are described by their frequency response $C(f)$, which may be expressed as

$$C(f) - A(f)e^{j\theta(f)} \tag{5.1}$$

where $A(f)$ is called the *amplitude response* and $\theta(f)$ is called the *phase response*. Another characteristic that is sometimes used in place of the phase response is the *envelope delay* or *group delay,* which is defined as

$$\tau(f) = -\frac{1}{2\pi}\frac{\mathrm{d}\theta(f)}{\mathrm{d}f} \tag{5.2}$$

A channel is said to be nondistorting or ideal if, within the bandwidth W occupied by the transmitted signal, $A(f) = \text{const}$ and $\theta(f)$ is a linear function of frequency [or the envelope delay $\tau(f) = \text{const}$]. On the other hand, if $A(f)$ and $\tau(f)$ are not constant within the bandwidth occupied by the transmitted signal, the channel distorts the signal. If $A(f)$ is not constant, the distortion is called *amplitude distortion* and if $\tau(f)$ is not constant, the distortion on the transmitted signal is called *delay distortion*.

As a result of the amplitude and delay distortion caused by the nonideal channel frequency response characteristic $C(f)$, a succession of pulses transmitted through the channel at rates comparable to the bandwidth W are smeared to the point that they are no longer distinguishable as well-defined pulses at the receiving terminal. Instead, they overlap and, thus, we have **intersymbol interference (ISI)**. As an example of the effect of delay distortion on a transmitted pulse, Fig. 5.1(a) illustrates a band-limited pulse having zeros periodically spaced in time at points labeled $\pm T$, $\pm 2T$, etc. If information is conveyed by the pulse amplitude, as in pulse amplitude modulation (PAM), for example, then one can transmit a sequence of pulses, each of which has a peak at the periodic zeros of the other pulses.   Transmission of the pulse through a channel modeled as having a linear envelope delay characteristic $\tau(f)$ [quadratic phase $\theta(f)$], however, results in the received pulse shown in Fig. 5.1(b) having zero crossings that are no longer periodically spaced. Consequently a sequence of successive pulses would be smeared into one another, and the peaks of the pulses would no longer be distinguishable. Thus, the channel delay distortion results in intersymbol interference. As will be discussed in this chapter, it is possible to compensate for the nonideal frequency response characteristic of the channel by use of a filter or equalizer at the demodulator. Figure 5.1(c) illustrates the output of a linear equalizer that compensates for the linear distortion in the channel.

The extent of the intersymbol interference on a telephone channel can be appreciated by observing a frequency response characteristic of the channel. Figure 5.2 illustrates the measured average amplitude and delay as a function of frequency for a medium-range (180– 725 mi) telephone channel of the switched telecommunications network as given by Duffy and Tratcher, 1971. We observe that the usable band of the channel extends from about 300 Hz to about 3000 Hz. The corresponding impulse response of the average channel is shown in Fig. 5.3. Its duration is about 10 ms. In comparison, the transmitted symbol rates on such a channel may be of the order of 2500 pulses or symbols per second. Hence, intersymbol interference might extend over 20–30 symbols.

Besides telephone channels, there are other physical channels that exhibit some form of time dispersion and, thus, introduce intersymbol interference. Radio channels, such as short-wave ionospheric propagation (HF), tropospheric scatter, and mobile cellular radio are three examples of time-dispersive wireless channels. In these channels, time dispersion and, hence, intersymbol interference is the result of multiple propagation paths with different path delays. The number of paths and the relative time delays among the paths vary with time and, for this reason, these radio channels are usually called time-variant multipath channels. The time-variant multipath conditions give rise to a wide variety of frequency response characteristics. Consequently, the frequency response characterization that is used for telephone channels is inappropriate for time-variant multipath channels. Instead, these radio channels are characterized statistically in terms of the scattering function, which, in brief, is a two-dimensional representation of the average received signal power as a function of relative time delay and Doppler frequency (see Proakis [4]).

For illustrative purposes, a scattering function measured on a medium-range (150 mi) tropospheric scatter channel is shown in Fig. 5.4. The total time duration (multipath spread) of the channel response is approximately 0.7 μs on the average, and the spread between half-power points in Doppler frequency is a little less than 1 Hz on the strongest path and somewhat larger on the other paths. Typically, if one is transmitting at a rate of 10^7 symbols/s over such a channel, the multipath spread of 0.7 μs will result in intersymbol interference that spans about seven symbols.

5.2   Characterization of Intersymbol Interference

In a digital communication system, channel distortion causes intersymbol interference, as illustrated in the preceding section. In this section, we shall present a model that characterizes the ISI. The digital modulation methods to which this treatment applies are PAM, phase-shift keying (PSK) and quadrature amplitude modulation (QAM). The transmitted

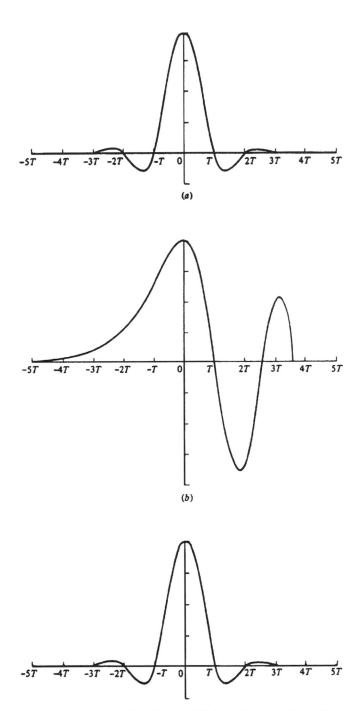

FIGURE 5.1   Effect of channel distortion: (a) channel input, (b) channel output, (c) equalizer output.

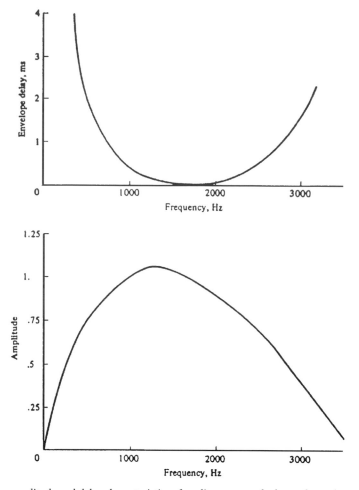

FIGURE 5.2 Average amplitude and delay characteristics of medium-range telephone channel.

signal for these three types of modulation may be expressed as

$$
\begin{aligned}
s(t) &= v_c(t)\cos 2\pi f_c t - v_s(t)\sin 2\pi f_c t \\
&= \operatorname{Re}\left[v(t)\,e^{j2\pi f_c t}\right]
\end{aligned}
\tag{5.3}
$$

where $v(t) = v_c(t) + jv_s(t)$ is called the *equivalent low-pass signal*, f_c is the carrier frequency, and Re[ ] denotes the real part of the quantity in brackets.

In general, the equivalent low-pass signal is expressed as

$$
v(t) = \sum_{n=0}^{\infty} I_n\, g_T(t - nT)
\tag{5.4}
$$

where $g_T(t)$ is the basic pulse shape that is selected to control the spectral characteristics of the transmitted signal, $\{I_n\}$ the sequence of transmitted information symbols selected from a signal constellation consisting of M points, and T the signal interval ($1/T$ is the symbol rate). For PAM, PSK, and QAM, the values of I_n are points from M-ary signal constellations. Figure 5.5 illustrates the signal constellations for the case of $M = 8$ signal points. Note that for PAM, the signal constellation is one dimensional. Hence, the equivalent low-pass signal $v(t)$ is real valued, i.e., $v_s(t) = 0$ and $v_c(t) = v(t)$. For M-ary ($M > 2$)

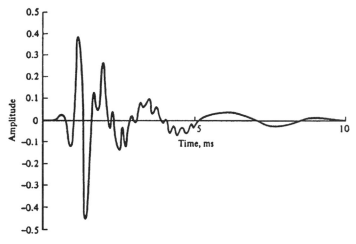

FIGURE 5.3   Impulse response of average channel with amplitude and delay shown in Fig.5.2.

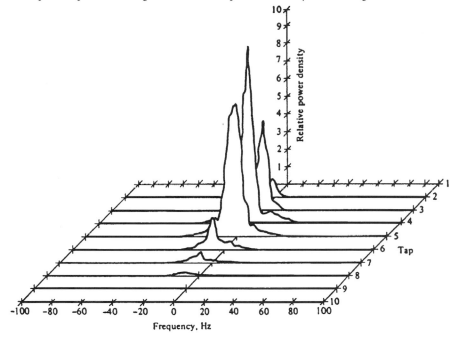

FIGURE 5.4   Scattering function of a medium-range tropospheric scatter channel.

PSK and QAM, the signal constellations are two dimensional and, hence, $v(t)$ is complex valued.

The signal $s(t)$ is transmitted over a bandpass channel that may be characterized by an equivalent low-pass frequency response $C(f)$. Consequently, the equivalent low-pass received signal can be represented as

$$r(t) = \sum_{n=0}^{\infty} I_n\, h(t - nT) + w(t) \tag{5.5}$$

where $h(t) = g_T(t) * c(t)$, and $c(t)$ is the impulse response of the equivalent low-pass channel, the asterisk denotes convolution, and $w(t)$ represents the additive noise in the channel.

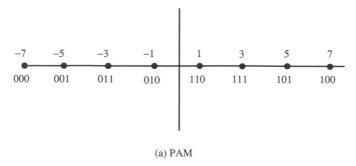

(a) PAM

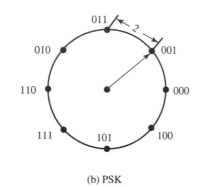

(b) PSK

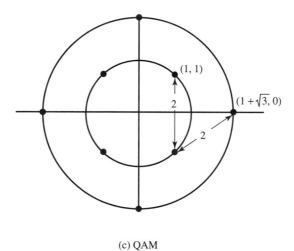

(c) QAM

FIGURE 5.5 $M = 8$ signal constellations for PAM, PSK, and QAM.

To characterize the ISI, suppose that the received signal is passed through a receiving filter and then sampled at the rate $1/T$ samples/s. In general, the optimum filter at the receiver is matched to the received signal pulse $h(t)$. Hence, the frequency response of this filter is $H^*(f)$. We denote its output as

$$y(t) = \sum_{n=0}^{\infty} I_n\, x(t - nT) + \nu(t) \tag{5.6}$$

where $x(t)$ is the signal pulse response of the receiving filter, i.e., $X(f) = H(f)H^*(f) = |H(f)|^2$, and $\nu(t)$ is the response of the receiving filter to the noise $w(t)$. Now, if $y(t)$ is sampled at times $t = kT$, $k = 0, 1, 2, \ldots$, we have

$$
\begin{aligned}
y(kT) \equiv y_k &= \sum_{n=0}^{\infty} I_n x(kT - nT) + \nu(kT) \\
&= \sum_{n=0}^{\infty} I_n x_{k-n} + \nu_k, \qquad k = 0, 1, \ldots
\end{aligned}
$$
(5.7)

The sample values $\{y_k\}$ can be expressed as

$$
y_k = x_0 \left(I_k + \frac{1}{x_0} \sum_{\substack{n=0 \\ n \neq k}}^{\infty} I_n x_{k-n} \right) + \nu_k, \qquad k = 0, 1, \ldots
$$
(5.8)

The term x_0 is an arbitrary scale factor, which we arbitrarily set equal to unity for convenience. Then

$$
y_k = I_k + \sum_{\substack{n=0 \\ n \neq k}}^{\infty} I_n x_{k-n} + \nu_k
$$
(5.9)

The term I_k represents the desired information symbol at the kth sampling instant, the term

$$
\sum_{\substack{n=0 \\ n \neq k}}^{\infty} I_n x_{k-n}
$$
(5.10)

represents the ISI, and ν_k is the additive noise variable at the kth sampling instant.

The amount of ISI, and noise in a digital communications system can be viewed on an oscilloscope. For PAM signals, we can display the received signal $y(t)$ on the vertical input with the horizontal sweep rate set at $1/T$. The resulting oscilloscope display is called an *eye pattern* because of its resemblance to the human eye. For example, Fig. 5.6 illustrates the eye patterns for binary and four-level PAM modulation. The effect of ISI is to cause the eye to close, thereby reducing the margin for additive noise to cause errors. Figure 5.7 graphically illustrates the effect of ISI in reducing the opening of a binary eye. Note that intersymbol interference distorts the position of the zero crossings and causes a reduction in the eye opening. Thus, it causes the system to be more sensitive to a synchronization error.

For PSK and QAM it is customary to display the eye pattern as a two-dimensional scatter diagram illustrating the sampled values $\{y_k\}$ that represent the decision variables at the sampling instants. Figure 5.8 illustrates such an eye pattern for an 8-PSK signal. In the absence of intersymbol interference and noise, the superimposed signals at the sampling instants would result in eight distinct points corresponding to the eight transmitted signal phases. Intersymbol interference and noise result in a deviation of the received samples $\{y_k\}$ from the desired 8-PSK signal. The larger the intersymbol interference and noise, the larger the scattering of the received signal samples relative to the transmitted signal points.

In practice, the transmitter and receiver filters are designed for zero ISI at the desired sampling times $t = kT$. Thus, if $G_T(f)$ is the frequency response of the transmitter filter and $G_R(f)$ is the frequency response of the receiver filter, then the product $G_T(f)\,G_R(f)$ is designed to yield zero ISI. For example, the product $G_T(f)\,G_R(f)$ may be selected as

$$
G_T(f)G_R(f) = X_{rc}(f)
$$
(5.11)

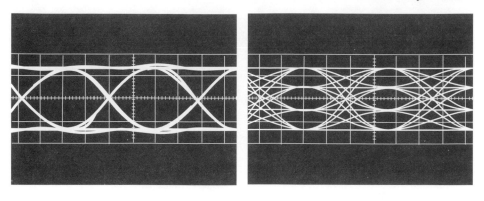

BINARY   QUATERNARY

FIGURE 5.6   Examples of eye patterns for binary and quaternary amplitude shift keying (or PAM).

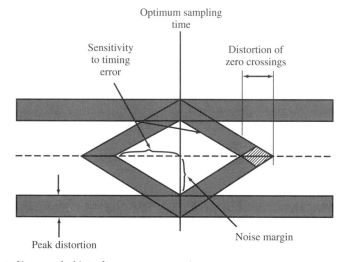

FIGURE 5.7   Effect of intersymbol interference on eye opening.

where $X_{rc}(f)$ is the raised-cosine frequency response characteristic, defined as

$$X_{rc}(f) = \begin{cases} T, & 0 \leq |f| \leq (1-\alpha)/2T \\ \dfrac{T}{2}\left[1 + \cos\dfrac{\pi T}{\alpha}\left(|f| - \dfrac{1-\alpha}{2T}\right), & \dfrac{1-\alpha}{2T} \leq |f| \leq \dfrac{1+\alpha}{2T} \\ 0, & |f| > \dfrac{1+\alpha}{2T} \end{cases} \qquad (5.12)$$

where α is called the *rolloff* factor, which takes values in the range $0 \leq \alpha \leq 1$, and $1/T$ is the symbol rate. The frequency response $X_{rc}(f)$ is illustrated in Fig. 5.9(a) for $\alpha = 0, 1/2$, and 1. Note that when $\alpha = 0$, $X_{rc}(f)$ reduces to an ideal brick wall physically nonrealizable frequency response with bandwidth occupancy $1/2T$. The frequency $1/2T$ is called the *Nyquist frequency.* For $\alpha > 0$, the bandwidth occupied by the desired signal $X_{rc}(f)$ beyond the Nyquist frequency $1/2T$ is called the *excess bandwidth,* and is usually expressed as a percentage of the Nyquist frequency. For example, when $\alpha = 1/2$, the excess bandwidth is 50% and when $\alpha = 1$, the excess bandwidth is 100%. The signal pulse $x_{rc}(t)$ having the

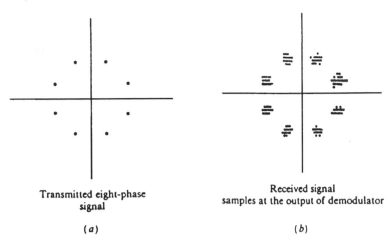

Transmitted eight-phase
signal

Received signal
samples at the output of demodulator

(*a*)

(*b*)

FIGURE 5.8   Two-dimensional digital eye patterns.

raised-cosine spectrum is

$$x_{rc}(t) = \frac{\sin \pi t/T}{\pi t/T} \frac{\cos (\pi \alpha t/T)}{1 - 4\alpha^2 t^2/T^2} \tag{5.13}$$

Figure 5.9(b) illustrates $x_{rc}(t)$ for $\alpha = 0$, $1/2$, and 1. Note that $x_{rc}(t) - 1$ at $t = 0$ and $x_{rc}(t) = 0$ at $t = kT$, $k = \pm 1, \pm 2, \ldots$. Consequently, at the sampling instants $t = kT$, $k \neq 0$, there is no ISI from adjacent symbols when there is no channel distortion. In the presence of channel distortion, however, the ISI given by Eq. (5.10) is no longer zero, and a **channel equalizer** is needed to minimize its effect on system performance.

5.3   Linear Equalizers

The most common type of channel equalizer used in practice to reduce SI is a linear transversal filter with adjustable coefficients $\{c_i\}$, as shown in Fig. 5.10.

On channels whose frequency response characteristics are unknown, but time invariant, we may measure the channel characteristics and adjust the parameters of the equalizer; once adjusted, the parameters remain fixed during the transmission of data. Such equalizers are called **preset equalizers.** On the other hand, **adaptive equalizers** update their parameters on a periodic basis during the transmission of data and, thus, they are capable of tracking a slowly time-varying channel response.

First, let us consider the design characteristics for a linear equalizer from a frequency domain viewpoint. Figure 5.11 shows a block diagram of a system that employs a linear filter as a **channel equalizer.**

The demodulator consists of a receiver filter with frequency response $G_R(f)$ in cascade with a channel equalizing filter that has a frequency response $G_E(f)$. As indicated in the preceding section, the receiver filter response $G_R(f)$ is matched to the transmitter response, i.e., $G_R(f) = G_T^*(f)$, and the product $G_R(f)G_T(f)$ is usually designed so that there is zero ISI at the sampling instants as, for example, when $G_R(t)G_T(f) = X_{rc}(f)$.

For the system shown in Fig. 5.11, in which the channel frequency response is not ideal, the desired condition for zero ISI is

$$G_T(f)C(f)G_R(f)G_E(f) = X_{rc}(f) \tag{5.14}$$

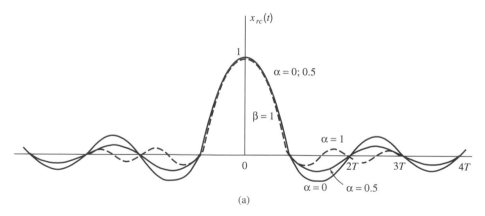

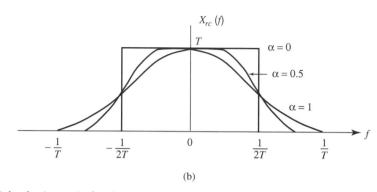

FIGURE 5.9   Pulses having a raised cosine spectrum.

where $X_{rc}(f)$ is the desired raised-cosine spectral characteristic. Since $G_T(f)G_R(f) = X_{rc}(f)$ by design, the frequency response of the equalizer that compensates for the channel distortion is

$$G_E(f) = \frac{1}{C(f)} = \frac{1}{|C(f)|}\, e^{-j\theta_c(f)} \tag{5.15}$$

Thus, the amplitude response of the equalizer is $|G_E(f)| = 1/|C(f)|$ and its phase response is $\theta_E(f) = -\theta_c(f)$. In this case, the equalizer is said to be the *inverse channel filter* to the channel response.

We note that the inverse channel filter completely eliminates ISI caused by the channel. Since it forces the ISI to be zero at the sampling instants $t = kT$, $k = 0, 1, \ldots$, the equalizer is called a **zero-forcing equalizer.** Hence, the input to the detector is simply

$$z_k = I_k + \eta_k, \qquad k = 0, 1, \ldots \tag{5.16}$$

where η_k represents the additive noise and I_k is the desired symbol.

In practice, the ISI caused by channel distortion is usually limited to a finite number of symbols on either side of the desired symbol. Hence, the number of terms that constitute the ISI in the summation given by Eq. (5.10) is finite. As a consequence, in practice the channel equalizer is implemented as a finite duration impulse response (FIR) filter, or transversal filter, with adjustable tap coefficients $\{c_n\}$, as illustrated in Fig. 5.10. The time delay τ between adjacent taps may be selected as large as T, the symbol interval, in which case the FIR equalizer is called a **symbol-spaced equalizer.** In this case, the input to the

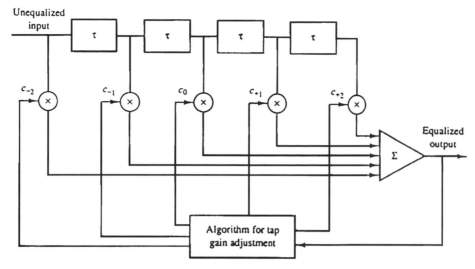

FIGURE 5.10 Linear transversal filter.

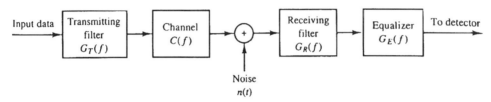

FIGURE 5.11 Block diagram of a system with an equalizer.

equalizer is the sampled sequence given by Eq. (5.7). We note that when the symbol rate $1/T < 2W$, however, frequencies in the received signal above the folding frequency $1/T$ are aliased into frequencies below $1/T$. In this case, the equalizer compensates for the aliased channel-distorted signal.

On the other hand, when the time delay τ between adjacent taps is selected such that $1/\tau \geq 2W > 1/T$, no aliasing occurs and, hence, the inverse channel equalizer compensates for the true channel distortion. Since $\tau < T$, the channel equalizer is said to have *fractionally spaced taps* and it is called a **fractionally spaced equalizer.** In practice, τ is often selected as $\tau = T/2$. Notice that, in this case, the sampling rate at the output of the filter $G_R(f)$ is $2/T$.

The impulse response of the FIR equalizer is

$$g_E(t) = \sum_{n=-N}^{N} c_n \delta(t - n\tau) \tag{5.17}$$

and the corresponding frequency response is

$$G_E(f) = \sum_{n=-N}^{N} c_n e^{-j2\pi f n\tau} \tag{5.18}$$

where $\{c_n\}$ are the $(2N+1)$ equalizer coefficients and N is chosen sufficiently large so that the equalizer spans the length of the ISI, i.e., $2N + 1 \geq L$, where L is the number of signal samples spanned by the ISI. Since $X(f) = G_T(f)C(f)G_R(f)$ and $x(t)$ is the signal pulse corresponding to $X(f)$, then the equalized output signal pulse is

$$q(t) = \sum_{n=-N}^{N} c_n x(t - n\tau) \tag{5.19}$$

The zero-forcing condition can now be applied to the samples of $q(t)$ taken at times $t = mT$. These samples are

$$q(mT) = \sum_{n=-N}^{N} c_n x(mT - n\tau), \qquad m = 0, \pm 1, \ldots, \pm N \tag{5.20}$$

Since there are $2N + 1$ equalizer coefficients, we can control only $2N + 1$ sampled values of $q(t)$. Specifically, we may force the conditions

$$q(mT) = \sum_{n=-N}^{N} c_n x(mT - n\tau) = \begin{cases} 1, & m = 0 \\ 0, & m = \pm 1, \pm 2, \ldots, \pm N \end{cases} \tag{5.21}$$

which may be expressed in matrix form as $\boldsymbol{X}\boldsymbol{c} = \boldsymbol{q}$, where $\boldsymbol{X}$ is a $(2N+1) \times (2N+1)$ matrix with elements $\{x(mT - n\tau)\}$, $\boldsymbol{c}$ is the $(2N + 1)$ coefficient vector and $\boldsymbol{q}$ is the $(2N + 1)$ column vector with one nonzero element. Thus, we obtain a set of $2N + 1$ linear equations for the coefficients of the zero-forcing equalizer.

We should emphasize that the FIR zero-forcing equalizer does not completely eliminate ISI because it has a finite length. As N is increased, however, the residual ISI can be reduced, and in the limit as $N \to \infty$, the ISI is completely eliminated.

EXAMPLE 5.1:

Consider a channel distorted pulse $x(t)$, at the input to the equalizer, given by the expression

$$x(t) = \frac{1}{1 + \left(\dfrac{2t}{T}\right)^2}$$

where $1/T$ is the symbol rate. The pulse is sampled at the rate $2/T$ and equalized by a zero-forcing equalizer. Determine the coefficients of a five-tap zero-forcing equalizer.

Solution 5.1   According to Eq. (5.21), the zero-forcing equalizer must satisfy the equations

$$q(mT) = \sum_{n=-2}^{2} c_n x\left(mT - nT/2\right) = \begin{cases} 1, & m = 0 \\ 0, & m = \pm 1, \pm 2 \end{cases}$$

The matrix $\boldsymbol{X}$ with elements $x(mT - nT/2)$ is given as

$$\boldsymbol{X} = \begin{bmatrix} \dfrac{1}{5} & \dfrac{1}{10} & \dfrac{1}{17} & \dfrac{1}{26} & \dfrac{1}{37} \\ 1 & \dfrac{1}{2} & \dfrac{1}{5} & \dfrac{1}{10} & \dfrac{1}{17} \\ \dfrac{1}{5} & \dfrac{1}{2} & 1 & \dfrac{1}{2} & \dfrac{1}{5} \\ \dfrac{1}{17} & \dfrac{1}{10} & \dfrac{1}{5} & \dfrac{1}{2} & 1 \\ \dfrac{1}{37} & \dfrac{1}{26} & \dfrac{1}{17} & \dfrac{1}{10} & \dfrac{1}{5} \end{bmatrix} \tag{5.22}$$

The coefficient vector c and the vector q are given as

$$c = \begin{bmatrix} c_{-2} \\ c_{-1} \\ c_0 \\ c_1 \\ c_2 \end{bmatrix} \qquad q = \begin{bmatrix} 0 \\ 0 \\ 1 \\ 0 \\ 0 \end{bmatrix} \tag{5.23}$$

Then, the linear equations $Xc = q$ can be solved by inverting the matrix X. Thus, we obtain

$$c_{\text{opt}} = X^{-1}q = \begin{bmatrix} -2.2 \\ 4.9 \\ -3 \\ 4.9 \\ -2.2 \end{bmatrix} \tag{5.24}$$

One drawback to the zero-forcing equalizer is that it ignores the presence of additive noise. As a consequence, its use may result in significant noise enhancement. This is easily seen by noting that in a frequency range where $C(f)$ is small, the channel equalizer $G_E(f) = 1/C(f)$ compensates by placing a large gain in that frequency range. Consequently, the noise in that frequency range is greatly enhanced. An alternative is to relax the zero ISI condition and select the channel equalizer characteristic such that the combined power in the residual ISI and the additive noise at the output of the equalizer is minimized. A channel equalizer that is optimized based on the minimum mean square error (MMSE) criterion accomplishes the desired goal.

To elaborate, let us consider the noise corrupted output of the FIR equalizer, which is

$$z(t) = \sum_{n=-N}^{N} c_n y(t - n\tau) \tag{5.25}$$

where $y(t)$ is the input to the equalizer, given by Eq. (5.6). The equalizer output is sampled at times $t = mT$. Thus, we obtain

$$z(mT) = \sum_{n=-N}^{N} c_n y(mT - n\tau) \tag{5.26}$$

The desired response at the output of the equalizer at $t = mT$ is the transmitted symbol I_m. The error is defined as the difference between I_m and $z(mT)$. Then, the mean square error (MSE) between the actual output sample $z(mT)$ and the desired values I_m is

$$\begin{aligned} MSE &= E\,|z(mT) - I_m|^2 \\ &= E\left[\left|\sum_{n=-N}^{N} c_n y(mT - n\tau) - I_m\right|^2\right] \\ &= \sum_{n=-N}^{N} \sum_{k=-N}^{N} c_n c_k R_Y(n-k) \\ &\quad - 2 \sum_{k=-N}^{N} c_k R_{IY}(k) + E\left(|I_m|^2\right) \end{aligned} \tag{5.27}$$

where the correlations are defined as

$$R_Y(n-k) = E\left[y^*(mT - n\tau)y(mT - k\tau)\right]$$

$$R_{IY}(k) = E\left[y(mT - k\tau)I_m^*\right] \tag{5.28}$$

and the expectation is taken with respect to the random information sequence $\{I_m\}$ and the additive noise.

The minimum *MSE* solution is obtained by differentiating Eq. (5.27) with respect to the equalizer coefficients $\{c_n\}$. Thus, we obtain the necessary conditions for the minimum *MSE* as

$$\sum_{n=-N}^{N} c_n R_Y(n-k) = R_{IY}(k), \qquad k = 0, \pm 1, 2, \ldots, \pm N \tag{5.29}$$

These are the $(2N + 1)$ linear equations for the equalizer coefficients. In contrast to the zero-forcing solution already described, these equations depend on the statistical properties (the autocorrelation) of the noise as well as the ISI through the autocorrelation $R_Y(n)$.

In practice, the autocorrelation matrix $R_Y(n)$ and the crosscorrelation vector $R_{IY}(n)$ are unknown a priori. These correlation sequences can be estimated, however, by transmitting a test signal over the channel and using the time-average estimates

$$\hat{R}_Y(n) = \frac{1}{K} \sum_{k=1}^{K} y^*(kT - n\tau)y(kT)$$

$$\hat{R}_{IY}(n) = \frac{1}{K} \sum_{k=1}^{K} y(kT - n\tau)I_k^* \tag{5.30}$$

in place of the ensemble averages to solve for the equalizer coefficients given by Eq. (5.29).

5.3.1 Adaptive Linear Equalizers

We have shown that the tap coefficients of a linear equalizer can be determined by solving a set of linear equations. In the zero-forcing optimization criterion, the linear equations are given by Eq. (5.21). On the other hand, if the optimization criterion is based on minimizing the *MSE,* the optimum equalizer coefficients are determined by solving the set of linear equations given by Eq. (5.29).

In both cases, we may express the set of linear equations in the general matrix form

$$\boldsymbol{B}\boldsymbol{c} = \boldsymbol{d} \tag{5.31}$$

where $\boldsymbol{B}$ is a $(2N+1)\times(2N+1)$ matrix, $\boldsymbol{c}$ is a column vector representing the $2N+1$ equalizer coefficients, and $\boldsymbol{d}$ a $(2N+1)$-dimensional column vector. The solution of Eq. (5.31) yields

$$\boldsymbol{c}_{\text{opt}} = \boldsymbol{B}^{-1}\boldsymbol{d} \tag{5.32}$$

In practical implementations of equalizers, the solution of Eq. (5.31) for the optimum coefficient vector is usually obtained by an iterative procedure that avoids the explicit computation of the inverse of the matrix $\boldsymbol{B}$. The simplest iterative procedure is the method of steepest descent, in which one begins by choosing arbitrarily the coefficient vector $\boldsymbol{c}$, say $\boldsymbol{c}_0$. This initial choice of coefficients corresponds to a point on the criterion function that is being optimized. For example, in the case of the *MSE* criterion, the initial guess $\boldsymbol{c}_0$

corresponds to a point on the quadratic *MSE* surface in the $(2N + 1)$-dimensional space of coefficients. The gradient vector, defined as $\boldsymbol{g}_0$, which is the derivative of the *MSE* with respect to the $2N + 1$ filter coefficients, is then computed at this point on the criterion surface, and each tap coefficient is changed in the direction opposite to its corresponding gradient component. The change in the jth tap coefficient is proportional to the size of the jth gradient component.

For example, the gradient vector denoted as $\boldsymbol{g}_k$, for the MSE criterion, found by taking the derivatives of the MSE with respect to each of the $2N + 1$ coefficients, is

$$\boldsymbol{g}_k = \boldsymbol{B}\boldsymbol{c}_k - \boldsymbol{d}, \qquad k = 0, 1, 2, \dots \tag{5.33}$$

Then the coefficient vector $\boldsymbol{c}_k$ is updated according to the relation

$$\boldsymbol{c}_{k+1} = \boldsymbol{c}_k - \Delta\boldsymbol{g}_k \tag{5.34}$$

where Δ is the *step-size parameter* for the iterative procedure. To ensure convergence of the iterative procedure, Δ is chosen to be a small positive number. In such a case, the gradient vector $\boldsymbol{g}_k$ converges toward zero, i.e., $\boldsymbol{g}_k \to \boldsymbol{0}$ as $k \to \infty$, and the coefficient vector $\boldsymbol{c}_k \to \boldsymbol{c}_{\mathrm{opt}}$ as illustrated in Fig. 5.12 based on two-dimensional optimization. In general, convergence of the equalizer tap coefficients to $\boldsymbol{c}_{\mathrm{opt}}$ cannot be attained in a finite number of iterations with the steepest-descent method. The optimum solution $\boldsymbol{c}_{\mathrm{opt}}$, however, can be approached as closely as desired in a few hundred iterations. In digital communication systems that employ channel equalizers, each iteration corresponds to a time interval for sending one symbol and, hence, a few hundred iterations to achieve convergence to $\boldsymbol{c}_{\mathrm{opt}}$ corresponds to a fraction of a second.

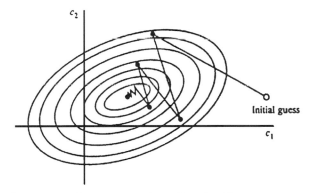

FIGURE 5.12   Examples of convergence characteristics of a gradient algorithm.

Adaptive channel equalization is required for channels whose characteristics change with time. In such a case, the ISI varies with time. The channel equalizer must track such time variations in the channel response and adapt its coefficients to reduce the ISI. In the context of the preceding discussion, the optimum coefficient vector $\boldsymbol{c}_{\mathrm{opt}}$ varies with time due to time variations in the matrix $\boldsymbol{B}$ and, for the case of the *MSE* criterion, time variations in the vector $\boldsymbol{d}$. Under these conditions, the iterative method described can be modified to use estimates of the gradient components. Thus, the algorithm for adjusting the equalizer tap coefficients may be expressed as

$$\hat{\boldsymbol{c}}_{k+1} = \hat{\boldsymbol{c}}_k - \Delta\hat{\boldsymbol{g}}_k \tag{5.35}$$

where $\hat{\boldsymbol{g}}_k$ denotes an estimate of the gradient vector $\boldsymbol{g}_k$ and $\hat{\boldsymbol{c}}_k$ denotes the estimate of the tap coefficient vector.

In the case of the *MSE* criterion, the gradient vector $\boldsymbol{g}_k$ given by Eq. (5.33) may also be expressed as

$$\boldsymbol{g}_k = -E\left(e_k \boldsymbol{y}_k^*\right)$$

An estimate $\hat{\boldsymbol{g}}_k$ of the gradient vector at the kth iteration is computed as

$$\hat{\boldsymbol{g}}_k = -e_k \boldsymbol{y}_k^* \tag{5.36}$$

where e_k denotes the difference between the desired output from the equalizer at the kth time instant and the actual output $z(kT)$, and $\boldsymbol{y}_k$ denotes the column vector of $2N + 1$ received signal values contained in the equalizer at time instant k. The *error signal* e_k is expressed as

$$e_k = I_k - z_k \tag{5.37}$$

where $z_k = z(kT)$ is the equalizer output given by Eq. (5.26) and I_k is the desired symbol. Hence, by substituting Eq. (5.36) into Eq. (5.35), we obtain the adaptive algorithm for optimizing the tap coefficients (based on the *MSE* criterion) as

$$\hat{\boldsymbol{c}}_{k+1} = \hat{\boldsymbol{c}}_k + \Delta e_k \boldsymbol{y}_k^* \tag{5.38}$$

Since an estimate of the gradient vector is used in Eq. (5.38) the algorithm is called a **stochastic gradient algorithm;** it is also known as the **LMS algorithm.**

A block diagram of an adaptive equalizer that adapts its tap coefficients according to Eq. (5.38) is illustrated in Fig. 5.13. Note that the difference between the desired output I_k and the actual output z_k from the equalizer is used to form the error signal e_k. This error is scaled by the step-size parameter Δ, and the scaled error signal Δe_k multiplies the received signal values $\{y(kT - n\tau)\}$ at the $2N + 1$ taps. The products $\Delta e_k y^*(kT - n\tau)$ at the $(2N + 1)$ taps are then added to the previous values of the tap coefficients to obtain the updated tap coefficients, according to Eq. (5.38). This computation is repeated as each new symbol is received. Thus, the equalizer coefficients are updated at the symbol rate.

Initially, the adaptive equalizer is trained by the transmission of a known pseudo-random sequence $\{I_m\}$ over the channel. At the demodulator, the equalizer employs the known sequence to adjust its coefficients. Upon initial adjustment, the adaptive equalizer switches from a **training mode** to a **decision-directed mode,** in which case the decisions at the output of the detector are sufficiently reliable so that the error signal is formed by computing the difference between the detector output and the equalizer output, i.e.,

$$e_k = \tilde{I}_k - z_k \tag{5.39}$$

where $\tilde{I}_k$ is the output of the detector. In general, decision errors at the output of the detector occur infrequently and, consequently, such errors have little effect on the performance of the tracking algorithm given by Eq. (5.38).

A rule of thumb for selecting the step-size parameter so as to ensure convergence and good tracking capabilities in slowly varying channels is

$$\Delta = \frac{1}{5(2N + 1)P_R} \tag{5.40}$$

where P_R denotes the received signal-plus-noise power, which can be estimated from the received signal (see Proakis [4]).

The convergence characteristic of the stochastic gradient algorithm in Eq. (5.38) is illustrated in Fig. 5.14. These graphs were obtained from a computer simulation of an 11-tap adaptive equalizer operating on a channel with a rather modest amount of ISI. The input

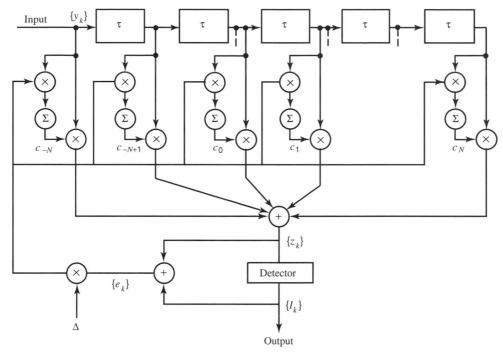

FIGURE 5.13   Linear adaptive equalizer based on the *MSE* criterion.

signal-plus-noise power P_R was normalized to unity. The rule of thumb given in Eq. (5.40) for selecting the step size gives $\Delta = 0.018$. The effect of making Δ too large is illustrated by the large jumps in *MSE* as shown for $\Delta = 0.115$. As Δ is decreased, the convergence is slowed somewhat, but a lower MSE is achieved, indicating that the estimated coefficients are closer to $\boldsymbol{c}_{\mathrm{opt}}$.

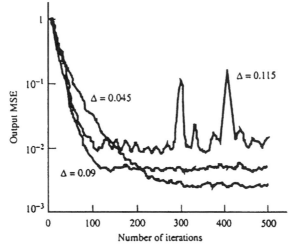

FIGURE 5.14   Initial convergence characteristics of the *LMS* algorithm with different step sizes.

Although we have described in some detail the operation of an adaptive equalizer that is optimized on the basis of the MSE criterion, the operation of an adaptive equalizer based on the zero-forcing method is very similar. The major difference lies in the method for estimating the gradient vectors g_k at each iteration. A block diagram of an adaptive zero-forcing equalizer is shown in Fig. 5.15.

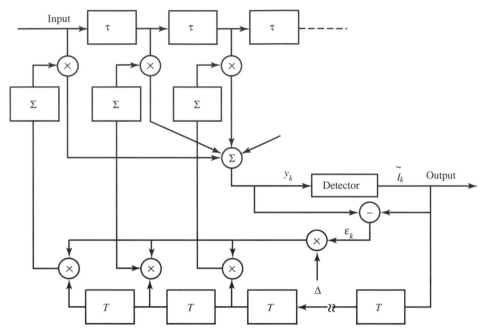

FIGURE 5.15   An adaptive zero-forcing equalizer.

For more details on the tap coefficient update method for a zero-forcing equalizer, the reader is referred to the papers by Lucky [2, 3], and the text by Proakis [4].

5.4   Decision-Feedback Equalizer

The linear filter equalizers described in the preceding section are very effective on channels, such as wire line telephone channels, where the ISI is not severe. The severity of the ISI is directly related to the spectral characteristics and not necessarily to the time span of the ISI. For example, consider the ISI resulting from the two channels that are illustrated in Fig. 5.16. The time span for the ISI in channel A is 5 symbol intervals on each side of the desired signal component, which has a value of 0.72. On the other hand, the time span for the ISI in channel B is one symbol interval on each side of the desired signal component, which has a value of 0.815. The energy of the total response is normalized to unity for both channels.

In spite of the shorter ISI span, channel B results in more severe ISI. This is evidenced in the frequency response characteristics of these channels, which are shown in Fig. 5.17. We observe that channel B has a spectral null [the frequency response $C(f) = 0$ for some frequencies in the band $|f| \leq W$] at $f = 1/2T$, whereas this does not occur in the case of channel A. Consequently, a linear equalizer will introduce a large gain in its frequency response to compensate for the channel null. Thus, the noise in channel B will be enhanced

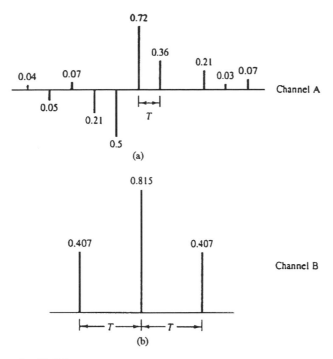

FIGURE 5.16 Two channels with ISI.

much more than in channel A. This implies that the performance of the linear equalizer for channel B will be sufficiently poorer than that for channel A. This fact is borne out by the computer simulation results for the performance of the two linear equalizers shown in Fig. 5.18. Hence, the basic limitation of a linear equalizer is that it performs poorly on channels having spectral nulls. Such channels are often encountered in radio communications, such as ionospheric transmission at frequencies below 30 MHz and mobile radio channels, such as those used for cellular radio communications.

A **decision-feedback equalizer** (**DFE**) is a nonlinear equalizer that employs previous decisions to eliminate the ISI caused by previously detected symbols on the current symbol to be detected. A simple block diagram for a DFE is shown in Fig. 5.19. The DFE consists of two filters. The first filter is called a *feedforward filter* and it is generally a fractionally spaced FIR filter with adjustable tap coefficients. This filter is identical in form to the linear equalizer already described. Its input is the received filtered signal $y(t)$ sampled at some rate that is a multiple of the symbol rate, e.g., at rate $2/T$. The second filter is a *feedback filter*. It is implemented as an FIR filter with symbol-spaced taps having adjustable coefficients. Its input is the set of previously detected symbols. The output of the feedback filter is subtracted from the output of the feedforward filter to form the input to the detector. Thus, we have

$$z_m = \sum_{n=-N_1}^{0} c_n y(mT - n\tau) - \sum_{n=1}^{N_2} b_n \tilde{I}_{m-n} \tag{5.41}$$

where $\{c_n\}$ and $\{b_n\}$ are the adjustable coefficients of the feedforward and feedback filters, respectively, $\tilde{I}_{m-n}, n = 1, 2, \ldots, N_2$ are the previously detected symbols, $N_1 + 1$ is the length of the feedforward filter, and N_2 is the length of the feedback filter. Based on the input z_m, the detector determines which of the possible transmitted symbols is closest in distance to the input signal I_m. Thus, it makes its decision and outputs $\tilde{I}_m$. What makes the

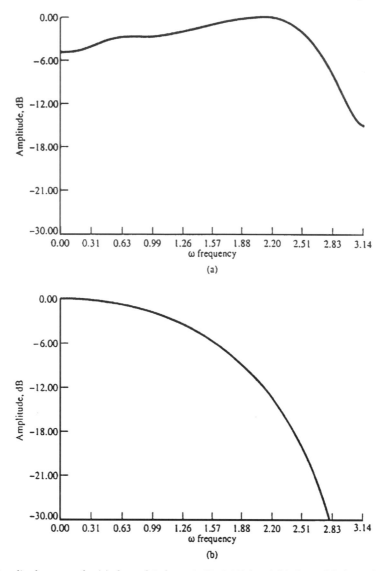

FIGURE 5.17   Amplitude spectra for (a) channel A shown in Fig.5.16(a) and (b) channel B shown in Fig.5.16(b).

DFE nonlinear is the nonlinear characteristic of the detector that provides the input to the feedback filter.

The tap coefficients of the feedforward and feedback filters are selected to optimize some desired performance measure. For mathematical simplicity, the *MSE* criterion is usually applied, and a stochastic gradient algorithm is commonly used to implement an adaptive DFE. Figure 5.20 illustrates the block diagram of an adaptive DFE whose tap coefficients are adjusted by means of the LMS stochastic gradient algorithm. Figure 5.21 illustrates the probability of error performance of the DFE, obtained by computer simulation, for binary PAM transmission over channel B. The gain in performance relative to that of a linear equalizer is clearly evident.

We should mention that decision errors from the detector that are fed to the feedback filter have a small effect on the performance of the DFE. In general, a small loss in performance

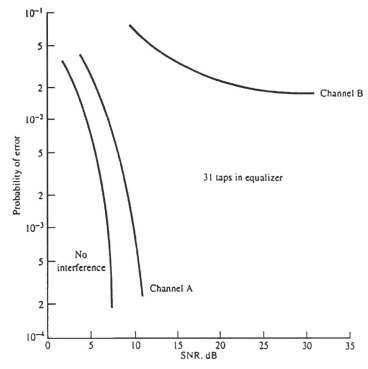

FIGURE 5.18 Error-rate performance of linear *MSE* equalizer.

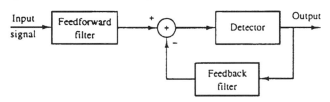

FIGURE 5.19 Block diagram of DFE.

of one to two decibels is possible at error rates below 10^{-2}, as illustrated in Fig. 5.21, but the decision errors in the feedback filters are not catastrophic.

5.5 Maximum-Likelihood Sequence Detection

Although the DFE outperforms a linear equalizer, it is not the optimum equalizer from the viewpoint of minimizing the probability of error in the detection of the information sequence $\{I_k\}$ from the received signal samples $\{y_k\}$ given in Eq. (5.5). In a digital communication system that transmits information over a channel that causes ISI, the optimum detector is a maximum-likelihood symbol sequence detector which produces at its output the most probable symbol sequence $\{\tilde{I}_k\}$ for the given received sampled sequence $\{y_k\}$. That is, the detector finds the sequence $\{\tilde{I}_k\}$ that maximizes the *likelihood function*

$$\Lambda\left(\{I_k\}\right) = \ln p\left(\{y_k\} \mid \{I_k\}\right) \tag{5.42}$$

where $p(\{y_k\} \mid \{I_k\})$ is the joint probability of the received sequence $\{y_k\}$ conditioned on $\{I_k\}$. The sequence of symbols $\{\tilde{I}_k\}$ that maximizes this joint conditional probability is called the **maximum-likelihood sequence detector.**

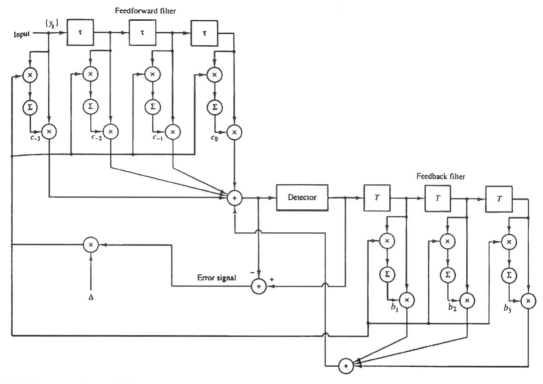

FIGURE 5.20   Adaptive DFE.

An algorithm that implements maximum-likelihood sequence detection (MLSD) is the Viterbi algorithm, which was originally devised for decoding convolutional codes. For a description of this algorithm in the context of sequence detection in the presence of ISI, the reader is referred to the paper by Forney [1] and the text by Proakis [4].

The major drawback of MLSD for channels with ISI is the exponential behavior in computational complexity as a function of the span of the ISI. Consequently, MLSD is practical only for channels where the ISI spans only a few symbols and the ISI is severe, in the sense that it causes a severe degradation in the performance of a linear equalizer or a decision-feedback equalizer. For example, Fig. 5.22 illustrates the error probability performance of the Viterbi algorithm for a binary PAM signal transmitted through channel B (see Fig. 5.16). For purposes of comparison, we also illustrate the probability of error for a DFE. Both results were obtained by computer simulation. We observe that the performance of the maximum likelihood sequence detector is about 4.5 dB better than that of the DFE at an error probability of 10^{-4}. Hence, this is one example where the ML sequence detector provides a significant performance gain on a channel with a relatively short ISI span.

5.6   Conclusions

Channel equalizers are widely used in digital communication systems to mitigate the effects of ISI caused by channel distortion. Linear equalizers are widely used for high-speed modems that transmit data over telephone channels. For wireless (radio) transmission, such as in mobile cellular communications and interoffice communications, the multipath propagation of the transmitted signal results in severe ISI. Such channels require more powerful equalizers to combat the severe ISI. The decision-feedback equalizer and the MLSD are two nonlinear channel equalizers that are suitable for radio channels with severe ISI.

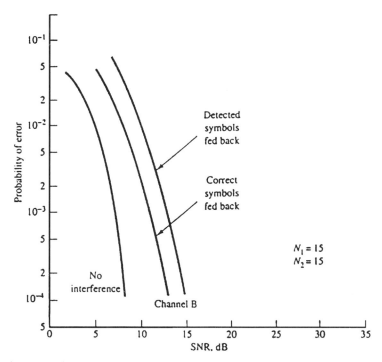

FIGURE 5.21   Performance of DFE with and without error propagation.

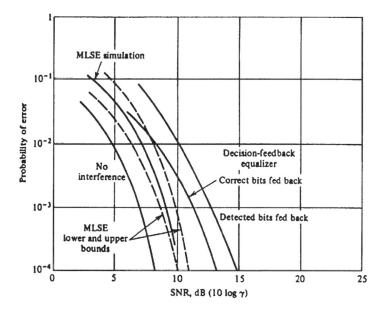

FIGURE 5.22   Comparison of performance between MLSE and decision-feedback equalization for channel B of Fig.5.16.

Defining Terms

Adaptive equalizer: A channel equalizer whose parameters are updated automatically and adaptively during transmission of data.

Channel equalizer: A device that is used to reduce the effects of channel distortion in a received signal.

Decision-directed mode: Mode for adjustment of the equalizer coefficient adaptively based on the use of the detected symbols at the output of the detector.

Decision-feedback equalizer (DFE): An adaptive equalizer that consists of a feedforward filter and a feedback filter, where the latter is fed with previously detected symbols that are used to eliminate the intersymbol interference due to the tail in the channel impulse response.

Fractionally spaced equalizer: A tapped-delay line channel equalizer in which the delay between adjacent taps is less than the duration of a transmitted symbol.

Intersymbol interference: Interference in a received symbol from adjacent (nearby) transmitted symbols caused by channel distortion in data transmission.

LMS algorithm: See stochastic gradient algorithm.

Maximum-likelihood sequence detector: A detector for estimating the most probable sequence of data symbols by maximizing the likelihood function of the received signal.

Preset equalizer: A channel equalizer whose parameters are fixed (time-invariant) during transmission of data.

Stochastic gradient algorithm: An algorithm for adaptively adjusting the coefficients of an equalizer based on the use of (noise-corrupted) estimates of the gradients.

Symbol-spaced equalizer: A tapped-delay line channel equalizer in which the delay between adjacent taps is equal to the duration of a transmitted symbol.

Training mode: Mode for adjustment of the equalizer coefficients based on the transmission of a known sequence of transmitted symbols.

Zero-forcing equalizer: A channel equalizer whose parameters are adjusted to completely eliminate intersymbol interference in a sequence of transmitted data symbols.

References

[1] Forney, G.D., Jr., Maximum-likelihood sequence estimation of digital sequences in the presence of intersymbol interference. *IEEE Trans. Inform. Theory*, IT-18, 363–378, May 1972.

[2] Lucky, R.W., Automatic equalization for digital communications. *Bell Syst. Tech. J.*, 44, 547–588, Apr. 1965.

[3] Lucky, R.W., Techniques for adaptive equalization of digital communication. *Bell Syst. Tech. J.*, 45, 255–286, Feb. 1966.

[4] Proakis, J.G., *Digital Communications*, 3rd ed., McGraw-Hill, New York, 1995.

Further Information

For a comprehensive treatment of adaptive equalization techniques and their performance characteristics, the reader may refer to the book by Proakis [4]. The two papers by

Lucky [2, 3], provide a treatment on linear equalizers based on the zero-forcing criterion. Additional information on decision-feedback equalizers may be found in the journal papers "An Adaptive Decision-Feedback Equalizer" by D.A. George, R.R. Bowen, and J.R. Storey, *IEEE Transactions on Communications Technology,* Vol. COM-19, pp. 281–293, June 1971, and "Feedback Equalization for Fading Dispersive Channels" by P. Monsen, *IEEE Transactions on Information Theory,* Vol. IT-17, pp. 56–64, January 1971. A through treatment of channel equalization based on maximum-likelihood sequence detection is given in the paper by Forney [1].

6

Line Coding

Joseph L. LoCicero
Illinois Institute of Technology

Bhasker P. Patel
Illinois Institute of Technology

6.1 Introduction

The terminology **line coding** originated in telephony with the need to transmit digital information across a copper telephone *line;* more specifically, binary data over a digital repeatered line. The concept of line coding, however, readily applies to any transmission line or channel. In a digital communication system, there exists a known set of symbols to be transmitted. These can be designated as $\{m_i\}$, $i = 1, 2, \ldots, N$, with a probability of occurrence $\{p_i\}$, $i = 1, 2, \ldots, N$, where the sequentially transmitted symbols are generally assumed to be statistically independent. The conversion or *coding* of these abstract symbols into real, temporal waveforms to be transmitted in baseband is the process of line coding. Since the most common type of line coding is for binary data, such a waveform can be succinctly termed a direct format for serial bits. The concentration in this section will be line coding for binary data.

Different channel characteristics, as well as different applications and performance requirements, have provided the impetus for the development and study of various types of line coding [1, 2]. For example, the channel might be ac coupled and, thus, could not support a line code with a dc component or large dc content. Synchronization or timing recovery requirements might necessitate a discrete component at the data rate. The channel bandwidth and **crosstalk** limitations might dictate the type of line coding employed. Even such factors as the complexity of the encoder and the economy of the decoder could determine the line code chosen. Each line code has its own distinct properties. Depending on the application, one property may be more important than the other. In what follows, we describe, in general, the most desirable features that are considered when choosing a line code.

It is commonly accepted [1, 2, 5, 8] that the dominant considerations effecting the choice of a line code are: 1) timing, 2) dc content, 3) power spectrum, 4) performance monitoring, 5) probability of error, and 6) transparency. Each of these are detailed in the following paragraphs.

1) *Timing:* The waveform produced by a line code should contain enough timing information such that the receiver can synchronize with the transmitter and decode the received signal properly. The timing content should be relatively independent of source statistics, i.e., a long string of **1**s or **0**s should not result in loss of timing or jitter at the receiver.

2) *DC content:* Since the repeaters used in telephony are ac coupled, it is desirable to have zero dc in the waveform produced by a given line code. If a signal with significant dc content is used in ac coupled lines, it will cause **dc wander** in the received waveform. That is, the received signal baseline will vary with time. Telephone lines do not pass dc due to ac coupling with transformers and capacitors to eliminate dc ground loops. Because of this, the telephone channel causes a droop in constant signals. This causes dc wander. It can be eliminated by dc restoration circuits, feedback systems, or with specially designed line codes.

3) *Power spectrum:* The power spectrum and bandwidth of the transmitted signal should be matched to the frequency response of the channel to avoid significant distortion. Also, the power spectrum should be such that most of the energy is contained in as small bandwidth as possible. The smaller is the bandwidth, the higher is the transmission efficiency.

4) *Performance monitoring:* It is very desirable to detect errors caused by a noisy transmission channel. The error detection capability in turn allows performance monitoring while the channel is in use (i.e., without elaborate testing procedures that require suspending use of the channel).

5) *Probability of error:* The average error probability should be as small as possible for a given transmitter power. This reflects the reliability of the line code.

6) *Transparency:* A line code should allow all the possible patterns of **1**s and **0**s. If a certain pattern is undesirable due to other considerations, it should be mapped to a unique alternative pattern.

6.2   Common Line Coding Formats

A line coding format consists of a formal definition of the line code that specifies how a string of binary digits are converted to a line code waveform. There are two major classes of binary line codes: **level codes** and **transition codes.** Level codes carry information in their voltage level, which may be high or low for a full bit period or part of the bit period. Level codes are usually instantaneous since they typically encode a binary digit into a distinct waveform, independent of any past binary data. However, some level codes do exhibit memory. Transition codes carry information in the change in level appearing in the line code waveform. Transition codes may be instantaneous, but they generally have memory, using past binary data to dictate the present waveform. There are two common forms of level line codes: one is called **return to zero (RZ)** and the other is called

nonreturn to zero (NRZ). In RZ coding, the level of the pulse returns to zero for a portion of the bit interval. In NRZ coding, the level of the pulse is maintained during the entire bit interval.

Line coding formats are further classified according to the polarity of the voltage levels used to represent the data. If only one polarity of voltage level is used, i.e., positive or negative (in addition to the zero level) then it is called **unipolar** signalling. If both positive and negative voltage levels are being used, with or without a zero voltage level, then it is called **polar** signalling. The term **bipolar** signalling is used by some authors to designate a specific line coding scheme with positive, negative, and zero voltage levels. This will be described in detail later in this section. The formal definition of five common line codes is given in the following along with a representative waveform, the *power spectral density* (PSD), the probability of error, and a discussion of advantages and disadvantages. In some cases specific applications are noted.

6.2.1 Unipolar NRZ (Binary On-Off Keying)

In this line code, a binary **1** is represented by a non-zero voltage level and a binary **0** is represented by a zero voltage level as shown in Fig. 6.1(a). This is an instantaneous level code. The PSD of this code with equally likely **1**s and **0**s is given by [5, 8]

$$S_1(f) = \frac{V^2 T}{4} \left(\frac{\sin \pi f T}{\pi f T} \right)^2 + \frac{V^2}{4} \delta(f) \tag{6.1}$$

where V is the binary **1** voltage level, $T = 1/R$ is the bit duration, and R is the bit rate in bits per second. The spectrum of unipolar NRZ is plotted in Fig. 6.2a. This PSD is a two-sided even spectrum, although only half of the plot is shown for efficiency of presentation. If the probability of a binary **1** is p, and the probability of a binary **0** is $(1 - p)$, then the PSD of this code, in the most general case, is $4p(1-p) S_1(f)$. Considering the frequency of the first spectral null as the bandwidth of the waveform, the bandwidth of unipolar NRZ is R in hertz. The error rate performance of this code, for equally likely data, with additive white Gaussian noise (AWGN) and optimum, i.e., matched filter, detection is given by [1, 5]

$$P_e = \frac{1}{2} \text{erfc} \left(\sqrt{\frac{E_b}{2N_0}} \right) \tag{6.2}$$

where E_b/N_0 is a measure of the signal-to-noise ratio (SNR) of the received signal. In general, E_b is the energy per bit and $N_0/2$ is the two-sided PSD of the AWGN. More specifically, for unipolar NRZ, E_b is the energy in a binary **1**, which is $V^2 T$. The performance of the unipolar NRZ code is plotted in Fig. 6.3

The principle advantages of unipolar NRZ are ease of generation, since it requires only a single power supply, and a relatively low bandwidth of R Hz. There are quite a few disadvantages of this line code. A loss of synchronization and timing jitter can result with a long sequence of **1**s or **0**s because no pulse transition is present. The code has no error detection capability and, hence, performance cannot be monitored. There is a significant dc component as well as a dc content. The error rate performance is not as good as that of polar line codes.

6.2.2 Unipolar RZ

In this line code, a binary **1** is represented by a nonzero voltage level during a portion of the bit duration, usually for half of the bit period, and a zero voltage level for rest of the bit

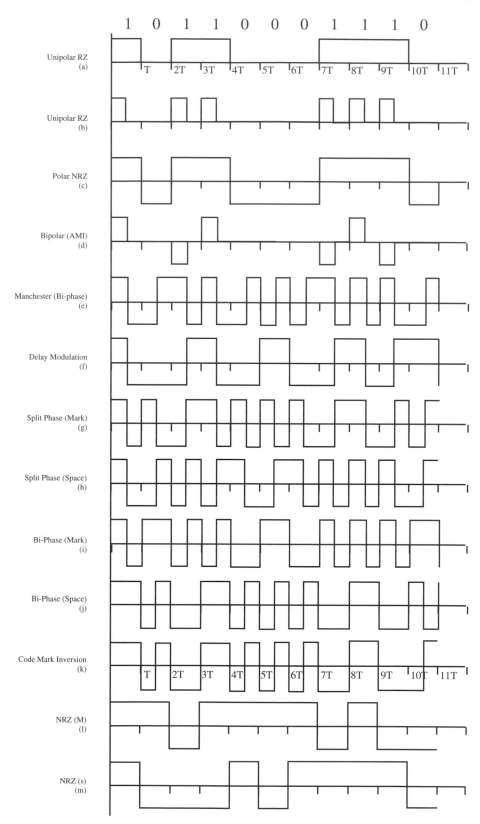

FIGURE 6.1   Waveforms for different line codes.

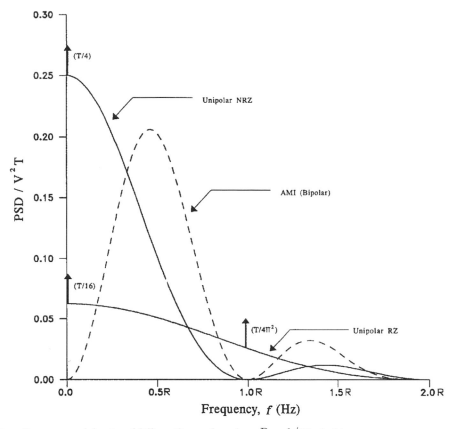

Figure 6.2a   Power spectral density of different line codes, where $R = 1/T$ is the bit rate.

duration. A binary **0** is represented by a zero voltage level during the entire bit duration. Thus, this is an instantaneous level code. Figure 6.1(b) illustrates a unipolar RZ waveform in which the **1** is represented by a nonzero voltage level for half the bit period. The PSD of this line code, with equally likely binary digits, is given by [5, 6, 8]

$$S_2(f) \;=\; \frac{V^2 T}{16} \left(\frac{\sin \pi fT/2}{\pi fT/2} \right)^2$$
$$+ \frac{V^2}{4\pi^2} \left[\frac{\pi^2}{4} \delta(f) + \sum_{n=-\infty}^{\infty} \frac{1}{(2n+1)^2} \delta(f - (2n+1)R) \right] \qquad (6.3)$$

where again V is the binary **1** voltage level, and $T = 1/R$ is the bit period. The spectrum of this code is drawn in Fig. 6.2a. In the most general case, when the probability of a **1** is p, the continuous portion of the PSD in Eq. (6.3) is scaled by the factor $4p(1-p)$ and the discrete portion is scaled by the factor $4p^2$. The first null bandwidth of unipolar RZ is $2R$ Hz. The error rate performance of this line code is the same as that of the unipolar NRZ provided we increase the voltage level of this code such that the energy in binary **1**, E_b, is the same for both codes. The probability of error is given by Eq. (6.2) and identified in Fig. 6.3. If the voltage level and bit period are the same for unipolar NRZ and unipolar RZ, then the energy in a binary **1** for unipolar RZ will be $V^2 T/2$ and the probability of error is worse by 3 dB.

The main advantages of unipolar RZ are, again, ease of generation since it requires a single power supply and the presence of a discrete spectral component at the symbol rate, which

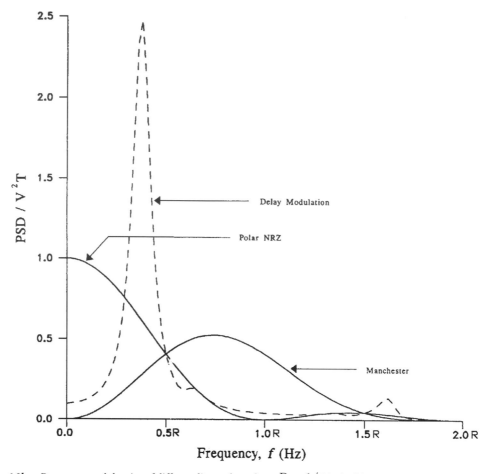

Figure 6.2b   Power spectral density of different line codes, where $R = 1/T$ is the bit rate.

allows simple timing recovery. A number of disadvantages exist for this line code. It has a nonzero dc component and nonzero dc content, which can lead to dc wander. A long string of **0**s will lack pulse transitions and could lead to loss of synchronization. There is no error detection capability and, hence, performance monitoring is not possible. The bandwidth requirement ($2R$ Hz) is higher than that of NRZ signals. The error rate performance is worse than that of polar line codes.

Unipolar NRZ as well as unipolar RZ are examples of pulse/no-pulse type of signalling. In this type of signalling, the pulse for a binary **0**, $g_2(t)$, is zero and the pulse for a binary **1** is specified generically as $g_1(t) = g(t)$. Using $G(f)$ as the Fourier transform of $g(t)$, the PSD of pulse/no-pulse signalling is given as [6, 7, 10]

$$S_{\text{PNP}}(f) = p(1-p)R|G(f)|^2 + p^2 R^2 \sum_{n=-\infty}^{\infty} |G(nR)|^2 \delta(f - nR) \qquad (6.4)$$

where p is the probability of a binary **1**, and R is the bit rate.

6.2.3 Polar NRZ

In this line code, a binary **1** is represented by a positive voltage $+V$ and a binary **0** is represented by a negative voltage $-V$ over the full bit period. This code is also referred to

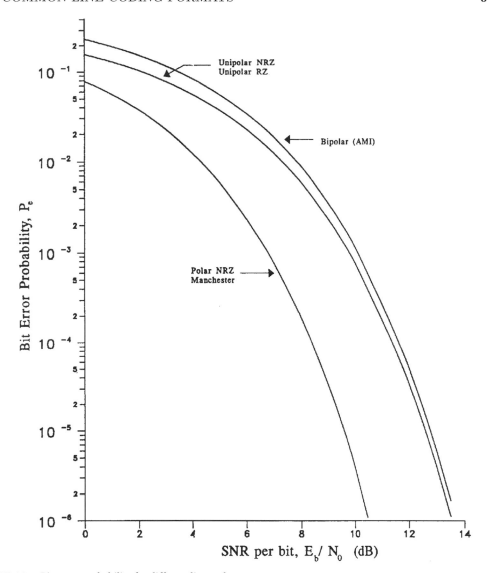

FIGURE 6.3 Bit error probability for different line codes.

as NRZ (L), since a bit is represented by maintaining a level (L) during its entire period. A polar NRZ waveform is shown in Fig. 6.1(c). This is again an instantaneous level code. Alternatively, a **1** may be represented by a $-V$ voltage level and a **0** by a $+V$ voltage level, without changing the spectral characteristics and performance of the line code. The PSD of this line code with equally likely bits is given by [5, 8]

$$S_3(f) = V^2 T \left(\frac{\sin \pi f T}{\pi f T} \right)^2 \tag{6.5}$$

This is plotted in Fig. 6.2b. When the probability of a **1** is p, and p is not 0.5, a dc component exists, and the PSD becomes [10]

$$S_{3p}(f) = 4V^2 T p(1-p) \left(\frac{\sin \pi f T}{\pi f T} \right)^2 + V^2 (1-2p)^2 \delta(f) \tag{6.6}$$

The first null bandwidth for this line code is again R Hz, independent of p. The probability of error of this line code when $p = 0.5$ is given by [1, 5]

$$P_e = \frac{1}{2}\text{erfc}\left(\sqrt{\frac{E_b}{N_0}}\right) \tag{6.7}$$

The performance of polar NRZ is plotted in Fig. 6.3. This is better than the error performance of the unipolar codes by 3 dB.

The advantages of polar NRZ include a low-bandwidth requirement, R Hz, comparable to unipolar NRZ, very good error probability, and greatly reduced dc because the waveform has a zero dc component when $p = 0.5$ even though the dc content is never zero. A few notable disadvantages are that there is no error detection capability, and that a long string of **1**s or **0**s could result in loss of synchronization, since there are no transitions during the string duration. Two power supplies are required to generate this code.

6.2.4   Polar RZ [Bipolar, Alternate Mark Inversion (AMI), or Pseudoternary]

In this scheme, a binary **1** is represented by alternating the positive and negative voltage levels, which return to zero for a portion of the bit duration, generally half the bit period. A binary **0** is represented by a zero voltage level during the entire bit duration. This line coding scheme is often called **alternate mark inversion (AMI)** since **1**s (marks) are represented by alternating positive and negative pulses. It is also called *pseudoternary* since three different voltage levels are used to represent binary data. Some authors designate this line code as bipolar RZ (BRZ). An AMI waveform is shown in Fig. 6.1(d). Note that this is a level code with memory. The AMI code is well known for its use in telephony. The PSD of this line code with memory is given by [1, 2, 7]

$$S_{4p}(f) = 2p(1-p)R|G(f)|^2 \left(\frac{1 - \cos 2\pi fT}{1 + (2p-1)^2 + 2(2p-1)\cos 2\pi fT}\right) \tag{6.8}$$

where $G(f)$ is the Fourier transform of the pulse used to represent a binary **1**, and p is the probability of a binary **1**. When $p = 0.5$ and square pulses with amplitude $\pm V$ and duration $T/2$ are used to represent binary **1**s, the PSD becomes

$$S_4(f) = \frac{V^2 T}{4}\left(\frac{\sin \pi fT/2}{\pi fT/2}\right)^2 \sin^2(\pi fT) \tag{6.9}$$

This PSD is plotted in Fig. 6.2a. The first null bandwidth of this waveform is R Hz. This is true for RZ rectangular pulses, independent of the value of p in Eq. (6.8). The error rate performance of this line code for equally likely binary data is given by [5]

$$P_e \approx \frac{3}{4}\text{erfc}\left(\sqrt{\frac{E_b}{2N_0}}\right), \qquad E_b/N_0 > 2 \tag{6.10}$$

This curve is plotted in Fig. 6.3 and is seen to be no more than 0.5 dB worse than the unipolar codes.

The advantages of polar RZ (or AMI, as it is most commonly called) outweigh the disadvantages. This code has no dc component and zero dc content, completely avoiding the dc wander problem. Timing recovery is rather easy since squaring, or full-wave rectifying, this type of signal yields a unipolar RZ waveform with a discrete component at the bit rate, R Hz. Because of the alternating polarity pulses for binary **1**s, this code has error detection and, hence, performance monitoring capability. It has a low-bandwidth requirement, R Hz, comparable to unipolar NRZ. The obvious disadvantage is that the error rate performance is worse than that of the unipolar and polar waveforms. A long string of **0**s could result in loss of synchronization, and two power supplies are required for this code.

6.2.5 Manchester Coding (Split Phase or Digital Biphase)

In this coding, a binary **1** is represented by a pulse that has positive voltage during the first-half of the bit duration and negative voltage during second-half of the bit duration. A binary **0** is represented by a pulse that is negative during the first-half of the bit duration and positive during the second-half of the bit duration. The negative or positive midbit transition indicates a binary **1** or binary **0**, respectively. Thus, a Manchester code is classified as an instantaneous transition code; it has no memory. The code is also called diphase because a square wave with a $0°$ phase is used to represent a binary **1** and a square wave with a phase of $180°$ used to represent a binary **0**; or vice versa. This line code is used in Ethernet local area networks (LANs). The waveform for Manchester coding is shown in Fig. 6.1(e). The PSD of a Manchester waveform with equally likely bits is given by [5, 8]

$$S_5(f) = V^2 T \left(\frac{\sin \pi f T/2}{\pi f T/2} \right)^2 \sin^2 (\pi f T/2) \tag{6.11}$$

where $\pm V$ are used as the positive/negative voltage levels for this code. Its spectrum is plotted in Fig. 6.2b. When the probability p of a binary **1**, is not equal to one-half, the continuous portion of the PSD is reduced in amplitude and discrete components appear at integer multiples of the bit rate, $R = 1/T$. The resulting PSD is [6, 10]

$$\begin{aligned} S_{5p}(f) &= V^2 T 4 p(1-p) \left(\frac{\sin \pi f T/2}{\pi f T/2} \right)^2 \sin^2 \frac{\pi f T}{2} \\ &+ V^2 (1-2p)^2 \sum_{n=-\infty, n\neq 0}^{\infty} \left(\frac{2}{n\pi} \right)^2 \delta(f - nR) \end{aligned} \tag{6.12}$$

The first null bandwidth of the waveform generated by a Manchester code is $2R$ Hz. The error rate performance of this waveform when $p = 0.5$ is the same as that of polar NRZ, given by Eq. (6.9), and plotted in Fig. 6.3.

The advantages of this code include a zero dc content on an individual pulse basis, so no pattern of bits can cause dc buildup; midbit transitions are always present making it is easy to extract timing information; and it has good error rate performance, identical to polar NRZ. The main disadvantage of this code is a larger bandwidth than any of the other common codes. Also, it has no error detection capability and, hence, performance monitoring is not possible.

Polar NRZ and Manchester coding are examples of the use of pure polar signalling where the pulse for a binary **0**, $g_2(t)$ is the negative of the pulse for a binary **1**, i.e., $g_2(t) = -g_1(t)$. This is also referred to as an antipodal signal set. For this broad type of polar binary line code, the PSD is given by [10]

$$S_{\text{BP}}(f) = 4p(1-p)R|G(f)|^2 + (2p-1)^2 R^2 \sum_{n=-\infty}^{\infty} |G(nR)|^2 \delta(f - nR) \tag{6.13}$$

where $|G(f)|$ is the magnitude of the Fourier transform of either $g_1(t)$ or $g_2(t)$.

A further generalization of the PSD of binary line codes can be given, wherein a continuous spectrum and a discrete spectrum is evident. Let a binary **1,** with probability p, be represented by $g_1(t)$ over the $T = 1/R$ second bit interval; and let a binary **0,** with probability $1 - p$, be represented by $g_2(t)$ over the same T second bit interval. The two-sided PSD for this general binary line code is [10]

$$S_{\text{GB}}(f) = p(1-p)R\,|G_1(f) - G_2(f)|^2$$
$$+ R^2 \sum_{n=-\infty}^{\infty} |pG_1(nR) + (1-p)G_2(nR)|^2\, \delta(f - nR) \qquad (6.14)$$

where the Fourier transform of $g_1(t)$ and $g_2(t)$ are given by $G_1(f)$ and $G_2(f)$, respectively.

6.3   Alternate Line Codes

Most of the line codes discussed thus far were instantaneous level codes. Only AMI had memory, and Manchester was an instantaneous transition code. The alternate line codes presented in this section all have memory. The first four are transition codes, where binary data is represented as the presence or absence of a transition, or by the direction of transition, i.e., positive to negative or vice versa. The last four codes described in this section are level line codes with memory.

6.3.1   Delay Modulation (Miller Code)

In this line code, a binary **1** is represented by a transition at the midbit position, and a binary **0** is represented by no transition at the midbit position. If a **0** is followed by another **0**, however, the signal transition also occurs at the end of the bit interval, that is, between the two **0**s. An example of delay modulation is shown in Fig. 6.1(f). It is clear that delay modulation is a transition code with memory. This code achieves the goal of providing good timing content without sacrificing bandwidth. The PSD of the Miller code for equally likely data is given by [10]

$$S_6(f) = \frac{V^2 T}{2(\pi fT)^2(17 + 8\cos 2\pi fT)}$$
$$\times (23 - 2\cos \pi fT - 22\cos 2\pi fT$$
$$- 12\cos 3\pi fT + 5\cos 4\pi fT + 12\cos 5\pi fT$$
$$+ 2\cos 6\pi fT - 8\cos 7\pi fT + 2\cos 8\pi fT) \qquad (6.15)$$

This spectrum is plotted in Fig. 6.2b. The advantages of this code are that it requires relatively low bandwidth, most of the energy is contained in less than $0.5R$. However, there is no distinct spectral null within the $2R$-Hz band. It has low dc content and no dc component. It has very good timing content, and carrier tracking is easier than Manchester coding. Error rate performance is comparable to that of the common line codes. One important disadvantage is that it has no error detection capability and, hence, performance cannot be monitored.

6.3.2   Split Phase (Mark)

This code is similar to Manchester in the sense that there are always midbit transitions. Hence, this code is relatively easy to synchronize and has no dc. Unlike Manchester, however, split phase (mark) encodes a binary digit into a midbit transition dependent on the midbit transition in the previous bit period [12]. Specifically, a binary **1** produces a reversal of midbit transition relative to the previous midbit transition. A binary **0** produces no reversal of the midbit transition. Certainly this is a transition code with memory. An example of a split phase (mark) coded waveform is shown in Fig. 6.1(g), where the waveform in the first bit period is chosen arbitrarily. Since this method encodes bits differentially, there is no 180°-phase ambiguity associated with some line codes. This phase ambiguity may not be an issue in most baseband links but is important if the line code is modulated. Split phase (space) is very similar to split phase (mark), where the role of the binary **1** and binary **0** are interchanged. An example of a split phase (space) coded waveform is given in Fig. 6.1(h); again, the first bit waveform is arbitrary.

6.3.3 Biphase (Mark)

This code, designated as Bi ϕ-M, is similar to a Miller code in that a binary **1** is represented by a midbit transition, and a binary **0** has no midbit transition. However, this code always has a transition at the beginning of a bit period [10]. Thus, the code is easy to synchronize and has no dc. An example of Bi ϕ-M is given in Fig. 6.1(i), where the direction of the transition at $t = 0$ is arbitrarily chosen. Biphase (space) or Bi ϕ-S is similar to Bi ϕ-M, except the role of the binary data is reversed. Here a binary **0** (space) produces a midbit transition, and a binary **1** does not have a midbit transition. A waveform example of Bi ϕ-S is shown in Fig. 6.1(j). Both Bi ϕ-S and Bi ϕ-M are transition codes with memory.

6.3.4 Code Mark Inversion (CMI)

This line code is used as the interface to a Consultative Committee on International Telegraphy and Telephony (CCITT) multiplexer and is very similar to Bi ϕ-S. A binary **1** is encoded as an NRZ pulse with alternate polarity, $+V$ or $-V$. A binary **0** is encoded with a definitive midbit transition (or square wave phase) [1]. An example of this waveform is shown in Fig. 6.1(k) where a negative to positive transition (or 180° phase) is used for a binary **0**. The voltage level of the first binary **1** in this example is chosen arbitrarily. This example waveform is identical to Bi ϕ-S shown in Fig. 6.1(j), except for the last bit. CMI has good synchronization properties and has no dc.

6.3.5 NRZ (I)

This type of line code uses an inversion (I) to designate binary digits, specifically, a change in level or no change in level. There are two variants of this code, NRZ mark (M) and NRZ space (S) [5, 12]. In NRZ (M), a change of level is used to indicate a binary **1,** and no change of level is used to indicate a binary **0.** In NRZ (S) a change of level is used to indicate a binary **0,** and no change of level is used to indicate a binary **1.** Waveforms for NRZ (M) and NRZ (S) are depicted in Fig. 6.1(l) and Fig. 6.1(m), respectively, where the voltage level of the first binary **1** in the example is chosen arbitrarily. These codes are level codes with memory. In general, line codes that use differential encoding, like NRZ (I), are insensitive to 180° phase ambiguity. Clock recovery with NRZ (I) is not particularly good, and dc wander is a problem as well. Its bandwidth is comparable to polar NRZ.

6.3.6 Binary *N* Zero Substitution (BNZS)

The common bipolar code AMI has many desirable properties of a line code. Its major limitation, however, is that a long string of zeros can lead to loss of synchronization and timing jitter because there are no pulses in the waveform for relatively long periods of time. **Binary N zero substitution (BNZS)** attempts to improve AMI by substituting a special code of length N for all strings of N zeros. This special code contains pulses that look like binary **1**s but purposely produce violations of the AMI pulse convention. Two consecutive pulses of the same polarity violate the AMI pulse convention, independent of the number of zeros between the two consecutive pulses. These violations can be detected at the receiver, and the special code replaced by N zeros. The special code contains pulses facilitating synchronization even when the original data has long string of zeros. The special code is chosen such that the desirable properties of AMI coding are retained despite the AMI pulse convention violations, i.e., dc balance and error detection capability. The only disadvantage of BNZS compared to AMI is a slight increase in crosstalk due to the increased number of pulses and, hence, an increase in the average energy in the code.

Choosing different values of N yields different BNZS codes. The value of N is chosen to meet the timing requirements of the application. In telephony, there are three commonly used BNZS codes: B6ZS, B3ZS, and B8ZS. All BNZS codes are level codes with memory.

In a B6ZS code, a string of six consecutive zeros is replaced by one of two the special codes according to the rule:

If the last pulse was positive $(+)$, the special code is: $0 \; + \; - \; 0 \; - \; +$.

If the last pulse was negative $(-)$, the special code is: $0 \; - \; + \; 0 \; + \; -$.

Here a zero indicates a zero voltage level for the bit period; a plus designates a positive pulse; and a minus indicates a negative pulse.

This special code causes two AMI pulse violations: in its second bit position and in its fifth bit position. These violations are easily detected at the receiver and zeros resubstituted. If the number of consecutive zeros is $12, 18, 24, \ldots$, the substitution is repeated $2, 3, 4, \ldots$ times. Since the number of violations is even, the B6ZS waveform is the same as the AMI waveform outside the special code, i.e., between special code sequences.

There are four pulses introduced by the special code that facilitates timing recovery. Also, note that the special code is dc balanced. An example of the B6ZS code is given as follows, where the special code is indicated by the bold characters.

Original data:	0	1	**0**	0	0	**0**	0	0	1	1	0	1	0	**0**	0	0	**0**	0	0	0	1	1
B6ZS format:	0	+	**0**	+	−	**0**	−	+	−	+	0	−	**0**	−	+	**0**	+	−	+	−

The computation of the PSD of a B6ZS code is tedious. Its shape is given in Fig. 6.4, for comparison purposes with AMI, for the case of equally likely data.

In a B3ZS code, a string of three consecutive zeros is replaced by either $B0V$ or **00V**, where B denotes a pulse obeying the AMI (bipolar) convention and V denotes a pulse violating the AMI convention. $B0V$ or **00V** is chosen such that the number of bipolar (B) pulses between the violations is odd. The B3ZS rules are summarized in Table 6.1.

TABLE 6.1   B3ZS Substitution Rules

Number of B Pulses Since Last Violation	Polarity of Last B Pulse	Substitution Code	Substitution Code Form
Odd	Negative $(-)$	**0 0 −**	**00V**
Odd	Positive $(+)$	**0 0 +**	**00V**
Even	Negative $(-)$	**+ 0 +**	*B0V*
Even	Positive $(+)$	**− 0 −**	*B0V*

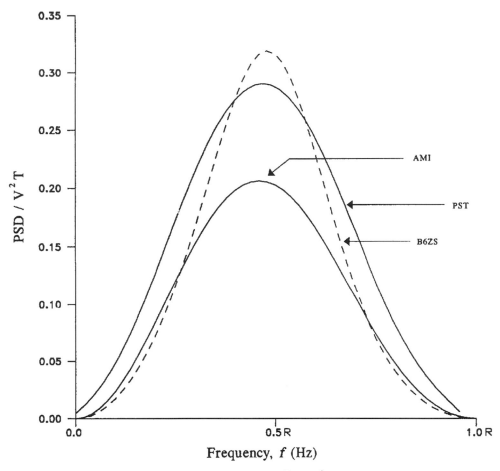

FIGURE 6.4 Power spectral density of different line codes, where $R = 1/\mathrm{T}$ is the bit rate.

Observe that the violation always occurs in the third bit position of the substitution code, and so it can be easily detected and zero replacement made at the receiver. Also, the substitution code selection maintains dc balance. There is either one or two pulses in the substitution code, facilitating synchronization. The error detection capability of AMI is retained in B3ZS because a single channel error would make the number of bipolar pulses between violations even instead of being odd. Unlike B6ZS, the B3ZS waveform between violations may not be the same as the AMI waveform. B3ZS is used in the digital signal-3 (DS-3) signal interface in North America and also in the long distance-4 (LD-4) coaxial transmission system in Canada. Next is an example of a B3ZS code, using the same symbol meaning as in the B6ZS code.

Original data:		1	0	0	1	0	0	0	1	1	0	0	0	0	1	0	0	0	1
B3ZS format:																			
Even No. of B pulses:		+	0	0	−	+	**0**	+	−	+	−	**0**	−	0	+	**0**	**0**	+	−
Odd No. of B pulses:		+	0	0	−	**0**	**0**	−	+	−	+	**0**	+	0	−	**0**	**0**	−	+

The last BNZS code considered here uses $N = 8$. A B8ZS code is used to provide transparent channels for the Integrated Services Digital Network (ISDN) on T1 lines and is similar to the B6ZS code. Here a string of eight consecutive zeros is replaced by one of two special codes according to the following rule:

If the last pulse was positive $(+)$, the special code is:	$0\ \ 0\ \ 0\ +\ -\ 0\ -\ +\ .$
If the last pulse was negative $(-)$, the special code is:	$0\ \ 0\ \ 0\ -\ +\ 0\ +\ -\ .$

There are two bipolar violations in the special codes, at the fourth and seventh bit positions. The code is dc balanced, and the error detection capability of AMI is retained. The waveform between substitutions is the same as that of AMI. If the number of consecutive zeros is $16, 24, \ldots$, then the substitution is repeated $2, 3, \ldots$, times.

6.3.7 High-Density Bipolar N (HDBN)

This coding algorithm is a CCITT standard recommended by the Conference of European Posts and Telecommunications Administrations (CEPT), a European standards body. It is quite similar to BNZS coding. It is thus a level code with memory. Whenever there is a string of $N + 1$ consecutive zeros, they are replaced by a special code of length $N + 1$ containing AMI violations. Specific codes can be constructed for different values of N. A specific **high-density bipolar N (HDBN)** code, HDB3, is implemented as a CEPT primary digital signal. It is very similar to the B3ZS code. In this code, a string of four consecutive zeros is replaced by either $B00V$ or $000V$. $B00V$ or $000V$ is chosen such that the number of bipolar (B) pulses between violations is odd. The HDB3 rules are summarized in Table 6.2.

TABLE 6.2   HDB3 Substitution Rules

Number of B Pulses Since Last Violation	Polarity of Last B Pulse	Substitution Code	Substitution Code Form
Odd	Negative $(-)$	$0\,0\,0\,-$	$000V$
Odd	Positive $(+)$	$0\,0\,0\,+$	$000V$
Even	Negative $(-)$	$+\,0\,0\,+$	$B00V$
Even	Positive $(+)$	$-\,0\,0\,-$	$B00V$

Here the violation always occurs in the fourth bit position of the substitution code, so that it can be easily detected and zero replacement made at the receiver. Also, the substitution code selection maintains dc balance. There is either one or two pulses in the substitution code facilitating synchronization. The error detection capability of AMI is retained in HDB3 because a single channel error would make the number of bipolar pulses between violations even instead of being odd.

6.3.8 Ternary Coding

Many line coding schemes employ three symbols or levels to represent only one bit of information, like AMI. Theoretically, it should be possible to transmit information more efficiently with three symbols, specifically the maximum efficiency is $\log_2 3 = 1.58$ bits per symbol. Alternatively, the redundancy in the code signal space can be used to provide better error control. Two examples of ternary coding are described next [1, 2]: **pair selected ternary (PST)** and **4 binary 3 ternary (4B3T).** The PST code has many of the desirable properties of line codes, but its transmission efficiency is still 1 bit per symbol. The 4B3T code also has many of the desirable properties of line codes, and it has increased transmission efficiency.

In the PST code, two consecutive bits, termed a binary pair, are grouped together to form a word. These binary pairs are assigned codewords consisting of two ternary symbols, where each ternary symbol can be $+$, $-$, or 0, just as in AMI. There are nine possible ternary codewords. Ternary codewords with identical elements, however, are avoided, i.e., $++$, $--$, and 00. The remaining six codewords are transmitted using two modes called $+$ mode and $-$ mode. The modes are switched whenever a codeword with a single pulse is transmitted. The PST code and mode switching rules are summarized in Table 6.3.

TABLE 6.3 PST Codeword Assignment and Mode Switching Rules

Binary Pair	Ternary Codewords + Mode	Ternary Codewords − Mode	Mode Switching
11	$+ -$	$+ -$	No
10	$+ 0$	$- 0$	Yes
01	$0 +$	$0 -$	Yes
00	$- +$	$- +$	No

PST is designed to maintain dc balance and include a strong timing component. One drawback of this code is that the bits must be framed into pairs. At the receiver, an *out-of-frame* condition is signalled when unused ternary codewords ($++$, $--$, and 00) are detected. The mode switching property of PST provides error detection capability. PST can be classified as a level code with memory.

If the original data for PST coding contains only **1**s or **0**s, an alternating sequence of $+- +- \cdots$ is transmitted. As a result, an out-of-frame condition can not be detected. This problem can be minimized by using the modified PST code as shown in Table 6.4.

TABLE 6.4 Modified PST Codeword Assignment and Mode Switching Rules

Binary Pair	Ternary Codewords + Mode	Ternary Codewords − Mode	Mode Switching
11	$+ 0$	$0 -$	Yes
10	$+ -$	$+ -$	No
01	$- +$	$- +$	No
00	$0 +$	$- 0$	Yes

It is tedious to derive the PSD of a PST coded waveform. Again, Fig. 6.4 shows the PSD of the PST code along with the PSD of AMI and B6ZS for comparison purposes, all for equally likely binary data. Observe that PST has more power than AMI and, thus, a larger amount of energy per bit, which translates into slightly increased crosstalk.

In 4B3T coding, words consisting of four binary digits are mapped into three ternary symbols. Four bits imply $2^4 = 16$ possible binary words, whereas three ternary symbols allow $3^3 = 27$ possible ternary codewords. The binary-to-ternary conversion in 4B3T insures dc balance and a strong timing component. The specific codeword assignment is as shown in Table 6.5.

TABLE 6.5   4B3T Codeword Assignment

Binary Words	Ternary Codewords		
	Column 1	Column 2	Column 3
0000	− − −		+ + +
0001	− − 0		+ + 0
0010	− 0 −		+ 0 +
0011	0 − −		0 + +
0100	− − +		+ + −
0101	− + −		+ − +
0110	+ − −		− + +
0111	− 0 0		+ 0 0
1000	0 − 0		0 + 0
1001	0 0 −		0 0 +
1010		0 + −	
1011		0 − +	
1100		+ 0 −	
1101		− 0 +	
1110		+ − 0	
1111		− + 0	

There are three types of codewords in Table 6.5, organized into three columns. The codewords in the first column have negative dc, codewords in the second column have zero dc, and those in the third column have positive dc. The encoder monitors the integer variable

$$I = N_p - N_n , \tag{6.16}$$

where N_p is the number of positive pulses transmitted and N_n are the number of negative pulses transmitted. Codewords are chosen according to following rule:

If $I < 0$,   choose the ternary codeword from columns 1 and 2.

If $I > 0$,   choose the ternary codeword from columns 2 and 3.

If $I = 0$,   choose the ternary word from column 2, and from column 1
if the previous $I > 0$ or from column 3 if the previous $I < 0$.

Note that the ternary codeword 000 is not used, but the remaining 26 codewords are used in a complementary manner. For example, the column 1 codeword for 0001 is −−0, whereas the column 3 codeword is ++0. The maximum transmission efficiency for the 4B3T code is 1.33 bits per symbol compared to 1 bit per symbol for the other line codes. The disadvantages of 4B3T are that framing is required and that performance monitoring is complicated. The 4B3T code is used in the T148 span line developed by ITT Telecommunications. This code allows transmission of 48 channels using only 50% more bandwidth than required by T1 lines, instead of 100% more bandwidth.

6.4   Multilevel Signalling, Partial Response Signalling, and Duobinary Coding

Ternary coding, such as 4B3T, is an example of the use of more than two levels to improve the transmission efficiency. To increase the transmission efficiency further, more levels and/or more signal processing is needed. Multilevel signalling allows an improvement in the transmission efficiency at the expense of an increase in the error rate, i.e., more transmitter power will be required to maintain a given probability of error. In partial response signalling, intersymbol interference is deliberately introduced by using pulses that are wider and, hence, require less bandwidth. The controlled amount of interference from each pulse can be removed at the receiver. This improves the transmission efficiency, at the expense of increased complexity. **Duobinary coding,** a special case of partial response signalling, requires only the minimum theoretical bandwidth of $0.5R$ Hz. In what follows these techniques are discussed in slightly more detail.

6.4.1 Multilevel Signalling

The number of levels that can be used for a line code is not restricted to two or three. Since more levels or symbols allow higher transmission efficiency, multilevel signalling can be considered in bandwidth-limited applications. Specifically, if the signalling rate or baud rate is R_s and the number of levels used is L, the equivalent transmission bit rate R_b is given by

$$R_b = R_s \log_2[L] \; . \tag{6.17}$$

Alternatively, multilevel signalling can be used to reduce the baud rate, which in turn can reduce crosstalk for the same equivalent bit rate. The penalty, however, is that the SNR must increase to achieve the same error rate. The T1G carrier system of AT&T uses multilevel signalling with $L = 4$ and a baud rate of 3.152 mega-symbols/s to double the capacity of the T1C system from 48 channels to 96 channels. Also, a four level signalling scheme at 80-kB is used to achieve 160 kb/s as a basic rate in a digital subscriber loop (DSL) for ISDN.

6.4.2 Partial Response Signalling and Duobinary Coding

This class of signalling is also called *correlative* coding because it purposely introduces a controlled or correlated amount of intersymbol interference in each symbol. At the receiver, the known amount of interference is effectively removed from each symbol. The advantage of this signalling is that wider pulses can be used requiring less bandwidth, but the SNR must be increased to realize a given error rate. Also, errors can propagate unless *precoding* is used.

There are many commonly used partial response signalling schemes, often described in terms of the delay operator D, which represents one signalling interval delay. For example, in $(1 + D)$ signalling the current pulse and the previous pulse are added. The T1D system of AT&T uses $(1 + D)$ signalling with precoding, referred to as duobinary signalling, to convert binary (two level) data into ternary (three level) data at the same rate. This requires the minimum theoretical channel bandwidth without the deleterious effects of intersymbol interference and avoids error propagation. Complete details regarding duobinary coding are found in Lender, 1963 and Schwartz, 1980. Some partial response signalling schemes, such as $(1 - D)$, are used to shape the bandwidth rather than control it. Another interesting example of duobinary coding is a $(1 - D^2)$, which can be analyzed as the product $(1 - D)$ $(1 + D)$. It is used by GTE in its modified T carrier system. AT&T also uses $(1 - D^2)$ with four input levels to achieve an equivalent data rate of 1.544 Mb/s in only a 0.5-MHz bandwidth.

6.5   Bandwidth Comparison

We have provided the PSD expressions for most of the commonly used line codes. The actual bandwidth requirement, however, depends on the pulse shape used and the definition of bandwidth itself. There are many ways to define bandwidth, for example, as a percentage of the total power or the sidelobe suppression relative to the main lobe. Using the first null of the PSD of the code as the definition of bandwidth, Table 6.6 provides a useful bandwidth comparison.

TABLE 6.6   First Null Bandwidth Comparison

Bandwidth	Codes	
	Unipolar NRZ	BNZS
R	Polar NRZ	HDBN
	Polar RZ (AMI)	PST
$2R$	Unipolar RZ	Split Phase
	Manchester	CMI

The notable omission in Table 6.6 is delay modulation (Miller code). It does not have a first null in the $2R$-Hz band, but most of its power is contained in less than $0.5R$ Hz.

6.6   Concluding Remarks

An in-depth presentation of line coding, particularly applicable to telephony, has been included in this chapter. The most desirable characteristics of line codes were discussed. We introduced five common line codes and eight alternate line codes. Each line code was illustrated by an example waveform. In most cases expressions for the PSD and the probability of error were given and plotted. Advantages and disadvantages of all codes were included in the discussion, and some specific applications were noted. Line codes for optical fiber channels and networks built around them, such as fiber distributed data interface (FDDI) were not included in this section. A discussion of line codes for optical fiber channels, and other new developments in this topic area can be found in [1, 3, 4].

Defining Terms

Alternate mark inversion (AMI): A popular name for bipolar line coding using three levels: zero, positive, and negative.

Binary N zero substitution (BNZS): A class of coding schemes that attempts to improve AMI line coding.

Bipolar: A particular line coding scheme using three levels: zero, positive, and negative.

Crosstalk: An unwanted signal from an adjacent channel.

DC wander: The dc level variation in the received signal due to a channel that cannot support dc.

Duobinary coding: A coding scheme with binary input and ternary output requiring the minimum theoretical channel bandwidth.

4 Binary 3 Ternary (4B3T): A line coding scheme that maps four binary digits into three ternary symbols.

High-density bipolar N (HDBN): A class of coding schemes that attempts to improve AMI.

Level codes: Line codes carrying information in their voltage levels.

Line coding: The process of converting abstract symbols into real, temporal waveforms to be transmitted through a baseband channel.

Nonreturn to zero (NRZ): A signal that stays at a nonzero level for the entire bit duration.

Pair selected ternary (PST): A coding scheme based on selecting a pair of three level symbols.

Polar: A line coding scheme using both polarity of voltages, with or without a zero level.

Return to zero (RZ): A signal that returns to zero for a portion of the bit duration.

Transition codes: Line codes carrying information in voltage level transitions.

Unipolar: A line coding scheme using only one polarity of voltage, in addition to a zero level.

References

[1] Bellamy, J., *Digital Telephony,* John Wiley & Sons, New York, NY, 1991.

[2] Bell Telephone Laboratories Technical Staff Members. *Transmission Systems for Communications,* 4th ed., Western Electric Company, Technical Publications, Winston-Salem, NC, 1970.

[3] Bic, J.C., Duponteil, D., and Imbeaux, J.C., *Elements of Digital Communication,* John Wiley & Sons, New York, NY, 1991.

[4] Bylanski, P., *Digital Transmission Systems,* Peter Peregrinus, Herts, England, 1976.

[5] Couch, L.W., *Modern Communication Systems: Principles and Applications,* Prentice-Hall, Englewood Cliffs, NJ, 1994.

[6] Feher, K., *Digital Modulation Techniques in an Interference Environment,* EMC Encyclopedia Series, Vol. IX. Don White Consultants, Germantown, MD, 1977.

[7] Gibson, J.D., *Principles of Analog and Digital Communications,* MacMillan Publishing, New York, NY, 1993.

[8] Lathi, B.P., *Modern Digital and Analog Communication Systems,* Holt, Rinehart and Winston, Philadelphia, PA, 1989.

[9] Lender, A., Duobinary Techniques for High Speed Data Transmission, *IEEE Trans. Commun. Electron.,* CE-82, 214–218, May 1963.

[10] Lindsey, W.C. and Simon, M.K., *Telecommunication Systems Engineering,* Prentice-Hall, Englewood Cliffs, NJ, 1973.

[11] Schwartz, M., *Information Transmission, Modulation, and Noise,* McGraw-Hill, New York, NY, 1980.

[12] Stremler, F.G., *Introduction to Communication Systems,* Addison-Wesley Publishing, Reading, MA, 1990.

7
Echo Cancellation

Giovanni Cherubini
IBM Zurich Research Laboratory

7.1 Introduction

Full-duplex data transmission over a single twisted-pair cable permits the simultaneous flow of information in two directions when the same frequency band is used. Examples of applications of this technique are found in digital communications systems that operate over the telephone network. In a digital subscriber loop, at each end of the full-duplex link, a circuit known as a hybrid separates the two directions of transmission. To avoid signal reflections at the near- and far-end hybrid, a precise knowledge of the line impedance would be required. Since the line impedance depends on line parameters that, in general, are not exactly known, an attenuated and distorted replica of the transmit signal leaks to the receiver input as an echo signal. Data-driven adaptive echo cancellation mitigates the effects of impedance mismatch.

A similar problem is caused by crosstalk in transmission systems over voice-grade unshielded twisted-pair cables for local-area network applications, where multipair cables are used to physically separate the two directions of transmission. Crosstalk is a statistical phenomenon due to randomly varying differential capacitive and inductive coupling between adjacent two-wire transmission lines. At the rates of several megabits per second that are usually considered for local-area network applications, near-end crosstalk (NEXT) represents the dominant disturbance; hence adaptive NEXT cancellation must be performed to ensure reliable communications.

In voiceband data modems, the model for the echo channel is considerably different from the echo model adopted in baseband transmission. The transmitted signal is a passband signal obtained by quadrature amplitude modulation (QAM), and the far-end echo may exhibit significant carrier-phase jitter and carrier-frequency shift, which are caused by signal processing at intermediate points in the telephone network. Therefore, a digital adaptive echo canceller for voiceband modems needs to embody algorithms that account for the presence of such additional impairments.

In this chapter, we describe the echo channel models and adaptive echo canceller structures that are obtained for various digital communications systems, which are classified according to the employed modulation techniques. We also address the tradeoffs between complexity, speed of adaptation, and accuracy of cancellation in adaptive echo cancellers.

7.2 Echo Cancellation for Pulse–Amplitude Modulation (PAM) Systems

The model of a full-duplex baseband data transmission system employing pulse–amplitude modulation (PAM) and adaptive echo cancellation is shown in Fig. 7.1. To describe system operations, we consider one end of the full-duplex link. The configuration of an echo canceller for a PAM transmission system is shown in Fig. 7.2. The transmitted data consist of a sequence $\{a_n\}$ of independent and identically distributed (i.i.d.) real-valued symbols from the M-ary alphabet $\mathcal{A} = \{\pm 1, \pm 3, \ldots, \pm (M-1)\}$. The sequence $\{a_n\}$ is converted into an analog signal by a digital-to-analog (D/A) converter. The conversion to a staircase signal by a zero-order hold D/A converter is described by the frequency response $H_{\mathrm{D/A}}(f) = T\sin(\pi fT)/(\pi fT)$, where T is the modulation interval. The D/A converter output is filtered by the analog transmit filter and is input to the channel through the hybrid.

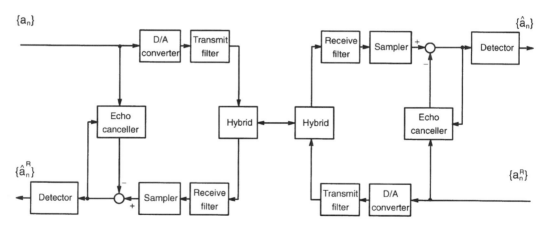

FIGURE 7.1   Model of a full-duplex PAM transmission system.

The signal $x(t)$ at the output of the low-pass analog receive filter has three components, namely, the signal from the far-end transmitter $r(t)$, the echo $u(t)$, and additive Gaussian noise $w(t)$. The signal $x(t)$ is given by

$$
\begin{aligned}
x(t) &= r(t) + u(t) + w(t) \\
&= \sum_{n=-\infty}^{\infty} a_n^R h(t - nT) + \sum_{n=-\infty}^{\infty} a_n h_E(t - nT) + w(t) \ ,
\end{aligned}
\tag{7.1}
$$

where $\{a_n^R\}$ is the sequence of symbols from the remote transmitter, and $h(t)$ and $h_E(t) = \{h_{\mathrm{D/A}} \otimes g_E\}(t)$ are the impulse responses of the overall channel and the echo channel, respectively. In the expression of $h_E(t)$, the function $h_{\mathrm{D/A}}(t)$ is the inverse Fourier transform of $H_{\mathrm{D/A}}(f)$, and the operator $\otimes$ denotes convolution. The signal obtained after echo cancellation is processed by a detector that outputs the sequence of estimated symbols $\{\hat{a}_n^R\}$. In

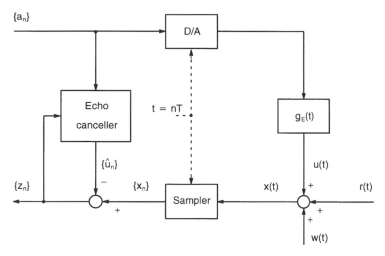

FIGURE 7.2 Configuration of an echo canceller for a PAM transmission system.

the case of full-duplex PAM data transmission over multi-pair cables for local-area network applications, where NEXT represents the main disturbance, the configuration of a digital NEXT canceller is obtained from Fig. 7.2, with the echo channel replaced by the crosstalk channel. For these applications, however, instead of *mono-duplex* transmission, where one pair is used to transmit only in one direction and the other pair to transmit only in the reverse direction, *dual-duplex* transmission may be adopted. Bi-directional transmission at rate ϱ over two pairs is then accomplished by full-duplex transmission of data streams at rate $\varrho/2$ over each of the two pairs. The lower modulation rate and/or spectral efficiency required per pair for achieving an aggregate rate equal to ϱ represents an advantage of dual-duplex over mono-duplex transmission. Dual-duplex transmission requires two transmitters and two receivers at each end of a link, as well as separation of the simultaneously transmitted and received signals on each pair, as illustrated in Fig. 7.3. In dual-duplex transceivers it is therefore necessary to suppress echoes returning from the hybrids and impedance discontinuities in the cable, as well as self NEXT, by adaptive digital echo and NEXT cancellation [3]. Although a dual-duplex scheme might appear to require higher implementation complexity than a mono-duplex scheme, it turns out that the two schemes are equivalent in terms of the number of multiply-and-add operations per second that are needed to perform the various filtering operations.

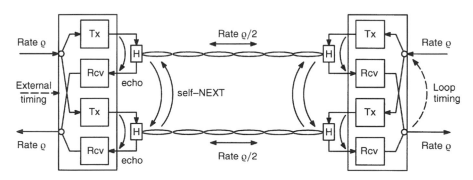

FIGURE 7.3 Model of a dual-duplex transmission system.

One of the transceivers in a full-duplex link will usually employ an externally provided reference clock for its transmit and receive operations. The other transceiver will extract timing from the received signal, and use this timing for its transmitter operations. This is known as *loop timing,* also illustrated in Fig. 7.3. If signals were transmitted in opposite directions with independent clocks, signals received from the remote transmitter would generally shift in phase relative to the also received echo signals. To cope with this effect, some form of interpolation would be required that can significantly increase the transceiver complexity [2].

In general, we consider baseband signalling techniques such that the signal at the output of the overall channel has nonnegligible excess bandwidth, i.e., nonnegligible spectral components at frequencies larger than half of the modulation rate, $|f| \geq 1/2T$. Therefore, to avoid aliasing, the signal $x(t)$ is sampled at twice the modulation rate or at a higher sampling rate. Assuming a sampling rate equal to $m/T, m > 1$, the ith sample during the nth modulation interval is given by

$$x\left[(nm+i)\frac{T}{m}\right] = x_{nm+i} = r_{nm+i} + u_{nm+i} + w_{nm+i}, \qquad i = 0, \ldots, m-1$$

$$= \sum_{k=-\infty}^{\infty} h_{km+i}a^{R}_{n-k} + \sum_{k=-\infty}^{\infty} h_{E,km+i}a_{n-k} + w_{nm+i} , \qquad (7.2)$$

where $\{h_{nm+i}, \ i = 0, \ldots, m-1\}$ and $\{h_{E,nm+i}, \ i = 0, \ldots, m-1\}$ are the discrete-time impulse responses of the overall channel and the echo channel, respectively, and $\{w_{nm+i}, i = 0, \ldots, m-1\}$ is a sequence of Gaussian noise samples with zero mean and variance σ^2_w. Equation (7.2) suggests that the sequence of samples $\{x_{nm+i}, \ i = 0, \ldots, m-1\}$ be regarded as a set of m interleaved sequences, each with a sampling rate equal to the modulation rate. Similarly, the sequence of echo samples $\{u_{nm+i}, \ i = 0, \ldots, m-1\}$ can be regarded as a set of m interleaved sequences that are output by m independent echo channels with discrete-time impulse responses $\{h_{E,nm+i}\}, \ i = 0, \ldots, m-1$, and an identical sequence $\{a_n\}$ of input symbols [7]. Hence, echo cancellation can be performed by m interleaved echo cancellers, as shown in Fig. 7.4. Since the performance of each canceller is independent of the other $m-1$ units, in the remaining part of this section we will consider the operations of a single echo canceller.

The echo canceller generates an estimate $\hat{u}_n$ of the echo signal. If we consider a transversal filter realization, $\hat{u}_n$ is obtained as the inner product of the vector of filter coefficients at time $t = nT$, $\boldsymbol{c}_n = (c_{n,0}, \ldots, c_{n,N-1})'$ and the vector of signals stored in the echo canceller delay line at the same instant, $\boldsymbol{a}_n = (a_n, \ldots, a_{n-N+1})'$, expressed by

$$\hat{u}_n = \boldsymbol{c}'_n\boldsymbol{a}_n = \sum_{k=0}^{N-1} c_{n,k}a_{n-k} \qquad (7.3)$$

where $\boldsymbol{c}'_n$ denotes the transpose of the vector $\boldsymbol{c}_n$. The estimate of the echo is subtracted from the received signal. The result is defined as the cancellation error signal

$$z_n = x_n - \hat{u}_n = x_n - \boldsymbol{c}'_n\boldsymbol{a}_n . \qquad (7.4)$$

The echo attenuation that must be provided by the echo canceller to achieve proper system operation depends on the application. For example, for the Integrated Services Digital Network (ISDN) U-Interface transceiver, the echo attenuation must be larger than 55 dB [10]. It is then required that the echo signals outside of the time span of the echo canceller delay line be negligible, i.e., $h_{E,n} \approx 0$ for $n < 0$ and $n > N-1$. As a measure

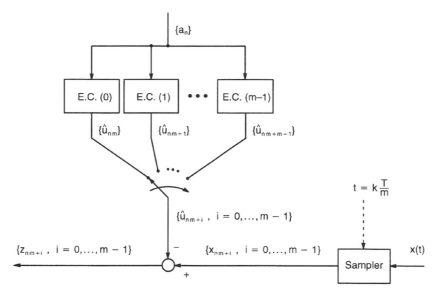

FIGURE 7.4 A set of m interleaved echo cancellers.

of system performance, we consider the mean square error ε_n^2 at the output of the echo canceller at time $t = nT$, defined by

$$\varepsilon_n^2 = E\left\{z_n^2\right\} , \tag{7.5}$$

where $\{z_n\}$ is the error sequence and $E\{\cdot\}$ denotes the expectation operator. For a particular coefficient vector $\boldsymbol{c}_n$, substitution of Eq. (7.4) into Eq. (7.5) yields

$$\varepsilon_n^2 = E\left\{x_n^2\right\} - 2\boldsymbol{c}_n'\boldsymbol{q} + \boldsymbol{c}_n'\boldsymbol{R}\boldsymbol{c}_n , \tag{7.6}$$

where $\boldsymbol{q} = E\{x_n \boldsymbol{a}_n\}$ and $\boldsymbol{R} = E\{\boldsymbol{a}_n \boldsymbol{a}_n'\}$. With the assumption of i.i.d. transmitted symbols, the correlation matrix $\boldsymbol{R}$ is diagonal. The elements on the diagonal are equal to the variance of the transmitted symbols, $\sigma_a^2 = (M^2 - 1)/3$. The minimum mean square error is given by

$$\varepsilon_{\min}^2 = E\left\{x_n^2\right\} - \boldsymbol{c}_{\text{opt}}'\boldsymbol{R}\boldsymbol{c}_{\text{opt}} , \tag{7.7}$$

where the optimum coefficient vector is $\boldsymbol{c}_{\text{opt}} = \boldsymbol{R}^{-1}\boldsymbol{q}$. We note that proper system operation is achieved only if the transmitted symbols are uncorrelated with the symbols from the remote transmitter. If this condition is satisfied, the optimum filter coefficients are given by the values of the discrete-time echo channel impulse response, i.e., $c_{\text{opt},k} = h_{E,k}, k = 0, \ldots, N - 1$.

By the decision-directed stochastic gradient algorithm, also known as the least mean square (LMS) algorithm, the coefficients of the echo canceller converge in the mean to $\boldsymbol{c}_{\text{opt}}$. The LMS algorithm for an N-tap adaptive linear transversal filter is formulated as follows:

$$\boldsymbol{c}_{n+1} = \boldsymbol{c}_n - \frac{1}{2}\alpha\nabla_{\boldsymbol{c}}\left\{z_n^2\right\} = \boldsymbol{c}_n + \alpha z_n \boldsymbol{a}_n , \tag{7.8}$$

where α is the adaptation gain and

$$\nabla_{\boldsymbol{c}}\left\{z_n^2\right\} = \left(\frac{\partial z_n^2}{\partial c_{n,0}}, \ldots, \frac{\partial z_n^2}{\partial c_{n,N-1}}\right)' = -2z_n \boldsymbol{a}_n$$

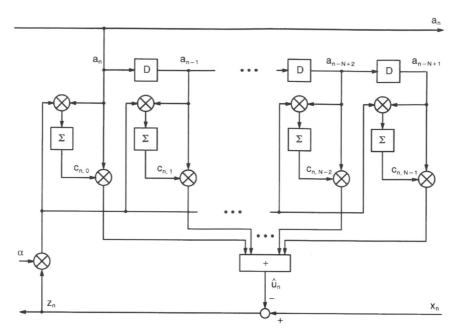

FIGURE 7.5   Block diagram of an adaptive transversal filter echo canceller.

is the gradient of the squared error with respect to the vector of coefficients. The block diagram of an adaptive transversal filter echo canceller is shown in Fig. 7.5.

If we define the vector $\boldsymbol{p}_n = \boldsymbol{c}_{\mathrm{opt}} - \boldsymbol{c}_n$, the mean square error can be expressed as

$$\varepsilon_n^2 = \varepsilon_{\min}^2 + \boldsymbol{p}_n' \boldsymbol{R} \boldsymbol{p}_n \;, \tag{7.9}$$

where the term $\boldsymbol{p}_n' \boldsymbol{R} \boldsymbol{p}_n$ represents an 'excess mean square distortion' due to the misadjustment of the filter settings. The analysis of the convergence behavior of the excess mean square distortion was first proposed for adaptive equalizers [13] and later extended to adaptive echo cancellers [9]. Under the assumption that the vectors $\boldsymbol{p}_n$ and $\boldsymbol{a}_n$ are statistically independent, the dynamics of the mean square error are given by

$$E\left\{\varepsilon_n^2\right\} = \varepsilon_0^2 \left[1 - \alpha\sigma_a^2 \left(2 - \alpha N \sigma_a^2\right)\right]^n + \frac{2\varepsilon_{\min}^2}{2 - \alpha N \sigma_a^2} \;, \tag{7.10}$$

where ε_0^2 is determined by the initial conditions. The mean square error converges to a finite steady-state value ε_∞^2 if the stability condition $0 < \alpha < 2/(N\sigma_a^2)$ is satisfied. The optimum adaptation gain that yields fastest convergence at the beginning of the adaptation process is $\alpha_{\mathrm{opt}} = 1/(N\sigma_a^2)$. The corresponding time constant and asymptotic mean square error are $\tau_{\mathrm{opt}} = N$ and $\varepsilon_\infty^2 = 2\varepsilon_{\min}^2$, respectively.

We note that a fixed adaptation gain equal to α_{opt} could not be adopted in practice, since after echo cancellation the signal from the remote transmitter would be embedded in a residual echo having approximately the same power. If the time constant of the convergence mode is not a critical system parameter, an adaptation gain smaller than α_{opt} will be adopted to achieve an asymptotic mean square error close to $\varepsilon_{\min}^2$. On the other hand, if fast convergence is required, a variable gain will be chosen.

Several techniques have been proposed to increase the speed of convergence of the LMS algorithm. In particular, for echo cancellation in data transmission, the speed of adaptation is reduced by the presence of the signal from the remote transmitter in the cancellation error.

To mitigate this problem, the data signal can be adaptively removed from the cancellation error by a decision-directed algorithm [5].

Modified versions of the LMS algorithm have been also proposed to reduce system complexity. For example, the sign algorithm suggests that only the sign of the error signal be used to compute an approximation of the stochastic gradient [4]. An alternative means to reduce the implementation complexity of an adaptive echo canceller consists in the choice of a filter structure with a lower computational complexity than the transversal filter.

At high data rates, very large scale integration (VLSI) technology is needed for the implementation of transceivers for full-duplex data transmission. High-speed echo cancellers and near-end crosstalk cancellers that do not require multiplications represent an attractive solution because of their low complexity. As an example of an architecture suitable for VLSI implementation, we consider echo cancellation by a distributed-arithmetic filter, where multiplications are replaced by table lookup and shift-and-add operations [12]. By segmenting the echo canceller into filter sections of shorter lengths, various tradeoffs concerning the number of operations per modulation interval and the number of memory locations needed to store the lookup tables are possible. Adaptivity is achieved by updating the values stored in the lookup tables by the LMS algorithm.

To describe the principles of operations of a distributed-arithmetic echo canceller, we assume that the number of elements in the alphabet of input symbols is a power of two, $M = 2^W$. Therefore, each symbol is represented by the vector $(a_n^{(0)}, \ldots, a_n^{(W-1)})$, where $a_n^{(i)}, i = 0, \ldots, W-1$, are independent binary random variables, i.e.,

$$a_n = \sum_{w=0}^{W-1} \left(2a_n^{(w)} - 1 \right) 2^w = \sum_{w=0}^{W-1} b_n^{(w)} 2^w \;, \tag{7.11}$$

where $b_n^{(w)} = (2a_n^{(w)} - 1) \in \{-1, +1\}$. By substituting Eq. (7.11) into Eq. (7.1) and segmenting the delay line of the echo canceller into L sections with $K = N/L$ delay elements each, we obtain

$$\hat{u}_n = \sum_{\ell=0}^{L-1} \sum_{w=0}^{W-1} 2^w \left[\sum_{k=0}^{K-1} b_{n-\ell K-k}^{(w)} c_{n,\ell K+k} \right] \;. \tag{7.12}$$

Equation (7.12) suggests that the filter output can be computed using a set of $L2^K$ values that are stored in L tables with 2^K memory locations each. The binary vectors $\boldsymbol{a}_{n,\ell}^{(w)}$ $= (a_{n-(\ell+1)K+1}^{(w)}, \ldots, a_{n-\ell K}^{(w)})$, $w = 0, \ldots, W-1$, $\ell = 0, \ldots, L-1$, determine the addresses of the memory locations where the values that are needed to compute the filter output are stored. The filter output is obtained by WL table lookup and shift-and-add operations.

We observe that $\boldsymbol{a}_{n,\ell}^{(w)}$ and its binary complement $\bar{\boldsymbol{a}}_{n,\ell}^{(w)}$ select two values that differ only in their sign. This symmetry is exploited to halve the number of values to be stored. To determine the output of a distributed-arithmetic filter with reduced memory size, we reformulate Eq. (7.12) as

$$\hat{u}_n = \sum_{\ell=0}^{L-1} \sum_{w=0}^{W-1} 2^w b_{n-\ell K-k_0}^{(w)} \left[c_{n,\ell K+k_0} + b_{n-\ell K-k_0}^{(w)} \sum_{\substack{k=0 \\ k \neq k_0}}^{K-1} b_{n-\ell K-k}^{(w)} c_{n,\ell K+k} \right] \;, \tag{7.13}$$

where k_0 can be any element of the set $\{0, \ldots, K-1\}$. In the following, we take $k_0 = 0$. Then the binary symbols $b_{n-\ell K}^{(w)}$ determine whether the selected values are to be added or subtracted. Each table has now 2^{K-1} memory locations, and the filter output is given by

$$\hat{u}_n = \sum_{\ell=0}^{L-1} \sum_{w=0}^{W-1} 2^w b_{n-\ell K}^{(w)} d_n \left(i_{n,\ell}^{(w)}, \ell \right) , \qquad (7.14)$$

where $d_n(k, \ell)$, $k = 0, \ldots, 2^{K-1} - 1$, $\ell = 0, \ldots, L - 1$, are the look up values, and $i_{n,\ell}^{(w)}$, $w = 0, \ldots, W - 1$, $\ell = 0, \ldots, L - 1$, are the look up indices computed as follows:

$$i_{n,\ell}^{(w)} = \begin{cases} \displaystyle\sum_{k=1}^{K-1} a_{n-\ell K-k}^{(w)} 2^{k-1} & \text{if } a_{n-\ell K}^{(w)} = 1 \\[4mm] \displaystyle\sum_{k=1}^{K-1} \bar{a}_{n-\ell K-k}^{(w)} 2^{k-1} & \text{if } a_{n-\ell K}^{(w)} = 0 \end{cases} . \qquad (7.15)$$

We note that, as long as Eqs. (7.12) and (7.13) hold for some coefficient vector $(c_{n,0}, \ldots, c_{n,N-1})$, the distributed-arithmetic filter emulates the operation of a linear transversal filter. For arbitrary values $d_n(k, \ell)$, however, a nonlinear filtering operation results.

The expression of the LMS algorithm to update the values of a distributed-arithmetic echo canceller takes the form

$$\boldsymbol{d}_{n+1} = \boldsymbol{d}_n - \frac{1}{2}\alpha \nabla_{\boldsymbol{d}} \left\{ z_n^2 \right\} = \boldsymbol{d}_n + \alpha z_n \boldsymbol{y}_n , \qquad (7.16)$$

where $\boldsymbol{d}_n' = [\boldsymbol{d}_n'(0), \ldots, \boldsymbol{d}_n'(L-1)]$, with $\boldsymbol{d}_n'(\ell) = [d_n(0, \ell), \ldots, d_n(2^{K-1} - 1, \ell)]$, and $\boldsymbol{y}_n' = [\boldsymbol{y}_n'(0), \ldots, \boldsymbol{y}_n'(L-1)]$, with

$$\boldsymbol{y}_n'(\ell) = \sum_{w=0}^{W-1} 2^w b_{n-\ell K}^{(w)} \left(\delta_{0,i_{n,\ell}^{(w)}}, \ldots, \delta_{2^{K-1}-1,i_{n,\ell}^{(w)}} \right) ,$$

are $L2^{K-1} \times 1$ vectors and where $\delta_{i,j}$ is the Kronecker delta. We note that at each iteration only those values that are selected to generate the filter output are updated. The block diagram of an adaptive distributed-arithmetic echo canceller with input symbols from a quaternary alphabet is shown in Fig. 7.6.

The analysis of the mean square error convergence behavior and steady-state performance has been extended to adaptive distributed-arithmetic echo cancellers [1]. The dynamics of the mean square error are given by

$$E \left\{ \varepsilon_n^2 \right\} = \varepsilon_0^2 \left[1 - \frac{\alpha \sigma_a^2}{2^{K-1}} \left(2 - \alpha L \sigma_a^2 \right) \right]^n + \frac{2\varepsilon_{\min}^2}{2 - \alpha L \sigma_a^2} . \qquad (7.17)$$

The stability condition for the echo canceller is $0 < \alpha < 2/(L\sigma_a^2)$. For a given adaptation gain, echo canceller stability depends on the number of tables and on the variance of the transmitted symbols. Therefore, the time span of the echo canceller can be increased without affecting system stability, provided that the number L of tables is kept constant. In that case, however, mean square error convergence will be slower. From Eq. (7.17), we find that the optimum adaptation gain that permits the fastest mean square error convergence at the beginning of the adaptation process is $\alpha_{\mathrm{opt}} = 1/(L\sigma_a^2)$. The time constant of the convergence mode is $\tau_{\mathrm{opt}} = L2^{K-1}$. The smallest achievable time constant is proportional to the total number of values. The realization of a distributed-arithmetic echo canceller can be further simplified by updating at each iteration only the values that are addressed by the most significant bits of the symbols stored in the delay line. The complexity required for adaptation can thus be reduced at the price of a slower rate of convergence.

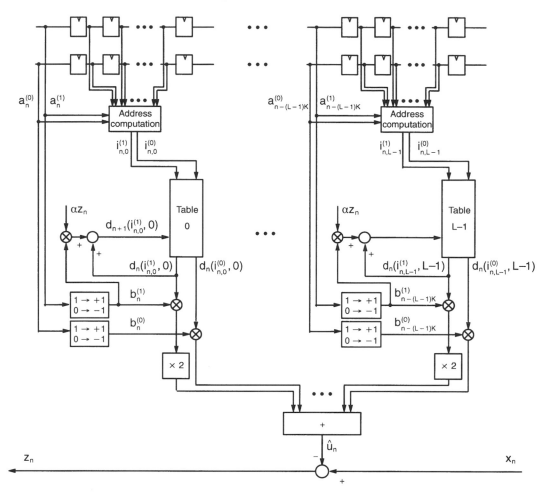

FIGURE 7.6 Block diagram of an adaptive distributed-arithmetic echo canceller.

7.3 Echo Cancellation for Quadrature Amplitude Modulation (QAM) Systems

Although most of the concepts presented in the preceding sections can be readily extended to echo cancellation for communications systems employing QAM, the case of full-duplex transmission over a voiceband data channel requires a specific discussion. We consider the system model shown in Fig. 7.7. The transmitter generates a sequence $\{a_n\}$ of i.i.d. complex-valued symbols from a two-dimensional constellation $\mathcal{A}$, which are modulated by the carrier $e^{j2\pi f_c nT}$, where T and f_c denote the modulation interval and the carrier frequency, respectively. The discrete-time signal at the output of the transmit Hilbert filter may be regarded as an analytic signal, which is generated at the rate of m/T samples/s, $m > 1$. The real part of the analytic signal is converted into an analog signal by a D/A converter and input to the channel. We note that by transmitting the real part of a complex-valued signal positive- and negative-frequency components become folded. The image band attenuation of the transmit Hilbert filter thus determines the achievable echo suppression. In fact, the receiver cannot extract aliasing image-band components from desired passband frequency components, and the echo canceller is able to suppress only echo arising from transmitted passband components.

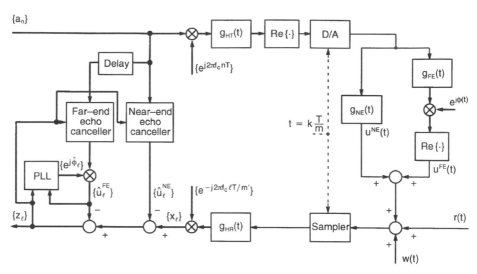

FIGURE 7.7   Configuration of an echo canceller for a QAM transmission system.

The output of the echo channel is represented as the sum of two contributions. The near-end echo $u^{\mathrm{NE}}(t)$ arises from the impedance mismatch between the hybrid and the transmission line, as in the case of baseband transmission. The far-end echo $u^{\mathrm{FE}}(t)$ represents the contribution due to echos that are generated at intermediate points in the telephone network. These echos are characterized by additional impairments, such as jitter and frequency shift, which are accounted for by introducing a carrier-phase rotation of an angle $\phi(t)$ in the model of the far-end echo.

At the receiver, samples of the signal at the channel output are obtained synchronously with the transmitter timing, at the sampling rate of m/T samples/s. The discrete-time received signal is converted to a complex-valued baseband signal $\{x_{nm'+i}, i = 0, \ldots, m'-1\}$, at the rate of m'/T samples/s, $1 < m' < m$, through filtering by the receive Hilbert filter, decimation, and demodulation. From delayed transmit symbols, estimates of the near- and far-end echo signals after demodulation, $\{\hat{u}^{\mathrm{NE}}_{nm'+i}, i = 0, \ldots, m'-1\}$ and $\{\hat{u}^{\mathrm{FE}}_{nm'+i}, i = 0, \ldots, m'-1\}$, respectively, are generated using m' interleaved near- and far-end echo cancellers. The cancellation error is given by

$$z_\ell = x_\ell - \hat{u}^{\mathrm{NE}}_\ell - \hat{u}^{\mathrm{FE}}_\ell \ . \tag{7.18}$$

A different model is obtained if echo cancellation is accomplished before demodulation. In this case, two equivalent configurations for the echo canceller may be considered. In one configuration, the modulated symbols are input to the transversal filter, which approximates the passband echo response. Alternatively, the modulator can be placed after the transversal filter, which is then called a baseband transversal filter [14].

In the considered realization, the estimates of the echo signals after demodulation are given by

$$\hat{u}^{\mathrm{NE}}_{nm'+i} = \sum_{k=0}^{N_{\mathrm{NE}}-1} c^{\mathrm{NE}}_{n,km'+i} a_{n-k}, \qquad i = 0, \ldots, m'-1 \ , \tag{7.19}$$

and

$$\hat{u}_{nm'+i}^{\mathrm{FE}} = \left[\sum_{k=0}^{N_{\mathrm{FE}}-1} c_{n,km'+i}^{\mathrm{FE}} a_{n-k-D_{\mathrm{FE}}} \right] e^{j\hat{\phi}_{nm'+i}}, \qquad i = 0, \ldots, m'-1 , \qquad (7.20)$$

where $(c_{n,0}^{\mathrm{NE}}, \ldots, c_{n,m'N_{\mathrm{NE}}-1}^{\mathrm{NE}})$ and $(c_{n,0}^{\mathrm{FE}}, \ldots, c_{n,m'N_{\mathrm{FE}}-1}^{\mathrm{FE}})$ are the coefficients of the m' interleaved near- and far-end echo cancellers, respectively, $\{\hat{\phi}_{nm'+i}, i = 0, \ldots, m'-1\}$ is the sequence of far-end echo phase estimates, and D_{FE} denotes the bulk delay accounting for the round-trip delay from the transmitter to the point of echo generation. To prevent overlap of the time span of the near-end echo canceller with the time span of the far-end echo canceller, the condition $D_{\mathrm{FE}} > N_{\mathrm{NE}}$ must be satisfied. We also note that, because of the different nature of near- and far-end echo generation, the time span of the far-end echo canceller needs to be larger than the time span of the near-end echo canceller, i.e., $N_{\mathrm{FE}} > N_{\mathrm{NE}}$.

Adaptation of the filter coefficients in the near- and far-end echo cancellers by the LMS algorithm leads to

$$\begin{aligned} c_{n+1,km'+i}^{\mathrm{NE}} &= c_{n,km'+i}^{\mathrm{NE}} + \alpha z_{nm'+i}(a_{n-k})^* \\ k &= 0, \ldots, N_{\mathrm{NE}} - 1, \qquad i = 0, \ldots, m'-1 , \end{aligned} \qquad (7.21)$$

and

$$\begin{aligned} c_{n+1,km'+i}^{\mathrm{FE}} &= c_{n,km'+i}^{\mathrm{FE}} + \alpha z_{nm'+i}(a_{n-k-D_{\mathrm{FE}}})^* e^{-j\hat{\phi}_{nm'+i}} \\ k &= 0, \ldots, N_{\mathrm{FE}} - 1, \qquad i = 0, \ldots, m'-1 , \end{aligned} \qquad (7.22)$$

respectively, where the asterisk denotes complex conjugation.

The far-end echo phase estimate is computed by a second-order phase-lock loop algorithm, where the following stochastic gradient approach is adopted:

$$\begin{cases} \hat{\phi}_{\ell+1} = \hat{\phi}_\ell - \frac{1}{2}\gamma_{\mathrm{FE}}\nabla_{\hat{\phi}}|z_\ell|^2 + \Delta\phi_\ell \qquad (\mathrm{mod}\ \ 2\pi) \\ \\ \Delta\phi_{\ell+1} = \Delta\phi_\ell - \frac{1}{2}\zeta_{\mathrm{FE}}\nabla_{\hat{\phi}}|z_\ell|^2 \end{cases}, \qquad (7.23)$$

where $\ell = nm' + i$, $i = 0, \ldots, m'-1, \gamma_{\mathrm{FE}}$ and ζ_{FE} are step-size parameters, and

$$\nabla_{\hat{\phi}}|z_\ell|^2 = \frac{\partial |z_\ell|^2}{\partial \hat{\phi}_\ell} = -2\mathrm{Im}\left\{ z_\ell \left(\hat{u}_\ell^{\mathrm{FE}} \right)^* \right\} . \qquad (7.24)$$

We note that algorithm (7.23) requires m' iterations per modulation interval, i.e., we cannot resort to interleaving to reduce the complexity of the computation of the far-end echo phase estimate.

7.4 Echo Cancellation for Orthogonal Frequency Division Multiplexing (OFDM) Systems

Orthogonal frequency division multiplexing (OFDM) is a modulation technique whereby blocks of M symbols are transmitted in parallel over M subchannels by employing M orthogonal subcarriers. We consider a real-valued discrete-time channel impulse response $\{h_i, i = 0, \ldots, L\}$ having length $L + 1 \ll M$. To illustrate the basic principles of OFDM systems, let us consider a noiseless ideal channel with impulse response given by $\{h_i\} = \{\delta_i\}$, where $\{\delta_i\}$ is defined as the discrete-time delta function. Modulation of the complex-valued input symbols at the n-th modulation interval, denoted by the vector $\boldsymbol{A}_n = \{A_n(i), i = 0, \ldots, M-1\}$, is performed by an inverse discrete Fourier transform (IDFT), as shown in Fig. 7.8. We assume that M is even, and that each block of symbols satisfies the Hermitian symmetry conditions, i.e., $A_n(0)$ and $A_n(M/2)$ are real valued, and $A_n(i) = A_n^*(M - i)$, $i = 1, \ldots, M/2 - 1$. Then the signals $\boldsymbol{a}_n = \{a_n(i), i = 0, \ldots, M-1\}$ obtained at the output of the IDFT are real valued. After parallel-to-serial conversion, the M signals are sent over the channel at the given transmission rate M/T, where T denotes the modulation interval. At the output of the channel, the noiseless signals are received without distortion. Serial-to-parallel conversion yields blocks of M elements, with boundaries placed such that each block obtained at the modulator output is also presented at the demodulator input. Then demodulation performed by a discrete Fourier transform (DFT) will reproduce the blocks of M input symbols. The overall input-output relationship is therefore equivalent to that of a bank of M parallel, independent subchannels.

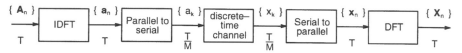

FIGURE 7.8   Block diagram of an OFDM system.

In the general case of a noisy channel with impulse response having length greater than one, M independent subchannels are obtained by a variant of OFDM that is also known as discrete multitone modulation (DMT) [11]. In a DMT system, modulation by the IDFT is performed at the rate $1/T' = M/(M + L)T < 1/T$. After modulation, each block of M signals is cyclically extended by copying the last L signals in front of the block, and converted from parallel to serial. The resulting $L + M$ signals are sent over the channel. At the receiver, blocks of samples with length $L + M$ are taken. Block boundaries are placed such that the last M samples depend only on the elements of one cyclically extended block of signals. The first L samples are discarded, and the vector $\boldsymbol{x}_n$ of the last M samples of the block received at the n-th modulation interval can be expressed as

$$\boldsymbol{x}_n = \Gamma_n \boldsymbol{h} + \boldsymbol{w}_n \,, \tag{7.25}$$

where $\boldsymbol{h}$ is the vector of the impulse response extended with $M - L - 1$ zeros, $\boldsymbol{w}_n$ is a vector of additive white Gaussian noise samples, and Γ_n is a $M \times M$ circulant matrix given by

$$\Gamma_n = \begin{bmatrix} a_n(0) & a_n(M-1) & \ldots & a_n(1) \\ a_n(1) & a_n(0) & \ldots & a_n(2) \\ \cdot & \cdot & & \cdot \\ \cdot & \cdot & & \cdot \\ \cdot & \cdot & & \cdot \\ a_n(M-1) & a_n(M-2) & \ldots & a_n(0) \end{bmatrix}. \tag{7.26}$$

Recalling that $\mathcal{F}_M \Gamma_n \mathcal{F}_M^{-1} = diag(\boldsymbol{A}_n)$, where $\mathcal{F}_M$ is the $M \times M$ DFT matrix defined as $\mathcal{F}_M = [(e^{-\frac{j2\pi}{M}})^{km}], k, m = 0, \ldots, M-1$, and $diag(\boldsymbol{A}_n)$ denotes the diagonal matrix with elements on the diagonal given by $\boldsymbol{A}_n$, we find that the output of the demodulator is given by

$$\boldsymbol{X}_n = diag(\boldsymbol{A}_n)\boldsymbol{H} + \boldsymbol{W}_n \,, \tag{7.27}$$

where $\boldsymbol{H}$ denotes the DFT of the vector $\boldsymbol{h}$, and $\boldsymbol{W}_n$ is a vector of independent Gaussian random variables. Equation (7.27) indicates that the sequence of transmitted symbol vectors can be detected by assuming a bank of M independent subchannels, at the price of a decrease in the data rate by a factor $(M + L)/M$. Note that in practice the computationally more efficient inverse fast Fourier transform and fast Fourier transform are used instead of IDFT and DFT.

We discuss echo cancellation for OFDM with reference to a DMT system [6], as shown in Fig. 7.9. The real-valued discrete-time echo impulse response is $\{h_{E,i}, i = 0, \ldots, N-1\}$,

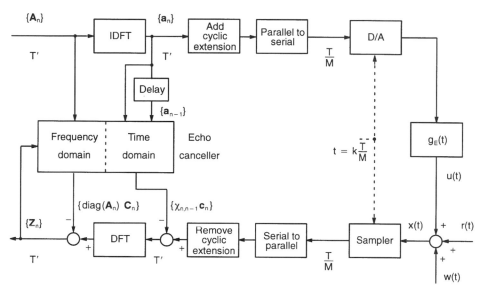

FIGURE 7.9 Configuration of an echo canceller for a DMT transmission system.

having length $N < M$. We initially assume $N \leq L + 1$. Furthermore, we assume that the boundaries of the received blocks are placed such that the last M samples of the n-th received block are expressed by the vector

$$x_n = \Gamma_n^R h + \Gamma_n h_E + w_n \,, \tag{7.28}$$

where Γ_n^R is the circulant matrix with elements given by the signals from the remote transmitter, and h_E is the vector of the echo impulse response extended with $M - N$ zeros. In the frequency domain, the echo is expressed as $U_n = diag(A_n)H_E$, where H_E denotes the DFT of the vector h_E. In this case, the echo canceller provides an echo estimate that is given by $\hat{U}_n = diag(A_n)C_n$, where C_n denotes the DFT of the vector c_n of the N coefficients of the echo canceller filter extended with $M - N$ zeros. In practice, however, we need to consider the case $N > L + 1$. The expression of the cancellation error is then given by

$$z_n = x_n - \Psi_{n,n-1} c_n \,, \tag{7.29}$$

where the vector of the last M elements of the n-th received block is now $x_n = \Gamma_n^R h + \Psi_{n,n-1} h_E + w_n$, and $\Psi_{n,n-1}$ is a $M \times M$ Toeplitz matrix given by

$$\Psi_{n,n-1} =$$

$$\begin{bmatrix} a_n(0) & a_n(M-1) & \cdots & a_n(M-L) & a_{n-1}(M-1) & \cdots & a_{n-1}(L+1) \\ a_n(1) & a_n(0) & \cdots & a_n(M-L+1) & a_n(M-L) & \cdots & a_{n-1}(L+2) \\ \vdots & & \ddots & & & & \vdots \\ a_n(M-1) & a_n(M-2) & \cdots & a_n(M-L-1) & a_n(M-L-2) & \cdots & a_n(0) \end{bmatrix} . \qquad (7.30)$$

In the frequency domain, the cancellation error can be expressed as

$$Z_n = \mathcal{F}_M \left(x_n - \chi_{n,n-1} c_n \right) - diag(A_n) C_n , \qquad (7.31)$$

where $\chi_{n,n-1} = \Psi_{n,n-1} - \Gamma_n$ is a $M \times M$ upper triangular Toeplitz matrix. Equation (7.31) suggests a computationally efficient, two-part echo cancellation technique. First, in the time domain, a short convolution is performed and the result subtracted from the received signals to compensate for the insufficient length of the cyclic extension. Second, in the frequency domain, cancellation of the residual echo is performed over a set of M independent echo subchannels. Observing that Eq. (7.31) is equivalent to $Z_n = X_n - \tilde{\Psi}_{n,n-1} C_n$, where $\tilde{\Psi}_{n,n-1} = \mathcal{F}_M \Psi_{n,n-1} \mathcal{F}_M^{-1}$, the echo canceller adaptation by the LMS algorithm in the frequency domain takes the form

$$C_{n+1} = C_n + \alpha \tilde{\Psi}_{n,n-1}^* Z_n , \qquad (7.32)$$

where α is the adaptation gain, and $\tilde{\Psi}_{n,n-1}^*$ denotes the transpose conjugate of $\tilde{\Psi}_{n,n-1}$. We note that, alternatively, echo canceller adaptation may also be performed by the algorithm $C_{n+1} = C_n + \alpha \, diag(A_n^*) Z_n$, which entails a substantially lower computational complexity than the LMS algorithm, at the price of a slower rate of convergence.

In DMT systems it is essential that the length of the channel impulse response be much less than the number of subchannels, so that the reduction in data rate due to the cyclic extension may be considered negligible. Therefore, equalization is adopted in practice to shorten the length of the channel impulse response. From Eq. (7.31), however, we observe that transceiver complexity depends on the relative lengths of the echo and of the channel impulse responses. To reduce the length of the cyclic extension as well as the computational complexity of the echo canceller, various methods have been proposed to shorten both the channel and the echo impulse responses jointly [8].

7.5   Summary and Conclusions

Digital signal processing techniques for echo cancellation provide large echo attenuation, and eliminate the need for additional line interfaces and digital-to-analog and analog-to-digital converters that are required by echo cancellation in the analog signal domain.

The realization of digital echo cancellers in transceivers for high-speed full-duplex data transmission today is possible at a low cost thanks to the advances in VLSI technology. Digital techniques for echo cancellation are also appropriate for near-end crosstalk cancellation in transceivers for transmission over voice-grade cables at rates of several megabits per second for local-area network applications.

In voiceband modems for data transmission over the telephone network, digital techniques for echo cancellation also allow a precise tracking of the carrier phase and frequency shift of far-end echos.

References

[1] Cherubini, G., Analysis of the convergence behavior of adaptive distributed-arithmetic echo cancellers. *IEEE Trans. Commun.*, 41(11), 1703–1714, 1993.

[2] Cherubini, G., Ölçer, S., and Ungerboeck, G., A quaternaty partial-response class-IV transceiver for 125 Mbit/s data transmission over unshielded twisted-pair cables: Principles of operation and VLSI realization. *IEEE J. Sel. Areas Commun.*, 13(9), 1656–1669, 1995.

[3] Cherubini, G., Creigh, J., Ölçer, S., Rao, S.K., and Ungerboeck, G., 100BASE-T2: A new standard for 100 Mb/s Ethernet transmission over voice-grade cables. *IEEE Commun. Mag.*, 35(11), 115–122, 1997.

[4] Duttweiler, D.L., Adaptive filter performance with nonlinearities in the correlation multiplier. *IEEE Trans. Acoust., Speech, Signal Processing*, 30(8), 578–586, 1982.

[5] Falconer, D.D., Adaptive reference echo-cancellation. *IEEE Trans. Commun.*, 30(9), 2083–2094, 1982.

[6] Ho, M., Cioffi, J.M. and Bingham, J.A.C., Discrete multitone echo cancellation. *IEEE Trans. Commun.*, 44(7), 817–825, 1996.

[7] Lee, E.A. and Messerschmitt, D.G., *Digital Communication*, 2nd ed., Kluwer Academic Publishers, Boston MA, 1994.

[8] Melsa, P.J.W., Younce, R.C., and Rohrs, C.E., Impulse response shortening for discrete multitone transceivers. *IEEE Trans. Commun.*, 44(12), 1662–1672, 1996.

[9] Messerschmitt, D.G., Echo cancellation in speech and data transmission. *IEEE J. Sel. Areas Commun.*, 2(2), 283–297, 1984.

[10] Messerschmitt, D.G., Design issues for the ISDN U-Interface transceiver. *IEEE J. Sel. Areas Commun.*, 4(8), 1281–1293, 1986.

[11] Ruiz, A., Cioffi, J.M., and Kasturia, S., Discrete multiple tone modulation with coset coding for the spectrally shaped channel. *IEEE Trans. Commun.*, 40(6), 1012–1029, 1992.

[12] Smith, M.J., Cowan, C.F.N., and Adams, P.F., Nonlinear echo cancellers based on transpose distributed arithmetic. *IEEE Trans. Circuits and Systems*, 35(1), 6–18, 1988.

[13] Ungerboeck, G., Theory on the speed of convergence in adaptive equalizers for digital communication. *IBM J. Res. Develop.*, 16(6), 546–555, 1972.

[14] Weinstein, S.B., A passband data-driven echo-canceller for full-duplex transmission on two-wire circuits. *IEEE Trans. Commun.*, 25(7), 654–666, 1977.

Further Information

For further information on adaptive transversal filters with application to echo cancellation, see *Adaptive Filters: Structures, Algorithms, and Applications*, M.L. Honig and D.G. Messerschmitt, Kluwer, 1984.

8

Pseudonoise Sequences

Tor Helleseth
University of Bergen

P. Vijay Kumar
University of Southern California

8.1   Introduction

Pseudonoise sequences (PN sequences), also referred to as pseudorandom sequences, are sequences that are deterministically generated and yet possess some properties that one would expect to find in randomly generated sequences. Applications of PN sequences include signal synchronization, navigation, radar ranging, random number generation, spread-spectrum communications, multipath resolution, cryptography, and signal identification in multiple-access communication systems. The *correlation* between two sequences $\{x(t)\}$ and $\{y(t)\}$ is the complex inner product of the first sequence with a shifted version of the second sequence. The correlation is called 1) an autocorrelation if the two sequences are the same, 2) a crosscorrelation if they are distinct, 3) a periodic correlation if the shift is a cyclic shift, 4) an aperiodic correlation if the shift is not cyclic, and 5) a partial-period correlation if the inner product involves only a partial segment of the two sequences. More precise definitions are given subsequently.

Binary m **sequences**, defined in the next section, are perhaps the best-known family of PN sequences. The balance, run-distribution, and autocorrelation properties of these sequences mimic those of random sequences. It is perhaps the random-like correlation properties of PN sequences that makes them most attractive in a communications system, and it is common to refer to any collection of low-correlation sequences as a family of PN sequences.

Section 8.2 begins by discussing m sequences. Thereafter, the discussion continues with a description of sequences satisfying various correlation constraints along the lines of the accompanying self-explanatory figure, Fig. 8.1. Expanded tutorial discussions on pseudo-random sequences may be found in [14], in [15, Chapter 5] and in [6].

8.2 m Sequences

A binary $\{0, 1\}$ **shift-register sequence** $\{s(t)\}$ is a sequence that satisfies a linear recurrence relation of the form

$$\sum_{i=0}^{r} f_i s(t + i) = 0 , \qquad \text{for all}\ \ t \geq 0 \tag{8.1}$$

where $r \geq 1$ is the *degree* of the recursion; the coefficients f_i belong to the finite field $GF(2) = \{0, 1\}$ where the leading coefficient $f_r = 1$. Thus, both sequences $\{a(t)\}$ and $\{b(t)\}$ appearing in Fig. 8.2 are shift-register sequences. A sequence satisfying a recursion of the form in Eq. (8.1) is said to have *characteristic polynomial* $f(x) = \sum_{i=0}^{r} f_i x^i$. Thus, $\{a(t)\}$ and $\{b(t)\}$ have characteristic polynomials given by $f(x) = x^3 + x + 1$ and $f(x) = x^3 + x^2 + 1$, respectively.

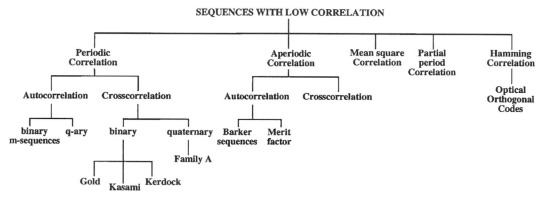

FIGURE 8.1   Overview of pseudonoise sequences.

Since an r-bit binary shift register can assume a maximum of 2^r different states, it follows that every shift-register sequence $\{s(t)\}$ is eventually periodic with period $n \leq 2^r$, i.e.,

$$s(t) = s(t + n), \qquad \text{for all}\ \ t \geq N$$

for some integer N. In fact, the maximum period of a shift-register sequence is $2^r - 1$, since a shift register that enters the all-zero state will remain forever in that state. The upper shift register in Fig. 8.2 when initialized with starting state $0\,0\,1$ generates the periodic sequence $\{a(t)\}$ given by

$$0010111 \quad 0010111 \quad 0010111 \quad \cdots \tag{8.2}$$

of period $n = 7$. It follows then that this shift register generates sequences of maximal period starting from any nonzero initial state.

An m sequence is simply a binary shift-register sequence having maximal period. For every $r \geq 1$, m sequences are known to exist. The periodic **autocorrelation** function θ_s of

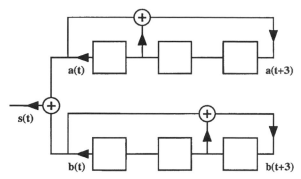

FIGURE 8.2 An example Gold sequence generator. Here $\{a(t)\}$ and $\{b(t)\}$ are m sequences of length 7.

a binary $\{0,1\}$ sequence $\{s(t)\}$ of period n is defined by

$$\theta_s(\tau) = \sum_{t=0}^{n-1} (-1)^{s(t+\tau)-s(t)} , \qquad 0 \le \tau \le n-1$$

An m sequence of length $2^r - 1$ has the following attributes. 1) *Balance property:* in each period of the m sequence there are 2^{r-1} ones and $2^{r-1} - 1$ zeros. 2) *Run property:* every nonzero binary s-tuple, $s \le r$ occurs 2^{r-s} times, the all-zero s-tuple occurs $2^{r-s} - 1$ times. 3) *Two-level autocorrelation function:*

$$\theta_s(\tau) = \left\{ \begin{array}{ll} n & \text{if } \tau = 0 \\ -1 & \text{if } \tau \ne 0 \end{array} \right. \tag{8.3}$$

The first two properties follow immediately from the observation that every nonzero r-tuple occurs precisely once in each period of the m sequence. For the third property, consider the difference sequence $\{s(t+\tau) - s(t)\}$ for $\tau \ne 0$. This sequence satisfies the same recursion as the m sequence $\{s(t)\}$ and is clearly not the all-zero sequence. It follows, therefore, that $\{s(t+\tau) - s(t)\} \equiv \{s(t+\tau')\}$ for some τ', $0 \le \tau' \le n-1$, i.e., is a different cyclic shift of the m sequence $\{s(t)\}$. The balance property of the sequence $\{s(t+\tau')\}$ then gives us attribute 3. The m sequence $\{a(t)\}$ in Eq. (8.2) can be seen to have the three listed properties.

If $\{s(t)\}$ is any sequence of period n and d is an integer, $1 \le d \le n$, then the mapping $\{s(t)\} \to \{s(dt)\}$ is referred to as a *decimation* of $\{s(t)\}$ by the integer d. If $\{s(t)\}$ is an m sequence of period $n = 2^r - 1$ and d is an integer relatively prime to $2^r - 1$, then the decimated sequence $\{s(dt)\}$ clearly also has period n. Interestingly, it turns out that the sequence $\{s(dt)\}$ is always also an m sequence of the same period. For example, when $\{a(t)\}$ is the sequence in Eq. (8.2), then

$$a(3t) = 0011101 \quad 0011101 \quad 0011101 \quad \cdots \tag{8.4}$$

and

$$a(2t) = 0111001 \quad 0111001 \quad 0111001 \quad \cdots \tag{8.5}$$

The sequence $\{a(3t)\}$ is also an m sequence of period 7, since it satisfies the recursion

$$s(t+3) + s(t+2) + s(t) = 0 \qquad \text{for all } t$$

of degree $r = 3$. In fact $\{a(3t)\}$ is precisely the sequence labeled $\{b(t)\}$ in Fig. 8.2. The sequence $\{a(2t)\}$ is simply a cyclically shifted version of $\{a(t)\}$ itself; this property holds in general. If $\{s(t)\}$ is any m sequence of period $2^r - 1$, then $\{s(2t)\}$ will always be a shifted version of the same m sequence. Clearly, the same is true for decimations by any power of 2.

Starting from an m sequence of period $2^r - 1$, it turns out that one can generate all m sequences of the same period through decimations by integers d relatively prime to $2^r - 1$. The set of integers d, $1 \leq d \leq 2^r - 1$ satisfying $(d, 2^r - 1) = 1$ forms a group under multiplication modulo $2^r - 1$, with the powers $\{2^i \mid 0 \leq i \leq r - 1\}$ of 2 forming a subgroup of order r. Since decimation by a power of 2 yields a shifted version of the same m sequence, it follows that the number of distinct m sequences of period $2^r - 1$ is $[\phi(2^r - 1)/r]$ where $\phi(n)$ denotes the number of integers d, $1 \leq d \leq n$, relatively prime to n. For example, when $r = 3$, there are just two cyclically distinct m sequences of period 7, and these are precisely the sequences $\{a(t)\}$ and $\{b(t)\}$ discussed in the preceding paragraph. Tables provided in [12] can be used to determine the characteristic polynomial of the various m sequences obtainable through the decimation of a single given m sequence. The classical reference on m sequences is [4].

If one obtains a sequence of some large length n by repeatedly tossing an unbiased coin, then such a sequence will very likely satisfy the balance, run, and autocorrelation properties of an m sequence of comparable length. For this reason, it is customary to regard the extent to which a given sequence possesses these properties as a measure of randomness of the sequence. Quite apart from this, in many applications such as signal synchronization and radar ranging, it is desirable to have sequences $\{s(t)\}$ with low autocorrelation sidelobes i.e., $|\theta_s(\tau)|$ is small for $\tau \neq 0$. Whereas m sequences are a prime example, there exist other methods of constructing binary sequences with low out-of-phase autocorrelation.

Sequences $\{s(t)\}$ of period n having an autocorrelation function identical to that of an m sequence, i.e., having θ_s satisfying Eq. (8.3) correspond to well-studied combinatorial objects known as *cyclic Hadamard difference sets*. Known infinite families fall into three classes 1) Singer and Gordon, Mills and Welch, 2) quadratic residue, and 3) twin-prime difference sets. These correspond, respectively, to sequences of period n of the form $n = 2^r - 1, r \geq 1$; n prime; and $n = p(p+2)$ with both p and $p+2$ being prime in the last case. For a detailed treatment of cyclic difference sets, see [2]. A recent observation by Maschietti in [9] provides additional families of cyclic Hadamard difference sets that also correspond to sequences of period $n = 2^r - 1$.

8.3 The q-ary Sequences with Low Autocorrelation

As defined earlier, the autocorrelation of a binary $\{0, 1\}$ sequence $\{s(t)\}$ leads to the computation of the inner product of an $\{-1, +1\}$ sequence $\{(-1)^{s(t)}\}$ with a cyclically shifted version $\{(-1)^{s(t+\tau)}\}$ of itself. The $\{-1, +1\}$ sequence is transmitted as a phase shift by either $0°$ and $180°$ of a radio-frequency carrier, i.e., using binary phase-shift keying (PSK) modulation. If the modulation is q-ary PSK, then one is led to consider sequences $\{s(t)\}$ with symbols in the set Z_q, i.e., the set of integers modulo q. The relevant autocorrelation function $\theta_s(\tau)$ is now defined by

$$\theta_s(\tau) = \sum_{t=0}^{n-1} \omega^{s(t+\tau)-s(t)}$$

where n is the period of $\{s(t)\}$ and ω is a complex primitive qth root of unity. It is possible to construct sequences $\{s(t)\}$ over Z_q whose autocorrelation function satisfies

$$\theta_s(\tau) = \begin{cases} n & \text{if } \tau = 0 \\ 0 & \text{if } \tau \neq 0 \end{cases}$$

For obvious reasons, such sequences are said to have an *ideal autocorrelation function.*

We provide without proof two sample constructions. The sequences in the first construction are given by

$$s(t) = \begin{cases} t^2/2 \pmod{n} & \text{when } n \text{ is even} \\ t(t+1)/2 \pmod{n} & \text{when } n \text{ is odd} \end{cases}$$

Thus, this construction provides sequences with ideal autocorrelation for any period n. Note that the size q of the sequence symbol alphabet equals n when n is odd and $2n$ when n is even.

The second construction also provides sequences over Z_q of period n but requires that n be a perfect square. Let $n = r^2$ and let π be an arbitrary permutation of the elements in the subset $\{0, 1, 2, \ldots, (r-1)\}$ of Z_n: Let g be an arbitrary function defined on the subset $\{0, 1, 2, \ldots, r-1\}$ of Z_n. Then any sequence of the form

$$s(t) = rt_1\pi(t_2) + g(t_2) \pmod{n}$$

where $t = rt_1 + t_2$ with $0 \le t_1, t_2 \le r - 1$ is the base-r decomposition of t, has an ideal autocorrelation function. When the alphabet size q equals or divides the period n of the sequence, ideal-autocorrelation sequences also go by the name *generalized bent functions.* For details, see [6].

8.4 Families of Sequences with Low Crosscorrelation

Given two sequences $\{s_1(t)\}$ and $\{s_2(t)\}$ over Z_q of period n, their **crosscorrelation** function $\theta_{1,2}(\tau)$ is defined by

$$\theta_{1,2}(\tau) = \sum_{t=0}^{n-1} \omega^{s_1(t+\tau)-s_2(t)}$$

where ω is a primitive qth root of unity. The crosscorrelation function is important in code-division multiple-access (CDMA) communication systems. Here, each user is assigned a distinct signature sequence and to minimize interference due to the other users, it is desirable that the signature sequences have pairwise, low values of crosscorrelation function. To provide the system in addition with a self-synchronizing capability, it is desirable that the signature sequences have low values of the autocorrelation function as well.

Let $\mathcal{F} = \{\{s_i(t)\} \mid 1 \le i \le M\}$ be a family of M sequences $\{s_i(t)\}$ over Z_q each of period n. Let $\theta_{i,j}(\tau)$ denote the crosscorrelation between the ith and jth sequence at shift τ, i.e.,

$$\theta_{i,j}(\tau) = \sum_{t=0}^{n-1} \omega^{s_i(t+\tau)-s_j(t)}, \qquad 0 \le \tau \le n - 1$$

The classical goal in sequence design for CDMA systems has been minimization of the parameter

$$\theta_{\max} = \max\{|\theta_{i,j}(\tau)| \mid \text{ either } i \ne j \text{ or } \tau \ne 0\}$$

for fixed n and M. It should be noted though that, in practice, because of data modulation the correlations that one runs into are typically of an aperiodic rather than a periodic nature (see Section 8.5). The problem of designing for low aperiodic correlation, however, is a more difficult one. A typical approach, therefore, has been to design based on periodic correlation, and then to analyze the resulting design for its aperiodic correlation properties.

Again, in many practical systems, the mean square correlation properties are of greater interest than the worst-case correlation represented by a parameter such as $\theta_{\max}$. The mean square correlation is discussed in Section 8.6.

Bounds on the minimum possible value of $\theta_{\max}$ for given period n, family size M, and alphabet size q are available that can be used to judge the merits of a particular sequence design. The most efficient bounds are those due to Welch, Sidelnikov, and Levenshtein, see [6]. In CDMA systems, there is greatest interest in designs in which the parameter $\theta_{\max}$ is in the range $\sqrt{n} \leq \theta_{\max} \leq 2\sqrt{n}$. Accordingly, Table 8.1 uses the Welch, Sidelnikov, and Levenshtein bounds to provide an order-of-magnitude upper bound on the family size M for certain $\theta_{\max}$ in the cited range.

Practical considerations dictate that q be small. The bit-oriented nature of electronic hardware makes it preferable to have q a power of 2. With this in mind, a description of some efficient sequence families having low auto- and crosscorrelation values and alphabet sizes $q = 2$ and $q = 4$ are described next.

TABLE 8.1   Bounds on Family Size M for Given $n, \theta_{\max}$

$\theta_{\max}$	Upper bound on M $q = 2$	Upper Bound on M $q > 2$
$\sqrt{n}$	$n/2$	n
$\sqrt{2n}$	n	$n^2/2$
$2\sqrt{n}$	$3n^2/10$	$n^3/2$

8.4.1   Gold and Kasami Sequences

Given the low autocorrelation sidelobes of an m sequence, it is natural to attempt to construct families of low correlation sequences starting from m sequences. Two of the better known constructions of this type are the families of Gold and Kasami sequences.

Let r be odd and $d = 2^k + 1$ where $k, 1 \leq k \leq r - 1$, is an integer satisfying $(k, r) = 1$. Let $\{s(t)\}$ be a cyclic shift of an m sequence of period $n = 2^r - 1$ that satisfies $S(dt) \not\equiv 0$ and let $\mathcal{G}$ be the *Gold* family of $2^r + 1$ sequences given by

$$\mathcal{G} = \{s(t)\} \cup \{s(dt)\} \cup \{\{s(t) + s(d[t + \tau])\} \mid 0 \leq \tau \leq n - 1\}$$

Then each sequence in $\mathcal{G}$ has period $2^r - 1$ and the maximum-correlation parameter $\theta_{\max}$ of $\mathcal{G}$ satisfies

$$\theta_{\max} \leq \sqrt{2^{r+1}} + 1$$

An application of the Sidelnikov bound coupled with the information that $\theta_{\max}$ must be an odd integer yields that for the family $\mathcal{G}$, $\theta_{\max}$ is as small as it can possibly be. In this sense the family $\mathcal{G}$ is an optimal family. We remark that these comments remain true even when d is replaced by the integer $d = 2^{2k} - 2^k + 1$ with the conditions on k remaining unchanged.

The Gold family remains the best-known family of m sequences having low crosscorrelation. Applications include the Navstar Global Positioning System whose signals are based on Gold sequences.

The family of Kasami sequences has a similar description. Let $r = 2v$ and $d = 2^v + 1$. Let $\{s(t)\}$ be a cyclic shift of an m sequence of period $n = 2^r - 1$ that satisfies $s(dt) \not\equiv 0$, and consider the family of Kasami sequences given by

$$\mathcal{K} = \{s(t)\} \cup \{\{s(t) + s(d[t + \tau])\} \mid 0 \leq \tau \leq 2^v - 2\}$$

Then the Kasami family $\mathcal{K}$ contains 2^v sequences of period $2^r - 1$. It can be shown that in this case

$$\theta_{\max} = 1 + 2^v$$

This time an application of the Welch bound and the fact that $\theta_{\max}$ is an integer shows that the Kasami family is optimal in terms of having the smallest possible value of $\theta_{\max}$ for given n and M.

8.4.2 Quaternary Sequences with Low Crosscorrelation

The entries in Table 8.1 suggest that nonbinary (i.e., $q > 2$) designs may be used for improved performance. A family of quaternary sequences that outperform the Gold and Kasami sequences is now discussed below.

Let $f(x)$ be the characteristic polynomial of a binary m sequence of length $2^r - 1$ for some integer r. The coefficients of $f(x)$ are either 0 or 1. Now, regard $f(x)$ as a polynomial over Z_4 and form the product $(-1)^r f(x) f(-x)$. This can be seen to be a polynomial in x^2. Define the polynomial $g(x)$ of degree r by setting $g(x^2) = (-1)^r f(x) f(-x)$. Let $g(x) = \sum_{i=0}^{r} g_i x^i$ and consider the set of all quaternary sequences $\{a(t)\}$ satisfying the recursion $\sum_{i=0}^{r} g_i a(t+i) = 0$ for all t.

It turns out that with the exception of the all-zero sequence, all of the sequences generated in this way have period $2^r - 1$. Thus, the recursion generates a family $\mathcal{A}$ of $2^r + 1$ cyclically distinct quaternary sequences. Closer study reveals that the maximum correlation parameter $\theta_{\max}$ of this family satisfies $\theta_{\max} \leq 1 + \sqrt{2^r}$. Thus, in comparison to the family of Gold sequences, the family $\mathcal{A}$ offers a lower value of $\theta_{\max}$ (by a factor of $\sqrt{2}$) for the same family size. In comparison to the set of Kasami sequences, it offers a much larger family size for the same bound on $\theta_{\max}$. Family $\mathcal{A}$ sequences may be found discussed in [16, 3].

We illustrate with an example. Let $f(x) = x^3 + x + 1$ be the characteristic polynomial of the m sequence $\{a(t)\}$ in Eq. (8.1). Then over Z_4

$$g\left(x^2\right) = (-1)^3 f(x) f(-x) = x^6 + 2x^4 + x^2 + 3$$

so that $g(x) = x^3 + 2x^2 + x + 3$. Thus, the sequences in family $\mathcal{A}$ are generated by the recursion $s(t+3) + 2s(t+2) + s(t+1) + 3s(t) = 0 \bmod 4$. The corresponding shift register is shown in Fig. 8.3. By varying initial conditions, this shift register can be made to generate nine cyclically distinct sequences, each of length 7. In this case $\theta_{\max} \leq 1 + \sqrt{8}$.

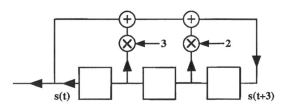

FIGURE 8.3 Shift register that generates family $\mathcal{A}$ quaternary sequences $\{s(t)\}$ of period 7.

8.4.3 Binary Kerdock Sequences

The Gold and Kasami families of sequences are closely related to binary linear cyclic codes. It is well known in coding theory that there exists nonlinear binary codes whose performance exceeds that of the best possible linear code. Surprisingly, some of these examples come from binary codes, which are images of *linear quaternary* ($q = 4$) *codes* under the Gray map: $0 \to 00, 1 \to 01, 2 \to 11, 3 \to 10$. A prime example of this is the Kerdock code, which recently has been shown to be the Gray image of a quaternary linear code. Thus, it is not surprising that the Kerdock code yields binary sequences that significantly outperform the family of Kasami sequences.

The Kerdock sequences may be constructed as follows: let $f(x)$ be the characteristic polynomial of an m sequence of period $2^r - 1, r$ odd. As before, regarding $f(x)$ as a polynomial over Z_4 (which happens to have $\{0, 1\}$ coefficients), let the polynomial $g(x)$ over Z_4 be defined via $g(x^2) = -f(x)f(-x)$. [Thus, $g(x)$ is the characteristic polynomial of a family $\mathcal{A}$ sequence set of period $2^r - 1$.] Set $h(x) = -g(-x) = \sum_{i=0}^{r} h_i x^i$, and let S be the set of all Z_4 sequences satisfying the recursion $\sum_{i=0}^{r} h_i s(t + i) = 0$. Then S contain 4^r-distinct sequences corresponding to all possible distinct initializations of the shift register.

Let T denote the subset S of size 2^r-consisting of those sequences corresponding to initializations of the shift register only using the symbols 0 and 2 in Z_4. Then the set $S - T$ of size $4^r - 2^r$ contains a set $\mathcal{U}$ of 2^{r-1} cyclically distinct sequences each of period $2(2^r - 1)$. Given $x = a + 2b \in Z_4$ with $a, b \in \{0, 1\}$, let μ denote the most significant bit (MSB) map $\mu(x) = b$. Let $\mathcal{K}_E$ denote the family of 2^{r-1} binary sequences obtained by applying the map μ to each sequence in $\mathcal{U}$. It turns out that each sequence in $\mathcal{U}$ also has period $2(2^r - 1)$ and that, furthermore, for the family $\mathcal{K}_E$, $\theta_{max} \leq 2 + \sqrt{2^{r+1}}$. Thus, $\mathcal{K}_E$ is a much larger family than the Kasami family, while having almost exactly the same value of θ_{max}.

For example, taking $r = 3$ and $f(x) = x^3 + x + 1$, we have from the previous family $\mathcal{A}$ example that $g(x) = x^3 + 2x^2 + x + 3$, so that $h(x) = -g(-x) = x^3 + 2x^2 + x + 1$. Applying the MSB map to the head of the shift register, and discarding initializations of the shift register involving only 0's and 2's yields a family of four cyclically distinct binary sequences of period 14. Kerdock sequences are discussed in [6, 11, 1, 17].

8.5   Aperiodic Correlation

Let $\{x(t)\}$ and $\{y(t)\}$ be complex-valued sequences of length (or period) n, not necessarily distinct. Their *aperiodic correlation* values $\{\rho_{x,y}(\tau)| - (n - 1) \leq \tau \leq n - 1\}$ are given by

$$\rho_{x,y}(\tau) = \sum_{t=\max\{0,-\tau\}}^{\min\{n-1, n-1-\tau\}} x(t + \tau) y^*(t)$$

where $y^*(t)$ denotes the complex conjugate of $y(t)$. When $x \equiv y$, we will abbreviate and write ρ_x in place of $\rho_{x,y}$. The sequences described next are perhaps the most famous example of sequences with low-aperiodic autocorrelation values.

8.5.1   Barker Sequences

A binary $\{-1, +1\}$ sequence $\{s(t)\}$ of length n is said to be a *Barker sequence* if the aperiodic autocorrelation values $\rho_s(\tau)$ satisfy $|\rho_s(\tau)| \leq 1$ for all $\tau, -(n - 1) \leq \tau \leq n - 1$. The Barker property is preserved under the following transformations:

$$s(t) \to -s(t), \qquad s(t) \to (-1)^t s(t) \quad \text{and} \quad s(t) \to s(n - 1 - t)$$

as well as under compositions of the preceding transformations. Only the following Barker sequences are known:

$$
\begin{array}{rl}
n = 2 & ++ \\
n = 3 & ++- \\
n = 4 & +++- \\
n = 5 & +++-+ \\
n = 7 & +++--+- \\
n = 11 & +++---+--+- \\
n = 13 & +++++--++-+-+
\end{array}
$$

where $+$ denotes $+1$ and $-$ denotes -1 and sequences are generated from these via the transformations already discussed. It is known that if any other Barker sequence exists, it must have length $n > 1,898,884$, that is a multiple of 4.

For an upper bound to the maximum out-of-phase aperiodic autocorrelation of an m sequence, see [13].

8.5.2 Sequences with High Merit Factor

The *merit factor* F of a $\{-1, +1\}$ sequence $\{s(t)\}$ is defined by

$$
F = \frac{n^2}{2 \sum_{\tau=1}^{n-1} \rho_s^2(\tau)}
$$

Since $\rho_s(\tau) = \rho_s(-\tau)$ for $1 \le |\tau| \le n - 1$ and $\rho_s(0) = n$, factor F may be regarded as the ratio of the square of the in-phase autocorrelation, to the sum of the squares of the out-of-phase aperiodic autocorrelation values. Thus, the merit factor is one measure of the aperiodic autocorrelation properties of a binary $\{-1, +1\}$ sequence. It is also closely connected with the signal to self-generated noise ratio of a communication system in which coded pulses are transmitted and received.

Let F_n denote the largest merit factor of any binary $\{-1, +1\}$ sequence of length n. For example, at length $n = 13$, the Barker sequence of length 13 has a merit factor $F = F_{13} = 14.08$. Assuming a certain ergodicity postulate it was established by Golay that $\lim_{n \to \infty} F_n = 12.32$. Exhaustive computer searches carried out for $n \le 40$ have revealed the following.

1. For $1 \le n \le 40$, $n \ne 11, 13$,

$$
3.3 \le F_n \le 9.85 \ ,
$$

2. $F_{11} = 12.1$, $F_{13} = 14.08$.

The value F_{11} is also achieved by a Barker sequence. From partial searches, for lengths up to 117, the highest known merit factor is between 8 and 9.56; for lengths from 118 to 200, the best-known factor is close to 6. For lengths > 200, statistical search methods have failed to yield a sequence having merit factor exceeding 5.

An *offset sequence* is one in which a fraction θ of the elements of a sequence of length n are chopped off at one end and appended to the other end, i.e., an offset sequence is a cyclic shift of the original sequence by $n\theta$ symbols. It turns out that the asymptotic merit factor of m sequences is equal to 3 and is independent of the particular offset of the m sequence. There exist offsets of sequences associated with quadratic-residue and twin-prime difference sets that achieve a larger merit factor of 6. Details may be found in [7].

8.5.3   Sequences with Low Aperiodic Crosscorrelation

If $\{u(t)\}$ and $\{v(t)\}$ are sequences of length $2n-1$ defined by

$$u(t) = \begin{cases} x(t) & \text{if } 0 \leq t \leq n-1 \\ 0 & \text{if } n \leq t \leq 2n-2 \end{cases}$$

and

$$v(t) = \begin{cases} y(t) & \text{if } 0 \leq t \leq n-1 \\ 0 & \text{if } n \leq t \leq 2n-2 \end{cases}$$

then

$$\{\rho_{x,y}(\tau) \mid -(n-1) \leq \tau \leq n-1\} = \{\theta_{u,v}(\tau) \mid 0 \leq \tau \leq 2n-2\} \tag{8.6}$$

Given a collection

$$U = \{\{x_i(t)\} \mid 1 \leq i \leq M\}$$

of sequences of length n over Z_q, let us define

$$\rho_{\max} = \max\{|\rho_{a,b}(\tau)| \mid a, b \in U, \quad \text{either } a \neq b \text{ or } \tau \neq 0\}$$

It is clear from Eq. (8.6) how bounds on the *periodic* correlation parameter $\theta_{\max}$ can be adapted to give bounds on $\rho_{\max}$. Translation of the Welch bound gives that for every integer $k \geq 1$,

$$\rho_{\max}^{2k} \geq \left(\frac{n^{2k}}{M(2n-1)-1}\right)\left\{\frac{M(2n-1)}{\binom{2n+k-2}{k}} - 1\right\}$$

Setting $k = 1$ in the preceding bound gives

$$\rho_{\max} \geq n\sqrt{\frac{M-1}{M(2n-1)-1}}$$

Thus, for fixed M and large n, Welch's bound gives

$$\rho_{\max} \geq \mathcal{O}\left(n^{1/2}\right)$$

There exist sequence families which asymptotically achieve $\rho_{\max} \approx \mathcal{O}(n^{1/2})$, [10].

8.6   Other Correlation Measures

8.6.1   Partial-Period Correlation

The *partial-period (p-p) correlation* between the sequences $\{u(t)\}$ and $\{v(t)\}$ is the collection $\{\Delta_{u,v}(l, \tau, t_0) \mid 1 \leq l \leq n, 0 \leq \tau \leq n-1, 0 \leq t_0 \leq n-1\}$ of inner products

$$\Delta_{u,v}(l, \tau, t_0) = \sum_{t=t_0}^{t=t_0+l-1} u(t+\tau)v^*(t)$$

where l is the length of the partial period and the sum $t + \tau$ is again computed modulo n.

In direct-sequence CDMA systems, the pseudorandom signature sequences used by the various users are often very long for reasons of data security. In such situations, to minimize receiver hardware complexity, correlation over a partial period of the signature sequence is often used to demodulate data, as well as to achieve synchronization. For this reason, the p-p correlation properties of a sequence are of interest.

Researchers have attempted to determine the moments of the p-p correlation. Here the main tool is the application of the Pless power-moment identities of coding theory [8]. The identities often allow the first and second p-p correlation moments to be completely determined. For example, this is true in the case of m sequences (the remaining moments turn out to depend upon the specific characteristic polynomial of the m sequence). Further details may be found in [15].

8.6.2 Mean Square Correlation

Frequently in practice, there is a greater interest in the mean-square correlation distribution of a sequence family than in the parameter $\theta_{\max}$. Quite often in sequence design, the sequence family is derived from a linear, binary cyclic code of length n by picking a set of cyclically distinct sequences of period n. The families of Gold and Kasami sequences are so constructed. In this case, as pointed out by Massey, the mean square correlation of the family can be shown to be either optimum or close to optimum, under certain easily satisfied conditions, imposed on the minimum distance of the dual code. A similar situation holds even when the sequence family does not come from a linear cyclic code. In this sense, mean square correlation is not a very discriminating measure of the correlation properties of a family of sequences. An expanded discussion of this issue may be found in [5].

8.6.3 Optical Orthogonal Codes

Given a pair of $\{0,1\}$ sequences $\{s_1(t)\}$ and $\{s_2(t)\}$ each having period n, we define the *Hamming correlation* function $\theta_{12}(\tau)$, $0 \leq \tau \leq n-1$, by

$$\theta_{12}(\tau) = \sum_{t=0}^{n-1} s_1(t+\tau)s_2(t)$$

Such correlations are of interest, for instance, in optical communication systems where the 1's and 0's in a sequence correspond to the presence or absence of pulses of transmitted light.

An (n, w, λ) optical orthogonal code (OOC) is a family $\mathcal{F} = \{\{s_i(t)\} \mid i = 1, 2, \ldots, M\}$, of M $\{0,1\}$ sequences of period n, constant Hamming weight w, where w is an integer lying between 1 and $n-1$ satisfying $\theta_{ij}(\tau) \leq \lambda$ whenever either $i \neq j$ or $\tau \neq 0$.

Note that the Hamming distance $d_{a,b}$ between a period of the corresponding codewords $\{a(t)\}$, $\{b(t)\}$, $0 \leq t \leq n-1$ in an (n, w, λ) OOC having Hamming correlation ρ, $0 \leq \rho \leq \lambda$, is given by $d_{a,b} = 2(w - \rho)$, and, thus, OOCs are closely related to constant-weight error correcting codes. Given an (n, w, λ) OOC, by enlarging the OOC to include every cyclic shift of each sequence in the code, one obtains a constant-weight, minimum distance $d_{\min} \geq 2(w - \lambda)$ code. Conversely, given a constant-weight cyclic code of length n, weight w and minimum distance $d_{\min}$, one can derive an (n, w, λ) OOC code with $\lambda \leq w - d_{\min}/2$ by partitioning the code into cyclic equivalence classes and then picking precisely one representative from each equivalence class of size n.

By making use of this connection, one can derive bounds on the size of an OOC from known bounds on the size of constant-weight codes. The bound given next follows directly from the Johnson bound for constant weight codes [8]. The number $M(n, w, \lambda)$ of codewords in a (n, w, λ) OOC satisfies

$$M(n, w, \lambda) \le \frac{1}{w} \left\lfloor \frac{n-1}{w-1} \cdots \left\lfloor \frac{n-\lambda+1}{w-\lambda+1} \left\lfloor \frac{n-\lambda}{w-\lambda} \right\rfloor \right\rfloor \cdots \right\rfloor$$

An OOC code that achieves the Johnson bound is said to be optimal. A family $\{\mathcal{F}_n\}$ of OOCs indexed by the parameter n and arising from a common construction is said to be asymptotically optimum if

$$\lim_{n \to \infty} \frac{|\mathcal{F}_n|}{M(n, w, \lambda)} = 1$$

Constructions for optical orthogonal codes are available for the cases when $\lambda = 1$ and $\lambda = 2$. For larger values of λ, there exist constructions which are asymptotically optimum. Further details may be found in [6].

Defining Terms

Autocorrelation of a sequence: The complex inner product of the sequence with a shifted version itself.

Crosscorrelation of two sequences: The complex inner product of the first sequence with a shifted version of the second sequence.

m Sequence: A periodic binary $\{0, 1\}$ sequence that is generated by a shift register with linear feedback and which has maximal possible period given the number of stages in the shift register.

Pseudonoise sequences: Also referred to as pseudorandom sequences (PN), these are sequences that are deterministically generated and yet possess some properties that one would expect to find in randomly generated sequences.

Shift-register sequence: A sequence with symbols drawn from a field, which satisfies a linear-recurrence relation and which can be implemented using a shift register.

References

[1] Barg, A. On small families of sequences with low periodic correlation, *Lecture Notes in Computer Science*, 781, 154–158, Berlin, Springer-Verlag, 1994.

[2] Baumert, L.D. *Cyclic Difference Sets,* Lecture Notes in Mathematics 182, Springer–Verlag, New York, 1971.

[3] Boztaş, S., Hammons, R., and Kumar, P.V. 4-phase sequences with near-optimum correlation properties, *IEEE Trans. Inform. Theory,* IT-38, 1101–1113, 1992.

[4] Golomb, S.W. *Shift Register Sequences,* Aegean Park Press, San Francisco, CA, 1982.

[5] Hammons, A.R., Jr. and Kumar, P.V. On a recent 4-phase sequence design for CDMA. *IEICE Trans. Commun.,* E76-B(8), 1993.

[6] Helleseth, T. and Kumar, P.V. (planned). Sequences with low correlation. In *Handbook of Coding Theory,* ed., V.S. Pless and W.C. Huffman, Elsevier Science Publishers, Amsterdam, 1998.

[7] Jensen, J.M., Jensen, H.E., and Høholdt, T. The merit factor of binary sequences related to difference sets. *IEEE Trans. Inform. Theory,* IT-37(May), 617–626, 1991.

[8] MacWilliams, F.J. and Sloane, N.J.A. *The Theory of Error-Correcting Codes,* North-Holland, Amsterdam, 1977.

[9] Maschietti, A. Difference sets and hyperovals, *Designs, Codes and Cryptography,* 14, 89–98, 1998.

[10] Mow, W.H. On McEliece's open problem on minimax aperiodic correlation. In *Proc. IEEE Intern. Symp. Inform. Theory,* 75, 1994.

[11] Nechaev, A. The Kerdock code in a cyclic form, *Discrete Math. Appl.,* 1, 365–384, 1991.

[12] Peterson, W.W. and Weldon, E.J., Jr. *Error-Correcting Codes,* 2nd ed. MIT Press, Cambridge, MA, 1972.

[13] Sarwate, D.V. An upper bound on the aperiodic autocorrelation function for a maximal-length sequence. *IEEE Trans. Inform. Theory,* IT-30(July), 685–687, 1984.

[14] Sarwate, D.V. and Pursley, M.B. Crosscorrelation properties of pseudorandom and related sequences. *Proc. IEEE,* 68(May), 593–619, 1980.

[15] Simon, M.K., Omura, J.K., Scholtz, R.A., and Levitt, B.K. *Spread Spectrum Communications Handbook,* revised ed., McGraw Hill, New York, 1994.

[16] Solé, P. A quaternary cyclic code and a family of quadriphase sequences with low correlation properties, *Coding Theory and Applications, Lecture Notes in Computer Science,* 388, 193–201, Berlin, Springer-Verlag, 1989.

[17] Udaya, P. and Siddiqi, M. Optimal biphase sequences with large linear complexity derived from sequences over Z_4, *IEEE Trans. Inform. Theory,* IT-42 (Jan), 206–216, 1996.

Further Information

A more in-depth treatment of pseudonoise sequences, may be found in the following.

[1] Golomb, S.W. *Shift Register Sequences,* Aegean Park Press, San Francisco, 1982.

[2] Helleseth, T. and Kumar, P.V. Sequences with Low Correlation, in *Handbook of Coding Theory,* edited by V.S. Pless and W.C. Huffman, Elsevier Science Publishers, Amsterdam, 1998 (planned).

[3] Sarwate, D.V. and Pursley, M.B. Crosscorrelation Properties of Pseudorandom and Related Sequences, *Proc. IEEE,* 68, May, 593–619, 1980.

[4] Simon, M.K., Omura, J.K., Scholtz, R.A., and Levitt, B.K. *Spread Spectrum Communications Handbook,* revised ed., McGraw Hill, New York, 1994.

9

Optimum Receivers

Geoffrey C. Orsak
Southern Methodist University

9.1 Introduction

Every engineer strives for optimality in design. This is particularly true for communications engineers since in many cases implementing suboptimal receivers and sources can result in dramatic losses in performance. As such, this chapter focuses on design principles leading to the implementation of optimum receivers for the most common communication environments.

The main objective in digital communications is to transmit a sequence of bits to a remote location with the highest degree of accuracy. This is accomplished by first representing bits (or more generally short bit sequences) by distinct waveforms of finite time duration. These time-limited waveforms are then transmitted (broadcasted) to the remote sites in accordance with the data sequence.

Unfortunately, because of the nature of the **communication channel,** the remote location receives a corrupted version of the concatenated signal waveforms. The most widely accepted model for the communication channel is the so-called **additive white Gaussian noise[1] channel (AWGN channel).** Mathematical arguments based upon the central

---

[1] For those unfamiliar with AWGN, a random process (waveform) is formally said to be white Gaussian noise if all collections of instantaneous observations of the process are jointly Gaussian and mutually independent. An important consequence of this property is that the power spectral density of the process

limit theorem [7], together with supporting empirical evidence, demonstrate that many common communication channels are accurately modeled by this abstraction. Moreover, from the design perspective, this is quite fortuitous since design and analysis with respect to this channel model is relatively straightforward.

9.2   Preliminaries

To better describe the digital communications process, we shall first elaborate on so-called binary communications. In this case, when the source wishes to transmit a bit value of 0, the transmitter broadcasts a specified waveform $s_0(t)$ over the **bit interval** $t \in [0, T]$. Conversely, if the source seeks to transmit the bit value of 1, the transmitter alternatively broadcasts the signal $s_1(t)$ over the same bit interval. The received waveform $R(t)$ corresponding to the first bit is then appropriately described by the following hypotheses testing problem:

$$\begin{aligned} H_0 &: R(t) = s_0(t) + \eta(t) \qquad 0 \le t \le T \\ H_1 &: R(t) = s_1(t) + \eta(t) \end{aligned} \tag{9.1}$$

where, as stated previously, $\eta(t)$ corresponds to AWGN with spectral height nominally given by $N_0/2$. It is the objective of the receiver to determine the bit value, i.e., the most accurate hypothesis from the received waveform $R(t)$.

The optimality criterion of choice in digital communication applications is the **total probability of error** normally denoted as P_e. This scalar quantity is expressed as

$$\begin{aligned} P_e &= Pr(\text{ declaring } 1\,|\,0 \text{ transmitted}) Pr(0 \text{ transmitted}) \\ &\quad + Pr(\text{ declaring } 0\,|\,1 \text{ transmitted}) Pr(1 \text{ transmitted}) \end{aligned} \tag{9.2}$$

The problem of determining the optimal binary receiver with respect to the probability of error is solved by applying stochastic representation theory [10] to detection theory [5, 9]. The specific waveform representation of relevance in this application is the **Karhunen–Loève (KL) expansion.**

9.3   Karhunen–Loève Expansion

The Karhunen–Loève expansion is a generalization of the Fourier series designed to represent a random process in terms of deterministic basis functions and uncorrelated random variables derived from the process. Whereas the Fourier series allows one to model or represent deterministic time-limited energy signals in terms of linear combinations of complex exponential waveforms, the Karhunen–Loève expansion allows us to represent a second-order random process in terms of a set of **orthonormal** basis functions scaled by a sequence of random variables. The objective in this representation is to choose the basis of time functions so that the coefficients in the expansion are mutually uncorrelated random variables.

To be more precise, if $R(t)$ is a zero mean second-order random process defined over $[0, T]$ with covariance function $K_R(t, s)$, then so long as the basis of deterministic functions satisfy

---

is a constant with respect to frequency variation (spectrally flat). For more on AWGN, see Papoulis [4].

certain integral constraints [9], one may write $R(t)$ as

$$R(t) = \sum_{i=1}^{\infty} R_i \phi_i(t) \qquad 0 \leq t \leq T \tag{9.3}$$

where

$$R_i = \int_0^T R(t)\phi_i(t)\ \mathrm{dt}$$

In this case the R_i will be mutually uncorrelated random variables with the ϕ_i being deterministic basis functions that are complete in the space of square integrable time functions over $[0, T]$. Importantly, in this case, equality is to be interpreted as **mean-square equivalence**, i.e.,

$$\lim_{N \to \infty} E \left[\left(R(t) - \sum_{i=1}^{N} R_i \phi_i(t) \right)^2 \right] = 0$$

for all $0 \leq t \leq T$.

FACT 9.1   *If $R(t)$ is AWGN, then any basis of the vector space of square integrable signals over $[0, T]$ results in uncorrelated and therefore independent Gaussian random variables.*

The use of Fact 9.1 allows for a conversion of a continuous time detection problem into a finite-dimensional detection problem. Proceeding, to derive the optimal binary receiver, we first construct our set of basis functions as the set of functions defined over $t \in [0, T]$ beginning with the signals of interest $s_0(t)$ and $s_1(t)$. That is,

$\{s_0(t), s_1(t),$ plus a countable number of functions which complete the basis$\}$

In order to insure that the basis is orthonormal, we must apply the Gramm–Schmidt procedure[2] [6] to the full set of functions beginning with $s_0(t)$ and $s_1(t)$ to arrive at our final choice of basis $\{\phi_i(t)\}$.

FACT 9.2   *Let $\{\phi_i(t)\}$ be the resultant set of basis functions.*
   Then for all $i > 2$, the $\phi_i(t)$ are orthogonal to $s_0(t)$ and $s_1(t)$. That is,

$$\int_0^T \phi_i(t) s_j(t)\, \mathrm{dt} = 0$$

for all $i > 2$ and $j = 0, 1$.

Using this fact in conjunction with Eq. (9.3), one may recognize that only the coefficients R_1 and R_2 are functions of our signals of interest. Moreover, since the R_i are mutually independent, the optimal receiver will, therefore, only be a function of these two values.

---

[2]The Gramm-Schmidt procedure is a deterministic algorithm that simply converts an arbitrary set of basis functions (vectors) into an equivalent set of orthonormal basis functions (vectors).

Thus, through the application of the KL expansion, we arrive at an equivalent hypothesis testing problem to that given in Eq. (9.1),

$$
\begin{aligned}
H_0 : \boldsymbol{R} &= \left[\begin{array}{c} \int_0^T \phi_1(t)s_0(t)\ \mathrm{dt} \\ \int_0^T \phi_2(t)s_0(t)\ \mathrm{dt} \end{array} \right] + \left[\begin{array}{c} \eta_1 \\ \eta_2 \end{array} \right] \\
H_1 : \boldsymbol{R} &= \left[\begin{array}{c} \int_0^T \phi_1(t)s_1(t)\ \mathrm{dt} \\ \int_0^T \phi_2(t)s_1(t)\ \mathrm{dt} \end{array} \right] + \left[\begin{array}{c} \eta_1 \\ \eta_2 \end{array} \right]
\end{aligned}
\tag{9.4}
$$

where it is easily shown that η_1 and η_2 are mutually independent, zero-mean, Gaussian random variables with variance given by $N_0/2$, and where ϕ_1 and ϕ_2 are the first two functions from our orthonormal set of basis functions. Thus, the design of the optimal binary receiver reduces to a simple two-dimensional detection problem that is readily solved through the application of detection theory.

9.4 Detection Theory

It is well known from detection theory [5] that under the minimum P_e criterion, the optimal detector is given by the *maximum a posteriori rule (MAP)*,

$$
\text{choose}_i \ \ \text{largest} \ \ p_{H_i|\boldsymbol{R}}\left(H_i \mid \boldsymbol{R} = \boldsymbol{r}\right)
\tag{9.5}
$$

i.e., determine the hypothesis that is most likely, given that our observation vector is $\boldsymbol{r}$. By a simple application of Bayes theorem [4], we immediately arrive at the central result in detection theory: the optimal binary detector is given by the likelihood ratio test (LRT),

$$
L(\boldsymbol{R}) = \frac{p_{\boldsymbol{R}|H_1}(\boldsymbol{R})}{p_{\boldsymbol{R}|H_0}(\boldsymbol{R})} \underset{H_0}{\overset{H_1}{\gtrless}} \frac{\pi_0}{\pi_1}
\tag{9.6}
$$

where the π_i are the a priori probabilities of the hypotheses H_i being true. Since in this case we have assumed that the noise is white and Gaussian, the LRT can be written as

$$
L(\boldsymbol{R}) = \frac{\prod_1^2 \frac{1}{\sqrt{\pi N_0}} \exp\left(-\frac{1}{2}\frac{(R_i - s_{1,i})^2}{N_0/2}\right)}{\prod_1^2 \frac{1}{\sqrt{\pi N_0}} \exp\left(-\frac{1}{2}\frac{(R_i - s_{0,i})^2}{N_0/2}\right)} \underset{H_0}{\overset{H_1}{\gtrless}} \frac{\pi_0}{\pi_1}
\tag{9.7}
$$

where

$$
s_{j,i} = \int_0^T \phi_i(t)s_j(t)\,\mathrm{dt}
$$

By taking the logarithm and cancelling common terms, it is easily shown that the optimum binary receiver can be written as

$$
\frac{2}{N_0}\sum_1^2 R_i\left(s_{1,i} - s_{0,i}\right) - \frac{1}{N_0}\sum_1^2 \left(s_{1,i}^2 - s_{0,i}^2\right) \underset{H_0}{\overset{H_1}{\gtrless}} \ln\frac{\pi_0}{\pi_1}
\tag{9.8}
$$

This finite-dimensional version of the optimal receiver can be converted back into a continuous time receiver by the direct application of Parseval's theorem [4] where it is easily shown that

$$
\sum_{i=1}^{2} R_i s_{k,i} = \int_0^T R(t) s_k(t) \, \mathrm{d}t
$$
$$
\sum_{i=1}^{2} s_{k,i}^2 = \int_0^T s_k^2(t) \, \mathrm{d}t
$$

(9.9)

By applying Eq. (9.9) to Eq. (9.8) the final receiver structure is then given by

$$
\int_0^T R(t) \left[s_1(t) - s_0(t) \right] \mathrm{d}t - \frac{1}{2} \left(E_1 - E_0 \right) \underset{H_0}{\overset{H_1}{\underset{<}{>}}} \frac{N_0}{2} \ln \frac{\pi_0}{\pi_1}
$$

(9.10)

where E_1 and E_0 are the energies of signals $s_1(t)$ and $s_0(t)$, respectively. (See Fig. 9.1 for a block diagram.) Importantly, if the signals are equally likely ($\pi_0 = \pi_1$), the optimal receiver is independent of the typically unknown spectral height of the background noise.

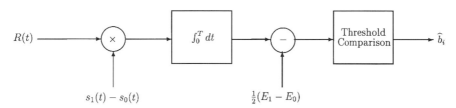

FIGURE 9.1 Optimal correlation receiver structure for binary communications.

One can readily observe that the optimal binary communication receiver correlates the received waveform with the difference signal $s_1(t) - s_0(t)$ and then compares the statistic to a threshold. This operation can be interpreted as identifying the signal waveform $s_i(t)$ that best correlates with the received signal $R(t)$. Based on this interpretation, the receiver is often referred to as the **correlation receiver.**

As an alternate means of implementing the correlation receiver, we may reformulate the computation of the left-hand side of Eq. (9.10) in terms of standard concepts in filtering. Let $h(t)$ be the impulse response of a linear, time-invariant (LTI) system. By letting $h(t) = s_1(T - t) - s_0(T - t)$, then it is easily verified that the output of $R(t)$ to a LTI system with impulse response given by $h(t)$ and then sampled at time $t = T$ gives the desired result. (See Fig. 9.2 for a block diagram.) Since the impulse response is matched to the signal waveforms, this implementation is often referred to as the **matched filter receiver.**

9.5 Performance

Because of the nature of the statistics of the channel and the relative simplicity of the receiver, performance analysis of the optimal binary receiver in AWGN is a straightforward task. Since the conditional statistics of the log likelihood ratio are Gaussian random vari-

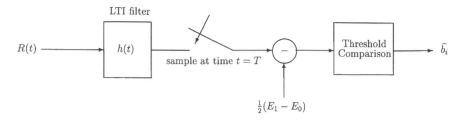

FIGURE 9.2 Optimal matched filter receiver structure for binary communications. In this case $h(t) = s_1(T - t) - s_0(t - t)$.

ables, the probability of error can be computed directly in terms of Marcum Q functions[3] as

$$P_e = Q\left(\frac{\|\boldsymbol{s}_0 - \boldsymbol{s}_1\|}{\sqrt{2\boldsymbol{N}_0}}\right)$$

where the $\boldsymbol{s}_i$ are the two-dimensional signal vectors obtained from Eq. (9.4), and where $\|\boldsymbol{x}\|$ denotes the Euclidean length of the vector $\boldsymbol{x}$. Thus, $\|\boldsymbol{s}_0 - \boldsymbol{s}_1\|$ is best interpreted as the distance between the respective signal representations. Since the Q function is monotonically decreasing with an increasing argument, one may recognize that the probability of error for the optimal receiver decreases with an increasing separation between the signal representations, i.e., the more dissimilar the signals, the lower the P_e.

9.6   Signal Space

The concept of a **signal space** allows one to view the signal classification problem (receiver design) within a geometrical framework. This offers two primary benefits: first it supplies an often more intuitive perspective on the receiver characteristics (e.g., performance) and second it allows for a straightforward generalization to standard M-ary signalling schemes.

To demonstrate this, in Fig. 9.3, we have plotted an arbitrary signal space for the binary signal classification problem. The axes are given in terms of the basis functions $\phi_1(t)$ and $\phi_2(t)$. Thus, every point in the signal space is a time function constructed as a linear combination of the two basis functions. By Fact 9.2, we recall that both signals $s_0(t)$ and $s_1(t)$ can be constructed as a linear combination of $\phi_1(t)$ and $\phi_2(t)$ and as such we may identify these two signals in this figure as two points.

Since the decision statistic given in Eq. (9.8) is a linear function of the observed vector $\boldsymbol{R}$ which is also located in the signal space, it is easily shown that the set of vectors under which the receiver declares hypothesis H_i is bounded by a line in the signal space. This so-called **decision boundary** is obtained by solving the equation $\ln[L(\boldsymbol{R})] = 0$. (Here again we have assumed equally likely hypotheses.) In the case under current discussion, this decision boundary is simply the hyperplane separating the two signals in signal space. Because of the generality of this formulation, many problems in communication system design are best cast in terms of the signal space, that is, signal locations and decision boundaries.

---

[3]The Q function is the probability that a standard normal random variable exceeds a specified constant, i.e., $Q(x) = \int_x^\infty 1/\sqrt{2\pi} \exp(-z^2/2)\,dz$.

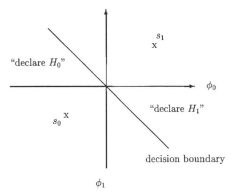

FIGURE 9.3   Signal space and decision boundary for optimal binary receiver.

9.7   Standard Binary Signalling Schemes

The framework just described allows us to readily analyze the most popular signalling schemes in binary communications: amplitude-shift keying (ASK), frequency-shift keying (FSK), and phase-shift keying (PSK). Each of these examples simply constitute a different selection for signals $s_0(t)$ and $s_1(t)$.

In the case of ASK, $s_0(t) = 0$, while $s_1(t) = \sqrt{2E/T}\sin(2\pi f_c t)$, where E denotes the energy of the waveform and f_c denotes the frequency of the carrier wave with $f_c T$ being an integer. Because $s_0(t)$ is the null signal, the signal space is a one-dimensional vector space with $\phi_1(t) = \sqrt{2/T}\sin(2\pi f_c t)$. This, in turn, implies that $\|s_0 - s_1\| = \sqrt{E}$. Thus, the corresponding probability of error for ASK is

$$P_e(\text{ ASK}) = Q\left(\sqrt{\frac{E}{2N_0}}\right)$$

For FSK, the signals are given by equal amplitude sinusoids with distinct center frequencies, that is, $s_i(t) = \sqrt{2E/T}\sin(2\pi f_i t)$ with $f_i T$ being two distinct integers. In this case, it is easily verified that the signal space is a two-dimensional vector space with $\phi_i(t) = \sqrt{2/T}\sin(2\pi f_i t)$ resulting in $\|s_0 - s_1\| = \sqrt{2E}$. The corresponding error rate is given to be

$$P_e(\text{FSK}) = Q\left(\sqrt{\frac{E}{N_0}}\right)$$

Finally, with regard to PSK signalling, the most frequently utilized binary PSK signal set is an example of an antipodal signal set. Specifically, the antipodal signal set results in the greatest separation between the signals in the signal space subject to an energy constraint on both signals. This, in turn, translates into the energy constrained signal set with the minimum P_e. In this case, the $s_i(t)$ are typically given by $\sqrt{2E/T}\sin[2\pi f_c t + \theta(i)]$, where $\theta(0) = 0$ and $\theta(1) = \pi$. As in the ASK case, this results in a one-dimensional signal space, however, in this case $\|s_0 - s_1\| = 2\sqrt{E}$ resulting in probability of error given by

$$P_e(\text{PSK}) = Q\left(\sqrt{\frac{2E}{N_0}}\right)$$

In all three of the described cases, one can readily observe that the resulting performance is a function of only the signal-to-noise ratio E/N_0. In the more general case, the performance will be a function of the intersignal energy to noise ratio. To gauge the relative difference in performance of the three signalling schemes, in Fig. 9.4, we have plotted the P_e as a function of the SNR. Please note the large variation in performance between the three schemes for even moderate values of SNR.

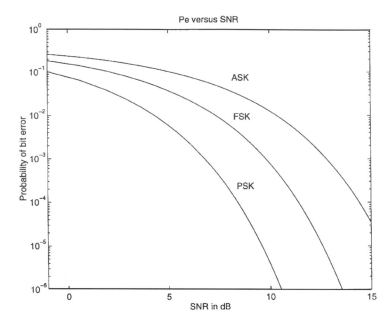

FIGURE 9.4 P_e vs. the signal to noise ratio in decibels [dB $= 10 \log(E/N_0)$] for amplitude-shift keying, frequency-shift keying, and phase-shift keying; note that there is a 3-dB difference in performance from ASK to FSK to PSK.

9.8 *M*-ary Optimal Receivers

In binary signalling schemes, one seeks to transmit a single bit over the bit interval $[0, T]$. This is to be contrasted with M-ary signalling schemes where one transmits multiple bits simultaneously over the so-called symbol interval $[0, T]$. For example, using a signal set with 16 separate waveforms will allow one to transmit a length four-bit sequence per symbol (waveform). Examples of M-ary waveforms are quadrature phase-shift keying (QPSK) and quadrature amplitude modulation (QAM).

The derivation of the optimum receiver structure for M-ary signalling requires the straightforward application of fundamental results in detection theory. As with binary signalling, the Karhunen–Loève expansion is the mechanism utilized to convert a hypotheses testing problem based on continuous waveforms into a vector classification problem. Depending on the complexity of the M waveforms, the signal space can be as large as an M-dimensional vector space.

By extending results from the binary signalling case, it is easily shown that the optimum M-ary receiver computes

$$\xi_i[R(t)] = \int_0^T s_i(t)R(t) \ dt - \frac{E_i}{2} + \frac{N_0}{2} \ln \pi_i \qquad i = 1, \ldots, M$$

where, as before, the $s_i(t)$ constitute the signal set with the π_i being the corresponding a priori probabilities. After computing M separate values of ξ_i, the minimum probability of error receiver simply chooses the largest amongst this set. Thus, the M-ary receiver is implemented with a bank of correlation or matched filters followed by choose-largest decision logic.

In many cases of practical importance, the signal sets are selected so that the resulting signal space is a two-dimensional vector space irrespective of the number of signals. This simplifies the receiver structure in that the sufficient statistics are obtained by implementing only two matched filters. Both QPSK and QAM signal sets fit into this category. As an example, in Fig. 9.5, we have depicted the signal locations for standard 16-QAM signalling with the associated decision boundaries. In this case we have assumed an equally likely signal set. As can be seen, the optimal decision rule selects the signal representation that is closest to the received signal representation in this two-dimensional signal space.

9.9 More Realistic Channels

As is unfortunately often the case, many channels of practical interest are not accurately modeled as simply an AWGN channel. It is often that these channels impose nonlinear effects on the transmitted signals. The best example of this are channels that impose a random phase and random amplitude onto the signal. This typically occurs in applications such as in mobile communications, where one often experiences rapidly changing path lengths from source to receiver.

Fortunately, by the judicious choice of signal waveforms, it can be shown that the selection of the ϕ_i in the Karhunen–Loève transformation is often independent of these unwanted parameters. In these situations, the random amplitude serves only to scale the signals in signal space, whereas the random phase simply imposes a rotation on the signals in signal space.

Since the Karhunen–Loève basis functions typically do not depend on the unknown parameters, we may again convert the continuous time classification problem to a vector channel problem where the received vector $\boldsymbol{R}$ is computed as in Eq. (9.3). Since this vector is a function of both the unknown parameters (i.e., in this case amplitude A and phase ν), to obtain a likelihood ratio test independent of A and ν, we simply apply Bayes theorem to obtain the following form for the LRT:

$$L(\boldsymbol{R}) = \frac{E\left[p_{\boldsymbol{R}|H_1,A,\nu}\left(\boldsymbol{R}\mid H_1,A,\nu\right)\right]}{E\left[p_{\boldsymbol{R}|H_0,A,\nu}\left(\boldsymbol{R}\mid H_0,A,\nu\right)\right]} \begin{matrix} H_1 \\ > \\ < \\ H_0 \end{matrix} \frac{\pi_0}{\pi_1}$$

where the expectations are taken with respect to A and ν, and where $p_{\boldsymbol{R}|H_i,A,\nu}$ are the conditional probability density functions of the signal representations. Assuming that the background noise is AWGN, it can be shown that the LRT simplifies to choosing the largest amongst

$$\xi_i[R(t)] = \pi_i \int_{A,\nu} \exp\left\{\frac{2}{N_0}\int_0^T R(t)s_i(t\mid A,\nu)\ \mathrm{d}t - \frac{E_i(A,\nu)}{N_0}\right\} p_{A,\nu}(A,\nu)\ \mathrm{d}A\,\mathrm{d}\nu$$

$$i = 1,\ldots,M \qquad (9.11)$$

It should be noted that in the Eq. (9.11) we have explicitly shown the dependence of the transmitted signals s_i on the parameters A and ν. The final receiver structures, together

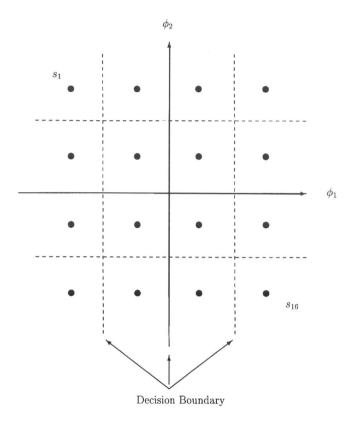

FIGURE 9.5 Signal space representation of 16-QAM signal set. Optimal decision regions for equally likely signals are also noted.

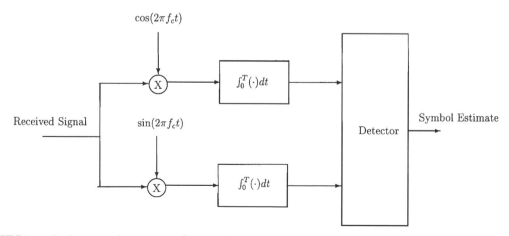

FIGURE 9.6 Optimum receiver structure for noncoherent (random or unknown phase) ASK demodulation.

with their corresponding performance are, thus, a function of both the choice of signal sets and the probability density functions of the random amplitude and random phase.

9.9.1 Random Phase Channels

If we consider first the special case where the channel simply imposes a uniform random phase on the signal, then it can be easily shown that the so-called in-phase and quadrature statistics obtained from the received signal $R(t)$ (denoted by R_I and R_Q, respectively), are sufficient statistics for the signal classification problem. These quantities are computed as

$$R_I(i) = \int_0^T R(t) \cos\left[2\pi f_c(i)t\right]\ \mathrm{d}t$$

and

$$R_Q(i) = \int_0^T R(t) \sin\left[2\pi f_c(i)t\right]\ \mathrm{d}t$$

where in this case the index i corresponds to the center frequencies of hypotheses H_i, (e.g., FSK signalling). The optimum binary receiver selects the largest from amongst

$$\xi_i[R(t)] = \pi_i \exp\left(-\frac{E_i}{N_0}\right) I_0 \left[\frac{2}{N_0}\sqrt{R_I^2(i) + R_Q^2(i)}\right] \qquad i = 1, \ldots, M$$

where I_0 is a zeroth-order, modified Bessel function of the first kind. If the signals have equal energy and are equally likely (e.g., FSK signalling), then the optimum receiver is given by

$$R_I^2(1) + R_Q^2(1) \underset{H_0}{\overset{H_1}{\underset{<}{\gtrless}}} R_I^2(0) + R_Q^2(0)$$

One may readily observe that the optimum receiver bases its decision on the values of the two envelopes of the received signal $\sqrt{R_I^2(i) + R_Q^2(i)}$ and, as a consequence, is often referred to as an envelope or square-law detector. Moreover, it should be observed that the computation of the envelope is independent of the underlying phase of the signal and is as such known as a noncoherent receiver.

The computation of the error rate for this detector is a relatively straightforward exercise resulting in

$$P_e(\text{ noncoherent}) = \frac{1}{2} \exp\left(-\frac{E}{2N_0}\right)$$

As before, note that the error rate for the noncoherent receiver is simply a function of the SNR.

9.9.2 Rayleigh Channel

As an important generalization of the described random phase channel, many communication systems are designed under the assumption that the channel introduces both a random amplitude and a random phase on the signal. Specifically, if the original signal sets are of the form $s_i(t) = m_i(t) \cos(2\pi f_c t)$ where $m_i(t)$ is the baseband version of the message (i.e., what distinguishes one signal from another), then the so-called **Rayleigh channel** introduces random distortion in the received signal of the following form:

$$s_i(t) = A m_i(t) \cos\left(2\pi f_c t + \nu\right)$$

where the amplitude A is a Rayleigh random variable[4] and where the random phase ν is a uniformly distributed between zero and 2π.

To determine the optimal receiver under this distortion, we must first construct an alternate statistical model for $s_i(t)$. To begin, it can be shown from the theory of random variables [4] that if X_I and X_Q are statistically independent, zero mean, Gaussian random variables with variance given by σ^2, then

$$Am_i(t)\cos(2\pi f_c t + \nu) = m_i(t)X_I\cos(2\pi f_c t) + m_i(t)X_Q\sin(2\pi f_c t)$$

Equality here is to be interpreted as implying that both A and ν will be the appropriate random variables. From this, we deduce that the combined uncertainty in the amplitude and phase of the signal is incorporated into the Gaussian random variables X_I and X_Q. The in-phase and quadrature components of the signal $s_i(t)$ are given by $s_{I_i}(t) = m_i(t)\cos(2\pi f_c t)$ and $s_{Q_i}(t) = m_i(t)\sin(2\pi f_c t)$, respectively. By appealing to Eq. (9.11), it can be shown that the optimum receiver selects the largest from

$$\xi_i[R(t)] = \frac{\pi_i}{1 + \frac{2E_i}{N_0}\sigma^2}\exp\left[\frac{\sigma^2}{\frac{1}{2} + \frac{E_i}{N_0}\sigma^2}\left(\langle R(t), s_{I_i}(t)\rangle^2 + \langle R(t), s_{Q_i}(t)\rangle^2\right)\right]$$

where the inner product

$$\langle R(t), S_i(t)\rangle = \int_0^T R(t)s_i(t)\ dt$$

Further, if we impose the conditions that the signals be equally likely with equal energy over the symbol interval, then optimum receiver selects the largest amongst

$$\xi_i[R(t)] = \sqrt{\langle R(t), s_{I_i}(t)\rangle^2 + \langle R(t), s_{Q_i}(t)\rangle^2}$$

Thus, much like for the random phase channel, the optimum receiver for the Rayleigh channel computes the projection of the received waveform onto the in-phase and quadrature components of the hypothetical signals. From a signal space perspective, this is akin to computing the length of the received vector in the subspace spanned by the hypothetical signal. The optimum receiver then chooses the largest amongst these lengths.

As with the random phase channel, computing the performance is a straightforward task resulting in (for the equally likely, equal energy case)

$$P_e(\text{ Rayleigh}) = \frac{\frac{1}{2}}{\left(1 + \frac{E\sigma^2}{N_0}\right)}$$

Interestingly, in this case the performance depends not only on the SNR, but also on the variance (spread) of the Rayleigh amplitude A. Thus, if the amplitude spread is large, we expect to often experience what is known as deep fades in the amplitude of the received waveform and as such expect a commensurate loss in performance.

---

[4]The density of a Rayleigh random variable is given by $p_A(a) = a/\sigma^2\exp(-a^2/2\sigma^2)$ for $a \geq 0$.

9.10   Dispersive Channels

The **dispersive channel** model assumes that the channel not only introduces AWGN but also distorts the signal through a filtering process. This model incorporates physical realities such as multipath effects and frequency selective fading. In particular, the standard model adopted is depicted in the block diagram given in Fig. 9.7. As can be seen, the receiver observes a filtered version of the signal plus AWGN. If the impulse response of the channel is known, then we arrive at the optimum receiver design by applying the previously presented theory. Unfortunately, the duration of the filtered signal can be a complicating factor. More often than not, the channel will increase the duration of the transmitted signals, hence, leading to the description, dispersive channel.

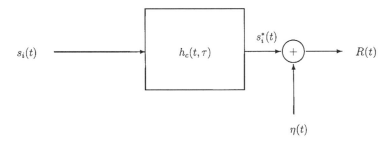

FIGURE 9.7   Standard model for dispersive channel. The time varying impulse response of the channel is denoted by $h_c(t, \tau)$.

However, if the designers take this into account by shortening the duration of $s_i(t)$ so that the duration of $s_i^*(t)$ is less than T, then the optimum receiver chooses the largest amongst

$$\xi_i(R(t)) = \frac{N_0}{2} \ln \pi_i + \langle R(t), s_i^*(t) \rangle - \frac{1}{2} E_i^*$$

If we limit our consideration to equally likely binary signal sets, then the minimum P_e matches the received waveform to the filtered versions of the signal waveforms. The resulting error rate is given by

$$P_e(\text{ dispersive}) = Q\left(\frac{\|s_0^* - s_0^*\|}{\sqrt{2N_0}} \right)$$

Thus, in this case the minimum P_e is a function of the separation of the filtered version of the signals in the signal space.

The problem becomes substantially more complex if we cannot insure that the filtered signal durations are less than the symbol lengths. In this case we experience what is known as **intersymbol interference (ISI)**. That is, observations over one symbol interval contain not only the symbol information of interest but also information from previous symbols. In this case we must appeal to optimum sequence estimation [5] to take full advantage of the information in the waveform. The basis for this procedure is the maximization of the joint likelihood function conditioned on the sequence of symbols. This procedure not only defines the structure of the optimum receiver under ISI but also is critical in the decoding of convolutional codes and coded modulation. Alternate adaptive techniques to solve this problem involve the use of channel equalization.

Defining Terms

Additive white Gaussian noise (AWGN) channel: The channel whose model is that of corrupting a transmitted waveform by the addition of white (i.e., spectrally flat) Gaussian noise.

Bit (symbol) interval: The period of time over which a single symbol is transmitted.

Communication channel: The medium over which communication signals are transmitted. Examples are fiber optic cables, free space, or telephone lines.

Correlation or matched filter receiver: The optimal receiver structure for digital communications in AWGN.

Decision boundary: The boundary in signal space between the various regions where the receiver declares H_i. Typically a hyperplane when dealing with AWGN channels.

Dispersive channel: A channel that elongates and distorts the transmitted signal. Normally modeled as a time-varying linear system.

Intersymbol interference: The ill-effect of one symbol smearing into adjacent symbols thus interfering with the detection process. This is a consequence of the channel filtering the transmitted signals and therefore elongating their duration, see dispersive channel.

Karhunen–Loève expansion: A representation for second-order random processes. Allows one to express a random process in terms of a superposition of deterministic waveforms. The scale values are uncorrelated random variables obtained from the waveform.

Mean-square equivalence: Two random vectors or time-limited waveforms are mean-square equivalent if and only if the expected value of their mean-square error is zero.

Orthonormal: The property of two or more vectors or time-limited waveforms being mutually orthogonal and individually having unit length. Orthogonality and length are typically measured by the standard Euclidean inner product.

Rayleigh channel: A channel that randomly scales the transmitted waveform by a Rayleigh random variable while adding an independent uniform phase to the carrier.

Signal space: An abstraction for representing a time limited waveform in a low-dimensional vector space. Usually arrived at through the application of the Karhunen–Loève transformation.

Total probability of error: The probability of classifying the received waveform into any of the symbols that were not transmitted over a particular bit interval.

References

[1] Gibson, J.D., *Principles of Digital and Analog Communications,* 2nd ed., MacMillan, New York, 1993.

[2] Haykin, S., *Communication Systems*, 3rd ed., John Wiley & Sons, New York, 1994.

[3] Lee, E.A. and Messerschmitt, D.G., *Digital Communication*, Kluwer Academic Publishers, Norwell, MA, 1988.

[4] Papoulis, A., *Probability, Random Variables, and Stochastic Processes,* 3rd ed., McGraw-Hill, New York, 1991.

[5] Poor, H.V., *An Introduction to Signal Detection and Estimation,* Springer-Verlag, New York, 1988.

[6] Proakis, J.G., *Digital Communications,* 2nd ed., McGraw-Hill, New York, 1989.

[7] Shiryayev, A.N., *Probability,* Springer-Verlag, New York, 1984.

[8] Sklar, B., *Digital Communications, Fundamentals and Applications,* Prentice Hall, Englewood Cliffs, NJ, 1988.

[9] Van Trees, H.L., *Detection, Estimation, and Modulation Theory, Part I,* John Wiley & Sons, New York, 1968.

[10] Wong, E. and Hajek, B., *Stochastic Processes in Engineering Systems,* Springer-Verlag, New York, 1985.

[11] Wozencraft, J.M. and Jacobs, I., *Principles of Communication Engineering,* reissue, Waveland Press, Prospect Heights, Illinois, 1990.

[12] Ziemer, R.E. and Peterson, R.L., *Introduction to Digital Communication,* Macmillan, New York, 1992.

Further Information

The fundamentals of receiver design were put in place by Wozencraft and Jacobs in their seminal book. Since that time, there have been many outstanding textbooks in this area. For a sampling see [1, 2, 3, 8, 12]. For a complete treatment on the use and application of detection theory in communications see [5, 9]. For deeper insights into the Karhunen–Loève expansion and its use in communications and signal processing see [10].

10

Forward Error Correction Coding

V.K. Bhargava
University of Victoria

I.J. Fair
University of Alberta

10.1 Introduction

In 1948, Claude Shannon issued a challenge to communications engineers by proving that communication systems could be made arbitrarily reliable as long as a fixed percentage of the transmitted signal was redundant [9]. He showed that limits exist only on the rate of communication and not its accuracy, and went on to prove that errorless transmission could be achieved in an additive white Gaussian noise (AWGN) environment with infinite bandwidth if the ratio of energy per data bit to noise power spectral density exceeds the **Shannon Limit**. He did not, however, indicate how this could be achieved. Subsequent research has led to a number of techniques that introduce redundancy to allow for correction of errors without retransmission. These techniques, collectively known as forward error correction (FEC) coding techniques, are used in systems where a reverse channel is not available for requesting retransmission, the delay with retransmission would be excessive, the expected number of errors would require a large number of retransmissions, or retransmission would be awkward to implement [10].

A simplified model of a digital communication system which incorporates FEC coding is shown in Fig. 10.1. The FEC code acts on a **discrete data channel** comprising all system elements between the encoder output and decoder input. The encoder maps the source data to q-ary code symbols which are modulated and transmitted. During transmission, this signal can be corrupted, causing errors to arise in the demodulated symbol sequence. The FEC decoder attempts to correct these errors and restore the original source data.

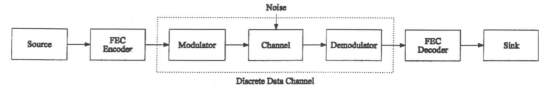

FIGURE 10.1   Block diagram of a digital communication system with forward error correction.

A demodulator which outputs only a value for the q-ary symbol received during each symbol interval is said to make **hard decisions.** In the **binary symmetric channel** (BSC), hard decisions are made on binary symbols and the probability of error is independent of the value of the symbol. One example of a BSC is the coherently demodulated binary phase-shift-keyed (BPSK) signal corrupted by AWGN. The conditional probability density functions which result with this system are depicted in Fig. 10.2. The probability of error is given by the area under the density functions that lies across the decision threshold, and is a function of the symbol energy E_s and the one-sided noise power spectral density N_0.

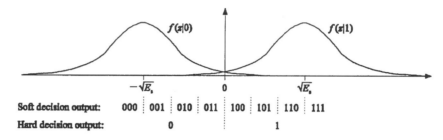

FIGURE 10.2   Hard and soft decision demodulation of a coherently demodulated BPSK signal corrupted by AWGN. $f(z \mid 1)$ and $f(z \mid 0)$ are the Gaussianly distributed conditional probability density functions at the threshold device.

Alternatively, the demodulator can make **soft decisions** or output an estimate of the symbol value along with an indication of its confidence in this estimate. For example, if the BPSK demodulator uses three-bit quantization, the two least significant bits can be taken as a confidence measure. Possible soft-decision thresholds for the BPSK signal are depicted in Fig. 10.2. In practice, there is little to be gained by using many soft-decision quantization levels.

Block and convolutional codes introduce redundancy by adding parity symbols to the message data. They map k source symbols to n code symbols and are said to have **code rate** $R = k/n$. With fixed information rates, this redundancy results in increased bandwidth and lower energy per transmitted symbol. At low signal-to-noise ratios, these codes cannot compensate for these impairments, and performance is degraded. At higher ratios of information symbol energy E_b to noise spectral density N_0, however, there is **coding gain** since the performance improvement offered by coding more than compensates for these impairments. Coding gain is usually defined as the reduction in required E_b/N_0 to achieve a specific error rate in an error-control coded system over one without coding. In contrast to block and convolutional codes, trellis-coded modulation introduces redundancy by expanding the size of the signal set rather than increasing the number of symbols transmitted, and so offers the advantages of coding to band-limited systems.

Each of these coding techniques is considered in turn. Following a discussion of **inter-**

leaving and concatenated coding, this chapter gives an overview of a recent and significant advance in coding, the development of Turbo codes, and concludes with a brief overview of FEC applications.

10.2 Fundamentals of Block Coding

In block codes there is a one-to-one mapping between k-symbol source words and n-symbol codewords. With q-ary signalling, q^k out of the q^n possible n-tuples are valid code vectors. The set of all n-tuples forms a **vector space** in which the q^k code vectors are distributed. The **Hamming distance** between any two code vectors is the number of symbols in which they differ; the **minimum distance** $d_{\min}$ of the code is the smallest Hamming distance between any two codewords.

There are two contradictory objectives of block codes. The first is to distribute the code vectors in the vector space such that the distance between them is maximized. Then, if the decoder receives a corrupted vector, by evaluating the nearest valid code vector it will decode the correct word with high probability. The second is to pack the vector space with as many code vectors as possible to reduce the redundancy in transmission.

When code vectors differ in at least $d_{\min}$ positions, a decoder which evaluates the nearest code vector to each received word is guaranteed to correct up to t random symbol errors per word if

$$d_{\min} \geq 2t + 1 \tag{10.1}$$

Alternatively, all $q^n - q^k$ illegal words can be detected, including all error patterns with $d_{\min} - 1$ or fewer errors. In general, a block code can correct all patterns of t or fewer errors and detect all patterns of u or fewer errors provided that $u \geq t$ and

$$d_{\min} \geq t + u + 1 \tag{10.2}$$

If $q = 2$, knowledge of the positions of the errors is sufficient for their correction; if $q > 2$, the decoder must determine both the positions and values of the errors. If the demodulator indicates positions in which the symbol values are unreliable, the decoder can assume their value unknown and has only to solve for the value of these symbols. These positions are called **erasures**. A block code can correct up to t errors and v erasures in each word if

$$d_{\min} \geq 2t + v + 1 \tag{10.3}$$

10.3 Structure and Decoding of Block Codes

Shannon showed that the performance limit of codes with fixed code rate improves as the block length increases. As n and k increase, however, practical implementation requires that the mapping from message to code vector not be arbitrary but that an underlying structure to the code exist. The structures developed to date limit the error correcting capability of these codes to below what Shannon proved possible, on average, for a code with random codeword assignments. Although Turbo codes have made significant strides towards approaching the Shannon Limit, the search for good constructive codes continues.

A property which simplifies implementation of the coding operations is that of code linearity. A code is **linear** if the addition of any two code vectors forms another code vector, which implies that the code vectors form a subspace of the vector space of n-tuples. This subspace, which contains the all-zero vector, is spanned by any set of k linearly independent code vectors. Encoding can be described as the multiplication of the information k-tuple by a **generator matrix G**, of dimension $k \times n$, which contains these basis vectors as rows. That is, a message vector m_i is mapped to a code vector c_i according to

$$c_i = m_i G, \qquad i = 0, 1, \ldots, q^k - 1 \tag{10.4}$$

where elementwise arithmetic is defined in the **finite field** $\mathrm{GF}(q)$. In general, this encoding procedure results in code vectors with nonsystematic form in that the values of the message symbols cannot be determined by inspection of the code vector. However, if G has the form $[I_k, P]$ where I_k is the $k \times k$ identity matrix and P is a $k \times (n-k)$ matrix of parity checks, then the k most significant symbols of each code vector are identical to the message vector and the code has **systematic** form. This notation assumes that vectors are written with their most significant or first symbols in time on the left, a convention used throughout this chapter.

For each generator matrix there is an $(n-k) \times k$ **parity check matrix H** whose rows are orthogonal to the rows in G, i.e., $GH^T = 0$. If the code is systematic, $H = [-P^T, I_{n-k}]$. Since all codewords are linear sums of the rows in G, it follows that $c_i H^T = 0$ for all $i, i = 0, 1, \ldots, q^k - 1$, and that the validity of the demodulated vectors can be checked by performing this multiplication. If a codeword c is corrupted during transmission so that the hard-decision demodulator outputs the vector $\hat{c} = c + e$, where e is a nonzero error pattern, the result of this multiplication is an $(n-k)$-tuple that is indicative of the validity of the sequence. This result, called the **syndrome s**, is dependent only on the error pattern since

$$s = \hat{c} H^T = (c + e) H^T = c H^T + e H^T = e H^T \tag{10.5}$$

If the error pattern is a code vector, the errors go undetected. For all other error patterns, however, the syndrome is nonzero. Since there are $q^{n-k} - 1$ nonzero syndromes, $q^{n-k} - 1$ error patterns can be corrected. When these patterns include all those with t or fewer errors and no others, the code is said to be a **perfect code.** Few codes are perfect; most codes are capable of correcting some patterns with more than t errors. **Standard array decoders** use lookup tables to associate each syndrome with an error pattern but become impractical as the block length and number of parity symbols increases. Algebraic decoding algorithms have been developed for codes with stronger structure. These algorithms are simplified with imperfect codes if the patterns corrected are limited to those with t or fewer errors, a simplification called **bounded distance decoding.**

Cyclic codes are a subclass of linear block codes with an algebraic structure that enables encoding to be implemented with a linear feedback shift register and decoding to be implemented without a lookup table. As a result, most block codes in use today are cyclic or are closely related to cyclic codes. These codes are best described if vectors are interpreted as polynomials and the arithmetic follows the rules for polynomials where the elementwise operations are defined in $\mathrm{GF}(q)$. In a cyclic code, all codeword polynomials are multiples of a **generator polynomial** $g(x)$ of degree $n-k$. This polynomial is chosen to be a divisor of $x^n - 1$ so that a cyclic shift of a code vector yields another code vector, giving this class of codes its name. A message polynomial $m_i(x)$ can be mapped to a codeword polynomial $c_i(x)$ in nonsystematic form as

$$c_i(x) = m_i(x) g(x), \qquad i = 0, 1, \ldots, q^k - 1 \tag{10.6}$$

In systematic form, codeword polynomials have the form

$$c_i(x) = m_i(x) x^{n-k} - r_i(x), \qquad i = 0, 1, \ldots, q^k - 1 \tag{10.7}$$

where $r_i(x)$ is the remainder of $m_i(x)x^{n-k}$ divided by $g(x)$. Polynomial multiplication and division can be easily implemented with shift registers [5].

The first step in decoding the demodulated word is to determine if the word is a multiple of $g(x)$. This is done by dividing it by $g(x)$ and examining the remainder. Since polynomial division is a linear operation, the resulting syndrome $s(x)$ depends only on the error pattern. If $s(x)$ is the all-zero polynomial, transmission is errorless or an undetectable error pattern has occurred. If $s(x)$ is nonzero, at least one error has occurred. This is the principle of the **cyclic redundancy check** (CRC). It remains to determine the most likely error pattern that could have generated this syndrome.

Single error correcting binary codes can use the syndrome to immediately locate the bit in error. More powerful codes use this information to determine the locations and values of multiple errors. The most prominent approach of doing so is with the iterative technique developed by Berlekamp. This technique, which involves computing an error-locator polynomial and solving for its roots, was subsequently interpreted by Massey in terms of the design of a minimum-length shift register. Once the location and values of the errors are known, Chien's search algorithm efficiently corrects them. The implementation complexity of these decoders increases only as the square of the number of errors to be corrected [4] but does not generalize easily to accommodate soft-decision information. Other decoding techniques, including Chase's algorithm and threshold decoding, are easier to implement with soft-decision input [6]. Berlekamp's algorithm can be used in conjunction with transform-domain decoding, which involves transforming the received block with a finite field Fourier-like transform and solving for errors in the transform domain. Since the implementation complexity of these decoders depends on the block length rather than the number of symbols corrected, this approach results in simpler circuitry for codes with high redundancy [13].

Other block codes have also been constructed, including codes that are based on transform-domain spectral properties, codes that are designed specifically for correction of burst errors, and codes that are decodable with straightforward threshold or majority logic decoders [5, 6, 7].

10.4 Important Classes of Block Codes

When errors occur independently, Bose–Chaudhuri–Hocquenghem (BCH) codes provide one of the best performances of known codes for a given block length and code rate. They are cyclic codes with $n = q^m - 1$, where m is any integer greater than 2. They are designed to correct up to t errors per word and so have **designed distance** $d = 2t + 1$; the minimum distance may be greater. Generator polynomials for these codes are listed in many texts, including [6]. These polynomials are of degree less than or equal to mt, and so $k \geq n - mt$. BCH codes can be shortened to accommodate system requirements by deleting positions for information symbols.

Some subclasses of these codes are of special interest. Hamming codes are perfect single error correcting binary BCH codes. Full length codes have $n = 2^m - 1$ and $k = n - m$ for any m greater than 2. The duals of these codes are maximal-length codes, with $n = 2^m - 1$, $k = m$, and $d_{\min} = 2^{m-1}$. All $2^m - 1$ nonzero code vectors in these codes are cyclic shifts of a single nonzero code vector. Reed–Solomon (RS) codes are nonbinary BCH codes defined over $\mathrm{GF}(q)$, where q is often taken as a power of two so that symbols can be represented by a sequence of bits. In these cases, correction of even a single symbol allows for correction of a burst of bit errors. The block length is $n = q - 1$, and the minimum distance $d_{\min} = 2t + 1$ is achieved using only $2t$ parity symbols. Since RS codes meet the Singleton bound of

$d_{\min} \leq n - k + 1$, they have the largest possible minimum distance for these values of n and k and are called **maximum distance separable** codes.

The Golay codes are the only nontrivial perfect codes that can correct more than one error. The $(11, 6)$ ternary Golay code has minimum distance 5. The $(23, 12)$ binary code is a triple error correcting BCH code with $d_{\min} = 7$. To simplify implementation, it is often extended to a $(24, 12)$ code through the addition of an extra parity bit. The extended code has $d_{\min} = 8$.

The $(23, 12)$ Golay code is also a binary quadratic residue code. These cyclic codes have prime length of the form $n = 8m \pm 1$, with $k = (n + 1)/2$ and $d_{\min} \geq \sqrt{n}$. Some of these codes are as good as the best codes known with these values of n and k, but it is unknown if there are good quadratic residue codes with large n [5].

Reed-Muller codes are equivalent to binary cyclic codes with an additional overall parity bit. For any m, the rth-order Reed-Muller code has $n = 2^m$, $k = \Sigma_{i=0}^{r} \binom{m}{i}$, and $d_{\min} = 2^{m-r}$. The rth-order and $(m - r - 1)$th-order codes are duals, and the first-order codes are similar to maximal-length codes. These codes, and the closely related Euclidean geometry and projective geometry codes, can be decoded with threshold decoding.

The performance of several of these block codes is shown in Fig. 10.3 in terms of decoded bit error probability vs. E_b/N_0 for systems using coherent, hard-decision demodulated BPSK signalling. Many other block codes have also been developed, including Goppa codes, quasicyclic codes, burst error correcting Fire codes, and other lesser known codes.

10.5   Principles of Convolutional Coding

Convolutional codes map successive information k-tuples to a series of n-tuples such that the sequence of n-tuples has distance properties that allow for detection and correction of errors. Although these codes can be defined over any alphabet, their implementation has largely been restricted to binary signals, and only binary convolutional codes are considered here.

In addition to the code rate $R = k/n$, the **constraint length** K is an important parameter for these codes. Definitions vary; we will use the definition that K equals the number of k-tuples that affect formation of each n-tuple during encoding. That is, the value of an n-tuple depends on the k-tuple that arrives at the encoder during that encoding interval as well as the $K - 1$ previous information k-tuples.

Binary convolutional encoders can be implemented with kK-stage shift registers and n modulo-2 adders, an example of which is given in Fig. 10.4(a) for a rate $1/2$, constraint length 3 code. The encoder shifts in a new k-tuple during each encoding interval and samples the outputs of the adders sequentially to form the coded output.

Although connection diagrams similar to that of Fig. 10.4(a) completely describe the code, a more concise description can be given by stating the values of n, k, and K and giving the adder connections in the form of vectors or polynomials. For instance, the rate $1/2$ code has the generator vectors $\boldsymbol{g}_1 = 111$ and $\boldsymbol{g}_2 = 101$, or equivalently, the generator polynomials $g_1(x) = x^2 + x + 1$ and $g_2(x) = x^2 + 1$. Alternatively, a convolutional code can be characterized by its impulse response, the coded sequence generated due to input of a single logic-1. It is straightforward to verify that the circuit in Fig. 10.4(a) has the impulse response 111011. Since modulo-2 addition is a linear operation, convolutional codes are linear, and the coded output can be viewed as the convolution of the input sequence with the impulse response, hence the name of this coding technique. Shifted versions of the impulse response or generator vectors can be combined to form an infinite-order generator matrix which also describes the code.

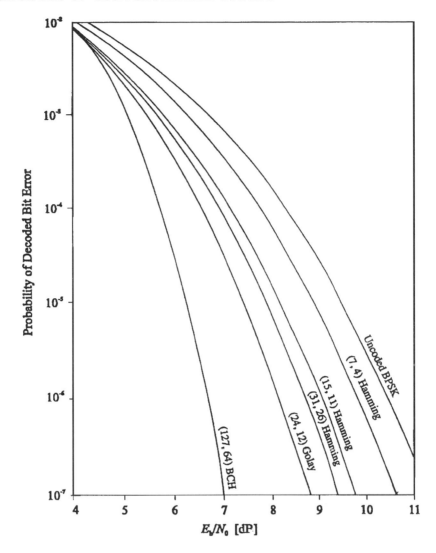

FIGURE 10.3 Block code performance. *Source:* Sklar, B., 1988, *Digital Communications: Fundamentals and Applications,* © 1988, p. 300. Reprinted by permission of Prentice-Hall, Inc., Englewood Cliffs, NJ.

Shift register circuits can be modeled as finite state machines. A Mealy machine description of a convolutional encoder requires $2^{k(K-1)}$ states, each describing a different value of the $K-1$ k-tuples which have most recently entered the shift register. Each state has 2^k exit paths which correspond to the value of the incoming k-tuple. A state machine description for the rate 1/2 encoder depicted in Fig. 10.4(a) is given in Fig. 10.4(b). States are labeled with the contents of the two leftmost register stages; edges are labeled with information bit values and their corresponding coded output.

The dimension of time is added to the description of the encoder with tree and trellis diagrams. The tree diagram for the rate 1/2 convolutional code is given in Fig. 10.4(c), assuming the shift register is initially clear. Each node represents an encoding interval, from which the upper branch is taken if the input bit is a 0 and the lower branch is taken if the input bit is a 1. Each branch is labeled with the corresponding output bit sequence. A drawback of the tree representation is that it grows without bound as the length of the

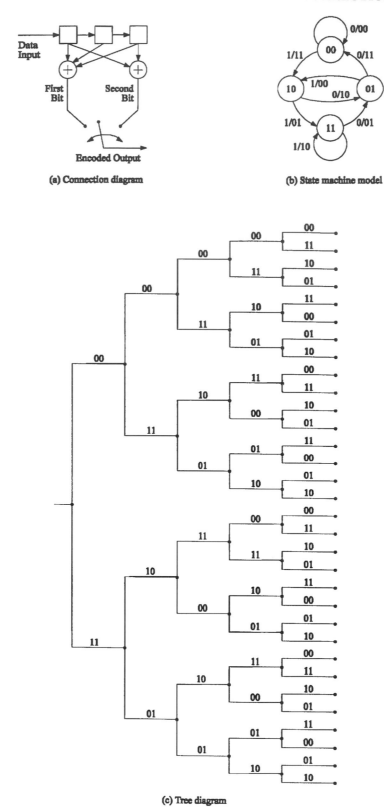

(a) Connection diagram

(b) State machine model

(c) Tree diagram

FIGURE 10.4   A rate 1/2, constraint length 3 convolutional code.

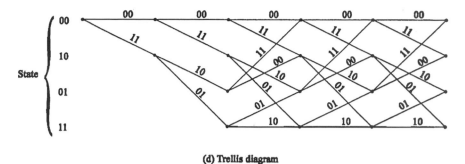

(d) Trellis diagram

FIGURE 10.4   (*Continued*).

input sequence increases. This is overcome with the trellis diagram depicted in Fig. 10.4(d),
Again, encoding results in left-to-right movement, where the upper of the two branches is
taken whenever the input is a 0, the lower branch is taken when the input is a 1, and the
output is the bit sequence which weights the branch taken. Each level of nodes corresponds
to a state of the encoder as shown on the left-hand side of the diagram.

If the received sequence contains errors, it may no longer depict a valid path through
the tree or trellis. It is the job of the decoder to determine the original path. In doing so,
the decoder does not so much correct errors as find the closest valid path to the received
sequence. As a result, the error correcting capability of a convolutional code is more difficult
to quantify than that of a block code; it depends on how valid paths differ. One measure of
this difference is the **column distance** $d_c(i)$, the minimum Hamming distance between all
coded sequences generated over i encoding intervals which differ in the first interval. The
nondecreasing sequence of column distance values is the **distance profile** of the code. The
column distance after K intervals is the minimum distance of the code and is important
for evaluating the performance of a code that uses threshold decoding. As i increases, $d_c(i)$
approaches the **free distance** of the code, d_{free}, which is the minimum Hamming distance
in the set of arbitrarily long paths that diverge and then remerge in the trellis.

With maximum likelihood decoding, convolutional codes can generally correct up to t
errors within three to five constraint lengths, depending on how the errors are distributed,
where

$$d_{\text{free}} \geq 2t + 1 \tag{10.8}$$

The free distance can be calculated by exhaustively searching for the minimum-weight path
that returns to the all-zero state, or evaluating the term of lowest degree in the generating
function of the code.

The objective of a convolutional code is to maximize these distance properties. They
generally improve as the constraint length of the code increases, and nonsystematic codes
generally have better properties than systematic ones. Good codes have been found by
computer search and are tabulated in many texts, including [6]. Convolutional codes with
high code rate can be constructed by **puncturing** or periodically deleting coded symbols
from a low rate code. A list of low rate codes and perforation matrices that result in good
high rate codes can be found in many sources, including [13]. The performance of good
punctured codes approaches that of the best convolutional codes known with similar rate,
and decoder implementation is significantly less complex.

Convolutional codes can be **catastrophic,** having the potential to generate an unlimited
number of decoded bit errors in response to a finite number of errors in the demodulated bit
sequence. Catastrophic error propagation is avoided if the code has generator polynomials
with a greatest common divisor of the form x^a for any a or, equivalently, if there are no

closed-loop paths in the state diagram with all-zero output other than the one taken with all-zero input. **Systematic codes** are not catastrophic.

10.6 Decoding of Convolutional Codes

In 1967, Viterbi developed a maximum likelihood decoding algorithm that takes advantage of the trellis structure to reduce the complexity of the evaluation. This algorithm has become known as the **Viterbi algorithm.** With each received n-tuple, the decoder computes a **metric** or measure of likelihood for all paths that could have been taken during that interval and discards all but the most likely to terminate on each node. An arbitrary decision is made if path metrics are equal. The metrics can be formed using either hard or soft decision information with little difference in implementation complexity.

If the message has finite length and the encoder is subsequently flushed with zeros, a single decoded path remains. With a BSC, this path corresponds to the valid code sequence with minimum Hamming distance from the demodulated sequence. Full-length decoding becomes impractical as the length of the message sequence increases. The most likely paths tend to have a common stem, however, and selecting the trace value four or five times the constraint length prior to the present decoding depth results in near-optimum performance. Since the number of paths examined during each interval increases exponentially with the constraint length, the Viterbi algorithm also becomes impractical for codes with large constraint length. To date, Viterbi decoding has been implemented for codes with constraint lengths up to ten. Other decoding techniques, such as sequential and threshold decoding, can be used with larger constraint lengths.

Sequential decoding was proposed by Wozencraft, and the most widely used algorithm was developed by Fano. Rather than tracking multiple paths through the trellis, the sequential decoder operates on a single path while searching the code tree for a path with high probability. It makes tentative decisions regarding the transmitted sequence, computes a metric between its proposed path and the demodulated sequence, and moves forward through the tree as long as the metric indicates that the path is likely. If the likelihood of the path becomes low, the decoder moves backward, searching other paths until it finds one with high probability. The number of computations involved in this procedure is almost independent of the constraint length and is typically quite small, but it can be highly variable, depending on the channel. Buffers must be provided to store incoming sequences as the decoder searches the tree. Their overflow is a significant limiting factor in the performance of these decoders.

Figure 10.5 compares the performance of the Viterbi and sequential decoding algorithms for several convolutional codes operating on coherently demodulated BPSK signals corrupted by AWGN. Other decoding algorithms have also been developed, including syndrome decoding methods such as table look-up feedback decoding and threshold decoding [6]. These algorithms are easily implemented but offer suboptimal performance. Techniques such as the one discussed by [1] have been developed to support both soft input and soft output, but these decoding techniques typically increase decoder complexity.

10.7 Trellis-Coded Modulation

Trellis-coded modulation (TCM) has received considerable attention since its development by Ungerboeck in the late 1970s [11]. Unlike block and convolutional codes, TCM schemes achieve coding gain by increasing the size of the signal alphabet and using multilevel/phase signalling. Like convolutional codes, sequences of coded symbols are restricted to certain

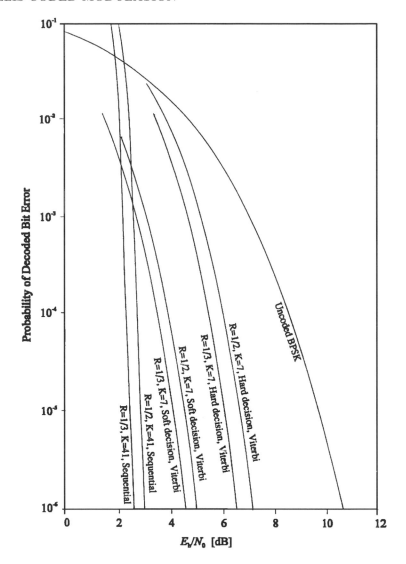

FIGURE 10.5   Convolutional code performance. *Source:* Omura, J.K. and Levitt, B.K., © 1982 IEEE, "Coded Error Probability Evaluation for Antijam Communication Systems," *IEEE Trans. Commun.,* vol. COM-30, no. 5, pp. 896–903. Reprinted by permission of IEEE.

valid patterns. In TCM, these patterns are chosen to have large Euclidean distance from one another so that a large number of corrupted sequences can be corrected. The Viterbi algorithm is often used to decode these sequences. Since the symbol transmission rate does not increase, coded and uncoded signals require the same transmission bandwidth. If transmission power is held constant, the signal constellation of the coded signal is denser. The loss in symbol separation, however, is more than overcome by the error correction capability of the code.

Ungerboeck investigated the increase in channel capacity that can be obtained by increasing the size of the signal set and restricting the pattern of transmitted symbols, and concluded that almost all of the additional capacity can be gained by doubling the number of points in the signal constellation. This is accomplished by encoding the binary data with a rate $R = k/(k+1)$ code and mapping sequences of $k+1$ coded bits to points in a constellation of 2^{k+1} symbols. For example, the rate 2/3 encoder of Fig. 10.6(a) encodes pairs of source bits to three coded bits. Figure 10.6(b) depicts one stage in the trellis of the coded output where, as with the convolutional code, the state of the encoder is defined by the values of the two most recent bits to enter the shift register. Note that unlike the trellis for the convolutional code, this trellis contains parallel paths between nodes.

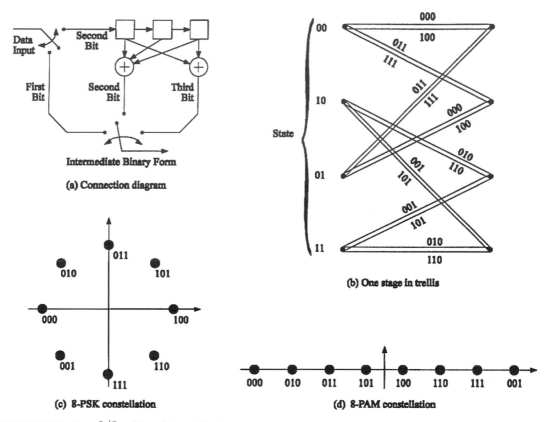

FIGURE 10.6   Rate 2/3 trellis-coded modulation.

The key to improving performance with TCM is to map the coded bits to points in the signal space such that the Euclidean distance between transmitted sequences is maximized. A method that ensures improved Euclidean distance is the method of **set partitioning.** This involves separating all parallel paths on the trellis with maximum distance and assigning the next greatest distance to paths that diverge from or merge onto the same node. Figures 10.6(c) and 10.6(d) give examples of mappings for the rate 2/3 code with 8-PSK and 8-PAM signal constellations, respectively.

As with convolutional codes, the free distance of a TCM code is defined as the minimum distance between paths through the trellis, where the distance of concern is now Euclidean distance rather than Hamming distance. The free distance of an uncoded signal is defined as the distance between the closest signal points. When coded and uncoded signals have the same average power, the coding gain of the TCM system is defined as

$$\text{coding gain} = 20 \log_{10} \left(\frac{d_{\text{free, coded}}}{d_{\text{free, uncoded}}} \right) \tag{10.9}$$

It can be shown that the simple, rate 2/3 8 phase-shift keying (PSK) and 8 pulse-amplitude modulation (PAM) TCM systems provide gains of 3 dB and 3.3 dB, respectively, [6]. More complex TCM systems yield gains up to 6 dB. Tables of good codes are given in [11].

10.8   Additional Measures

When the demodulated sequence contains bursts of errors, the performance of codes designed to correct independent errors improves if coded sequences are **interleaved** prior to transmission and deinterleaved prior to decoding. Deinterleaving separates the burst errors, making them appear more random and increasing the likelihood of accurate decoding. It is generally sufficient to interleave several block lengths of a block coded signal or several constraint lengths of a convolutionally encoded signal. Block interleaving is the most straightforward approach, but delay and memory requirements are halved with convolutional and helical interleaving techniques. Periodicity in the way sequences are combined is avoided with pseudorandom interleaving.

Serially **concatenated codes,** first investigated by Forney, use two levels of coding to achieve a level of performance with less complexity than a single coding stage would require. The inner code interfaces with the modulator and demodulator and corrects the majority of the errors; the outer code corrects errors that appear at the output of the inner-code decoder. A convolutional code with Viterbi decoding is usually chosen as the inner code, and an RS code is often chosen as the outer code due to its ability to correct the bursts of bit errors which can result with incorrect decoding of trellis-coded sequences. Interleaving and deinterleaving outer-code symbols between coding stages offers further protection against the burst error output of the inner code.

Product codes effectively place the data in a two dimensional array and use FEC techniques over both the rows and columns of this array. Not only do these codes result in error protection in two dimensions, but the manner in which the array is constructed can offer advantages similar to those achieved through interleaving.

10.9   Turbo Codes

The most recent significant achievement in FEC coding is the development of **Turbo codes** [3]. The principle of this coding technique is to encode the data with two or more **constituent codes** concatenated in parallel form. The received sequence is decoded in an iterative, serial approach using soft-input, soft-output decoders. This iterative decoding approach involves feedback of information in a manner similar to processes within the turbo engine, giving this coding technique its name.

Turbo codes effectively result in the construction of relatively long codewords with few codewords being close in terms of Hamming distance, while at the same time constraining the implementation complexity of the decoder to practical limits. The first Turbo codes developed used recursive systematic convolutional codes as the constituent codes, and punctured them to improve the code rate. The use of other constituent codes has since been considered. Two or more of these codes are concatenated in parallel,where code concatenation is combined with interleaving in order to increase the independence of the data sequences encoded by the constituent encoders. This apparent increase in randomness, implemented with simple interleavers, is an important contributing factor to the excellent performance of the decoders.

As in other multi-stage coding techniques, the complexity of the decoder is limited through use of separate decoding stages for each constituent code. The input to the first stage is the soft output of the demodulator for a finite-length received symbol sequence. Subsequent stages use both the demodulator output and an output of the previous decoding stage which is indicative of the reliability of the symbols. This information, gleaned from soft-output decoders, is called **extrinsic information**.

Decoding proceeds by iterating through constituent decoders, each forwarding updated extrinsic information to the next decoder, until a predefined number of iterations has been completed or the extrinsic information indicates that high reliability has been achieved. This approach results in very good performance at low values of E_b/N_0. Simulations have demonstrated error rates of 10^{-5} at signal-to-noise ratios appreciably less than 1 dB. At higher values of E_b/N_0, however, the performance curves can exhibit flattening if constituent codes are chosen in a manner that results in an overall small Hamming distance for the code.

Although this coding technique has shown great promise, there remains considerable work with regard to optimizing code parameters. Great strides have been made over the last few years in understanding the structure of these codes and relating them to serially concatenated and product codes, but many researchers are still examining these codes in order to advance their development. With this research will come optimization of the Turbo code process and application of these codes in various communication systems.

10.10 Applications

FEC coding remained of theoretical interest until advances in digital technology and improvements in decoding algorithms made their implementation possible. It has since become an attractive alternative to improving other system components or boosting transmission power. FEC codes are commonly used in digital storage systems, deep-space and satellite communication systems, terrestrial radio and band limited wireline systems, and have also been proposed for fiber optic transmission. Accordingly, the theory and practice of error correcting codes now occupies a prominent position in the field of communications engineering.

Deep-space systems began using forward error correction in the early 1970s to reduce transmission power requirements, and used multiple error correcting RS codes for the first time in 1977 to protect against corruption of compressed image data in the Voyager missions [12]. The Consultative Committee for Space Data Systems (CCSDS) has since recommended use of a concatenated coding system which uses a rate 1/2, constraint length 7 convolutional inner code and a (255, 223) RS outer code.

Coding is now commonly used in satellite systems to reduce power requirements and overall hardware costs and to allow closer orbital spacing of geosynchronous satellites [2]. FEC codes play integral roles in the VSAT, MSAT, INTELSAT, and INMARSAT systems [13]. Further, a (31, 15) RS code is used in the joint tactical information distribution system (JTIDS), a (7, 2) RS code is used in the air force satellite communication system (AFSAT-COM), and a (204, 192) RS code has been designed specifically for satellite time division multiple access (TDMA) systems. Another code designed for military applications involves concatenation of a Golay and RS code with interleaving to ensure an imbalance of 1's and 0's in the transmitted symbol sequence and enhance signal recovery under severe noise and interference [2].

TCM has become commonplace in transmission of data over voiceband telephone channels. Modems developed since 1984 use trellis coded QAM modulation to provide robust communication at rates above 9.6 kb/s. Various coding techniques are used in the new digital cellular and personal communication standards, with an emphasis on convolutional and cyclic redundancy check codes [8].

FEC codes have also been widely used in digital recording systems, most prominently in the compact disc digital audio system. This system uses two levels of coding and interleaving in the cross-interleaved RS coding (CIRC) system to correct errors that result from disc imperfections and dirt and scratches which accumulate during use. Steps are also taken to mute uncorrectable sequences [12].

Defining Terms

Binary symmetric channel: A memoryless discrete data channel with binary signalling, hard-decision demodulation, and channel impairments that do not depend on the value of the symbol transmitted.

Bounded distance decoding: Limiting the error patterns which are corrected in an imperfect code to those with t or fewer errors.

Catastrophic code: A convolutional code in which a finite number of code symbol errors can cause an unlimited number of decoded bit errors.

Code rate: The ratio of source word length to codeword length, indicative of the amount of information transmitted per encoded symbol.

Coding gain: The reduction in signal-to-noise ratio required for specified error performance in a block or convolutional coded system over an uncoded system with the same information rate, channel impairments, and modulation and demodulation techniques. In TCM, the ratio of the squared free distance in the coded system to that of the uncoded system.

Column distance: The minimum Hamming distance between convolutionally encoded sequences of a specified length with different leading n-tuples.

Constituent codes: Two or more FEC codes that are combined in concatenated coding techniques.

Cyclic code: A block code in which cyclic shifts of code vectors are also code vectors.

Cyclic redundancy check: When the syndrome of a cyclic block code is used to detect errors.

Designed distance: The guaranteed minimum distance of a BCH code designed to correct up to t errors.

Discrete data channel: The concatenation of all system elements between FEC encoder output and decoder input.

Distance profile: The minimum Hamming distance after each encoding interval of convolutionally encoded sequences which differ in the first interval.

Erasure: A position in the demodulated sequence where the symbol value is unknown.

Extrinsic information: The output of a constituent soft decision decoder that is forwarded as input to the next decoding stage in iterative decoding of Turbo codes.

Finite field: A finite set of elements and operations of addition and multiplication that satisfy specific properties. Often called Galois fields and denoted $GF(q)$, where q is the number of elements in the field. Finite fields exist for all q which are prime or the power of a prime.

Free distance: The minimum Hamming weight of convolutionally encoded sequences that diverge and remerge in the trellis. Equals the maximum column distance and the limiting value of the distance profile.

Generator matrix: A matrix used to describe a linear code. Code vectors equal the information vectors multiplied by this matrix.

Generator polynomial: The polynomial that is a divisor of all codeword polynomials in a cyclic block code; a polynomial that describes circuit connections in a convolutional encoder.

Hamming distance: The number of symbols in which codewords differ.

Hard decision: Demodulation that outputs only a value for each received symbol.

Interleaving: Shuffling the coded bit sequence prior to modulation and reversing this operation following demodulation. Used to separate and redistribute burst errors over several codewords (block codes) or constraint lengths (trellis codes) for higher probability of correct decoding by codes designed to correct random errors.

Linear code: A code whose code vectors form a vector space. Equivalently, a code where the addition of any two code vectors forms another code vector.

Maximum distance separable: A code with the largest possible minimum distance given the block length and code rate. These codes meet the Singleton bound of $d_{\min} \leq n - k + 1$.

Metric: A measure of goodness against which items are judged. In the Viterbi algorithm, an indication of the probability of a path being taken given the demodulated symbol sequence.

Minimum distance: In a block code, the smallest Hamming distance between any two codewords. In a convolutional code, the column distance after K intervals.

Parity check matrix: A matrix whose rows are orthogonal to the rows in the generator matrix of a linear code. Errors can be detected by multiplying the received vector by this matrix.

Perfect code: A t error correcting (n, k) block code in which $q^{n-k} - 1 = \Sigma_{i=1}^{t} \binom{n}{i}$.

Puncturing: Periodic deletion of code symbols from the sequence generated by a convolutional encoder for purposes of constructing a higher rate code. Also, deletion of parity bits in a block code.

Set partitioning: Rules for mapping coded sequences to points in the signal constellation that always result in a larger Euclidean distance for a TCM system than an uncoded system, given appropriate construction of the trellis.

Shannon Limit: The ratio of energy per data bit E_b to one-sided noise power spectral density N_0 in an AWGN channel above which errorless transmission is possible when bandwidth limitations are not placed on the signal and transmission is at channel capacity. This limit has the value $\ln 2 = 0.693 = -1.6$ dB.

Soft decision: Demodulation that outputs an estimate of the received symbol value along with an indication of the reliability of this value. Usually implemented by quantizing the received signal to more levels than there are symbol values.

Standard array decoding: Association of an error pattern with each syndrome by way of a lookup table.

Syndrome: An indication of whether or not errors are present in the demodulated symbol sequence.

Systematic code: A code in which the values of the message symbols can be identified by inspection of the code vector.

Vector space: An algebraic structure comprised of a set of elements in which operations of vector addition and scalar multiplication are defined. For our purposes, a set of n-tuples consisting of symbols from $GF(q)$ with addition and multiplication defined in terms of elementwise operations from this finite field.

Viterbi algorithm: A maximum-likelihood decoding algorithm for trellis codes that discards low-probability paths at each stage of the trellis, thereby reducing the total number of paths that must be considered.

References

[1] Bahl, L.R., Cocke, J., Jelinek, F., and Raviv, J., Optimal Decoding of Linear Codes for Minimizing Symbol Error Rate. *IEEE Transactions on Information Theory*, 20, 248–287, 1974.

[2] Berlekamp, E.R., Peile, R.E., and Pope, S.P., The application of error control to communications. *IEEE Commun. Mag.*, 25(4), 44–57, 1987.

[3] Berrou, C., Glavieux, A., and Thitimajshima, P., Near Shannon Limit Error-Correcting Coding and Decoding: Turbo Codes. *Proceedings of ICC'93*, Geneva, Switzerland, 1064–1070, 1993. Later expanded and published as: Berrou, C., Glavieux, A., 1996. Near Optimum Error Correcting Coding and Decoding. *IEEE Transactions on Communications*, 44(10), 1261–1271, 1996.

[4] Bhargava, V.K., Forward error correction schemes for digital communications. *IEEE Commun. Mag.*, 21(1), 11–19, 1983.

[5] Blahut, R.E., *Theory and Practice of Error Control Codes*, Addison-Wesley, Reading, MA, 1983.

[6] Clark, G.C. Jr. and Cain, J.B., *Error Correction Coding for Digital Communications*, Plenum Press, New York, 1981.

[7] Lin, S. and Costello, D.J. Jr., *Error Control Coding: Fundamentals and Applications*, Prentice-Hall, Englewood Cliffs, NJ, 1983.

[8] Rappaport, T.S., *Wireless Communications, Principles and Practice*, Prentice-Hall and IEEE Press, NJ, 1996.

[9] Shannon, C.E., A mathematical theory of communication. *Bell Syst. Tech. J.*, 27(3), 379–423 and 623–656, 1948.

[10] Sklar, B., *Digital Communications: Fundamentals and Applications*, Prentice-Hall, Englewood Cliffs, NJ, 1988.

[11] Ungerboeck, G., Trellis-coded modulation with redundant signal sets. *IEEE Commun. Mag.*, 25(2), 5–11 and 12–21, 1987.

[12] Wicker, S.B. and Bhargava, V.K., *Reed-Solomon Codes and Their Applications*, IEEE Press, NJ, 1994.

[13] Wu, W.W., Haccoun, D., Peile, R., and Hirata, Y., Coding for satellite communication. *IEEE J. Selected Areas in Commun.*, SAC-5(4), 724–748, 1987.

Further Information

There is now a large amount of literature on the subject of FEC coding. An introduction to the philosophy and limitations of these codes can be found in the second chapter of Lucky's book *Silicon Dreams: Information, Man, and Machine*, St. Martin's Press, New York, 1989. More practical introductions can be found in overview chapters of many communications texts. The number of texts devoted entirely to this subject also continues to grow. Although these texts summarize the algebra underlying block codes, more in-depth treatments can be found in mathematical texts. Survey papers appear occasionally in the literature, but the interested reader is directed to the seminal papers by Shannon, Hamming, Reed and Solomon, Bose and Chaudhuri, Hocquenghem, Wozencraft, Fano, Forney, Berlekamp, Massey, Viterbi, Ungerboeck, Berrou and Glavieux, among others. The most recent advances in the theory and implementation of error control codes are published in *IEEE Transactions on Information Theory, IEEE Transactions on Communications*, and special issues of *IEEE Journal on Selected Areas in Communications*.

Spread Spectrum Communications

Laurence B. Milstein
University of California

Marvin K. Simon
Jet Propulsion Laboratory

11.1   A Brief History

Spread spectrum (SS) has its origin in the military arena where the friendly communicator is 1) susceptible to detection/interception by the enemy and 2) vulnerable to intentionally introduced unfriendly interference (jamming). Communication systems that employ spread spectrum to reduce the communicator's detectability and combat the enemy-introduced interference are respectively referred to as **low probability of intercept (LPI)** and **antijam (AJ) communication systems**. With the change in the current world political situation wherein the U.S. Department of Defense (DOD) has reduced its emphasis on the development and acquisition of new communication systems for the original purposes, a host of new commercial applications for SS has evolved, particularly in the area of cellular mobile communications. This shift from military to commercial applications of SS has demonstrated that the basic concepts that make SS techniques so useful in the military can also be put to practical peacetime use. In the next section, we give a simple description of these basic concepts using the original military application as the basis of explanation. The extension of these concepts to the mentioned commercial applications will be treated later on in the chapter.

11.2   Why Spread Spectrum?

Spread spectrum is a communication technique wherein the transmitted modulation is *spread* (increased) in bandwidth prior to transmission over the channel and then *despread* (decreased) in bandwidth by the same amount at the receiver. If it were not for the fact that the communication channel introduces some form of narrowband (relative to the spread bandwidth) interference, the receiver performance would be transparent to the spreading and despreading operations (assuming that they are identical inverses of each other). That is, after **despreading** the received signal would be identical to the transmitted signal prior to **spreading**. In the presence of narrowband interference, however, there is a significant advantage to employing the spreading/despreading procedure described. The reason for this is as follows. Since the interference is introduced after the transmitted signal is spread, then, whereas the despreading operation at the receiver shrinks the desired signal back to its original bandwidth, at the same time it spreads the undesired signal (interference) in bandwidth by the same amount, thus reducing its power spectral density. This, in turn, serves to diminish the effect of the interference on the receiver performance, which depends on the amount of interference power in the despread bandwidth. It is indeed this very simple explanation, which is at the heart of all spread spectrum techniques.

11.3　Basic Concepts and Terminology

To describe this process analytically and at the same time introduce some terminology that is common in spread spectrum parlance, we proceed as follows. Consider a communicator that desires to send a message using a transmitted power S Watts (W) at an information rate R_b bits/s (bps). By introducing a SS modulation, the bandwidth of the transmitted signal is increased from R_b Hz to W_{SS} Hz where $W_{SS} \gg R_b$ denotes the **spread spectrum bandwidth**. Assume that the channel introduces, in addition to the usual thermal noise (assumed to have a single-sided power spectral density (PSD) equal to N_0 W/Hz), an additive interference (jamming) having power J distributed over some bandwidth W_J. After despreading, the desired signal bandwidth is once again now equal to R_b Hz and the interference PSD is now $N_J = J/W_{SS}$. Note that since the thermal noise is assumed to be white, i.e., it is uniformly distributed over all frequencies, its PSD is unchanged by the despreading operation and, thus, remains equal to N_0. Regardless of the signal and interferer waveforms, the equivalent bit energy-to-total noise spectral density ratio is, in terms of the given parameters,

$$\frac{E_b}{N_t} = \frac{E_b}{N_0 + N_J} = \frac{S/R_b}{N_0 + J/W_{SS}} \tag{11.1}$$

For most practical scenarios, the jammer limits performance and, thus, the effects of receiver noise in the channel can be ignored. Thus, assuming $N_J \gg N_0$, we can rewrite Eq. (11.1) as

$$\frac{E_b}{N_t} \cong \frac{E_b}{N_J} = \frac{S/R_b}{J/W_{SS}} = \frac{S}{J}\frac{W_{SS}}{R_b} \tag{11.2}$$

where the ratio J/S is the *jammer-to-signal power ratio* and the ratio W_{SS}/R_b is the **spreading ratio** and is defined as the **processing gain** of the system. Since the ultimate error probability performance of the communication receiver depends on the ratio E_b/N_J, we see that from the communicator's viewpoint his goal should be to minimize J/S (by choice of S) and maximize the processing gain (by choice of W_{SS} for a given desired information rate). The possible strategies for the jammer will be discussed in the section on military applications dealing with AJ communications.

11.4　Spread Spectrum Techniques

By far the two most popular spreading techniques are **direct sequence (DS) modulation** and **frequency hopping (FH) modulation**. In the following subsections, we present a brief description of each.

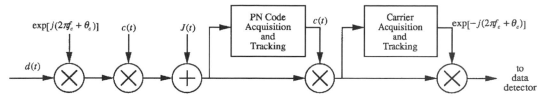

FIGURE 11.1   A DS-BPSK system (complex form).

11.4.1   Direct Sequence Modulation

A direct sequence modulation $c(t)$ is formed by linearly modulating the output sequence $\{c_n\}$ of a pseudorandom number generator onto a train of pulses, each having a duration T_c called the **chip time**. In mathematical form,

$$c(t) = \sum_{n=-\infty}^{\infty} c_n p\,(t - nT_c) \tag{11.3}$$

where $p(t)$ is the basic pulse shape and is assumed to be of rectangular form. This type of modulation is usually used with binary phase-shift-keyed (BPSK) information signals, which have the complex form $d(t)\exp\{j(2\pi f_c t + \theta_c)\}$, where $d(t)$ is a binary-valued data waveform of rate $1/T_b$ bits/s and f_c and θ_c are the frequency and phase of the data-modulated carrier, respectively. As such, a DS/BPSK signal is formed by multiplying the BPSK signal by $c(t)$ (see Fig. 11.1), resulting in the real transmitted signal

$$x(t) = \mathrm{Re}\left\{c(t)d(t)\exp\left[j\left(2\pi f_c t + \theta_c\right)\right]\right\} \tag{11.4}$$

Since T_c is chosen so that $T_b \gg T_c$, then relative to the bandwidth of the BPSK information signal, the bandwidth of the DS/BPSK signal[1] is effectively increased by the ratio $T_b/T_c = W_{\mathrm{SS}}/2R_b$, which is one-half the spreading factor or processing gain of the system. At the receiver, the sum of the transmitted DS/BPSK signal and the channel interference $I(t)$ (as discussed before, we ignore the presence of the additive thermal noise) are ideally multiplied by the identical DS modulation (this operation is known as despreading), which returns the DS/BPSK signal to its original BPSK form whereas the real interference signal is now the real wideband signal $\mathrm{Re}\{I(t)c(t)\}$. In the previous sentence, we used the word ideally, which implies that the PN waveform used for despreading at the receiver is identical to that used for spreading at the transmitter. This simple implication covers up a multitude of tasks that a practical DS receiver must perform. In particular, the receiver must first acquire the PN waveform. That is, the local PN random generator that generates the PN waveform at the receiver used for despreading must be aligned (synchronized) to within one chip of the PN waveform of the received DS/BPSK signal. This is accomplished by employing some sort of **search algorithm** which typically steps the local PN waveform sequentially in time by a fraction of a chip (e.g., half a chip) and at each position searches for a high degree of correlation between the received and local PN reference waveforms. The search terminates when the correlation exceeds a given threshold, which is an indication that the alignment has been achieved. After bringing the two PN waveforms into **coarse alignment**, a **tracking algorithm** is employed to maintain **fine alignment**. The most

---

[1]For the usual case of a rectangular spreading pulse $p(t)$, the PSD of the DS/BPSK modulation will have $(\sin x/x)^2$ form with first zero crossing at $1/T_c$, which is nominally taken as one-half the spread spectrum bandwidth W_{SS}.

popular forms of tracking loops are the continuous time **delay-locked loop** and its time-multiplexed version the **tau–dither loop**. It is the difficulty in synchronizing the receiver PN generator to subnanosecond accuracy that limits PN chip rates to values on the order of hundreds of Mchips/s, which implies the same limitation on the DS spread spectrum bandwidth W_{SS}.

11.4.2   Frequency Hopping Modulation

A **frequency hopping (FH) modulation** $c(t)$ is formed by nonlinearly modulating a train of pulses with a sequence of pseudorandomly generated frequency shifts $\{f_n\}$. In mathematical terms, $c(t)$ has the complex form

$$c(t) = \sum_{n=-\infty}^{\infty} \exp\left\{j\left(2\pi f_n + \phi_n\right)\right\} p\left(t - nT_h\right) \tag{11.5}$$

where $p(t)$ is again the basic pulse shape having a duration T_h, called the **hop time** and $\{\phi_n\}$ is a sequence of random phases associated with the generation of the hops. FH modulation is traditionally used with multiple-frequency-shift-keyed (MFSK) information signals, which have the complex form $\exp\{j[2\pi(f_c+d(t))t]\}$, where $d(t)$ is an M-level digital waveform (M denotes the symbol alphabet size) representing the information frequency modulation at a rate $1/T_s$ symbols/s (sps). As such, an FH/MFSK signal is formed by complex multiplying the MFSK signal by $c(t)$ resulting in the real transmitted signal

$$x(t) = \mathrm{Re}\left\{c(t)\exp\left\{j\left[2\pi(f_c+d(t))t\right]\right\}\right\} \tag{11.6}$$

In reality, $c(t)$ is never generated in the transmitter. Rather, $x(t)$ is obtained by applying the sequence of pseudorandom frequency shifts $\{f_n\}$ directly to the frequency synthesizer that generates the carrier frequency f_c (see Fig. 11.2). In terms of the actual implementation,

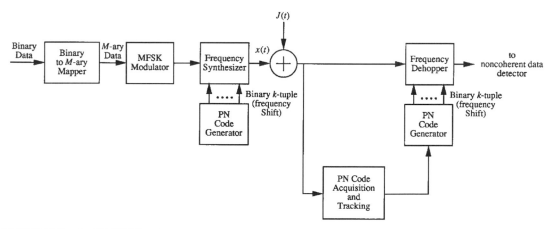

FIGURE 11.2   An FH-MFSK system.

successive (not necessarily disjoint) k-chip segments of a PN sequence drive a frequency synthesizer, which hops the carrier over 2^k frequencies. In view of the large bandwidths over which the frequency synthesizer must operate, it is difficult to maintain phase coherence from hop to hop, which explains the inclusion of the sequence $\{\phi_n\}$ in the Eq. (11.5) model for $c(t)$. On a short term basis, e.g., within a given hop, the signal bandwidth is identical to that of the MFSK information modulation, which is typically much smaller than W_{ss}. On the other hand, when averaged over many hops, the signal bandwidth is equal to W_{ss}, which can be on the order of several GHz, i.e., an order of magnitude larger than that of implementable DS bandwidths. The exact relation between W_{SS}, T_h, T_s and the number of frequency shifts in the set $\{f_n\}$ will be discussed shortly.

At the receiver, the sum of the transmitted FH/MFSK signal and the channel interference $I(t)$ is ideally complex multiplied by the identical FH modulation (this operation is known as **dehopping**), which returns the FH/MFSK signal to its original MFSK form, whereas the real interference signal is now the wideband (in the average sense) signal $\text{Re}\{I(t)c(t)\}$. Analogous to the DS case, the receiver must acquire and track the FH signal so that the dehopping waveform is as close to the hopping waveform $c(t)$ as possible.

FH systems are traditionally classified in accordance with the relationship between T_h and T_s. **Fast frequency-hopped (FFH)** systems are ones in which there exists one or more hops per data symbol, that is, $T_s = NT_h$ (N an integer) whereas **slow frequency-hopped (SFH)** systems are ones in which there exists more than one symbol per hop, that is, $T_h = NT_s$. It is customary in SS parlance to refer to the FH/MFSK tone of shortest duration as a "chip", despite the same usage for the PN chips associated with the code generator that drives the frequency synthesizer. Keeping this distinction in mind, in an FFH system where, as already stated, there are multiple hops per data symbol, a chip is equal to a hop. For SFH, where there are multiple data symbols per hop, a chip is equal to an MFSK symbol. Combining these two statements, the chip rate R_c in an FH system is given by the larger of $R_h = 1/T_h$ and $R_s = 1/T_s$ and, as such, is the highest system clock rate.

The frequency spacing between the FH/MFSK tones is governed by the chip rate R_c and is, thus, dependent on whether the FH modulation is FFH or SFH. In particular, for SFH where $R_c = R_s$, the spacing between FH/MFSK tones is equal to the spacing between the MFSK tones themselves. For noncoherent detection (the most commonly encountered in FH/MFSK systems), the separation of the MFSK symbols necessary to provide orthogonality[2] is an integer multiple of R_s. Assuming the minimum spacing, i.e., R_s, the entire spread spectrum band is then partitioned into a total of $N_t = W_{SS}/R_s = W_{SS}/R_c$ equally spaced FH tones. One arrangement, which is by far the most common, is to group these N_t tones into $N_b = N_t/M$ contiguous, nonoverlapping bands, each with bandwidth $MR_s = MR_c$; see Fig. 11.3a. Assuming symmetric MFSK modulation around the carrier frequency, then the center frequencies of the $N_b = 2^k$ bands represent the set of hop carriers, each of which is assigned to a given k-tuple of the PN code generator. In this fixed arrangement, each of the N_t FH/MFSK tones corresponds to the combination of a unique hop carrier (PN code k-tuple) and a unique MFSK symbol. Another arrangement, which provides more protection against the sophisticated interferer (jammer), is to overlap adjacent M-ary bands by an amount equal to R_c; see Fig. 11.3b. Assuming again that the center frequency of each band corresponds to a possible hop carrier, then since all but $M - 1$ of the N_t tones are available as center frequencies, the number of hop carriers has been increased from N_t/M to $N_t - (M-1)$, which for $N_t \gg M$ is approximately an increase in randomness by a factor of M.

---

[2] An optimum noncoherent MFSK detector consists of a bank of energy detectors each matched to one of the M frequencies in the MFSK set. In terms of this structure, the notion of *orthogonality* implies that for a given transmitted frequency there will be no crosstalk (energy spillover) in any of the other $M - 1$ energy detectors.

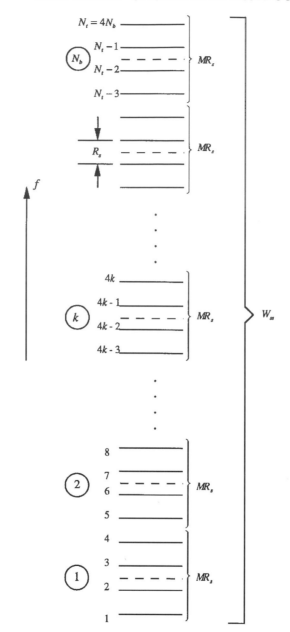

Figure 11.3a   Frequency distribution for FH-4FSK—nonoverlapping bands. Dashed lines indicate location of hop frequencies.

For FFH, where $R_c = R_h$, the spacing between FH/MFSK tones is equal to the hop rate. Thus, the entire spread spectrum band is partitioned into a total of $N_t = W_{SS}/R_h = W_{SS}/R_c$ equally spaced FH tones, each of which is assigned to a unique k-tuple of the PN code generator that drives the frequency synthesizer. Since for FFH there are R_h/R_s hops per symbol, then the metric used to make a noncoherent decision on a particular symbol is obtained by summing up R_h/R_s detected chip (hop) energies, resulting in a so-called *noncoherent combining loss*.

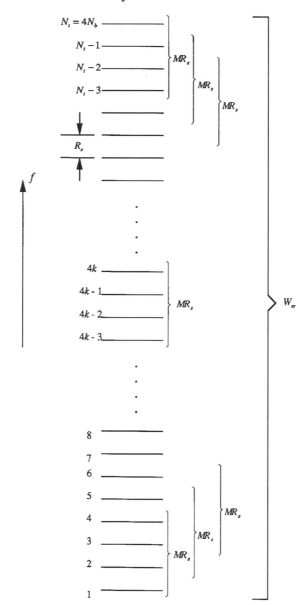

Figure 11.3b   Frequency distribution for FH-4FSK—over-lapping bands.

11.4.3   Time Hopping Modulation

Time hopping (TH) is to spread spectrum modulation what pulse position modulation (PPM) is to information modulation. In particular, consider segmenting time into intervals of T_f seconds and further segment each T_f interval into M_T increments of width T_f/M_T. Assuming a pulse of maximum duration equal to T_f/M_T, then a **time hopping spread spectrum** modulation would take the form

$$c(t) = \sum_{n=-\infty}^{\infty} p\left[t - \left(n + \frac{a_n}{M_T}\right)T_f\right] \tag{11.7}$$

where a_n denotes the pseudorandom position (one of M_T uniformly spaced locations) of the pulse within the T_f-second interval.

For DS and FH, we saw that *multiplicative* modulation, that is the transmitted signal is the product of the SS and information signals, was the natural choice. For TH, *delay* modulation is the natural choice. In particular, a TH-SS modulation takes the form

$$x(t) = \text{Re}\left\{c(t - d(t)) \exp\left[j\left(2\pi f_c + \phi_T\right)\right]\right\} \tag{11.8}$$

where $d(t)$ is a digital information modulation at a rate $1/T_s$ sps. Finally, the dehopping procedure at the receiver consists of removing the sequence of delays introduced by $c(t)$, which restores the information signal back to its original form and spreads the interferer.

11.4.4 Hybrid Modulations

By blending together several of the previous types of SS modulation, one can form **hybrid** modulations that, depending on the system design objectives, can achieve a better performance against the interferer than can any of the SS modulations acting alone. One possibility is to multiply several of the $c(t)$ wideband waveforms [now denoted by $c^{(i)}(t)$ to distinguish them from one another] resulting in a SS modulation of the form

$$c(t) = \prod_i c^{(i)}(t) \tag{11.9}$$

Such a modulation may embrace the advantages of the various $c^{(i)}(t)$, while at the same time mitigating their individual disadvantages.

11.5 Applications of Spread Spectrum

11.5.1 Military

Antijam (AJ) Communications

As already noted, one of the key applications of spread spectrum is for antijam communications in a hostile environment. The basic mechanism by which a **direct sequence spread spectrum** receiver attenuates a noise jammer was illustrated in Section 11.3. Therefore, in this section, we will concentrate on tone jamming.

Assume the received signal, denoted $r(t)$, is given by

$$r(t) = Ax(t) + I(t) + n_w(t) \tag{11.10}$$

where $x(t)$ is given in Eq. (11.4), A is a constant amplitude,

$$I(t) = \alpha \cos\left(2\pi f_c t + \theta\right) \tag{11.11}$$

and $n_w(t)$ is additive white Gaussian noise (AWGN) having two-sided spectral density $N_0/2$. In Eq. (11.11), α is the amplitude of the tone jammer and θ is a random phase uniformly distributed in $[0, 2\pi]$.

If we employ the standard correlation receiver of Fig. 11.4, it is straightforward to show that the final test statistic out of the receiver is given by

$$g(T_b) = AT_b + \alpha \cos\theta \int_0^{T_b} c(t)\,\mathrm{d}t + N\left(T_b\right) \tag{11.12}$$

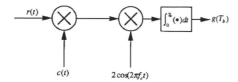

FIGURE 11.4 Standard correlation receiver.

where $N(T_b)$ is the contribution to the test statistic due to the AWGN. Noting that, for rectangular chips, we can express

$$\int_0^{T_b} c(t)\,\mathrm{d}t = T_c \sum_{i=1}^{M} c_i \tag{11.13}$$

where

$$M \triangleq \frac{T_b}{T_c} \tag{11.14}$$

is one-half of the processing gain, it is straightforward to show that, for a given value of θ, the signal-to-noise-plus-interference ratio, denoted by S/N_{total}, is given by

$$\frac{S}{N_{\text{total}}} = \frac{1}{\frac{N_0}{2E_b} + \left(\frac{J}{MS}\right)\cos^2\theta} \tag{11.15}$$

In Eq. (11.15), the jammer power is

$$J \triangleq \frac{\alpha^2}{2} \tag{11.16}$$

and the signal power is

$$S \triangleq \frac{A^2}{2} \tag{11.17}$$

If we look at the second term in the denominator of Eq. (11.15), we see that the ratio J/S is divided by M. Realizing that J/S is the ratio of the jammer power to the signal power before despreading, and J/MS is the ratio of the same quantity after despreading, we see that, as was the case for noise jamming, the benefit of employing direct sequence spread spectrum signalling in the presence of tone jamming is to reduce the effect of the jammer by an amount on the order of the processing gain.

Finally, one can show that an estimate of the average probability of error of a system of this type is given by

$$P_e = \frac{1}{2\pi} \int_0^{2\pi} \phi\left(-\sqrt{\frac{S}{N_{\text{total}}}}\right) \mathrm{d}\theta \tag{11.18}$$

where

$$\phi(x) \triangleq \frac{1}{\sqrt{2\pi}} \int_{-\infty}^{x} e^{-y^2/2}\,\mathrm{d}y \tag{11.19}$$

If Eq. (11.18) is evaluated numerically and plotted, the results are as shown in Fig. 11.5. It is clear from this figure that a large initial power advantage of the jammer can be overcome by a sufficiently large value of the processing gain.

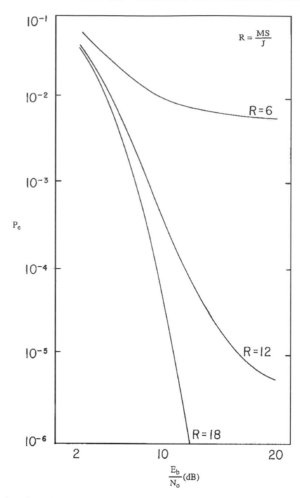

FIGURE 11.5   Plotted results of Eq. (11.18).

Low-Probability of Intercept (LPI)

The opposite side of the AJ problem is that of LPI, that is, the desire to hide your signal from detection by an intelligent adversary so that your transmissions will remain unnoticed and, thus, neither jammed nor exploited in any manner. This idea of designing an LPI system is achieved in a variety of ways, including transmitting at the smallest possible power level, and limiting the transmission time to as short an interval in time as is possible. The choice of signal design is also important, however, and it is here that spread spectrum techniques become relevant.

The basic mechanism is reasonably straightforward; if we start with a conventional narrowband signal, say a BPSK waveform having a spectrum as shown in Fig. 11.6a, and then spread it so that its new spectrum is as shown in Fig. 11.6b, the peak amplitude of the spectrum after spreading has been reduced by an amount on the order of the processing gain relative to what it was before spreading. Indeed, a sufficiently large processing gain will result in the spectrum of the signal after spreading falling below the ambient thermal noise level. Thus, there is no easy way for an unintended listener to determine that a transmission is taking place.

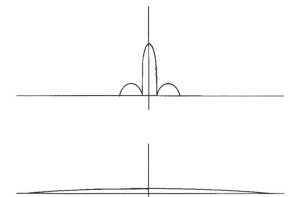

Figure 11.6a

Figure 11.6b

That is not to say the spread signal cannot be detected, however, merely that it is more difficult for an adversary to learn of the transmission. Indeed, there are many forms of so-called intercept receivers that are specifically designed to accomplish this very task. By way of example, probably the best known and simplest to implement is a **radiometer**, which is just a device that measures the total power present in the received signal. In the case of our intercept problem, even though we have lowered the power spectral density of the transmitted signal so that it falls below the noise floor, we have not lowered its power (i.e., we have merely spread its power over a wider frequency range). Thus, if the radiometer integrates over a sufficiently long period of time, it will eventually determine the presence of the transmitted signal buried in the noise. The key point, of course, is that the use of the spreading makes the interceptor's task much more difficult, since he has no knowledge of the spreading code and, thus, cannot despread the signal.

11.5.2 Commercial

Multiple Access Communications

From the perspective of commercial applications, probably the most important use of spread spectrum communications is as a multiple accessing technique. When used in this manner, it becomes an alternative to either frequency division multiple access (FDMA) or time division multiple access (TDMA) and is typically referred to as either code division multiple access (CDMA) or spread spectrum multiple access (SSMA). When using CDMA, each signal in the set is given its own spreading sequence. As opposed to either FDMA, wherein all users occupy disjoint frequency bands but are transmitted simultaneously in time, or TDMA, whereby all users occupy the same bandwidth but transmit in disjoint intervals of time, in CDMA, all signals occupy the same bandwidth and are transmitted simultaneously in time; the different waveforms in CDMA are distinguished from one another at the receiver by the specific spreading codes they employ.

Since most CDMA detectors are correlation receivers, it is important when deploying such a system to have a set of spreading sequences that have relatively low-pairwise cross-correlation between any two sequences in the set. Further, there are two fundamental types of operation in CDMA, synchronous and asynchronous. In the former case, the symbol transition times of all of the users are aligned; this allows for orthogonal sequences to be used as the spreading sequences and, thus, eliminates interference from one user to another.

Alternately, if no effort is made to align the sequences, the system operates asychronously; in this latter mode, multiple access interference limits the ultimate channel capacity, but the system design exhibits much more flexibility.

CDMA has been of particular interest recently for applications in wireless communications. These applications include cellular communications, personal communications services (PCS), and wireless local area networks. The reason for this popularity is primarily due to the performance that spread spectrum waveforms display when transmitted over a multipath fading channel.

To illustrate this idea, consider DS signalling. As long as the duration of a single chip of the spreading sequence is less than the multipath delay spread, the use of DS waveforms provides the system designer with one of two options. First, the multipath can be treated as a form of interference, which means the receiver should attempt to attenuate it as much as possible. Indeed, under this condition, all of the multipath returns that arrive at the receiver with a time delay greater than a chip duration from the multipath return to which the receiver is synchronized (usually the first return) will be attenuated because of the processing gain of the system.

Alternately, the multipath returns that are separated by more than a chip duration from the main path represent independent "looks" at the received signal and can be used constructively to enhance the overall performance of the receiver. That is, because all of the multipath returns contain information regarding the data that is being sent, that information can be extracted by an appropriately designed receiver. Such a receiver, typically referred to as a RAKE receiver, attempts to resolve as many individual multipath returns as possible and then to sum them coherently. This results in an *implicit* diversity gain, comparable to the use of *explicit* diversity, such as receiving the signal with multiple antennas.

The condition under which the two options are available can be stated in an alternate manner. If one envisions what is taking place in the frequency domain, it is straightforward to show that the condition of the chip duration being smaller than the multipath delay spread is equivalent to requiring that the spread bandwidth of the transmitted waveform exceed what is called the coherence bandwidth of the channel. This latter quantity is simply the inverse of the multipath delay spread and is a measure of the range of frequencies that fade in a highly correlated manner. Indeed, anytime the coherence bandwidth of the channel is less than the spread bandwidth of the signal, the channel is said to be *frequency selective* with respect to the signal. Thus, we see that to take advantage of DS signalling when used over a multipath fading channel, that signal should be designed such that it makes the channel appear frequency selective.

In addition to the desirable properties that spread spectrum signals display over multipath channels, there are two other reasons why such signals are of interest in cellular-type applications. The first has to do with a concept known as the reuse factor. In conventional cellular systems, either analog or digital, in order to avoid excessive interference from one cell to its neighbor cells, the frequencies used by a given cell are not used by its immediate neighbors (i.e., the system is designed so that there is a certain spatial separation between cells that use the same carrier frequencies). For CDMA, however, such spatial isolation is typically not needed, so that so-called *universal reuse* is possible.

Further, because CDMA systems tend to be interference limited, for those applications involving voice transmission, an additional gain in the capacity of the system can be achieved by the use of *voice activity detection*. That is, in any given two-way telephone conversation, each user is typically talking only about 50% of the time. During the time when a user is quiet, he is not contributing to the instantaneous interference. Thus, if a sufficiently large number of users can be supported by the system, statistically only about one-half of them will be active simultaneously, and the effective capacity can be doubled.

Interference Rejection

In addition to providing multiple accessing capability, spread spectrum techniques are of interest in the commercial sector for basically the same reasons they are in the military community, namely their AJ and LPI characteristics. However, the motivations for such interest differ. For example, whereas the military is interested in ensuring that systems they deploy are robust to interference generated by an intelligent adversary (i.e., exhibit jamming resistance), the interference of concern in commercial applications is unintentional. It is sometimes referred to as cochannel interference (CCI) and arises naturally as the result of many services using the same frequency band at the same time. And while such scenarios almost always allow for some type of spatial isolation between the interfering waveforms, such as the use of narrow-beam antenna patterns, at times the use of the inherent interference suppression property of a spread spectrum signal is also desired. Similarly, whereas the military is very much interested in the LPI property of a spread spectrum waveform, as indicated in Section 11.3, there are applications in the commercial segment where the same characteristic can be used to advantage.

To illustrate these two ideas, consider a scenario whereby a given band of frequencies is somewhat sparsely occupied by a set of conventional (i.e., nonspread) signals. To increase the overall spectral efficiency of the band, a set of spread spectrum waveforms can be overlaid on the same frequency band, thus forcing the two sets of users to share common spectrum. Clearly, this scheme is feasible only if the mutual interference that one set of users imposes on the other is within tolerable limits. Because of the interference suppression properties of spread spectrum waveforms, the despreading process at each spread spectrum receiver will attenuate the components of the final test statistic due to the overlaid narrowband signals. Similarly, because of the LPI characteristics of spread spectrum waveforms, the increase in the overall noise level as seen by any of the conventional signals, due to the overlay, can be kept relatively small.

Defining Terms

Antijam communication system: A communication system designed to resist intentional jamming by the enemy.

Chip time (interval): The duration of a single pulse in a direct sequence modulation; typically much smaller than the information symbol interval.

Coarse alignment: The process whereby the received signal and the despreading signal are aligned to within a single chip interval.

Dehopping: Despreading using a frequency-hopping modulation.

Delay-locked loop: A particular implementation of a closed-loop technique for maintaining fine alignment.

Despreading: The notion of decreasing the bandwidth of the received (spread) signal back to its information bandwidth.

Direct sequence modulation: A signal formed by linearly modulating the output sequence of a pseudorandom number generator onto a train of pulses.

Direct sequence spread spectrum: A spreading technique achieved by multiplying the information signal by a direct sequence modulation.

Fast frequency-hopping: A spread spectrum technique wherein the hop time is less than or equal to the information symbol interval, i.e., there exist one or more hops per data symbol.

Fine alignment: The state of the system wherein the received signal and the despreading signal are aligned to within a small fraction of a single chip interval.

Frequency-hopping modulation: A signal formed by nonlinearly modulating a train of pulses with a sequence of pseudorandomly generated frequency shifts.

Hop time (interval): The duration of a single pulse in a frequency-hopping modulation.

Hybrid spread spectrum: A spreading technique formed by blending together several spread spectrum techniques, e.g., direct sequence, frequency-hopping, etc.

Low-probability-of-intercept communication system: A communication system designed to operate in a hostile environment wherein the enemy tries to detect the presence and perhaps characteristics of the friendly communicator's transmission.

Processing gain (spreading ratio): The ratio of the spread spectrum bandwidth to the information data rate.

Radiometer: A device used to measure the total energy in the received signal.

Search algorithm: A means for coarse aligning (synchronizing) the despreading signal with the received spread spectrum signal.

Slow frequency-hopping: A spread spectrum technique wherein the hop time is greater than the information symbol interval, i.e., there exists more than one data symbol per hop.

Spread spectrum bandwidth: The bandwidth of the transmitted signal after spreading.

Spreading: The notion of increasing the bandwidth of the transmitted signal by a factor far in excess of its information bandwidth.

Tau–dither loop: A particular implementation of a closed-loop technique for maintaining fine alignment.

Time-hopping spread spectrum: A spreading technique that is analogous to pulse position modulation.

Tracking algorithm: An algorithm (typically closed loop) for maintaining fine alignment.

References

[1] Cook, C.F., Ellersick, F.W., Milstein, L.B., and Schilling, D.L., *Spread Spectrum Communications*, IEEE Press, 1983.

[2] Dixon, R.C., *Spread Spectrum Systems*, 3rd ed., John Wiley and Sons, Inc. 1994.

[3] Holmes, J.K., *Coherent Spread Spectrum Systems*, John Wiley and Sons, Inc. 1982.

[4] Simon, M.K., Omura, J.K., Scholtz, R.A., and Levitt, B.K., *Spread Spectrum Communications Handbook*, McGraw Hill, 1994 (previously published as *Spread Spectrum Communications*, Computer Science Press, 1985).

[5] Ziemer, R.E. and Peterson, R.L., *Digital Communications and Spread Spectrum Techniques*, Macmillan, 1985.

12

Diversity

Arogyaswami J. Paulraj
Stanford University

12.1 Introduction

Diversity is a commonly used technique in mobile radio systems to combat signal **fading.**
The basic principle of diversity is as follows. If several replicas of the same information-
carrying signal are received over multiple channels with comparable strengths, which exhibit
independent fading, then there is a good likelihood that at least one or more of these
received signals will not be in a fade at any given instant in time, thus making it possible to
deliver adequate signal level to the receiver. Without diversity techniques, in noise limited
conditions, the transmitter would have to deliver a much higher power level to protect the
link during the short intervals when the channel is severely faded. In mobile radio, the
power available on the reverse link is severely limited by the battery capacity of hand-held
subscriber units. Diversity methods play a crucial role in reducing transmit power needs.
Also, cellular communication networks are mostly interference limited and, once again,
mitigation of channel fading through use of diversity can translate into reduced variability
of carrier-to-interference ratio (C/I), which in turn means lower C/I margin and hence
better reuse factors and higher system capacity.

    The basic principles of diversity have been known since 1927 when the first experiments
in space diversity were reported. There are many techniques for obtaining independently
fading branches, and these can be subdivided into two main classes. The first are explicit
techniques where explicit redundant signal transmission is used to exploit diversity channels.
Use of dual polarized signal transmission and reception in many point-to-point radios is an
example of explicit diversity. Clearly such redundant signal transmission involves a penalty
in frequency spectrum or additional power. In the second class are implicit diversity tech-

0-8493-8597-0/99/$0.00+$.50

niques: the signal is transmitted only once, but the decorrelating effects in the propagation medium such as multipaths are exploited to receive signals over multiple diversity channels. A good example of implicit diversity is the **RAKE receiver** in code division multiple access (CDMA) systems, which uses independent fading of resolvable multipaths to achieve diversity gain. Figure 12.1 illustrates the principle of diversity where two independently fading signals are shown along with the selection diversity output signal which selects the stronger signal. The fades in the resulting signal have been substantially smoothed out while also yielding higher average power.

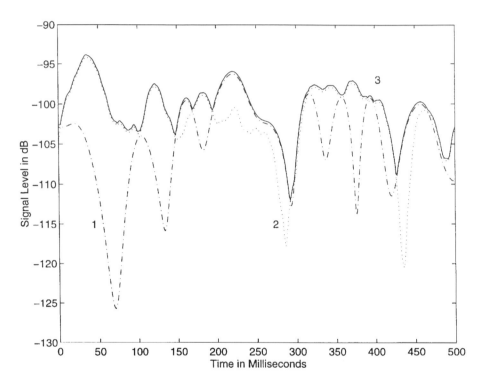

FIGURE 12.1   Example of diversity combining. Two independently fading signals 1 and 2. The signal 3 is the result of selecting the strongest signal.

If antennas are used in transmit, they can be exploited for diversity. If the transmit channel is known, the antennas can be driven with complex conjugate channel weighting to co-phase the signals at the receive antenna. If the forward channel is not known, we have several methods to convert space selective fading at the transmit antennas to other forms of diversity exploitable in the receiver.

Exploiting diversity needs careful design of the communication link. In explicit diversity, multiple copies of the same signal are transmitted in channels using either a frequency, time, or polarization dimension. At the receiver end we need arrangements to receive the different diversity branches (this is true for both explicit and implicit diversity). The different diversity branches are then combined to reduce signal **outage probability** or bit error rate.

In practice, the signals in the diversity branches may not show completely independent fading. The envelope cross correlation ρ between these signals is a measure of their independence.

$$\rho = \frac{E\left[\left[r_1 - \bar{r}_1\right]\left[r_2 - \bar{r}_2\right]\right]}{\sqrt{E\left|r_1 - \bar{r}_1\right|^2 E\left|r_2 - \bar{r}_2\right|^2}}$$

where r_1 and r_2 represent the instantaneous envelope levels of the normalized signals at the two receivers and $\bar{r}_1$ and $\bar{r}_2$ are their respective means. It has been shown that a cross correlation of 0.7 [3] between signal envelopes is sufficient to provide a reasonable degree of diversity gain. Depending on the type of diversity employed, these diversity channels must be sufficiently *separated* along the appropriate diversity dimension. For spatial diversity, the antennas should be separated by more than the *coherence distance* to ensure a cross correlation of less than 0.7. Likewise in frequency diversity, the frequency separation must be larger than the *coherence bandwidth,* and in time diversity the separation between channel reuse in time should be longer than the *coherence time.* These coherence factors in turn depend on the channel characteristics. The coherence distance, coherence bandwidth and coherence time vary inversely as the angle spread, delay spread, and Doppler spread, respectively.

If the receiver has a number of diversity branches, it has to combine these branches to maximize the signal level. Several techniques have been studied for diversity combining. We will describe three main techniques: selection combining, equal gain combining, and maximal ratio combining.

Finally, we should note that diversity is primarily used to combat fading and if the signal does not show significant fading in the first place, for example when there is a direct path component, diversity combining may not provide significant diversity gain. In the case of antenna diversity, array gain proportional to the number of antennas will still be available.

12.2 Diversity Schemes

There are several techniques for obtaining diversity branches, sometimes also known as diversity dimensions. The most important of these are discussed in the following sections.

12.2.1 Space Diversity

This has historically been the most common form of diversity in mobile radio base stations. It is easy to implement and does not require additional frequency spectrum resources. Space diversity is exploited on the reverse link at the base station receiver by spacing antennas apart so as to obtain sufficient decorrelation. The key for obtaining minimum uncorrelated fading of antenna outputs is adequate spacing of the antennas. The required spacing depends on the degree of multipath angle spread. For example if the multipath signals arrive from all directions in the azimuth, as is usually the case at the mobile, antenna spacing (coherence distance) of the order of 0.5λ to 0.8λ is quite adequate [5]. On the other hand if the multipath angle spread is small, as in the case of base stations, the coherence distance is much larger. Also empirical measurements show a strong coupling between antenna height and spatial correlation. Larger antenna heights imply larger coherence distances. Typically 10λ to 20λ separation is adequate to achieve $\rho = 0.7$ at base stations in suburban settings when the signals arrive from the broadside direction. The coherence distance can be 3 to 4 times larger for endfire arrivals. The endfire problem is averted in base stations with trisectored antennas as each sector needs to handle only signals arriving $\pm 60°$ off the broadside. The coherence distance depends strongly on the terrain. Large multipath angle spread means smaller coherence distance. Base stations normally use space diversity in the horizontal plane only. Separation in the vertical plane can also be used, and the necessary spacing depends upon vertical multipath angle spread. This can be small for distant mobiles making vertical plane diversity less attractive in most applications.

Space diversity is also exploitable at the transmitter. If the forward channel is known, it works much like receive space diversity. If it is not known, then space diversity can be transformed to another form of diversity exploitable at the receiver. (See Section 12.2.7 below).

If antennas are used at transmit and receive, the M transmit and N receive antennas both contribute to diversity. It can be shown that if simple weighting is used without additional bandwidth or time/memory processing, then maximum diversity gain is obtained if the transmitter and receiver use the left and right singular vectors of the $M \times N$ channel matrix, respectively. However, to approach the maximum $M \times N$ order diversity order will require the use of additional bandwidth or time/memory-based methods.

12.2.2   Polarization Diversity

In mobile radio environments, signals transmitted on orthogonal polarizations exhibit low fade correlation, and therefore, offer potential for diversity combining. Polarization diversity can be obtained either by explicit or implicit techniques. Note that with polarization only two diversity branches are available as against space diversity where several branches can be obtained using multiple antennas. In explicit polarization diversity, the signal is transmitted and received in two orthogonal polarizations. For a fixed total transmit power, the power in each branch will be 3 dB lower than if single polarization is used. In the implicit polarization technique, the signal is launched in a single polarization, but is received with cross-polarized antennas. The propagation medium couples some energy into the cross-polarization plane. The observed cross-polarization coupling factor lies between 8 to 12 dB in mobile radio [8, 1]. The cross-polarization envelope decorrelation has been found to be adequate. However, the large branch imbalance reduces the available diversity gain.

With hand-held phones, the handset can be held at random orientations during a call. This results in energy being launched with varying polarization angles ranging from vertical to horizontal. This further increases the advantage of cross-polarized antennas at the base station since the two antennas can be combined to match the received signal polarization. This makes polarization diversity even more attractive. Recent work [4] has shown that with variable launch polarization, a cross-polarized antenna can give comparable overall (matching plus diversity) performance to a vertically polarized space diversity antenna.

Finally, we should note that cross-polarized antennas can be deployed in a compact antenna assembly and do not need large physical separation needed in space diversity antennas. This is an important advantage in the **PCS** base stations where low profile antennas are needed.

12.2.3   Angle Diversity

In situations where the angle spread is very high, such as indoors or at the mobile unit in urban locations, signals collected from multiple nonoverlapping beams offer low fade correlation with balanced power in the diversity branches. Clearly, since directional beams imply use of antenna aperture, angle diversity is closely related to space diversity. Angle diversity has been utilized in indoor wireless LANs, where its use allows substantial increase in LAN throughputs [2].

12.2.4   Frequency Diversity

Another technique to obtain decorrelated diversity branches is to transmit the same signal over different frequencies. The frequency separation between carriers should be larger than the coherence bandwidth. The coherence bandwidth, of course, depends on the multipath delay spread of the channel. The larger the delay spread, the smaller the coherence bandwidth and the more closely we can space the frequency diversity channels. Clearly, frequency diversity is an explicit diversity technique and needs additional frequency spectrum.

A common form of frequency diversity is multicarrier (also known as multitone) modulation. This technique involves sending redundant data over a number of closely spaced carriers to benefit from frequency diversity, which is then exploited by applying **interleaving** and **channel coding/forward error correction** across the carriers. Another technique is to use **frequency hopping** wherein the interleaved and channel coded data stream is transmitted with widely separated frequencies from burst to burst. The wide frequency separation is chosen to guarantee independent fading from burst to burst.

12.2.5   Path Diversity

This implicit diversity is available if the signal bandwidth is much larger than the channel coherence bandwidth. The basis for this method is that when the multipath arrivals can be resolved in the receiver and since the paths fade independently, diversity gain can be obtained. In CDMA systems, the multipath arrivals must be separated by more than one *chip* period and the RAKE receiver provides the diversity [9]. In TDMA systems, the multipath arrivals must be separated by more than one *symbol* period and the MLSE receiver provides the diversity.

12.2.6   Time Diversity

In mobile communications channels, the mobile motion together with scattering in the vicinity of the mobile causes time selective fading of the signal with Rayleigh fading statistics for the signal envelope. Signal fade levels separated by the *coherence time* show low correlation and can be used as diversity branches if the same signal can be transmitted at multiple instants separated by the coherence time. The coherence time depends on the Doppler spread of the signal, which in turn is a function of the mobile speed and the carrier frequency.

Time diversity is usually exploited via interleaving, forward-error correction (FEC) coding, and **automatic request for repeat** (ARQ). These are sophisticated techniques to exploit channel coding and time diversity. One fundamental drawback with time diversity approaches is the delay needed to collect the repeated or interleaved transmissions. If the coherence time is large, as for example when the vehicle is slow moving, the required delay becomes too large to be acceptable for interactive voice conversation.

The statistical properties of fading signals depend on the field component used by the antenna, the vehicular speed, and the carrier frequency. For an idealized case of a mobile surrounded by scatterers in all directions, the autocorrelation function of the received signal $x(t)$ (note this is not the envelope $r(t)$) can be shown to be

$$E\left[x(t)x(t+\tau)\right] = J_0\left(2\pi\tau v/\lambda\right)$$

where J_0 is a Bessel function of the 0th order and v is the mobile velocity.

12.2.7 Transformed Diversity

In transformed diversity, the space diversity branches at the transmitter are transformed into other forms of diversity branches exploitable at the receiver. This is used when the forward channel is not known and shifts the responsibility of diversity combining to the receiver which has the necessary channel knowledge.

Space to Frequency

- *Antenna-delay.* Here the signal is transmitted from two or more antennas with delays of the order of a chip or symbol period in CDMA or TDMA, respectively. The different transmissions simulate resolved path arrivals that can be used as diversity branches by the RAKE or MLSE equalizer.
- *Multicarrier modulation.* The data stream after interleaving and coding is modulated as a multicarrier output using an inverse DFT. The carriers are then mapped to the different antennas. The space selective fading at the antennas is now transformed to frequency selective fading and diversity is obtained during decoding.

Space to Time

- *Antenna hopping/phase rolling.* In this method the data stream after coding and interleaving is switched randomly from antenna to antenna. The space selective fading at the transmitter is converted into a time selective fading at the receiver. This is a form of "active" fading.
- *Space-time coding.* The approach in space-time coding is to split the encoded data into multiple data streams each of which is modulated and simultaneously transmitted from different antennas. The received signal is a superposition of the multiple transmitted signals. Channel decoding can be used to recover the data sequence. Since the encoded data arrive over uncorrelated fade branches, diversity gain can be realized.

12.3 Diversity Combining Techniques

Several diversity combining methods are known. We describe three main techniques: selection, maximal ratio, and equal gain. They can be used with each of the diversity schemes discussed above.

12.3.1 Selection Combining

This is the simplest and perhaps the most frequently used form of diversity combining. In this technique, one of the two diversity branches with the highest carrier-to-noise ratio (C/N) is connected to the output. See Fig. 12.2(a).

The performance improvement due to selection diversity can be seen as follows. Let the signal in each branch exhibit Rayleigh fading with mean power σ^2. The density function of the envelope is given by

$$p\left(r_i\right) = \frac{r_i}{\sigma^2} e^{\frac{-r_i^2}{2\sigma^2}} \tag{12.1}$$

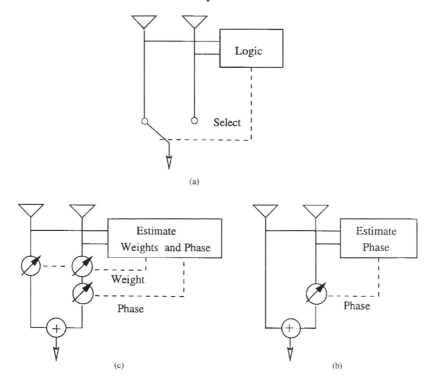

FIGURE 12.2 Diversity combining methods for two diversity branches.

where r_i is the signal envelope in each branch. If we define two new variables

$$\gamma_i = \frac{\text{Instantaneous signal power in each branch}}{\text{Mean noise power}}$$

$$\Gamma = \frac{\text{Mean signal power in each branch}}{\text{Mean noise power}}$$

then the probability that the C/N is less than or equal to some specified value γ_s is

$$\text{Prob}\left[\gamma_i \leq \gamma_s\right] = 1 - e^{-\gamma_s/\Gamma} \tag{12.2}$$

The probability that γ_i in all branches with independent fading will be simultaneously less than or equal to γ_s is then

$$\text{Prob}\left[\gamma_1, \gamma_2, \ldots \gamma_M \leq \gamma_s\right] = \left(1 - e^{-\gamma_s/\Gamma}\right)^M \tag{12.3}$$

This is the distribution of the best signal envelope from the two diversity branches. Figure 12.3 shows the distribution of the combiner output C/N for $M = 1,2,3$, and 4 branches. The improvement in signal quality is significant. For example at 99% reliability level, the improvement in C/N is 10 dB for two branches and 16 dB for four branches.

Selection combining also increases the mean C/N of the combiner output and can be shown to be [3]

$$\text{Mean}\left(\gamma_s\right) = \Gamma \sum_{k=1}^{M} \frac{1}{k} \tag{12.4}$$

This indicates that with 4 branches, for example, the mean C/N of the selected branch is 2.08 better than the mean C/N in any one branch.

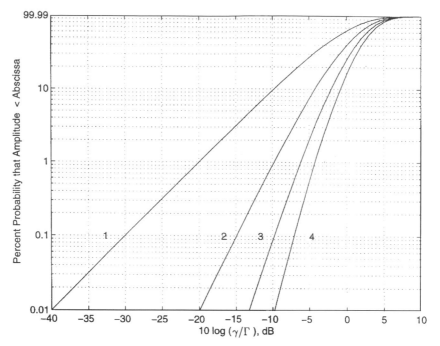

FIGURE 12.3   Probability distribution of signal envelope for selection combining.

12.3.2   Maximal Ratio Combining

In this technique the M diversity branches are first co-phased and then weighted proportionally to their signal level before summing. See Fig. 12.2(b). The distribution of the maximal ratio combiner has been shown to be [5]

$$\text{Prob}\left[\gamma \leq \gamma_m\right] = 1 - e^{(-\gamma_m/\Gamma)} \sum_{k=1}^{M} \frac{(\gamma_m/\Gamma)^{k-1}}{(k-1)!} \tag{12.5}$$

The distribution of output of a maximal ratio combiner is shown in Fig. 12.4. Maximal ratio combining is known to be optimal in the sense that it yields the best statistical reduction of fading of any linear diversity combiner. In comparison to the selection combiner, at 99% reliability level, the maximal ratio combiner provides a 11.5 dB gain for two branches and a 19 dB gain for four branches, an improvement of 1.5 and 3 dB, respectively, over the selection diversity combiner.

The mean C/N of the combined signal may be easily shown to be

$$\text{Mean}\left(\gamma_m\right) = M\Gamma \tag{12.6}$$

Therefore, combiner output mean varies linearly with M. This confirms the intuitive result that the output C/N averaged over fades should provide gain proportional to the number of diversity branches. This is a situation similar to conventional beamforming.

12.3.3   Equal Gain Combining

In some applications, it may be difficult to estimate the amplitude accurately, the combining gains may all be set to unity, and the diversity branches merely summed after co-phasing. [See Fig. 12.2(c)].

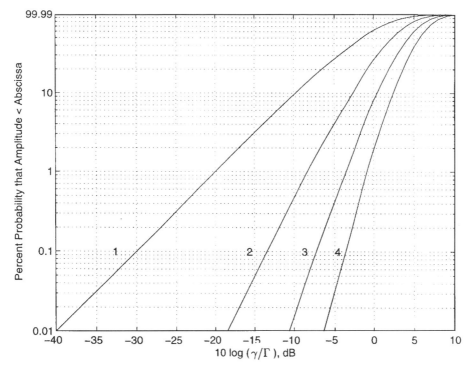

FIGURE 12.4   Probability distribution for signal envelope for maximal ratio combining.

The distribution of equal gain combiner does not have a neat expression and has been computed by numerical evaluation. Its performance has been shown to be very close to within a decibel to maximal ratio combining. The mean C/N can be shown to be [3]

$$\text{Mean}\,(\gamma_e) = \Gamma\left[1 + \frac{\pi}{4}(M-1)\right] \tag{12.7}$$

Like maximal ratio combining, the mean C/N for equal gain combining grows almost linearly with M and is approximately only one decibel poorer than maximal ratio combiner even with an infinite number of branches.

12.3.4   Loss of Diversity Gain Due to Branch Correlation and Unequal Branch Powers

The above analysis assumed that the fading signals in the diversity branches were all uncorrelated and of equal power. In practice, this may be difficult to achieve and as we saw earlier, the branch cross-correlation coefficient $\rho = 0.7$ is considered to be acceptable. Also, equal mean powers in diversity branches are rarely available. In such cases we can expect a certain loss of diversity gain. However, since most of the damage in fading is due to deep fades, and also since the chance of coincidental deep fades is small even for moderate branch correlation, one can expect a reasonable tolerance to branch correlation.

The distribution of the output signal envelope of maximal ratio combiner has been shown to be [6]:

$$\text{Prob}\,[\gamma_m] = \sum_{n=1}^{M} \frac{A_n}{2\lambda_n} e^{-\gamma_m/2\lambda_n} \tag{12.8}$$

where λ_n are the eigenvalues of the M $\times$ M branch envelope covariance matrix whose elements are defined by

$$\mathbf{R}_{ij} = E\left[r_i r_j^*\right] \tag{12.9}$$

and A_n is defined by

$$A_n = \prod_{\substack{k=1 \\ k \neq n}}^{M} \frac{1}{1 - \lambda_k/\lambda_n} \tag{12.10}$$

12.4 Effect of Diversity Combining on Bit Error Rate

So far we have studied the distribution of the instantaneous envelope or C/N after diversity combining. We will now briefly survey how diversity combining affects BER performance in digital radio links; we assume maximal ratio combining.

To begin let us first examine the effect of Rayleigh fading on the BER performance of digital transmission links. This has been studied by several authors and is summarized in [7]. Table 12.1 gives the BER expressions in the large E_b/N_0 case for coherent binary PSK and coherent binary orthogonal FSK for unfaded and Rayleigh faded AWGN (additive white Gaussian noise channels) channels. $\bar{E}_b/N_0$ represents the average E_b/N_0 for the fading channel.

TABLE 12.1 Comparison of BER Performance for Unfaded and Rayleigh Faded Signals

Modulaton	Unfaded BER	Faded BER
Coh BPSK	$\frac{1}{2}\mathrm{erfc}\left(\sqrt{E_b/N_0}\right)$	$\dfrac{1}{4\left(\bar{E}_b/N_0\right)}$
Coh FSK	$\frac{1}{2}\mathrm{erfc}\left(\sqrt{\frac{1}{2}E_b/N_0}\right)$	$\dfrac{1}{2\left(\bar{E}_b/N_0\right)}$

Observe that error rates decrease only inversely with SNR as against exponential decreases for the unfaded channel. Also note that for fading channels, coherent binary PSK is 3 dB better than coherent binary FSK, exactly the same advantage as in unfaded case. Even for modest target BER of 10^{-2} that is usually needed in mobile communications, the loss due to fading can be very high—17.2 dB.

To obtain the BER with maximal ratio diversity combining we have to average the BER expression for the unfaded BER with the distribution obtained for the maximal ratio combiner given in (12.5). Analytical expressions have been derived for these in [7]. For a branch SNR greater than 10 dB, the BER after maximal ratio diversity combining is given in Table 12.2.

We observe that the probability of error varies as $1/\bar{E}_b/N_0$ raised to the Lth power. Thus, diversity reduces the error rate exponentially as the number of independent branches increases.

TABLE 12.2 BER Performance for Coherent BPSK and FSK with Diversity

Modulaton	Post Diversity BER
Coherent BPSK	$\left(\dfrac{1}{4\ \bar{E}_b/N_0}\right)^L \left(\begin{array}{c} 2L-1 \\ L \end{array}\right)$
Coherent FSK	$\left(\dfrac{1}{2\bar{E}_b/N_0}\right)^L \left(\begin{array}{c} 2L-1 \\ L \end{array}\right)$

12.5 Concluding Remarks

Diversity provides a powerful technique for combating fading in mobile communication systems. Diversity techniques seek to generate and exploit multiple branches over which the signal shows low fade correlation. To obtain the best diversity performance, the multiple access, modulation, coding and antenna design of the wireless link must all be carefully chosen so as to provide a rich and reliable level of well-balanced, low-correlation diversity branches in the target propagation environment. Successful diversity exploitation can impact a mobile network in several ways. Reduced power requirements can result in increased coverage or improved battery life. Low signal outage improves voice quality and handoff performance. Finally, reduced fade margins directly translate to better reuse factors and, hence, increased system capacity.

Defining Terms

Automatic request for repeat: An error control mechanism in which received packets that cannot be corrected are retransmitted.

Channel coding/Forward error correction: A technique that inserts redundant bits during transmission to help detect and correct bit errors during reception.

Fading: Fluctuation in the signal level due to shadowing and multipath effects.

Frequency hopping: A technique where the signal bursts are transmitted at different frequencies separated by random spacing that are multiples of signal bandwidth.

Interleaving: A form of data scrambling that spreads burst of bit errors evenly over the received data allowing efficient forward error correction.

Outage probability: The probability that the signal level falls below a specified minimum level.

PCS: Personal Communications Services.

RAKE receiver: A receiver used in direct sequence spread spectrum signals. The receiver extracts energy in each path and then adds them together with appropriate weighting and delay.

References

[1] Adachi, F., Feeney, M.T., Williason, A.G., and Parsons, J.D., Crosscorrelation between the envelopes of 900 MHz signals received at a mobile radio base station site. *Proc. IEE*, 133(6), 506–512, 1986.

[2] Freeburg, T.A., Enabling technologies for in-building network communications—four technical challenges and four solutions. *IEEE Trans. Veh. Tech.*, 29(4), 58–64, 1991.

[3] Jakes, W.C., *Microwave Mobile Communications*, John Wiley & Sons, New York, 1974.

[4] Jefford, P.A., Turkmani, A.M.D., Arowojulu, A.A., and Kellet, C.J., An experimental evaluation of the performance of the two branch space and polarization schemes at 1800 MHz. *IEEE Trans. Veh. Tech.*, VT-44(2), 318–326, 1995.

[5] Lee, W.C.Y., *Mobile Communications Engineering*, McGraw-Hill, New York, 1982.

[6] Pahlavan, K. and Levesque, A.H., *Wireless Information Networks*, John Wiley & Sons, New York, 1995.

[7] Proakis, J.G., *Digital Communications*, McGraw-Hill, New York, 1989.

[8] Vaughan, R.G., Polarization diversity system in mobile communications. *IEEE Trans. Veh. Tech.*, VT-39(3), 177–186, 1990.

[9] Viterbi, A.J., *CDMA: Principle of Spread Spectrum Communications*, Addison-Wesley, Reading, MA, 1995.

13

Digital Communication System Performance[1]

Bernard Sklar
Communications Engineering Services

13.1 Introduction

In this section we examine some fundamental tradeoffs among bandwidth, power, and error performance of digital communication systems. The criteria for choosing modulation and coding schemes, based on whether a system is bandwidth limited or power limited, are reviewed for several system examples. Emphasis is placed on the subtle but straightforward relationships we encounter when transforming from data-bits to channel-bits to symbols to chips.

---

[1]A version of this chapter has appeared as a paper in the *IEEE Communications Magazine,* November 1993, under the title "Defining, Designing, and Evaluating Digital Communication Systems."

0-8493-8597-0/99/$0.00+$.50
© 1999 by CRC Press LLC

The design or definition of any digital communication system begins with a description of the communication link. The *link* is the name given to the communication transmission path from the modulator and transmitter, through the channel, and up to and including the receiver and demodulator. The *channel* is the name given to the propagating medium between the transmitter and receiver. A link description quantifies the average signal power that is received, the available bandwidth, the noise statistics, and other impairments, such as fading. Also needed to define the system are basic requirements, such as the data rate to be supported and the error performance.

13.1.1   The Channel

For radio communications, the concept of *free space* assumes a channel region free of all objects that might affect radio frequency (RF) propagation by absorption, reflection, or refraction. It further assumes that the atmosphere in the channel is perfectly uniform and nonabsorbing, and that the earth is infinitely far away or its reflection coefficient is negligible. The RF energy arriving at the receiver is assumed to be a function of distance from the transmitter (simply following the inverse-square law of optics). In practice, of course, propagation in the atmosphere and near the ground results in refraction, reflection, and absorption, which modify the free space transmission.

13.1.2   The Link

A radio transmitter is characterized by its average output signal power P_t and the gain of its transmitting antenna G_t. The name given to the product $P_t G_t$, with reference to an isotropic antenna is *effective radiated power* (*EIRP*) in watts (or dBW). The predetection average signal power S arriving at the output of the receiver antenna can be described as a function of the *EIRP,* the gain of the receiving antenna G_r, the path loss (or space loss) L_s, and other losses, L_o, as follows [14, 15]:

$$S = \frac{EIRP\ G_r}{L_s L_o} \tag{13.1}$$

The path loss L_s can be written as follows [15]:

$$L_s = \left(\frac{4\pi d}{\lambda}\right)^2 \tag{13.2}$$

where d is the distance between the transmitter and receiver and λ is the wavelength.

We restrict our discussion to those links distorted by the mechanism of additive white Gaussian noise (AWGN) only. Such a noise assumption is a very useful model for a large class of communication systems. A valid approximation for average received noise power N that this model introduces is written as follows [5, 9]:

$$N \cong kT^\circ W \tag{13.3}$$

where k is Boltzmann's constant (1.38×10^{-23} joule/K), T° is effective temperature in kelvin, and W is bandwidth in hertz. Dividing Eq. (13.3) by bandwidth, enables us to write the received noise-power spectral density N_0 as follows:

$$N_0 = \frac{N}{W} = kT^\circ \tag{13.4}$$

Dividing Eq. (13.1) by N_0 yields the received average signal-power to noise-power spectral density S/N_0 as

$$\frac{S}{N_0} = \frac{EIRP\ G_r/T^\circ}{kL_sL_o} \tag{13.5}$$

where G_r/T° is often referred to as the receiver figure of merit. A link budget analysis is a compilation of the power gains and losses throughout the link; it is generally computed in decibels, and thus takes on the bookkeeping appearance of a business enterprise, highlighting the assets and liabilities of the link. Once the value of S/N_0 is specified or calculated from the link parameters, we then shift our attention to optimizing the choice of signalling types for meeting system bandwidth and error performance requirements.

Given the received S/N_0, we can write the received bit-energy to noise-power spectral density E_b/N_0, for any desired data rate R, as follows:

$$\frac{E_b}{N_0} = \frac{ST_b}{N_0} = \frac{S}{N_0}\left(\frac{1}{R}\right) \tag{13.6}$$

Equation (13.6) follows from the basic definitions that received bit energy is equal to received average signal power times the bit duration and that bit rate is the reciprocal of bit duration. Received E_b/N_0 is a key parameter in defining a digital communication system. Its value indicates the apportionment of the received waveform energy among the bits that the waveform represents. At first glance, one might think that a system specification should entail the symbol-energy to noise-power spectral density E_s/N_0 associated with the arriving waveforms. We will show, however, that for a given S/N_0 the value of E_s/N_0 is a function of the modulation and coding. The reason for defining systems in terms of E_b/N_0 stems from the fact that E_b/N_0 depends only on S/N_0 and R and is unaffected by any system design choices, such as modulation and coding.

13.2   Bandwidth and Power Considerations

Two primary communications resources are the received power and the available transmission bandwidth. In many communication systems, one of these resources may be more precious than the other and, hence, most systems can be classified as either bandwidth limited or power limited. In bandwidth-limited systems, spectrally efficient modulation techniques can be used to save bandwidth at the expense of power; in power-limited systems, power efficient modulation techniques can be used to save power at the expense of bandwidth. In both bandwidth- and power-limited systems, error-correction coding (often called channel coding) can be used to save power or to improve error performance at the expense of bandwidth. Recently, trellis-coded modulation (TCM) schemes have been used to improve the error performance of bandwidth-limited channels without any increase in bandwidth [17], but these methods are beyond the scope of this chapter.

13.2.1   The Bandwidth Efficiency Plane

Figure 13.1 shows the abscissa as the ratio of bit-energy to noise-power spectral density E_b/N_0 (in decibels) and the ordinate as the ratio of throughput, R (in bits per second), that can be transmitted per hertz in a given bandwidth W. The ratio R/W is called bandwidth efficiency, since it reflects how efficiently the bandwidth resource is utilized. The plot stems from the Shannon–Hartley capacity theorem [12, 13, 15], which can be stated as

$$C = W \log_2\left(1 + \frac{S}{N}\right) \tag{13.7}$$

where S/N is the ratio of received average signal power to noise power. When the logarithm is taken to the base 2, the capacity C, is given in bits per second. The capacity of a channel defines the maximum number of bits that can be reliably sent per second over the channel. For the case where the data (information) rate R is equal to C, the curve separates a region of practical communication systems from a region where such communication systems cannot operate reliably [12, 15].

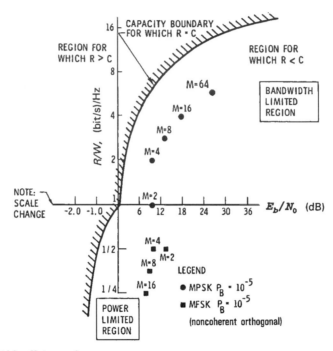

FIGURE 13.1   Bandwidth-efficiency plane.

13.2.2   *M*-ary Signalling

Each symbol in an M-ary alphabet can be related to a unique sequence of m bits, expressed as

$$M = 2^m \qquad \text{or} \qquad m = \log_2 M \tag{13.8}$$

where M is the size of the alphabet. In the case of digital transmission, the term symbol refers to the member of the M-ary alphabet that is transmitted during each symbol duration T_s. To transmit the symbol, it must be mapped onto an electrical voltage or current waveform. Because the waveform represents the symbol, the terms symbol and waveform are sometimes used interchangeably. Since one of M symbols or waveforms is transmitted during each symbol duration T_s, the data rate R in bits per second can be expressed as

$$R = \frac{m}{T_s} = \frac{\log_2 M}{T_s} \tag{13.9}$$

Data-bit-time duration is the reciprocal of data rate. Similarly, symbol-time duration is the reciprocal of symbol rate. Therefore, from Eq. (13.9), we write that the effective time duration T_b of each bit in terms of the symbol duration T_s or the symbol rate R_s is

$$T_b = \frac{1}{R} = \frac{T_s}{m} = \frac{1}{mR_s} \tag{13.10}$$

Then, using Eqs. (13.8) and (13.10) we can express the symbol rate R_s in terms of the bit rate R as follows:

$$R_s = \frac{R}{\log_2 M} \tag{13.11}$$

From Eqs. (13.9) and (13.10), any digital scheme that transmits $m = \log_2 M$ bits in T_s seconds, using a bandwidth of W hertz, operates at a bandwidth efficiency of

$$\frac{R}{W} = \frac{\log_2 M}{WT_s} = \frac{1}{WT_b} \qquad \text{(b/s)/Hz} \tag{13.12}$$

where T_b is the effective time duration of each data bit.

13.2.3 Bandwidth-Limited Systems

From Eq. (13.12), the smaller the WT_b product, the more bandwidth efficient will be any digital communication system. Thus, signals with small WT_b products are often used with bandwidth-limited systems. For example, the European digital mobile telephone system known as Global System for Mobile Communications (GSM) uses Gaussian minimum shift keying (GMSK) modulation having a WT_b product equal to 0.3 Hz/(b/s), where W is the 3-dB bandwidth of a Gaussian filter [4].

For uncoded bandwidth-limited systems, the objective is to maximize the transmitted information rate within the allowable bandwidth, at the expense of E_b/N_0 (while maintaining a specified value of bit-error probability P_B). The operating points for coherent M-ary phase-shift keying (MPSK) at $P_B = 10^{-5}$ are plotted on the bandwidth-efficiency plane of Fig. 13.1. We assume Nyquist (ideal rectangular) filtering at baseband [10]. Thus, for MPSK, the required double-sideband (DSB) bandwidth at an intermediate frequency (IF) is related to the symbol rate as follows:

$$W = \frac{1}{T_s} = R_s \tag{13.13}$$

where T_s is the symbol duration and R_s is the symbol rate. The use of Nyquist filtering results in the minimum required transmission bandwidth that yields zero intersymbol interference; such ideal filtering gives rise to the name Nyquist minimum bandwidth.

From Eqs. (13.12) and (13.13), the bandwidth efficiency of MPSK modulated signals using Nyquist filtering can be expressed as

$$R/W = \log_2 M \qquad \text{(b/s)/Hz} \tag{13.14}$$

The MPSK points in Fig. 13.1 confirm the relationship shown in Eq. (13.14). Note that MPSK modulation is a bandwidth-efficient scheme. As M increases in value, R/W also increases. MPSK modulation can be used for realizing an improvement in bandwidth efficiency at the cost of increased E_b/N_0. Although beyond the scope of this chapter, many highly bandwidth-efficient modulation schemes are under investigation [1].

13.2.4   Power-Limited Systems

Operating points for noncoherent orthogonal M-ary FSK (MFSK) modulation at $P_B = 10^{-5}$ are also plotted on Fig. 13.1. For MFSK, the IF minimum bandwidth is as follows [15]

$$W = \frac{M}{T_s} = MR_s \tag{13.15}$$

where T_s is the symbol duration and R_s is the symbol rate. With MFSK, the required transmission bandwidth is expanded M-fold over binary FSK since there are M different orthogonal waveforms, each requiring a bandwidth of $1/T_s$. Thus, from Eqs. (13.12) and (13.15), the bandwidth efficiency of noncoherent orthogonal MFSK signals can be expressed as

$$\frac{R}{W} = \frac{\log_2 M}{M} \qquad \text{(b/s)/Hz} \tag{13.16}$$

The MFSK points plotted in Fig. 13.1 confirm the relationship shown in Eq. (13.16). Note that MFSK modulation is a bandwidth-expansive scheme. As M increases, R/W decreases. MFSK modulation can be used for realizing a reduction in required E_b/N_0 at the cost of increased bandwidth.

In Eqs. (13.13) and (13.14) for MPSK, and Eqs. (13.15) and (13.16) for MFSK, and for all the points plotted in Fig. 13.1, ideal filtering has been assumed. Such filters are not realizable! For realistic channels and waveforms, the required transmission bandwidth must be increased in order to account for realizable filters.

In the examples that follow, we will consider radio channels that are disturbed only by additive white Gaussian noise (AWGN) and have no other impairments, and for simplicity, we will limit the modulation choice to constant-envelope types, i.e., either MPSK or noncoherent orthogonal MFSK. For an uncoded system, MPSK is selected if the channel is bandwidth limited, and MFSK is selected if the channel is power limited. When error-correction coding is considered, modulation selection is not as simple, because coding techniques can provide power-bandwidth tradeoffs more effectively than would be possible through the use of any M-ary modulation scheme considered in this chapter [3].

In the most general sense, M-ary signalling can be regarded as a waveform-coding procedure, i.e., when we select an M-ary modulation technique instead of a binary one, we in effect have replaced the binary waveforms with better waveforms—either better for bandwidth performance (MPSK) or better for power performance (MFSK). Even though orthogonal MFSK signalling can be thought of as being a coded system, i.e., a first-order Reed-Muller code [8], we restrict our use of the term coded system to those traditional error-correction codes using redundancies, e.g., block codes or convolutional codes.

13.2.5   Minimum Bandwidth Requirements for MPSK and MFSK Signalling

The basic relationship between the symbol (or waveform) transmission rate R_s and the data rate R was shown in Eq. (13.11). Using this relationship together with Eqs. (13.13–13.16) and $R = 9600$ b/s, a summary of symbol rate, minimum bandwidth, and bandwidth efficiency for MPSK and noncoherent orthogonal MFSK was compiled for $M = 2, 4, 8, 16$, and 32 (Table 13.1). Values of E_b/N_0 required to achieve a bit-error probability of 10^{-5} for MPSK and MFSK are also given for each value of M. These entries (which were computed using relationships that are presented later in this chapter) corroborate the tradeoffs shown in Fig. 13.1. As M increases, MPSK signalling provides more bandwidth efficiency at the cost of increased E_b/N_0, whereas MFSK signalling allows for a reduction in E_b/N_0 at the cost of increased bandwidth.

TABLE 13.1 Symbol Rate, Minimum Bandwidth, Bandwidth Efficiency, and Required E_b/N_0 for MPSK and Noncoherent Orthogonal MFSK Signalling at 9600 bit/s

M	m	R (b/s)	R_s (symb/s)	MPSK Minimum Bandwidth (Hz)	MPSK R/W	MPSK E_b/N_0 (dB) $P_B = 10^{-5}$	Noncoherent Orthog MFSK Min Bandwidth (Hz)	MFSK R/W	MFSK E_b/N_0 (dB) $P_B = 10^{-5}$
2	1	9600	9600	9600	1	9.6	19,200	1/2	13.4
4	2	9600	4800	4800	2	9.6	19,200	1/2	10.6
8	3	9600	3200	3200	3	13.0	25,600	3/8	9.1
16	4	9600	2400	2400	4	17.5	38,400	1/4	8.1
32	5	9600	1920	1920	5	22.4	61,440	5/32	7.4

13.3 Example 1: Bandwidth-Limited Uncoded System

Suppose we are given a bandwidth-limited AWGN radio channel with an available bandwidth of $W = 4000$ Hz. Also, suppose that the link constraints (transmitter power, antenna gains, path loss, etc.) result in the ratio of received average signal-power to noise-power spectral density S/N_0 being equal to 53 dB-Hz. Let the required data rate R be equal to 9600 b/s, and let the required bit-error performance P_B be at most 10^{-5}. The goal is to choose a modulation scheme that meets the required performance. In general, an error-correction coding scheme may be needed if none of the allowable modulation schemes can meet the requirements. In this example, however, we shall find that the use of error-correction coding is not necessary.

13.3.1 Solution to Example 1

For any digital communication system, the relationship between received S/N_0 and received bit-energy to noise-power spectral density, E_b/N_0 was given in Eq. (13.6) and is briefly rewritten as

$$\frac{S}{N_0} = \frac{E_b}{N_0} R \qquad (13.17)$$

Solving for E_b/N_0 in decibels, we obtain

$$\begin{aligned} \frac{E_b}{N_0} \text{ (dB)} &= \frac{S}{N_0} \text{ (dB-Hz)} - R \text{ (dB-b/s)} \\ &= 53 \text{ dB-Hz} - (10 \times \log_{10} 9600) \text{ dB-b/s} \\ &= 13.2 \text{ dB (or 20.89)} \qquad (13.18) \end{aligned}$$

Since the required data rate of 9600 b/s is much larger than the available bandwidth of 4000 Hz, the channel is bandwidth limited. We therefore select MPSK as our modulation scheme. We have confined the possible modulation choices to be constant-envelope types; without such a restriction, we would be able to select a modulation type with greater bandwidth efficiency. To conserve power, we compute the *smallest possible* value of M such that the MPSK minimum bandwidth does not exceed the available bandwidth of 4000 Hz.

Table 13.1 shows that the smallest value of M meeting this requirement is $M = 8$. Next we determine whether the required bit-error performance of $P_B \leq 10^{-5}$ can be met by using 8-PSK modulation alone or whether it is necessary to use an error-correction coding scheme. Table 13.1 shows that 8-PSK alone will meet the requirements, since the required E_b/N_0 listed for 8-PSK is less than the received E_b/N_0 derived in Eq. (13.18). Let us imagine that we do not have Table 13.1, however, and evaluate whether or not error-correction coding is necessary.

Figure 13.2 shows the basic modulator/demodulator (MODEM) block diagram summarizing the functional details of this design. At the modulator, the transformation from data bits to symbols yields an output symbol rate R_s, that is, a factor $\log_2 M$ smaller than the input data-bit rate R, as is seen in Eq. (13.11). Similarly, at the input to the demodulator, the symbol-energy to noise-power spectral density E_S/N_0 is a factor $\log_2 M$ larger than E_b/N_0, since each symbol is made up of $\log_2 M$ bits. Because E_S/N_0 is larger than E_b/N_0 by the same factor that R_s is smaller than R, we can expand Eq. (13.17), as follows:

$$\frac{S}{N_0} = \frac{E_b}{N_0} R = \frac{E_s}{N_0} R_s \qquad (13.19)$$

The demodulator receives a waveform (in this example, one of $M = 8$ possible phase shifts) during each time interval T_s. The probability that the demodulator makes a symbol error $P_E(M)$ is well approximated by the following equation for $M > 2$ [6]:

$$P_E(M) \cong 2Q\left[\sqrt{\frac{2E_s}{N_0}} \sin\left(\frac{\pi}{M}\right)\right] \qquad (13.20)$$

where $Q(x)$, sometimes called the complementary error function, represents the probability under the tail of a zero-mean unit-variance Gaussian density function. It is defined as follows [18]:

$$Q(x) = \frac{1}{\sqrt{2\pi}} \int_x^\infty \exp\left(-\frac{u^2}{2}\right) du \qquad (13.21)$$

A good approximation for $Q(x)$, valid for $x > 3$, is given by the following equation [2]

$$Q(x) \cong \frac{1}{x\sqrt{2\pi}} \exp\left(-\frac{x^2}{2}\right) \qquad (13.22)$$

In Fig. 13.2 and all of the figures that follow, rather than show explicit probability relationships, the generalized notation $f(x)$ has been used to indicate some functional dependence on x.

A traditional way of characterizing communication efficiency in digital systems is in terms of the received E_b/N_0 in decibels. This E_b/N_0 description has become standard practice, but recall that there are no bits at the input to the demodulator; there are only waveforms that have been assigned bit meanings. The received E_b/N_0 represents a bit-apportionment of the arriving waveform energy.

To solve for $P_E(M)$ in Eq. (13.20), we first need to compute the ratio of received symbol-energy to noise-power spectral density E_s/N_0. Since from Eq. (13.18)

$$\frac{E_b}{N_0} = 13.2 \text{ dB (or } 20.89)$$

and because each symbol is made up of $\log_2 M$ bits, we compute the following using $M = 8$.

$$\frac{E_s}{N_0} = (\log_2 M) \frac{E_b}{N_0} = 3 \times 20.89 = 62.67 \qquad (13.23)$$

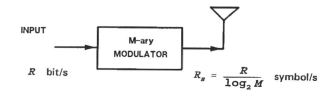

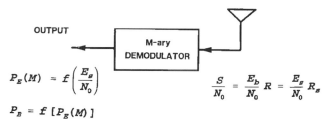

FIGURE 13.2 Basic modulator/demodulator (MODEM) without channel coding.

Using the results of Eq. (13.23) in Eq. (13.20), yields the symbol-error probability $P_E = 2.2 \times 10^{-5}$. To transform this to bit-error probability, we use the relationship between bit-error probability P_B and symbol-error probability P_E, for multiple-phase signalling [8] for $P_E \ll 1$ as follows:

$$P_B \cong \frac{P_E}{\log_2 M} = \frac{P_E}{m} \tag{13.24}$$

which is a good approximation when Gray coding is used for the bit-to-symbol assignment [6]. This last computation yields $P_B = 7.3 \times 10^{-6}$, which meets the required bit-error performance. No error-correction coding is necessary, and 8-PSK modulation represents the design choice to meet the requirements of the bandwidth-limited channel, which we had predicted by examining the required E_b/N_0 values in Table 13.1.

13.4 Example 2: Power-Limited Uncoded System

Now, suppose that we have exactly the same data rate and bit-error probability requirements as in Example 1, but let the available bandwidth W be equal to 45 kHz, and the available S/N_0 be equal to 48 dB-Hz. The goal is to choose a modulation or modulation/coding scheme that yields the required performance. We shall again find that error-correction coding is not required.

13.4.1 Solution to Example 2

The channel is clearly not bandwidth limited since the available bandwidth of 45 kHz is more than adequate for supporting the required data rate of 9600 bit/s. We find the received E_b/N_0 from Eq. (13.18), as follows:

$$\frac{E_b}{N_0} \ \text{(dB)} = 48 \ \text{dB-Hz} - (10 \times \log_{10} 9600) \ \text{dB-b/s} = 8.2 \ \text{dB (or 6.61)} \tag{13.25}$$

Since there is abundant bandwidth but a relatively small E_b/N_0 for the required bit-error probability, we consider that this channel is power limited and choose MFSK as the modulation scheme. To conserve power, we search for the *largest possible M* such that the MFSK minimum bandwidth is not expanded beyond our available bandwidth of 45 kHz. A search results in the choice of $M = 16$ (Table 13.1). Next, we determine whether the required error performance of $P_B \leq 10^{-5}$ can be met by using 16-FSK alone, i.e., without error-correction coding. Table 13.1 shows that 16-FSK alone meets the requirements, since the required E_b/N_0 listed for 16-FSK is less than the received E_b/N_0 derived in Eq. (13.25). Let us imagine again that we do not have Table 13.1, and evaluate whether or not error-correction coding is necessary.

The block diagram in Fig. 13.2 summarizes the relationships between symbol rate R_s, and bit rate R, and between E_s/N_0 and E_b/N_0, which is identical to each of the respective relationships in Example 1. The 16-FSK demodulator receives a waveform (one of 16 possible frequencies) during each symbol time interval T_s. For noncoherent orthogonal MFSK, the probability that the demodulator makes a symbol error $P_E(M)$ is approximated by the following upper bound [20]:

$$P_E(M) \leq \frac{M-1}{2} \exp\left(-\frac{E_s}{2N_0}\right) \tag{13.26}$$

To solve for $P_E(M)$ in Eq. (13.26), we compute E_S/N_0, as in Example 1. Using the results of Eq. (13.25) in Eq. (13.23), with $M = 16$, we get

$$\frac{E_s}{N_0} = (\log_2 M) \frac{E_b}{N_0} = 4 \times 6.61 = 26.44 \tag{13.27}$$

Next, using the results of Eq. (13.27) in Eq. (13.26), yields the symbol-error probability $P_E = 1.4 \times 10^{-5}$. To transform this to bit-error probability, P_B, we use the relationship between P_B and P_E for orthogonal signalling [20], given by

$$P_B = \frac{2^{m-1}}{(2^m - 1)} P_E \tag{13.28}$$

This last computation yields $P_B = 7.3 \times 10^{-6}$, which meets the required bit-error performance. Thus, we can meet the given specifications for this power-limited channel by using 16-FSK modulation, without any need for error-correction coding, as we had predicted by examining the required E_b/N_0 values in Table 13.1.

13.5 Example 3: Bandwidth-Limited and Power-Limited Coded System

We start with the same channel parameters as in Example 1 ($W = 4000$ Hz, $S/N_0 = 53$ dB-Hz, and $R = 9600$ b/s), with one exception.

In this example, we specify that P_B must be at most 10^{-9}. Table 13.1 shows that the system is both bandwidth limited and power limited, based on the available bandwidth of 4000 Hz and the available E_b/N_0 of 20.2 dB, from Eq. (13.18); 8-PSK is the only possible choice to meet the bandwidth constraint; however, the available E_b/N_0 of 20.2 dB is certainly insufficient to meet the required P_B of 10^{-9}. For this small value of P_B, we need to consider the performance improvement that error-correction coding can provide within the available bandwidth. In general, one can use convolutional codes or block codes.

The Bose–Chaudhuri–Hocquenghem (BCH) codes form a large class of powerful error-correcting cyclic (block) codes [7]. To simplify the explanation, we shall choose a block code from the BCH family. Table 13.2 presents a partial catalog of the available BCH codes in terms of n, k, and t, where k represents the number of information (or data) bits that the code transforms into a longer block of n coded bits (or channel bits), and t represents the largest number of incorrect channel bits that the code can correct within each n-sized block. The rate of a code is defined as the ratio k/n; its inverse represents a measure of the code's redundancy [7].

TABLE 13.2

BCH Codes (Partial Catalog)

n	k	t
7	4	1
15	11	1
	7	2
	5	3
31	26	1
	21	2
	16	3
	11	5
63	57	1
	51	2
	45	3
	39	4
	36	5
	30	6
127	120	1
	113	2
	106	3
	99	4
	92	5
	85	6
	78	7
	71	9
	64	10

13.5.1 Solution to Example 3

Since this example has the same bandwidth-limited parameters given in Example 1, we start with the same 8-PSK modulation used to meet the stated bandwidth constraint. We now employ error-correction coding, however, so that the bit-error probability can be lowered to $P_B \leq 10^{-9}$.

To make the optimum code selection from Table 13.2, we are guided by the following goals.

1. The output bit-error probability of the combined modulation/coding system must meet the system error requirement.

2. The rate of the code must not expand the required transmission bandwidth beyond the available channel bandwidth.

3. The code should be as simple as possible. Generally, the shorter the code, the simpler will be its implementation.

The uncoded 8-PSK minimum bandwidth requirement is 3200 Hz (Table 13.1) and the allowable channel bandwidth is 4000 Hz, and so the uncoded signal bandwidth can be increased by no more than a factor of 1.25 (i.e., an expansion of 25%). The very first step in this (simplified) code selection example is to eliminate the candidates in Table 13.2 that would expand the bandwidth by more than 25%. The remaining entries form a much reduced set of bandwidth-compatible codes (Table 13.3).

In Table 13.3, a column designated Coding Gain G (for MPSK at $P_B = 10^{-9}$) has been added. Coding gain in decibels is defined as follows:

$$G = \left(\frac{E_b}{N_0}\right)_{\text{uncoded}} - \left(\frac{E_b}{N_0}s\right)_{\text{coded}} \tag{13.29}$$

G can be described as the reduction in the required E_b/N_0 (in decibels) that is needed due to the error-performance properties of the channel coding. G is a function of the

TABLE 13.3   Bandwidth-Compatible BCH
Codes

n	k	t	Coding Gain, G (dB) MPSK, $P_B = 10^{-9}$
31	26	1	2.0
63	57	1	2.2
	51	2	3.1
127	120	1	2.2
	113	2	3.3
	106	3	3.9

modulation type and bit-error probability, and it has been computed for MPSK at $P_B = 10^{-9}$ (Table 13.3). For MPSK modulation, G is relatively independent of the value of M. Thus, for a particular bit-error probability, a given code will provide about the same coding gain when used with any of the MPSK modulation schemes. Coding gains were calculated using a procedure outlined in the subsequent Calculating Coding Gain section.

A block diagram summarizes this system, which contains both modulation and coding (Fig. 13.3). The introduction of encoder/decoder blocks brings about additional transformations. The relationships that exist when transforming from R b/s to R_c channel-b/s to R_s symbol/s are shown at the encoder/modulator. Regarding the channel-bit rate R_c, some authors prefer to use the units of channel-symbol/s (or code-symbol/s). The benefit is that error-correction coding is often described more efficiently with nonbinary digits. We reserve the term symbol for that group of bits mapped onto an electrical waveform for transmission, and we designate the units of R_c to be channel-b/s (or coded-b/s).

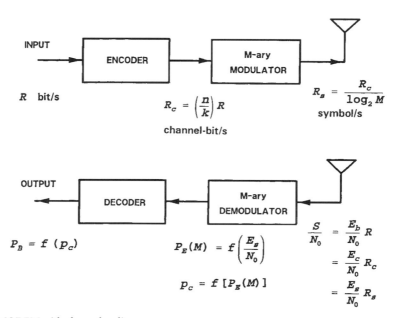

FIGURE 13.3   MODEM with channel coding.

We assume that our communication system cannot tolerate any message delay, so that the channel-bit rate R_c must exceed the data-bit rate R by the factor n/k. Further, each symbol is made up of $\log_2 M$ channel bits, and so the symbol rate R_s is less than R_c by the factor $\log_2 M$. For a system containing both modulation and coding, we summarize the rate transformations as follows:

$$R_c = \left(\frac{n}{k}\right) R \tag{13.30}$$

$$R_s = \frac{R_c}{\log_2 M} \tag{13.31}$$

At the demodulator/decoder in Fig.13.3, the transformations among data-bit energy, channel- bit energy, and symbol energy are related (in a reciprocal fashion) by the same factors as shown among the rate transformations in Eqs. (13.30) and (13.31). Since the encoding transformation has replaced k data bits with n channel bits, then the ratio of channel-bit energy to noise-power spectral density E_c/N_0 is computed by decrementing the value of E_b/N_0 by the factor k/n. Also, since each transmission symbol is made up of $\log_2 M$ channel bits, then E_S/N_0, which is needed in Eq. (13.20) to solve for P_E, is computed by incrementing E_c/N_0 by the factor $\log_2 M$. For a system containing both modulation and coding, we summarize the energy to noise-power spectral density transformations as follows:

$$\frac{E_c}{N_0} = \left(\frac{k}{n}\right) \frac{E_b}{N_0} \tag{13.32}$$

$$\frac{E_s}{N_0} = (\log_2 M) \frac{E_c}{N_0} \tag{13.33}$$

Using Eqs. (13.30) and (13.31), we can now expand the expression for S/N_0 in Eq. (13.19), as follows (Appendix).

$$\frac{S}{N_0} = \frac{E_b}{N_0} R = \frac{E_c}{N_0} R_c = \frac{E_s}{N_0} R_s \tag{13.34}$$

As before, a standard way of describing the link is in terms of the received E_b/N_0 in decibels. However, there are no data bits at the input to the demodulator, and there are no channel bits; there are only waveforms that have bit meanings and, thus, the waveforms can be described in terms of bit-energy apportionments.

Since S/N_0 and R were given as 53 dB-Hz and 9600 b/s, respectively, we find as before, from Eq. (13.18), that the received $E_b/N_0 = 13.2$ dB. The received E_b/N_0 is fixed and independent of n, k, and t (Appendix). As we search, in Table 13.3 for the ideal code to meet the specifications, we can iteratively repeat the computations suggested in Fig. 13.3. It might be useful to program on a personal computer (or calculator) the following four steps as a function of n, k, and t. Step 1 starts by combining Eqs. (13.32) and (13.33), as follows.

Step 1:

$$\frac{E_s}{N_0} = (\log_2 M) \frac{E_c}{N_0} = (\log_2 M) \left(\frac{k}{n}\right) \frac{E_b}{N_0} \tag{13.35}$$

Step 2:

$$P_E(M) \cong 2Q \left[\sqrt{\frac{2E_s}{N_0}} \sin\left(\frac{\pi}{M}\right) \right] \tag{13.36}$$

which is the approximation for symbol-error probability P_E rewritten from Eq. (13.20). At each symbol-time interval, the demodulator makes a symbol decision, but it delivers a channel-bit sequence representing that symbol to the decoder. When the channel-bit output of the demodulator is quantized to two levels, 1 and 0, the demodulator is said to make hard decisions. When the output is quantized to more than two levels, the demodulator is said to make soft decisions [15]. Throughout this paper, we shall assume hard-decision demodulation.

Now that we have a decoder block in the system, we designate the channel-bit-error probability out of the demodulator and into the decoder as p_c, and we reserve the notation P_B for the bit-error probability out of the decoder. We rewrite Eq. (13.24) in terms of p_c for $P_E \ll 1$ as follows.

Step 3:

$$p_c \cong \frac{P_E}{\log_2 M} = \frac{P_E}{m} \tag{13.37}$$

relating the channel-bit-error probability to the symbol-error probability out of the demodulator, assuming Gray coding, as referenced in Eq. (13.24).

For traditional channel-coding schemes and a given value of received S/N_0, the value of E_s/N_0 with coding will always be less than the value of E_s/N_0 without coding. Since the demodulator with coding receives less E_s/N_0, it makes more errors! When coding is used, however, the system error-performance does not only depend on the performance of the demodulator, it also depends on the performance of the decoder. For error-performance improvement due to coding, the decoder must provide enough error correction to more than compensate for the poor performance of the demodulator.

The final output decoded bit-error probability P_B depends on the particular code, the decoder, and the channel-bit-error probability p_c. It can be expressed by the following approximation [11].

Step 4:

$$P_B \cong \frac{1}{n} \sum_{j=t+1}^{n} j \binom{n}{j} p_c^j (1 - p_c)^{n-j} \tag{13.38}$$

where t is the largest number of channel bits that the code can correct within each block of n bits. Using Eqs. (13.35–13.38) in the four steps, we can compute the decoded bit-error probability P_B as a function of n, k, and t for each of the codes listed in Table 13.3. The entry that meets the stated error requirement with the largest possible code rate and the smallest value of n is the double-error correcting (63, 51) code. The computations are as follows.

Step 1:

$$\frac{E_s}{N_0} = 3 \left(\frac{51}{63} \right) 20.89 = 50.73$$

where $M = 8$, and the received $E_b/N_0 = 13.2$ dB (or 20.89).

Step 2:

$$P_E \cong 2Q \left[\sqrt{101.5} \times \sin \left(\frac{\pi}{8} \right) \right] = 2Q(3.86) = 1.2 \times 10^{-4}$$

Step 3:

$$p_c \cong \frac{1.2 \times 10^{-4}}{3} = 4 \times 10^{-5}$$

Step 4:

$$P_B \cong \frac{3}{63} \binom{63}{3} \left(4 \times 10^{-5} \right)^3 \left(1 - 4 \times 10^{-5} \right)^{60}$$

$$+ \frac{4}{63} \binom{63}{4} \left(4 \times 10^{-5} \right)^4 \left(1 - 4 \times 10^{-5} \right)^{59} + \cdots$$

$$= 1.2 \times 10^{-10}$$

where the bit-error-correcting capability of the code is $t = 2$. For the computation of P_B in step 4, we need only consider the first two terms in the summation of Eq. (13.38) since the other terms have a vanishingly small effect on the result. Now that we have selected the (63, 51) code, we can compute the values of channel-bit rate R_c and symbol rate R_s using Eqs. (13.30) and (13.31), with $M = 8$,

$$R_c = \left(\frac{n}{k}\right) R = \left(\frac{63}{51}\right) 9600 \approx 11{,}859 \text{ channel-b/s}$$

$$R_s = \frac{R_c}{\log_2 M} = \frac{11859}{3} = 3953 \text{ symbol/s}$$

13.5.2 Calculating Coding Gain

Perhaps a more direct way of finding the simplest code that meets the specified error performance is to first compute how much coding gain G is required in order to yield $P_B = 10^{-9}$ when using 8-PSK modulation alone; then, from Table 13.3, we can simply choose the code that provides this performance improvement. First, we find the uncoded E_s/N_0 that yields an error probability of $P_B = 10^{-9}$, by writing from Eqs. (13.24) and (13.36), the following:

$$P_B \cong \frac{P_E}{\log_2 M} \cong \frac{2Q\left[\sqrt{\frac{2E_s}{N_0}}\sin\left(\frac{\pi}{M}\right)\right]}{\log_2 M} = 10^{-9} \tag{13.39}$$

At this low value of bit-error probability, it is valid to use Eq. (13.22) to approximate $Q(x)$ in Eq. (13.39) By trial and error (on a programmable calculator), we find that the uncoded $E_s/N_0 = 120.67 = 20.8$ dB, and since each symbol is made up of $\log_2 8 = 3$ bits, the required $(E_b/N_0)_{\text{uncoded}} = 120.67/3 = 40.22 = 16$ dB. From the given parameters and Eq. (13.18), we know that the received $(E_b/N_0)_{\text{coded}} = 13.2$ dB. Using Eq. (13.29), the required coding gain to meet the bit-error performance of $P_B = 10^{-9}$ in decibels is

$$G = \left(\frac{E_b}{N_0}\right)_{\text{uncoded}} - \left(\frac{E_b}{N_0}\right)_{\text{coded}} = 16 - 13.2 = 2.8$$

To be precise, each of the E_b/N_0 values in the preceding computation must correspond to exactly the same value of bit-error probability (which they do not). They correspond to $P_B = 10^{-9}$ and $P_B = 1.2 \times 10^{-10}$, respectively. At these low probability values, however, even with such a discrepancy, this computation still provides a good approximation of the required coding gain. In searching Table 13.3 for the simplest code that will yield a coding gain of at least 2.8 dB, we see that the choice is the (63, 51) code, which corresponds to the same code choice that we made earlier.

13.6   Example 4: Direct-Sequence (DS) Spread-Spectrum Coded System

Spread-spectrum systems are not usually classified as being bandwidth- or power-limited. They are generally perceived to be power-limited systems, however, because the bandwidth occupancy of the information is much larger than the bandwidth that is intrinsically needed for the information transmission. In a direct-sequence spread-spectrum (DS/SS) system, spreading the signal bandwidth by some factor permits lowering the signal-power spectral density by the same factor (the total average signal power is the same as before spreading).

The bandwidth spreading is typically accomplished by multiplying a relatively narrowband data signal by a wideband spreading signal. The spreading signal or spreading code is often referred to as a pseudorandom code or PN code.

13.6.1   Processing Gain

A typical DS/SS radio system is often described as a two-step BPSK modulation process. In the first step, the carrier wave is modulated by a bipolar data waveform having a value $+1$ or -1 during each data-bit duration; in the second step, the output of the first step is multiplied (modulated) by a bipolar PN-code waveform having a value $+1$ or -1 during each PN-code-bit duration. In reality, DS/SS systems are usually implemented by first multiplying the data waveform by the PN-code waveform and then making a single pass through a BPSK modulator. For this example, however, it is useful to characterize the modulation process in two separate steps—the outer modulator/demodulator for the data, and the inner modulator/demodulator for the PN code (Fig. 13.4).

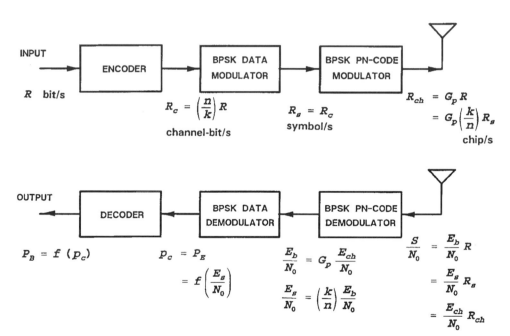

FIGURE 13.4   Direct-sequence spread-spectrum MODEM with channel coding.

A spread-spectrum system is characterized by a processing gain G_p, that is defined in terms of the spread-spectrum bandwidth W_{ss} and the data rate R as follows [20]:

$$G_p = \frac{W_{\text{ss}}}{R} \tag{13.40}$$

For a DS/SS system, the PN-code bit has been given the name chip, and the spread-spectrum signal bandwidth can be shown to be about equal to the chip rate R_{ch} as follows:

$$G_p = \frac{R_{\text{ch}}}{R} \tag{13.41}$$

Some authors define processing gain to be the ratio of the spread-spectrum bandwidth to the symbol rate. This definition separates the system performance that is due to bandwidth spreading from the performance that is due to error-correction coding. Since we ultimately want to relate all of the coding mechanisms relative to the information source, we shall conform to the most usually accepted definition for processing gain, as expressed in Eqs. (13.40) and (13.41).

A spread-spectrum system can be used for interference rejection and for multiple access (allowing multiple users to access a communications resource simultaneously). The benefits of DS/SS signals are best achieved when the processing gain is very large; in other words, the chip rate of the spreading (or PN) code is much larger than the data rate. In such systems, the large value of G_p allows the signalling chips to be transmitted at a power level well below that of the thermal noise. We will use a value of $G_p = 1000$. At the receiver, the despreading operation correlates the incoming signal with a synchronized copy of the PN code and, thus, accumulates the energy from multiple (G_p) chips to yield the energy per data bit. The value of G_p has a major influence on the performance of the spread-spectrum system application. We shall see, however, that the value of G_p has no effect on the received E_b/N_0. In other words, spread spectrum techniques offer no error-performance advantage over thermal noise. For DS/SS systems, there is no disadvantage either! Sometimes such spread-spectrum radio systems are employed only to enable the transmission of very small power-spectral densities and thus avoid the need for FCC licensing [16].

13.6.2   Channel Parameters for Example 13.4

Consider a DS/SS radio system that uses the same (63, 51) code as in the previous example. Instead of using MPSK for the data modulation, we shall use BPSK. Also, we shall use BPSK for modulating the PN-code chips. Let the received $S/N_0 = 48$ dB-Hz, the data rate $R = 9600$ b/s, and the required $P_B \leq 10^{-6}$. For simplicity, assume that there are no bandwidth constraints. Our task is simply to determine whether or not the required error performance can be achieved using the given system architecture and design parameters. In evaluating the system, we will use the same type of transformations used in the previous examples.

13.6.3   Solution to Example 13.4

A typical DS/SS system can be implemented more simply than the one shown in Fig. 13.4. The data and the PN code would be combined at baseband, followed by a single pass through a BPSK modulator. We will, however, assume the existence of the individual blocks in Fig. 13.4 because they enhance our understanding of the transformation process. The relationships in transforming from data bits, to channel bits, to symbols, and to chips Fig. 13.4 have the same pattern of subtle but straightforward transformations in rates and energies as previous relationships (Figs. 13.2 and 13.3). The values of R_c, R_s, and R_{ch} can now be calculated immediately since the (63, 51) BCH code has already been selected. From Eq. (13.30) we write

$$R_c = \left(\frac{n}{k}\right) R = \left(\frac{63}{51}\right) 9600 \approx 11{,}859 \text{ channel-b/s}$$

Since the data modulation considered here is BPSK, then from Eq. (13.31) we write

$$R_s = R_c \approx 11{,}859 \text{ symbol/s}$$

and from Eq. (13.41), with an assumed value of $G_p = 1000$

$$R_{\text{ch}} = G_p R = 1000 \times 9600 = 9.6 \times 10^6 \text{ chip/s}$$

Since we have been given the same S/N_0 and the same data rate as in Example 2, we find the value of received E_b/N_0 from Eq. (13.25) to be 8.2 dB (or 6.61). At the demodulator, we can now expand the expression for S/N_0 in Eq. (13.34) and the Appendix as follows:

$$\frac{S}{N_0} = \frac{E_b}{N_0}R = \frac{E_c}{N_0}R_c = \frac{E_s}{N_0}R_s = \frac{E_{\text{ch}}}{N_0}R_{\text{ch}} \tag{13.42}$$

Corresponding to each transformed entity (data bit, channel bit, symbol, or chip) there is a change in rate and, similarly, a reciprocal change in energy-to-noise spectral density for that received entity. Equation (13.42) is valid for any such transformation when the rate and energy are modified in a reciprocal way. There is a kind of *conservation of power* (or energy) phenomenon that exists in the transformations. The total received average power (or total received energy per symbol duration) is fixed regardless of how it is computed, on the basis of data bits, channel bits, symbols, or chips.

The ratio E_{ch}/N_0 is much lower in value than E_b/N_0. This can be seen from Eqs. (13.42) and (13.41), as follows:

$$\frac{E_{\text{ch}}}{N_0} = \frac{S}{N_0}\left(\frac{1}{R_{\text{ch}}}\right) = \frac{S}{N_0}\left(\frac{1}{G_p R}\right) = \left(\frac{1}{G_p}\right)\frac{E_b}{N_0} \tag{13.43}$$

But, even so, the despreading function (when properly synchronized) accumulates the energy contained in a quantity G_p of the chips, yielding the same value $E_b/N_0 = 8.2$ dB, as was computed earlier from Eq. (13.25). Thus, the DS spreading transformation has no effect on the error performance of an AWGN channel [15], and the value of G_p has no bearing on the value of P_B in this example.

From Eq. (13.43), we can compute, in decibels,

$$\begin{aligned}
\frac{E_{\text{ch}}}{N_0} &= E_b/N_0 - G_p \\
&= 8.2 - (10 \times \log_{10} 1000) \\
&= -21.8 \tag{13.44}
\end{aligned}$$

The chosen value of processing gain ($G_p = 1000$) enables the DS/SS system to operate at a value of chip energy well below the thermal noise, with the same error performance as without spreading.

Since BPSK is the data modulation selected in this example, each message symbol therefore corresponds to a single channel bit, and we can write

$$\frac{E_s}{N_0} = \frac{E_c}{N_0} = \left(\frac{k}{n}\right)\frac{E_b}{N_0} = \left(\frac{51}{63}\right) \times 6.61 = 5.35 \tag{13.45}$$

where the received $E_b/N_0 = 8.2$ dB (or 6.61). Out of the BPSK data demodulator, the symbol-error probability P_E (and the channel-bit error probability p_c) is computed as follows [15]:

$$p_c = P_E = Q\left(\sqrt{\frac{2E_c}{N_0}}\right) \tag{13.46}$$

Using the results of Eq. (13.45) in Eq. (13.46) yields

$$p_c = Q(3.27) = 5.8 \times 10^{-4}$$

Finally, using this value of p_c in Eq. (13.38) for the (63,51) double-error correcting code yields the output bit-error probability of $P_B = 3.6 \times 10^{-7}$. We can, therefore, verify that for the given architecture and design parameters of this example the system does, in fact, achieve the required error performance.

13.7 Conclusion

The goal of this section has been to review fundamental relationships used in evaluating the performance of digital communication systems. First, we described the concept of a link and a channel and examined a radio system from its transmitting segment up through the output of the receiving antenna. We then examined the concept of bandwidth-limited and power-limited systems and how such conditions influence the system design when the choices are confined to MPSK and MFSK modulation. Most important, we focused on the definitions and computations involved in transforming from data bits to channel bits to symbols to chips. In general, most digital communication systems share these concepts; thus, understanding them should enable one to evaluate other such systems in a similar way.

Appendix: Received E_b/N_0 Is Independent of the Code Parameters

Starting with the basic concept that the received average signal power S is equal to the received symbol or waveform energy, E_s, divided by the symbol-time duration, T_s (or multiplied by the symbol rate, R_s), we write

$$\frac{S}{N_0} = \frac{E_s/T_s}{N_0} = \frac{E_s}{N_0}R_s \tag{A13.1}$$

where N_0 is noise-power spectral density.

Using Eqs. (13.27) and (13.25), rewritten as

$$\frac{E_s}{N_0} = (\log_2 M)\frac{E_c}{N_0} \quad \text{and} \quad R_s = \frac{R_c}{\log_2 M}$$

let us make substitutions into Eq. (A13.1), which yields

$$\frac{S}{N_0} = \frac{E_c}{N_0}R_c \tag{A13.2}$$

Next, using Eqs. (13.26) and (13.24), rewritten as

$$\frac{E_c}{N_0} = \left(\frac{k}{n}\right)\frac{E_b}{N_0} \quad \text{and} \quad R_c = \left(\frac{n}{k}\right)R$$

let us now make substitutions into Eq. (A13.2), which yields the relationship expressed in Eq. (13.11)

$$\frac{S}{N_0} = \frac{E_b}{N_0}R \tag{A13.3}$$

Hence, the received E_b/N_0 is only a function of the received S/N_0 and the data rate R. It is independent of the code parameters, $n, k,$ and t. These results are summarized in Fig. 13.3.

References

[1] Anderson, J.B. and Sundberg, C.-E.W., Advances in constant envelope coded modulation, *IEEE Commun., Mag.*, 29(12), 36–45, 1991.

[2] Borjesson, P.O. and Sundberg, C.E., Simple approximations of the error function $Q(x)$ for communications applications, *IEEE Trans. Comm.*, COM-27, 639–642, Mar. 1979.

[3] Clark Jr., G.C. and Cain, J.B., *Error-Correction Coding for Digital Communications*, Plenum Press, New York, 1981.

[4] Hodges, M.R.L., The GSM radio interface, *British Telecom Technol. J.*, 8(1), 31–43, 1990.

[5] Johnson, J.B., Thermal agitation of electricity in conductors, *Phys. Rev.*, 32, 97–109, Jul. 1928.

[6] Korn, I., *Digital Communications*, Van Nostrand Reinhold Co., New York, 1985.

[7] Lin, S. and Costello Jr., D.J., *Error Control Coding: Fundamentals and Applications*, Prentice-Hall, Englewood Cliffs, NJ, 1983.

[8] Lindsey, W.C. and Simon, M.K., *Telecommunication Systems Engineering*, Prentice-Hall, Englewood Cliffs, NJ, 1973.

[9] Nyquist, H., Thermal agitation of electric charge in conductors, *Phys. Rev.*, 32, 110–113, Jul. 1928.

[10] Nyquist, H., Certain topics on telegraph transmission theory, *Trans. AIEE*, 47, 617–644, Apr. 1928.

[11] Odenwalder, J.P., *Error Control Coding Handbook*. Linkabit Corp., San Diego, CA, Jul. 15, 1976.

[12] Shannon, C.E., A mathematical theory of communication, *BSTJ.* 27, 379–423, 623–657, 1948.

[13] Shannon, C.E., Communication in the presence of noise, *Proc. IRE.* 37(1), 10–21, 1949.

[14] Sklar, B., What the system link budget tells the system engineer or how I learned to count in decibels, *Proc. of the Intl. Telemetering Conf.*, San Diego, CA, Nov. 1979.

[15] Sklar, B., *Digital Communications: Fundamentals and Applications*, Prentice-Hall, Englewood Cliffs, NJ, 1988.

[16] Title 47, *Code of Federal Regulations,* Part 15 Radio Frequency Devices.

[17] Ungerboeck, G., Trellis-coded modulation with redundant signal sets, Pt. I and II, *IEEE Comm. Mag.*, 25, 5–21. Feb. 1987.

[18] Van Trees, H.L., *Detection, Estimation, and Modulation Theory*, Pt. I, John Wiley & Sons, New York, 1968.

[19] Viterbi, A.J., *Principles of Coherent Communication*, McGraw-Hill, New York, 1966.

[20] Viterbi, A.J., Spread spectrum communications—myths and realities, *IEEE Comm. Mag.*, 11–18, May, 1979.

Further Information

A useful compilation of selected papers can be found in: *Cellular Radio & Personal Communications–A Book of Selected Readings,* edited by Theodore S. Rappaport, Institute of Electrical and Electronics Engineers, Inc., Piscataway, New Jersey, 1995. Fundamental design issues, such as propagation, modulation, channel coding, speech coding, multiple-accessing and networking, are well represented in this volume.

Another useful sourcebook that covers the fundamentals of mobile communications in great detail is: *Mobile Radio Communications,* edited by Raymond Steele, Pentech Press, London 1992. This volume is also available through the Institute of Electrical and Electronics Engineers, Inc., Piscataway, New Jersey.

For spread spectrum systems, an excellent reference is: *Spread Spectrum Communications Handbook,* by Marvin K. Simon, Jim K. Omura, Robert A. Scholtz, and Barry K. Levitt, McGraw-Hill Inc., New York, 1994.

14

Telecommunications Standardization

Spiros Dimolitsas
Lawrence Livermore National Laboratory

Michael Onufry
COMSAT Laboratories

14.1 Introduction

National economies are increasingly becoming information based, where networking and information transport provide a foundation for productivity and economic growth. Concurrently, many countries are rapidly adopting deregulation policies that are resulting in a telecommunications industry that is increasingly multicarrier and multivendor based, and where interconnectivity and compatibility between different networks is emerging as key to the success of this technological and regulatory transition. The communications industry has, consequently, become more interested in standardization; **standards** give manufacturers, service providers, and users freedom of choice at reasonable cost.

In this chapter, a review is provided of the primary telecommunications standards setting bodies. As will be seen, these bodies are often driven by slightly different underlying philosophies, but the output of their activities, i.e., the standards, possess essentially the same characteristics. An all-encompassing review of standardization bodies is not attempted here; this would clearly take many volumes to describe. Furthermore, as country after country increasingly deregulates its telecommunication industry, new standards setting bodies emerge to fill in the void of the de-facto (but no longer existing) standards setting bodies: the national telecommunications administration.

The principal communications standards bodies that will be covered are the following: the International Telecommunications Union (**ITU**); the United States **ANSI** Committee

T1 on Telecommunications; the Telecommunications Industry Association (**TIA**); the European Telecommunications Standards Institute (**ETSI**); the Inter-American Telecommunications Commission (**CITEL**); the Japanese Telecommunications Technology Committee (**TTC**); and the Institute of Electrical and Electronics Engineers (**IEEE**). Not addressed explicitly are other standards setting bodies that are either national or regional in character, even though it is recognized that sometimes there is overlap in scope with the bodies explicitly covered here.

Most notably, standards setting bodies that are not covered, but that are worth noting, include: the United States ANSI Committee X3; the International Standards Organization (**ISO**), the International Electrotechnical Commission (**IEC**) [except ISO/IEC joint technical committee (JTC) 1], the Telecommunications Standards Advisory Council of Canada (TSACC), the Australian Telecommunications Standardization Committee (ATSC), the Telecommunication Technology Association (TTA) in Korea, and several forums (whose scope is, in principle, somewhat different) such as the asynchronous transfer mode (ATM) forum, the frame relay forum, the integrated digital services network (ISDN) users' forum, and telocator. As will be described later, many of these bodies operate in a coherent fashion through a mechanism developed by the Interregional Telecommunications Standards Conference (ITSC) and its successor, the Global Standards Collaboration (**GSC**).

14.2   Global Standardization

When it comes to setting global communications standards, the ITU comes to the forefront. The ITU is an intergovernmental organization, whereby each sovereign state that is a member of the United Nations may become a member of the ITU. Member governments (in most cases represented by their telecommunications administrations) are constitutional members with a right to vote. Other organizations, such as network and service providers, manufacturers, and scientific and industrial organizations also participate in ITU activities but with a lower legal status.

ITU traces its history back to 1865 in the era of telegraphy. The supreme organ of the ITU is the plenipotentiary conference, which is held not less than every five years and plays a major role in the management of ITU. In 1993 the ITU as a U.N.-specialized agency was reorganized into three sectors (see Fig. 14.1): The *telecommunications standardization* sector (**ITU-T**), the *radiocommunications* sector (**ITU-R**), and the *development* sector (BDT). These sectors' activities are, respectively, standardization of telecommunications, including radio communications; regulation of telecommunications (mainly for radio communications); and development of telecommunications.

It should be noted that, in general, the ITU-T is the successor of the international telephone and telegraph consultative committee (**CCITT**) of the ITU with additional responsibilities for standardization of network-related radio communications. Similarly, the ITU-R is the successor of the international radio consultative committee (**CCIR**) and the international frequency registration bureau (IFRB) of the ITU (after transferring some of its standardization activities to the ITU-T). The BDT is a new sector, which became operational in 1989.

14.2.1   ITU-T

Within the ITU structure, standardization work is undertaken by a number of study groups (SG) dealing with specific areas of communications. There are currently 14 study groups, as shown in Table 14.1.

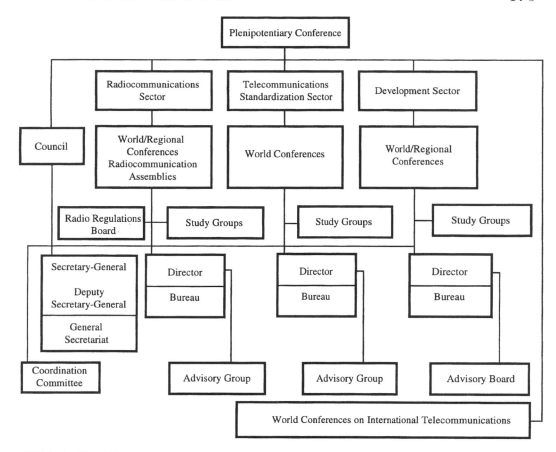

FIGURE 14.1 The ITU structure.

Study groups develop standards for their respective work areas, which then have to be agreed upon by consensus—a process that for the time being is reserved to administrations only. The standards so developed are called **recommendations** to indicate their legal nonbinding nature. Technically, however, there is no distinction between recommendations developed by the ITU and standards developed by other standards setting bodies.

The study groups' work is undertaken by delegation members, sent or sponsored by their national administrations, and delegates from recognized private operating organizations (RPOA), scientific and industrial organizations (SIO) or international organizations. Because an ITU-T study group can typically have from 100 to more than 500 participating members and deal with 20–50 project standards, the work of each study group is often divided among working parties (WP). Such working parties are usually split further into experts' groups led by a chair or "rapporteur" with responsibility for leading the work defined in an approved active question or subelement of a question.

To coordinate standardization work that spans several study groups, two joint coordination groups (**JCG**) have also been established (not shown in Fig. 14.1): International Mobile Communications (IMT-2000) and Satellite Matters.

Such groups do not have executive powers but are merely there to coordinate work of pervasive interest within the ITU-T sector.

Also part of the ITU-T structure is the telecommunications standardization bureau (TSB) or, as it was formerly called, the CCITT secretariat. The TSB is responsible for the organization of numerous meetings held by the sector each year as well as all other support

TABLE 14.1 ITU-T Study Group Structure

SG 2	Network and service operation Lead SG on Service definition, Numbering, Routing and Global Mobility
SG 3	Tariff and accounting principles
SG 4	TMN and network maintenance Lead SG on Telecommunication management network (TMN) studies
SG 5	Protection against electromagnetic environmental effects
SG 6	Outside plant
SG 7	Data networks and open systems communications Lead SG on Open Distributed Processing (ODP), Frame Relay and for Communications System Security
SG 8	Characteristics of telematic services Lead SG on Facsimile
SG 9	Television and sound transmission
SG 10	Languages and general software aspects for telecommunications systems
SG 11	Signalling requirements and protocols Lead SG on Intelligent Network and IMT-2000
SG 12	End-to-end transmission performance of networks and terminals
SG 13	General network aspects Lead SG on General network aspects Global Information Infrastructures and Broadband ISDN
SG 15	Transport networks, systems and equipment Lead SG on Access Network Transport
SG 16	Transmission systems and equipment Lead SG on Multimedia services and systems

services required to ensure the smooth and efficient operation of the sector (including, but not limited to, document production and distribution). The TSB is headed by a director, who holds the executive power and, in collaboration with the study groups, bears full responsibility for the ITU-T activities. In this structure, unlike other U.N. organizations, the secretary general is the legal representative of the ITU, with the executive powers being vested in the director.

Finally, the ITU-T is supported by an advisory group, i.e., the telecommunications standardization advisory group (TSAG), which together with interested ITU members, the ITU-T Director, and ITU-T SG chairman, guides standardization activities.

14.2.2 ITU-R

The radiocommunications sector emphasizes the regulatory and pure radio-interface aspects. The functional structure of the ITU-R currently includes eight study groups, (shown in Table 14.2) a radiocommunications bureau, and an advisory board. The role of the latter two elements is very similar to the ITU-T and, thus, need not be repeated here.

As within the ITU-T, there are areas of pervasive interest, and so areas of common interest can be found between the ITU-T and ITU-R where activities need to be coordinated. To achieve this objective, two intersector coordination groups (**ICG**) have been established (not shown in Fig. 14.1) dealing with international mobile telecommunications (IMT-2000), and satellite matters.

Three major special activities have been organized within ITU-R:

- IMT-2000 (formerly known as Future Public Land Mobile Telecommunications Systems FPLMTS). The objective of the International Mobile Telecommunications (IMT)-2000 activity is to provide seamless satellite and terrestrial operation

TABLE 14.2   ITU-R Study Group Structure

SG 1	Spectrum management
SG 3	Radio wave propagation
SG 4	Fixed satellite service
SG 7	Science services
SG 8	Mobile, radio determination, amateur and related satellite services
SG 9	Fixed service
SG 10	Broadcasting services: sound
SG 11	Broadcasting services: television

of mobile terminals throughout the world—anywhere, anytime—where communication coverage requires interoperation of satellite and terrestrial networks. This is to be accomplished using technology available around the year 2000.

- Mobile-satellite and radionavigation-satellite service (MSS-RNSS). The rapid growth of service in these areas has created a need to focus attention on interference and spectrum allocation.
- Wireless Access Systems (WAS). This is an application of radio technology and personal communications systems directed toward lowering the installation and maintenance cost of the local access network. The traditional high cost has prevented penetration of basic telephone service in evolving and developing countries of the world. Overcoming this barrier is an objective of the BDT, described next.

14.2.3   BDT

Unlike the ITU-T (and to some extent ITU-R), which deals with standardization, the BDT deals with aspects that promote the integration and deployment of communications in developing countries. Typical outputs from this sector include implementation guides that expand the utility of ITU recommendations and ensure their expeditious implementation. Communications has been recognized as a necessary element for economic growth. The BDT also seeks to arrange special financing involving communication suppliers and governments or authorized carriers within developing countries to enable provision of basic communications service where otherwise it would not be possible.

14.2.4   ISO/IEC JTC 1

Two global organizations are active in the information processing systems area, the International Standards Organization (ISO) and the International Electrotechnical Commission (IEC), particularly through the Joint Technical Committee 1 (JTC 1).

The ISO comprises national standards bodies, which have the responsibility for promoting and distributing ISO standards within their own countries. ISO technical work is carried out by some 200 technical committees (TC). Technical committees are established by the ISO council and their work program is approved by the technical board on behalf of the council.

The IEC comprises national committees (one from each country) and deals with almost all spheres of electrotechnology, including power, electronics, telecommunications, and nuclear energy. IEC technical work is performed by some 200 TCs set up by its council and some 700 working groups. Part of this organization, a President's Advisory Committee on future technology (PACT) advises the IEC president on new technologies which require preliminary or immediate standardization work. PACT is designed to form a direct link with private and public research and development activities, keeping the IEC abreast of accelerating technological changes and the accompanying demand for new standards. Small industrial project teams examine new work initiatives which can be introduced into the regular IEC working structure.

In 1987 a joint technical committee was established incorporating ISO TC97, IEC TC83, and subcommittee 47B to deal with generic information technology. The international standards developed by JTC1 are published under the ISO and IEC logos. The activities of ISO/IEC/JTC 1 are listed in Table 14.3 expressed in terms of its subcommittees (SC).

TABLE 14.3   ISO/IEC/JTC1 Subcommittees

SC 1	Vocabulary
SC 2	Coded character sets
SC 6	Telecommunications information exchange between systems
SC 7	Software engineering
SC 11	Flexible magnetic media for digital data interchange
SC 17	Identification cards and related devices
SC 22	Programming languages, their environments and systems software interfaces
SC 23	Optical disk cartridges for information interchange
SC 24	Computer graphics and image processing
SC 25	Interconnection information technology management
SC 26	Microprocessor systems
SC 27	IT security techniques
SC 28	Office equipment
SC 29	Coding of audio, picture, multimedia and hypermedia information
SC 31	Automatic data capture
SC 32	Data management services
SC 33	Distributed application services

The ISO and IEC jointly issue directions for the work of the technical committees. The scope (or area of activity) of each technical committee (TC)/subcommittee (SC) is defined by the TC/SC itself, and then submitted to the Committee of Action (CA)/parent TC for approval. The TCs/SCs prepare technical documents on specific subjects within their respective scopes, which area then submitted to the National Committees for voting with a view to their approval as international standards.

14.3   Regional Standardization

Today the ETSI comes closest to being a true regional standards setting body, together with CITEL, the regional (Latin-American) standardization body.

ETSI is the result of the Single Act of the European community and the EC commission green paper in 1987 that analyzed the consequences of the Single Act and recommended that a European telecommunications standards body be created to develop common standards for telecommunications equipment and networks. Out of this recommendation, the Committee for Harmonization (CCH) and the European Conference for Post and Telecommunications (**CEPT**) evolved into ETSI, which formally came into being in March 1988. It should be noted, however, that even though ETSI attributes at least part of its existence to the European Community, its membership is wider than just the European Union Nations.

Because of the way ETSI came into being, ETSI is characterized by a unique aspect, namely, it is often called upon by the European Commission to develop standards that are necessary to implement legislation. Such standards, which are referred to as technical basis reports (TBR) and whose application is usually mandatory, are often needed in public procurements, as well as in provisioning for open network interconnection as national telecommunications administrations are being deregulated. Like ITU, however, ETSI also develops voluntary standards in accordance with common international understanding against which industry is not obliged to produce conforming products. These standards fall into either the European technical standard (**ETS**) class when fully approved, or into the interim-ETS class, when not fully stable or proven.

ETSI standards are typically sought when either the subject matter is not studied at the global level (such as when it may be required to support some piece of legislation), or the development of the standard is justified by market needs that exist in Europe and not in other parts of the world. In some cases, it may be necessary to adapt ITU standards for the European continent, although a simple endorsement of an ITU standard as a European standard is also possible. A more delicate case arises when both the ITU and ETSI are pursuing parallel standards activities, in which case close coordination with the ITU is sought either through member countries that may input ETSI standards to the ITU for consideration or through the global standards collaboration process.

The highest authority of ETSI is the general assembly, which determines ETSI's policy, appoints its director and deputy, adopts the budget, and approves the audited accounts. The more technical issues are addressed by the technical assembly, which approves technical standards, advises on the work to be undertaken, and sets priorities. The ETSI technical committees are listed in Table 14.4.

TABLE 14.4 ETSI Technical Committees

TCEE	Environmental engineering
TCHF	Human factors
TCMTS	Methods for testing and specification
TCSEC	Security
TCSPS	Signalling protocols and switching
TCTM	Transmission and multiplexing
TCERM	EMC and radio spectrum matters
TCICC	Integrated circuit cards
TCNA	Network aspects
TCSES	Satellite earth stations and systems
TCSTQ	Speech processing, transmission and quality
TCTMN	Telecommunications management networks
ECMA TC32	Communication, networks and systems interconnection
EBU/CENELEC/ETSI JTC	Joint technical committee

It can be seen that ETSI currently comprises 14 technical committees reporting to the technical assembly. These committees are responsible for the development of technical standards. In addition, these committees are responsible for prestandardization activities, that is, activities lead to ETSI technical reports (ETR) that eventually become the basis for future standards.

In addition to the technical assembly, a strategic review committee (SRC) is responsible for prospective examination of a single technical domain, whereas an intellectual property rights committee defines ETSI's policy in the area of intellectual property. Although by no means unique to ETSI, the rapid pace of technological progress has resulted in more standards being adopted that embrace technologies that are still under patent protection. This creates a fundamental conflict between the private, exclusive nature of industrial property rights, and the open, public nature of standards. Harmonizing those conflicting claims has emerged as a thorny issue in all standards organizations; ETSI has established a formal function for this purpose. Finally, the ETS/EBU technical committee coordinates activities with the European broadcasting union (EBU), whereas the ISDN committee is in charge of managing and coordinating the standardization process for narrowband ISDN.

14.3.1   CITEL

On June 11, 1993, the Organization of American States (OAS) General Assembly revised the existing Inter-American Telecommunication Commission (CITEL) strengthening and reorganizing the activities of CITEL, creating a position for the executive secretariat of CITEL and opening the doors, as associate members, to enterprises, organisms, and private telecommunication organizations, to act as observers of the permanent consultative committees of CITEL and its working groups.

CITEL's objectives include facilitating and promoting the continuous development of telecommunications in the hemisphere. It serves as the organization's principal advisory body on matters related to telecommunications. The commission represents all the members states. It has a permanent executive committee consisting of 11 members, and three permanent consultative committees. The permanent consultative committees, whose members are all member states of the organization, also have associate members that represent various private telecommunications agencies or companies.

The general assembly of CITEL, through resolution CITEL Res.8(I-94) established the following specific mandates for the three permanent consultative committees and the steering committee.

Permanent Consultative Committee I: Public Telecommunication Services. To promote and watch over the integration and strengthening of networks and public telecommunication services operating in the countries of the Americas, taking into account the need for modernization of networks and promotion of universal telephone basic services, as well as for increasing the public availability of specialized services and the promotion of the use of international ITU standards and radio regulations.

Permanent Consultative Committee II: Broadcasting. To stimulate and encourage the regional presence of broadcasting services, promoting the use of modern technologies and improving the public availability of such communication media, including audio and video systems, and the promotion of the use of international ITU standards and radio regulations.

Permanent Consultative Committee III: Radiocommunications. To promote the harmonization of radiocommunication services bearing especially in mind the need for a reduction to the minimum of those factors that may cause harmful interferences in the performance and operation of networks and services. To promote the use of modern technologies and the application of the ITU radio regulations and standards.

Steering Committee. The Steering Committee shall be formed by the chairman and vice-chairman of COM/CITEL and the chairman of the PCCs. The committee will be responsible for the revision and proposal to COM/CITEL of the continuous updating of the regulations, mandates and work programs of CITEL bodies; the executive secretary of CITEL will act as the secretary of said committee.

14.4 National Standardization

As standardization moves from global to regional and then to national levels, the number of actual participating entities rapidly grows. Here, the function of two national standards bodies are reviewed, primarily because these have been in existence the longest and secondarily because they also represent major markets for commercial communications.

14.4.1 ANSI T1

Unlike the ETSI, which came into being partly as a consequence of legislative recommendations, the ANSI Committee T1 on telecommunications came into being as a result of the realization that with the breakup of the Bell System, de-facto standards could no longer be expected. In fact, T1 came into being the very same year (1984) that the breakup of the Bell System came into effect.

The T1 membership comprises four types of interest groups: users and general interest groups, manufacturers, interexchange carriers, and exchange carriers. This rather broad membership is reflected, to some extent, by the scope to which T1 standards are being applied; this means that nontraditional telecommunications service providers are utilizing the technologies standardized by committee T1. This situation is the result of the rapid evolution and convergence of the telecommunications, computer, and cable television industries in the United States, and advances in wireless technology.

Committee T1 currently addresses approximately 150 approved projects, which led to the establishment of six, primarily functionally oriented, technical subcommittees (TSC), as shown in Table 14.5 and Fig. 14.2 [although not evident from Table 14.3, subcommittee T1P1 has primary responsibility for management of activities on personal communications systems (PCS)]. In-turn, each of these six subcommittees is divided into a number of subtending working groups, and subworking groups.

TABLE 14.5 T1 Subcommittee Structure

TSC: T1A1	Performance and signal processing
TSC: T1E1	Network interfaces and environmental considerations
TSC: T1M1	Interwork operations, administration, maintenance, and provisioning
TSC: T1P1	Systems engineering, standards planning, and program management
TSC: T1S1	Services, architecture, and signalling
TSC: T1X1	Digital hierarchy and synchronization

Committee T1 also has an advisory group (**T1AG**) made up of elected representatives from each of the four interest groups to carry out committee T1 directives and to develop proposals for consideration by the T1 membership.

In parallel to serving as the forum that establishes ANSI telecommunications network standards, committee T1 technical subcommittees draft candidate U.S. technical contributions to the ITU. These contributions are submitted to the U.S. Department of State National Committee for the ITU, which administers U.S. participation and contributions to the ITU (see Fig. 14.3). In this manner, activities within T1 are coordinated with

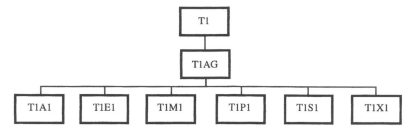

FIGURE 14.2   T1 committee structure.

those of the ITU. This coordination with other standards setting bodies is also reflected in T1's involvement with Latin-American standards, through the formation of an ad hoc group with CITEL's permanent technical committee 1 (PTC 1/T1). Further coordination with ETSI and other standards setting bodies is accomplished through the global standards collaboration process.

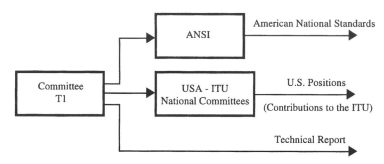

FIGURE 14.3   Committee T1 output.

14.4.2   TIA

The TIA is a full-service trade organization that provides its members with numerous services including government relations, market support activities, educational programs, and standards setting activities.

TIA is a member-driven organization. Policy is formulated by 25 board members selected from member companies, and is carried out by a permanent professional staff located in Washington D.C. TIA comprises six issue-oriented standing committees, each of which is chaired by a board member. The six committees are membership scope and development, international, marketing and trade shows, public policy and government relations, and technical. It is this last committee that in 1992 was accredited by ANSI in the United States to standardize telecommunications products. Technology standardization activities are reflected by TIA's four product-oriented divisions, namely, user premises equipment, network equipment, mobile and personal communications equipment, and fiber optics.

In these divisions the legislative and regulatory concerns of product manufacturers and the preparation of standards dealing with performance testing and compatibility are addressed. For example, modem and telematic standards, as well as much of the cellular standards technology, has been standardized in the United States under the mandate of TIA.

14.4.3 TTC

The third national committee to be addressed is the TTC in Japan. TTC was established in October 1985 to develop and disseminate Japanese domestic standards for deregulated technical items and protocols. It is a nongovernmental, nonprofit standards setting organization established to ensure fair and transparent standardization procedures.

TTC's primary emphasis is to develop, conduct studies and research, and disseminate protocols and standards for the connection of telecommunications networks. TTC is organized along six technical subcommittees that report to a board of directors through a technical assembly (see Fig. 14.4).

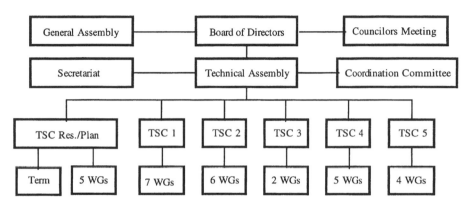

FIGURE 14.4 Organization of TTC.

The TTC organization comprises a general assembly, which is in charge of matters such as business plans and budgets. The councilors meeting examines standards development procedures in order to assure impartiality and clarity. The secretariat provides overall support to the organization; the technical assembly develops standards and handles technical matters including surveys and research. Each technical subcommittee is partitioned into two or more working groups (WG). The coordination committee handles all issues in or between the TSCs and WGs, and it assures the smooth running of all technical committee meetings.

Under the coordination committee, a subcommittee examines users' requests and studies their applicability to the five-year standardization-project plan. This subcommittee also conducts user-request surveys. The areas of involvement of each of the five subcommittees are shown in Table 14.6.

TTC membership is divided into four categories. Type I telecommunications carriers, that is, those carriers that own telecommunications circuits and facilities; type II telecom-

TABLE 14.6 TTC SubCommittees

Strategic Research and Planning Committee: Technical Survey and International Collaboration
TSC 1 Network-to-network interfaces, mobile communications
TSC 2 User-network interfaces
TSC 3 PBX, LAN
TSC 4 Higher level protocols
TSC 5 Voice and video signal coding scheme and systems

munications carriers, that is, those with telecommunications circuits leased from type I carriers; related equipment manufacturers; and others, including users.

Underlying objectives that guide TTC's approach to standards development are 1) to conform to international recommendations or standards; 2) standardize items, where either international recommendations or standards are not clear, or where national standards need to be set, and where a consensus is achieved; and 3) to conduct further studies into any of the items just mentioned whenever the technical assembly is unable to arrive at a consensus.

These objectives, which give highest priority in developing standards that are compatible with international recommendations or standards, have often driven TTC to adapt international standards for national use through the use of supplements that:

- Give guidelines for users of TTC standards on how to apply them
- Help clarify the contents of standards
- Help with the implementation of standards in terminal equipment and adaptors
- Assure interconnection between terminal equipment and adaptors
- Provide background information regarding the content of standards
- Assure interconnection.

These supplements also include questions and answers that help in implementing the standards, including encoding examples of various parameters and explanation of the practical meaning of a standard.

14.5   Intellectual Property

In the deregulating telecommunication arena patents have become increasingly more important. New ideas that are incorporated in standards often have global market potential and patent holders are seeking to obtain an income from their intellectual property as well as from products. In addition, the general effort to develop standards quickly places them closer to the leading edge of technology. There are some cases, for example speech encoding algorithms, where terms of reference for performance are typically set as objectives that no one can meet when the objectives are defined. The state of the art is being pushed by goals of the standards development organization. In this environment, incorporation of some intellectual property in standards is practically unavoidable.

With regard to intellectual property rights in the ITU, the TSB has developed a "code of practice" which may be summarized as follows.

The TSB requests members putting forth standards to draw the attention of the TSB to any known patent or patent pending application relevant to the developing standard. Where such information has been declared to the TSB, a log of registered patent holders for each affected recommendation is maintained for the convenience of users of ITU standards. If a recommendation, which is a nonbinding international standard, is developed and contains patented intellectual property there are three situations that may arise.

- The patent holder waives the rights and the recommendation is freely accessible to everybody.
- The patent holder will not waive the rights but is willing to negotiate licenses with other parties on a nondiscriminatory basis and on reasonable terms and conditions. What is reasonable is not defined, and the ITU-T will not participate in such negotiations.

- The patent holder is not willing to comply with either of the above two situations, in which case the ITU-T will not approve a recommendation containing such intellectual property.

The patent policy of the American National Standards Institute (ANSI), which governs all standards development organizations accredited by ANSI, is defined in ANSI procedures 1.2.11. It is similar to that of the ITU in that it requires a statement from patent holders or identified parties to indicate granting of a royalty-free license, willingness to license on reasonable and nondiscriminatory terms and conditions, or a disclaimer of no patent. Unlike the ITU, ANSI advises that is prepared to get involved in resolving disputes of what is considered "nondiscriminatory" and "reasonable." Additional information on ANSI patent guidelines can be found at `http://web.ansi.org/public/library/guides/ppguide.html`.

As mentioned earlier ETSI produces a combination of mandatory and voluntary standards. This can create additional complications when intellectual property issues are encapsulated within the standards. To formally address these issues an intellectual property rights committee defines ETSI's policy in the area of intellectual property.

Given the different patent policies adopted by various standards organizations, it is recommended that companies developing products based on standards investigate and understand the patent policy of the associated standards body and the patent statements filed regarding the standard being implemented.

14.6   Standards Coordination

The pace of technological advancements coupled with deregulation has given rise to increased global telecommunications standards activities. At the same time a growth of regional standards bodies has occurred which has increased the potential for duplication of work, wasting resources, and creating conflicting standards. This potentially adverse situation was addressed by a number of interregional telecommunications standardization conferences (ITSCs) that were held in the early 1990s. A global standards collaboration (GSC) group was established to oversee collaborative activities including electronic document handling (EDH) and five high-interest standards subjects:

- Broadband integrated services digital network (B-ISDN)
- Intelligent Networks (IN)
- Transmission management network (TMN)
- Universal personal telecommunications (UPT)
- Synchronous digital hierarchy/synchronous optical network (SDH/SONET)

This early activity was successful in avoiding duplication of effort and coordinating activities on these major standardization efforts. Today the level of cooperative activities, again driven by the pressure to avoid wasting valuable resources and reaching agreed standards more rapidly, are being driven to lower levels through the use of liaison statements between regional standards groups and permitting "documents of information" to flow between standards development organizations. The processes for this information flow are evolving and the electronic addresses provided at the end of this chapter should be consulted for the current interstandards organization communication mechanisms.

14.7   Scientific

Another global, scientifically based organization that has been particularly active in standards development (more recently emphasizing information processing) is the IEEE. Responsibility for standards adoption within the IEEE lies with the IEEE standards board. The board is supported by nine standing committees (see Fig. 14.5).

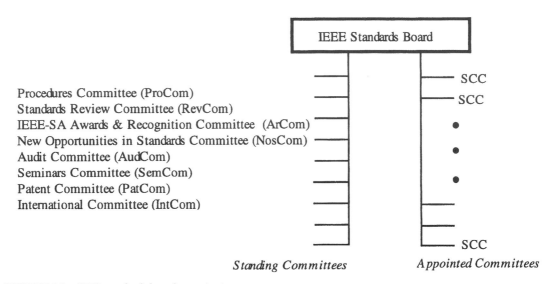

Procedures Committee (ProCom)
Standards Review Committee (RevCom)
IEEE-SA Awards & Recognition Committee (ArCom)
New Opportunities in Standards Committee (NosCom)
Audit Committee (AudCom)
Seminars Committee (SemCom)
Patent Committee (PatCom)
International Committee (IntCom)

Standing Committees   *Appointed Committees*

FIGURE 14.5   IEEE standards board organization.

Proposed standards are normally developed in the technical committees of the IEEE societies. There are occasions, however, when the scope of activity is too broad to be encompassed by a single society or where the societies are not able to do so for other reasons. In this case the standards board establishes its own standards developing committees, namely, the standards coordinating committees (**SCC**), to perform this function.

The adoption of IEEE standards is based on projects that have been approved by the IEEE standards board, while each project is the responsibility of a sponsor. Sponsors need not be an SCC, but can also include technical committees of IEEE societies; a standards, or standards coordinating committee of an IEEE Society; an accredited standards committee; or another organization approved by the IEEE standards board.

14.8   Standards Development Cycle

Although the manner in which standards are developed and approved somewhat varies between standards organizations, there are common characteristics to be found.

For most standards, first a set of requirements is defined. This may be done either by the standards committee actually developing the standard or by another entity in collaboration with such a committee. Subsequently, the technical details of a standard are developed. The actual entity developing a standard may be a member of the standards committee, or the actual standards committee itself. Outsiders may also contribute to standards development but, typically, only if sponsored by a committee member. Membership in the standards committee and the right to contribute technical information towards the development of the standard differs among the various standards' organizations, as indicated. This process is illustrated in Fig. 14.6.

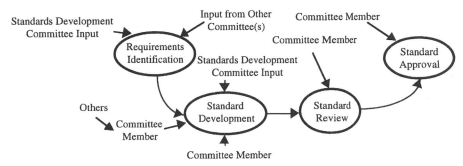

FIGURE 14.6 Typical standards development and approval process.

Finally, once the standard has been fully developed, it is placed under an approval cycle. Each standards setting body typically has precisely defined and often complex procedures for reviewing and then approving proposed standards, which although different in detail, are typically consensus driven.

Defining Terms

ANSI: The American National Standards Institute.

CCIR: The International Radio Consultative Committee, the predecessor of the ITU-R.

CCITT: The International Telephone and Telegraph Consultative Committee, the predecessor of the ITU-T.

CEPT: The European Conference for Post and Telecommunications, a predecessor of ETSI.

CITEL: Inter-American Telecommunications Commission, a standards setting body for the Americas.

ETS: A European (ETSI) technical standard.

ETSI: The European Telecommunications Standards Institute.

GSC: The Global Standards Collaboration group.

ICG: Intersector Coordination Group, a group which coordinates activities between the ITU-T and ITU-R.

IEC: The International Electrotechnical Commission.

IEEE: The Institute of Electrical and Electronics Engineers.

ISO: The International Standards Organization.

ITU: The International Telecommunications Union, an international treaty organization, which is part of the United Nations.

ITU-R: The radio communications sector of the ITU, the successor of the CCIR.

ITU-T: The standardization sector of the ITU, the successor of the CCITT.

JCG: The Joint Coordination Group, which oversees the coordination of common work between ITU-T study groups.

Recommendation: An ITU technical standard.

SCC: A standard's coordinating committee within the IEEE organization.

Standard: A publicly approved technical specification.

T1: An ANSI-approved standards body, which develops telecommunications standards in the United States.

T1AG: The primary advisory group within ANSI Committee T1 on Telecommunications.

TIA: The Telecommunications Industry Association, which is an ANSI-approved standards body that develops terminal equipment standards.

TTC: The Telecommunications Technology Committee, a Japanese standards setting body.

Further Information

[1] Irmer, T., Shaping future telecommunications: the challenge of global standardization, *IEEE Comm. Mag.*, 32(1), 20–28, 1994.

[2] Matute, M.A., CITEL: formulating telecommunications in the Americas. *IEEE Comm. Mag.*, 32(1), 38–39, 1994.

[3] Robin, G., The European perspective for telecommunications standards. *IEEE Comm. Mag.*, 32(1), 40–50, 1994.

[4] Reilly, A.K., A U.S. perspective on standards development. *IEEE Comm. Mag.*, 32(1), 30–36, 1994.

[5] Iida, T., Domestic standards in a changing world. *IEEE Comm. Mag.*, 32(1), 46–50, 1994.

[6] Habara, K., Cooperation in standardization. *IEEE Comm. Mag.*, 32(1), 78–84, 1994.

[7] *IEEE Standards Board Bylaws.* Institute of Electrical and Electronics Engineers. Dec. 1993.

[8] Chiarottino, W. and Pirani, G., International telecommunications standards organizations, *CSELT Tech. Repts.*, XXI(2), 207–236, 1993.

[9] ITU, Book No. 1. Resolutions; Recommendations on the organization of the work of ITU-T (series A); study groups and other groups; list of study questions (1993-1996). World Standardization Conf. Helsinki, 1–12, Mar. 1993.

[10] Standards Committee T1., *Telecommunications.* Procedures Manual. 7th Iss. Jun. 1992.

The standards' organizations often undergo structural and substantive changes. It is recommended that the following web sites be visited for the most updated information.

ANSI	http://www.ansi.org/
CITEL	http://www.oas.org
ETSI	http://www.etsi.org
IEC	http://www.iec.ch
IEEE	http://www.ieee.org
ISO	http://www.iso.ch
ITU	http://www.itu.ch
T1	http://www.t1.org
TIA	http://www.tia.org
TTC	http://www.ttc.or.jp

II

Wireless

15

Wireless Personal Communications: A Perspective

Donald C. Cox
Stanford University

{ This chapter has been updated using { } as indicators of inserts into the text of the original chapter of the same title that appeared in the first edition of this Handbook in 1996. }

15.1 Introduction

Wireless personal communications has captured the attention of the media and with it, the imagination of the public. Hardly a week goes by without one seeing an article on the subject appearing in a popular U.S. newspaper or magazine. Articles ranging from a short paragraph to many pages regularly appear in local newspapers, as well as in nationwide print media, e.g., *The Wall Street Journal, The New York Times, Business Week,* and *U.S. News and World Report.* Countless marketing surveys continue to project enormous demand, often projecting that at least half of the households, or half of the people, want wireless personal communications. Trade magazines, newsletters, conferences, and seminars on the subject by many different names have become too numerous to keep track of, and technical journals, magazines, conferences, and symposia continue to proliferate and to have ever increasing attendance and numbers of papers presented. It is clear that wireless personal communications is, by any measure, the fastest growing segment of telecommunications. { The explosive growth of wireless personal communications has continued unabated worldwide. Cellular and high-tier PCS pocketphones, pagers, and cordless telephones have become so common in many countries that few people even notice them anymore. These items have become an expected part of everyday life in most developed countries and in many developing countries around the world. }

If you look carefully at the seemingly endless discussions of the topic, however, you cannot help but note that they are often describing different things, i.e., different versions of wireless personal communications [29, 50]. Some discuss pagers, or messaging, or data systems, or access to the national information infrastructure, whereas others emphasize cellular radio, or cordless telephones, or dense systems of satellites. Many make reference to popular fiction entities such as Dick Tracy, Maxwell Smart, or *Star Trek.* { In addition to the things noted above, the topic of wireless loops [24], [30], [32] has also become popular in the widespread discussions of wireless communications. As discussed in [30], this topic includes several fixed wireless applications as well as the low-tier PCS application that was discussed originally under the wireless loop designation [24, 32]. The fixed wireless applications are aimed at reducing the cost of wireline loop-ends, i.e., the so-called "last mile" or "last km" of wireline telecommunications. }

Thus, it appears that almost everyone wants wireless personal communications, but *What is it?* There are many different ways to segment the complex topic into different communications applications, modes, functions, extent of coverage, or mobility [29, 30, 50]. The complexity of the issues has resulted in considerable confusion in the industry, as evidenced by the many different wireless systems, technologies, and services being offered, planned, or proposed. Many different industry groups and regulatory entities are becoming involved. The confusion is a natural consequence of the massive dislocations that are occurring, and will continue to occur, as we progress along this large change in the paradigm of the way we communicate. Among the different changes that are occurring in our communications paradigm, perhaps the major constituent is the change from wired fixed place-to-place communications to wireless mobile person-to-person communications. Within this major change are also many other changes, e.g., an increase in the significance of data and message communications, a perception of possible changes in video applications, and changes in the regulatory and political climates. { The fixed wireless loop applications noted earlier do not fit the new mobile communications paradigm. After many years of decline of fixed wireless communications applications, e.g., intercontinental HF radio and later satellites, point-to-point terrestrial microwave radio, and tropospheric scatter, it is interesting to see this rebirth of interest in fixed wireless applications. This rebirth is riding on the gigantic "wireless wave" resulting from the rapid public acceptance of mobile wireless communications. It will be interesting to observe this rebirth to see if communications history repeats; certainly mobility is wireless, but there is also considerable historical evidence that wireless is also mobility. }

This chapter attempts to identify different issues and to put many of the activities in wireless into a framework that can provide perspective on what is driving them, and perhaps even to yield some indication of where they appear to be going in the future. Like any attempt to categorize many complex interrelated issues, however, there are some that do not quite fit into neat categories, and so there will remain some dangling loose ends. Like any major paradigm shift, there will continue to be considerable confusion as many entities attempt to interpret the different needs and expectations associated with the new paradigm.

15.2 Background and Issues

15.2.1 Mobility and Freedom from Tethers

Perhaps the clearest constituents in all of the wireless personal communications activity are the desire for mobility in communications and the companion desire to be free from tethers, i.e., from physical connections to communications networks. These desires are clear from the very rapid growth of mobile technologies that provide primarily two-way voice services, even though economical wireline voice services are readily available. For example, cellular mobile radio has experienced rapid growth. Growth rates have been between 35 and 60% per year in the United States for a decade, with the total number of subscribers reaching 20 million by year-end 1994. The often neglected wireless companions to cellular radio, i.e., cordless telephones, have experienced even more rapid, but harder to quantify, growth with sales rates often exceeding 10 million sets a year in the United States, and with an estimated usage significantly exceeding 50 million in 1994. Telephones in airlines have also become commonplace. Similar or even greater growth in these wireless technologies has been experienced throughout the world. { The explosive growth in cellular and its identical companion, high-tier PCS, has continued to about 55 million subscribers in the U.S. at year-end 1997 and a similar number worldwide. In Sweden the penetration of cellular subscribers by 1997 was over one-third of the total population, i.e., the total including every man, woman, and child! And the growth has continued since. Similar penetrations of mobile wireless services are seen in some other developed nations, e.g., Japan. The growth in users of cordless telephones also has continued to the point that they have become the dominant subscriber terminal on wireline telephone loops in the U.S. It would appear that, taking into account cordless telephones and cellular and high-tier PCS phones, half of all telephone calls in the U.S. terminate with at least one end on a wireless device. }

{ Perhaps the most significant event in wireless personal communications since the writing of this original chapter was the widespread deployment and start of commercial service of personal handphone (PHS) in Japan in July of 1995 and its very rapid early acceptance by the consumer market [53]. By year-end 1996 there were 5 million PHS subscribers in Japan with the growth rate exceeding one-half million/month for some months. The PHS "phenomena" was one of the fastest adoptions of a new technology ever experienced. However, the PHS success story [41] peaked at a little over 7 million subscribers in 1997 and has declined slightly to a little under 7 million in mid-1998. This was the first mass deployment of a low-tier-like PCS technology (see later sections of this chapter), but PHS has some significant limitations. Perhaps the most significant limitation is the inability to successfully handoff at vehicular speeds. This handoff limitation is a result of the cumbersome radio link structure and control algorithms used to implement dynamic channel allocation (DCA) in PHS. DCA significantly increases channel occupancy (base station capacity) but incurs considerable complexity in implementing handoff. Another significant limitation of the PHS standard has been insufficient receiver sensitivity to permit "adequate" coverage from a "reasonably" dense deployment of base stations. These technology deficiencies coupled with heavy price cutting by the cellular service providers to compete with the rapid advancing of the PHS market were significant contributors to the leveling out of PHS growth. It is again evident, as with CT-2 phone point discussed in a later section, that low-tier PCS has very attractive features that can attract many subscribers, but it must also provide vehicle speed handoff and widespread coverage of highways as well as populated regions.

Others might point out the deployment and start of service of CDMA systems as a significant event since the first edition. However, the major significance of this CDMA activity is that it confirmed that CDMA performance was no better than other less-complex technologies and that those, including this author, who had been branded as "unbelieving skeptics" were correct in their assessments of the shortcomings of this technology. The overwhelming failure of CDMA technology to live up to the early claims for it can hardly be seen as a significant positive event in the evolution of wireless communication. It was, of course, a significant negative event. After years of struggling with the problems of this technology, service providers still have significantly fewer subscribers on CDMA worldwide than there are PHS subscribers in Japan alone! CDMA issues are discussed more in later sections dealing with technology issues. }

Paging and associated messaging, although not providing two-way voice, do provide a form of tetherless mobile communications to many subscribers worldwide. These services have also experienced significant growth { and have continued to grow since 1996. } There is even a glimmer of a market in the many different specialized wireless data applications evident in the many wireless local area network (WLAN) products on the market, the several wide area data services being offered, and the specialized satellite-based message services being provided to trucks on highways. { Wireless data technologies still have many supporters, but they still have fallen far short of the rapid deployment and growth of the more voice oriented wireless technologies. However, hope appears to be eternal in the wireless data arena. }

The topics discussed in the preceding two paragraphs indicate a dominant issue separating the different evolutions of wireless personal communications. That issue is the voice versus data communications issue that permeates all of communications today; this division also is very evident in fixed networks. The packet-oriented computer communications community and the circuit-oriented voice telecommunications (telephone) community hardly talk to each other and often speak different languages in addressing similar issues. Although they often converge to similar overall solutions at large scales (e.g., hierarchical routing with exceptions for embedded high-usage routes), the small-scale initial solutions are frequently quite different. Asynchronous transfer mode (ATM-) based networks are an attempt to integrate, at least partially, the needs of both the packet-data and circuit-oriented communities.

Superimposed on the voice-data issue is an issue of competing modes of communications that exist in both fixed and mobile forms. These different modes include the following.

Messaging is where the communication is not real time but is by way of message transmission, storage, and retrieval. This mode is represented by voice mail, electronic facsimile (fax), and electronic mail (e-mail), the latter of which appears to be a modern automated version of an evolution that includes telegraph and telex. Radio paging systems often provide limited one-way messaging, ranging from transmitting only the number of a calling party to longer alpha-numeric text messages.

Real-time two-way communications are represented by the telephone, cellular mobile radio telephone, and interactive text (and graphics) exchange over data networks. Two-way video phone always captures significant attention and fits into this mode; however, its benefit/cost ratio has yet to exceed a value that customers are willing to pay.

Paging, i.e., broadcast with no return channel, alerts a paged party that someone wants to communicate with him/her. Paging is like the ringer on a telephone without having the capability for completing the communications.

Agents are new high-level software applications or entities being incorporated into some computer networks. When launched into a data network, an agent is aimed at finding information by some title or characteristic and returning the information to the point from which the agent was launched. { The rapid growth of the worldwide web is based on this mode of communications. }

There are still other ways in which wireless communications have been segmented in attempts to optimize a technology to satisfy the needs of some particular group. Examples include 1) user location, which can be differentiated by indoors or outdoors, or on an airplane or a train and 2) degree of mobility, which can be differentiated either by speed, e.g., vehicular, pedestrian, or stationary, or by size of area throughout which communications are provided. { As noted earlier, wireless local loop with stationary terminals has become a major segment in the pursuit of wireless technology. }

At this point one should again ask; wireless personal communications—*What is it?* The evidence suggests that what is being sought by users, and produced by providers, can be categorized according to the following two main characteristics.

Communications portability and mobility on many different scales:

- Within a house or building [cordless telephone, (WLANs)]
- Within a campus, a town, or a city (cellular radio, WLANs, wide area wireless data, radio paging, extended cordless telephone)
- Throughout a state or region (cellular radio, wide area wireless data, radio paging, satellite-based wireless)
- Throughout a large country or continent (cellular radio, paging, satellite-based wireless)
- Throughout the world?

Communications by many different modes for many different applications:

- Two-way voice
- Data
- Messaging
- Video?

Thus, it is clear why wireless personal communications today is not one technology, not one system, and not one service but encompasses many technologies, systems, and services optimized for different applications.

15.3 Evolution of Technologies, Systems, and Services

Technologies and systems [27, 29, 30, 39, 50, 59, 67, 87], that are currently providing, or are proposed to provide, wireless communications services can be grouped into about seven relatively distinct groups, { the seven previous groups are still evident in the technology but with the addition of the fixed point-to-multipoint wireless loops there are now eight, } although there may be some disagreement on the group definitions, and in what group some particular technology or system belongs. All of the technologies and systems are evolving as technology advances and perceived needs change. Some trends are becoming evident in the evolutions. In this section, different groups and evolutionary trends are explored along with factors that influence the characteristics of members of the groups. The grouping is generally with respect to scale of mobility and communications applications or modes.

15.3.1 Cordless Telephones

Cordless telephones [29, 39, 50] generally can be categorized as providing low-mobility, low-power, two-way tetherless voice communications, with low mobility applying both to the range and the user's speed. Cordless telephones using analog radio technologies appeared in the late 1970s, and have experienced spectacular growth. They have evolved to digital radio technologies in the forms of second-generation cordless telephone (CT-2), and digital European cordless telephone (DECT) standards in Europe, and several different industrial scientific medical (ISM) band technologies in the United States.[1]

{ Personal handyphone (PHS) noted earlier and discussed in later sections and inserts can be considered either as a quite advanced digital cordless telephone similar to DECT or as a somewhat limited low-tier PCS technology. It has most of the attributes of similarity of the digital cordless telephones listed later in this section except that PHS uses $\pi/4$ QPSK modulation. }

Cordless telephones were originally aimed at providing economical, tetherless voice communications inside residences, i.e., at using a short wireless link to replace the cord between a telephone base unit and its handset. The most significant considerations in design compromises made for these technologies are to minimize total cost, while maximizing the talk time away from the battery charger. For digital cordless phones intended to be carried away from home in a pocket, e.g., CT-2 or DECT, handset weight and size are also major factors. These considerations drive designs toward minimizing complexity and minimizing the power used for signal processing and for transmitting.

Cordless telephones compete with wireline telephones. Therefore, high circuit quality has become a requirement. Early cordless sets had marginal quality. They were purchased by the millions, and discarded by the millions, until manufacturers produced higher-quality sets. Cordless telephones sales then exploded. Their usage has become commonplace, approaching, and perhaps exceeding, usage of corded telephones.

The compromises accepted in cordless telephone design in order to meet the cost, weight, and talk-time objectives are the following.

- Few users per megahertz
- Few users per base unit (many link together a particular handset and base unit)
- Large number of base units per unit area; one or more base units per wireline access line (in high-rise apartment buildings the density of base units is very large)
- Short transmission range

There is no added network complexity since a base unit looks to a telephone network like a wireline telephone. These issues are also discussed in [29, 50].

Digital cordless telephones in Europe have been evolving for a few years to extend their domain of use beyond the limits of inside residences. Cordless telephone, second generation, (CT-2) has evolved to provide telepoint or phone-point services. Base units are located in places where people congregate. e.g., along city streets and in shopping malls, train stations, etc. Handsets registered with the phone-point provider can place calls when within range of a telepoint. CT-2 does not provide capability for transferring (handing off) active wireless calls from one phone point to another if a user moves out of range of the one to which the

---

[1]These ISM technologies either use spread spectrum techniques (direct sequence or frequency hopping) or very low-transmitter power ($< \sim 1$ mW) as required by the ISM band regulations.

call was initiated. A CT-2+ technology, evolved from CT-2 and providing limited handoff capability, is being deployed in Canada. { CT-2+ deployment was never completed. } Phone-point service was introduced in the United Kingdom twice, but failed to attract enough customers to become a viable service. In Singapore and Hong Kong, however, CT-2 phone point has grown rapidly, reaching over 150,000 subscribers in Hong Kong [75] in mid-1994. The reasons for success in some places and failure in others are still being debated, but it is clear that the compactness of the Hong Kong and Singapore populations make the service more widely available, using fewer base stations than in more spreadout cities. Complaints of CT-2 phone-point users in trials have been that the radio coverage was not complete enough, and/or they could not tell whether there was coverage at a particular place, and the lack of handoff was inconvenient. In order to provide the alerting or ringing function for phone-point service, conventional radio pagers have been built into some CT-2 handsets. (The telephone network to which a CT-2 phone point is attached has no way of knowing from which base units to send a ringing message, even though the CT-2 handsets can be rung from a home base unit). { CT-2 phone points in Hong Kong peaked at about 170,000 subscribers. There was then a precipitous decline as these subscribers abandoned the limited service CT-2 in favor of high-tier PCS and cellular services which had reduced their prices to compete with phone points. Phone points have now been removed from Hong Kong. CT-2 phone points were also deployed in Korea, again with initial success followed by decline. The message is clear from the CT-2 and PHS experiences. The attributes of cordless phone-like low-tier PCS are very attractive, but need widespread coverage and vehicle speed handoff in order to be long term attractive and viable in the market. }

Another European evolution of cordless telephones is DECT, which was optimized for use inside buildings. Base units are attached through a controller to private branch exchanges (PBXs), key telephone systems, or phone company CENTREX telephone lines. DECT controllers can hand off active calls from one base unit to another as users move, and can page or ring handsets as a user walks through areas covered by different base units. { DECT has increased in deployment in some countries, notably Italy, for several applications including cordless telephone and wireless loop. There are reported to be perhaps 7 million or more DECT users in Europe. }

These cordless telephone evolutions to more widespread usage outside and inside with telepoints and to usage inside large buildings are illustrated in Fig. 15.1, along with the integration of paging into handsets to provide alerting for phone-point services. They represent the first attempts to increase the service area of mobility for low-power cordless telephones.

Some of the characteristics of the digital cordless telephone technologies, CT-2 and DECT, are listed in Table 15.1. Additional information can be found in [32, 50]. Even though there are significant differences between these technologies, e.g., multiple access technology [frequency division multiple access (FDMA) or time division multiple access (TDMA)/FDMA], and channel bit rate, there are many similarities that are fundamental to the design objectives discussed earlier and to a user's perception of them. These similarities and their implications are as follows.

32-kb/s Adaptive Differential Pulse Code Modulation (ADPCM) Digital Speech Encoding: This is a low-complexity (low-signal processing power) speech encoding process that provides wireline speech quality and is an international standard.

Average Transmitter Power ≤ 10 mW: This permits many hours of talk time with small, low-cost, lightweight batteries, but provides limited radio range.

Low-Complexity Radio Signal Processing: There is no forward error correction and no complex multipath mitigation (i.e., no equalization or spread spectrum).

TABLE 15.1   Wireless PCS Technologies

	High-Power Systems				Low-Power Systems			
	Digital Cellular (High-Tier PCS)				Low-Tier PCS		Digital Cordless	
System	IS-54	IS-95 (DS)	GSM	DCS-1800	WACS/PACS	Handi-Phone	DECT	CT-2
Multiple Access	TDMA/FDMA	CDMA/FDMA	TDMA/FDMA	TDMA/FDMA	TDMA/FDMA	TDMA/FDMA	TDMA/FDMA	FDMA
Freq. band, MHz Uplink, MHz Downlink, MHz	869–894 824–849 (USA)	869–894 824–849 (USA)	935–960 890–915 (Eur.)	1710–1785 1805–1880 (UK)	Emerg. Tech.* (USA)	1895–1907 (Japan)	1880–1990 (Eur.)	864–868 (Eur. and Asia)
RF ch. spacing Downlink, KHz Uplink, KHz	30 30	1250 1250	200 200	200 200	300 300	300	1728	100
Modulation	π/4 DQPSK	BPSK/QPSK	GMSK	GMSK	π/4 QPSK	π/4 DQPSK	GFSK	GFSK
Portable txmit Power, max./avg.	600 mW/ 200 mW	600 mW	1 W/ 125 mW	1 W/ 125 mW	200 mW/ 25 mW	80 mW/ 10 mW	250 mW/ 10 mW	10 mW/ 5 mW
Speech coding	VSELP	QCELP	RPE-LTP	RPE-LTP	ADPCM	ADPCM	ADPCM	ADPCM
Speech rate, kb/s	7.95	8 (var.)	13	13	32/16/8	32	32	32
Speech ch./RF ch.	3	—	8	8	8/16/32	4	12	1
Ch. Bit rate, kb/s Uplink, kb/s Downlink, kb/s	48.6 48.6		270.833 270.833	270.833 270.833	384 384	384	1152	72
Ch. coding	1/2 rate conv.	1/2 rate fwd. 1/3 rate rev.	1/2 rate conv.	1/2 rate conv.	CRC	CRC	CRC (control)	None
Frame, ms	40	20	4.615	4.615	2.5	5	10	2

* Spectrum is 1.85–2.2 GHz allocated by the FCC for emerging technologies; DS is direct sequence.

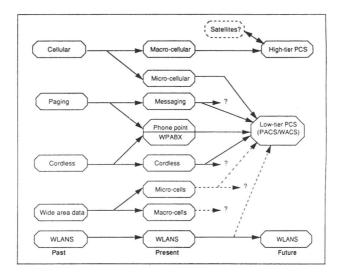

FIGURE 15.1

Low Transmission Delay, e.g., < 50 ms, and for CT-2 < 10-ms Round Trip: This is a speech-quality and network-complexity issue. A maximum of 10 ms should be allowed, taking into account additional inevitable delay in long-distance networks. Echo cancellation is generally required for delays > 10 ms.

Simple Frequency-Shift Modulation and Noncoherent Detection: Although still being low in complexity, the slightly more complex 4QPSK modulation with coherent detection provides significantly more spectrum efficiency, range, and interference immunity.

Dynamic Channel Allocation: Although this technique has potential for improved system capacity, the cordless-telephone implementations do not take full advantage of this feature for handoff and, thus, cannot reap the full benefit for moving users [15, 19].

Time Division Duplex (TDD): This technique permits the use of a single contiguous frequency band and implementation of diversity from one end of a radio link. Unless all base station transmissions are synchronized in time, however, it can incur severe cochannel interference penalties in outside environments [15, 16]. Of course, for cordless telephones used inside with base stations not having a propagation advantage, this is not a problem. Also, for small indoor PBX networks, synchronization of base station transmission is easier than is synchronization throughout a widespread outdoor network, which can have many adjacent base stations connected to different geographic locations for central control and switching.

15.3.2   Cellular Mobile Radio Systems

Cellular mobile radio systems are becoming known in the United States as high-tier personal communications service (PCS), particularly when implemented in the new 1.9-GHz PCS bands [20]. These systems generally can be categorized as providing high-mobility, wide-ranging, two-way tetherless voice communications. In these systems, high mobility refers to vehicular speeds, and also to widespread regional to nationwide coverage [27, 29, 50]. Mobile radio has been evolving for over 50 years. Cellular radio integrates wireless access with large-scale networks having sophisticated intelligence to manage mobility of users.

Cellular radio was designed to provide voice service to wide-ranging vehicles on streets and highways [29, 39, 50, 82], and generally uses transmitter power on the order of 100 times that of cordless telephones ($\approx$ 2 W for cellular). Thus, cellular systems can only provide reduced service to handheld sets that are disadvantaged by using somewhat lower transmitter power ($<$ 0.5 W) and less efficient antennas than vehicular sets. Handheld sets used inside buildings have the further disadvantage of attenuation through walls that is not taken into account in system design.

Cellular radio or high-tier PCS has experienced large growth as noted earlier. In spite of the limitations on usage of handheld sets already noted, handheld cellular sets have become very popular, with their sales becoming comparable to the sales of vehicular sets. Frequent complaints from handheld cellular users are that batteries are too large and heavy, and both talk time and standby time are inadequate. { Cellular and high-tier PCS pocket handsets have continued to decrease in size and weight and more efficient lithium batteries have been incorporated. This has increased their attractiveness (more on this in the later section "Reality Check"). For several years there have been many more pocket handsets sold than vehicular mounted sets every year. However, despite the improvements in these handsets and batteries, the complaints of weight and limited talk time still persist. The electronics have become essentially weightless compared to the batteries required for these high-tier PCS and cellular handsets. }

Cellular radio at 800 MHz has evolved to digital radio technologies [29, 39, 50] in the forms of the deployed systems standards

- Global Standard for Mobile (GSM) in Europe
- Japanese or personal digital cellular (JDC or PDC) in Japan
- U.S. TDMA digital cellular known as USDC or IS-54.

and in the form of the code division multiple access (CDMA) standard, IS-95, which is under development but not yet deployed. { Since the first edition was published, CDMA systems have been deployed in the U.S., Korea, Hong Kong, and other countries after many months (years) of redesign, reprogramming, and adjustment. These CDMA issues are discussed later in the section "New Technology." }

The most significant consideration in the design compromises made for the U.S. digital cellular or high-tier PCS systems was the high cost of cell sites (base stations). A figure often quoted is U.S. $1 million for a cell site. This consideration drove digital system designs to maximize users per megahertz and to maximize the users per cell site.

Because of the need to cover highways running through low-population-density regions between cities, the relatively high transmitter power requirement was retained to provide maximum range from high antenna locations.

Compromises that were accepted while maximizing the two just cited parameters are as follows.

- High transmitter power consumption.
- High user-set complexity, and thus high signal-processing power consumption.
- Low circuit quality.
- High network complexity, e.g., the new IS-95 technology will require complex new switching and control equipment in the network, as well as high-complexity wireless-access technology.

Cellular radio or high-tier PCS has also been evolving for a few years in a different direction, toward very small coverage areas or microcells. This evolution provides increased capacity in areas having high user density, as well as improved coverage of shadowed areas. Some microcell base stations are being installed inside, in conference center lobbies and similar places of high user concentrations. Of course, microcells, also permit lower transmitter power that conserves battery power when power control is implemented, and base stations inside buildings circumvent the outside wall attenuation. Low-complexity microcell base stations also are considerably less expensive than conventional cell sites, perhaps two orders of magnitude less expensive. Thus, the use of microcell base stations provides large increases in overall system capacity, while also reducing the cost per available radio channel and the battery drain on portable subscriber equipment. This microcell evolution, illustrated in Fig. 15.1, moves handheld cellular sets in a direction similar to that of the expanded-coverage evolution of cordless telephones to phone points and wireless PBX.

Some of the characteristics of digital-cellular or high-tier PCS technologies are listed in Table 15.1 for IS-54, IS-95, and GSM at 900 MHz, and DCS-1800, which is GSM at 1800 MHz. { The technology listed here as IS-54 has also become known as IS-136 having more sophisticated digital control channels. These technologies, IS-54/IS-136 are also sometimes known as DAMPS (i.e., Digital AMPS), as U.S. TDMA or North American TDMA, or sometimes just as "TDMA." } Additional information can be found in [29, 39, 50]. The JDC or PDC technology, not listed, is similar to IS-54. As with the digital cordless technologies, there are significant differences among these cellular technologies, e.g., modulation type, multiple access technology, and channel bit rate. There are also many similarities, however, that are fundamental to the design objectives discussed earlier. These similarities and their implications are as follows.

Low Bit-Rate Speech Coding $\leq$13 kb/s with Some $\leq$8 kb/s: Low bit-rate speech coding obviously increases the number of users per megahertz and per cell site. However, it also significantly reduces speech quality [29], and does not permit speech encodings in tandem while traversing a network; see also the section on Other Issues later in this chapter.

Some Implementations Make Use of Speech Inactivity: This further increases the number of users per cell site, i.e., the cell-site capacity. It also further reduces speech quality [29], however, because of the difficulty of detecting the onset of speech. This problem is even worse in an acoustically noisy environment like an automobile.

High Transmission Delay; $\approx$200-ms Round Trip: This is another important circuit-quality issue. Such large delay is about the same as one-way transmission through a synchronous-orbit communications satellite. A voice circuit with digital cellular technology on both ends will experience the delay of a full satellite circuit. It should be recalled that one reason long-distance circuits have been removed from satellites and put onto fiber-optic cable is because customers find the delay to be objectionable. This delay in digital cellular technology results from both computation for speech bit-rate reduction and from complex signal processing, e.g., bit interleaving, error correction decoding, and multipath mitigation [equalization or spread spectrum code division multiple access (CDMA)].

High-Complexity Signal Processing, Both for Speech Encoding and for Demodulation: Signal processing has been allowed to grow without bound and is about a factor of 10 greater than that used in the low-complexity digital cordless telephones [29]. Since several watts are required from a battery to produce the high transmitter power in a cellular or high-tier PCS set, signal-processing power is not as significant as it is in the low-power cordless telephones; see also the section on Complexity/Coverage Area Comparisons later in this chapter.

Fixed Channel Allocation: The difficulties associated with implementing capacity-increasing dynamic channel allocation to work with handoff [15, 19] have impeded its adoption in systems requiring reliable and frequent handoff.

Frequency Division Duplex (FDD): Cellular systems have already been allocated paired-frequency bands suitable for FDD. Thus, the network or system complexity required for providing synchronized transmissions [15, 16] from all cell sites for TDD has not been embraced in these digital cellular systems. Note that TDD has not been employed in IS-95 even though such synchronization is required for other reasons.

Mobile/Portable Set Power Control: The benefits of increased capacity from lower over-all cochannel interference and reduced battery drain have been sought by incorporating power control in the digital cellular technologies.

15.3.3   Wide-Area Wireless Data Systems

Existing wide area data systems generally can be categorized as providing high mobility, wide-ranging, low-data-rate digital data communications to both vehicles and pedestrians [29, 50]. These systems have not experienced the rapid growth that the two-way voice technologies have, even though they have been deployed in many cities for a few years and have established a base of customers in several countries. Examples of these packet data systems are shown in Table 15.2.

TABLE 15.2   Wide-Area Wireless Packet Data Systems

	CDPD[1]	RAM Mobile (Mobitex)	ARDIS[2] (KDT)	Metricom (MDN)[3]
Data rate, kb/s	19.2	8 (19.2)	4.8 (19.2)	76
Modulation	GMSK BT = 0.5	GMSK	GMSK	GMSK
Frequency, MHz	800	900	800	915
Chan. spacing, kHz	30	12.5	25	160
Status	1994 service	Full service	Full service	In service
Access means	Unused AMPS channels	Slotted Aloha CSMA		FH SS (ISM)
Transmit power, W			40	1

Note: Data in parentheses (   ) indicates proposed.
[1]Cellular Digital Packet Data
[2]Advanced Radio Data Information Service
[3]Microcellular Data Network

The earliest and best known of these systems in the United States are the ARDIS network developed and run by Motorola, and the RAM mobile data network based on Ericsson Mobitex Technology. These technologies were designed to make use of standard, two-way voice, land mobile-radio channels, with 12.5- or 25-kHz channel spacing. In the United States these are specialized mobile radio services (SMRS) allocations around 450 MHz and 900 MHz. Initially, the data rates were low: 4.8 kb/s for ARDIS and 8 kb/s for RAM. The systems use high transmitter power (several tens of watts) to cover large regions from a few base stations having high antennas. The relatively low data capacity of a relatively expensive base station has resulted in economics that have not favored rapid growth.

The wide-area mobile data systems also are evolving in several different directions in an attempt to improve base station capacity, economics, and the attractiveness of the service. The technologies used in both the ARDIS and RAM networks are evolving to higher channel bit rates of 19.2 kb/s.

The cellular carriers and several manufacturers in the United States are developing and deploying a new wide area packet data network as an overlay to the cellular radio networks. This cellular digital packet data (CDPD) technology shares the 30-kHz spaced 800-MHz voice channels used by the analog FM advanced mobile phone service (AMPS) systems. Data rate is 19.2 kb/s. The CDPD base station equipment also shares cell sites with the voice cellular radio system. The aim is to reduce the cost of providing packet data service by sharing the costs of base stations with the better established and higher cell-site capacity cellular systems. This is a strategy similar to that used by nationwide fixed wireline packet data networks that could not provide an economically viable data service if they did not share costs by leasing a small amount of the capacity of the interexchange networks that are paid for largely by voice traffic. { CDPD has been deployed in many U.S. cities for several years. However, it has not lived up to early expectations and has become "just another" wireless data service with some subscribers, but not with the large growth envisioned earlier. }

Another evolutionary path in wide-area wireless packet data networks is toward smaller coverage areas or microcells. This evolutionary path also is indicated on Fig. 15.1. The microcell data networks are aimed at stationary or low-speed users. The design compromises are aimed at reducing service costs by making very small and inexpensive base stations that can be attached to utility poles, the sides of buildings and inside buildings and can be widely distributed throughout a region. Base-station-to-base-station wireless links are used to reduce the cost of the interconnecting data network. In one network this decreases the overall capacity to serve users, since it uses the same radio channels that are used to provide service. Capacity is expected to be made up by increasing the number of base stations that have connections to a fixed-distribution network as service demand increases. Another such network uses other dedicated radio channels to interconnect base stations. In the high-capacity limit, these networks will look more like a conventional cellular network architecture, with closely spaced, small, inexpensive base stations, i.e., microcells, connected to a fixed infrastructure. Specialized wireless data networks have been built to provide metering and control of electric power distributions, e.g., Celldata and Metricom in California.

A large microcell network of small inexpensive base stations has been installed in the lower San Francisco Bay Area by Metricom, and public packet-data service was offered during early 1994. Most of the small (shoe-box size) base stations are mounted on street light poles. Reliable data rates are about 75 kb/s. The technology is based on slow frequency-hopped spread spectrum in the 902–928 MHz U.S. ISM band. Transmitter power is 1 W maximum, and power control is used to minimize interference and maximize battery life time. { The metricom network has been improved and significantly expanded in the San Francisco Bay Area and has been deployed in Washington, D.C. and a few other places in the U.S. However, like all wireless data services so far, it has failed to grow as rapidly or to attract as many subscribers as was originally expected. Wireless data overall has had only very limited success compared to that of the more voice-oriented technologies, systems, and services. }

15.3.4   High-Speed Wireless Local-Area Networks (WLANs)

Wireless local-area data networks can be categorized as providing low-mobility high-data-rate data communications within a confined region, e.g., a campus or a large building. Coverage range from a wireless data terminal is short, tens to hundreds of feet, like cordless telephones. Coverage is limited to within a room or to several rooms in a building. WLANs have been evolving for a few years, but overall the situation is chaotic, with many different products being offered by many different vendors [29, 59]. There is no stable definition of the needs or design objectives for WLANs, with data rates ranging from hundreds of kb/s to more than 10 Mb/s, and with several products providing one or two Mb/s wireless link rates. The best description of the WLAN evolutionary process is: having severe birth pains. An IEEE standards committee, 802.11, has been attempting to put some order into this topic, but their success has been somewhat limited. A partial list of some advertised products is given in Table 15.3. Users of WLANs are not nearly as numerous as the users of more voice-oriented wireless systems. Part of the difficulty stems from these systems being driven by the computer industry that views the wireless system as just another plug-in interface card, without giving sufficient consideration to the vagaries and needs of a reliable radio system. { This section still describes the WLAN situation in spite of some attempts at standards in the U.S. and Europe, and continuing industry efforts. Some of the products in Table 15.3 have been discontinued because of lack of market and some new products have been offered, but the manufacturers still continue to struggle to find enough customers to support their efforts. Optimism remains high in the WLAN community that "eventually" they will find the "right" technology, service, or application to make WLANs "take off" — but the world still waits. Success is still quite limited. }

There are two overall network architectures pursued by WLAN designers. One is a centrally coordinated and controlled network that resembles other wireless systems. There are base stations in these networks that exercise overall control over channel access [44].

The other type of network architecture is the self-organizing and distributed controlled network where every terminal has the same function as every other terminal, and networks are formed ad hoc by communications exchanges among terminals. Such ad hoc networks are more like citizen band (CB) radio networks, with similar expected limitations if they were ever to become very widespread.

Nearly all WLANs in the United States have attempted to use one of the ISM frequency bands for unlicensed operation under part 15 of the FCC rules. These bands are 902–928 MHz, 2400–2483.5 MHz, and 5725–5850 MHz, and they require users to accept interference from any interfering source that may also be using the frequency. The use of ISM bands has further handicapped WLAN development because of the requirement for use of either frequency hopping or direct sequence spread spectrum as an access technology, if transmitter power is to be adequate to cover more than a few feet. One exception to the ISM band implementations is the Motorola ALTAIR, which operates in a licensed band at 18 GHz. { It appears that ALTAIR has been discontinued because of the limited market. } The technical and economic challenges of operation at 18 GHz have hampered the adoption of this 10–15 Mb/s technology. The frequency-spectrum constraints have been improved in the United States with the recent FCC allocation of spectrum from 1910–1930 MHz for unlicensed data PCS applications. Use of this new spectrum requires implementation of an access etiquette incorporating listen before transmit in an attempt to provide some coordination of an otherwise potentially chaotic, uncontrolled environment [68]. Also, since spread spectrum is not a requirement, access technologies and multipath mitigation techniques more compatible with the needs of packet-data transmission [59], e.g., multipath equalization or multicarrier transmission can be incorporated into new WLAN designs. { The FCC is allocating spectrum at 5 GHz for wideband wireless data for internet and next generation data network access, BUT it remains to be seen whether this initiative is any more successful than past wireless data attempts. Optimism is again high, BUT... }

Three other widely different WLAN activities also need mentioning. One is a large European Telecommunications Standards Institute (ETSI) activity to produce a standard for high performance radio local area network (HIPERLAN), a 20-Mb/s WLAN technology to operate near 5 GHz. Other activities are large U.S. Advance Research Projects Agency- (ARPA-) sponsored, WLAN research projects at the Universities of California at Berkeley (UCB), and at Los Angeles (UCLA). The UCB Infopad project is based on a coordinated network architecture with fixed coordinating nodes and direct-sequence spread spectrum

TABLE 15.3 Partial List of WLAN Products

Product Company Location	Freq., MHz	Link Rate, Mb/s	User Rate	Protocol(s)	Access	No. of chan. or Spread Factor	Mod./ Coding	Power, mW	Network Topology
Altair Plus Motorola Arlington Hts, IL	18–19 GHz	15	5.7 Mb/s	Ethernet			4-level FSK	25 peak	Eight devices/radio; radio to base to ethernet
WaveLAN NCR/AT&T Dayton, OH	902–928	2	1.6 Mb/s	Ethernet-like	DS SS		DQPSK	250	Peer-to-peer
AirLan Solectek San Diego, CA	902–928		2 Mb/s	Ethernet	DS SS		DQPSK	250	PCMCIA w/ant.; radio to hub
Freeport Windata Inc. Northboro, MA	902–928	16	5.7 Mb/s	Ethernet	DS SS	32 chips/bit	16 PSK trellis coding	650	Hub
Intersect Persoft Inc. Madison, WI	902–928		2 Mb/s	Ethernet token ring	DS SS		DQPSK	250	Hub
LAWN O'Neill Comm. Horsham, PA	902–928		38.4 kb/s	AX.25	SS	20 users/chan.; max. 4 chan.		20	Peer-to-peer
WILAN Wi-LAN Inc. Calgary, Alberta	902–928	20	1.5 Mb/s/ chan.	Ethernet, token ring	CDMA/ TDMA	3 chan. 10–15 links each	unconventional	30	Peer-to-peer
RadioPort ALPS Electric USA	902–928		242 kb/s	Ethernet	SS	7/3 channels		100	Peer-to-peer
ArLAN 600 Telesys. SLW Don Mills, Ont.	902–928; 2.4 GHz		1.35 Mb/s	Ethernet	SS			1 W max	PCs with ant.; radio to hub
Radio Link Cal. Microwave Sunnyvale, CA	902–928; 2.4 GHz	250 kb/s	64 kb/s		FH SS	250 ms/hop 500 kHz space			Hub
Range LAN Proxim, Inc. Mountain View, CA	902–928		242 kb/s	Ethernet, token ring	DS SS	3 chan.		100	
RangeLAN 2 Proxim, Inc. Mountain View, CA	2.4 GHz	1.6	50 kb/s max.	Ethernet, token ring	FH SS	10 chan. at 5 kb/s; 15 sub-ch. each		100	Peer-to-peer bridge
Netwave Xircom Calabasas, CA	2.4 GHz	1/adaptor		Ethernet, token ring	FH SS	82 1-MHz chn. or "hops"			Hub
Freelink Cabletron Sys. Rochester, NH	2.4 and 5.8 GHz		5.7 Mb/s	Ethernet	DS SS	32 chips/bit	16 PSK trellis coding	100	Hub

(CDMA), whereas, the UCLA project is aimed at peer-to-peer networks and uses frequency hopping. Both ARPA sponsored projects are concentrated on the 900-MHz ISM band.

As computers shrink in size from desktop to laptop to palmtop, mobility in data network access is becoming more important to the user. This fact, coupled with the availability of more usable frequency spectrum, and perhaps some progress on standards, may speed the evolution and adoption of wireless mobile access to WLANs. From the large number of companies making products, it is obvious that many believe in the future of this market. { It should be noted that the objective for 10 MB/s data service with widespread coverage from a sparse distribution of widely separated base stations equivalent to cellular is unrealistic and unrealizable. This can be readily seen by considering a simple example. Consider a cellular coverage area that requires full cellular power of 0.5 watt to cover from a handset. Consider the handset to use a typical digital cellular bit rate of about 10 kb/s (perhaps 8 kb/s speech coding + overhead). With all else in the system the same, e.g., antennas, antenna height, receiver noise figure, detection sensitivity, etc., the 10 MB/s data would require 10 MB/s ÷ 10 kb/s = 1000 times as much power as the 10 kb/s cellular. Thus, it would require $0.5 \times 1000 = 500$ watts for the wireless data transmitter. This is a totally unrealistic situation. If the data system operates at a higher frequency (e.g., 5 GHz) than the cellular system (e.g., 1 or 2 GHz) then there will be even more power required to overcome the additional loss at a higher frequency. The sometimes expressed desire by the wireless data community for a system to provide network access to users in and around buildings and to provide 10 MB/s over 10 miles with 10 milliwatts of transmitter power and costing $10.00 is totally impossible. It requires violation of the "laws of physics." }

15.3.5 Paging/Messaging Systems

Radio paging began many years ago as a one-bit messaging system. The one bit was: some one wants to communicate with you. More generally, paging can be categorized as one-way messaging over wide areas. The one-way radio link is optimized to take advantage of the asymmetry. High transmitter power (hundreds of watts to kilowatts), and high antennas at the fixed base stations permit low-complexity, very low-power-consumption, pocket paging receivers that provide long usage time from small batteries. This combination provides the large radio-link margins needed to penetrate walls of buildings without burdening the user set battery. Paging has experienced steady rapid growth for many years and serves about 15 million subscribers in the United States.

Paging also has evolved in several different directions. It has changed from analog tone coding for user identification to digitally encoded messages. It has evolved from the 1-b message, someone wants you, to multibit messages from, first, the calling party's telephone number to, now, short e-mail text messages. This evolution is noted in Fig. 15.1.

The region over which a page is transmitted has also increased from 1) local, around one transmitting antenna; to 2) regional, from multiple widely-dispersed antennas; to 3) nationwide, from large networks of interconnected paging transmitters. The integration of paging with CT-2 user sets for phone-point call alerting was noted previously.

Another evolutionary paging route sometimes proposed is two-way paging. This is an ambiguous and unrealizable concept, however, since the requirement for two-way communications destroys the asymmetrical link advantage so well exploited by paging. Two-way paging puts a transmitter in the user's set and brings along with it all of the design compromises that must be faced in such a two-way radio system. Thus, the word paging is not appropriate to describe a system that provides two-way communications. { The two-way paging situation is as unrealistic as that noted earlier for wide-area, high-speed, low-power wireless data. This can be seen by looking at the asymmetry situation in paging. In order to achieve comparable coverage uplink and downlink, a 500-watt paging transmitter downlink advantage must be overcome in the uplink. Even considering the relatively high cellular handset transmit power levels on the order of 0.5 watt results in a factor of 1000 disadvantage, and 0.5 watt is completely incompatible with the low power paging receiver power requirements. If the same uplink and downlink coverage is required for an equivalent set of system parameters, then the only variable left to work with is bandwidth. If the paging link bit rate is taken to be 10 kb/sec (much higher than many paging systems), then the usable uplink rate is 10 kb/s/1000 = 10 B/s, an unusably low rate. Of course, some uplink benefit can be gained because of better base station receiver noise figure and by using forward error correction and perhaps ARQ. However, this is unlikely to raise the allowable rate to greater than 100 B/s which even though likely overoptimistic is still unrealistically low and we have assumed an unrealistically high transmit power in the two-way "pager!" }

15.3.6   Satellite-Based Mobile Systems

Satellite-based mobile systems are the epitome of wide-area coverage, expensive base station systems. They generally can be categorized as providing two-way (or one-way) limited quality voice and/or very limited data or messaging to very wide-ranging vehicles (or fixed locations). These systems can provide very widespread, often global, coverage, e.g., to ships at sea by INMARSAT. There are a few messaging systems in operation, e.g., to trucks on highways in the United States by Qualcomm's Omnitracs system.

A few large-scale mobile satellite systems have been proposed and are being pursued: perhaps the best known is Motorola's Iridium; others include Odyssey, Globalstar, and Teledesic. The strength of satellite systems is their ability to provide large regional or global coverage to users outside buildings. However, it is very difficult to provide adequate link margin to cover inside buildings, or even to cover locations shadowed by buildings, trees, or mountains. A satellite system's weakness is also its large coverage area. It is very difficult to provide from Earth orbit the small coverage cells that are necessary for providing high overall systems capacity from frequency reuse. This fact, coupled with the high cost of the orbital base stations, results in low capacity along with the wide overall coverage but also in expensive service. Thus, satellite systems are not likely to compete favorably with terrestrial systems in populated areas or even along well-traveled highways. They can complement terrestrial cellular or PCS systems in low-population-density areas. It remains to be seen whether there will be enough users with enough money in low-population-density regions of the world to make satellite mobile systems economically viable. { Some of the mobile satellite systems have been withdrawn, e.g., Odyssey. Some satellites in the Iridium and Globalstar systems have been launched. The industry will soon find out whether these systems are economically viable. }

Proposed satellite systems range from 1) low-Earth-orbit systems (LEOS) having tens to hundreds of satellites through 2) intermediate- or medium-height systems (MEOS) to 3) geostationary or geosynchronous orbit systems (GEOS) having fewer than ten satellites. LEOS require more, but less expensive, satellites to cover the Earth, but they can more easily produce smaller coverage areas and, thus, provide higher capacity within a given spectrum allocation. Also, their transmission delay is significantly less (perhaps two orders of magnitude!), providing higher quality voice links, as discussed previously. On the other hand, GEOS require only a few, somewhat more expensive, satellites (perhaps only three) and are likely to provide lower capacity within a given spectrum allocation and suffer severe transmission-delay impairment on the order of 0.5 s. Of course, MEOS fall in between these extremes. The possible evolution of satellite systems to complement high-tier PCS is indicated in Fig. 15.1.

15.3.7 {Fixed Point-to-Multipoint Wireless Loops

Wideband point-to-multipoint wireless loop technologies sometimes have been referred to earlier as "wireless cable" when they were proposed as an approach for providing interactive video services to homes [30]. However, as the video application started to appear less attractive, the application emphasis shifted to providing wideband data access for the internet, the worldwide web, and future wideband data networks. Potentially lower costs are the motivation for this wireless application. As such, these technologies will have to compete with existing coaxial cable and fiber/coax distribution by CATV companies, with satellites, and with fiber and fiber/coax systems being installed or proposed by telephone companies and other entities [30]. Another competitor is asymmetric digital subscriber line technology, which uses advanced digital signal processing to provide high-bandwidth digital distribution over twisted copper wire pairs.

In the U.S. two widely different frequency bands are being pursued for fixed point-to-multipoint wireless loops. These bands are at 28 GHz for local multipoint distribution systems or services (LMDS) [52] and 2.5 to 2.7 GHz for microwave or metropolitan distribution systems (MMDS) [74]. The goal of low-cost fixed wireless loops is based on the low cost of point-to-multipoint line-of-sight wireless technology. However, significant challenges are presented by the inevitable blockage by trees, terrain, and houses, and by buildings in heavily built-up residential areas. Attenuation in rainstorms presents an additional problem at 28 GHz in some localities. Even at the 2.5-GHz MMDS frequencies, the large bandwidth required for distribution of many video channels presents a challenge to provide adequate radio-link margin over obstructed paths. From mobile satellite investigations it is known that trees can often produce over 15 dB additional path attenuation [38]. Studies of blockage by buildings in cities have shown that it is difficult to have line-of-sight access to more than 60% of the buildings from a single base station [55]. Measurements in a region in Brooklyn, NY [60], suggest that access from a single base station can range from 25% to 85% for subscriber antenna heights of 10 to 35 ft and a base station height of about 290 ft. While less blockage by houses could be expected in residential areas, such numbers would suggest that greater than 90% access to houses could be difficult, even from multiple elevated locations, when mixes of one- and two-story houses, trees, and hills are present. In regions where tree cover is heavy, e.g., the eastern half of the U.S., tree cover in many places will present a significant obstacle. Heavy rainfall is an additional problem at 28 GHz in some regions. In spite of these challenges, the lure of low-cost wireless loops is attracting many participants, both service providers and equipment manufacturers. }

15.3.8 Reality Check

Before we go on to consider other applications and compromises, perhaps it would be helpful to see if there is any indication that the previous discussion is valid. For this check, we could look at cordless telephones for telepoint use (i.e., pocketphones) and at pocket cellular telephones that existed in the 1993 time frame.

Two products from one United States manufacturer are good for this comparison. One is a third-generation hand-portable analog FM cellular phone from this manufacturer that represents their second generation of pocketphones. The other is a first-generation digital cordless phone built to the United Kingdom CT-2 common air interface (CAI) standard. Both units are of flip phone type with the earpiece on the main handset body and the mouthpiece formed by or on the flip-down part. Both operate near 900 MHz and have 1/4 wavelength pull-out antennas. Both are fully functional within their class of operation (i.e., full number of U.S. cellular channels, full number of CT-2 channels, automatic channel setup, etc.) Table 15.4 compares characteristics of these two wireless access pocketphones from the same manufacturer.

TABLE 15.4   Comparison of CT-2 and Cellular Pocket Size Flip-Phones from the Same Manufacturer

Characteristics/Parameter	CT-2	Cellular
Weight, oz		
Flip phone only	5.2	4.2
Battery[1] only	1.9	3.6
Total unit	7.1	7.8
Size (max.dimensions), in		
Flip phone only	$5.9 \times 2.2 \times 0.95$ 8.5 in^3	$5.5 \times 2.4 \times 0.9$ —
Battery[1] only	$1.9 \times 1.3 \times 0.5$ internal	$4.7 \times 2.3 \times 0.4$ external
Total unit	$5.9 \times 2.2 \times 0.95$ 8.5 in^3	$5.9 \times 2.4 \times 1.1$ 11.6 in^3
Talk-time, min (h)		
Rechargeable battery[2]	180 (3)	45
Nonrechargeable battery	600 (10)	N/A
Standby time, h		
Rechargeable battery	30	8
Nonrechargeable battery	100	N/A
Speech quality	32 kb/s telephone quality	30 kHz FM depends on channel quality
Transmit power avg., W	0.005	0.5

[1] Rechargeable battery.
[2] Ni-cad battery.

The following are the most important items to note in the Table 15.4 comparison.

1. The talk time of the low-power pocketphone is four times that of the high-power pocket-phone.

2. The battery inside the low-power pocketphone is about one-half the weight and size of the battery attached to the high-power pocketphone.

3. The battery-usage ratio, talk time/weight of battery, is eight times greater, almost an order of magnitude, for the low-power pocketphone compared to the high-power pocketphone!

4. Additionally, the lower power (5 mW) digital cordless pocketphone is slightly smaller and lighter than the high-power (500 mW) analog FM cellular mobile pocketphone.

{ Similar comparisons can be made between PHS advanced cordless/low-tier PCS phones and advanced cellular/high-tier PCS pocketphones. New lithium batteries have permitted increased talk time in pocketphones. Digital control/paging channels facilitate significantly extended standby time. Advances in solid-state circuits have reduced the size and weight of cellular pocketphone electronics so that they are almost insignificant compared to the battery required for the high power transmitter and complex digital signal processing. However, even with all these changes, there is still a very significant weight and talk time benefit in the low complexity PHS handsets compared to the most advanced cellular/high-tier PCS handsets. Picking typical minimum size and weight handsets for both technologies results in the following comparisons.

	PHS	Cellular
weight, oz		
total unit	3	4.5
size		
total unit	4.2 in^3	—
talk-time, h	8	3
standby time, h	600	48

From the table, the battery usage ratio has been reduced to a factor of about 4 from a factor of 8, but this is based on total weight, not battery weight alone as used for the earlier CT-2 and cellular comparison.

Thus, there is still a large talk time and weight benefit for low-power low-complexity low-tier PCS compared to higher power high-complexity, high tier PCS and cellular. }

The following should also be noted.

1. The room for technology improvement of the CT-2 cordless phone is greater since it is first generation and the cellular phone is second/third generation.

2. A digital cellular phone built to the IS-54, GSM, or JDC standard, or in the proposed United States CDMA technology, would either have less talk time or be heavier and larger than the analog FM phone, because: a) the low-bit-rate digital speech coder is more complex and will consume more power than the analog speech processing circuits; b) the digital units have complex digital signal-processing circuits for forward error correction—either for delay dispersion equalizing or for spread-spectrum processing—that will consume significant amounts of power and that have no equivalents in the analog FM unit; and c) power amplifiers for the shaped-pulse nonconstant-envelope digital signals will be less efficient than the amplifiers for constant-envelope analog FM. Although it may be suggested that transmitter power control will reduce the weight and size of a CDMA handset and battery, if that handset is to be capable of operating at full power in fringe areas, it will have to have capabilities similar to other cellular sets. Similar power control applied to a CT-2-like low-maximum-power set would also reduce its power consumption and thus also its weight and size.

The major difference in size, weight, and talk time between the two pocketphones is directly attributable to the two orders of magnitude difference in average transmitter power. The generation of transmitter power dominates power consumption in the analog cellular phone. Power consumption in the digital CT-2 phone is more evenly divided between transmitter-power generation and digital signal processing. Therefore, power consumption in complex digital signal processing would have more impact on talk time in small low-power personal communicators than in cellular handsets where the transmitter-power generation is so large. Other than reducing power consumption for both functions, the only alternative for increasing talk time and reducing battery weight is to invent new battery technology having greater density; see section on Other Issues later in this chapter.

In contrast, lowering the transmitter power requirement, modestly applying digital signal processing, and shifting some of the radio coverage burden to a higher density of small, low-power, low-complexity, low-cost fixed radio ports has the effect of shifting some of the talk time, weight, and cost constraints from battery technology to solid state electronics technology, which continues to experience orders-of-magnitude improvements in the span of several years. Digital signal-processing complexity, however, cannot be permitted to overwhelm power consumption in low-power handsets; whereas small differences in complexity will not matter much, orders-of-magnitude differences in complexity will continue to be significant.

Thus, it can be seen from Table 15.4 that the size, weight, and quality arguments in the preceding sections generally hold for these examples. It also is evident from the preceding paragraphs that they will be even more notable when comparing digital cordless pocketphones with digital cellular pocketphones of the same development generations.

15.4 Evolution Toward the Future and to Low-Tier Personal Communications Services

After looking at the evolution of several wireless technologies and systems in the preceding sections it appears appropriate to ask again: wireless personal communications, What is it? All of the technologies in the preceding sections claim to provide wireless personal communications, and all do to some extent. All have significant limitations, however, and all are evolving in attempts to overcome the limitations. It seems appropriate to ask, what are the likely endpoints? Perhaps some hint of the endpoints can be found by exploring what users see as limitations of existing technologies and systems and by looking at the evolutionary trends.

In order to do so, we summarize some important clues from the preceding sections and project them, along with some U.S. standards activity, toward the future.

Digital Cordless Telephones

- Strengths: good circuit quality; long talk time; small lightweight battery; low-cost sets and service.
- Limitations: limited range; limited usage regions.
- Evolutionary trends: phone points in public places; wireless PBX in business.
- Remaining limitations and issues: limited usage regions and coverage holes; limited or no handoff; limited range.

{ Experience with PHS and CT-2 phone point have provided more emphasis on the need for vehicle speed handoff and continuous widespread coverage of populated areas and of highways in between. }

Digital Cellular Pocket Handsets

- Strength: widespread service availability.
- Limitations: limited talk time; large heavy batteries; high-cost sets and service; marginal circuit quality; holes in coverage and poor in-building coverage; limited data capabilities; complex technologies.
- Evolutionary trends: microcells to increase capacity and in-building coverage and to reduce battery drain; satellite systems to extend coverage.
- Remaining limitations and issues: limited talk time and large battery; marginal circuit quality; complex technologies.

Wide Area Data

- Strength: digital messages.
- Limitations: no voice, limited data rate; high cost.
- Evolutionary trends: microcells to increase capacity and reduce cost; share facilities with voice systems to reduce cost.

- Remaining limitations and issues: no voice; limited capacity.

Wireless Local Area Networks (WLANs)

- Strength: high data rate.
- Limitations: insufficient capacity for voice, limited coverage; no standards; chaos.
- Evolutionary trends: hard to discern from all of the churning.

Paging/messaging

- Strengths: widespread coverage; long battery life; small lightweight sets and batteries; economical.
- Limitations: one-way message only; limited capacity.
- Evolutionary desire: two-way messaging and/or voice; capacity.
- Limitations and issues: two-way link cannot exploit the advantages of one-way link asymmetry.

{Fixed Wireless Loops

- Strength: High data rates.
- Limitations: no mobility. }

There is a strong trajectory evident in these systems and technologies aimed at providing the following features.

High Quality Voice and Data

- To small, lightweight, pocket carried communicators.
- Having small lightweight batteries.
- Having long talk time and long standby battery life.
- Providing service over large coverage regions.
- For pedestrians in populated areas (but not requiring high population density).
- Including low to moderate speed mobility with handoff. { It has become evident from the experience with PHS and CT-2 phone point that vehicle speed handoff is essential so that handsets can be used in vehicles also. }

Economical Service

- Low subscriber-set cost.
- Low network-service cost.

Privacy and Security of Communications

- Encrypted radio links.

This trajectory is evident in all of the evolving technologies but can only be partially satisfied by any of the existing and evolving systems and technologies! Trajectories from all of the evolving technologies and systems are illustrated in Fig. 15.1 as being aimed at low-tier personal communications systems or services, i.e., low-tier PCS. Taking characteristics from cordless, cellular, wide-area data and, at least moderate-rate, WLANs, suggests the following attributes for this low-tier PCS.

1. 32 kb/s ADPCM speech encoding in the near future to take advantage of the low complexity and low power consumption, and to provide low-delay high-quality speech.

2. Flexible radio link architecture that will support multiple data rates from several kilobits per second. This is needed to permit evolution in the future to lower bit rate speech as technology improvements permit high quality without excessive power consumption or transmission delay and to provide multiple data rates for data transmission and messaging.

3. Low transmitter power (≤ 25 mW average) with adaptive power control to maximize talk time and data transmission time. This incurs short radio range that requires many base stations to cover a large region. Thus, base stations must be small and inexpensive, like cordless telephone phone points or the Metricom wireless data base stations. { The lower power will require somewhat closer spacing of base stations in cluttered environments with many buildings, etc. This issue is dealt with in more detail in Section 15.5. The issues associated with greater base station spacing along highways are also considered in Section 15.5. }

4. Low-complexity signal processing to minimize power consumption. Complexity one-tenth that of digital cellular or high-tier PCS technologies is required [29]. With only several tens of milliwatts (or less under power control) required for transmitter power, signal processing power becomes significant.

5. Low cochannel interference and high coverage area design criteria. In order to provide high-quality service over a large region, at least 99% of any covered area must receive good or better coverage and be below acceptable cochannel interference limits. This implies less than 1% of a region will receive marginal service. This is an order-of-magnitude higher service requirement than the 10% of a region permitted to receive marginal service in vehicular cellular system (high-tier PCS) design criteria.

6. Four-level phase modulation with coherent detection to maximize radio link performance and capacity with low complexity.

7. Frequency division duplexing to relax the requirement for synchronizing base station transmissions over a large region. { PHS uses time division duplexing and requires base station synchronization. In first deployments, one provider did not implement this synchronization. The expected serious performance degradation prompted system upgrades to provide the needed synchronization. While this is not a big issue, it does add complexity to the system and decreases the overall robustness. }

8. { As noted previously, experience with PHS and CT-2 phone point have emphasized the need for vehicular speed handoff in these low-tier PCS systems. Such handoff is readily implemented in PACS and has been demonstrated in the field [51]. This issue is discussed in more detail later in this section. }

Such technologies and systems have been designed, prototyped, and laboratory and field tested and evaluated for several years [7, 23, 24, 25, 26, 27, 28, 29, 31, 32, 50]. The viewpoint expressed here is consistent with the progress in the Joint Technical Committee (JTC) of the U.S. standards bodies, Telecommunications Industry Association (TIA) and Committee T1 of the Alliance for Telecommunications Industry Solutions (ATIS). Many technologies and systems were submitted to the JTC for consideration for wireless PCS in the new 1.9-GHz frequency bands for use in the United States [20]. Essentially all of the technologies and systems listed in Table 15.1, and some others, were submitted in late 1993. It was evident that there were at least two and perhaps three distinctly different classes of submissions. No systems optimized for packet data were submitted, but some of the technologies are optimized for voice.

One class of submissions was the group labeled high-power systems, digital cellular (high-tier PCS) in Table 15.1. These are the technologies discussed previously in this chapter. They are highly optimized for low-bit-rate voice and, therefore, have somewhat limited capability for serving packet-data applications. Since it is clear that wireless services to wide ranging high-speed mobiles will continue to be needed, and that the technology already described for low-tier PCS may not be optimum for such services, Fig. 15.1 shows a continuing evolution and need in the future for high-tier PCS systems that are the equivalent of today's cellular radio. There are more than 100 million vehicles in the United States alone. In the future, most, if not all, of these will be equipped with high-tier cellular mobile phones. Therefore, there will be a continuing and rapidly expanding market for high-tier systems.

Another class of submissions to the JTC [20] included the Japanese personal handyphone system (PHS) and a technology and system originally developed at Bellcore but carried forward to prototypes and submitted to the JTC by Motorola and Hughes Network Systems. This system was known as wireless access communications systems (WACS).[2] These two submissions were so similar in their design objectives and system characteristics that, with the agreement of the delegations from Japan and the United States, the PHS and WACS submissions were combined under a new name, personal access communication systems (PACS), that was to incorporate the best features of both. This advanced, low-power wireless access system, PACS, was to be known as low-tier PCS. Both WACS/PACS and Handyphone (PHS) are shown in Table 15.1 as low-tier PCS and represent the evolution to low-tier PCS in Fig. 15.1. The WACS/PACS/ UDPC system and technology are discussed in [7, 23, 24, 25, 26, 28, 29, 31, 32, 50].

In the JTC, submissions for PCS of DECT and CT-2 and their variations were also lumped under the class of low-tier PCS, even though these advanced digital cordless telephone technologies were somewhat more limited in their ability to serve all of the low-tier PCS needs. They are included under digital cordless technologies in Table 15.1. Other technologies and systems were also submitted to the JTC for high-tier and low-tier applications, but they have not received widespread industry support.

One wireless access application discussed earlier that is not addressed by either high-tier or low-tier PCS is the high-speed WLAN application. Specialized high-speed WLANs also are likely to find a place in the future. Therefore, their evolution is also continued in Fig. 15.1. The figure also recognizes that widespread low-tier PCS can support data at several hundred kilobits per second and, thus, can satisfy many of the needs of WLAN users.

It is not clear what the future roles are for paging/messaging, cordless telephone appliances, or wide-area packet-data networks in an environment with widespread contiguous coverage by low-tier and high-tier PCS. Thus, their extensions into the future are indicated with a question mark in Fig. 15.1.

Those who may object to the separation of wireless PCS into high-tier and low-tier should review this section again, and note that we have two tiers of PCS now. On the voice side there is cellular radio, i.e., high-tier PCS, and cordless telephone, i.e., an early form of low-tier PCS. On the data side there is wide-area data, i.e., high-tier data PCS, and WLANs,

---

[2]WACS was known previously as Universal Digital Portable Communications (UDPC).

i.e., perhaps a form of low-tier data PCS. In their evolutions, these all have the trajectories discussed and shown in Fig. 15.1 that point surely toward low-tier PCS. It is this low-tier PCS that marketing studies continue to project is wanted by more than half of the U.S. households or by half of the people, a potential market of over 100 million subscribers in the United States alone. Similar projections have been made worldwide.

{ PACS technology [6] has been prototyped by several manufacturers. In 1995 field demonstrations were run in Boulder, CO at a U.S. West test site using radio ports (base stations) and radio port control units made by NEC. "Handset" prototypes made by Motorola and Panasonic were trialed. The handsets and ports were brought together for the first time in Boulder. The highly successful trial demonstrated the ease of integrating the subsystems of the low-complexity PACS technology and the overall advantages of PACS from a user's perspective as noted throughout this chapter. Effective vehicular speed operation was demonstrated in these tests. Also, Hughes Network Systems (HNS) has developed and tested many sets of PACS infrastructure technology with different handsets in several settings and has many times demonstrated highly reliable high vehicular speed (in excess of 70 mi/hr) operation and handoff among several radio ports. Motorola also has demonstrated PACS equipment in several settings at vehicular speeds as well as for wireless loop applications. Highly successful demonstrations of PACS prototypes have been conducted from Alaska to Florida, from New York to California, and in China and elsewhere.

A PACS deployment in China using NEC equipment started to provide service in 1998. The U.S. Service Provider, 21st Century Telesis, is poised to begin a PACS deployment in several states in the U.S. using infrastructure equipment from IINS and handsets and switching equipment from different suppliers. Perhaps, with a little more support of these deployments, the public will finally be able to obtain the benefits of low-tier PCS. }

15.5 Comparisons with Other Technologies

15.5.1 Complexity/Coverage Area Comparisons

Experimental research prototypes of radio ports and subscriber sets [64, 66] have been constructed to demonstrate the technical feasibility of the radio link requirements in [7]. These WACS prototypes generally have the characteristics and parameters previously noted, with the exceptions that 1) the portable transmitter power is lower (10 mW average, 100 mW peak), 2) dynamic power control and automatic time slot transfer are not implemented, and 3) a rudimentary automatic link-transfer implementation is based only on received power. The experimental base stations transmit near 2.17 GHz; the experimental subscriber sets transmit near 2.12 GHz. Both operated under a Bellcore experimental license. The experimental prototypes incorporate application-specific, very large-scale integrated circuits[3] fabricated to demonstrate the feasibility of the low-complexity high-performance digital signal-processing techniques [63, 64] for symbol timing and coherent bit detection. These techniques permit the efficient short TDMA bursts having only 100 b that are necessary for low-delay TDMA implementations. Other digital signal-processing functions in the prototypes are implemented in programmable logic devices. All of the digital signal-processing functions combined require about 1/10 of the logic gates that are required for digital signal processing in vehicular digital cellular mobile implementations [42, 62, 63]; that is, this low-complexity PCS implementation having no delay-dispersion-compensating circuits and

---

[3]Applications specific integrated circuits (ASIC), very large-scale integration (VLSI).

no forward error-correction decoding and is about 1/10 as complex as the digital cellular implementations that include these functions.[4] The 32 kb/s ADPCM speech-encoding in the low-complexity PCS implementation is also about 1/10 as complex as the less than 10-kb/s speech encoding used in digital cellular implementations. This significantly lower complexity will continue to translate into lower power consumption and cost. It is particularly important for low-power pocket personal communicators with power control in which the DC power expended for radio frequency transmitting can be only tens of milliwatts for significant lengths of time.

The experimental radio links have been tested in the laboratory for detection sensitivity [bit error rate (BER) vs SNR] [18, 61, 66] and for performance against cochannel interference [1] and intersymbol interference caused by multipath delay spread [66]. These laboratory tests confirm the performance of the radio-link techniques. In addition to the laboratory tests, qualitative tests have been made in several PCS environments to compare these experimental prototypes with several United States CT-1 cordless telephones at 50 MHz, with CT-2 cordless telephones at 900 MHz, and with DCT-900 cordless telephones at 900 MHz. Some of these comparisons have been reported [8, 71, 84, 85]. In general, depending on the criteria, e.g., either no degradation or limited degradation of circuit quality, these WACS experimental prototypes covered areas inside buildings that ranged from 1.4 to 4 times the areas covered by the other technologies. The coverage areas for the experimental prototypes were always substantially limited in two or three directions by the outside walls of the buildings. These area factors could be expected to be even larger if the coverage were not limited by walls, i.e., once all of a building is covered in one direction, no more area can be covered no matter what the radio link margin. The earlier comparisons [8, 84, 85] were made with only two-branch uplink diversity before subscriber-set transmitting antenna switching was implemented and, with only one radio port before automatic radio-link transfer was implemented. The later tests [71] included these implementations. These reported comparisons agree with similar unreported comparisons made in a Bellcore Laboratory building. Similar coverage comparison results have been noted for a 900-MHz ISM-band cordless telephone compared to the 2-GHz experimental prototype. The area coverage factors (e.g., ×1.4 to ×4) could be expected to be even greater if the cordless technologies had also been operated at 2 GHz since attenuation inside buildings between similar small antennas is about 7 dB greater at 2 GHz than at 900 MHz [35, 36] and the 900 MHz handsets transmitted only 3 dB less average power than the 2-GHz experimental prototypes. The greater area coverage demonstrated for this technology is expected because of the different compromises noted earlier; the following, in particular.

1. Coherent detection of QAM provides more detection sensitivity than noncoherent detection of frequency-shift modulations [17].

2. Antenna diversity mitigates bursts of errors from multipath fading [66].

3. Error detection and blanking of TDMA bursts having errors significantly improves perceived speech quality [72]. (Undetected errors in the most significant bit cause sharp audio pops that seriously degrade perceived speech quality.)

4. Robust symbol timing and burst and frame synchronization reduce the number of frames in error due to imperfect timing and synchronization [66].

---

[4]Some indication of VLSI complexity can be seen by the number of people required to design the circuits. For the low-complexity TDMA ASIC set, only one person part time plus a student part time were required; the complex CDMA ASIC has six authors on the paper alone.

5. Transmitting more power from the radio port compared to the subscriber set offsets the less sensitive subscriber set receiver compared to the port receiver that results from power and complexity compromises made in a portable set.

Of course, as expected, the low-power (10-mW) radio links cover less area than high-power (0.5-W) cellular mobile pocketphone radio links because of the 17-dB transmitter power difference resulting from the compromises discussed previously. In the case of vehicular mounted sets, even more radio-link advantage accrues to the mobile set because of the higher gain of vehicle-mounted antennas and higher transmitter power (3 W).

{ The power difference between a low-tier PACS handset and a high-tier PCS or cellular pocket handset is not as significant in limiting range as is often portrayed. Other differences in deployment scenarios for low-tier and high-tier systems are as large or even larger factors, e.g., base station antenna height and antenna gain. This can be seen by considering using the same antennas and receiver noise figures at base stations and looking at the range of high-tier and low-tier handsets. High-tier handsets typically transmit a maximum average power of 0.5 watt. The PACS handset average transmit power is 25 milliwatts (peak power is higher for TDMA, but equal comparisons can be made considering average power and equivalent receiver sensitivities). This power ratio of $\times 20$ translates to approximately a range reduction of a factor of about 0.5 for an environment with a distance dependence of $1/(d)^4$ or a factor of about 0.4 for a $1/(d)^{3.5}$ environment. These represent typical values of distance dependence for PCS and cellular environments. Thus, if the high-tier handset would provide a range of 5 miles in some environment, the low-tier handset would provide a range of 2 to 2.5 miles in the same environment, if the base station antennas and receiver noise figures were the same. This difference in range is no greater than the difference in range between high-tier PCS handsets used in 1.9 GHz systems and cellular handsets with the same power used in 800 MHz systems. Range factors are discussed further in the next section. }

15.5.2 {Coverage, Range, Speed, and Environments

Interest has been expressed in having greater range for low-tier PCS technology for low-population-density areas. One should first note that the range of a wireless link is highly dependent on the amount of clutter or obstructions in the environment in which it is operated. For example, radio link calculations that result in a 1400-ft base station (radio-port) separation at 1.9 GHz contain over 50-dB margin for shadowing from obstructions and multipath effects [25, 37]. Thus, in an environment without obstructions, e.g., along a highway, the base station separation can be increased at least by a factor of 4 to over a mile, i.e., 25 dB for an attenuation characteristic of d^{-4}, while providing the same quality of service, without any changes to the base station or subscriber transceivers, and while still allowing over 25-dB margin for multipath and some shadowing. This remaining margin allows for operation of a handset inside an automobile. In such an unobstructed environment, multipath RMS delay spread [21, 33] will still be less than the 0.5 μs in which PACS was designed to operate [28].

Operation at still greater range along unobstructed highways or at a range of a mile along more obstructed streets can be obtained in several ways. Additional link gain of 6 dB can be obtained by going from omnidirectional antennas at base stations to 90° sectored antennas (four sectors). Another 6 dB can be obtained by raising base station antennas by a factor of 2 from 27 ft to 55 ft in height. This additional 12 dB will allow another factor of 2 increase in range to 2-mile base station separation along highways, or to about 3000-ft separation in residential areas. Even higher-gain and taller antennas could be used to concentrate coverage along highways, particularly in rural areas. Of course, range could be further increased by increasing the power transmitted.

As the range of the low-tier PACS technology is extended in cluttered areas by increasing link gain, increased RMS delay spread is likely to be encountered. This will require increasing complexity in receivers. A factor of 2 in tolerance of delay spread can be obtained by interference-canceling signal combining [76, 77, 78, 79, 80] from two antennas instead of the simpler selection diversity combining originally used in PACS. This will provide adequate delay-spread tolerance for most suburban environments [21, 33].

The PACS downlink contains synchronization words that could be used to train a conventional delay-spread equalizer in subscriber set receivers. Constant-modulus (blind) equalization will provide greater tolerance to delay spread in base station receivers on the uplink [45, 46, 47, 48] than can be obtained by interference-cancellation combining from only two antennas. The use of more base-station antennas and receivers can also help mitigate uplink delay spread. Thus, with some added complexity, the low-tier PACS technology can work effectively in the RMS delay spreads expected in cluttered environments for base station separations of 2 miles or so.

The guard time in the PACS TDMA uplink is adequate for 1-mile range, i.e., 2-mile separation between base station and subscriber transceivers. A separation of up to 3 miles between transceivers could be allowed if some statistical outage were accepted for the few times when adjacent uplink timeslots are occupied by subscribers at the extremes of range (near–far). With some added complexity in assigning timeslots, the assignment of subscribers at very different ranges to adjacent timeslots could be avoided, and the base station separation could be increased to several miles without incurring adjacent slot interference. A simple alternative in low-density (rural) areas, where lower capacity could be acceptable and greater range could be desirable, would be to use every other timeslot to ensure adequate guard time for range differences of many tens of miles. Also, the capability of transmitter time advance has been added to PACS standard in order to increase the range of operation. Such time advance is applied in the cellular TDMA technologies.

The synchronization, carrier recovery, and detection in the low-complexity PACS transceivers will perform well at highway speeds. The two-receiver diversity used in uplink transceivers also will perform well at highway speeds. The performance of the single-receiver selection diversity used in the low-complexity PACS downlink transceivers begins to deteriorate at speeds above about 30 mi/h. However, at any speed, the performance is always at least as good as that of a single transceiver without the low-complexity diversity. Also, fading in the relatively uncluttered environment of a highway is likely to have a less severe Ricean distribution, so diversity will be less needed for mitigating the fading. Cellular handsets do not have diversity. Of course, more complex two-receiver diversity could be added to downlink transceivers to provide two-branch diversity performance at highway speeds. It should be noted that the very short 2.5-ms TDMA frames incorporated into PACS to provide low transmission delay (for high speech quality) also make the technology significantly less sensitive to high-speed fading than the longer-frame-period cellular technologies. The short frame also facilitates the rapid coordination needed to make reliable high-speed handoffs between base stations. Measurements on radio links to potential handoff base stations can be made rapidly, i.e., a measurement on at least one radio link every 2.5 ms. Once a handoff decision is made, signalling exchanges every 2.5 ms ensure that the radio link handoff is completed quickly. In contrast, the long frame periods in the high-tier (cellular) technologies prolong the time it takes to complete a handoff. As noted earlier, high speed handoff has been demonstrated many times with PACS technology and at speeds over 70 mi/hr. }

15.6 Quality, Capacity, and Economic Issues

Although the several trajectories toward low-tier PCS discussed in the preceding section are clear, it does not fit the existing wireless communications paradigms. Thus, low-tier PCS has attracted less attention than the systems and technologies that are compatible with the existing paradigms. Some examples are cited in the following paragraphs.

The need for intense interaction with an intelligent network infrastructure in order to manage mobility is not compatible with the cordless telephone appliance paradigm. In that paradigm, independence of network intelligence and base units that mimic wireline telephones are paramount.

Wireless data systems often do not admit to the dominance of wireless voice communications and, thus, do not take advantage of the economics of sharing network infrastructure and base station equipment. Also, wireless voice systems often do not recognize the importance of data and messaging and, thus, only add them in as bandaids to systems.

The need for a dense collection of many low-complexity, low-cost, low-tier PCS base stations interconnected with inexpensive fixed-network facilities (copper or fiber based) does not fit the cellular high-tier paradigm that expects sparsely distributed $1 million cell sites. Also, the need for high transmission quality to compete with wireline telephones is not compatible with the drive toward maximizing users-per-cell-site and per megahertz to minimize the number of expensive cell sites. These concerns, of course, ignore the hallmark of frequency-reusing cellular systems. That hallmark is the production of almost unlimited overall system capacity by reducing the separation between base stations. The cellular paradigm does not recognize the fact that almost all houses in the U.S. have inexpensive copper wires connecting telephones to the telephone network. The use of low-tier PCS base stations that concentrate individual user services before backhauling in the network will result in less fixed interconnecting facilities than exist now for wireline telephones. Thus, inexpensive techniques for interconnecting many low-tier base stations are already deployed to provide wireline telephones to almost all houses. { The cost of backhaul to many base stations (radio ports) in a low-tier system is often cited as an economic disadvantage that cannot be overcome. However, this perception is based on existing tariffs for T1 digital lines which are excessive considering current digital subscriber line technology. These tariffs were established many years ago when digital subscriber line electronics were very expensive. With modern low-cost high-rate digital subscriber line (HDSL) electronics, the cost of backhaul could be greatly reduced. If efforts were made to revise tariffs for digital line backhaul based on low cost electronics and copper loops like residential loops, the resulting backhaul costs would more nearly approach the cost of residential telephone lines. As it is now, backhaul costs are calculated based on antiquated high T1 line tariffs that were established for "antique" high cost electronics. }

This list could be extended, but the preceding examples are sufficient, along with the earlier sections of the paper, to indicate the many complex interactions among circuit quality, spectrum utilization, complexity (circuit and network), system capacity, and economics that are involved in the design compromises for a large, high-capacity wireless-access system. Unfortunately, the tendency has been to ignore many of the issues and focus on only one, e.g., the focus on cell site capacity that drove the development of digital-cellular high-tier systems in the United States. Interactions among circuit quality, complexity, capacity, and economics are considered in the following sections.

15.6.1 Capacity, Quality, and Complexity

Although capacity comparisons frequently are made without regard to circuit quality, complexity, or cost per base station, such comparisons are not meaningful. An example in Table 15.5 compares capacity factors for U.S. cellular or high-tier PCS technologies with the low-tier PCS technology, PACS/WACS. The mean opinion scores (MOS) (noted in Table 15.5) for speech coding are discussed later. Detection of speech activity and turning off the transmitter during times of no activity is implemented in IS-95. Its impact on MOS also is noted later. A similar technique has been proposed as E-TDMA for use with IS-54 and is discussed with respect to TDMA system in [29]. Note that the use of low-bit-rate speech coding combined with speech activity degrades the high-tier system's quality by nearly one full MOS point on the five-point MOS scale when compared to 32 kb/s ADPCM. Tandem encoding is discussed in a later section. These speech quality degrading factors alone provide a base station capacity increasing factor of $\times 4 \times 2.5 = \times 10$ over the high-speech-quality low-tier system! Speech coding, of course, directly affects base station capacity and, thus, overall system capacity by its effect on the number of speech channels that can fit into a given bandwidth.

TABLE 15.5 Comparison of Cellular (IS-54/IS-95) and Low-Tier PCS (WACS/PACS). Capacity Comparisons Made without Regard to Quality Factors, Complexity, and Cost per Base Station Are not Meaningful

Parameter	Cellular (High-Tier)	Low-Tier PCS	Capacity Factor
Speech Coding, kb/s	8 (MOS 3.4) No tandem coding	32 (MOS 4.1) 3 or 4 tandem	$\times 4$
Speech activity	Yes (MOS 3.2)	No (MOS 4.1)	$\times 2.5$
Percentage of good areas, %	90	99	$\times 2$
Propagation σ, dB	8	10	$\times 1.5$
Total: trading quality for capacity			$\times 30$

The allowance of extra system margin to provide coverage of 99% of an area for low-tier PCS versus 90% coverage for high-tier is discussed in the previous section and [29]. This additional quality factor costs a capacity factor of $\times 2$. The last item in Table 15.5 does not change the actual system, but only changes the way that frequency reuse is calculated. The additional 2-dB margin in standard deviation σ, allowed for coverage into houses and small buildings for low-tier PCS, costs yet another factor of $\times 1.5$ in calculation only. Frequency reuse factors affect the number of sets of frequencies required and, thus, the bandwidth available for use at each base station. Thus, these factors also affect the base station capacity and the overall system capacity.

For the example in Table 15.5, significant speech and coverage quality has been traded for a factor of $\times 30$ in base station capacity! Whereas base station capacity affects overall system capacity directly, it should be remembered that overall system capacity can be increased arbitrarily by decreasing the spacing between base stations. Thus, if the PACS low-tier PCS technology were to start with a base station capacity of $\times 0.5$ of AMPS cellular[5] (a

---

[5]Note that the $\times 0.5$ factor is an arbitrary factor taken for illustrating this example. The so-called $\times$AMPS factors are only with regard to base station capacity, although they are often misused as system capacity.

much lower figure than the ×0.8 sometimes quoted [20]), and then were degraded in quality as described above to yield the ×30 capacity factor, it would have a resulting capacity of ×15 of AMPS! Thus, it is obvious that making such a base station capacity comparison without including quality is not meaningful.

15.6.2 Economics, System Capacity, and Coverage Area Size

Claims are sometimes made that low-tier PCS cannot be provided economically, even though it is what the user wants. These claims are often made based on economic estimates from the cellular paradigm. These include the following.

- Very low estimates of market penetration, much less than cordless telephones, and often even less than cellular.
- High estimates of base station costs more appropriate to high-complexity, high-cost cellular technology than to low-complexity, low-cost, low-tier technology.
- Very low estimates of circuit usage time more appropriate to cellular than to cordless/wireline telephone usage, which is more likely for low-tier PCS.
- { Backhaul costs based on existing T1 line tariffs that are based on "antique" high cost digital loop electronics. (See discussion in fourth paragraph at start of Section 15.6.) }

Such economic estimates are often done by making absolute economic calculations based on very uncertain input data. The resulting estimates for low-tier and high-tier are often closer together than the large uncertainties in the input data. A perhaps more realistic approach for comparing such systems is to vary only one or two parameters while holding all others fixed and then looking at relative economics between high-tier and low-tier systems. This is the approach used in the following examples.

EXAMPLE 15.1:

In the first example (see Table 15.6), the number of channels per megahertz is held constant for cellular and for low-tier PCS. Only the spacing is varied between base stations, e.g., cell sites for cellular and radio ports for low-tier PCS, to account for the differences in transmitter power, antenna height, etc. In this example, overall system capacity varies directly as the square of base station spacing, but base station capacity is the same for both cellular and low-tier PCS. For the typical values in the example, the resulting low-tier system capacity is ×400 greater, only because of the closer base station spacing. If the two systems were to cost the same, the equivalent low-tier PCS base stations would have to cost less than $2,500.

This cost is well within the range of estimates for such base stations, including equivalent infrastructure. These low-tier PCS base stations are of comparable or lower complexity than cellular vehicular subscriber sets, and large-scale manufacture will be needed to produce the millions that will be required. Also, land, building, antenna tower and legal fees for zoning approval, or rental of expensive space on top of commercial buildings, represent large expenses for cellular cell sites. Low-tier PCS base stations that are mounted on utility poles and sides of buildings will not incur such large additional expenses. Therefore, costs of the order of magnitude indicated seem reasonable in large quantities. Note that, with these estimates, the per-wireless-circuit cost of the low-tier PCS circuits would be only $14/circuit compared to $5,555/circuit for the high-tier circuits. Even if there were a factor

TABLE 15.6   System Capacity/Coverage Area Size/Economics

Example 15.1

Assume channels/MHz are the same for cellular and PCS
   Cell site: spacing = 20.000 ft   cost \$ = 1 M
   PCS port: spacing = 1,000 ft
   PCS system capacity is $(20000/1000)^2 = 400 \times$ cellular capacity

Then, for the system costs to be the same
   Port cost = (\$ 1 M/400) \$2,500 a reasonable figure

If, cell site and port each have 180 channels
   Cellular cost/circuit = \$ 1 M/180 = \$5,555/circuit
   PCS cost/circuit = \$2500/180 = \$14/circuit

Example 15.2

Assume equal cellular and PCS system capacity
   Cell site: spacing = 20,000 ft
   PCS port: spacing = 1,000 ft

If, a cell site has 180 channels
   then, for equal system capacity, a PCS port needs 180/400 < 1 channel/port

Example 15.3

Quality/cost trade
   Cell site: Spacing = 20,000 ft   cost = \$1 M   channels = 180
   PCS port: Spacing = 1,000 ft   cost = \$2,500

Cellular to PCS, base station spacing capacity factor = $\times$ 400
   PCS to cellular quality reduction factors:
      32 to 8 kb/s speech $\times 4$
      Voice activity (buying) $\times 2$
      99–90% good areas $\times 2$
      Both in same environment (same σ) $\underline{\times 1}$
      Capacity factor traded $\overline{\times 16}$

180 ch/16 = 11.25 channels/port   then, \$2500/11.25 = \$222/circuit
   and remaining is $\times 400/16 = \times 25$ system capacity of PCS over cellular

---

of 10 error in cost estimates, or a reduction of channels per radio port of a factor of 10, the per-circuit cost of low-tier PCS would still be only \$140/circuit, which is still much less than the per-circuit cost of high-tier.

EXAMPLE 15.2:

In the second example (see Table 15.6), the overall system capacity is held constant, and the number of channels/port, i.e., channels/(base station) is varied. In this example, less than 1/2 channel/port is needed, again indicating the tremendous capacity that can be produced with close-spaced low-complexity base stations.

EXAMPLE 15.3:

Since the first two examples are somewhat extreme, the third example (see Table 15.6) uses a more moderate, intermediate approach. In this example, some of the cellular high-tier channels/(base station) are traded to yield higher quality low-tier PCS as in the previous subsection. This reduces the channels/port to 11+, with an accompanying increase in cost/circuit up to \$222/circuit, which is still much less than the \$5,555/circuit for the high-tier system. Note, also, that the low-tier system still has $\times 25$ the capacity of the high-tier system!

Low-tier base station (Port) cost would have to exceed \$62,500 for the low-tier per-circuit cost to exceed that of the high-tier cellular system. Such a high port cost far exceeds any existing realistic estimate of low-tier system costs.

It can be seen from these examples, and particularly Example 15.3, that the circuit economics of low-tier PCS are significantly better than for high-tier PCS, if the user demand and density is sufficient to make use of the large system capacity. Considering the high penetration of cordless telephones, the rapid growth of cellular handsets, and the enormous market projections for wireless PCS noted earlier in this chapter, filling such high capacity in the future would appear to be certain. The major problem is providing rapidly the widespread coverage (buildout) required by the FCC in the United States. If this unrealistic regulatory demand can be overcome, low-tier wireless PCS promises to provide the wireless personal communications that everyone wants.

15.6.3 {Loop Evolution and Economics

It is interesting to note that several wireless loop applications are aimed at reducing cost by replacing parts of wireline or CATV loops with wireless links between transceivers. The economics of these applications are driven by the replacing of labor-intensive wireline and cable technologies with mass-produced solid-state electronics in transceivers.

Consider first a cordless telephone base unit. The cordless base-unit transceiver usually serves one or, at most, two handsets at the end of one wireline loop. Now consider moving such a base unit back along the copper-wire-pair loop end a distance that can be reliably covered by a low-power wireless link [25, 31], i.e., several hundred to a thousand feet or so, and mounting it on a utility pole or a street light pole. This replaces the copper loop end with the wireless link. Many additional copper loop ends to other subscribers will be contained within a circle around the pole having a maximum usable radius of this wireless link. Replace all of the copper loop ends within the circle with cordless base units on the same pole. Note that this process replaces the most expensive parts of these many loops, i.e., the many individual loop ends, with the wireless links from cordless handsets to "equivalent" cordless base units on a pole. Of course, being mounted outside will require somewhat stronger enclosures and means of powering the base units, but these additional costs are considerably more than offset by eliminating the many copper wire drops.

It is instructive to consider how many subscribers could be collected at a pole containing base units. Consider, as an example, a coverage square of 1400 ft on a side (PACS will provide good coverage over this range, i.e., for base unit pole separations of about 1400 ft, at 1.9 GHz). Within this square will be 45 houses for a 1 house/acre density typical of low-housing-density areas, or 180 houses for 4 house/acre density more typical of high-density single-family housing areas. These represent significant concentration of traffic at a pole.

Because of the trunking advantage of the significant number of subscribers concentrated at a pole, they can share a smaller number of base unit, i.e., wireless base unit transceivers, than there are wireless subscriber sets. Therefore, the total cost compared with having a cordless base unit per subscriber also is reduced by the concentration of users.

A single PACS transceiver will support simultaneously eight TDMA channels or circuits at 32 kb/s (or 16 at 16 kb/s or 32 at 8 kb/s) [56]. Of these, one channel is reserved for system control. The cost of such moderate-rate transceivers is relatively insensitive to the number of channels supported; i.e., the cost of such an 8-channel (or 16 or 32) transceiver will be significantly less than twice the cost of a similar one-channel transceiver. Thus, another economic advantage accrues to this wireless loop approach from using time-multiplexed (TDMA) transceivers instead of single-channel-per-transceiver cordless telephone base units.

For an offered traffic of 0.06 Erlang, a typical busy-hour value for a wireline subscriber, a seven-channel transceiver could serve about 40 subscribers at 1% blocking, based on the Erlang B queuing discipline. From the earlier example, such a transceiver could serve most of the 45 houses within a 1400-ft square. Considering partial penetration, the transceiver capacity is more than adequate for the low-density housing.[6]

Considering the high-density example of 4 houses/acre, a seven-channel transceiver could serve only about 20% of the subscribers within a 1400-ft square. If the penetration became greater than about 20%, either additional transceivers, perhaps those of other service providers, or closer transceiver spacing would be required.

Another advantageous economic factor for wireless loops results when considering time-multiplexed transmission in the fixed distribution facilities. For copper or fiber digital subscriber loop carrier (SLC), e.g., T1 or high-rate digital subscriber line (HDSL), a de-multiplexing/multiplexing terminal and drop interface are required at the end of the time-multiplexed SLC line to provide the individual circuits for each subscriber loop-end circuit, i.e., for each drop. The most expensive part of such an SLC terminating unit is the subscriber line cards that provide per-line interfaces for each subscriber drop. Terminating a T1 or HDSL line on a wireless loop transceiver eliminates all per-line interfaces, i.e., all line cards, the most expensive part of a SLC line termination. Thus, the greatly simplified SLC termination can be incorporated within a TDMA wireless loop transceiver, resulting in another cost savings over the conventional copper-wire-pair telephone loop end.

The purpose of the previous discussions is not to give an exact system design or economic analysis, but to illustrate the inherent economic advantages of low-power wireless loops over copper loop ends and over copper loop ends with cordless telephone base units. Some economic analyses have found wireless loop ends to be more economical than copper loop ends when subscribers use low-power wireless handsets. Rizzo and Sollenberger [56] have also discussed the advantageous economics of PACS wireless loop technology in the context of low-tier PCS.

The discussions in this section can be briefly summarized as follows. Replacing copper wire telephone loop ends with low-complexity wireless loop technology like PACS can produce economic benefits in at least four ways. These are.

1. Replacing the most expensive part of a loop, the per-subscriber loop-end, with a wireless link.

2. Taking advantage of trunking in concentrating many wireless subscriber loops into a smaller number of wireless transceiver channels.

3. Reducing the cost of wireless transceivers by time multiplexing (TDMA) a few (7, 15, or 31) wireless loop circuits (channels) in each transceiver.

4. Eliminating per-line interface cards in digital subscriber line terminations by terminating time-multiplexed subscriber lines in the wireless loop transceivers. }

15.7   Other Issues

Several issues in addition to those addressed in the previous two sections continue to be raised with respect to low-tier PCS. These are treated in this section.

---

[6]The range could be extended by using higher base unit antennas, by using higher-gain directional (sectored) antennas, and/or by increasing the maximum power that can be transmitted.

15.7.1 Improvement of Batteries

Frequently, the suggestion is made that battery technology will improve so that high-power handsets will be able to provide the desired 5 or 6 hours of talk time in addition to 10 or 12 hours of standby time, and still weigh less than one-fourth of the weight of today's smallest cellular handset batteries. This hope does not take into account the maturity of battery technology, and the long history (many decades) of concerted attempts to improve it. Increases in battery capacity have come in small increments, a few percent, and very slowly over many years, and the shortfall is well over a factor of 10. In contrast, integrated electronics and radio frequency devices needed for low-power low-tier PCS continue to improve and to decrease in cost by factors of greater than 2 in time spans on the order of a year or so. It also should be noted that, as the energy density of a battery is increased, the energy release rate per volume must also increase in order to supply the same amount of power. If energy storage density and release rate are increased significantly, the difference between a battery and a bomb become indistinguishable! The likelihood of a $\times 10$ improvement in battery capacity appears to be essentially zero. If even a modest improvement in battery capacity were possible, many people would be driving electric vehicles.

{ As noted in the addition to the "Reality Check" section, new lithium batteries have become the batteries of choice for the smallest cellular/high-tier PCS handsets. While these lithium batteries have higher energy density than earlier nickel cadmium batteries, they still fall far short of the factor of 10 improvement that was needed to make long talk time, small size, and low weight possible. With the much larger advance in electronics, the battery is even more dominant in the size and weight of the newest cellular handsets. The introduction of these batteries incurred considerable startup pain because of the greater fire and explosive hazard associated with lithium materials, i.e., closer approach to a bomb. Further attempts in this direction will be even more hazardous. }

15.7.2 People Only Want One Handset

This issue is often raised in support of high-tier cellular handsets over low-tier handsets. Whereas the statement is likely true, the assumption that the handset must work with high-tier cellular is not. Such a statement follows from the current large usage of cellular handsets; but such usage results because that is the only form of widespread wireless service currently available, not because it is what people want. The statement assumes inadequate coverage of a region by low-tier PCS, and that low-tier handsets will not work in vehicles. The only way that high-tier handsets could serve the desires of people discussed earlier would be for an unlikely breakthrough in battery technology to occur. A low-tier system, however, can cover economically any large region having some people in it. (It will not cover rural or isolated areas but, by definition, there is essentially no one there to want communications anyway.)

Low-tier handsets will work in vehicles on village and city streets at speeds up to 30 or 40 mi/h, and the required handoffs make use of computer technology that is rapidly becoming inexpensive. { As noted earlier, vehicular speed handoff is readily accomplished with PACS. Reliable handoff has been demonstrated for PACS at speeds in excess of 70 mi/hr. } Highways between populated areas, and also streets within them, will need to be covered by high-tier cellular PCS, but users are likely to use vehicular sets in these cellular systems. Frequently the vehicular mobile user will want a different communications device anyway, e.g., a hands-free phone. The use of hands-free phones in vehicles is becoming a legal requirement in some places now and is likely to become a requirement in many more places in the future. Thus, handsets may not be legally usable in vehicles anyway.

With widespread deployment of low-tier PCS systems, the one handset of choice will be the low-power, low-tier PCS pocket handset or voice/data communicator.

{ As discussed in earlier sections, it is quite feasible economically to cover highways between cities with low-tier systems, if the low-tier base stations have antennas with the same height and gain as used for cellular and high-tier PCS systems. (The range penalty for the lower power was noted earlier to be only on the order of 1/2, or about the same as the range penalty in going from 800 MHz cellular to 1.9 GHz high-tier PCS.) }

There are approaches for integrating low-tier pocket phones or pocket communicators with high-tier vehicular cellular mobile telephones. The user's identity could be contained either in memory in the low-tier set or in a small smart card inserted into the set, as is a feature of the European GSM system. When entering an automobile, the small low-tier communicator or card could be inserted into a receptacle in a high-tier vehicular cellular set installed in the automobile.[7] The user's identity would then be transferred to the mobile set. { "Car adapters" that have a cradle for a small cellular handset providing battery charging and connection to an outside antenna are quite common — e.g., in Sweden use of such adapters is commonplace. Thus, this concept has already evolved significantly, even for the disadvantaged cellular handsets when they are used in vehicles. } The mobile set could then initiate a data exchange with the high-tier system, indicating that the user could now receive calls at that mobile set. This information about the user's location would then be exchanged between the network intelligence so that calls to the user could be correctly routed.[8] In this approach the radio sets are optimized for their specific environments, high-power, high-tier vehicular or low-power, low-tier pedestrian, as discussed earlier, and the network access and call routing is coordinated by the interworking of network intelligence. This approach does not compromise the design of either radio set or radio system. It places the burden on network intelligence technology that benefits from the large and rapid advances in computer technology.

The approach of using different communications devices for pedestrians than for vehicles is consistent with what has actually happened in other applications of technology in similarly different environments. For example, consider the case of audio cassette tape players. Pedestrians often carry and listen to small portable tape players with lightweight headsets (e.g., a Walkman).[9] When one of these people enters an automobile, he or she often removes the tape from the Walkman and inserts it into a tape player installed in the automobile. The automobile player has speakers that fill the car with sound. The Walkman is optimized for a pedestrian, whereas the vehicular-mounted player is optimized for an automobile. Both use the same tape, but they have separate tape heads, tape transports, audio preamps, etc. They do not attempt to share electronics. In this example, the tape cassette is the information-carrying entity similar to the user identification in the personal communications example discussed earlier. The main points are that the information is shared among different devices but that the devices are optimized for their environments and do not share electronics.

Similarly, a high-tier vehicular-cellular set does not need to share oscillators, synthesizers, signal processing, or even frequency bands or protocols with a low-tier pocket-size communicator. Only the information identifying the user and where he or she can be reached

---

[7]Inserting the small personal communicator in the vehicular set would also facilitate charging the personal communicator's battery.

[8]This is a feature proposed for FPLMTS in CCIR Rec. 687.

[9]Walkman is a registered trademark of Sony Corporation.

needs to be shared among the intelligence elements, e.g., routing logic, databases, and common channel signalling [26, 29] of the infrastructure networks. This information exchange between network intelligence functions can be standardized and coordinated among infrastructure subnetworks owned and operated by different business entities (e.g., vehicular cellular mobile radio networks and intelligent low-tier PCS networks). Such standardization and coordination are the same as are required today to pass intelligence among local exchange networks and interexchange carrier networks.

15.7.3 Other Environments

Low-tier personal communications can be provided to occupants of airplanes, trains, and buses by installing compatible low-tier radio access ports inside these vehicles. The ports can be connected to high-power, high-tier vehicular cellular mobile sets or to special air-ground or satellite-based mobile communications sets. Intelligence between the internal ports and mobile sets could interact with cellular mobile, air-ground, or satellite networks in one direction, using protocols and spectrum allocated for that purpose, and with low-tier personal communicators in the other direction to exchange user identification and route calls to and from users inside these large vehicles. Radio isolation between the low-power units inside the large metal vehicles and low-power systems outside the vehicles can be ensured by using windows that are opaque to the radio frequencies. Such an approach also has been considered for automobiles, i.e., a radio port for low-tier personal communications connected to a cellular mobile set in a vehicle so that the low-tier personal communicator can access a high-tier cellular network. (This could be done in the United States using unlicensed PCS frequencies within the vehicle.)

15.7.4 Speech Quality Issues

All of the PCS and cordless telephone technologies that use CCITT standardized 32-kb/s ADPCM speech encoding can provide similar error-free speech distortion quality. This quality often is rated on a five-point subjective mean opinion score (MOS) with 5 excellent, 4 good, 3 fair, 2 poor, and 1 very poor. The error-free MOS of 32-kb/s ADPCM is about 4.1 and degrades very slightly with tandem encodings. Tandem encodings could be expected in going from a digital-radio PCS access link, through a network using analog transmission or 64-kb/s PCM, and back to another digital-radio PCS access link on the other end of the circuit. In contrast, a low-bit-rate (<10-kb/s) vocoder proposed for a digital cellular system was recently reported [54] to yield an MOS of 3.4 on an error-free link without speech-activity detection. This score dropped to 3.2 when speech-activity detection was implemented to increase system capacity. This nearly one full point decrease on the five-point MOS score indicates significant degradation below accepted CCITT wireline speech distortion quality. Either almost half of the population must have rated it as poor or most of the population must have rated it as only fair. It should also be noted that these MOS scores may not reflect additional degradation that may occur in low-bit-rate speech encoding when the speech being encoded is combined with acoustical noise in a mobile environment, e.g., tire, wind, and engine noise in automobiles and street noise, background talking, etc., in handheld phone environments along streets and in buildings. Comments from actual users of low-bit-rate speech technology in acoustically noisy environments suggest that the MOS scores just quoted are significantly degraded in these real world environments. Waveform coders, e.g., ADPCM are not likely to experience degradation from such background noise. In addition, the low-bit-rate speech encoding is not at all tolerant of the tandem speech encodings that will inevitably occur for PCS for many years. That is, when low-bit-rate speech is transcoded to a different encoding format, e.g., to 64 kb/s as is used in many networks or from an IS-54 phone on one end to a GSM or IS-95 phone on the other end, the speech quality deteriorates precipitously. Although this may not be a serious issue for a vehicular mobile user who has no choice other than not to communicate at all, it is likely to be a serious issue in an environment where a wireline telephone is available as an alternative. It is also less serious when there are few mobile-to-mobile calls through the network, but as wireless usage increases and digital mobile-to-mobile calls become commonplace, the marginal transcoded speech quality is likely to become a serious issue. These comments in this paragraph are generally applicable to speech encoding at rates of 13 kb/s or less. { The predictions above about the seriousness of the speech coding quality issues have been proven to be true in virtually all cellular technologies [2, 5]. The much touted speech quality of the original IS-95 CDMA 8 kb/s QCELP speech coder proved to be unacceptable to users.

The 8 kb/s speech coder was replaced with a 13 kb/s speech coder when IS-95 CDMA technology was deployed for high-tier PCS at 1.9 GHz. Many of the 800 MHz CDMA systems are converting or considering converting to the higher bit rate coder to improve their speech quality. This change was accompanied by a base station capacity reduction of 1/3, but the CDMA advocates seldom acknowledge this fact. The GSM effort to create a "half-rate" speech coder was redirected to develop an improved speech quality coder at 13 kb/s that would be less sensitive to acoustic background noise, bit errors, tandem encoding, etc. The quality of the new coder is better under some conditions. The 8 kb/s North American TDMA/IS-136 speech coder was redesigned to improve speech quality. Some improvement has been achieved. In all these revised coder cases, the issues discussed above in the original article still apply. }

In the arena of transmission delay, the short-frame (2-ms) FDMA/TDD and TDMA technologies (e.g., CT-2 and WACS noted earlier) can readily provide single-radio-link round-trip delays of <10 ms and, perhaps, even <5 ms. The longer frame (10 ms and greater) cordless-phone TDMA technologies, e.g., DCT-900/CT-3/DECT and some ISM-band implementations, inherently have a single-link round-trip delay of at least 20 ms and can range 30-40 ms or more in some implementations. As mentioned earlier, the digital vehicular-cellular technologies with low-bit-rate speech encoding, bit interleaving, forward error-correction decoding, and relatively long frame time (~16–20 ms) result in single-link round-trip delays on the order of 200 ms, well over an order of magnitude greater than the short-frame technologies, and on the same order of magnitude as a single-direction synchronous satellite link. { Even with the somewhat improved low bit rate speech coders noted previously, the transmission delay remains excessive for all of the cellular/high-tier PCS digital technologies. } It should be noted that almost all United States domestic long-distance telephone circuits have been removed from such satellite links, and many international satellite links also are being replaced by undersea fiber links. These long-distance-circuit technology changes are made partially to reduce the perceptual impairment of long transmission delay.

15.7.5 New Technology

New technology, e.g., spread spectrum or CDMA, is sometimes offered as a solution to both the higher-tier cell site capacity and transmitter power issues. As these new technologies are pursued vigorously, however, it becomes increasingly evident that the early projections were considerably overoptimistic, that the base station capacity will be about the same as other technologies [29], and that the high complexity will result in more, not less, power consumption.

With the continuing problems and delays in initial deployments, there is increasing concern throughout the industry as to whether CDMA is a viable technology for high-capacity cellular applications. With the passage of time, it is becoming more obvious that Viterbi was correct in his 1985 paper in which he questioned the use of spread spectrum for commercial communications [73].

The IS-95 proposal is considerably more technically sophisticated than earlier spread spectrum proposals. It includes fast feedback control of mobile transmitter power, heavy forward error correction, speech detection and speech-encoding rate adjustment to take advantage of speech inactivity, and multiple receiver correlators to latch onto and track resolvable multipath maxima [57]. The spreading sequence rate is 1.23 MHz.

The near–far problem is addressed directly and elegantly on the uplink by a combination of the fast-feedback power control and a technique called soft handoff that permits the instantaneous selection of the best paths between a mobile and two cell sites. Path selection is done on a frame-by-frame basis when paths between a mobile and the two cell sites are within a specified average level (perhaps 6–10 dB) for each other. This soft handoff provides a form of macroscopic diversity [10] between pairs of cell sites when it is advantageous. Increasing capacity by soft handoff requires precise time synchronization (on the order of a microsecond) among all cell sites in a system. An advantage of this proposal is that frequency coordination is not needed among cell sites since all sites can share a frequency channel. Coordination of the absolute time delays of spreading sequences among cell sites is required, however, since these sequence delays are used to distinguish different cell sites for initial access and for soft handoff. Also, handoff from one frequency to another is complicated.

Initially, the projected cell site capacity of this CDMA system, determined by mathematical analysis and computer simulation of simplified versions of the system, was ×20 to ×40 that of the analog AMPS, with a coverage criterion of 99% of the covered area [11]. However, some other early estimates [83] suggested that the factors were more likely to be ×6 to ×8 of AMPS.

A limited experiment was run in San Diego, California, during the fourth quarter of 1991 under the observation of cellular equipment vendors and service providers. This experiment had 42–62 mobile units in fewer than that many vehicles,[10] and four or five cell sites, one with three sectors. Well over half of the mobiles needed to provide the interference environment for system capacity tests were simulated by hardware noise simulation by a method not yet revealed for technical assessment. Estimates of the cell site capacity from this CDMA experiment center around ×10 that of AMPS [12, 54][11] with coverage criteria <99%, perhaps 90-95%, and with other capacity estimates ranging between ×8 and ×15.

This experiment did not exercise several potential capacity-reducing factors; the following, for example.

1. Only four cells participated in capacity tests. The test mobiles were all located in a relatively limited area and had limited choices of cell sites with which to communicate for soft handoffs. This excludes the effects of selecting a strong cell-site downlink for soft handoff that does not have the lowest uplink attenuation because of uncorrelated uplink and downlink multipath fading at slow vehicle speeds [3].

---

[10]Some vehicles contained more than one mobile unit.
[11]AT&T stated that the San Diego data supported a capacity improvement over analog cellular of at least ×10.

2. The distribution of time-dispersed energy in hilly environments like San Diego usually is more concentrated around one or two delays than is the dispersed energy scattered about in heavily built-up urban areas like downtown Manhattan [13, 14], or Chicago. Energy concentrated at one or two delays is more fully captured by the limited number of receiver correlators than is energy more evenly dispersed in time.

3. Network delay in setting up soft handoff channels can result in stronger paths to other cell sites than to the one controlling uplink transmitter power. This effect can be more pronounced when coming out of shadows of tall buildings at intersections in heavily built-up areas. The effect will not occur as frequently in a system with four or five cell sites as it will in a large, many cell site system.

All of these effects and others [4] will increase the interference in a large system, similarly to the increase in interference that results from additional mobiles and, thus, will decrease cell site capacity significantly below that estimated in the San Diego trial. Factors like these have been shown to reduce the San Diego estimate of ×10 to an expected CDMA capacity of ×5 or ×6 of analog AMPS [3, 4, 13]. This further reduction in going from a limited experiment to a large-scale system in a large metropolitan area is consistent with the reduction already experienced in going from simplified theoretical estimates to the limited experiment in a restricted environment. { There are many "small" effects that decrease the performance and, thus, the base station capacity of actual mobile CDMA systems compared to calculations of performance based on ideal assumptions — e.g., stationary channel and additive white Gaussian noise analysis. Some of these effects, in addition to those noted above, included inaccuracy in assigning correlators ("fingers") to time varying statistically nonstationary multiple "paths" at different delays when many "paths" exist, and lack of diversity because many environments have delay spreads less than the resolution of correlators. }

The San Diego trial also indicated a higher rate of soft(er) handoffs [54] between antenna sectors at a single cell site than expected for sectors well isolated by antenna patterns. This result suggests a lower realizable sectorization gain because of reflected energy than would be expected from more idealized antennas and locations. This could further reduce the estimated cell site capacity of a large-scale system.

Even considering the aforementioned factors, capacity increases of ×5 or ×6 are significant. These estimates, however, are consistent with the factor of ×3 obtained from low-bit-rate (<10-kb/s) speech coding and the ×2 to ×2.5 obtained by taking advantage of speech pauses. These factors result in an expected increase of ×6–7.5, with none of these speech-processing-related contributions being directly attributable to the spread-spectrum processing in CDMA. These results are consistent with the factor of ×6–8 estimate made earlier [83] and are not far from the factor of ×8 quoted recently [86].

Thus, it is clear that new high-complexity, high-tier technology will not be a substitute for low-complexity, low-power, low-tier PCS.

{ Since the writing for the first edition of this handbook, IS-95 based CDMA systems have been deployed and are providing service in several parts of the world, e.g., the U.S.A., Korea, and Hong Kong. The "high priests" and the "disciples" of CDMA continue to espouse their "belief" in the superiority of their CDMA technology. Good hard data on system performance is still scarce and of questionable reliability. However, there is sufficient evidence from some "usually reliable sources" and from "conversations in the halls" to confirm that, not only does CDMA not perform as well as was originally promised, but also in many places it is likely to be providing performance that is inferior to the other high-tier digital technologies. In spite of the continuing dismal failures of CDMA technology to meet the early claims for it, a significant fraction of the wireless industry appears to be committed to pursuing variations of CDMA for future applications. At this time it is not obvious how many more redesign and reprogramming failures, or perhaps even failures of manufacturing companies and service providers will be needed to overcome the religiously fanatic attachment of a part of the industry to CDMA. It appears that some are heavily committed again to "sweep on to the grand fallacy." Viterbi may even have been over optimistic about CDMA in 1985 [73], but he was correct in the way it attracts advocates.

IS-95 CDMA has failed to satisfy any of the major claims made for it in the late 1980s and early 1990s. The listing in Table 15.7 and the following discussion cites specific claims that have NOT been met now that reality has set in.

TABLE 15.7   CDMA Scoresheet

Early Claim	Reality
Easy to install and expand (no frequency planning)	Many "parameters" to adjust; adjustment is very difficult; many software revisions required
"Capacity" of AMPS ×20 (with high expectation of AMPS ×40)	AMPS ×3 or 4; AMPS × <3? (at unspecified but likely high call dropping rate) (continuously revised downward over time)
No more dropped calls	More dropped calls than other technologies (some reports say up to 40% dropped calls when system is loaded!)
No problem with interference (CDMA "loves" interference)	Can't live with interference from AMPS or other sources
Greater range and coverage than other technologies	System coverage "breathes" as loading increases; range often is less than other technologies
"Toll quality" 8 kb/s QCELP speech	Original 8 kb/s QCELP speech quality worse than VSELP in TDMA; required change to 13 kb/s coder (change reduced "capacity")

Much of the original CDMA hype was focused on undefined cell site capacity (see text from original Handbook, particularly this section and the section on "Capacity, Quality, and Complexity.") Information from many sources (e.g., [9], Benjamin and Lusignan, 1997 — Benjamin is with Lucent Technologies; Lusignan with Stanford University; [40, 49, 69], etc.) suggest that with 13 kb/s speech coding, "fully loaded" systems may provide base station capacities on the order of AMPS ×3 or perhaps ×4. However, the meaning of fully loaded is not well defined. If it is taken to be at a very high call dropping rate (sometimes said to be as high as 40%!), perhaps even these low factors are again unrealistically inflated. A usable system loading must provide an acceptable call dropping rate, perhaps 1% or less. If loading is reduced to provide sufficient operating margin to prevent too many dropped calls, perhaps the usable base station capacity may be even significantly less than AMPS ×3. (Note: Publicly presented CDMA claims have been consistently inflated in the past. Why should we be any more inclined to believe them now?) The capacity situation is so bad that the CDMA crowd doesn't even want to discuss this topic anymore. They now attempt to shift attention to the fact that there are operating CDMA systems, and try to brush aside the fact CDMA performance is so very poor. Low capacity has a significant negative impact on the economics of systems; high call dropping rates have a significant negative impact on service quality.

Another big early claim for CDMA was that it was easy to install and expand, that is, that it needed little engineering and planning for installation because it didn't require frequency reuse and power coordination among base stations. In reality, many parameters require adjustment and critical thresholds require setting for a system to provide even the low capacity numbers cited above. Many software revisions have been made in attempts to cope with fundamental problems that keep appearing and reappearing in new deployments and expansions. Deployments and service starts have been consistently late by many months and even years. [40, 69, and others]. A partial list of late CDMA deployments that required many months or even years of "parameter adjustment" and/or "software revisions" and/or "redesign" include those in Seattle, WA, U.S.A. (800 MHz); Los Angeles, CA, U.S.A. (800 MHz); Hong Kong, China (800 MHz); Trenton, NJ, U.S.A.; Seoul, Korea (800 MHz); Houston and Dallas, TX, U.S.A. (1.9 GHz); etc. As these systems were deployed a consistent pattern developed. Announcements would concentrate on a projected future service date, e.g., Seattle, 1Q 1994. When that date approached, news focused on yet another service date elsewhere, e.g., after Seattle was Los Angeles, to divert attention from the deployment problems causing large slips in earlier announcements. Sometimes when service date slips became excessive, sizes of deployed areas were revised downward, and "trials" were substituted for service starts, e.g., Los Angeles. This pattern continued through the early deployments and service eventually was started on systems, even though their performance was significantly below earlier claims. Excuses for slips were often vague, for example: software problems (e.g., Seattle, Hong Kong, and Korea); adjusting parameters (e.g., Seattle, Houston, and Dallas); too much interference from AMPS (e.g., Los Angeles); not enough multipath! (e.g., Seattle). These sound like descriptions of attempts to overcome some of the fundamental problems in mobile CDMA systems. In any event it is obvious that CDMA is not easy to install and is likely more difficult to adjust than other technologies.

The CDMA speech quality issue was discussed earlier in the "Speech Quality Issues" section. It is just another example of unrealistic initial claims.

Throughout the CDMA debate over almost 10 years, when unrealistic claims were questioned and even when sound technical analysis was cited by more conservative members of the wireless technical community, including this author, these members were immediately branded as heretics and their concerns were dismissed with rhetoric by the high priests and their disciples. (In the technical world, "never have so many been misled for so long by so few!") However, as deployments have proceeded and reality set in, even the more conservative initial assessments have, unfortunately, turned out to be somewhat optimistic. Thus, even the conservative heretics in the community now appear to be in the unfortunate position of having also been overoptimistic in their initial estimates of the performance of IS-95 CDMA. }

Statistical Multiplexing, Speech Activity, CDMA, and TDMA

Factors of $\times 2$–2.5 have been projected for the capacity increase possible by taking advantage of pauses in speech. It has been suggested that implementing statistical multiplexing is easier for CDMA systems because it is sometimes thought to be time consuming to negotiate channels for speech spurts for implementation in TDMA systems. The most negative quality-impacting factor in implementing statistical multiplexing for speech, however, is not in obtaining a channel when needed but is in the detection of the onset of speech, particularly in an acoustically noisy environment. The effect of clipping at the onset of speech is evident in the MOS scores noted for the speech-activity implementation in the United States cellular CDMA proposal discussed earlier (i.e., an MOS of 3.4 without statistical multiplexing and of 3.2 with it). The degradation in MOS can be expected to be even greater for encoding that starts with a higher MOS, e.g., 32-kb/s ADPCM.

It was noted earlier that the proposed cellular CDMA implementation was $\times 10$ as complex as the proposed WACS wireless access for personal communications TDMA implementation. From earlier discussion, the CDMA round-trip delay approaches 200 ms, whereas the short 2-ms-frame TDMA delay is $<$10-ms round trip. It should be noted that the TDMA architecture could permit negotiation for time slots when speech activity is detected. Since the TDMA frames already have capability for exchange of signalling data, added complexity for statistical multiplexing of voice could readily be added within less than 200 ms of delay and less than $\times 10$ in complexity. That TDMA implementation supports 8 circuits at 32 kb/s or 16 circuits at 16 kb/s for each frequency. These are enough circuits to gain benefit from statistical multiplexing. Even more gain could be obtained at radio ports that support two or three frequencies and, thus, have 16–48 circuits over which to multiplex.

A statistical multiplexing protocol for speech and data has been researched at Rutgers WINLAB [39]. The Rutgers packet reservation multiple access (PRMA) protocol has been used to demonstrate the feasibility of increasing capacity on TDMA radio links. These PRMA TDMA radio links are equivalent to slotted ALOHA packet-data networks. Transmission delays of less than 50 ms are realizable. The capacity increase achievable depends on the acceptable packet-dropping ratio. This increase is soft in that a small increase in users causes a small increase in packet-dropping ratio. This is analogous to the soft capacity claimed for CDMA.

Thus, for similar complexity and speech quality, there appears to be no inherent advantage of either CDMA or TDMA for the incorporation of statistical multiplexing. It is not included in the personal communications proposal but is included in cellular proposals because of the different speech-quality/complexity design compromises discussed throughout this chapter, not because of any inherent ease of incorporating it in any particular access technology.

15.7.6   High-Tier to Low-Tier or Low-Tier to High-Tier Dual Mode

Industry and the FCC in the United States appear willing to embrace multimode handsets for operating in very different high-tier cellular systems, e.g., analog FM AMPS, TDMA IS-54, and CDMA IS-95. Such sets incur significant penalties for dual mode operation with dissimilar air interface standards and, of course, incur the high-tier complexity penalties.

It has been suggested that multimode high-tier and low-tier handsets could be built around one air-interface standard, for example, TDMA IS-54 or GSM. When closely spaced low-power base stations were available, the handset could turn off unneeded power-consuming circuitry, e.g., the multipath equalizer. The problem with this approach is that the handset is still encumbered with power-consuming and quality-reducing signal processing inherent in the high-tier technology, e.g., error correction decoding and low-bit-rate speech encoding and decoding.

{ With widespread and successful deployment of GSM, perhaps an initial dual-mode strategy of GSM and PACS could be desirable. Of all the high-tier and low-tier technologies, GSM and PACS appear most compatible for dual-mode handset and base station implementation. They are considerably more compatible than the dual-mode combinations, e.g., AMPS/IS-54/IS-136 and AMPS/IS-95, already implemented. }

An alternative dual-mode low-tier, high-tier system based on a common air-interface standard can be configured around the low-tier PACS/WACS system, if such a dual-mode system is deemed desirable in spite of the discussion in this chapter. The range of PACS can readily be extended by increasing transmitter power and/or the height and gain of base station antennas. With increased range, the multipath delay spread will be more severe in some locations [21, 22, 33]. Two different solutions to the increased delay spread can be employed, one for the downlink and another for the uplink. The PACS radio-link architecture has a specified bit sequence, i.e., a unique word, between each data word on the TDM downlink [7, 28]. This unique word can be used as a training sequence for setting the tap weights of a conventional equalizer added to subscriber sets for use in a high-tier PACS mode. Since received data can be stored digitally [62, 65], tap weights can be trimmed, if necessary, by additional passes through an adaptive equalizer algorithm, e.g., a decision feedback equalizer algorithm.

The PACS TDMA uplink has no unique word. The high-tier uplink, however, will terminate on a base station that can support greater complexity but still be no more complex than the high-tier cellular technologies. Research at Stanford University has indicated that blind equalization, using constant-modulus algorithms (CMA), [58, 70], can be effective for equalizing the PACS uplink. Techniques have been developed for converging the CMA equalizer on the short TDMA data burst.

{ See earlier added section on "Coverage, Range, Speed and Environments." }

The advantages of building a dual-mode high-tier, low-tier PCS system around the low-tier PACS air-interface standard follow.

1. The interface can still support small low-complexity, low-power, high-speech-quality low-tier handsets.

2. Both data and voice can be supported in a PACS personal communicator.

3. In high-tier low-tier dual mode PACS sets, circuits used for low-tier operation will also be used for high-tier operation, with additional circuits being activated only for high-tier operation.

4. The flexibility built into the PACS radio link to handle different data rates from 8 kb/s to several hundred kb/s will be available to both modes of operation.

15.8   Infrastructure Networks

It is beyond the scope of this chapter to consider the details of PCS network infrastructures. There are, however, perhaps as many network issues as there are wireless access issues discussed herein [26, 32, 43, 81]. With the possible exception of the self-organizing WLANS, wireless PCS technologies serve as access technologies to large integrated intelligent fixed communications infrastructure networks.

These infrastructure networks must incorporate intelligence, i.e., database storage, signalling, processing and protocols, to handle both small-scale mobility, i.e., handoff from base station to base station as users move, and large-scale mobility, i.e., providing service to users who roam over large distances, and perhaps from one network to another. The fixed infrastructure networks also must provide the interconnection among base stations and other network entities, e.g., switches, databases, and control processors. Of course, existing cellular mobile networks now contain or are incorporating these infrastructure network capabilities. Existing cellular networks, however, are small compared to the expected size of future high-tier and low-tier PCS networks, e.g., 20 million cellular users in the United States compared with perhaps 100 million users or more each in the future for high-tier and low-tier PCS.

Several other existing networks have some of the capabilities needed to serve as access networks for PCS. Existing networks that could provide fixed base station interconnection include:

- Local exchange networks that could provide interconnection using copper or glass-fiber distribution facilities

- Cable TV networks that could provide interconnection using new glass-fiber and coaxial-cable distribution facilities
- Metropolitan fiber digital networks that could provide interconnection in some cities in which they are being deployed

Networks that contain intelligence, e.g., databases, control processors, and signalling that is suitable or could be readily adapted to support PCS access include:

- Local exchange networks that are equipped with signalling system 7 common channel signalling (SS7 CCS), databases, and digital control processors
- Interexchange networks that are similarly equipped

Data networks, e.g., the internet, could perhaps be adapted to provide the needed intelligence for wireless data access, but they do not have the capacity needed to support large voice/data wireless low-tier PCS access.

Many entities and standards bodies worldwide are working on the access network aspects of wireless PCS. The signalling, control processing, and database interactions required for wireless access PCS are considerably greater than those required for fixed place-to-place networks, but that fact must be accepted when considering such networks.

Low-tier PCS, when viewed from a cellular high-tier paradigm, requires much greater fixed interconnection for the much closer spaced base stations. When viewed from a cordless telephone paradigm of a base unit for every handset and, perhaps, several base units per wireline, however, the requirement is much less fixed interconnection because of the concentration of users and trunking that occurs at the multiuser base stations. One should remember that there are economical fixed wireline connections to almost all houses and business offices in the United States now. If wireless access displaces some of the wireline connections, as expected, the overall need for fixed interconnection could decrease!

15.9 Conclusion

Wireless personal communications embraces about seven relatively distinct groups of tetherless voice and data applications or services having different degrees of mobility for operation in different environments. Many different technologies and systems are evolving to provide the different perceived needs of different groups. Different design compromises are evident in the different technologies and systems. The evidence suggests that the evolutionary trajectories are aimed toward at least three large groups of applications or services, namely, high-tier PCS (current cellular radio), high-speed wireless local-area networks (WLANS), and low-tier PCS (an evolution from several of the current groups). It is not clear to what extent several groups, e.g., cordless telephones, paging, and wide-area data, will remain after some merging with the three large groups. Major considerations that separate current cellular technologies from evolving low-tier low-power PCS technologies are speech quality, complexity, flexibility of radio-link architecture, economics for serving high-user-density or low-user-density areas, and power consumption in pocket carried handsets or communicators. High-tier technologies make use of large complex expensive cell sites and have attempted to increase capacity and reduce circuit costs by increasing the capacity of the expensive cell sites. Low-tier technologies increase capacity by reducing the spacing between base stations, and achieve low circuit cost by using low-complexity, low-cost base stations. The differences between these approaches result in significantly different compromises in circuit quality and power consumption in pocket-sized handsets or communicators. These kinds of differences also can be seen in evolving wireless systems optimized for data. Advantages of the low-tier PACS/WACS technology are reviewed in the chapter, along with techniques for using that technology in high-tier PCS systems.

References

[1] Afrashteh, A., Sollenberg, N.R., and Chukurov, D.D., Signal to interference performance for a TDMA portable radio link. *IEEE VTC'91*, St. Louis, MO, 19–22, May, 1991.

[2] {Andrews, E.L., "When Digital Doesn't Always Mean Clear," *New York Times*, p. C1, 4, Jun. 26, 1995.}

[3] Ariyavisitakul, S., SIR-based power control in a CDMA system. *IEEE GLOBECOM'92*, Orlando, FL, Paper 26.3, Dec. 6–9, to be published in *IEEE Trans. Comm.*, 1992.

[4] Ariyavisitakul, S., et al., private communications.

[5] {Baugh, C.R., Wireless quality: a ticking time bomb, *Telephony*, 232(9), 50–58, Mar. 3, 1997.}

[6] {Baugh, C.R., Laborde, E., Pandey, V., and Varma, V., Personal access communications system: fixed wireless local loop and mobile configurations and services, *Proc. IEEE*, 1498–1506, July, 1998. }

[7] Bellcore Technical Advisories. Generic framework criteria for universal digital personal communications systems (PCS). FA-TSY-001013(1) March and FA-NWT-001013(2) Dec., and Tech. Ref. 1993. Generic criteria for version 0.1 wireless access communications systems (WACS). (1) Oct. rev. 1, Jun. 1994.

[8] BellSouth Services Inc., Quarterly progress report number 3 for experimental licenses KF2XFO and KF2XFN. To the Federal Communications Commission, Nov. 25, 1991.

[9] { Benjamin, B. and Lusignan, B., CDMA vs. TDMA, Stanford Univ. Center for Telecom. Symposium, The Acceleration of World Wide Wireless Communication, May 21, 1997.}

[10] Bernhardt, R.C., Macroscopic diversity in frequency reuse radio systems. *IEEE J. Sel. Areas in Comm.*, SAC-5, 862–870, Jun. 1987.

[11] Cellular Telecommunications Industries Association. CDMA digital cellular technology open forum. Jun. 6, 1989.

[12] Cellular Telecommunications Industries Association. CTIA Meeting, Washington, D.C., 5–6, Dec. 1991.

[13] Chang, L.F., Dispersive fading effects in CDMA radio systems. *Elect. Lett.*, 28(19), 1801–1802, 1992.

[14] Chang, L.F. and Ariyavisitakul, S., Performance of a CDMA radio communications system with feed-back power control and multipath dispersion. *IEEE GLOBECOM'91*, Phoenix, AZ, 1017–1021, Dec. 1991.

[15] Chuang, J.C.-I., Performance issues and algorithms for dynamic channel assignment. *IEEE JSAC*, Aug. 1993.

[16] Chuang, J.C.-I., Performance limitations of TDD wireless personal communications with asynchronous radio ports. *Electron. Lett.*, 28, 532–533, Mar. 1992.

[17] Chuang, J.C.-I., Comparison of coherent and differential detection of BPSK and QPSK in a quasistatic fading channel. *IEEE Trans. Comm.*, 565–576, May 1990.

[18] Chuang, J.C.-I. and Sollenberger, N.R., Burst coherent detection with robust frequency and timing estimation for portable radio communications. *IEEE GLOBECOM'88*, Hollywood, FL, 28–30, Nov. 1988.

[19] Chuang, J.C.-I., Sollenberger, N.R., and Cox, D.C., A pilot based dynamic channel assignment scheme for wireless access TDMA/FDMA systems. *Proc. IEEE ICUPC'93*, Ottawa, Canada, Oct. 12–15, 706–712, 1994. *Intl. J. Wireless Inf. Net.*, 1(1), 37–48, 1993.

[20] Cook, C.I., Development of air interface standards for PCS. *IEEE Personal Comm.*, 4th Quarter, 30–34, 1994.

[21] Cox, D.C., Delay-Doppler characteristics of multipath propagation at 910 MHz in a suburban mobile radio environment. *IEEE Trans. on Antennas and Propagation,* 625–635, Sep. 1972.

[22] Cox, D.C., Multipath delay spread and path loss correlation for 910 MHz urban mobile radio propagation. *IEEE Trans. on Veh. Tech.,* 340–344, Nov. 1977.

[23] Cox, D.C., Universal portable radio communications. *IEEE Trans. on Veh. Tech.,* 117–121, Aug. 1985.

[24] Cox, D.C., Research toward a wireless digital loop. *Bellcore Exchange,* 2, 2–7, Nov./Dec. 1986.

[25] Cox, D.C., Universal digital portable radio communications. *Proc. IEEE,* 75, 436–477, Apr. 1987.

[26] Cox, D.C., Portable digital radio communications—An approach to tetherless access. *IEEE Comm. Mag.,* 30–40, Jul. 1989.

[27] Cox, D.C., Personal communications—A viewpoint. *IEEE Comm. Mag.,* 8–20, Nov. 1990.

[28] Cox, D.C., A radio system proposal for widespread low-power tetherless communications. *IEEE Trans. on Comm.,* 324–335, Feb. 1991.

[29] Cox, D.C., Wireless network access for personal communications. *IEEE Comm. Mag.,* 96–115, Dec. 1992.

[30] {Cox, D.C., Wireless loops: what are they. *Intl. J. Wireless Inf. Net.,* 3(3), Plenum Press, 1996.}

[31] Cox, D.C., Arnold, H.W., and Porter, P.T., Universal digital portable communications—a system perspective. *IEEE JSAC,* JSAC-5, 764–773, Jun. 1987.

[32] Cox, D.C., Gifford, W.G., and Sherry, H., Low-power digital radio as a ubiquitous subscriber loop. *IEEE Comm. Mag.,* 92–95, Mar. 1991.

[33] {Devasirvatham, D.M.J., Radio propagation studies in a small city for universal portable communications. *IEEE VTC'88,* Conference Record. Philadelphia, PA, 100–104, Jun. 15–17, 1988.}

[34] {Devasirvatham, D.M.J., Radio propagation studies in a small city for universal portable communications, *IEEE VTC'88,* Philadelphia, PA, 100–104, Jun. 15–17, 1988.}

[35] Devasirvatham, D.M.J., et al., Radio propagation measurements at 850 MHz, 1.7 GHz and 4 GHz inside two dissimilar office buildings. *Elect. Lett.,* 26(7), 445–447, 1990a.

[36] Devasirvatham, D.M.J., et al., Multi-frequency radiowave propagation measurements in the portable radio environment. *IEEE ICC'90,* 1334–1340, Apr. 1990b.

[37] {Devasirvatham, D.M.J., Murray, R.R., Arnold, H.W., and Cox, D.C., Four-frequency CW measurements in residential environments for personal communications, symposium on personal indoor and mobile radio communications (PIMRC'93), Yokohama, Japan, 201–205, Sep. 9–11, 1993.}

[38] {Goldhirsh, J. and Vogel, W., Mobile satellite system fade and statistics for shadowing and multipath from roadside trees at UFH and L-band, *IEEE Transactions on Antennas and Propagation,* 37, Apr. 1989.}

[39] Goodman, D.J., Trends in cellular and cordless communications. *IEEE Comm. Mag.,* 31–40, Jun. 1991.

[40] {Hardy, Q., Jacob's Patter, The Wall Street Journal, Friday, Sep. 6, 1996.}

[41] {Hattori, T., PHS system deployment success story, Stanford Univ. Center for Telecom. Symposium, May 21, 1997.}

[42] Hinderling, J., et al., CDMA mobile station modem ASIC. *IEEE CICC 92,* Boston, MA, May. *Intl. J. Wireless Info. Net.,* 1(1), 37–48, Jan. 1994.

[43] Jabbari, B., et al., Network issues for wireless personal communications. *IEEE Comm. Mag.,* 88–98, Jan. 1995.

[44] Katz, R.H., Adaptation and mobility in wireless information systems. *IEEE Personal Comm.*, 1st Quarter, 6–17, 1994.

[45] {Kim, B.-J. and Cox, D.C., Blind equalization for short burst communications over frequency selective wireless channels, 1997 *IEEE VTC'97*, Phoenix, AZ, 2, 544–548, May 4–7, 1997a.}

[46] {Kim, B.-J. and Cox, D.C., Blind sequence estimation for short burst communications over time-varying wireless channels, *IEEE ICUPC'97*, San Diego, CA, 2, 713–717, Oct. 12–16, 1997b.}

[47] {Kim, B.-J. and Cox, D.C., Blind diversity combining equalization for short burst wireless communications, *IEEE Globecom'97*, Phoenix, AZ, 3, 1163–1167, Nov. 3–8, 1997c.}

[48] {Kim, B.-J., Blind equalization for short burst wireless communications, Stanford University Ph.D. Thesis, Jan. 1998.}

[49] {Korea Mobile Telecom, Article on CDMA, Mobile Communications International, Feb. 1997.}

[50] Padgett, J.E., Hattori, T., and Gunther, C., Overview of wireless personal communications. *IEEE Comm. Mag.*, 28–41, Jan. 1995.

[51] {Patel, V., Personal Access Communications System, Stanford Univ. Center for Telecom. Symposium, May 21, 1997.}

[52] {Phillips, B.W., Broadband in the Local Loop, Telecommunications, 37–42, Nov. 1994.}

[53] {PHS, World Congress, Singapore, Nov. 18–20, 1996.}

[54] Qualcomm Technology Forum, Open meeting on status of CDMA technology and review of San Diego experiment. San Diego, CA, Jan. 16–17, 1992.

[55] {Ranade, A., Local access radio interference due to building reflections, *IEEE Transaction on Communications*, 70–74, Jan. 1989.}

[56] {Rizzo, J.F. and Sollenberger, N.R., Multiuser Wireless Access, *IEEE Personal Communications Magazine*, 18–30, Jun. 1995.}

[57] Salmasi, A. and Gilhousen, K.S., On the system design aspects of code division multiple access (CDMA) applied to digital cellular and personal communications networks. *IEEE VTC'91*, St. Louis, MO, 57–62, May 1991.

[58] Sato, Y., A method of self-recovering equalization for multilevel amplitude modulation systems. *IEEE Trans. on Comm.*, 679–682, Jun. 1975.

[59] Schneideman, R., Spread spectrum gains wireless applications. *Microwaves and RF*, 31–42, May 1992.

[60] {Seidel, S.Y. and Arnold, H.W., Propagation measurements of 28 GHz to investigate the performance of local multipoint distribution services (LMDS), *IEEE Globecom'95*, 1995.}

[61] Sollenberger, N.R. and Chuang, J.C-I., Low overhead symbol timing and carrier recovery for TDMA portable radio systems. Third nordic sem. digital land mobile radio comm., Copenhagen, Denmark, 13–15, Sep. 1988.

[62] Sollenberger, N.R., An experimental VLSI implementation of low-overhead symbol timing and frequency offset estimation for TDMA portable radio applications. *IEEE Globecom'90*, San Diego, CA, 1701–1711, Dec. 1990.

[63] Sollenberger, N.R., An experimental TDMA modulation/demodulation CMOS VLSI chip-set. *IEEE CICC'91*, San Diego, CA, May 12–15, 1991.

[64] Sollenberger, N.R. and Afrashteh, A., An experimental low-delay TDMA portable radio link. *Wireless'91*, Calgary, Canada, Jul. 8–10, 1991.

[65] Sollenberger, N.R. and Chuang, J.C.-I., Low-overhead symbol timing and carrier recovery for TDMA portable radio systems. *IEEE Trans. on Comm.*, 1886–1892, Oct. 1990.

[66] Sollenberger, N.R., et al., Architecture and implementation of an efficient and robust TDMA frame structure for digital portable communications. *IEEE Trans. Veh. Tech.*, 40, 250–260, Feb. 1991.

[67] Steele, R., Deploying personal communications networks. *IEEE Comm. Mag.*, 12–15, Sep. 1990.

[68] Steer, D.G., Coexistence and access etiquette in the United States unlicensed PCS band. *IEEE Personal Comm. Mag.*, 4th Quarter, 36–43, 1994.

[69] {Titch, S., Blind Faith, Telephony, Sep. 8, 1997 and Global Telephony, Intertec Publishing Corp., Oct. 30, 1997.}

[70] Treichler, J.R. and Agee, B.G., A new approach to multipath correction of constant modulus signals, *IEEE Trans. on Acoustics, Speech and Signal Processing,* 459–472, Apr. 1983.

[71] Tuthill, J.P., Granger, B.S., and Wurtz, J.L., Request for a pioneer's preference. Before the Federal Communications Commission, Pacific Bell submission for FCC General Docket No. 90–314, RM-7140, and RM-7175, May 4, 1992.

[72] Varma, V.K., et al., Performance of sub-band and RPE coders in the portable communication environment. *Fourth Intl. Conf. on Land Mobile Radio,* Coventry, UK, 221–227, Dec. 14–17, 1987.

[73] Viterbi, A.J., When not to spread spectrum—A sequel. *IEEE Comm. Mag.*, 12–17, Apr. 1985.

[74] {Weseloh, C., Solutions for Wireless Communications, Stanford University Center for Telecommunications, Symposium, Stanford, CA, May 16, 1995.}

[75] Wong, A., Regulating public wireless networks. Workshop on lightwave, wireless and networking technologies, Chinese Univ. of Hong Kong, Hong Kong, Aug. 24, 1994.

[76] {Wong, P., Low-power, low-complexity diversity combining algorithm and VLSI circuit chip for hand-held PCS receivers, Stanford University Ph.D. Thesis, Apr. 1997.}

[77] {Wong, P. and Cox, D.C., Low-complexity cochannel interference cancellation and macroscopic diversity for high capacity PCS, *IEEE ICC'95 Digest,* Seattle, WA, 852–857, Jun. 19–22, 1995.}

[78] {Wong, P. and Cox, D.C., Low-complexity diversity combining algorithms and circuit architectures for cochannel interference cancellation and frequency selective fading mitigation, *IEEE ICC'96 Digest,* Dallas TX, Jun. 23–27, 1996a.}

[79] {Wong, P. and Cox, D.C., Low-complexity diversity combining algorithms and circuit architectures for cochannel interference cancellation and frequency-selective fading mitigation, *IEEE Trans. on Comm.,* 44(9), 1107–1116, Sep. 1996b.}

[80] {Wong, P. and Cox, D.C., Low-complexity interference cancellation and macroscopic diversity for high capacity PCS, *IEEE Trans. on Veh. Tech.,* 124–132, Feb. 1998.}

[81] Zaid, M., Personal mobility in PCS. *IEEE Personal Comm.,* 4th Quarter, 12–16, 1994.

[82] Special issue on advanced mobile phone service (AMPS). *Bell System Tech. J.,* (BSTS). 58, Jan. 1979.

[83] CDMA capacity seen as less than advertised. *Adv. Wireless Comm.,* (AWC). 1–2, Feb. 6, 1991.

[84] Bellcore's Wireless Prototype. *Microcell News,* (MN). 5–6, Jan. 25, 1992.

[85] Bellcore PCS phone shines in Bellsouth test; others fall short. *Adv. Wireless Comm.,* (AWC). 2–3, Feb. 19, 1992a.

[86] TDMA accelerates, but CDMA could still be second standard. *Adv. Wireless Comm.,* (AWC). 6, Mar. 4, 1992b.

[87] Special issue on wireless personal communications. *IEEE Comm. Mag.,* Jan. 1995.

16

Modulation Methods

Gordon L. Stüber
Georgia Institute of Technology

16.1 Introduction

Modulation is the process where the message information is added to the radio carrier. Most first generation cellular systems such as the advanced mobile telephone system (AMPS) use analog **frequency modulation** (**FM**), because analog technology was very mature when these systems were first introduced. Digital modulation schemes, however, are the obvious choice for future wireless systems, especially if data services such as wireless multimedia are to be supported. Digital modulation can also improve spectral efficiency, because digital signals are more robust against channel impairments. Spectral efficiency is a key attribute of wireless systems that must operate in a crowded radio frequency spectrum.

To achieve high spectral efficiency, modulation schemes must be selected that have a high **bandwidth efficiency** as measured in units of bits per second per Hertz of bandwidth. Many wireless communication systems, such as cellular telephones, operate on the principle of frequency reuse, where the carrier frequencies are reused at geographically separated locations. The link quality in these systems is limited by cochannel interference. Hence, modulation schemes must be identified that are both bandwidth efficient and capable of tolerating high levels of cochannel interference. More specifically, digital modulation techniques are chosen for wireless systems that satisfy the following properties.

Compact Power Density Spectrum: To minimize the effect of adjacent channel interference, it is desirable that the power radiated into the adjacent channel be 60–80 dB below that in the desired channel. Hence, modulation techniques with a narrow main lobe and fast rolloff of sidelobes are desirable.

0-8493-8597-0/99/$0.00+$.50

Good Bit-Error-Rate Performance: A low-bit-error probability should be achieved in the presence of cochannel interference, adjacent channel interference, thermal noise, and other channel impairments, such as fading and intersymbol interference.

Envelope Properties: Portable and mobile applications typically employ nonlinear (class C) power amplifiers to minimize battery drain. Nonlinear amplification may degrade the bit-error-rate performance of modulation schemes that transmit information in the amplitude of the carrier. Also, spectral shaping is usually performed prior to up-conversion and nonlinear amplification. To prevent the regrowth of spectral sidelobes during nonlinear amplification, the input signal must have a relatively constant envelope.

A variety of digital modulation techniques are currently being used in wireless communication systems. Two of the more widely used digital modulation techniques for cellular mobile radio are $\pi/4$ phase-shifted quadrature **phase shift keying** ($\pi/4$-QPSK) and **Gaussian minimum shift keying** (**GMSK**). The former is used in the North American IS-54 digital cellular system and Japanese Personal Digital Cellular (PDC), whereas the latter is used in the global system for mobile communications (GSM system). This chapter provides a discussion of these and other modulation techniques that are employed in wireless communication systems.

16.2   Basic Description of Modulated Signals

With any modulation technique, the bandpass signal can be expressed in the form

$$s(t) = \text{Re}\left\{v(t)e^{j2\pi f_c t}\right\} \tag{16.1}$$

where $v(t)$ is the complex envelope, f_c is the carrier frequency, and $\text{Re}\{z\}$ denotes the real part of z. For digital modulation schemes $v(t)$ has the general form

$$v(t) = A\sum_k b\left(t - kT, \boldsymbol{x}_k\right) \tag{16.2}$$

where A is the amplitude of the carrier $\boldsymbol{x}_k = (x_k, x_{k-1}, \ldots, x_{k-K})$ is the data sequence, T is the symbol or baud duration, and $b(t, \boldsymbol{x}_i)$ is an equivalent shaping function usually of duration T. The precise form of $b(t, \boldsymbol{x}_i)$ and the memory length K depends on the type of modulation that is employed. Several examples are provided in this chapter where information is transmitted in the amplitude, phase, or frequency of the bandpass signal.

The **power spectral density** of the bandpass signal $S_{ss}(f)$ is related to the power spectral density of the complex envelope $S_{vv}(f)$ by

$$S_{ss}(f) = \frac{1}{2}\left[S_{vv}\left(f - f_c\right) + S_{vv}\left(f + f_c\right)\right] \tag{16.3}$$

The power density spectrum of the complex envelope for a digital modulation scheme has the general form

$$S_{vv}(f) = \frac{A^2}{T}\sum_m S_{b,m}(f)e^{-j2\pi f m T} \tag{16.4}$$

where

$$S_{b,m}(f) = \frac{1}{2}E\left[B\left(f, \boldsymbol{x}_m\right)B^*\left(f, \boldsymbol{x}_0\right)\right] \tag{16.5}$$

$B(f, \boldsymbol{x}_m)$ is the Fourier transform of $b(t, \boldsymbol{x}_m)$, and $E[\cdot]$ denotes the expectation operator. Usually symmetric signal sets are chosen so that the complex envelope has zero mean, i.e.,

$E[b(t, \boldsymbol{x}_0)] = 0$. This implies that the power density spectrum has no discrete components. If, in addition, $\boldsymbol{x}_m$ and $\boldsymbol{x}_0$ are independent for $|m| > K$, then

$$S_{vv}(f) = \frac{A^2}{T} \sum_{|m|<K} S_{b,m}(f) e^{-j2\pi fmT} \tag{16.6}$$

16.3 Analog Frequency Modulation

With analog frequency modulation the complex envelope is

$$v(t) = A \exp \left[j2\pi k_f \int_0^t m(\tau) \, d\tau \right] \tag{16.7}$$

where $m(t)$ is the modulating waveform and k_f in Hz/v is the frequency sensitivity of the FM modulator. The bandpass signal is

$$s(t) = A \cos \left[2\pi f_c t + 2\pi k_f \int_0^t m(t) \, dt \right] . \tag{16.8}$$

The instantaneous frequency of the carrier $f_i(t) = f_c + k_f m(t)$ varies linearly with the waveform $m(t)$, hence, the name frequency modulation. Notice that FM has a constant envelope making it suitable for nonlinear amplification. However, the complex envelope is a nonlinear function of the modulating waveform $m(t)$ and, therefore, the spectral characteristics of $v(t)$ cannot be obtained directly from the spectral characteristics of $m(t)$.

With the sinusoidal modulating waveform $m(t) = A_m \cos(2\pi f_m t)$ the instantaneous carrier frequency is

$$f_i(t) = f_c + \Delta_f \cos (2\pi f_m t) \tag{16.9}$$

where $\Delta_f = k_f A_m$ is the peak frequency deviation. The complex envelope becomes

$$\begin{aligned} v(t) &= \exp \left[2\pi \int_0^t f_i(t) \, dt \right] \\ &= \exp \left[2\pi f_c t + \beta \sin (2\pi f_m t) \right] \end{aligned} \tag{16.10}$$

where $\beta = \Delta_f / f_m$ is called the modulation index. The bandwidth of $v(t)$ depends on the value of β. If $\beta < 1$, then narrowband FM is generated, where the spectral widths of $v(t)$ and $m(t)$ are about the same, i.e., $2f_m$. If $\beta \gg 1$, then wideband FM is generated, where the spectral occupancy of $v(t)$ is slightly greater than $2\Delta_f$. In general, the approximate bandwidth of an FM signal is

$$W \approx 2\Delta_f + 2f_m = 2\Delta_f \left(1 + \frac{1}{\beta} \right) \tag{16.11}$$

which is a relation known as Carson's rule. Unfortunately, typical analog cellular radio systems use a modulation index in the range $1 \lesssim \beta \lesssim 3$ where Carson's rule is not accurate. Furthermore, the message waveform $m(t)$ is not a pure sinusoid so that Carson's rule does not directly apply.

In analog cellular systems the waveform $m(t)$ is obtained by first companding the speech waveform and then hard limiting the resulting signal. The purpose of the limiter is to control the peak frequency deviation Δ_f. The limiter introduces high-frequency components that must be removed with a low-pass filter prior to modulation. To estimate the bandwidth occupancy, we first determine the ratio of the frequency deviation Δ_f corresponding to the maximum amplitude of $m(t)$, and the highest frequency component B that is present in $m(t)$. These two conditions are the most extreme cases, and the resulting ratio, $D = \Delta_f/B$, is called the *deviation ratio*. Then replace β by D and f_m by B in Carson's rule, giving

$$W \approx 2\Delta_f + 2B = 2\Delta_f \left(1 + \frac{1}{D} \right) \tag{16.12}$$

This approximation will overestimate the bandwidth requirements. A more accurate estimate of the bandwidth requirements must be obtained from simulation or measurements.

16.4   Phase Shift Keying (PSK) and $\pi/4$-QPSK

With **phase shift keying** (PSK), the equivalent shaping function in Eq. (16.2) has the form

$$b(t, \boldsymbol{x}_k) = \psi_T(t)\exp\left[j\frac{\pi}{M} x_k h_s(t) \right], \qquad \boldsymbol{x}_k = x_k \tag{16.13}$$

where $h_s(t)$ is a phase shaping pulse, $\psi_T(t)$ an amplitude shaping pulse, and M the size of the modulation alphabet. Notice that the phase varies linearly with the symbol sequence $\{x_k\}$, hence the name phase shift keying. For a modulation alphabet size of M, $x_k \in \{\pm 1, \pm 3, \ldots, \pm(M-1)\}$. Each symbol x_k is mapped onto $\log_2 M$ source bits. A QPSK signal is obtained by using $M = 4$, resulting in a transmission rate of 2 b/symbol.

Usually, the phase shaping pulse is chosen to be the rectangular pulse $h_s(t) = u_T(t) \triangleq u(t) - u(t - T)$, where $u(t)$ is the unit step function. The amplitude shaping pulse is very often chosen to be a square root raised cosine pulse, where the Fourier transform of $\psi_T(t)$ is

$$\Psi_T(f) = \begin{cases} \sqrt{T} & 0 \leq |f| \leq (1-\beta)/2T \\ \sqrt{\dfrac{T}{2}\left[1 - \sin\dfrac{\pi T}{\beta}\left(f - \dfrac{1}{2T} \right) \right]} & (1-\beta)/2T \leq |f| \leq (1+\beta)/2T \end{cases} \tag{16.14}$$

The receiver implements the same filter $\Psi_R(f) = \Psi_T(f)$ so that the overall pulse has the raised cosine spectrum $\Psi(f) = \Psi_R(f)\Psi_T(f) = |\Psi_T(f)|^2$. If the channel is affected by flat fading and additive white Gaussian noise, then this partitioning of the filtering operations between the transmitter and receiver will optimize the signal to noise ratio at the output of the receiver filter at the sampling instants. The rolloff factor β usually lies between 0 and 1 and defines the **excess bandwidth** $100\beta\%$. Using a smaller β results in a more compact power density spectrum, but the link performance becomes more sensitive to errors in the symbol timing. The IS-54 system uses $\beta = 0.35$, while PDC uses $\beta = 0.5$.

The time domain pulse corresponding to Eq. (16.14) can be obtained by taking the inverse Fourier transform, resulting in

$$\psi_T(t) = 4\beta\frac{\cos\left[(1+\beta)\,\pi t/T \right] + \sin\left[(1-\beta)\,\pi t/T \right](4\beta t/T)^{-1}}{\pi\sqrt{T}\left[1 - 16\beta^2 t^2/T^2 \right]} \tag{16.15}$$

A typical square root raised cosine pulse with a rolloff factor of $\beta = 0.5$ is shown in Fig. 16.1. Strictly speaking the pulse $\psi_T(t)$ is noncausal, but in practice a truncated time domain pulse is used. For example, in Fig. 16.1 the pulse is truncated to $6T$ and time shifted by $3T$ to yield a causal pulse.

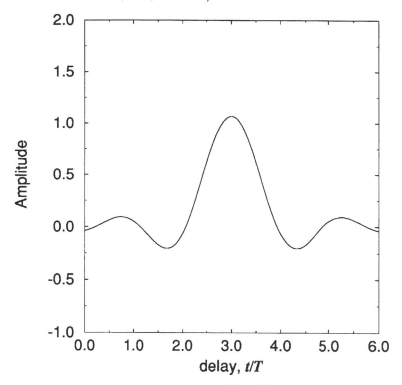

FIGURE 16.1   Square root raised cosine pulse with rolloff factor $\beta = 0.5$.

Unlike conventional QPSK that has four possible transmitted phases, $\pi/4$-QPSK has eight possible transmitted phases. Let $\theta(n)$ be the transmitted carrier phase for the nth epoch, and let $\Delta\theta(n) = \theta(n) - \theta(n-1)$ be the differential carrier phase between epochs n and $n-1$. With $\pi/4$-QPSK, the transmission rate is 2 b/symbol and the differential phase is related to the symbol sequence $\{x_n\}$ through the mapping

$$\Delta\theta(n) = \begin{cases} -3\pi/4, & x_n = -3 \\ -\pi/4, & x_n = -1 \\ \pi/4, & x_n = +1 \\ 3\pi/4, & x_n = +3 \end{cases} \qquad (16.16)$$

Since the symbol sequence $\{x_n\}$ is random, the mapping in Eq. (16.16) is arbitrary, except that the phase differences must be $\pm\pi/4$ and $\pm3\pi/4$. The phase difference with the given mapping can be written in the convenient algebraic form

$$\Delta\theta(n) = x_n \frac{\pi}{4} \qquad (16.17)$$

which allows us to write the equivalent shaping function of the $\pi/4$-QPSK signal as

$$\begin{aligned} b\left(t, \underline{x}_k\right) &= \psi(t)\exp\left\{j\left[\theta(k-1) + x_k\frac{\pi}{4}\right]\right\} \\ &= \psi_T(t)\exp\left[j\frac{\pi}{4}\left(\sum_{n=-\infty}^{k-1} x_n + x_k\right)\right] \end{aligned} \qquad (16.18)$$

The summation in the exponent represents the accumulated carrier phase, whereas the last term is the phase change due to the kth symbol. Observe that the phase shaping function is the rectangular pulse $u_T(t)$. The amplitude shaping function $\psi_T(t)$ is usually the square root raised cosine pulse in Eq. (16.15).

The phase states of QPSK and $\pi/4$-QPSK signals can be summarized by the signal space diagram in Fig. 16.2 that shows the phase states and allowable transitions between the phase states. However, it does not describe the actual phase trajectories. A typical diagram showing phase trajectories with square root raised cosine pulse shaping is shown in Fig. 16.3. Note that the phase trajectories do not pass through the origin. This reduces the envelope fluctuations of the signal making it less susceptible to amplifier nonlinearities and reduces the dynamic range required of the power amplifier.

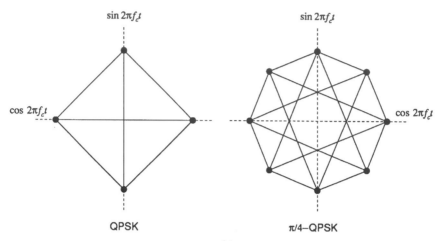

FIGURE 16.2   Signal-space constellations for QPSK and $\pi/4$-DQPSK.

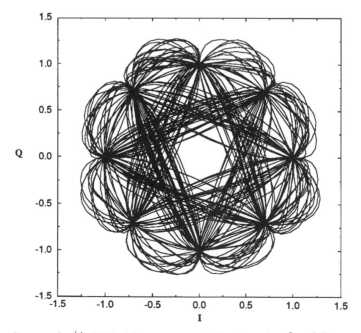

FIGURE 16.3   Phase diagram of $\pi/4$-QPSK with square root raised cosine pulse; $\beta = 0.5$.

The power density spectrum of QPSK and $\pi/4$-QPSK depends on both the amplitude and phase shaping pulses. For the rectangular phase shaping pulse $h_s(t) = u_T(t)$, the power density spectrum of the complex envelope is

$$S_{vv}(f) = \frac{A^2}{T}\left|\Psi_T(f)\right|^2 \tag{16.19}$$

With square root raised cosine pulse shaping, $\Psi_T(f)$ has the form defined in Eq. (16.14). The power density spectrum of a pulse $\tilde{\psi}_T(t)$ that is obtained by truncating $\psi_T(t)$ to length τ can be obtained by writing $\tilde{\psi}_T(t) = \psi_T(t)\text{rect}(t/\tau)$. Then $\tilde{\Psi}_T(f) = \Psi_T(f)*\tau\text{sinc}(f\tau)$, where $*$ denotes the operation of convolution, and the power density spectrum is again obtained by applying Eq. (16.19). Truncation of the pulse will regenerate some side lobes, thus causing adjacent channel interference. Figure 16.4 illustrates the power density spectrum of a truncated square root raised cosine pulse for various truncation lengths τ.

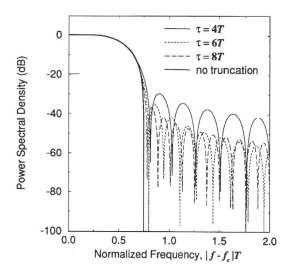

FIGURE 16.4 Power density spectrum of truncated square root raised cosine pulse with various truncation lengths; $\beta = 0.5$.

16.5 Continuous Phase Modulation (CPM) and MSK

Continuous phase modulation (CPM) refers to a broad class of frequency modulation techniques where the carrier phase varies in a continuous manner. A comprehensive treatment of CPM is provided in [1]. CPM schemes are attractive because they have constant envelope and excellent spectral characteristics. The complex envelope of any CPM signal is

$$v(t) = A\exp\left[j2\pi k_f \int_{-\infty}^{t} \sum_n x_n h_s(\tau - nT)\,\mathrm{d}\tau\right] \tag{16.20}$$

The instantaneous frequency deviation from the carrier is

$$f_{\text{dev}}(t) = k_f \sum_n x_n h_s(t - nT) \tag{16.21}$$

where k_f is the peak frequency deviation. If the frequency shaping pulse $h_s(t)$ has duration T, then the equivalent shaping function in Eq. (16.2) has the form

$$b(t, \boldsymbol{x}_k) = \exp\left\{ j\left[\beta(T) \sum_{n=-\infty}^{k-1} x_n + x_k \beta(t) \right] \right\} u_T(t) \tag{16.22}$$

where

$$\beta(t) = \begin{cases} 0, & t < 0 \\ \dfrac{\pi h}{\int_0^T h_s(\tau)\,\mathrm{d}\tau} \displaystyle\int_0^t h_s(\tau)\,\mathrm{d}\tau, & 0 \le t \le T \\ \pi h, & t \ge T \end{cases} \tag{16.23}$$

is the phase shaping pulse, and $h = \beta(T)/\pi$ is called the modulation index.

Minimum shift keying (MSK) is a special form of binary CPM ($x_k \in \{-1, +1\}$) that is defined by a rectangular frequency shaping pulse $h_s(t) = u_T(t)$, and a modulation index $h = 1/2$ so that

$$\beta(t) = \begin{cases} 0, & t < 0 \\ \pi t/2T, & 0 \le t \le T \\ \pi/2, & t \ge T \end{cases} \tag{16.24}$$

Therefore, the complex envelope is

$$v(t) = A \exp\left(j\frac{\pi}{2} \sum_{n=-\infty}^{k-1} x_n + \frac{\pi}{2} x_k \frac{t - kT}{T} \right) \tag{16.25}$$

A MSK signal can be described by the phase trellis diagram shown in Fig. 16.5 which plots the time behavior of the phase

$$\theta(t) = \frac{\pi}{2} \sum_{n=-\infty}^{k-1} x_n + \frac{\pi}{2} x_k \frac{t - kT}{T} \tag{16.26}$$

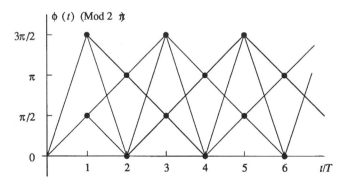

FIGURE 16.5   Phase-trellis diagram for MSK.

The MSK bandpass signal is

$$\begin{aligned} s(t) &= A \cos\left(2\pi f_c t + \frac{\pi}{2} \sum_{n=-\infty}^{k-1} x_n + \frac{\pi}{2} x_k \frac{t - kT}{T} \right) \\ &= A \cos\left[2\pi\left(f_c + \frac{x_k}{4T} \right) t - \frac{k\pi}{2} x_k + \frac{\pi}{2} \sum_{n=-\infty}^{k-1} x_n \right] \quad kT \le t \le (k+1)T \end{aligned} \tag{16.27}$$

From Eq. (16.27) we observe that the MSK signal has one of two possible frequencies $f_L = f_c - 1/4T$ or $f_U = f_c + 1/4T$ during each symbol interval. The difference between these frequencies is $f_U - f_L = 1/2T$. This is the minimum frequency difference between two sinusoids of duration T that will ensure orthogonality with coherent demodulation [7], hence, the name minimum shift keying. By applying various trigonometric identities to Eq. (16.27) we can write

$$s(t) = A\left[x_k^I \psi(t - k2T)\cos\left(2\pi f_c t\right) - x_k^Q \psi(t - k2T - T)\sin\left(2\pi f_c t\right)\right] ,$$
$$kT \leq t \leq (k+1)T \qquad (16.28)$$

where

$$x_k^I = -x_{k-1}^Q x_{2k-1}$$
$$x_k^Q = x_k^I x_{2k}$$
$$\psi(t) = \cos\left(\frac{\pi t}{2T}\right), \qquad -T \leq t \leq T$$

Note that the x_k^I and x_k^Q are independent binary symbols that take on elements from the set $\{-1, +1\}$, and the half-sinusoid amplitude shaping pulse $\psi(t)$ has duration $2T$ and $\psi(t - T) = \sin(\pi t/2T), 0 \leq t \leq 2T$. Therefore, MSK is equivalent to offset quadrature amplitude shift keying (OQASK) with a half-sinusoid amplitude shaping pulse.

To obtain the power density spectrum of MSK, we observe from Eq. (16.28) that the equivalent shaping function of MSK has the form

$$b\left(t, \boldsymbol{x}_k\right) = x_k^I \psi(t) + j x_k^Q \psi(t - T) \qquad (16.29)$$

The Fourier transform of Eq. (16.29) is

$$B\left(f, \boldsymbol{x}_k\right) = \left(x_k^I + j x_k^Q e^{-j2\pi fT}\right) \Psi(f) \qquad (16.30)$$

Since the symbols x_k^I and x_k^Q are independent and zero mean, it follows from Eqs. (16.5) and (16.6) that

$$S_{vv}(f) = \frac{A^2 |\Psi(f)|^2}{2T} \qquad (16.31)$$

Therefore, the power density spectrum of MSK is determined solely by the Fourier transform of the half-sinusoid amplitude shaping pulse $\psi(t)$, resulting in

$$S_{vv}(f) = \frac{16A^2 T}{\pi^2}\left[\frac{\cos 2\pi fT}{1 - 16f^2 T^2}\right]^2 \qquad (16.32)$$

The power spectral density of MSK is plotted in Fig. 16.8. Observe that an MSK signal has fairly large sidelobes compared to $\pi/4$-QPSK with a truncated square root raised cosine pulse (c.f., Fig. 16.4).

16.6 Gaussian Minimum Shift Keying

MSK signals have all of the desirable attributes for mobile radio, except for a compact power density spectrum. This can be alleviated by filtering the modulating signal $x(t) = \Sigma_n x_n u_T(t - nT)$ with a low-pass filter prior to frequency modulation, as shown in Fig. 16.6.

Such filtering removes the higher frequency components in $x(t)$ and, therefore, yields a more compact spectrum. The low-pass filter is chosen to have 1) narrow bandwidth and a sharp transition band, 2) low-overshoot impulse response, and 3) preservation of the output pulse area to ensure a phase shift of $\pi/2$.

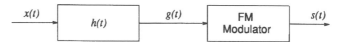

FIGURE 16.6   Premodulation filtered MSK.

GMSK uses a low-pass filter with the following transfer function:

$$H(f) = A \exp\left\{ -\left(\frac{f}{B}\right)^2 \frac{\ln 2}{2} \right\} \tag{16.33}$$

where B is the 3-dB bandwidth of the filter and A a constant. It is apparent that $H(f)$ is bell shaped about $f = 0$, hence the name Gaussian MSK. A rectangular pulse $\text{rect}(t/T) = u_T(t + T/2)$ transmitted through this filter yields the frequency shaping pulse

$$h_s(t) = A\sqrt{\frac{2\pi}{\ln 2}}(BT) \int_{t/T-1/2}^{t/T+1/2} \exp\left\{ -\frac{2\pi^2(BT)^2 x^2}{\ln 2} \right\} dx \tag{16.34}$$

The phase change over the time interval from $-T/2 \leq t \leq T/2$ is

$$\theta\left(\frac{T}{2}\right) - \theta\left(\frac{-T}{2}\right) = x_0\beta_0(T) + \sum_{\substack{n=-\infty \\ n\neq 0}}^{\infty} x_n\beta_n(T) \tag{16.35}$$

where

$$\beta_n(T) = \frac{\pi h}{\int_{-\infty}^{\infty} h_s(\nu)\, d\nu} \int_{-T/2-nT}^{T/2-nT} h_s(\nu)\, d\nu \tag{16.36}$$

The first term in Eq. (16.35) is the desired term, and the second term is the intersymbol interference (ISI) introduced by the premodulation filter. Once again, with GMSK $h = 1/2$ so that a total phase shift of $\pi/2$ is maintained.

Notice that the pulse $h_s(t)$ is noncausal so that a truncated pulse must be used in practice. Figure 16.7 plots a GMSK frequency shaping pulse that is truncated to $\tau = 5T$ and time shifted by $2.5T$, for various normalized filter bandwidths BT. Notice that the frequency shaping pulse has a duration greater than T so that ISI is introduced. As BT decreases, the induced ISI is increased. Thus, whereas a smaller value of BT results in a more compact power density spectrum, the induced ISI will degrade the bit-error-rate performance. Hence, there is a tradeoff in the choice of BT. Some studies have indicated that $BT = 0.25$ is a good choice for cellular radio systems [6].

The power density spectrum of GMSK is quite difficult to obtain, but can be computed by using published methods [3]. Figure 16.8 plots the power density spectrum for $BT = 0.2$, 0.25, and 0.3, obtained from Wesolowski, [8]. Observe that the spectral sidelobes are greatly reduced by the Gaussian low-pass filter.

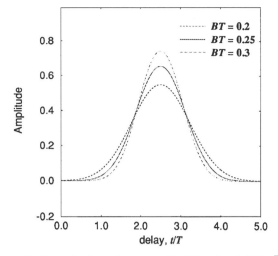

FIGURE 16.7   GMSK frequency shaping pulse for various normalized filter bandwidths BT.

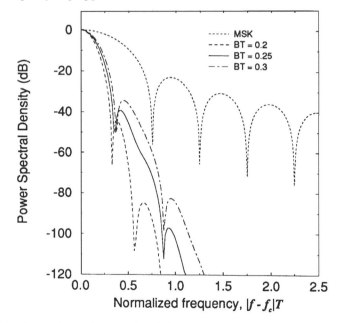

FIGURE 16.8   Power density spectrum of MSK and GMSK.

16.7   Orthogonal Frequency Division Multiplexing (OFDM)

Orthogonal frequency division multiplexing (OFDM) is a modulation technique that has been recently suggested for use in cellular radio [2], digital audio broadcasting [4], and digital video broadcasting. The basic idea of OFDM is to transmit blocks of symbols in parallel by employing a (large) number of orthogonal subcarriers. With block transmission, N serial source symbols each with period T_s are converted into a block of N parallel modulated symbols each with period $T = NT_s$. The block length N is chosen so that $NT_s \gg \sigma_\tau$, where σ_τ is the rms delay spread of the channel. Since the symbol rate on each subcarrier is much less than the serial source rate, the effects of delay spread are greatly reduced. This has practical advantages because it may reduce or even eliminate the need for equalization.

Although the block length N is chosen so that $NT_s \gg \sigma_\tau$, the channel dispersion will still cause consecutive blocks to overlap. This results in some residual ISI that will degrade the performance. This residual ISI can be eliminated at the expense of channel capacity by using guard intervals between the blocks that are at least as long as the effective channel impulse response.

The complex envelope of an OFDM signal is described by

$$v(t) = A \sum_k \sum_{n=0}^{N-1} x_{k,n} \phi_n(t - kT) \tag{16.37}$$

where

$$\phi_n(t) = \exp\left\{ j \frac{2\pi \left(n - \dfrac{N-1}{2}\right) t}{T} \right\} U_T(t), \qquad n = 0, 1, \ldots, N-1 \tag{16.38}$$

are orthogonal waveforms and $U_T(t)$ is a rectangular shaping function. The frequency separation of the subcarriers, $1/T$, ensures that the subcarriers are orthogonal and phase continuity is maintained from one symbol to the next, but is twice the minimum required for orthogonality with coherent detection. At epoch k, N-data symbols are transmitted by using the N distinct pulses. The data symbols $x_{k,n}$ are often chosen from an M-ary **quadrature amplitude modulation** (M-QAM) constellation, where $x_{k,n} = x_{k,n}^I + jx_{k,n}^Q$ with $x_{k,n}^I, x_{k,n}^Q \in \{\pm 1, \pm 3, \ldots, \pm(N-1)\}$ and $N = \sqrt{M}$.

A key advantage of using OFDM is that the modulation can be achieved in the discrete domain by using either an inverse discrete Fourier transform (IDFT) or the more computationally efficient inverse fast Fourier transform (IFFT). Considering the data block at epoch $k = 0$ and ignoring the frequency offset $\exp\{-j[2\pi(N-1)t/2T]\}$, the complex low-pass OFDM signal has the form

$$v(t) = \sum_{n=0}^{N-1} x_{0,n} \exp\left\{ \frac{j2\pi nt}{NT_s} \right\}, \qquad 0 \le t \le T \tag{16.39}$$

If this signal is sampled at epochs $t = kT_s$, then

$$v^k = v\left(kT_s\right) = \sum_{n=0}^{N-1} x_{0,n} \exp\left\{ \frac{j2\pi nk}{N} \right\}, \qquad k = 0, 1, \ldots, N-1 \tag{16.40}$$

Observe that the sampled OFDM signal has duration N and the samples $v^0, v^1, \ldots, v^{N-1}$ are just the IDFT of the data block $x_{0,0}, x_{0,1}, \ldots, x_{0,N-1}$. A block diagram of an OFDM transmitter is shown in Fig. 16.9.

The power spectral density of an OFDM signal can be obtained by treating OFDM as independent modulation on subcarriers that are separated in frequency by $1/T$. Because the subcarriers are only separated by $1/T$, significant spectral overlap results. Because the subcarriers are orthogonal, however, the overlap improves the spectral efficiency of the scheme. For a signal constellation with zero mean and the waveforms in Eq. (16.38), the power density spectrum of the complex envelope is

$$S_{vv}(f) = \frac{A^2}{T} \sigma_x^2 \sum_{n=0}^{N-1} \left| \mathrm{sinc}\left[fT - \left(n - \frac{N-1}{2}\right) \right] \right|^2 \tag{16.41}$$

where $\sigma_x^2 = \frac{1}{2} E[|x_{k,n}|^2]$ is the variance of the signal constellation. For example, the complex envelope power spectrum of OFDM with $N = 32$ subcarriers is shown in Fig. 16.10.

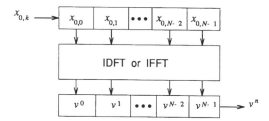

FIGURE 16.9 Block diagram of OFDM transmitter using IDFT or IFFT.

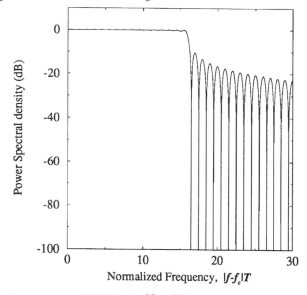

FIGURE 16.10 Power density spectrum of OFDM with $N - 32$.

16.8 Conclusions

A variety of modulation schemes are employed in wireless communication systems. Wireless modulation schemes must have a compact power density spectrum, while at the same time providing a good bit-error-rate performance in the presence of channel impairments such as cochannel interference and fading. The most popular digital modulation techniques employed in wireless systems are GMSK in the European GSM system, $\pi/4$-QPSK in the North American IS-54 and Japanese PDC systems, and OFDM in digital audio broadcasting systems.

Defining Terms

Bandwidth efficiency: Transmission efficiency of a digital modulation scheme measured in units of bits per second per Hertz of bandwidth.

Continuous phase modulation: Frequency modulation where the phase varies in a continuous manner.

Excess bandwidth: Percentage of bandwidth that is in excess of the minimum of $1/2T$ (T is the baud or symbol duration) required for data communication.

Frequency modulation: Modulation where the instantaneous frequency of the carrier varies linearly with the data signal.

Gaussian minimum shift keying: MSK where the data signal is prefiltered with a Gaussian filter prior to frequency modulation.

Minimum shift keying: A special form of continuous phase modulation having linear phase trajectories and a modulation index of 1/2.

Orthogonal frequency division multiplexing: Modulation by using a collection of low-bit-rate orthogonal subcarriers.

Phase shift keying: Modulation where the instantaneous phase of the carrier varies linearly with the data signal.

Power spectral density: Relative power in a modulated signal as a function of frequency.

Quadrature amplitude modulation: Modulation where information is transmitted in the amplitude of the cosine and sine components of the carrier.

References

[1] Anderson, J.B., Aulin, T., and Sundberg, C.-E., *Digital Phase Modulation*, Plenum Press, New York, 1986.

[2] Birchler, M.A. and Jasper, S.C., A 64 kbps digital land mobile radio system employing M-16QAM. *Proc. 5th Nordic Sem. Dig. Mobile Radio Commun.*, 237–241, Dec. 1992.

[3] Garrison, G.J., A power spectral density analysis for digital FM, *IEEE Trans. Commun.*, COM-23, 1228–1243, Nov. 1975.

[4] Le Floch, B., Halbert-Lassalle, R., and Castelain, D., Digital sound broadcasting to mobile receivers, *IEEE Trans. Consum. Elec.*, 35, Aug. 1989.

[5] Murota, K. and Hirade, K., GMSK modulation for digital mobile radio telephony, *IEEE Trans. Commun.*, COM-29, 1044–1050, Jul. 1981.

[6] Murota, K., Kinoshita, K., and Hirade, K., Spectral efficiency of GMSK land mobile radio. *Proc. ICC'81*, 23.8.1, Jun. 1981.

[7] Proakis, J.G., *Digital Communications*, 2nd ed., McGraw-Hill, New York, 1989.

[8] Wesolowski, K., Private Communication, 1994.

Further Information

A good discussion of digital modem techniques is presented in *Advanced Digital Communications,* edited by K. Feher, Prentice-Hall, 1987.

Proceedings of various IEEE conferences such as the Vehicular Technology Conference, International Conference on Communications, and Global Telecommunications Conference, document the lastest development in the field of wireless communications each year.

Journals such as the *IEEE Transactions on Communications* and *IEEE Transactions on Vehicular Technology* report advances in wireless modulation.

17

Access Methods

Bernd-Peter Paris
George Mason University

17.1 Introduction

The radio channel is fundamentally a broadcast communication medium. Therefore, signals transmitted by one user can potentially be received by all other users within range of the transmitter. Although this high connectivity is very useful in some applications, like broadcast radio or television, it requires stringent access control in wireless communication systems to avoid, or at least to limit, interference between transmissions. Throughout, the term wireless communication systems is taken to mean communication systems that facilitate two-way communication between a portable radio communication terminal and the fixed network infrastructure. Such systems range from mobile cellular systems through personal communication systems (PCS) to cordless telephones.

The objective of wireless communication systems is to provide communication channels on demand between a portable radio station and a radio port or base station that connects the user to the fixed network infrastructure. Design criteria for such systems include **capacity**, cost of implementation, and quality of service. All of these measures are influenced by the method used for providing multiple-access capabilities. However, the opposite is also true: the access method should be chosen carefully in light of the relative importance of design criteria as well as the system characteristics.

Multiple access in wireless radio systems is based on insulating signals used in different connections from each other. The support of parallel transmissions on the uplink and downlink, respectively, is called multiple access, whereas the exchange of information in both directions of a connection is referred to as **duplexing**. Hence, multiple access and duplexing are methods that facilitate the sharing of the broadcast communication medium. The necessary insulation is achieved by assigning to each transmission different components of the domains that contain the signals. The signal domains commonly used to provide multiple access capabilities include the following.

Spatial domain: All wireless communication systems exploit the fact that radio signals experience rapid attenuation during propagation. The propagation exponent ρ on typical radio channels lies between $\rho = 2$ and $\rho = 6$ with $\rho = 4$ a typical value. As signal strength decays inversely proportional to the ρth power of the distance, far away transmitters introduce interference that is negligible compared to the strength of the desired signal. The cellular design principle is based on the ability to reuse signals safely if a minimum reuse distance is maintained. Directional antennas can be used to enhance the insulation between signals. We will not focus further on the spatial domain in this treatment of access methods.

Frequency domain: Signals which occupy nonoverlapping frequency bands can be easily separated using appropriate bandpass filters. Hence, signals can be transmitted simultaneously without interfering with each other. This method of providing multiple access capabilities is called **frequency-division multiple access (FDMA)**.

Time domain: Signals can be transmitted in nonoverlapping time slots in a round-robin fashion. Thus, signals occupy the same frequency band but are easily separated based on their time of arrival. This multiple access method is called **time-division multiple access (TDMA)**.

Code domain: In **code-division multiple access (CDMA)** different users employ signals that have very small cross-correlation. Thus, correlators can be used to extract individual signals from a mixture of signals even though they are transmitted simultaneously and in the same frequency band. The term code-division multiple-access is used to denote this form of channel sharing. Two forms of CDMA are most widely employed and will be described in detail subsequently, frequency hopping (FH) and direct sequence (DS).

System designers have to decide in favor of one, or a combination, of the latter three domains to facilitate multiple access. The three access methods are illustrated in Fig. 17.1. The principal idea in all three of these access methods is to employ signals that are orthogonal or nearly orthogonal. Then, correlators that project the received signal into the subspace of the desired signal can be employed to extract a signal without interference from other transmissions.

Preference for one access method over another depends largely on overall system characteristics, as we will see in the sequel. No single access method is universally preferable, and system considerations should be carefully weighed before the design decision is made. Before going into the detailed description of the different access methods, we will discuss briefly the salient features of some wireless communication systems. This will allow us later to assess the relative merits of the access methods in different scenarios.

17.2   Relevant Wireless Communication System Characteristics

Modern wireless radio systems range from relatively simple cordless telephones to mobile cellular systems and the emerging personal communication systems (PCS). It is useful to consider such diverse systems as cordless telephone and mobile cellular radio to illustrate some of the fundamental characteristics of wireless communication systems [2].

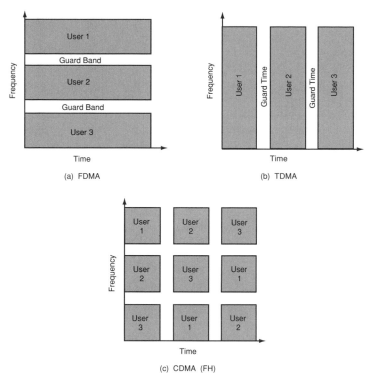

FIGURE 17.1 Multiple-access methods for wireless communication systems.

A summary of the relevant parameters and characteristics for cordless telephone and cellular radio is given in Table 17.1 As evident from that table, the fundamental differences between the two systems are speech quality and the area covered by a base station. The high speech quality requirement in the cordless application is the consequence of the availability of tethered access in the home and office and the resulting direct competition with wire-line telephone services. In the mobile cellular application, the user has no alternative to the wireless access and may be satisfied with lower, but still acceptable, quality of service.

TABLE 17.1 Summary of Relevant Characteristics of Cordless Telephone and Cellular Mobile Radio

Characteristic or Parameter	Cordless Telephone	Cellular Radio
Speech quality	Toll quality	Varying with channelquality; possibly decreased by speech pause exploitation
Transmission range	<100 m	100 m–30 km
Transmit power	Milliwatts	Approx. 1 W
Base station antenna height	Approx. 1 m	Tens of meters
Delay spread	Approx. 1 μs	Approx. 10 μs
Complexity of base station	Low	High
Complexity of user set	Low	High

In cordless telephone applications the transmission range is short because the base station can simply be moved to a conveniently located wire-line access point (wall jack) to provide wireless network access where desired. In contrast, the mobile cellular base station must provide access for users throughout a large geographical area of up to approximately 30 km

(20 mi) around the base station. This large coverage area is necessary to economically meet the promise of uninterrupted service to roaming users.

The different range requirements directly affect the transmit power and antenna height for the two systems. High-power transmitters used in mobile cellular user sets consume far more power than even complex signal processing hardware. Hence, sophisticated signal processing, including speech compression, voice activity detection, error correction and detection, and adaptive equalization, can be employed without substantial impact on the battery life in portable hand sets. Furthermore, such techniques are consistent with the goals of increased range and support of large numbers of users with a single, expensive base station. On the other hand, the high mobile cellular base station antennas introduce delay spreads that are one or two orders of magnitude larger than those commonly observed in cordless telephone applications.

Clearly, the two systems just considered are at extreme ends of the spectrum of wireless communications systems. Most notably, the emerging PCS systems fall somewhere between the two. However, the comparison above highlights some of the system characteristics that should be considered when discussing access methods for wireless communication systems.

17.3   Frequency Division Multiple Access

As mentioned in Section 17.1, in FDMA nonoverlapping frequency bands are allocated to different users on a continuous time basis. Hence, signals assigned to different users are clearly orthogonal, at least ideally. In practice, out-of-band spectral components can not be completely suppressed leaving signals not quite orthogonal. This necessitates the introduction of guard bands between frequency bands to reduce adjacent channel interference, i.e., inference from signals transmitted in adjacent frequency bands; see also Fig. 17.1(a).

It is advantageous to combine FDMA with time-division duplexing (TDD) to avoid simultaneous reception and transmission that would require insulation between receive and transmit antennas. In this scenario, the base station and portable take turns using the same frequency band for transmission. Nevertheless, combining FDMA and frequency division duplex is possible in principle, as is evident from the analog FM-based systems deployed throughout the world since the early 1980s.

17.3.1   Channel Considerations

In principle there exists the well-known duality between TDMA and FDMA; see [1, p. 113]. In the wireless environment, however, propagation related factors have a strong influence on the comparison between FDMA and TDMA. Specifically, the duration of a transmitted symbol is much longer in FDMA than in TDMA. As an immediate consequence, an equalizer is typically not required in an FDMA-based system because the delay spread is small compared to the symbol duration.

To illustrate this point, consider a hypothetical system that transmits information at a constant rate of 50 kb/s. This rate would be sufficient to support 32-kb/s adaptive differential pulse code modulation (ADPCM) speech encoding, some coding for error protection, and control overhead. If we assume further that some form of QPSK modulation is employed, the resulting symbol duration is 40 μs. In relation to delay spreads of approximately 1 μs in the cordless application and 10 μs in cellular systems, this duration is large enough that only little intersymbol interference is introduced. In other words, the channel is frequency nonselective, i.e., all spectral components of the signal are affected equally by the channel. In the cordless application an equalizer is certainly not required; cellular receivers may require equalizers capable of removing intersymbol interference between adjacent bits. Furthermore, it is well known that intersymbol interference between adjacent bits can be removed without loss in SNR by using maximum-likelihood sequence estimation; e.g., [8, p. 622].

Hence, rather simple receivers can be employed in FDMA systems at these data rates. However, there is a flip side to the argument. Recall that the Doppler spread, which characterizes the rate at which the channel impulse response changes, is given approximately by $B_d = v/cf_c$, where v denotes the speed of the mobile user, c is the propagation speed of the electromagnetic waves carrying the signal, and f_c is the carrier frequency. Thus, for systems operating in the vicinity of 1 GHz, B_d will be less than 1 Hz in the cordless application and typically about 100 Hz for a mobile traveling on a highway. In either case, the signal bandwidth is much larger than the Doppler spread B_d, and the channel can be characterized as slowly fading. Whereas this allows tracking of the carrier phase and the use of coherent receivers, it also means that fade durations are long in comparison to the symbol duration and can cause long sequences of bits to be subject to poor channel conditions. The problem is compounded by the fact that the channel is frequency nonselective because it implies that the entire signal is affected by a fade.

To overcome these problems either time diversity, frequency diversity, or spatial diversity could be employed. Time diversity can be accomplished by a combination of coding and interleaving if the fading rate is sufficiently large. For very slowly fading channels, such as the cordless application, the necessary interleaving depth would introduce too much delay to be practical. Frequency diversity can be introduced simply by slow frequency hopping, a technique that prescribes users to change the carrier frequency periodically. Frequency hopping is a form of spectrum spreading because the bandwidth occupied by the resulting signal is much larger than the symbol rate. In contrast to direct sequence spread spectrum discussed subsequently, however, the instantaneous bandwidth is not increased. The jumps between different frequency bands effectively emulate the movement of the portable and, thus, should be combined with the just described time-diversity methods. Spatial diversity is provided by the use of several receive or transmit antennas. At carrier frequencies exceeding 1 GHz, antennas are small and two or more antennas can be accommodated even in the hand set. Furthermore, if FDMA is combined with time-division duplexing, multiple antennas at the base station can provide diversity on both uplink and downlink. This is possible because the channels for the two links are virtually identical, and the base station, using channel information gained from observing the portable's signal, can transmit signals at each antenna such that they combine coherently at the portable's antenna. Thus, signal processing complexity is moved to the base station extending the portable's battery life.

17.3.2 Influence of Antenna Height

In the cellular mobile environment base station antennas are raised considerably to increase the coverage area. Antennas mounted on towers and rooftops are a common sight, and antenna heights of 50 m above ground are no exceptions. Besides increasing the coverage area, this has the additional effect that frequently there exists a better propagation path between two base station antennas than between a mobile and the base station; see Fig. 17.2.

Assuming that FDMA is used in conjunction with TDD as specified at the beginning of this section, then base stations and mobiles transmit on the same frequency. Now, unless there is tight synchronization between all base stations, signals from other base stations will interfere with the reception of signals from portables at the base station. To keep the interference at acceptable levels, it is necessary to increase the reuse distance, i.e., the distance between cells using the same frequencies. In other words, sufficient insulation in

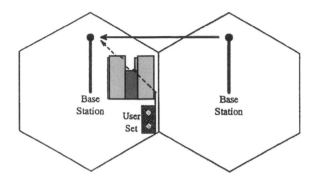

FIGURE 17.2   High base station antennas lead to stronger propagation paths between base stations than between a user set and its base stations.

the spatial domain must be provided to facilitate the separation of signals. Note that these comments apply equally to cochannel and adjacent channel interference.

This problem does not arise in cordless applications. Base station antennas are generally of the same height as user sets. Hence, interference created by base stations is subject to the same propagation conditions as signals from user sets. Furthermore, in cordless telephone applications there are frequently attenuating obstacles, such as walls, between base stations that reduce intracell interference further. Note that this reduction is vital for the proper functioning of cordless telephones since there is typically no network planning associated with installing a cordless telephone. As a safety feature, to overcome intercell interference, adaptive channel management strategies based on sensing interference levels can be employed.

17.3.3   Example 17.1: CT2

The CT2 standard was originally adopted in 1987 in Great Britain and improved with a common air interface (CAI) in 1989. The CAI facilitates interoperability between equipment from different vendors whereas the original standard only guarantees noninterference. The CT2 standard is used in home and office cordless telephone equipment and has been used for telepoint applications [5].

CT2 operates in the frequency band 864–868 MHz and uses carriers spaced at 100 kHz. FDMA with time division duplexing is employed. The combined gross bit rate is 72 kb/s, transmitted in frames of 2-ms duration of which the first-half carries downlink and the second-half carries uplink information. This setup supports a net bit rate of 32 kb/s of user data (32-kb/s ADPCM encoded speech) and 2-kb/s control information in each direction. The CT2 modulation technique is binary frequency shift keying.

17.3.4   Further Remarks

From the preceding discussion it is obvious that FDMA is a good candidate for applications like cordless telephone. In particular, the simple signal processing makes it a good choice for inexpensive implementation in the benign cordless environment. The possibility of concentration of signal processing functions in the base station strengthens this aspect.

In the cellular application, on the other hand, FDMA is inappropriate because of the lack of built-in diversity and the potential for severe intercell interference between base stations. A further complication arises from the difficulty of performing handovers if base-stations are not tightly synchronized.

For PCS the decision is not as obvious. Depending on whether the envisioned PCS application resembles more a cordless private branch exchange (PBX) than a cellular system, FDMA may be an appropriate choice. We will see later that it is probably better to opt for a combined TDMA/FDMA or a CDMA-based system to avoid the pitfalls of pure FDMA systems and still achieve moderate equipment complexities.

Finally, there is the problem of channel assignment. Clearly, it is not reasonable to assign a unique frequency to each user as there are not sufficient frequencies and the spectral resource would be unused whenever the user is idle. Instead, methods that allocate channels on demand can make much more efficient use of the spectrum. Such methods will be discussed further during the description of TDMA systems.

17.4 Time Division Multiple Access

In TDMA systems users share the same frequency band by accessing the channel in non-overlapping time intervals in a round-robin fashion [3]. Since the signals do not overlap, they are clearly orthogonal, and the signal of interest is easily extracted by switching the receiver on only during the transmission of the desired signal. Hence, the receiver filters are simply windows instead of the bandpass filters required in FDMA. As a consequence, the guard time between transmissions can be made as small as the synchronization of the network permits. Guard times of 30–50 μs between time slots are commonly used in TDMA-based systems. As a consequence, all users must be synchronized with the base station to within a fraction of the guard time. This is achievable by distributing a master clock signal on one of the base station's broadcast channels.

TDMA can be combined with TDD or frequency-division duplexing (FDD). The former duplexing scheme is used, for example, in the Digital European Cordless Telephone (DECT) standard and is well suited for systems in which base-to-base and mobile-to-base propagation paths are similar, i.e., systems without extremely high base station antennas. Since both the portable and the base station transmit on the same frequency, some signal processing functions for the downlink can be implemented in the base station, as discussed earlier for FDMA/TDD systems.

In the cellular application, the high base station antennas make FDD the more appropriate choice. In these systems, separate frequency bands are provided for uplink and downlink communication. Note that it is still possible and advisable to stagger the uplink and downlink transmission intervals such that they do not overlap, to avoid the situation that the portable must transmit and receive at the same time. With FDD the uplink and downlink channel are not identical and, hence, signal processing functions can not be implemented in the base-station; antenna diversity and equalization have to be realized in the portable.

17.4.1 Propagation Considerations

In comparison to a FDMA system supporting the same user data rate, the transmitted data rate in a TDMA system is larger by a factor equal to the number of users sharing the frequency band. This factor is eight in the pan-European global system for mobile communications (GSM) and three in the advanced mobile phone service (D-AMPS) system. Thus, the symbol duration is reduced by the same factor and severe intersymbol interference results, at least in the cellular environment.

To illustrate, consider the earlier example where each user transmits 25 K symbols per second. Assuming eight users per frequency band leads to a symbol duration of 5 μs. Even in the cordless application with delay spreads of up to 1 μs, an equalizer may be useful to combat the resulting interference between adjacent symbols. In cellular systems, however, the delay spread of up to 20 μs introduces severe intersymbol interference spanning up to 5 symbol periods. As the delay spread often exceeds the symbol duration, the channel can be classified as frequency selective, emphasizing the observation that the channel affects different spectral components differently.

The intersymbol interference in cellular TDMA systems can be so severe that linear equalizers are insufficient to overcome its negative effects. Instead, more powerful, nonlinear decision feedback or maximum-likelihood sequence estimation equalizers must be employed [9]. Furthermore, all of these equalizers require some information about the channel impulse response that must be estimated from the received signal by means of an embedded training sequence. Clearly, the training sequence carries no user data and, thus, wastes valuable bandwidth.

In general, receivers for cellular TDMA systems will be fairly complex. On the positive side of the argument, however, the frequency selective nature of the channel provides some built-in diversity that makes transmission more robust to channel fading. The diversity stems from the fact that the multipath components of the received signal can be resolved at a resolution roughly equal to the symbol duration, and the different multipath components can be combined by the equalizer during the demodulation of the signal. To further improve robustness to channel fading, coding and interleaving, slow frequency hopping and antenna diversity can be employed as discussed in connection with FDMA.

17.4.2   Initial Channel Assignment

In both FDMA and TDMA systems, channels should not be assigned to a mobile on a permanent basis. A fixed assignment strategy would either be extremely wasteful of precious bandwidth or highly susceptible to cochannel interference. Instead, channels must be assigned on demand. Clearly, this implies the existence of a separate uplink channel on which mobiles can notify the base station of their need for a traffic channel. This uplink channel is referred to as the **random-access channel** because of the type of strategy used to regulate access to it.

The successful procedure for establishing a call that originates from the mobile station is outlined in Fig. 17.3. The mobile initiates the procedure by transmitting a request on the random-access channel. Since this channel is shared by all users in range of the base station, a random access protocol, like the ALOHA protocol, has to be employed to resolve possible collisions. Once the base station has received the mobile's request, it responds with an immediate assignment message that directs the mobile to tune to a dedicated control channel for the ensuing call setup. Upon completion of the call setup negotiation, a traffic channel, i.e., a frequency in FDMA systems or a time slot in TDMA systems, is assigned by the base station and all future communication takes place on that channel. In the case of a mobile-terminating call request, the sequence of events is preceded by a paging message alerting the base station of the call request.

17.4.3   Example 17.2: GSM

Named after the organization that created the system standards (Groupe Speciale Mobile) this pan-European digital cellular system has been deployed in Europe since the early 1990s [6]. GSM uses combined TDMA and FDMA with frequency-division duplex for access. Carriers are spaced at 200 kHz and support eight TDMA time slots each. For the uplink the frequency band 890–915 MHz is allocated, whereas the downlink uses the band 935–960 MHz. Each time slot is of duration 577 μs, which corresponds to 156.26-b periods,

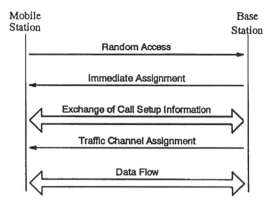

FIGURE 17.3   Mobile-originating call establishment.

including a guard time of 8.25-b periods. Eight consecutive time slots form a GSM frame of duration 4.62 ms.

The GSM modulation is Gaussian minimum shift keying with time-bandwidth product of 0.3, i.e., the modulator bandpass has a cutoff frequency of 0.3 times the bit rate. At the bit rate of 270.8 kb/s, severe intersymbol interference arises in the cellular environment. To facilitate coherent detection, a 26-b training sequence is embedded into every time slot. Time diversity is achieved by interleaving over 8 frames for speech signals and 20 frames for data communication. Sophisticated error-correction coding with varying levels of protection for different outputs of the speech coder is provided. Note that the round-trip delay introduced by the interleaver is on the order of 80 ms for speech signals. GSM provides slow frequency hopping as a further mechanism to improve the efficiency of the interleaver.

17.4.4   Further Remarks

In cellular systems, such as GSM or the North-American D-AMPS, TDMA is combined with FDMA. Different frequencies are used in neighboring cells to provide orthogonal signalling without the need for tight synchronization of base stations. Furthermore, channel assignment can then be performed in each cell individually. Within a cell, one or more frequencies are shared by users in the time domain.

From an implementation standpoint TDMA systems have the advantage that common radio and signal processing equipment at the base station can be shared by users communicating on the same frequency. A somewhat more subtle advantage of TDMA systems arises from the possibility of monitoring surrounding base stations and frequencies for signal quality to support mobile assisted handovers.

17.5   Code Division Multiple Access

CDMA systems employ wideband signals with good cross-correlation properties [7]. That means the output of a filter matched to one user's signal is small when a different user's signal is input. A large body of work exists on spreading sequences that lead to signal sets with small cross correlations [10]. Because of their noise-like appearance such sequences are often referred to as pseudonoise (PN) sequences, and because of their wideband nature CDMA systems are often called spread-spectrum systems.

Spectrum spreading can be achieved mainly in two ways: through frequency hopping as explained earlier or through direct sequence spreading. In direct sequence spread spectrum, a high-rate, antipodal pseudorandom spreading sequence modulates the transmitted signal such that the bandwidth of the resulting signal is roughly equal to the rate of the spreading sequence. The cross correlation of the signals is then largely determined by the cross-correlation properties of the spreading signals. Clearly, CDMA signals overlap in both time and frequency domains but are separable based on their spreading waveforms.

An immediate consequence of this observation is that CDMA systems do not require tight synchronization between users as do TDMA systems. By the same token, frequency planning and management are not required as frequencies are reused throughout the coverage area.

17.5.1   Propagation Considerations

Spread spectrum is well suited for wireless communication systems because of its built-in frequency diversity. As discussed before, in cellular systems the delay spread measures several microseconds and, hence, the coherence bandwidth of the channel is smaller than 1 MHz. Spreading rates can be chosen to exceed the coherence bandwidth such that the channel becomes frequency selective, i.e., different spectral components are affected unequally by the channel and only parts of the signal are affected by fades. Expressing the same observation in time domain terms, multipath components are resolvable at a resolution equal to the chip period and can be combined coherently, for example, by means of a RAKE receiver [8]. An estimate of the channel impulse response is required for the coherent combination of multipath components. This estimate can be gained from a training sequence or by means of a so-called pilot signal.

Even for cordless telephone systems, operating in environments with submicrosecond delay spread and corresponding coherence bandwidths of a few megahertz, the spreading rate can be chosen large enough to facilitate multipath diversity. If the combination of multipath components already described is deemed too complex, a simpler, but less powerful, form of diversity can be used that decorrelates only the strongest received multipath component and relies on the suppression of other path components by the matched filter.

17.5.2   Multiple-Access Interference

If it is possible to control the relative timing of the transmitted signals, such as on the down-link, the transmitted signals can be made perfectly orthogonal, and if the channel only adds white Gaussian noise, matched filter receivers are optimal for extracting a signal from the superposition of waveforms. If the channel is dispersive because of multipath, the signals arriving at the receiver will no longer be orthogonal and will introduce some multiple-access interference, i.e., signal components from other signals that are not rejected by the matched filter.

On the uplink, extremely tight synchronization between users to within a fraction of a chip period, which is defined as the inverse of the spreading rate, is generally not possible, and measures to control the impact of multiple-access interference must be taken. Otherwise, the near–far problem, i.e., the problem of very strong undesired users' signals overwhelming the weaker signal of the desired user, can severely decrease performance. Two approaches are proposed to overcome the near–far problem: power control with soft handovers and multiuser detection.

Power control attempts to ensure that signals from all mobiles in a cell arrive at the base station with approximately equal power levels. To be effective, power control must be accurate to within about 1 dB and fast enough to compensate for channel fading. For a mobile moving at 55 mph and transmitting at 1 GHz, the Doppler bandwidth is approximately 100 Hz. Hence, the channel changes its characteristic drastically about 100 times per second and on the order of 1000 b/s must be sent from base station to mobile for power control purposes. As different mobiles may be subject to vastly different fading and shadowing conditions, a large dynamic range of about 80 dB must be covered by power control. Notice, that power control on the downlink is really only necessary for mobiles that are about equidistant from two base stations, and even then neither the update rate nor the dynamic range of the uplink is required.

The interference problem that arises at the cell boundaries where mobiles are within range of two or more base stations can be turned into an advantage through the idea of soft handover. On the downlink, all base stations within range can transmit to the mobile, which in turn can combine the received signals to achieve some gain from the antenna diversity. On the uplink, a similar effect can be obtained by selecting the strongest received signal from all base stations that received a user's signal. The base station that receives the strongest signal will also issue power control commands to minimize the transmit power of the mobile. Note, however, that soft handover requires fairly tight synchronization between base stations, and one of the advantages of CDMA over TDMA is lost.

Multiuser detection is still an emerging technique. It is probably best used in conjunction with power control. The fundamental idea behind this technique is to model multiple-access interference explicitly and devise receivers that reject or cancel the undesired signals. A variety of techniques have been proposed ranging from optimum maximum-likelihood sequence estimation via multistage schemes, reminiscent of decision feedback algorithms, to linear decorrelating receivers. An excellent survey of the theory and practice of multiuser detection is given by [11].

17.5.3 Further Remarks

CDMA systems work well in conjunction with frequency division duplexing. This arrangement decouples the power control problem on the uplink and downlink, respectively.

Signal quality enhancing methods, such as time diversity through coding and interleaving, can be applied just as with the other access methods. In spread spectrum systems, however, coding can be built into the spreading process, avoiding the loss of bandwidth associated with error protection. Additionally, CDMA lends itself naturally to the exploitation of speech pauses that make up more than half the time of a connection. If no signals are transmitted during such pauses, then the instantaneous interference level is reduced and the total number of users supportable by the system can be approximately doubled.

17.6 Comparison and Outlook

The question of which of the access methods is best does not have a single answer. Based on the preceding discussion FDMA is only suited for applications such as cordless telephone with very small cells and submicrosecond delay spreads. In cellular systems and for most versions of personal communication systems, the choice reduces to TDMA vs. CDMA.

In terms of complexity, TDMA receivers require adaptive, nonlinear equalizers when operating in environments with large delay spreads. CDMA systems, in turn, need RAKE receivers and sophisticated power control algorithms. In the future, some form of multiple-access interference rejection is likely to be implemented as well. Time synchronization is required in both systems, albeit for different reasons. The additional complexity for coding and interleaving is comparable for both access methods.

An often quoted advantage of CDMA systems is the fact that the performance will degrade gracefully as the load increases. In TDMA systems, in turn, requests will have to be blocked once all channels in a cell are in use. Hence, there is a hard limit on the number of channels per cell. There are proposals for extended TDMA systems, however, that incorporate reassignment of channels during speech pauses. Not only would such extended TDMA systems match the advantage of the exploitation of speech pauses of CDMA systems, they would also lead to a soft limit on the system capacity. The extended TDMA proposals would implement the statistical multiplexing of the user data, e.g., by means of the packet reservation multiple access protocol [4]. The increase in capacity depends on the acceptable packet loss rate; in other words, small increases in the load lead to small increases in the packet loss probability.

Many comparisons in terms of capacity between TDMA and CDMA can be found in the recent literature. Such comparisons, however, are often invalidated by making assumptions that favor one access method over the other. An important exception constitutes the recent paper by Wyner [12]. Under a simplified model that nevertheless captures the essence of cellular systems, he computes the Shannon capacity. Highlights of his results include the following.

- TDMA is distinctly suboptimal in cellular systems.
- When the signal-to-noise-ratio is large, CDMA appears to achieve twice the capacity of TDMA.
- Multiuser detectors are essential to realize near-optimum performance in CDMA systems.
- Intercell interference in CDMA systems has a detrimental effect when the signal-to-noise ratio is large, but it can be exploited via diversity combining to increase capacity when the signal-to-noise ratio is small.

More research along this avenue is necessary to confirm the validity of the results. In particular, incorporation of realistic channel models into the analysis is required. However, this work represents a substantial step towards quantifying capacity increases achievable with CDMA.

Defining Terms

Capacity: Shannon originally defined capacity as the maximum data rate which permits error-free communication in a given environment. A looser interpretation is normally employed in wireless communication systems. Here capacity denotes the traffic density supported by the system under consideration normalized with respect to bandwidth and coverage area.

Code-division multiple access (CDMA): Systems use signals with very small cross-correlations to facilitate sharing of the broadcast radio channel. Correlators are used to extract the desired user's signal while simultaneously suppressing interfering, parallel transmissions.

Duplexing: Refers to the exchange of messages in both directions of a connection.

Frequency-division multiple access (FDMA): Simultaneous access to the radio channel is facilitated by assigning nonoverlapping frequency bands to different users.

Multiple access: Denotes the support of simultaneous transmissions over a shared communication channel.

Random-access channel: This uplink control channel is used by mobiles to request assignment of a traffic channel. A random access protocol is employed to arbitrate access to this channel.

Time-division multiple access (TDMA): Systems assign nonoverlapping time slots to different users in a round-robin fashion.

References

[1] Bertsekas, D. and Gallager, R., *Data Networks,* Prentice-Hall, Englewood Cliffs, NJ, 1987.

[2] Cox, D.C., Wireless network access for personal communications, *IEEE Comm. Mag.,* 96–115, 1992.

[3] Falconer, D.D., Adachi, F., and Gudmundson, B. Time division multiple access methods for wireless personal communications. *IEEE Comm. Mag.,* 33(1), 50–57, 1995.

[4] Goodman, D., Trends in cellular and cordless communications. *IEEE Comm. Mag.,* 31–40, 1991a.

[5] Goodman, D.J., Second generation wireless information networks. *IEEE Trans. on Vehicular Tech.,* 40(2), 366–374, 1991b.

[6] Hodges, M.R.L., The GSM radio interface. *Br. Telecom Tech. J.,* 8(1), 31–43, 1990.

[7] Kohno, R., Meidan, R., and Milstein, L.B., Spread spectrum access methods for wireless communications. *IEEE Comm. Mag.,* 33(1), 58, 1995.

[8] Proakis, J.G., *Digital Communications.* 2nd ed., McGraw-Hill, New York, 1989.

[9] Proakis, J.G., Adaptive equalization for TDMA digital mobile radio. *IEEE Trans. on Vehicular Tech.,* 40(2), 333–341, 1991.

[10] Sarwate, D.V. and Pursley, M.B., Crosscorrelation properties of pseudorandom and related sequences. *Proceedings of the IEEE,* 68(5), 593–619, 1980.

[11] Verdu, S., Multi-user detection. In *Advances in Statistical Signal Processing—Vol. 2: Signal Detection,* JAI Press, Greenwich, CT, 1992.

[12] Wyner, A.D., Shannon-theoretic approach to a Gaussian cellular multiple-access channel. *IEEE Trans. on Information Theory,* 40(6), 1713–1727, 1994.

Further Information

Several of the IEEE publications, including the *Transactions on Communications, Journal on Selected Areas in Communications, Transactions on Vehicular Technology, Communications Magazine,* and *Personal Communications Magazine* contain articles the on subject of access methods on a regular basis.

18

Rayleigh Fading Channels[1]

Bernard Sklar
Communications Engineering Services

0-8493-8597-0/99/$0.00+$.50
© 1999 by CRC Press LLC

18.1   Introduction

When the mechanisms of fading channels were first modeled in the 1950s and 1960s, the ideas were primarily applied to over-the-horizon communications covering a wide range of frequency bands. The 3–30 MHz high-frequency (HF) band is used for ionospheric communications, and the 300 MHz–3 GHz ultra-high-frequency (UHF) and 3–30 GHz super-high-frequency (SHF) bands are used for tropospheric scatter. Although the fading effects in a mobile radio system are somewhat different from those in ionospheric and tropospheric channels, the early models are still quite useful to help characterize fading effects in mobile digital communication systems. This chapter addresses Rayleigh fading, primarily in the UHF band, that affects mobile systems such as cellular and personal communication systems (PCS). The chapter itemizes the fundamental fading manifestations, types of degradation, and methods to mitigate the degradation. Two particular mitigation techniques are examined: the Viterbi equalizer implemented in the Global System for Mobile Communication (GSM), and the Rake receiver used in CDMA systems built to meet Interim Standard-95 (IS-95).

18.2   The Challenge of a Fading Channel

In the study of communication systems, the classical (ideal) additive-white-Gaussian-noise (AWGN) channel, with statistically independent Gaussian noise samples corrupting data samples free of intersymbol interference (ISI), is the usual starting point for understanding basic performance relationships. The primary source of performance degradation is thermal noise generated in the receiver. Often, external interference received by the antenna is more significant than the thermal noise. This external interference can sometimes be characterized as having a broadband spectrum and quantified by a parameter called antenna temperature [1]. The thermal noise usually has a flat power spectral density over the signal band and a zero-mean Gaussian voltage probability density function (pdf). When modeling practical systems, the next step is the introduction of bandlimiting filters. The filter in the transmitter usually serves to satisfy some regulatory requirement on spectral containment. The filter in the receiver often serves the purpose of a classical "matched filter" [2] to the signal bandwidth. Due to the bandlimiting and phase-distortion properties of filters, special signal design and equalization techniques may be required to mitigate the filter-induced ISI.

If a radio channel's propagating characteristics are not specified, one usually infers that the signal attenuation vs. distance behaves as if propagation takes place over ideal free space. The model of free space treats the region between the transmit and receive antennas as being free of all objects that might absorb or reflect radio frequency (RF) energy. It also assumes that, within this region, the atmosphere behaves as a perfectly uniform and nonabsorbing medium. Furthermore, the earth is treated as being infinitely far away from the propagating signal (or, equivalently, as having a reflection coefficient that is negligible). Basically, in this idealized free-space model, the attenuation of RF energy between the transmitter and receiver behaves according to an inverse-square law. The received power expressed in terms of transmitted power is attenuated by a factor, $L_s(d)$, where this factor is called **path loss** or **free space loss.** When the receiving antenna is isotropic, this factor

---

[1]A version of this chapter has appeared as two papers in the *IEEE Communications Magazine,* September 1997, under the titles "Rayleigh Fading Channels in Mobile Digital Communication Systems, Part I: Characterization" and "Part II: Mitigation."

is expressed as [1]:

$$L_s(d) = \left(\frac{4\pi d}{\lambda}\right)^2 \tag{18.1}$$

In Eq. (18.1), d is the distance between the transmitter and the receiver, and λ is the wavelength of the propagating signal. For this case of idealized propagation, received signal power is very predictable.

For most practical channels, where signal propagation takes place in the atmosphere and near the ground, the free-space propagation model is inadequate to describe the channel and predict system performance. In a wireless mobile communication system, a signal can travel from transmitter to receiver over multiple reflective paths; this phenomenon is referred to as **multipath propagation.** The effect can cause fluctuations in the received signal's amplitude, phase, and angle of arrival, giving rise to the terminology **multipath fading.** Another name, **scintillation,** having originated in radio astronomy, is used to describe the multipath fading caused by physical changes in the propagating medium, such as variations in the density of ions in the ionospheric layers that reflect high frequency (HF) radio signals. Both names, fading and scintillation, refer to a signal's random fluctuations or fading due to multipath propagation. The main difference is that scintillation involves mechanisms (e.g., ions) that are much smaller than a wavelength. The end-to-end modeling and design of systems that mitigate the effects of fading are usually more challenging than those whose sole source of performance degradation is AWGN.

18.3 Mobile-Radio Propagation: Large-Scale Fading and Small-Scale Fading

Figure 18.1 represents an overview of fading channel manifestations. It starts with two types of fading effects that characterize mobile communications: large-scale fading and small-scale fading. Large-scale fading represents the average signal power attenuation or the path loss due to motion over large areas. In Fig. 18.1, the large-scale fading manifestation is shown in blocks 1, 2, and 3. This phenomenon is affected by prominent terrain contours (e.g., hills, forests, billboards, clumps of buildings, etc.) between the transmitter and receiver. The receiver is often represented as being "shadowed" by such prominences. The statistics of large-scale fading provide a way of computing an estimate of path loss as a function of distance. This is described in terms of a mean-path loss (nth-power law) and a log-normally distributed variation about the mean. Small-scale fading refers to the dramatic changes in signal amplitude and phase that can be experienced as a result of small changes (as small as a half-wavelength) in the spatial separation between a receiver and transmitter. As indicated in Fig. 18.1, blocks 4, 5, and 6, small-scale fading manifests itself in two mechanisms, namely, time-spreading of the signal (or signal dispersion) and time-variant behavior of the channel. For mobile-radio applications, the channel is time-variant because motion between the transmitter and receiver results in propagation path changes. The rate of change of these propagation conditions accounts for the fading rapidity (rate of change of the fading impairments). Small-scale fading is also called **Rayleigh fading** because if the multiple reflective paths are large in number and there is no line-of-sight signal component, the envelope of the received signal is statistically described by a Rayleigh pdf. When there is a dominant nonfading signal component present, such as a line-of-sight propagation path, the small-scale fading envelope is described by a Rician pdf [3]. A mobile radio roaming over a large area must process signals that experience both types of fading: small-scale fading superimposed on large-scale fading.

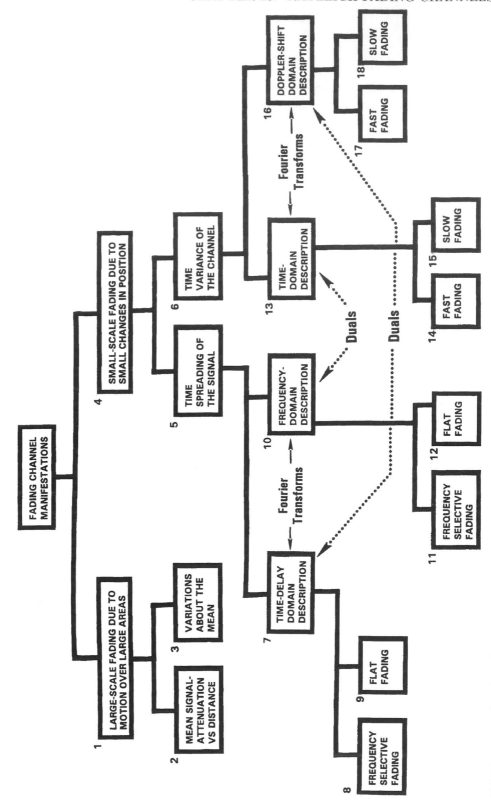

FIGURE 18.1   Fading channel manifestations.

There are three basic mechanisms that impact signal propagation in a mobile communication system. They are reflection, diffraction, and scattering [3].

- Reflection occurs when a propagating electromagnetic wave impinges upon a smooth surface with very large dimensions compared to the RF signal wavelength (λ).

- Diffraction occurs when the radio path between the transmitter and receiver is obstructed by a dense body with large dimensions compared to λ, causing secondary waves to be formed behind the obstructing body. Diffraction is a phenomenon that accounts for RF energy travelling from transmitter to receiver without a line-of-sight path between the two. It is often termed **shadowing** because the diffracted field can reach the receiver even when shadowed by an impenetrable obstruction.

- Scattering occurs when a radio wave impinges on either a large rough surface or any surface whose dimensions are on the order of λ or less, causing the reflected energy to spread out (scatter) in all directions. In an urban environment, typical signal obstructions that yield scattering are lampposts, street signs, and foliage.

Figure 18.1 may serve as a table of contents for the sections that follow. We will examine the two manifestations of small-scale fading: signal time-spreading (signal dispersion) and the time-variant nature of the channel. These examinations will take place in two domains: time and frequency, as indicated in Fig. 18.1, blocks 7, 10, 13, and 16. For signal dispersion, we categorize the fading degradation types as being frequency-selective or frequency-nonselective (flat), as listed in blocks 8, 9, 11, and 12. For the time-variant manifestation, we categorize the fading degradation types as fast-fading or slow-fading, as listed in blocks 14, 15, 17, and 18. The labels indicating Fourier transforms and duals will be explained later.

Figure 18.2 illustrates the various contributions that must be considered when estimating path loss for a link budget analysis in a cellular application [4]. These contributions are:

- Mean path loss as a function of distance, due to large-scale fading
- Near-worst-case variations about the mean path loss (typically 6–10 dB) or large-scale fading margin
- Near-worst-case Rayleigh or small-scale fading margin (typically 20–30 dB)

In Fig. 18.2, the annotations " $\approx$ 1–2% " indicate a suggested area (probability) under the tail of each pdf as a design goal. Hence, the amount of margin indicated is intended to provide adequate received signal power for approximately 98–99% of each type of fading variation (large- and small-scale).

A received signal, is generally described in terms of a transmitted signal $s(t)$ convolved with the impulse response of the channel $h_c(t)$. Neglecting the degradation due to noise, we write:

$$r(t) = s(t) * h_c(t) \qquad (18.2)$$

where $*$ denotes convolution. In the case of mobile radios, $r(t)$ can be partitioned in terms of two component random variables, as follows [5]:

$$r(t) = m(t) \times r_0(t) \qquad (18.3)$$

where $m(t)$ is called the large-scale-fading component, and $r_0(t)$ is called the small-scale-fading component. $m(t)$ is sometimes referred to as the **local mean** or **log-normal fading**

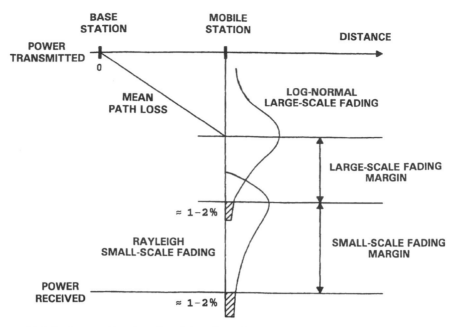

FIGURE 18.2   Link-budget considerations for a fading channel.

because the magnitude of $m(t)$ is described by a log-normal pdf (or, equivalently, the magnitude measured in decibels has a Gaussian pdf). $r_0(t)$ is sometimes referred to as multipath or Rayleigh fading. Figure 18.3 illustrates the relationship between large-scale and small-scale fading. In Fig. 18.3(a), received signal power $r(t)$ vs. antenna displacement (typically in units of wavelength) is plotted for the case of a mobile radio. Small-scale fading superimposed on large-scale fading can be readily identified. The typical antenna displacement between the small-scale signal nulls is approximately a half wavelength. In Fig. 18.3(b), the large-scale fading or local mean, $m(t)$, has been removed in order to view the small-scale fading, $r_0(t)$, about some average constant power.

In the sections that follow, we enumerate some of the details regarding the statistics and mechanisms of large-scale and small-scale fading.

18.3.1   Large-Scale Fading: Path-Loss Mean and Standard Deviation

For the mobile radio application, Okumura [6] made some of the earlier comprehensive path-loss measurements for a wide range of antenna heights and coverage distances. Hata [7] transformed Okumura's data into parametric formulas. For the mobile radio application, the mean path loss, $\overline{L_p}(d)$, as a function of distance, d, between the transmitter and receiver is proportional to an nth-power of d relative to a reference distance d_0 [3].

$$\overline{L_p}(d) \propto \left(\frac{d}{d_0}\right)^n \tag{18.4}$$

$\overline{L_p}(d)$ is often stated in decibels, as shown below.

$$\overline{L_p}(d) \text{ (dB) } = L_s(d_0) \text{ (dB) } + 10\,n\,\log\left(\frac{d}{d_0}\right) \tag{18.5}$$

The reference distance d_0, corresponds to a point located in the far field of the antenna. Typically, the value of d_0 is taken to be 1 km for large cells, 100 m for microcells, and 1 m

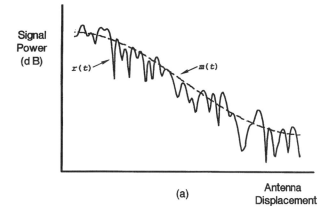

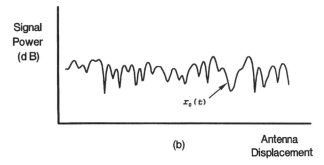

FIGURE 18.3   Large-scale fading and small-scale fading.

for indoor channels. $\overline{L_p}(d)$ is the average path loss (over a multitude of different sites) for a given value of d. Linear regression for a minimum mean-squared estimate (MMSE) fit of $\overline{L_p}(d)$ vs. d on a log-log scale (for distances greater than d_0) yields a straight line with a slope equal to $10\,n$ dB/decade. The value of the exponent n depends on the frequency, antenna heights, and propagation environment. In free space, $n = 2$, as seen in Eq. (18.1). In the presence of a very strong guided wave phenomenon (like urban streets), n can be lower than 2. When obstructions are present, n is larger. The path loss $L_s(d_0)$ to the reference point at a distance d_0 from the transmitter is typically found through field measurements or is calculated using the free-space path loss given by Eq. (18.1). Figure 18.4 shows a scatter plot of path loss vs. distance for measurements made in several German cities [8]. Here, the path loss has been measured relative to the free-space reference measurement at $d_0 = 100$ m. Also shown are straight-line fits to various exponent values.

   The path loss vs. distance expressed in Eq. (18.5) is an average, and therefore not adequate to describe any particular setting or signal path. It is necessary to provide for variations about the mean since the environment of different sites may be quite different for similar transmitter-receiver separations. Figure 18.4 illustrates that path-loss variations can be quite large. Measurements have shown that for any value of d, the path loss $L_p(d)$ is a random variable having a log-normal distribution about the mean distant-dependent value $\overline{L_p}(d)$ [9]. Thus, path loss $L_p(d)$ can be expressed in terms of $\overline{L_p}(d)$ plus a random variable X_σ, as follows [3].

$$L_p(d)\ (\text{dB})\ = L_s\,(d_0)\ (\text{dB})\ + 10\,n\,\log_{10}\left(\frac{d}{d_0}\right) + X_\sigma\ (\text{dB}) \qquad (18.6)$$

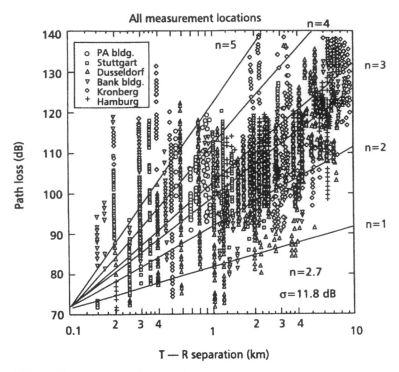

FIGURE 18.4   Path loss vs. distance measured in several German cities.

where X_σ denotes a zero-mean, Gaussian random variable (in decibels) with standard deviation σ (also in decibels). X_σ is site and distance dependent. The choice of a value for X_σ is often based on measurements; it is not unusual for X_σ to take on values as high as 6–10 dB or greater. Thus, the parameters needed to statistically describe path loss due to large-scale fading for an arbitrary location with a specific transmitter-receiver separation are:

- The reference distance d_0
- The path-loss exponent n
- The standard deviation σ of X_σ

There are several good references dealing with the measurement and estimation of propagation path loss for many different applications and configurations [3], [7]–[11].

18.3.2   Small-Scale Fading: Statistics and Mechanisms

When the received signal is made up of multiple reflective rays plus a significant line-of-sight (nonfaded) component, the envelope amplitude due to small-scale fading has a Rician pdf, and is referred to as **Rician fading** [3]. The nonfaded component is called the **specular component.** As the amplitude of the specular component approaches zero, the Rician pdf approaches a Rayleigh pdf, expressed as:

$$
p(r) = \left\{ \begin{array}{ll} \dfrac{r}{\sigma^2} \exp\left[-\dfrac{r^2}{2\sigma^2} \right] & \text{for } r \geq 0 \\[2mm] 0 & \text{otherwise} \end{array} \right\} \tag{18.7}
$$

where r is the envelope amplitude of the received signal, and $2\sigma^2$ is the predetection mean power of the multipath signal. The Rayleigh faded component is sometimes called the **random, scatter,** or **diffuse component.** The Rayleigh pdf results from having no specular component of the signal; thus for a single link it represents the pdf associated with the worst case of fading per mean received signal power. For the remainder of this chapter, it will be assumed that loss of signal-to-noise ratio (SNR) due to fading follows the Rayleigh model described. It will also be assumed that the propagating signal is in the UHF band, encompassing present-day cellular and personal communications services (PCS) frequency allocations—nominally 1 GHz and 2 GHz, respectively.

As indicated in Fig. 18.1, blocks 4, 5, and 6, small-scale fading manifests itself in two mechanisms:

- Time-spreading of the underlying digital pulses within the signal
- A time-variant behavior of the channel due to motion (e.g., a receive antenna on a moving platform).

Figure 18.5 illustrates the consequences of both manifestations by showing the response of a multipath channel to a narrow pulse vs. delay, as a function of antenna position (or time, assuming a constant velocity of motion). In Fig. 18.5, we distinguish between

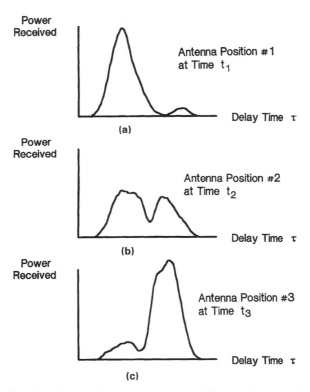

FIGURE 18.5 Response of a multipath channel to a narrow pulse vs. delay, as a function of antenna position.

two different time references—delay time τ and transmission or observation time t. Delay time refers to the time-spreading manifestation which results from the fading channel's nonoptimum impulse response. The transmission time, however, is related to the antenna's motion or spatial changes, accounting for propagation path changes that are perceived as the channel's time-variant behavior. Note that, for constant velocity, as is assumed in Fig. 18.5, either antenna position or transmission time can be used to illustrate this time-variant behavior. Figures 18.5(a)–(c) show the sequence of received pulse-power profiles as the antenna moves through a succession of equally spaced positions. Here, the interval between antenna positions is 0.4 λ, where λ is the wavelength of the carrier frequency. For each of the three cases shown, the response-pattern differs significantly in the delay time of the largest signal component, the number of signal copies, their magnitudes, and the total received power (area) in the received power profile. Figure 18.6 summarizes these two small-scale fading mechanisms, the two domains (time or time-delay and frequency or Doppler shift) for viewing each mechanism and the degradation categories each mechanism can exhibit. Note that any mechanism characterized in the time domain can be characterized equally well in the frequency domain. Hence, as outlined in Fig. 18.6, the time-spreading mechanism will be characterized in the time-delay domain as a multipath delay spread and in the frequency domain as a channel coherence bandwidth. Similarly, the time-variant mechanism will be characterized in the time domain as a channel coherence time and in the Doppler-shift (frequency) domain as a channel fading rate or Doppler spread. These mechanisms and their associated degradation categories will be examined in greater detail in the sections that follow.

18.4   Signal Time-Spreading Viewed in the Time-Delay Domain: Figure 18.1, Block 7—The Multipath Intensity Profile

A simple way to model the fading phenomenon was introduced by Bello [13] in 1963; he proposed the notion of wide-sense stationary uncorrelated scattering (WSSUS). The model treats signal variations arriving with different delays as uncorrelated. It can be shown [4, 13] that such a channel is effectively WSS in both the time and frequency domains. With such a model of a fading channel, Bello was able to define functions that apply for all time and all frequencies. For the mobile channel, Fig. 18.7 contains four functions that make up this model [4], [13]–[16]. We will examine these functions, starting with Fig. 18.7(a) and proceeding counter-clockwise toward Fig. 18.7(d).

In Fig. 18.7(a), a **multipath-intensity profile,** $S(\tau)$ vs. time delay τ is plotted. Knowledge of $S(\tau)$ helps answer the question, "For a transmitted impulse, how does the average received power vary as a function of time delay, τ?" The term "time delay" is used to refer to the excess delay. It represents the signal's propagation delay that exceeds the delay of the first signal arrival at the receiver. For a typical wireless radio channel, the received signal usually consists of several discrete multipath components, sometimes referred to as fingers. For some channels, such as the tropospheric scatter channel, received signals are often seen as a continuum of multipath components [14, 16]. For making measurements of the multipath intensity profile, wideband signals (impulses or spread spectrum) need to be used [16]. For a single transmitted impulse, the time, T_m, between the first and last received component represents the **maximum excess delay,** during which the multipath signal power falls to some threshold level below that of the strongest component. The threshold level might be chosen at 10 dB or 20 dB below the level of the strongest component. Note, that for an ideal system (zero excess delay), the function $S(\tau)$ would consist of an ideal impulse with weight equal to the total average received signal power.

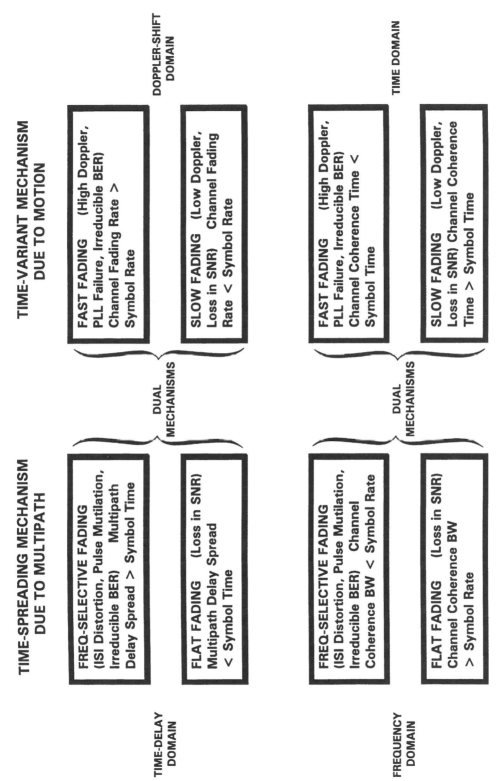

FIGURE 18.6 Small-scale fading; mechanisms, degradation categories, and effects.

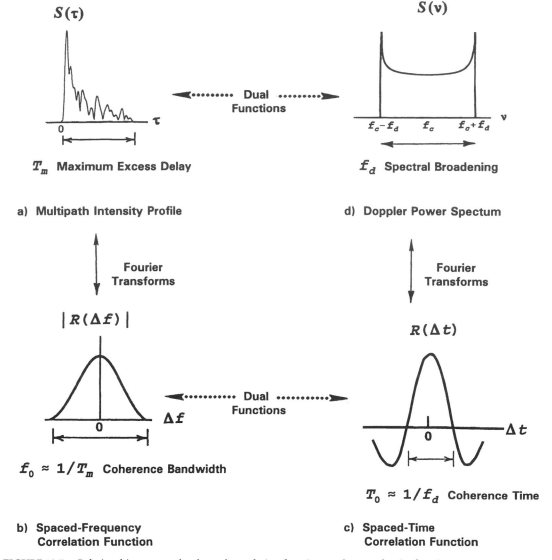

FIGURE 18.7   Relationships among the channel correlation functions and power density functions.

18.4.1   Degradation Categories due to Signal Time-Spreading Viewed in the Time-Delay Domain

In a fading channel, the relationship between maximum excess delay time, T_m, and symbol time, T_s, can be viewed in terms of two different degradation categories, **frequency-selective fading** and **frequency nonselective** or **flat fading,** as indicated in Fig. 18.1, blocks 8 and 9, and Fig. 18.6. A channel is said to exhibit frequency-selective fading if $T_m > T_s$. This condition occurs whenever the received multipath components of a symbol extend beyond the symbol's time duration. Such multipath dispersion of the signal yields the same kind of ISI distortion that is caused by an electronic filter. In fact, another name for this category of fading degradation is **channel-induced ISI.** In the case of frequency-selective fading, mitigating the distortion is possible because many of the multipath components are resolvable by the receiver. Later, several such mitigation techniques are described.

A channel is said to exhibit frequency nonselective or flat fading if $T_m < T_s$. In this case, all of the received multipath components of a symbol arrive within the symbol time duration; hence, the components are not resolvable. Here, there is no channel-induced ISI distortion, since the signal time spreading does not result in significant overlap among neighboring received symbols. There is still performance degradation since the unresolvable phasor components can add up destructively to yield a substantial reduction in SNR. Also, signals that are classified as exhibiting flat fading can sometimes experience frequency-selective distortion. This will be explained later when viewing degradation in the frequency domain, where the phenomenon is more easily described. For loss in SNR due to flat fading, the mitigation technique called for is to improve the received SNR (or reduce the required SNR). For digital systems, introducing some form of signal diversity and using error-correction coding is the most efficient way to accomplish this.

18.5 Signal Time-Spreading Viewed in the Frequency Domain: Figure 18.1, Block 10—The Spaced-Frequency Correlation Function

A completely analogous characterization of signal dispersion can begin in the frequency domain. In Fig. 18.7(b), the function $|R(\Delta f)|$ is seen, designated a **spaced-frequency correlation function;** it is the Fourier transform of $S(\tau)$. $R(\Delta f)$ represents the correlation between the channel's response to two signals as a function of the frequency difference between the two signals. It can be thought of as the channel's frequency transfer function. Therefore, the time-spreading manifestation can be viewed as if it were the result of a filtering process. Knowledge of $R(\Delta f)$ helps answer the question, "What is the correlation between received signals that are spaced in frequency $\Delta f = f_1 - f_2$?" $R(\Delta f)$ can be measured by transmitting a pair of sinusoids separated in frequency by Δf, cross-correlating the two separately received signals, and repeating the process many times with ever-larger separation Δf. Therefore, the measurement of $R(\Delta f)$ can be made with a sinusoid that is swept in frequency across the band of interest (a wideband signal). The **coherence bandwidth,** f_0 , is a statistical measure of the range of frequencies over which the channel passes all spectral components with approximately equal gain and linear phase. Thus, the coherence bandwidth represents a frequency range over which frequency components have a strong potential for amplitude correlation. That is, a signal's spectral components in that range are affected by the channel in a similar manner, as for example, exhibiting fading or no fading. Note that f_0 and T_m are reciprocally related (within a multiplicative constant). As an approximation, it is possible to say that

$$f_0 \approx \frac{1}{T_m} \tag{18.8}$$

The maximum excess delay, T_m, is not necessarily the best indicator of how any given system will perform on a channel because different channels with the same value of T_m can exhibit very different profiles of signal intensity over the delay span. A more useful measurement of delay spread is most often characterized in terms of the root mean squared (rms) delay spread, σ_τ, where

$$\sigma_\tau = \sqrt{\overline{\tau^2} - (\overline{\tau})^2} \tag{18.9}$$

$\overline{\tau}$ is the mean excess delay, $(\overline{\tau})^2$ is the mean squared, $\overline{\tau^2}$ is the second moment, and σ_τ is the square root of the second central moment of $S(\tau)$ [3].

An exact relationship between coherence bandwidth and delay spread does not exist, and must be derived from signal analysis (usually using Fourier techniques) of actual signal dispersion measurements in particular channels. Several approximate relationships have been described. If coherence bandwidth is defined as the frequency interval over which the channel's complex frequency transfer function has a correlation of at least 0.9, the coherence bandwidth is approximately [17]

$$f_0 \approx \frac{1}{50\sigma_\tau} \tag{18.10}$$

For the case of a mobile radio, an array of radially uniformly spaced scatterers, all with equal-magnitude reflection coefficients but independent, randomly occurring reflection phase angles [18, 19] is generally accepted as a useful model for urban surroundings. This model is referred to as the **dense-scatterer channel model.** With the use of such a model, coherence bandwidth has similarly been defined [18] for a bandwidth interval over which the channel's complex frequency transfer function has a correlation of at least 0.5 to be

$$f_0 = \frac{0.276}{\sigma_\tau} \tag{18.11}$$

The ionospheric-effects community employs the following definition

$$f_0 = \frac{1}{2\pi\sigma_\tau} \tag{18.12}$$

A more popular approximation of f_0 corresponding to a bandwidth interval having a correlation of at least 0.5 is [3]

$$f_0 \approx \frac{1}{5\sigma_\tau} \tag{18.13}$$

18.5.1 Degradation Categories due to Signal Time-Spreading Viewed in the Frequency Domain

A channel is referred to as frequency-selective if $f_0 < 1/T_s \approx W$, where the symbol rate $1/T_s$ is nominally taken to be equal to the signal bandwidth W. In practice, W may differ from $1/T_s$ due to system filtering or data modulation type (quaternary phase shift keying, QPSK, minimum shift keying, MSK, etc.) [21]. Frequency-selective fading distortion occurs whenever a signal's spectral components are not all affected equally by the channel. Some of the signal's spectral components, falling outside the coherence bandwidth, will be affected differently (independently) compared to those components contained within the coherence bandwidth. This occurs whenever $f_0 < W$ and is illustrated in Fig. 18.8(a).

Frequency-nonselective or flat fading degradation occurs whenever $f_0 > W$. Hence, all of the signal's spectral components will be affected by the channel in a similar manner (e.g., fading or no fading); this is illustrated in Fig. 18.8(b). Flat-fading does not introduce channel-induced ISI distortion, but performance degradation can still be expected due to the loss in SNR whenever the signal is fading. In order to avoid channel-induced ISI distortion, the channel is required to exhibit flat fading by insuring that

$$f_0 > W \approx \frac{1}{T_s} \tag{18.14}$$

Hence, the channel coherence bandwidth f_0 sets an upper limit on the transmission rate that can be used without incorporating an equalizer in the receiver.

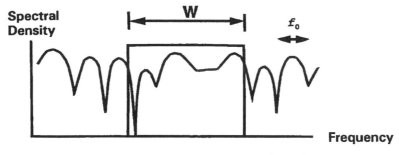

(a) Typical Frequency-Selective Fading Case ($f_0 < W$)

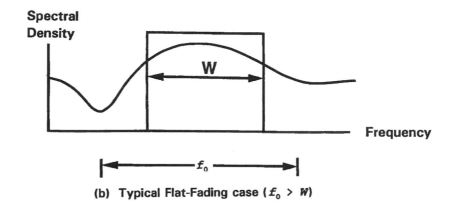

(b) Typical Flat-Fading case ($f_0 > W$)

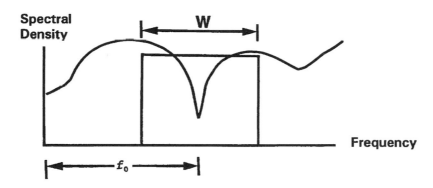

(c) Null of Channel Frequency-Transfer Function occurs at Signal Band Center ($f_0 > W$)

FIGURE 18.8   Relationships between the channel frequency-transfer function and a signal with bandwidth W.

For the flat-fading case, where $f_0 > W$ (or $T_m < T_s$), Fig. 18.8(b) shows the usual flat-fading pictorial representation. However, as a mobile radio changes its position, there will be times when the received signal experiences frequency-selective distortion even though $f_0 > W$. This is seen in Fig. 18.8(c), where the null of the channel's frequency transfer function occurs at the center of the signal band. Whenever this occurs, the baseband pulse will be especially mutilated by deprivation of its DC component. One consequence of the loss of DC (zero mean value) is the absence of a reliable pulse peak on which to establish the timing synchronization, or from which to sample the carrier phase carried by the pulse [18].

Thus, even though a channel is categorized as flat fading (based on rms relationships), it can still manifest frequency-selective fading on occasions. It is fair to say that a mobile radio channel, classified as having flat-fading degradation, cannot exhibit flat fading all of the time. As f_0 becomes much larger than W (or T_m becomes much smaller than T_s), less time will be spent in conditions approximating Fig. 18.8(c). By comparison, it should be clear that in Fig. 18.8(a) the fading is independent of the position of the signal band, and frequency-selective fading occurs all the time, not just occasionally.

18.6   Typical Examples of Flat Fading and Frequency-Selective Fading Manifestations

Figure 18.9 shows some examples of flat fading and frequency-selective fading for a direct-sequence spread-spectrum (DS/SS) system [20, 22]. In Fig. 18.9, there are three plots of the

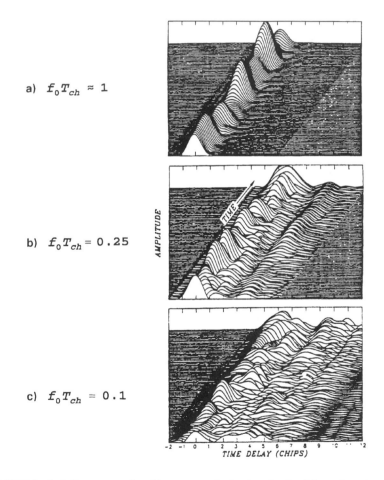

a) $f_0 T_{ch} \approx 1$

b) $f_0 T_{ch} = 0.25$

c) $f_0 T_{ch} = 0.1$

FIGURE 18.9   DS/SS Matched-filter output time-history examples for three levels of channel conditions, where T_{ch} is the time duration of a chip.

output of a pseudonoise (PN) code correlator vs. delay as a function of time (transmission or observation time). Each amplitude vs. delay plot is akin to $S(\tau)$ vs. τ shown in Fig. 18.7(a).

The key difference is that the amplitudes shown in Fig. 18.9 represent the output of a correlator; hence, the waveshapes are a function not only of the impulse response of the channel, but also of the impulse response of the correlator. The delay time is expressed in units of chip durations (chips), where the chip is defined as the spread-spectrum minimal-duration keying element. For each plot, the observation time is shown on an axis perpendicular to the amplitude vs. time-delay plane. Figure 18.9 is drawn from a satellite-to-ground communications link exhibiting scintillation because of atmospheric disturbances. However, Fig. 18.9 is still a useful illustration of three different channel conditions that might apply to a mobile radio situation. A mobile radio that moves along the observation-time axis is affected by changing multipath profiles along the route, as seen in the figure. The scale along the observation-time axis is also in units of chips. In Fig. 18.9(a), the signal dispersion (one "finger" of return) is on the order of a chip time duration, T_{ch}. In a typical DS/SS system, the spread-spectrum signal bandwidth is approximately equal to $1/T_{ch}$; hence, the normalized coherence bandwidth $f_0 T_{ch}$ of approximately unity in Fig. 18.9(a) implies that the coherence bandwidth is about equal to the spread-spectrum bandwidth. This describes a channel that can be called frequency-nonselective or slightly frequency-selective. In Fig. 18.9(b), where $f_0 T_{ch} = 0.25$, the signal dispersion is more pronounced. There is definite interchip interference, and the coherence bandwidth is approximately equal to 25% of the spread-spectrum bandwidth. In Fig. 18.9(c), where $f_0 T_{ch} = 0.1$, the signal dispersion is even more pronounced, with greater interchip-interference effects, and the coherence bandwidth is approximately equal to 10% of the spread-spectrum bandwidth. The channels of Figs. 18.9(b) and (c) can be categorized as moderately and highly frequency-selective, respectively, with respect to the basic signalling element, the chip. Later, we show that a DS/SS system operating over a frequency-selective channel at the chip level does not necessarily experience frequency-selective distortion at the symbol level.

18.7 Time Variance Viewed in the Time Domain: Figure 18.1, Block 13—The Spaced-Time Correlation Function

Until now, we have described signal dispersion and coherence bandwidth, parameters that describe the channel's time-spreading properties in a local area. However, they do not offer information about the time-varying nature of the channel caused by relative motion between a transmitter and receiver, or by movement of objects within the channel. For mobile-radio applications, the channel is time variant because motion between the transmitter and receiver results in propagation-path changes. Thus, for a transmitted continuous wave (CW) signal, as a result of such motion, the radio receiver sees variations in the signal's amplitude and phase. Assuming that all scatterers making up the channel are stationary, then whenever motion ceases, the amplitude and phase of the received signal remain constant; that is, the channel appears to be time invariant. Whenever motion begins again, the channel appears time variant. Since the channel characteristics are dependent on the positions of the transmitter and receiver, time variance in this case is equivalent to spatial variance.

Figure 18.7(c) shows the function $R(\Delta t)$, designated the **spaced-time correlation function;** it is the autocorrelation function of the channel's response to a sinusoid. This function specifies the extent to which there is correlation between the channel's response to a sinusoid sent at time t_1 and the response to a similar sinusoid sent at time t_2, where $\Delta t = t_2 - t_1$. The **coherence time,** T_0, is a measure of the expected time duration over which the channel's response is essentially invariant. Earlier, we made measurements of signal dispersion and coherence bandwidth by using wideband signals. Now, to measure the time-variant nature of the channel, we use a narrowband signal. To measure $R(\Delta t)$ we can transmit a single sinusoid ($\Delta f = 0$) and determine the autocorrelation function of the received signal. The function $R(\Delta t)$ and the parameter T_0 provide us with knowledge about the fading rapidity of the channel. Note that for an ideal **time-invariant channel** (e.g., a mobile radio exhibiting no motion at all), the channel's response would be highly correlated for all values of Δt, and $R(\Delta t)$ would be a constant function. When using the dense-scatterer channel model described earlier, with constant velocity of motion, and an unmodulated CW signal, the normalized $R(\Delta t)$ is described as

$$R(\Delta t) = J_0 \left(kV \Delta t \right) \tag{18.15}$$

where $J_0(\cdot)$ is the zero-order Bessel function of the first kind, V is velocity, $V \Delta t$ is distance traversed, and $k = 2\pi/\lambda$ is the free-space phase constant (transforming distance to radians of phase). Coherence time can be measured in terms of either time or distance traversed (assuming some fixed velocity of motion). Amoroso described such a measurement using a CW signal and a dense-scatterer channel model [18]. He measured the statistical correlation between the combination of received magnitude and phase sampled at a particular antenna location x_0, and the corresponding combination sampled at some displaced location $x_0 + \zeta$, with displacement measured in units of wavelength λ. For a displacement ζ of 0.38λ between two antenna locations, the combined magnitudes and phases of the received CW are statistically uncorrelated. In other words, the state of the signal at x_0 says nothing about the state of the signal at $x_0 + \zeta$. For a given velocity of motion, this displacement is readily transformed into units of time (coherence time).

18.7.1   The Concept of Duality

Two operators (functions, elements, or systems) are dual when the behavior of one with reference to a time-related domain (time or time-delay) is identical to the behavior of the other with reference to the corresponding frequency-related domain (frequency or Doppler shift).

In Fig. 18.7, we can identify functions that exhibit similar behavior across domains. For understanding the fading channel model, it is useful to refer to such functions as duals. For example, $R(\Delta f)$ in Fig. 18.7(b), characterizing signal dispersion in the frequency domain, yields knowledge about the range of frequency over which two spectral components of a received signal have a strong potential for amplitude and phase correlation. $R(\Delta t)$ in Fig. 18.7(c), characterizing fading rapidity in the time domain, yields knowledge about the span of time over which two received signals have a strong potential for amplitude and phase correlation. We have labeled these two correlation functions as duals. This is also noted in Fig. 18.1 as the duality between blocks 10 and 13, and in Fig. 18.6 as the duality between the time-spreading mechanism in the frequency domain and the time-variant mechanism in the time domain.

18.7.2   Degradation Categories due to Time Variance Viewed in the Time Domain

The time-variant nature of the channel or fading rapidity mechanism can be viewed in terms of two degradation categories as listed in Fig. 18.6: **fast fading** and **slow fading.** The terminology "fast fading" is used for describing channels in which $T_0 < T_s$, where T_0 is the channel coherence time and T_s is the time duration of a transmission symbol. Fast fading describes a condition where the time duration in which the channel behaves in a correlated manner is short compared to the time duration of a symbol. Therefore, it can be expected that the fading character of the channel will change several times during the time that a symbol is propagating, leading to distortion of the baseband pulse shape. Analogous to the distortion previously described as channel-induced ISI, here distortion takes place because the received signal's components are not all highly correlated throughout time. Hence, fast fading can cause the baseband pulse to be distorted, resulting in a loss of SNR that often yields an irreducible error rate. Such distorted pulses cause synchronization problems (failure of phase-locked-loop receivers), in addition to difficulties in adequately defining a matched filter.

A channel is generally referred to as introducing slow fading if $T_0 > T_s$. Here, the time duration that the channel behaves in a correlated manner is long compared to the time duration of a transmission symbol. Thus, one can expect the channel state to virtually remain unchanged during the time in which a symbol is transmitted. The propagating symbols will likely not suffer from the pulse distortion described above. The primary degradation in a slow-fading channel, as with flat fading, is loss in SNR.

18.8 Time Variance Viewed in the Doppler-Shift Domain: Figure 18.1, Block 16—The Doppler Power Spectrum

A completely analogous characterization of the time-variant nature of the channel can begin in the Doppler-shift (frequency) domain. Figure 18.7(d) shows a **Doppler power spectral density**, $S(v)$, plotted as a function of Doppler-frequency shift, v. For the case of the dense-scatterer model, a vertical receive antenna with constant azimuthal gain, a uniform distribution of signals arriving at all arrival angles throughout the range $(0, 2\pi)$, and an unmodulated CW signal, the signal spectrum at the antenna terminals is [19]

$$S(v) = \frac{1}{\pi f_d \sqrt{1 - \left(\frac{v - f_c}{f_d}\right)^2}} \tag{18.16}$$

The equality holds for frequency shifts of v that are in the range $\pm f_d$ about the carrier frequency f_c and would be zero outside that range. The shape of the RF Doppler spectrum described by Eq. (18.16) is classically bowl-shaped, as seen in Fig. 18.7(d). Note that the spectral shape is a result of the dense-scatterer channel model. Equation (18.16) has been shown to match experimental data gathered for mobile radio channels [23]; however, different applications yield different spectral shapes. For example, the dense-scatterer model does not hold for the indoor radio channel; the channel model for an indoor area assumes $S(v)$ to be a flat spectrum [24].

In Fig. 18.7(d), the sharpness and steepness of the boundaries of the Doppler spectrum are due to the sharp upper limit on the Doppler shift produced by a vehicular antenna traveling among the stationary scatterers of the dense scatterer model. The largest magnitude (infinite) of $S(v)$ occurs when the scatterer is directly ahead of the moving antenna platform or directly behind it. In that case the magnitude of the frequency shift is given by

$$f_d = \frac{V}{\lambda} \tag{18.17}$$

where V is relative velocity and λ is the signal wavelength. f_d is positive when the transmitter and receiver move toward each other and negative when moving away from each other. For scatterers directly broadside of the moving platform, the magnitude of the frequency shift is zero. The fact that Doppler components arriving at exactly $0°$ and $180°$ have an infinite power spectral density is not a problem, since the angle of arrival is continuously distributed and the probability of components arriving at exactly these angles is zero [3, 19].

$S(v)$ is the Fourier transform of $R(\Delta t)$. We know that the Fourier transform of the autocorrelation function of a time series is the magnitude squared of the Fourier transform of the original time series. Therefore, measurements can be made by simply transmitting a sinusoid (narrowband signal) and using Fourier analysis to generate the power spectrum of the received amplitude [16]. This Doppler power spectrum of the channel yields knowledge about the spectral spreading of a transmitted sinusoid (impulse in frequency) in the Doppler-shift domain. As indicated in Fig. 18.7, $S(v)$ can be regarded as the dual of the multipath intensity profile, $S(\tau)$, since the latter yields knowledge about the time spreading of a transmitted impulse in the time-delay domain. This is also noted in Fig. 18.1 as the duality between blocks 7 and 16, and in Fig. 18.6 as the duality between the time-spreading mechanism in the time-delay domain and the time-variant mechanism in the Doppler-shift domain.

Knowledge of $S(v)$ allows us to glean how much spectral broadening is imposed on the signal as a function of the rate of change in the channel state. The width of the Doppler power spectrum is referred to as the **spectral broadening** or **Doppler spread,** denoted by f_d, and sometimes called the **fading bandwidth** of the channel. Equation (18.16) describes the Doppler frequency shift. In a typical multipath environment, the received signal arrives from several reflected paths with different path distances and different angles of arrival, and the Doppler shift of each arriving path is generally different from that of another path. The effect on the received signal is seen as a Doppler spreading or spectral broadening of the transmitted signal frequency, rather than a shift. Note that the Doppler spread, f_d, and the coherence time, T_0, are reciprocally related (within a multiplicative constant). Therefore, we show the approximate relationship between the two parameters as

$$T_0 \approx \frac{1}{f_d} \tag{18.18}$$

Hence, the Doppler spread f_d or $1/T_0$ is regarded as the typical **fading rate** of the channel. Earlier, T_0 was described as the expected time duration over which the channel's response to a sinusoid is essentially invariant. When T_0 is defined more precisely as the time duration over which the channel's response to a sinusoid has a correlation of at least 0.5, the relationship between T_0 and f_d is approximately [4]

$$T_0 \approx \frac{9}{16\pi f_d} \tag{18.19}$$

A popular "rule of thumb" is to define T_0 as the geometric mean of Eqs. (18.18) and (18.19). This yields

$$T_0 = \sqrt{\frac{9}{16\pi f_d^2}} = \frac{0.423}{f_d} \tag{18.20}$$

For the case of a 900 MHz mobile radio, Fig. 18.10 illustrates the typical effect of Rayleigh fading on a signal's envelope amplitude vs. time [3]. The figure shows that the distance traveled by the mobile in the time interval corresponding to two adjacent nulls (small-scale fades) is on the order of a half-wavelength $(\lambda/2)$ [3]. Thus, from Fig. 18.10 and Eq. (18.17), the time (approximately, the coherence time) required to traverse a distance $\lambda/2$ when traveling at a constant velocity, V, is:

$$T_0 \approx \frac{\lambda/2}{V} = \frac{0.5}{f_d} \tag{18.21}$$

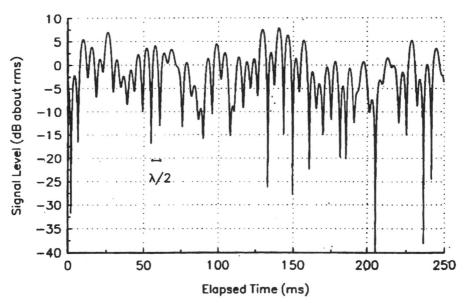

FIGURE 18.10   A typical Rayleigh fading envelope at 900 MHz.

Thus, when the interval between fades is taken to be $\lambda/2$, as in Fig. 18.10, the resulting expression for T_0 in Eq. (18.21) is quite close to the rule-of-thumb shown in Eq. (18.20). Using Eq. (18.21), with the parameters shown in Fig. 18.10 (velocity = 120 km/hr, and carrier frequency = 900 MHz), it is straightforward to compute that the coherence time is approximately 5 ms and the Doppler spread (channel fading rate) is approximately 100 Hz. Therefore, if this example represents a voice-grade channel with a typical transmission rate of 10^4 symbols/s, the fading rate is considerably less than the symbol rate. Under such conditions, the channel would manifest slow-fading effects. Note that if the abscissa of Fig. 18.10 were labeled in units of wavelength instead of time, the figure would look the same for any radio frequency and any antenna speed.

18.9   Analogy Between Spectral Broadening in Fading Channels and Spectral Broadening in Digital Signal Keying

Help is often needed in understanding why spectral broadening of the signal is a function of fading rate of the channel. Figure 18.11 uses the keying of a digital signal (such as amplitude-shift-keying or frequency-shift-keying) to illustrate an analogous case. Figure 18.11(a) shows that a single tone, $\cos 2\pi f_c t$ $(-\infty < t < \infty)$ that exists for all time is characterized in the frequency domain in terms of impulses (at $\pm f_c$). This frequency domain representation is ideal (i.e., zero bandwidth), since the tone is pure and neverending. In practical applications, digital signalling involves switching (keying) signals on and off at a required rate. The keying operation can be viewed as multiplying the infinite-duration tone in Fig. 18.11(a) by an ideal rectangular (switching) function in Fig. 18.11(b). The frequency-domain description of the ideal rectangular function is of the form $(\sin f)/f$. In Fig. 18.11(c), the result of the multiplication yields a tone, $\cos 2\pi f_c t$, that is time-duration limited in the interval

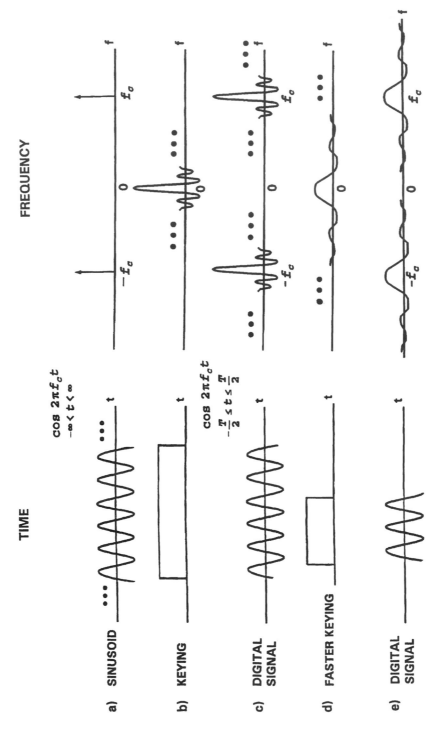

FIGURE 18.11   Analogy between spectral broadening in fading and spectral broadening in keying a digital signal.

$-T/2 < t < T/2$. The resulting spectrum is obtained by convolving the spectral impulses in part (a) with the $(\sin f)/f$ function in part (b), yielding the broadened spectrum in part (c). It is further seen that, if the signalling occurs at a faster rate characterized by the rectangle of shorter duration in part (d), the resulting spectrum of the signal in part (e) exhibits greater spectral broadening. The changing state of a fading channel is somewhat analogous to the keying on and off of digital signals. The channel behaves like a switch, turning the signal "on" and "off." The greater the rapidity of the change in the channel state, the greater the spectral broadening of the received signals. The analogy is not exact because the on and off switching of signals may result in phase discontinuities, but the typical multipath-scatterer environment induces phase-continuous effects.

18.10 Degradation Categories due to Time Variance, Viewed in the Doppler-Shift Domain

A channel is referred to as fast fading if the symbol rate, $1/T_s$ (approximately equal to the signalling rate or bandwidth W) is less than the fading rate, $1/T_0$ (approximately equal to f_d); that is, fast fading is characterized by

$$W < f_d \tag{18.22a}$$

or

$$T_s > T_0 \tag{18.22b}$$

Conversely, a channel is referred to as slow fading if the signalling rate is greater than the fading rate. Thus, in order to avoid signal distortion caused by fast fading, the channel must be made to exhibit slow fading by insuring that the signalling rate must exceed the channel fading rate. That is

$$W > f_d \tag{18.23a}$$

or

$$T_s < T_0 \tag{18.23b}$$

In Eq. (18.14), it was shown that due to signal dispersion, the coherence bandwidth, f_0, sets an upper limit on the signalling rate which can be used without suffering frequency-selective distortion. Similarly, Eq. (18.23) shows that due to Doppler spreading, the channel fading rate, f_d, sets a lower limit on the signalling rate that can be used without suffering fast-fading distortion. For HF communicating systems, when teletype or Morse-coded messages were transmitted at a low data rate, the channels were often fast fading. However, most present-day terrestrial mobile-radio channels can generally be characterized as slow fading.

Equation (18.23) doesn't go far enough in describing what we desire of the channel. A better way to state the requirement for mitigating the effects of fast fading would be that we desire $W \gg f_d$ (or $T_s \ll T_0$). If this condition is not satisfied, the random frequency modulation (FM) due to varying Doppler shifts will limit the system performance significantly. The Doppler effect yields an irreducible error rate that cannot be overcome by simply increasing E_b/N_0 [25]. This irreducible error rate is most pronounced for any modulation that involves switching the carrier phase. A single specular Doppler path, without scatterers, registers an instantaneous frequency shift, classically calculated as $f_d = V/\lambda$. However, a combination of specular and multipath components yields a rather complex time dependence of instantaneous frequency which can cause much larger frequency swings than $\pm V/\lambda$ when detected by an instantaneous frequency detector (a nonlinear device) [26]. Ideally, coherent demodulators that lock onto and track the information signal should suppress the effect of this FM noise and thus cancel the impact of Doppler shift. However, for large values of f_d, carrier recovery becomes a problem because very wideband (relative to the data rate) phase-lock loops (PLLs) need to be designed. For voice-grade applications with bit-error rates of 10^{-3} to 10^{-4}, a large value of Doppler shift is considered to be on the order of $0.01 \times W$. Therefore, to avoid fast-fading distortion and the Doppler-induced irreducible error rate, the signalling rate should exceed the fading rate by a factor of 100 to 200 [27]. The exact factor depends on the signal modulation, receiver design, and required error-rate [3], [26]–[29]. Davarian [29] showed that a frequency-tracking loop can help lower, but not completely remove, the irreducible error rate in a mobile system when using differential minimum-shift keyed (DMSK) modulation.

18.11   Mitigation Methods

Figure 18.12, subtitled "The Good, The Bad, and The Awful," highlights three major performance categories in terms of bit-error probability, P_B, vs. E_b/N_0. The leftmost exponentially-shaped curve represents the performance that can be expected when using any nominal modulation type in AWGN. Observe that with a reasonable amount of E_b/N_0, good performance results. The middle curve, referred to as the **Rayleigh limit,** shows the performance degradation resulting from a loss in SNR that is characteristic of flat fading or slow fading when there is no line-of-sight signal component present. The curve is a function of the reciprocal of E_b/N_0 (an inverse-linear function), so for reasonable values of SNR, performance will generally be "bad." In the case of Rayleigh fading, parameters with overbars are often introduced to indicate that a mean is being taken over the "ups" and "downs" of the fading experience. Therefore, one often sees such bit-error probability plots with mean parameters denoted by $\overline{P_B}$ and $\overline{E_b}/N_0$. The curve that reaches an irreducible level, sometimes called an **error floor,** represents "awful" performance, where the bit-error probability can approach the value of 0.5. This shows the severe distorting effects of frequency-selective fading or fast fading.

If the channel introduces signal distortion as a result of fading, the system performance can exhibit an irreducible error rate; when larger than the desired error rate, no amount of E_b/N_0 will help achieve the desired level of performance. In such cases, the general approach for improving performance is to use some form of mitigation to remove or reduce the distortion. The mitigation method depends on whether the distortion is caused by frequency-selective fading or fast fading. Once the distortion has been mitigated, the P_B vs. E_b/N_0 performance should have transitioned from the "awful" bottoming out curve to the merely "bad" Rayleigh limit curve. Next, we can further ameliorate the effects of fading and strive to approach AWGN performance by using some form of diversity to provide the receiver with a collection of uncorrelated samples of the signal, and by using a powerful error-correction code.

In Fig. 18.13, several mitigation techniques for combating the effects of both signal distortion and loss in SNR are listed. Just as Figs. 18.1 and 18.6 serve as a guide for characterizing fading phenomena and their effects, Fig. 18.13 can similarly serve to describe mitigation

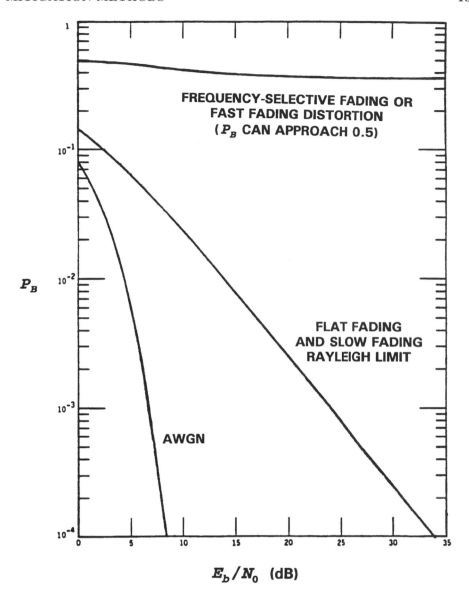

FIGURE 18.12 Error performance: The good, the bad, and the awful.

methods that can be used to ameliorate the effects of fading. The mitigation approach to be used should follow two basic steps: first, provide distortion mitigation; second, provide diversity.

18.11.1 Mitigation to Combat Frequency-Selective Distortion

- Equalization can compensate for the channel-induced ISI that is seen in frequency-selective fading. That is, it can help move the operating point from the error-performance curve that is "awful" in Fig. 18.12 to the one that is "bad." The process of equalizing the ISI involves some method of gathering the dispersed symbol energy back together into its original time interval. In effect, equaliza-

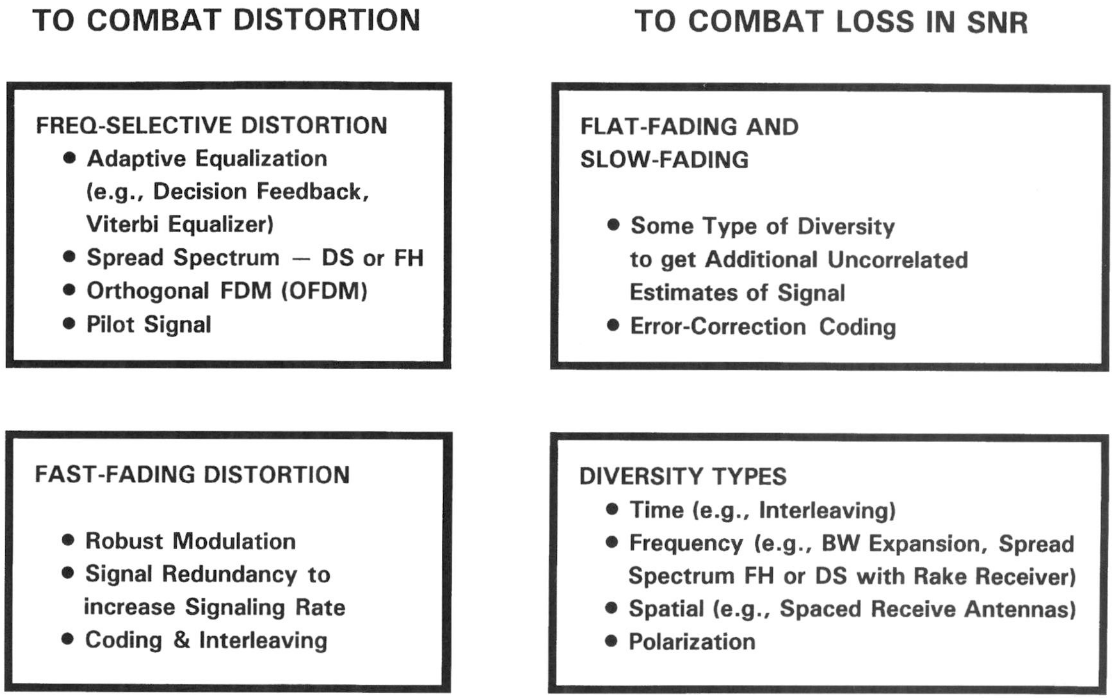

TO COMBAT LOSS IN SNR

FLAT-FADING AND
SLOW-FADING

- Some Type of Diversity
  to get Additional Uncorrelated
  Estimates of Signal
- Error-Correction Coding

DIVERSITY TYPES

- Time (e.g., Interleaving)
- Frequency (e.g., BW Expansion, Spread
  Spectrum FH or DS with Rake Receiver)
- Spatial (e.g., Spaced Receive Antennas)
- Polarization

TO COMBAT DISTORTION

FREQ-SELECTIVE DISTORTION

- Adaptive Equalization
  (e.g., Decision Feedback,
  Viterbi Equalizer)
- Spread Spectrum — DS or FH
- Orthogonal FDM (OFDM)
- Pilot Signal

FAST-FADING DISTORTION

- Robust Modulation
- Signal Redundancy to
  increase Signaling Rate
- Coding & Interleaving

FIGURE 18.13   Basic mitigation types.

tion involves insertion of a filter to make the combination of channel and filter yield a flat response with linear phase. The phase linearity is achieved by making the equalizer filter the complex conjugate of the time reverse of the dispersed pulse [30]. Because in a mobile system the channel response varies with time, the equalizer filter must also change or adapt to the time-varying channel. Such equalizer filters are, therefore, called adaptive equalizers. An equalizer accomplishes more than distortion mitigation; it also provides diversity. Since distortion mitigation is achieved by gathering the dispersed symbol's energy back into the symbol's original time interval so that it doesn't hamper the detection of other symbols, the equalizer is simultaneously providing each received symbol with energy that would otherwise be lost.

- The decision feedback equalizer (DFE) has a feedforward section that is a linear transversal filter [30] whose length and tap weights are selected to coherently combine virtually all of the current symbol's energy. The DFE also has a feedback section which removes energy that remains from previously detected symbols [14], [30]–[32]. The basic idea behind the DFE is that once an information symbol has been detected, the ISI that it induces on future symbols can be estimated and subtracted before the detection of subsequent symbols.

- The maximum-likelihood sequence estimation (MLSE) equalizer tests all possible data sequences (rather than decoding each received symbol by itself) and chooses the data sequence that is the most probable of the candidates. The MLSE equalizer was first proposed by Forney [33] when he implemented the equalizer using the Viterbi decoding algorithm [34]. The MLSE is optimal in the sense that it minimizes the probability of a sequence error. Because the Viterbi decoding algorithm is the way in which the MLSE equalizer is typically implemented, the equalizer is often referred to as the **Viterbi equalizer.** Later in this chapter, we illustrate the adaptive equalization performed in the Global System for Mobile Communications (GSM) using the Viterbi equalizer.

- Spread-spectrum techniques can be used to mitigate frequency-selective ISI distortion because the hallmark of any spread-spectrum system is its capability to reject interference, and ISI is a type of interference. Consider a direct-sequence spread-spectrum (DS/SS) binary phase shift keying (PSK) communication channel comprising one direct path and one reflected path. Assume that the propagation from transmitter to receiver results in a multipath wave that is delayed by τ_k compared to the direct wave. If the receiver is synchronized to the waveform arriving via the direct path, the received signal, $r(t)$, neglecting noise, can be expressed as

$$r(t) = Ax(t)g(t) \cos\left(2\pi f_c t\right) + \alpha Ax\left(t - \tau_k\right) g\left(t - \tau_k\right) \cos\left(2\pi f_c t + \Theta\right) \quad (18.24)$$

where $x(t)$ is the data signal, $g(t)$ is the pseudonoise (PN) spreading code, and τ_k is the differential time delay between the two paths. The angle Θ is a random phase, assumed to be uniformly distributed in the range $(0, 2\pi)$, and α is the attenuation of the multipath signal relative to the direct path signal. The receiver multiplies the incoming $r(t)$ by the code $g(t)$. If the receiver is synchronized to the direct path signal, multiplication by the code signal yields

$$Ax(t)g^2(t) \cos\left(2\pi f_c t\right) + \alpha Ax\left(t - \tau_k\right) g(t)g\left(t - \tau_k\right) \cos\left(2\pi f_c t + \Theta\right) \quad (18.25)$$

where $g^2(t) = 1$, and if τ_k is greater than the chip duration, then,

$$\left| \int g^*(t) g\left(t - \tau_k\right) dt \right| \ll \int g^*(t) g(t) dt \tag{18.26}$$

over some appropriate interval of integration (correlation), where $*$ indicates complex conjugate, and τ_k is equal to or larger than the PN chip duration. Thus, the spread spectrum system effectively eliminates the multipath interference by virtue of its code-correlation receiver. Even though channel-induced ISI is typically transparent to DS/SS systems, such systems suffer from the loss in energy contained in all the multipath components not seen by the receiver. The need to gather up this lost energy belonging to the received chip was the motivation for developing the Rake receiver [35]–[37]. The Rake receiver dedicates a separate correlator to each multipath component (finger). It is able to coherently add the energy from each finger by selectively delaying them (the earliest component gets the longest delay) so that they can all be coherently combined.

- Earlier, we described a channel that could be classified as flat fading, but occasionally exhibits frequency-selective distortion when the null of the channel's frequency transfer function occurs at the center of the signal band. The use of DS/SS is a good way to mitigate such distortion because the wideband SS signal would span many lobes of the selectively faded frequency response. Hence, a great deal of pulse energy would then be passed by the scatterer medium, in contrast to the nulling effect on a relatively narrowband signal [see Fig. 18.8(c)] [18].

- Frequency-hopping spread-spectrum (FH/SS) can be used to mitigate the distortion due to frequency-selective fading, provided the hopping rate is at least equal to the symbol rate. Compared to DS/SS, mitigation takes place through a different mechanism. FH receivers avoid multipath losses by rapid changes in the transmitter frequency band, thus avoiding the interference by changing the receiver band position before the arrival of the multipath signal.

- Orthogonal frequency-division multiplexing (OFDM) can be used in frequency-selective fading channels to avoid the use of an equalizer by lengthening the symbol duration. The signal band is partitioned into multiple subbands, each one exhibiting a lower symbol rate than the original band. The subbands are then transmitted on multiple orthogonal carriers. The goal is to reduce the symbol rate (signalling rate), $W \approx 1/T_s$, on each carrier to be less than the channel's coherence bandwidth f_0. OFDM was originally referred to as Kineplex. The technique has been implemented in the U.S. in mobile radio systems [38], and has been chosen by the European community under the name Coded OFDM (COFDM), for high-definition television (HDTV) broadcasting [39].

- Pilot signal is the name given to a signal intended to facilitate the coherent detection of waveforms. Pilot signals can be implemented in the frequency domain as an in-band tone [40], or in the time domain as a pilot sequence, which can also provide information about the channel state and thus improve performance in fading [41].

18.11.2   Mitigation to Combat Fast-Fading Distortion

- For fast fading distortion, use a robust modulation (noncoherent or differentially coherent) that does not require phase tracking, and reduce the detector integration time [20].

- Increase the symbol rate, $W \approx 1/T_s$, to be greater than the fading rate, $f_d \approx 1/T_0$, by adding signal redundancy.

- Error-correction coding and interleaving can provide mitigation because instead of providing more signal energy, a code reduces the required E_b/N_0. For a given E_b/N_0, with coding present, the error floor will be lowered compared to the uncoded case.

- An interesting filtering technique can provide mitigation in the event of fast-fading distortion and frequency-selective distortion occurring simultaneously. The frequency-selective distortion can be mitigated by the use of an OFDM signal set. Fast fading, however, will typically degrade conventional OFDM because the Doppler spreading corrupts the orthogonality of the OFDM subcarriers. A polyphase filtering technique [42] is used to provide time-domain shaping and duration extension to reduce the spectral sidelobes of the signal set and thus help preserve its orthogonality. The process introduces known ISI and adjacent channel interference (ACI) which are then removed by a post-processing equalizer and canceling filter [43].

18.11.3 Mitigation to Combat Loss in SNR

After implementing some form of mitigation to combat the possible distortion (frequency-selective or fast fading), the next step is to use some form of diversity to move the operating point from the error-performance curve labeled as "bad" in Fig. 18.12 to a curve that approaches AWGN performance. The term "diversity" is used to denote the various methods available for providing the receiver with uncorrelated renditions of the signal. Uncorrelated is the important feature here, since it would not help the receiver to have additional copies of the signal if the copies were all equally poor. Listed below are some of the ways in which diversity can be implemented.

- Time diversity—Transmit the signal on L different time slots with time separation of at least T_0. Interleaving, often used with error-correction coding, is a form of time diversity.

- Frequency diversity—Transmit the signal on L different carriers with frequency separation of at least f_0. Bandwidth expansion is a form of frequency diversity. The signal bandwidth, W, is expanded to be greater than f_0, thus providing the receiver with several independently fading signal replicas. This achieves frequency diversity of the order $L = W/f_0$. Whenever W is made larger than f_0, there is the potential for frequency-selective distortion unless we further provide some mitigation such as equalization. Thus, an expanded bandwidth can improve system performance (via diversity) only if the frequency-selective distortion the diversity may have introduced is mitigated.

- Spread spectrum is a form of bandwidth expansion that excels at rejecting interfering signals. In the case of direct-sequence spread-spectrum (DS/SS), it was shown earlier that multipath components are rejected if they are delayed by more than one chip duration. However, in order to approach AWGN performance, it is necessary to compensate for the loss in energy contained in those rejected components. The Rake receiver (described later) makes it possible to coherently combine the energy from each of the multipath components arriving along different paths. Thus, used with a Rake receiver, DS/SS modulation can be said to achieve path diversity. The Rake receiver is needed in phase-coherent reception, but in differentially coherent bit detection, a simple delay line (one bit long) with complex conjugation will do the trick [44].

- Frequency-hopping spread-spectrum (FH/SS) is sometimes used as a diversity mechanism. The GSM system uses slow FH (217 hops/s) to compensate for those cases where the mobile user is moving very slowly (or not at all) and happens to be in a spectral null.

- Spatial diversity is usually accomplished through the use of multiple receive antennas, separated by a distance of at least 10 wavelengths for a base station (much less for a mobile station). Signal processing must be employed to choose the best antenna output or to coherently combine all the outputs. Systems have also been implemented with multiple spaced transmitters; an example is the Global Positioning System (GPS).

- Polarization diversity [45] is yet another way to achieve additional uncorrelated samples of the signal.

- Any diversity scheme may be viewed as a trivial form of repetition coding in space or time. However, there exist techniques for improving the loss in SNR in a fading channel that are more efficient and more powerful than repetition coding. Error-correction coding represents a unique mitigation technique, because instead of providing more signal energy it reduces the required E_b/N_0 in order to accomplish the desired error performance. Error-correction coding coupled with interleaving [20], [46]–[51] is probably the most prevalent of the mitigation schemes used to provide improved performance in a fading environment.

18.12 Summary of the Key Parameters Characterizing Fading Channels

We summarize the conditions that must be met so that the channel does not introduce frequency-selective distortion and fast-fading distortion. Combining the inequalities of Eqs. (18.14) and (18.23), we obtain

$$f_0 > W > f_d \tag{18.27a}$$

or

$$T_m < T_s < T_0 \tag{18.27b}$$

In other words, we want the channel coherence bandwidth to exceed our signalling rate, which in turn should exceed the fading rate of the channel. Recall that without distortion mitigation, f_0 sets an upper limit on signalling rate, and f_d sets a lower limit on it.

18.12.1 Fast-Fading Distortion: Example #1

If the inequalities of Eq. (18.27) are not met and distortion mitigation is not provided, distortion will result. Consider the fast-fading case where the signalling rate is less than the channel fading rate, that is,

$$f_0 > W < f_d \tag{18.28}$$

Mitigation consists of using one or more of the following methods. (See Fig. 18.13).

- Choose a modulation/demodulation technique that is most robust under fast-fading conditions. That means, for example, avoiding carrier recovery with PLLs since the fast fading could keep a PLL from achieving lock conditions.
- Incorporate sufficient redundancy so that the transmission symbol rate exceeds the channel fading rate. As long as the transmission symbol rate does not exceed the coherence bandwidth, the channel can be classified as flat fading. However, even flat-fading channels will experience frequency-selective distortion whenever a channel null appears at the band center.

Since this happens only occasionally, mitigation might be accomplished by adequate error-correction coding and interleaving.

- The above two mitigation approaches should result in the demodulator operating at the Rayleigh limit [20] (see Fig. 18.12). However, there may be an irreducible floor in the error-performance vs. E_b/N_0 curve due to the FM noise that results from the random Doppler spreading. The use of an in-band pilot tone and a frequency-control loop can lower this irreducible performance level.
- To avoid this error floor caused by random Doppler spreading, increase the signalling rate above the fading rate still further (100–200 × fading rate) [27]. This is one architectural motive behind time-division multiple access (TDMA) mobile systems.
- Incorporate error-correction coding and interleaving to lower the floor and approach AWGN performance.

18.12.2   Frequency-Selective Fading Distortion: Example #2

Consider the frequency-selective case where the coherence bandwidth is less than the symbol rate; that is,

$$f_0 < W > f_d \qquad (18.29)$$

Mitigation consists of using one or more of the following methods. (See Fig. 18.13).

- Since the transmission symbol rate exceeds the channel-fading rate, there is no fast-fading distortion. Mitigation of frequency-selective effects is necessary. One or more of the following techniques may be considered:
- Adaptive equalization, spread spectrum (DS or FH), OFDM, pilot signal. The European GSM system uses a midamble training sequence in each transmission time slot so that the receiver can learn the impulse response of the channel. It then uses a Viterbi equalizer (explained later) for mitigating the frequency-selective distortion.
- Once the distortion effects have been reduced, introduce some form of diversity and error-correction coding and interleaving in order to approach AWGN performance. For direct-sequence spread-spectrum (DS/SS) signalling, the use of a Rake receiver (explained later) may be used for providing diversity by coherently combining multipath components that would otherwise be lost.

18.12.3   Fast-Fading and Frequency-Selective Fading Distortion: Example #3

Consider the case where the coherence bandwidth is less than the signalling rate, which in turn is less than the fading rate. The channel exhibits both fast-fading and frequency-selective fading which is expressed as

$$f_0 < W < f_d \qquad\qquad (18.30a)$$

or

$$f_0 < f_d \qquad\qquad (18.30b)$$

Recalling from Eq. (18.27) that f_0 sets an upper limit on signalling rate and f_d sets a lower limit on it, this is a difficult design problem because, unless distortion mitigation is provided, the maximum allowable signalling rate is (in the strict terms of the above discussion) less than the minimum allowable signalling rate. Mitigation in this case is similar to the initial approach outlined in example #1.

- Choose a modulation/demodulation technique that is most robust under fast-fading conditions.
- Use transmission redundancy in order to increase the transmitted symbol rate.
- Provide some form of frequency-selective mitigation in a manner similar to that outlined in example #2.
- Once the distortion effects have been reduced, introduce some form of diversity and error-correction coding and interleaving in order to approach AWGN performance.

18.13   The Viterbi Equalizer as Applied to GSM

Figure 18.14 shows the GSM time-division multiple access (TDMA) frame, having a duration of 4.615 ms and comprising 8 slots, one assigned to each active mobile user. A normal transmission burst occupying one slot of time contains 57 message bits on each side

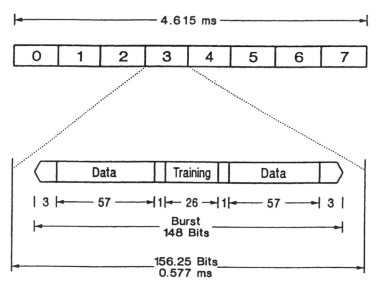

FIGURE 18.14   The GSM TDMA frame and time-slot containing a normal burst.

of a 26-bit midamble called a **training** or **sounding sequence.** The slot-time duration is 0.577 ms (or the slot rate is 1733 slots/s). The purpose of the midamble is to assist the receiver in estimating the impulse response of the channel in an adaptive way (during the time duration of each 0.577 ms slot). In order for the technique to be effective, the fading behavior of the channel should not change appreciably during the time interval of one slot. In other words, there should not be any fast-fading degradation during a slot time when the receiver is using knowledge from the midamble to compensate for the channel's fading behavior. Consider the example of a GSM receiver used aboard a high-speed train, traveling at a constant velocity of 200 km/hr (55.56 m/s). Assume the carrier frequency to be 900 MHz, (the wavelength is $\lambda = 0.33$ m). From Eq. (18.21), we can calculate that a half-wavelength is traversed in approximately the time (coherence time)

$$T_0 \approx \frac{\lambda/2}{V} \approx 3 \text{ ms} \tag{18.31}$$

Therefore, the channel coherence time is over 5 times greater than the slot time of 0.577 ms. The time needed for a significant change in fading behavior is relatively long compared to the time duration of one slot. Note, that the choices made in the design of the GSM TDMA slot time and midamble were undoubtedly influenced by the need to preclude fast fading with respect to a slot-time duration, as in this example.

The GSM symbol rate (or bit rate, since the modulation is binary) is 271 kilosymbols/s and the bandwidth is $W = 200$ kHz. If we consider that the typical rms delay spread in an urban environment is on the order of $\sigma_\tau = 2\mu s$, then using Eq. (18.13) the resulting coherence bandwidth is $f_0 \approx 100$ kHz. It should therefore be apparent that since $f_0 < W$, the GSM receiver must utilize some form of mitigation to combat frequency-selective distortion. To accomplish this goal, the Viterbi equalizer is typically implemented.

Figure 18.15 illustrates the basic functional blocks used in a GSM receiver for estimating the channel impulse response, which is then used to provide the detector with channel-corrected reference waveforms [52]. In the final step, the Viterbi algorithm is used to

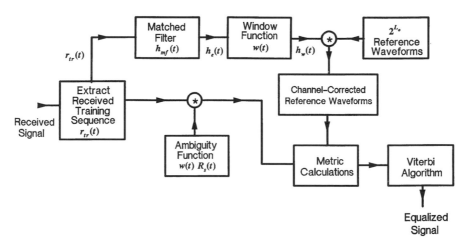

FIGURE 18.15 The Viterbi equalizer as applied to GSM.

compute the MLSE of the message. As stated in Eq. (18.2), a received signal can be described in terms of the transmitted signal convolved with the impulse response of the channel, $h_c(t)$. We show this below, using the notation of a received training sequence,

$r_{tr}(t)$, and the transmitted training sequence, $s_{tr}(t)$, as follows:

$$r_{tr}(t) = s_{tr}(t) * h_c(t) \tag{18.32}$$

where * denotes convolution. At the receiver, $r_{tr}(t)$ is extracted from the normal burst and sent to a filter having impulse response, $h_{mf}(t)$, that is matched to $s_{tr}(t)$. This matched filter yields at its output an estimate of $h_c(t)$, denoted $h_e(t)$, developed from Eq. (18.32) as follows.

$$
\begin{aligned}
h_e(t) &= r_{tr}(t) * h_{mf}(t) \\
       &= s_{tr}(t) * h_c(t) * h_{mf}(t) \\
       &= R_s(t) * h_c(t)
\end{aligned}
\tag{18.33}
$$

where $R_s(t)$ is the autocorrelation function of $s_{tr}(t)$. If $R_s(t)$ is a highly peaked (impulse-like) function, then $h_e(t) \approx h_c(t)$.

Next, using a windowing function, $w(t)$, we truncate $h_e(t)$ to form a computationally affordable function, $h_w(t)$. The window length must be large enough to compensate for the effect of typical channel-induced ISI. The required observation interval L_0 for the window can be expressed as the sum of two contributions. The interval of length L_{CISI} is due to the controlled ISI caused by Gaussian filtering of the baseband pulses, which are then MSK modulated. The interval of length L_C is due to the channel-induced ISI caused by multipath propagation; therefore, L_0 can be written as

$$L_0 = L_{CISI} + L_C \tag{18.34}$$

The GSM system is required to provide mitigation for distortion due to signal dispersions of approximately 15–20 μs. The bit duration is 3.69 μs. Thus, the Viterbi equalizer used in GSM has a memory of 4–6 bit intervals. For each L_0-bit interval in the message, the function of the Viterbi equalizer is to find the most likely L_0-bit sequence out of the 2^{L_0} possible sequences that might have been transmitted. Determining the most likely L_0-bit sequence requires that 2^{L_0} meaningful reference waveforms be created by modifying (or disturbing) the 2^{L_0} ideal waveforms in the same way that the channel has disturbed the transmitted message. Therefore, the 2^{L_0} reference waveforms are convolved with the windowed estimate of the channel impulse response, $h_w(t)$ in order to derive the disturbed or channel-corrected reference waveforms. Next, the channel-corrected reference waveforms are compared against the received data waveforms to yield metric calculations. However, before the comparison takes place, the received data waveforms are convolved with the known windowed autocorrelation function $w(t)R_s(t)$, transforming them in a manner comparable to that applied to the reference waveforms. This filtered message signal is compared to all possible 2^{L_0} channel-corrected reference signals, and metrics are computed as required by the Viterbi decoding algorithm (VDA). The VDA yields the maximum likelihood estimate of the transmitted sequence [34].

18.14   The Rake Receiver Applied to Direct-Sequence Spread-Spectrum (DS/SS) Systems

Interim Specification 95 (IS-95) describes a DS/SS cellular system that uses a Rake receiver [35]–[37] to provide path diversity. In Fig. 18.16, five instances of chip transmissions corresponding to the code sequence 1 0 1 1 1 are shown, with the transmission or observation times labeled t_{-4} for the earliest transmission and t_0 for the latest. Each abscissa shows three "fingers" of a signal that arrive at the receiver with delay times τ_1, τ_2, and τ_3. Assume that the intervals between the t_i transmission times and the intervals between the τ_i delay times are each one chip long. From this, one can conclude that the finger arriving at the receiver at time t_{-4}, with delay τ_3, is time coincident with two other fingers, namely the fingers arriving at times t_{-3} and t_{-2} with delays τ_2 and τ_1, respectively. Since, in this example, the delayed components are separated by exactly one chip time, they are *just* resolvable. At the receiver, there must be a sounding device that is dedicated to estimating the τ_i delay times. Note that for a terrestrial mobile radio system, the fading rate is relatively slow (milliseconds) or the channel coherence time large compared to the chip time ($T_0 > T_{ch}$). Hence, the changes in τ_i occur slowly enough so that the receiver can readily adapt to them.

Once the τ_i delays are estimated, a separate correlator is dedicated to processing each finger. In this example, there would be three such dedicated correlators, each one processing a delayed version of the same chip sequence 1 0 1 1 1. In Fig. 18.16, each correlator receives chips with power profiles represented by the sequence of fingers shown along a diagonal line. Each correlator attempts to match these arriving chips with the same PN code, similarly delayed in time. At the end of a symbol interval (typically there may be hundreds or thousands of chips per symbol), the outputs of the correlators are coherently combined, and a symbol detection is made. At the chip level, the Rake receiver resembles an equalizer, but its real function is to provide diversity.

The interference-suppression nature of DS/SS systems stems from the fact that a code sequence arriving at the receiver merely one chip time late, will be approximately orthogonal to the particular PN code with which the sequence is correlated. Therefore, any code chips that are delayed by one or more chip times will be suppressed by the correlator. The delayed chips only contribute to raising the noise floor (correlation sidelobes). The mitigation provided by the Rake receiver can be termed path diversity, since it allows the energy of a chip that arrives via multiple paths to be combined coherently. Without the Rake receiver, this energy would be transparent and therefore lost to the DS/SS system. In Fig. 18.16, looking vertically above point τ_3, it is clear that there is interchip interference due to different fingers arriving simultaneously. The spread-spectrum processing gain allows the system to endure such interference at the chip level. No other equalization is deemed necessary in IS-95.

18.15   Conclusion

In this chapter, the major elements that contribute to fading in a communication channel have been characterized. Figure 18.1 was presented as a guide for the characterization of fading phenomena. Two types of fading, large-scale and small-scale, were described. Two manifestations of small-scale fading (signal dispersion and fading rapidity) were examined, and the examination involved two views, time and frequency. Two degradation categories were defined for dispersion: frequency-selective fading and flat-fading. Two degradation categories were defined for fading rapidity: fast and slow. The small-scale fading degradation categories were summarized in Fig. 18.6. A mathematical model using correlation and power density functions was presented in Fig. 18.7. This model yields a nice symmetry, a kind of "poetry" to help us view the Fourier transform and duality relationships that describe the fading phenomena. Further, mitigation techniques for ameliorating the effects of each degradation category were treated, and these techniques were summarized in Fig. 18.13. Finally, mitigation methods that have been implemented in two system types, GSM and CDMA systems meeting IS-95, were described.

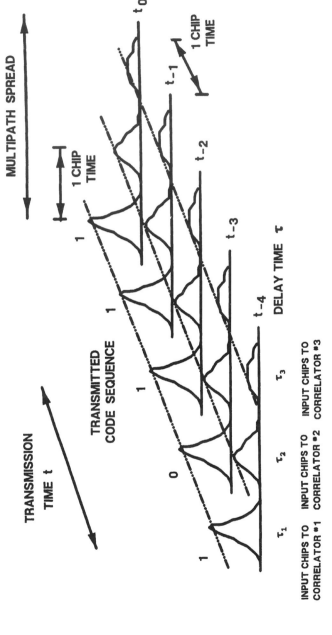

FIGURE 18.16   Example of received chips seen by a 3-finger rake receiver.

References

[1] Sklar, B., *Digital Communications: Fundamentals and Applications*, Prentice-Hall, Englewood Cliffs, NJ, Ch. 4, 1988.

[2] Van Trees, H.L., *Detection, Estimation, and Modulation Theory, Part I*, John Wiley & Sons, New York, Ch. 4, 1968.

[3] Rappaport, T.S., *Wireless Communications*, Prentice-Hall, Upper Saddle River, New Jersey, Chs. 3 and 4, 1996.

[4] Greenwood, D. and Hanzo, L., Characterisation of Mobile Radio Channels, *Mobile Radio Communications*, Steele, R., Ed., Pentech Press, London, Ch. 2, 1994.

[5] Lee, W.C.Y., Elements of cellular mobile radio systems, *IEEE Trans. Vehicular Technol.*, V-35(2), 48–56, May 1986.

[6] Okumura, Y. et al., Field strength and its variability in VHF and UHF land mobile radio service, *Rev. Elec. Comm. Lab.*, 16(9-10), 825–873, 1968.

[7] Hata, M., Empirical formulæ for propagation loss in land mobile radio services, *IEEE Trans. Vehicular Technol.*, VT-29(3), 317–325, 1980.

[8] Seidel, S.Y. et al., Path loss, scattering and multipath delay statistics in four European cities for digital cellular and microcellular radiotelephone, *IEEE Trans. Vehicular Technol.*, 40(4), 721–730, Nov. 1991.

[9] Cox, D.C., Murray, R., and Norris, A., 800 MHz Attenuation measured in and around suburban houses, *AT&T Bell Laboratory Technical Journal*, 673(6), 921–954, Jul.-Aug. 1984.

[10] Schilling, D.L. et al., Broadband CDMA for personal communications systems, *IEEE Commun. Mag.*, 29(11), 86–93, Nov. 1991.

[11] Andersen, J.B., Rappaport, T.S., and Yoshida, S., Propagation measurements and models for wireless communications channels, *IEEE Commun. Mag.*, 33(1), 42–49, Jan. 1995.

[12] Amoroso, F., Investigation of signal variance, bit error rates and pulse dispersion for DSPN signalling in a mobile dense scatterer ray tracing model, *Intl. J. Satellite Commun.*, 12, 579–588, 1994.

[13] Bello, P.A., Characterization of randomly time-variant linear channels, *IEEE Trans. Commun. Syst.*, 360–393, Dec. 1963.

[14] Proakis, J.G., *Digital Communications*, McGraw-Hill, New York, Ch. 7, 1983.

[15] Green, P.E., Jr., Radar astronomy measurement techniques, *MIT Lincoln Laboratory*, Lexington, MA, Tech. Report No. 282, Dec. 1962.

[16] Pahlavan, K. and Levesque, A.H., *Wireless Information Networks*, John Wiley & Sons, New York, Chs. 3 and 4, 1995.

[17] Lee, W.Y.C., *Mobile Cellular Communications*, McGraw-Hill, New York, 1989.

[18] Amoroso, F., Use of DS/SS signalling to mitigate Rayleigh fading in a dense scatterer environment, *IEEE Personal Commun.*, 3(2), 52–61, Apr. 1996.

[19] Clarke, R.H., A statistical theory of mobile radio reception, *Bell Syst. Tech. J.*, 47(6), 957–1000, Jul.-Aug. 1968.

[20] Bogusch, R.L., *Digital Communications in Fading Channels: Modulation and Coding*, Mission Research Corp., Santa Barbara, California, Report No. MRC-R-1043, Mar. 11, 1987.

[21] Amoroso, F., The bandwidth of digital data signals, *IEEE Commun. Mag.*, 18(6), 13–24, Nov. 1980.

[22] Bogusch, R.L. et al., Frequency selective propagation effects on spread-spectrum receiver tracking, *Proc. IEEE*, 69(7), 787–796, Jul. 1981.

[23] Jakes, W.C., Ed., *Microwave Mobile Communications*, John Wiley & Sons, New York, 1974.

[24] *Joint Technical Committee of Committee T1 R1P1.4 and TIA TR46.3.3/TR45.4.4 on Wireless Access,* Draft Final Report on RF Channel Characterization, Paper No. JTC(AIR)/94.01.17-238R4, Jan. 17, 1994.

[25] Bello, P.A. and Nelin, B.D., The influence of fading spectrum on the binary error probabilities of incoherent and differentially coherent matched filter receivers, *IRE Trans. Commun. Syst.,* CS-10, 160–168, Jun. 1962.

[26] Amoroso, F., Instantaneous frequency effects in a Doppler scattering environment, *IEEE International Conference on Communications,* 1458–1466, Jun. 7–10, 1987.

[27] Bateman, A.J. and McGeehan, J.P., Data transmission over UHF fading mobile radio channels, *IEEE Proc.,* 131, Pt. F(4), 364–374, Jul. 1984.

[28] Feher, K., *Wireless Digital Communications,* Prentice-Hall, Upper Saddle River, NJ, 1995.

[29] Davarian, F., Simon, M., and Sumida, J., DMSK: A Practical 2400-bps Receiver for the Mobile Satellite Service, Jet Propulsion Laboratory Publication 85-51 (MSAT-X Report No. 111), Jun. 15, 1985.

[30] Rappaport, T.S., *Wireless Communications,* Prentice-Hall, Upper Saddle River, NJ, Ch. 6, 1996.

[31] Bogusch, R.L., Guigliano, F.W., and Knepp, D.L., Frequency-selective scintillation effects and decision feedback equalization in high data-rate satellite links, *Proc. IEEE,* 71(6), 754–767, Jun. 1983.

[32] Qureshi, S.U.H., Adaptive equalization, *Proc. IEEE,* 73(9), 1340–1387, Sept. 1985.

[33] Forney, G.D., The Viterbi algorithm, *Proc. IEEE,* 61(3), 268–278, Mar. 1978.

[34] Sklar, B., *Digital Communications: Fundamentals and Applications,* Prentice-Hall, Englewood Cliffs, NJ, Ch. 6, 1988.

[35] Price, R. and Green, P.E., Jr., A communication technique for multipath channels, *Proc. IRE,* 555–570, Mar. 1958.

[36] Turin, G.L., Introduction to spread-spectrum antimultipath techniques and their application to urban digital radio, *Proc. IEEE,* 68(3), 328–353, Mar. 1980.

[37] Simon, M.K., Omura, J.K., Scholtz, R.A., and Levitt, B.K., *Spread Spectrum Communications Handbook,* McGraw-Hill, New York, 1994.

[38] Birchler, M.A. and Jasper, S.C., A 64 kbps Digital Land Mobile Radio System Employing M-16QAM, *Proceedings of the 1992 IEEE Intl. Conference on Selected Topics in Wireless Communications,* Vancouver, British Columbia, 158–162, Jun. 25–26, 1992.

[39] Sari, H., Karam, G., and Jeanclaude, I., Transmission techniques for digital terrestrial TV broadcasting, *IEEE Commun. Mag.,* 33(2), 100–109, Feb. 1995.

[40] Cavers, J.K., The performance of phase locked transparent tone-in-band with symmetric phase detection, *IEEE Trans. Commun.,* 39(9), 1389–1399, Sept. 1991.

[41] Moher, M.L. and Lodge, J.H., TCMP—A modulation and coding strategy for Rician fading channel, *IEEE J. Selected Areas Commun.,* 7(9), 1347–1355, Dec. 1989.

[42] Harris, F., On the Relationship Between Multirate Polyphase FIR Filters and Windowed, Overlapped FFT Processing, *Proceedings of the Twenty Third Annual Asilomar Conference on Signals, Systems, and Computers,* Pacific Grove, California, 485–488, Oct. 30 to Nov. 1, 1989.

[43] Lowdermilk, R.W. and Harris, F., Design and Performance of Fading Insensitive Orthogonal Frequency Division Multiplexing (OFDM) using Polyphase Filtering Techniques, *Proceedings of the Thirtieth Annual Asilomar Conference on Signals, Systems, and Computers,* Pacific Grove, California, Nov. 3–6, 1996.

[44] Kavehrad, M. and Bodeep, G.E., Design and experimental results for a direct-sequence spread-spectrum radio using differential phase-shift keying modulation for indoor wireless communications, *IEEE JSAC,* SAC-5(5), 815–823, Jun. 1987.

[45] Hess, G.C., *Land-Mobile Radio System Engineering,* Artech House, Boston, 1993.

[46] Hagenauer, J. and Lutz, E., Forward error correction coding for fading compensation in mobile satellite channels, *IEEE JSAC*, SAC-5(2), 215–225, Feb. 1987.

[47] McLane, P.I. et al., PSK and DPSK trellis codes for fast fading, shadowed mobile satellite communication channels, *IEEE Trans. Commun.*, 36(11), 1242–1246, Nov. 1988.

[48] Schlegel, C. and Costello, D.J., Jr., Bandwidth efficient coding for fading channels: code construction and performance analysis, *IEEE JSAC*, 7(9), 1356–1368, Dec. 1989.

[49] Edbauer, F., Performance of interleaved trellis-coded differential 8–PSK modulation over fading channels, *IEEE J. Selected Areas Commun.*, 7(9), 1340–1346, Dec. 1989.

[50] Soliman, S. and Mokrani, K., Performance of coded systems over fading dispersive channels, *IEEE Trans. Commun.*, 40(1), 51–59, Jan. 1992.

[51] Divsalar, D. and Pollara, F., Turbo Codes for PCS Applications, *Proc. ICC'95*, Seattle, Washington, 54–59, Jun. 18–22, 1995.

[52] Hanzo, L. and Stefanov, J., The Pan-European Digital Cellular Mobile Radio System—known as GSM, *Mobile Radio Communications*. Steele, R., Ed., Pentech Press, London, Ch. 8, 1992.

19

Space-Time Processing

Arogyaswami J. Paulraj
Stanford University

19.1 Introduction

Mobile radio signal processing includes modulation and demodulation, channel coding and decoding, equalization and diversity. Current cellular modems mainly use temporal signal processing. Use of spatio-temporal signal processing can improve average signal power, mitigate fading, and reduce cochannel and intersymbol interference. This can significantly improve the capacity, coverage, and quality of wireless networks.

A space-time processing radio operates simultaneously on multiple antennas by processing signal samples both in space and time. In receive, space-time (ST) processing can increase array gain, spatial and temporal diversity and reduce cochannel interference and intersymbol interference. In transmit, the spatial dimension can enhance array gain, improve diversity, reduce generation of cochannel and inter-symbol interference.

19.2 The Space-Time Wireless Channel

19.2.1 Multipath Propagation

Multipath scattering gives rise to a number of propagation effects described below.

0-8493-8597-0/99/$0.00+$.50

Scatterers Local to Mobile

Scattering local to the mobile is caused by buildings/other scatterers in the vicinity of the mobile (a few tens of meters). Mobile motion and local scattering give rise to Doppler spread which causes time-selective fading. For a mobile traveling at 65 mph, the Doppler spread is about 200 Hz in the 1900 MHz band. While local scatterers contribute to Doppler spread, the delay spread they contribute is usually insignificant because of the small scattering radius. Likewise, the angle spread induced at the base station is also small.

Remote Scatterers

The emerging wavefront from the local scatterers may then travel directly to the base or may be scattered toward the base by remote dominant scatterers, giving rise to specular multipaths. These remote scatterers can be either terrain features or high-rise building complexes. Remote scattering can cause significant delay and angle spreads.

Scatterers Local to Base

Once these multiple wavefronts reach the base station, they may be scattered further by local structures such as buildings or other structures in the vicinity of the base. Such scattering will be more pronounced for low elevation and below roof-top antennas. Scattering local to the base can cause severe angle spread which in turn causes space-selective fading. See Figure 19.1 for a depiction of different types of scattering.

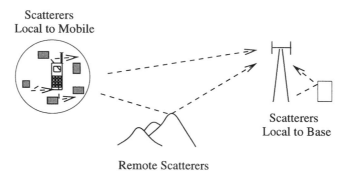

FIGURE 19.1   Multipath propagation in macrocells.

The forward link channel is affected in similar ways by these scatterers, but in a reverse order.

19.2.2   Space-Time Channel Model

The effect of delay, Doppler and angle spreads makes the channel selective in frequency, time, and space. Figure 19.2 shows plots of the frequency response at each branch of a four-antenna receiver operating with a 200 Khz bandwidth. We can see that the channel is highly frequency-selective since the delay spread reaches 10 to 15 μs. Also, an angle spread of 30° causes variations in the channel from antenna to antenna. The channel variation in time depends upon the Doppler spread. As expected, the plots show negligible channel variation between adjacent time slots, despite the high velocity of the mobile (100 kph). Use of longer time slots such as in IS-136 will result in significant channel variations over the slot period. Therefore, space-time processing should address the effect of the three spreads on the signal.

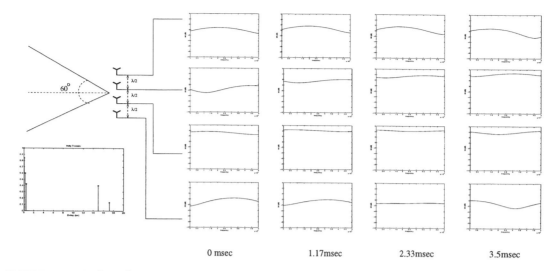

| 0 msec | 1.17msec | 2.33msec | 3.5msec |

FIGURE 19.2 ST channel.

19.3 Signal Models

We develop signal models for nonspread modulation used in time division multiple access (TDMA) systems.

19.3.1 Signal Model at Base Station (Reverse Link)

We assume that antenna arrays are used at the base station only and that the mobile has a single omni antenna. The mobile transmits a channel coded and modulated signal which does not incorporate any spatial (or indeed any special temporal) processing. See Figure 19.3.

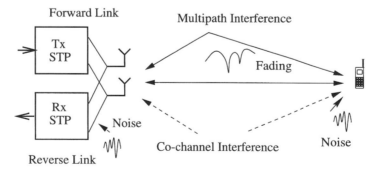

FIGURE 19.3 ST Processing Model.

The baseband signal $x_i(t)$ received by the base station at the ith element of an m element antenna array is given by

$$x_i(t) = \sum_{l=1}^{L} a_i(\theta_l)\,\alpha_l^R(t)u(t - \tau_l) + n_i(t) \tag{19.1}$$

where L is the number of multipaths, $a_i(\theta_l)$ is the response of the ith element for the lth path from direction θ_l, $\alpha_l^R(t)$ is the complex path fading, τ_l is the path delay, $n_i(t)$ is the additive noise and $u(\cdot)$ is the transmitted signal that depends on the modulation waveform and the information data stream.

For a linear modulation, the baseband transmitted signal is given by

$$u(t) = \sum_k g(t - kT)s(k) \tag{19.2}$$

where $g(\cdot)$ is the pulse shaping waveform and $s(k)$ represents the information bits.

In the above model we have assumed that the inverse signal bandwidth is large compared to the travel time across the array. Therefore, the complex envelopes of the signals received by different antennas from a given path are identical except for phase and amplitude differences that depend on the path angle-of-arrival, array geometry and the element pattern. This angle-of-arrival dependent phase and amplitude response at the ith element is $a_i(\theta_l)$.

We collect all the element responses to a path arriving from angle θ_l into an m-dimensional vector, called the *array response vector* defined as

$$\mathbf{a}(\theta_l) = [a_1(\theta_l)\, a_2(\theta_l) \ldots a_m(\theta_l)]^T$$

$$\mathbf{x}(t) = \sum_{l=1}^{L} \mathbf{a}(\theta_l)\, \alpha_l^R(t)u(t - \tau_l) + \mathbf{n}(t) \tag{19.3}$$

where $\mathbf{x}(t)$ and $\mathbf{n}(t)$ are m-dimensional complex vectors. The fading $|\alpha^R(t)|$ is Rayleigh or Rician distributed depending on the propagation model.

19.3.2   Signal Model at Mobile (Forward Link)

In this model, the base station transmits different signals from each antenna with a defined relationship between them. In the case of a two element array, some examples of transmitted signals $u_i(t), i = 1, 2$ can be: (a) delay diversity: $u_2(t) = u_1(t - T)$ where T is the symbol period; (b) Doppler diversity: $u_2(t) = u_1(t)e^{j\omega t}$ where ω is differential carrier offset; (c) beamforming: $u_2(t) = w_2 u_1(t)$ where w_2 is complex scalar; and (d) space-time coding: $u_1(t) = \sum_k g(t - kT)s^1(k), u_2(t) = \sum_k g(t - kT)s^2(k)$ where $s^1(k)$ and $s^2(k)$ are related to the symbol sequence $s(k)$ through coding.

The received signal at the mobile is then given by

$$x(t) = \sum_{i=1}^{m}\sum_{l=1}^{L} a_i(\theta_l)\, \alpha_l^F(t)u_i(t - \tau_l) + n(t) \tag{19.4}$$

where the path delay τ_l and angle parameters θ_l are the same as those of the reverse link. $\alpha_l^F(t)$ is the complex fading on the forward link. In (fast) **TDD** systems $\alpha_l^F(t)$ will be identical to the reverse link complex fading $\alpha_l^R(t)$. In a **FDD** system $\alpha_l^F(t)$ and $\alpha_l^R(t)$ will usually have the same statistics but will in general be uncorrelated with each other. We assume $a_i(\theta_l)$ is the same for both links. This is only approximately true in FDD systems.

If simple beamforming alone is used in transmit, the signals radiated from the antennas are related by a complex scalar and result in a directional transmit beam which may selectively couple into the multipath environment and differentially scale the power in each path.

The signal received by the mobile in this case can be written as

$$x(t) = \sum_{l=1}^{L} \mathbf{w}^H \mathbf{a}(\theta_l)\, \alpha_l^F(t)u(t - \tau_l) + n(t) \tag{19.5}$$

where $\mathbf{w}$ is the beamforming vector.

19.3.3 Discrete Time Signal Model

The channel model described above uses physical path parameters such as path gain, delay, and angle of arrival. In practice these are not known and the discrete time received signal uses a more convenient discretized "symbol response" channel model.

We derive a discrete-time signal model at the base station antenna array. Let the continuous-time output from the receive antenna array $\mathbf{x}(t)$ be sampled at the symbol rate at instants $t = t_o + kT$. Then the vector array output may be written as

$$\mathbf{x}(k) = \mathbf{H}^R \mathbf{s}(k) + n(k) \tag{19.6}$$

where $\mathbf{H}^R$ is the reverse link symbol response channel (a $m \times N$ matrix) that captures the effects of the array response, symbol waveform and path fading. m is the number of antennas, N is the channel length in symbol periods and $\mathbf{n}(k)$ the sampled vector of additive noise. Note that $\mathbf{n}(k)$ may be colored in space and time, as discussed later. $\mathbf{H}^R$ is assumed to be time invariant. $\mathbf{s}(k)$ is a vector of N consecutive elements of the data sequence and is defined as

$$\mathbf{s}(k) = \begin{bmatrix} s(k) \\ \vdots \\ s(k-N+1) \end{bmatrix} \tag{19.7}$$

Note that we have assumed a sampling rate of one sample per symbol. Higher sampling rates may be used. Also, $\mathbf{H}^R$ is given by

$$\mathbf{H}^R = \sum_{l=1}^{L} \mathbf{a}\left(\theta_l\right) \alpha_l^R \mathbf{g}^T\left(\tau_l\right) \tag{19.8}$$

where $\mathbf{g}(\tau_l)$ is a vector defined by T spaced sampling of the pulse shaping function $g(\cdot)$ with an offset of τ_l.

Likewise the forward discrete signal model at the mobile is given by

$$x(k) = \sum_{i=1}^{m} \mathbf{h}_i^F \mathbf{s}(k) + n(k) \tag{19.9}$$

where $\mathbf{h}_i^F$ is a $1 \times N$ composite channel from the symbol sequence via the ith antenna to the mobile receiver which includes the effect transmit ST processing at the base station.

In the case of two antenna delay diversity, $\mathbf{h}_i^F$ is given by

$$\mathbf{h}_1^F = \sum_{l=1}^{L} a_1\left(\theta_l\right) \alpha_l^F \mathbf{g}\left(\tau_l\right) \tag{19.10}$$

and

$$\mathbf{h}_2^F = \sum_{l=1}^{L} a_2\left(\theta_l\right) \alpha_l^F \mathbf{g}\left(\tau_l - T\right) \tag{19.11}$$

If spatial beamforming alone is used, the signal model becomes

$$x(k) = \sum_{l=1}^{L} \mathbf{w}^H \mathbf{H}^F \mathbf{s}(k) + n(k) \tag{19.12}$$

where $\mathbf{H}^F$ is the *intrinsic* forward (F) channel given by

$$\mathbf{H}^F = \begin{bmatrix} \mathbf{h}_1^F \\ \mathbf{h}_2^F \end{bmatrix} = \sum_{l=1}^{L} \mathbf{a}\left(\theta_l\right) \alpha_l^F \mathbf{g}^T\left(\tau_l\right) \tag{19.13}$$

19.3.4 Signal-Plus-Interference Model

The overall received signal-plus-interference-and-noise model at the base station antenna array can be written as

$$\mathbf{x}(k) = \mathbf{H}_s^R \mathbf{s}_s(k) + \sum_{q=1}^{Q-1} \mathbf{H}_q^R \mathbf{s}_q(k) + \mathbf{n}(k) \tag{19.14}$$

where $\mathbf{H}_s^R$ and $\mathbf{H}_q^R$ are channels for signal and **CCI**, respectively, while $\mathbf{s}_s$ and $\mathbf{s}_q$ are the corresponding data sequences. Note that Eq. (19.14) appears to suggest that the signal and interference are baud synchronous. However, this can be relaxed and the time offsets can be absorbed into the channel $\mathbf{H}_q^R$.

Similarly, the signal at the mobile can also be extended to include CCI. Note that in this case, the source of interference is from other base stations (in TDMA) and the channel is between the interfering base station and the desired mobile. It is often convenient to handle signals in blocks. Therefore, we may collect M consecutive snapshots of $\mathbf{x}(\cdot)$ corresponding to time instants $k, \ldots, k + M - 1$, (and dropping subscripts for a moment), we get

$$\mathbf{X}(k) = \mathbf{H}^R \mathbf{S}(k) + \mathbf{N}(k) \tag{19.15}$$

where $\mathbf{X}(k)$, $\mathbf{S}(k)$ and $\mathbf{N}(k)$ are defined appropriately. Similarly the received signal at the mobile in the forward link has a block representation using a row vector.

19.4   ST Receive Processing (Base)

The base station receiver receives the signals from the array antenna which consist of the signals from the desired mobile and the cochannel signals along with associated intersymbol interference and fading. The task of the receiver is to maximize signal power and mitigate fading, CCI and **ISI**. There are two broad approaches for doing this—one is multiuser detection wherein we demodulate both the cochannel and desired signals jointly, the other is to cancel CCI. The structure of the receiver depends on the nature of the channel estimates available and the tolerable receiver complexity. There are a number of options and we discuss only a few salient cases. Before discussing the receiver processing, we discuss how receiver channel is estimated.

19.4.1   Receive Channel Estimation (Base)

In many mobile communications standards, such as GSM and IS-54, explicit training signals are inserted inside the TDMA data bursts.

Let $\mathbf{T}$ be the training sequence arranged in a matrix form ($\mathbf{T}$ is arranged to be a Toeplitz matrix). Then, during the training burst, the received data is given by

$$\mathbf{X} = \mathbf{H}^R \mathbf{T} + \mathbf{N} \tag{19.16}$$

Clearly $\mathbf{H}^R$ can be estimated using least squares as

$$\mathbf{H}^R = \mathbf{X}\mathbf{T}^\dagger \tag{19.17}$$

where $\mathbf{T}^\dagger = \mathbf{T}^H \left(\mathbf{T}\mathbf{T}^H\right)^{-1}$.

The use of training consumes spectrum resource. In GSM, for example, about 20% of the bits are dedicated to training. Moreover, in rapidly varying mobile channels, we may have to retrain frequently, resulting in even poorer spectral efficiency. There is, therefore, increased interest in blind methods that can estimate a channel without an explicit training signal.

19.4.2 Multiuser ST Receive Algorithms

In multiuser (MU) algorithms, we address the problem of jointly demodulating the multiple signals. Recall the received signal is given by

$$\mathbf{X} = \mathbf{H}^R\mathbf{S} + \mathbf{N} \tag{19.18}$$

where $\mathbf{H}^R$ and $\mathbf{S}$ are suitably defined to include multiple users and are of dimensions $m \times NQ$ and $NQ \times M$, respectively.

If the channels for all the arriving signals are known, then we jointly demodulate all the user data sequences using multiuser **maximum likelihood sequence estimation** (MLSE). Starting with the data model in Eq. (19.18), we can then search for multiple user data sequences that minimize the ML cost function

$$\min_{\mathbf{S}} \left\| \mathbf{X} - \mathbf{H}^R\mathbf{S} \right\|_F^2 \tag{19.19}$$

The multiuser MLSE will have a large number of states in the trellis. Efficient techniques for implementing this complex receiver are needed. Multiuser MLSE detection schemes outperform all other receivers.

19.4.3 Single-User ST Receive Algorithms

In this scheme we only demodulate the desired user and cancel the CCI. Therefore, after CCI cancellation we can use MLSE receivers to handle diversity and ISI. In this method there is potential conflict between CCI mitigation and diversity maximization. We are forced to allocate the available degrees of freedom (antennas) to the competing requirements.

One approach is to cancel CCI by a space-time filter followed by an MLSE receiver to handle ISI. We do this by reformulating the MLSE criterion to arrive at a joint solution for the ST-MMSE filter and the effective channel for the scalar MLSE.

Another approach is to use a ST-MMSE receiver to handle both CCI and ISI. In a space-time filter (equalizer-beamformer), $\mathbf{W}$ has the following form

$$\mathbf{W}(k) = \begin{bmatrix} w_{11}(k) & \cdots & w_{1M}(k) \\ \vdots & \cdots & \vdots \\ w_{m1}(k) & \cdots & w_{mM}(k) \end{bmatrix} \tag{19.20}$$

In order to obtain a convenient formulation for the space-time filter output, we introduce the quantities $W(k)$ and $X(k)$ as follows

$$\begin{aligned} X(k) &= vec\left(\mathbf{X}(k)\right) & (mM \times 1) \\ W(k) &= vec\left(\mathbf{W}(k)\right) & (mM \times 1) \end{aligned} \tag{19.21}$$

where the operator $vec(\cdot)$ is defined as:

$$vec\left([\mathbf{v}_1 \cdots \mathbf{v}_M]\right) = \begin{bmatrix} \mathbf{v}_1 \\ \vdots \\ \mathbf{v}_M \end{bmatrix}$$

The ST-MMSE filter chooses the filter weights to achieve the minimum mean square error. The ST-MMSE filter takes the familiar form

$$W = \mathbf{R}_{XX}^{-1} \overline{H^R} \tag{19.22}$$

where $\overline{H^R}$ is one column of $vec\,(\mathbf{H}^R)$. In ST-MMSE the CCI and spatial diversity conflict for the spatial degrees of freedom. Likewise, temporal diversity and ISI cancellation conflict for the temporal degrees of freedom.

19.5   ST Transmit Processing (Base)

The goal in ST transmit processing is to maximize the average signal power and diversity at the receiver as well as minimize cochannel generation to other mobiles. Note that the base station transmission cannot directly affect the CCI seen by its intended mobile. In transmit the space-time processing needs channel knowledge, but since it is carried out prior to transmission and, therefore, before the signal encounters the channel, this is different from the reverse link where the space-time processing is carried out after the channel has affected the signal. Note that the mobile receiver will, of course, need to know the channel for signal demodulation, but since it sees the signal after transmission through the channel, it can estimate the forward link channel using training signals transmitted from the individual transmitter antennas.

19.5.1   Transmit Channel Estimation (Base)

The transmit channel estimation at the base of the vector forward channel can be done via feedback by use of reciprocity principles. In a TDD system, if the duplexing time is small compared to the coherence time of the channel, both channels are the same and the base-station can use its estimate of the reverse channel as the forward channel; i.e., $\mathbf{H}^F = \mathbf{H}^R$, where $\mathbf{H}^R$ is the reverse channel (we have added superscript R to emphasize the receive channel). In FDD systems, the forward and reverse channels can potentially be very different. This arises from differences in instantaneous complex path gains $\alpha^R \neq \alpha^F$. The other channel components $\mathbf{a}(\theta_l)$ and $\mathbf{g}(\tau_l)$ are very nearly equal.

A direct approach to estimating the forward channel is to feed back the signal from the mobile unit and then estimate the channel. We can do this by transmitting orthogonal training signals through each base station antenna. We can feed back from the mobile to the base the received signal for each transmitted signal and thus estimate the channel.

19.5.2   ST Transmit Processing

The primary goals at the transmitter are to maximize diversity in the link and to reduce CCI generation to other mobiles. The diversity maximization depends on the inherent diversity at the antenna array and cannot be created at the transmitter. The role of ST processing is limited to maximizing the exploitability of this diversity at the receiver. This usually leads to use of orthogonal or near orthogonal signalling at each antenna: $\int u_1(t)\, u_2(t)\, dt \approx 0$.

Orthogonality ensures that the transmitted signals are separable at the mobile which can now combine these signals after appropriate weighting to attain maximum diversity.

In order to minimize CCI, our goal is to use the beamforming vector $\mathbf{w}$ to steer the radiated energy and therefore minimize the interference at the other mobiles while maximizing the signal level at one's own mobile. Note that the CCI at the reference mobile is not controlled by its own base station but is generated by other base stations. Reducing CCI at one's own mobile requires the cooperation of the other base stations.

Therefore we choose $\mathbf{w}$ such that

$$\max_{\mathbf{w}} \frac{E(\mathbf{w}^H \mathbf{H}^F \mathbf{s}(k) \mathbf{s}(k)^H \mathbf{H}^{FH} \mathbf{w})}{\displaystyle\sum_{q=1}^{Q-1} \mathbf{w}^H \mathbf{H}_q^F \mathbf{H}_q^{FH} \mathbf{w}} \tag{19.23}$$

where $Q-1$ is the number of susceptible outer cell mobiles. $\mathbf{H}_q^F$ is the channel from the base station to the qth outer cell mobile. In order to solve the above equation, we need to know the forward link channel $\mathbf{H}^F$ to the reference mobile and $\mathbf{H}_q^F$ to cochannel mobiles. In general, such complete channel knowledge may not be available and suboptimum receivers must be designed. Furthermore, we need to find a receiver that harmonizes maximization of diversity and reduction of CCI. Use of transmit ST processing affects $\mathbf{H}^F$ and thus can be incorporated.

19.5.3 Forward Link Processing at the Mobile

The mobile will receive the composite signal from all the base station transmit antennas and will need to demodulate the signal to estimate the symbol sequence. In doing so it usually needs to estimate the individual channels from each base station antenna to itself. This is usually done via the use of training signals on each transmit antenna. Note that as the number of transmit antennas increases, there is a greater burden of training requirements. The use of transmit ST processing reduces the CCI power observed by the mobile as well enhances the diversity available.

19.6   Summary

Use of space-time processing can significantly improve average signal power, mitigate fading, and reduce cochannel and intersymbol interference in wireless networks. This can in turn result in significantly improved capacity, coverage, and quality of wireless networks.

In this chapter we have discussed applications of ST processing to TDMA systems. The applications to CDMA systems follow similar principles, but differences arise due to the nature of the signal and interference models.

Defining Terms

ISI: Intersymbol intereference is caused by multipath propagation where one symbol interferes with other symbols.

CCI: Cochannel interference arises from neighboring cells where the frequency channel is reused.

Maximum Likelihood Sequence Estimation: A technique for channel equalization based on determining the best symbol sequence that matches the received signal.

References

[1] Lindskog, E. and Paulraj, A., A taxonomy of space-time signal processing, *IEE Trans. Radar and Sonar*, 25–31, Feb. 1998.

[2] Ng, B.C. and Paulraj, A., Space-time processing for PCS, *IEEE PCS Magazine*, 5(1), 36–48, Feb. 1998.

[3] Paulraj, A. and Papadias, C.B., Space-time processing for wireless communications, *IEEE Signal Processing Magazine*, 14(5), 49–83, Nov. 1997.

[4] Paulraj, A., Papadias, C., Reddy, V.U., and Van der Veen, A., *A Review of Space-Time Signal Processing for Wireless Communications*, in *Signal Processing for Wireless Communications*, V. Poor, Ed., Prentice Hall, 179–210, Dec. 1997.

20

Location Strategies for Personal Communications Services

Ravi Jain
Bell Communications Research

Yi-Bing Lin
Bell Communications Research

Seshadri Mohan[1]
Bell Communications Research

[1] Address correspondence to: Seshadri Mohan, MCC-1A216B, Bellcore, 445 South St, Morristown, NJ 07960; Phone: 973-829-5160, Fax: 973-829-5888, e-mail: smohan@bellcore.com.

20.1   Introduction

The vision of nomadic personal communications is the ubiquitous availability of services to facilitate exchange of information (voice, data, video, image, etc.) between nomadic end users independent of time, location, or access arrangements. To realize this vision, it is necessary to locate users that move from place to place. The strategies commonly proposed are two-level hierarchical strategies, which maintain a system of mobility databases, home location registers (HLR) and visitor location resisters (VLR), to keep track of user locations. Two standards exist for carrying out two-level hierarchical strategies using HLRs and VLRs. The standard commonly used in North America is the EIA/TIA Interim Standard 41 (IS 41) [6] and in Europe the Global System for Mobile Communications (GSM) [15, 18]. In this chapter, we refer to these two strategies as *basic* location strategies.

We introduce these two strategies for locating users and provide a tutorial on their usage. We then analyze and compare these basic location strategies with respect to load on mobility databases and signalling network. Next we propose an auxiliary strategy, called the *per-user caching* or, simply, the *caching* strategy, that augments the basic location strategies to reduce the signalling and database loads.

The outline of this chapter is as follows. In Section 20.2 we discuss different forms of mobility in the context of personal communications services (PCS) and describe a reference model for a PCS architecture. In Sections 20.3 and 20.4, we describe the user location strategies specified in the IS-41 and GSM standards, respectively, and in Section 20.5, using a simple example, we present a simplified analysis of the database loads generated by each strategy. In Section 20.6, we briefly discuss possible modifications to these protocols that are likely to result in significant benefits by either reducing query and update rate to databases or reducing the signalling traffic or both. Section 20.7 introduces the caching strategy followed by an analysis in the next two sections. This idea attempts to exploit the spatial and temporal locality in calls received by users, similar to the idea of exploiting locality of file access in computer systems [20]. A feature of the caching location strategy is that it is useful only for certain classes of PCS users, those meeting certain call and mobility criteria. We encapsulate this notion in the definition of the user's call-to-mobility ratio (CMR), and local CMR (LCMR), in Section 20.8. We then use this definition and our PCS network reference architecture to quantify the costs and benefits of caching and the threshold LCMR for which caching is beneficial, thus characterizing the classes of users for which caching should be applied. In Section 20.9 we describe two methods for estimating users' LCMR and compare their effectiveness when call and mobility patterns are fairly stable, as well as when they may be variable. In Section 20.10, we briefly discuss alternative architectures and implementation issues of the strategy proposed and mention other auxiliary strategies that can be designed. Section 20.11 provides some conclusions and discussion of future work.

The choice of platforms on which to realize the two location strategies (IS-41 and GSM) may vary from one service provider to another. In this paper, we describe a possible realization of these protocols based on the advanced intelligent network (AIN) architecture (see [2, 5]), and signalling system 7 (SS7). It is also worthwhile to point out that several strategies have been proposed in the literature for locating users, many of which attempt to reduce the signalling traffic and database loads imposed by the need to locate users in PCS.

20.2 An Overview of PCS

This section explains different aspects of mobility in PCS using an example of two nomadic users who wish to communicate with each other. It also describes a reference model for PCS.

20.2.1 Aspects of Mobility—Example 20.1

PCS can involve two possible types of mobility, terminal mobility and personal mobility, that are explained next.

Terminal Mobility: This type of mobility allows a terminal to be identified by a unique terminal identifier independent of the point of attachment to the network. Calls intended for that terminal can therefore be delivered to that terminal regardless of its network point of attachment. To facilitate terminal mobility, a network must provide several functions, which include those that locate, identify, and validate a terminal and provide services (e.g., deliver calls) to the terminal based on the location information. This implies that the network must store and maintain the location information of the terminal based on a unique identifier assigned to that terminal. An example of a terminal identifier is the IS-41 EIA/TIA cellular industry term mobile identification number (MIN), which is a North American Numbering Plan (NANP) number that is stored in the terminal at the time of manufacture and cannot be changed. A similar notion exists in GSM (see Section 20.4).

Personal Mobility: This type of mobility allows a PCS user to make and receive calls independent of both the network point of attachment and a specific PCS terminal. This implies that the services that a user has subscribed to (stored in that user's service profile) are available to the user even if the user moves or changes terminal equipment. Functions needed to provide personal mobility include those that identify (authenticate) the end user and provide services to an end user independent of both the terminal and the location of the user. An example of a functionality needed to provide personal mobility for voice calls is the need to maintain a user's location information based on a unique number, called the universal personal telecommunications (UPT) number, assigned to that user. UPT numbers are also NANP numbers. Another example is one that allows end users to define and manage their service profiles to enable users to tailor services to suit their needs. In Section 20.4, we describe how GSM caters to personal mobility via smart cards.

For the purposes of the example that follows, the terminal identifiers (TID) and UPT numbers are NANP numbers, the distinction being TIDs address terminal mobility and UPT numbers address personal mobility. Though we have assigned two different numbers to address personal and terminal mobility concerns, the same effect could be achieved by a single identifier assigned to the terminal that varies depending on the user that is currently utilizing the terminal. For simplicity we assume that two different numbers are assigned.

Figure 20.1 illustrates the terminal and personal mobility aspects of PCS, which will be explained via an example. Let us assume that users Kate and Al have, respectively, subscribed to PCS services from PCS service provider (PSP) A and PSP B. Kate receives the UPT number, say, 500 111 4711, from PSP A. She also owns a PCS terminal with TID 200 777 9760. Al too receives his UPT number 500 222 4712 from PSP B, and he owns a PCS terminal with TID 200 888 5760. Each has been provided a personal identification number (PIN) by their respective PSP when subscription began. We assume that the two PSPs have subscribed to PCS access services from a certain network provider such as, for example, a local exchange carrier (LEC). (Depending on the capabilities of the PSPs, the access services provided may vary. Examples of access services include translation of UPT number to a routing number, terminal and personal registration, and call delivery. Refer to Bellcore, [3], for further details). When Kate plugs in her terminal to the network, or when she activates it, the terminal registers itself with the network by providing its TID to the network. The network creates an entry for the terminal in an appropriate database, which, in this example, is entered in the terminal mobility database (TMDB) A. The entry provides a mapping of her terminal's TID, 200 777 9760, to a routing number (RN), RN1. All of these activities happen without Kate being aware of them. After activating her terminal, Kate registers herself at that terminal by entering her UPT number (500 111 4711) to inform the network that all calls to her UPT number are to be delivered to her at the terminal. For security reasons, the network may want to authenticate her and she may be prompted to enter her PIN number into her terminal. (Alternatively, if the terminal is equipped with a smart card reader, she may enter her smart card into the reader. Other techniques, such as, for example, voice recognition, may be employed). Assuming that she is authenticated, Kate has now registered herself. As a result of personal registration by Kate, the network creates an entry for her in the personal mobility database (PMDB) A that maps her UPT number to the TID of the terminal at which she registered. Similarly, when Al activates his terminal and registers himself, appropriate entries are created in TMDB B and PMDB B. Now Al wishes to call Kate and, hence, he dials Kate's UPT number (500 111 4711). The network carries out the following tasks.

1. The switch analyzes the dialed digits and recognizes the need for AIN service, determines that the dialed UPT number needs to be translated to a RN by querying PMDB A and, hence, it queries PMDB A.
2. PMDB A searches its database and determines that the person with UPT number 500 111 4711 is currently registered at terminal with TID 200 777 9760.
3. PMDB A then queries TMDB A for the RN of the terminal with TID 200 777 9760. TMDB A returns the RN (RN1).
4. PMDB A returns the RN (RN1) to the originating switch.
5. The originating switch directs the call to the switch RN1, which then alerts Kate's terminal. The call is completed when Kate picks up her terminal.

Kate may take her terminal wherever she goes and perform registration at her new location. From then on, the network will deliver all calls for her UPT number to her terminal at the new location. In fact, she may actually register on someone else's terminal too. For example, suppose that Kate and Al agree to meet at Al's place to discuss a school project they are working on together. Kate may register herself on Al's terminal (TID 200 888 9534). The network will now modify the entry corresponding to 4711 in PMDB A to point to B 9534. Subsequent calls to Kate will be delivered to Al's terminal.

The scenario given here is used only to illustrate the key aspects of terminal and personal mobility; an actual deployment of these services may be implemented in ways different from those suggested here. We will not discuss personal registration further. The analyses that follow consider only terminal mobility but may easily be modified to include personal mobility.

20.2.2 A Model for PCS

Figure 20.2 illustrates the reference model used for the comparative analysis. The model assumes that the HLR resides in a service control point (SCP) connected to a regional signal transfer point (RSTP). The SCP is a storehouse of the AIN service logic, i.e., functionality used to perform the processing required to provide advanced services, such as speed call-

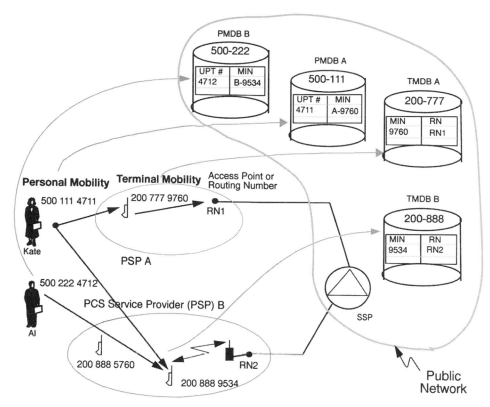

FIGURE 20.1   Illustrating terminal and personal mobility.

ing, outgoing call screening, etc., in the AIN architecture (see Bellcore, [2] and Berman and Brewster, [5]). The RSTP and the local STP (LSTP) are packet switches, connected together by various links such A links or D links, that perform the signalling functions of the SS7 network. Such functions include, for example, global title translation for routing messages between the AIN switching system, which is also referred to as the service switching point (SSP), and SCP and IS-41 messages [6]. Several SSPs may be connected to an LSTP.

The reference model in Fig. 20.2 introduces several terms which are explained next. We have tried to keep the terms and discussions fairly general. Wherever possible, however, we point to equivalent cellular terms from IS-41 or GSM.

For our purposes, the geographical area served by a PCS system is partitioned into a number of radio port coverage areas (or cells, in cellular terms) each of which is served by a radio port (or, equivalently, base station) that communicates with PCS terminals in that cell. A registration area (also known in the cellular world as location area) is composed of a number of cells. The base stations of all cells in a registration area are connected by wireline links to a mobile switching center (MSC). We assume that each registration area is served by a single VLR. The MSC of a registration area is responsible for maintaining and accessing the VLR and for switching between radio ports. The VLR associated with a registration area is responsible for maintaining a subset of the user information contained in the HLR.

Terminal registration process is initiated by terminals whenever they move into a new registration area. The base stations of a registration area periodically broadcast an identifier associated with that area. The terminals periodically compare an identifier they have stored

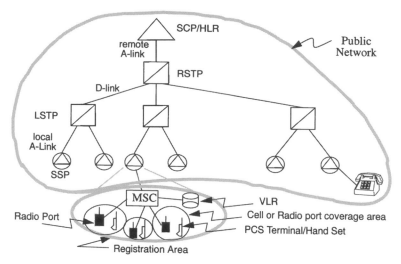

FIGURE 20.2   Example of a reference model for a PCS.

with the identifier to the registration area being broadcast. If the two identifiers differ, the terminal recognizes that it has moved from one registration area to another and will, therefore, generate a registration message. It also replaces the previous registration area identifier with that of the new one. Movement of a terminal within the same registration area will not generate registration messages. Registration messages may also be generated when the terminals are switched on. Similarly, messages are generated to deregister them when they are switched off.

PCS services may be provided by different types of commercial service vendors. Bellcore, [3] describes three different types of PSPs and the different access services that a public network may provide to them. For example, a PSP may have full network capabilities with its own switching, radio management, and radio port capabilities. Certain others may not have switching capabilities, and others may have only radio port capabilities. The model in Fig. 20.2 assumes full PSP capabilities. The analysis in Section 20.5 is based on this model and modifications may be necessary for other types of PSPs.

It is also quite possible that one or more registration areas may be served by a single PSP. The PSP may have one or more HLRs for serving its service area. In such a situation users that move within the PSP's serving area may generate traffic to the PSP's HLR (not shown in Fig. 20.2) but not to the network's HLR (shown in Fig. 20.2). In the interest of keeping the discussions simple, we have assumed that there is one-to-one correspondence between SSPs and MSCs and also between MSCs, registration areas, and VLRs. One impact of locating the SSP, MSC, and VLR in separate physical sites connected by SS7 signalling links would be to increase the required signalling message volume on the SS7 network. Our model assumes that the messages between the SSP and the associated MSC and VLR do not add to signalling load on the public network. Other configurations and assumptions could be studied for which the analysis may need to be suitably modified. The underlying analysis techniques will not, however, differ significantly.

20.3   IS-41 Preliminaries

We now describe the message flow for call origination, call delivery, and terminal registration, sometimes called location registration, based on the IS-41 protocol. This protocol is described in detail in EIA/TIA, [6]. Only an outline is provided here.

20.3.1 Terminal/Location Registration

During IS-41 registration, signalling is performed between the following pairs of network elements:

- New serving MSC and the associated database (or VLR)
- New database (VLR) in the visited area and the HLR in the public network
- HLR and the VLR in former visited registration area or the old MSC serving area.

Figure 20.3 shows the signalling message flow diagram for IS-41 registration activity, focusing only on the essential elements of the message flow relating to registration; for details of variations from the basic registration procedure, see Bellcore, [3].

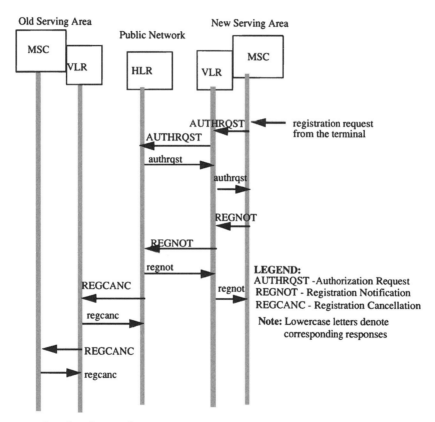

FIGURE 20.3   Signalling flow diagram for registration in IS-41.

The following steps describe the activities that take place during registration.

1. Once a terminal enters a new registration area, the terminal sends a registration request to the MSC of that area.
2. The MSC sends an authentication request (AUTHRQST) message to its VLR to authenticate the terminal, which in turn sends the request to the HLR. The HLR sends its response in the authrqst message.

3. Assuming the terminal is authenticated, the MSC sends a registration notification (REGNOT) message to its VLR.

4. The VLR in turn sends a REGNOT message to the HLR serving the terminal. The HLR updates the location entry corresponding to the terminal to point to the new serving MSC/VLR. The HLR sends a response back to the VLR, which may contain relevant parts of the user's service profile. The VLR stores the service profile in its database and also responds to the serving MSC.

5. If the user/terminal was registered previously in a different registration area, the HLR sends a registration cancellation (REGCANC) message to the previously visited VLR. On receiving this message, the VLR erases all entries for the terminal from the record and sends a REGCANC message to the previously visited MSC, which then erases all entries for the terminal from its memory.

The protocol shows authentication request and registration notification as separate messages. If the two messages can be packaged into one message, then the rate of queries to HLR may be cut in half. This does not necessarily mean that the total number of messages are cut in half.

20.3.2   Call Delivery

The signalling message flow diagram for IS-41 call delivery is shown in Fig. 20.4. The following steps describe the activities that take place during call delivery.

1. A call origination is detected and the number of the called terminal (for example, MIN) is received by the serving MSC. Observe that the call could have originated from within the public network from a wireline phone or from a wireless terminal in an MSC/VLR serving area. (If the call originated within the public network, the AIN SSP analyzes the dialed digits and sends a query to the SCP.)

2. The MSC determines the associated HLR serving the called terminal and sends a location request (LOCREQ) message to the HLR.

3. The HLR determines the serving VLR for that called terminal and sends a routing address request (ROUTEREQ) to the VLR, which forwards it to the MSC currently serving the terminal.

4. Assuming that the terminal is idle, the serving MSC allocates a temporary identifier, called a temporary local directory number (TLDN), to the terminal and returns a response to the HLR containing this information. The HLR forwards this information to the originating SSP/MSC in response to its LOCREQ message.

5. The originating SSP requests call setup to the serving MSC of the called terminal via the SS7 signalling network using the usual call setup protocols.

Similar to the considerations for reducing signalling traffic for location registration, the VLR and HLR functions could be united in a single logical database for a given serving area and collocated; further, the database and switch can be integrated into the same piece of physical equipment or be collocated. In this manner, a significant portion of the messages exchanged between the switch, HLR and VLR as shown in Fig. 20.4 will not contribute to signalling traffic.

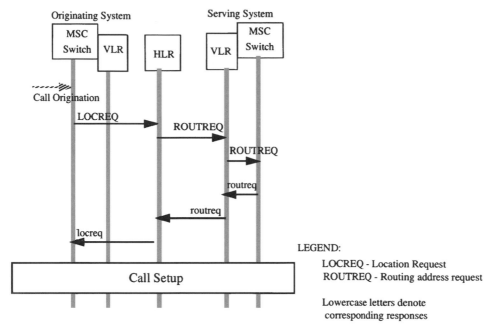

FIGURE 20.4 Signalling flow diagram for call delivery in IS-41.

20.4 Global System for Mobile Communications

In this section we describe the user location strategy proposed in the European Global System for Mobile Communications (GSM) standard and its offshoot, digital cellular system 1800 (DCS1800). There has recently been increased interest in GSM in North America, since it is possible that early deployment of PCS will be facilitated by using the communication equipment already available from European manufacturers who use the GSM standard. Since the GSM standard is relatively unfamiliar to North American readers, we first give some background and introduce the various abbreviations. The reader will find additional details in Mouley and Pautet, [18]. For an overview on GSM, refer to Lycksell, [15].

The abbreviation GSM originally stood for Groupe Special Mobile, a committee created within the pan-European standardization body Conference Europeenne des Posts et Telecommunications (CEPT) in 1982. There were numerous national cellular communication systems and standards in Europe at the time, and the aim of GSM was to specify a uniform standard around the newly reserved 900-MHz frequency band with a bandwidth of twice 25 MHz. The phase 1 specifications of this standard were frozen in 1990. Also in 1990, at the request of the United Kingdom, specification of a version of GSM adapted to the 1800-MHz frequency, with bandwidth of twice 75 MHz, was begun. This variant is referred to as DCS1800; the abbreviation GSM900 is sometimes used to distinguish between the two variations, with the abbreviation GSM being used to encompass both GSM900 and DSC1800. The motivation for DCS1800 is to provide higher capacities in densely populated urban areas, particularly for PCS. The DCS1800 specifications were frozen in 1991, and by 1992 all major GSM900 European operators began operation.

At the end of 1991, activities concerning the post-GSM generation of mobile communications were begun by the standardization committee, using the name universal mobile telecommunications system (UMTS) for this effort. In 1992, the name of the standardization committee was changed from GSM to special mobile group (SMG) to distinguish it from the 900-MHz system itself, and the term GSM was chosen as the commercial trademark

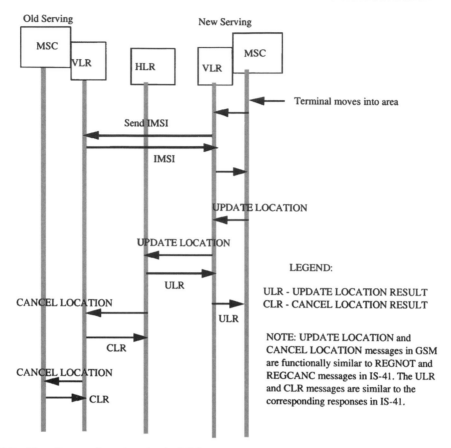

FIGURE 20.5   Flow diagram for registration in GSM.

of the European 900-MHz system, where GSM now stands for global system for mobile communications.

The GSM standard has now been widely adopted in Europe and is under consideration in several other non-European countries, including the United Arab Emirates, Hong Kong, and New Zealand. In 1992, Australian operators officially adopted GSM.

20.4.1   Architecture

In this section we describe the GSM architecture, focusing on those aspects that differ from the architecture assumed in the IS-41 standard.

A major goal of the GSM standard was to enable users to move across national boundaries and still be able to communicate. It was considered desirable, however, that the operational network within each country be operated independently. Each of the operational networks is called a public land mobile network (PLMN) and its commercial coverage area is confined to the borders of one country (although some radio coverage overlap at national boundaries may occur), and each country may have several competing PLMNs.

A GSM customer subscribes to a single PLMN called the home PLMN, and subscription information includes the services the customer subscribes to. During normal operation, a user may elect to choose other PLMNs as their service becomes available (either as the user moves or as new operators enter the marketplace). The user's terminal [GSM calls the terminal a mobile station (MS)] assists the user in choosing a PLMN in this case, either presenting a list of possible PLMNs to the user using explicit names (e.g., DK Sonofon for the Danish PLMN) or choosing automatically based on a list of preferred PLMNs stored in the terminal's memory. This PLMN selection process allows users to choose between the services and tariffs of several competing PLMNs. Note that the PLMN selection process differs from the cell selection and handoff process that a terminal carries out automatically without any possibility of user intervention, typically based on received radio signal strengths and, thus, requires additional intelligence and functionality in the terminal.

The geographical area covered by a PLMN is partitioned into MSC serving areas, and a registration area is constrained to be a subset of a single MSC serving area. The PLMN operator has complete freedom to allocate cells to registration areas. Each PLMN has, logically speaking, a single HLR, although this may be implemented as several physically distributed databases, as for IS-41. Each MSC also has a VLR, and a VLR may serve one or several MSCs. As for IS-41, it is interesting to consider how the VLR should be viewed in this context. The VLR can be viewed as simply a database off loading the query and signalling load on the HLR and, hence, logically tightly coupled to the HLR or as an ancillary processor to the MSC. This distinction is not academic; in the first view, it would be natural to implement a VLR as serving several MSCs, whereas in the second each VLR would serve one MSC and be physically closely coupled to it. For GSM, the MSC implements most of the signalling protocols, and at present all switch manufacturers implement a combined MSC and VLR, with one VLR per MSC [18].

A GSM mobile station is split in two parts, one containing the hardware and software for the radio interface and the other containing subscribers-specific and location information, called the subscriber identity module (SIM), which can be removed from the terminal and is the size of a credit card or smaller. The SIM is assigned a unique identity within the GSM system, called the international mobile subscriber identity (IMSI), which is used by the user location strategy as described the next subsection. The SIM also stores authentication information, services lists, PLMN selection lists, etc., and can itself be protected by password or PIN.

The SIM can be used to implement a form of large-scale mobility called SIM roaming. The GSM specifications standardize the interface between the SIM and the terminal, so that a user carrying his or her SIM can move between different terminals and use the SIM to personalize the terminal. This capability is particularly useful for users who move between PLMNs which have different radio interfaces. The user can use the appropriate terminal for each PLMN coverage area while obtaining the personalized facilities specified in his or her SIM. Thus, SIMs address personal mobility. In the European context, the usage of two closely related standards at different frequencies, namely, GSM900 and DCS1800, makes this capability an especially important one and facilitates interworking between the two systems.

20.4.2 User Location Strategy

We present a synopsis of the user location strategy in GSM using call flow diagrams similar to those used to describe the strategy in IS-41.

In order to describe the registration procedure, it is first useful to clarify the different identifiers used in this procedure. The SIM of the terminal is assigned a unique identity, called the IMSI, as already mentioned. To increase confidentiality and make more efficient use of the radio bandwidth, however, the IMSI is not normally transmitted over the radio link. Instead, the terminal is assigned a temporary mobile subscriber identity (TMSI) by the VLR when it enters a new registration area. The TMSI is valid only within a given registration area and is shorter than the IMSI. The IMSI and TMSI are identifiers that are internal to the system and assigned to a terminal or SIM and should not be confused with the user's number that would be dialed by a calling party; the latter is a separate number called the mobile subscriber integrated service digital network (ISDN) number (MSISDN), and is similar to the usual telephone number in a fixed network.

We now describe the procedure during registration. The terminal can detect when it has moved into the cell of a new registration area from the system information broadcast by the base station in the new cell. The terminal initiates a registration update request to the new base station; this request includes the identity of the old registration area and the TMSI of the terminal in the old area. The request is forwarded to the MSC, which, in turn, forwards it to the new VLR. Since the new VLR cannot translate the TMSI to the IMSI of the terminal, it sends a request to the old VLR to send the IMSI of the terminal corresponding to that TMSI. In its response, the old VLR also provides the required authentication information. The new VLR then initiates procedures to authenticate the terminal. If the authentication succeeds, the VLR uses the IMSI to determine the address of the terminal's HLR.

The ensuing protocol is then very similar to that in IS-41, except for the following differences. When the new VLR receives the registration affirmation (similar to regnot in IS-41) from the HLR, it assigns a new TMSI to the terminal for the new registration area. The HLR also provides the new VLR with all relevant subscriber profile information required for call handling (e.g., call screening lists, etc.) as part of the affirmation message. Thus, in contrast with IS-41, authentication and subscriber profile information are obtained from both the HLR and old VLR and not just the HLR.

The procedure for delivering calls to mobile users in GSM is very similar to that in IS-41. The sequence of messages between the caller and called party's MSC/VLRs and the HLR is identical to that shown in the call flow diagrams for IS-41, although the names, contents and lengths of messages may be different and, hence, the details are left out. The interested reader is referred to Mouly and Pautet, [18], or Lycksell, [15], for further details.

20.5   Analysis of Database Traffic Rate for IS-41 and GSM

In the two subsections that follow, we state the common set of assumptions on which we base our comparison of the two strategies.

20.5.1   The Mobility Model for PCS Users

In the analysis that follows in the IS-41 analysis subsection, we assume a simple mobility model for the PCS users. The model, which is described in [23], assumes that PCS users carrying terminals are moving at an average velocity of v and their direction of movement is uniformly distributed over $[0, 2\pi]$. Assuming that the PCS users are uniformly populated with a density of ρ and the registration area boundary is of length L, it has been shown that the rate of registration area crossing R is given by

$$R = \frac{\rho v L}{\pi} \tag{20.1}$$

Using Eq. (20.1), we can calculate the signalling traffic due to registration, call origination, and delivery. We now need a set of assumptions so that we may proceed to derive the traffic rate to the databases using the model in Fig. 20.2.

20.5.2 Additional Assumptions

The following assumptions are made in performing the analysis.

- 128 total registration areas
- Square registration area size: $(7.575 \text{ km})^2 = 57.5 \text{ km}^2$, with border length $L = 30.3$ km
- Average call origination rate = average call termination (delivery) rate = 1.4/h/terminal
- Mean density of mobile terminals = $\rho = 390/\text{km}^2$
- Total number of mobile terminals = $128 \times 57.4 \times 390 = 2.87 \times 10^6$
- Average call origination rate = average call termination (delivery) rate = 1.4/h/terminal
- Average speed of a mobile, $v = 5.6$ km/h
- Fluid flow mobility model

The assumptions regarding the total number of terminals may also be obtained by assuming that a certain public network provider serves 19.15×10^6 users and that 15% (or 2.87×10^6) of the users also subscribe to PCS services from various PSPs.

Note that we have adopted a simplified model that ignores situations where PCS users may turn their handsets on and off that will generate additional registration and deregistration traffic. The model also ignores wireline registrations. These activities will increase the total number of queries and updates to HLR and VLRs.

20.5.3 Analysis of IS-41

Using Eq. (20.1) and the parameter values assumed in the preceding subsection, we can compute the traffic due to registration. The registration traffic is generated by mobile terminals moving into a new registration area, and this must equal the mobile terminals moving out of the registration area, which per second is

$$R_{\text{reg, VLR}} = \frac{390 \times 30.3 \times 5.6}{3600\pi} = 5.85$$

This must also be equal to the number of deregistrations (registration cancellations),

$$R_{\text{dereg, VLR}} = 5.85$$

The total number of registration messages per second arriving at the HLR will be

$$R_{\text{reg, HLR}} = R_{\text{reg, VLR}} \times \text{total No. of registration areas} = 749$$

The HLR should, therefore, be able to handle, roughly, 750 updates per second. We observe from Fig. 20.3 that authenticating terminals generate as many queries to VLR and HLR as the respective number of updates generated due to registration notification messages.

The number of queries that the HLR must handle during call origination and delivery can be similarly calculated. Queries to HLR are generated when a call is made to a PCS user. The SSP that receives the request for a call, generates a location request (LOCREQ) query to the SCP controlling the HLR. The rate per second of such queries must be equal to the rate of calls made to PCS users.

This is calculated as

$$
\begin{aligned}
R_{\text{CallDeliv, HLR}} &= \text{call rate per user} \times \text{total number of users} \\
&= \frac{1.4 \times 2.87 \times 10^5}{3600} \\
&= 1116
\end{aligned}
$$

For calls originated from a mobile terminal by PCS users, the switch authenticates the terminal by querying the VLR. The rate per second of such queries is determined by the rate of calls originating in an SSP serving area, which is also a registration area (RA). This is given by

$$
R_{\text{CallOrig, VLR}} = \frac{1116}{128} = 8.7
$$

This is also the number of queries per second needed to authenticate terminals of PCS users to which calls are delivered:

$$
R_{\text{CallDeliv, VLR}} = 8.7
$$

Table 20.1 summarizes the calculations.

TABLE 20.1   IS-41 Query and Update Rates to HLR and VLR

Activity	HLR Updates/s	VLR Updates/s	HLR Queries/s	VLR queries/s
Mobility-related activities at registration	749	5.85	749	5.85
Mobility-related activities at deregistration		5.85		
Call origination				8.7
Call delivery			1116	8.7
Total (per RA)	5.85	11.7	14.57	23.25
Total (Network)	749	1497.6	1865	2976

20.5.4   Analysis of GSM

Calculations for query and update rates for GSM may be performed in the same manner as for IS-41, and they are summarized in Table 20.2. The difference between this table and Table 20.1 is that in GSM the new serving VLR does not query the HLR separately in order to authenticate the terminal during registration and, hence, there are no HLR queries during registration. Instead, the entry (749 queries) under HLR queries in Table 20.1, corresponding to mobility-related authentication activity at registration, gets equally divided between the 128 VLRs. Observe that with either protocol the total database traffic rates are conserved, where the total database traffic for the entire network is given by the sum of all of the entries in the last row total (Network), i.e.,

HLR updates + VLR updates + HLR queries + VLR queries

From Tables 20.1 and 20.2 we see that this quantity equals 7087.

The conclusion is independent of any variations we may provide to the assumptions in earlier in the section. For example, if the PCS penetration (the percentage of the total users

TABLE 20.2 GSM Query and Update Rates to HLR and VLR

Activity	HLR Updates/s	VLR Updates/s	HLR Queries/s	VLR Queries/s
Mobility-related activities at registration	749	5.85		11.7
Mobility-related activities at deregistration		5.85		
Call origination				8.7
Call delivery			1116	8.7
Total (per VLR)	749	11.7	1116	29.1
Total (Network)	749	1497.6	1116	3724.8

subscribing to PCS services) were to increase from 15 to 30%, all of the entries in the two tables will double and, hence, the total database traffic generated by the two protocols will still be equal.

20.6 Reducing Signalling During Call Delivery

In the preceding section, we provided a simplified analysis of some scenarios associated with user location strategies and the associated database queries and updates required. Previous studies [13, 16] indicate that the signalling traffic and database queries associated with PCS due to user mobility are likely to grow to levels well in excess of that associated with a conventional call. It is, therefore, desirable to study modifications to the two protocols that would result in reduced signalling and database traffic. We now provide some suggestions.

For both GSM and IS-41, delivery of calls to a mobile user involves four messages: from the caller's VLR to the called party's HLR, from the HLR to the called party's VLR, from the called party's VLR to the HLR, and from the HLR to the caller's VLR. The last two of these messages involve the HLR, whose role is to simply relay the routing information provided by the called party's VLR to the caller's VLR. An obvious modification to the protocol would be to have the called VLR directly send the routing information to the calling VLR. This would reduce the total load on the HLR and on signalling network links substantially. Such a modification to the protocol may not be easy, of course, due to administrative, billing, legal, or security concerns. Besides, this would violate the query/response model adopted in IS-41, requiring further analysis.

A related question which arises is whether the routing information obtained from the called party's VLR could instead be stored in the HLR. This routing information could be provided to the HLR, for example, whenever a terminal registers in a new registration area. If this were possible, two of the four messages involved in call delivery could be eliminated. This point was discussed at length by the GSM standards body, and the present strategy was arrived at. The reason for this decision was to reduce the number of temporary routing numbers allocated by VLRs to terminals in their registration area. If a temporary routing number (TLDN in IS-41 or MSRN in GSM) is allocated to a terminal for the whole duration of its stay in a registration area, the quantity of numbers required is much greater than if a number is assigned on a per-call basis. Other strategies may be employed to reduce signalling and database traffic via intelligent paging or by storing user's mobility behavior in user profiles (see, for example, Tabbane, [22]). A discussion of these techniques is beyond the scope of the paper.

20.7   Per-User Location Caching

The basic idea behind per-user location caching is that the volume of SS7 message traffic and database accesses required in locating a called subscriber can be reduced by maintaining local storage, or cache, of user location information at a switch. At any switch, location caching for a given user should be employed only if a large number of calls originate for that user from that switch, relative to the user's mobility. Note that the cached information is kept at the switch from which calls originate, which may or may not be the switch where the user is currently registered.

Location caching involves the storage of location pointers at the originating switch; these point to the VLR (and the associated switch) where the user is currently registered. We refer to the procedure of locating a PCS user a *FIND* operation, borrowing the terminology from Awerbuch and Peleg, [1]. We define a basic *FIND*, or *BasicFIND*( ), as one where the following sequence of steps takes place.

1. The incoming call to a PCS user is directed to the nearest switch.
2. Assuming that the called party is not located within the immediate RA, the switch queries the HLR for routing information.
3. The HLR contains a pointer to the VLR in whose associated RA the subscriber is currently situated and launches a query to that VLR.
4. The VLR, in turn, queries the MSC to determine whether the user terminal is capable of receiving the call (i.e., is idle) and, if so, the MSC returns a routable address (TLDN in IS-41) to the VLR.
5. The VLR relays the routing address back to the originating switch via the HLR.

At this point, the originating switch can route the call to the destination switch. Alternately, *BasicFIND*( ) can be described by pseudocode as follows. (We observe that a more formal method of specifying PCS protocols may be desirable).

> *BasicFIND*( ){
>
>> Call to PCS user is detected at local switch;
>> *if* called party is in same RA *then* return;
>> Switch queries called party's HLR;
>> Called party's HLR queries called party's current VLR, V;
>> V returns called party's location to HLR;
>> HLR returns location to calling switch;
>
> }

In the *FIND* procedure involving the use of location caching, or *CacheFIND*( ), each switch contains a local memory (cache) that stores location information for subscribers. When the switch receives a call origination (from either a wire-line or wireless caller) directed to a PCS subscriber, it first checks its cache to see if location information for the called party is maintained. If so, a query is launched to the pointed VLR; if not, *BasicFIND*( ), as just described, is followed. If a cache entry exists and the pointed VLR is queried, two situations are possible. If the user is still registered at the RA of the pointed VLR (i.e., we have a *cache hit*), the pointed VLR returns the user's routing address. Otherwise, the pointed VLR returns a *cache miss*.

> *CacheFIND*( ){
>
>> Call to PCS user is detected at local switch;

> *if* called is in same RA *then* return;
> *if* there is no cache entry for called user
> *then* invoke *BasicFIND*( ) and return;
> Switch queries the VLR, V, specified in the cache entry;
> *if* called is at V, *then*
> > V returns called party's location to calling switch;
> *else* {
> > V returns "miss" to calling switch;
> > Calling switch invokes *BasicFIND*( );
> }
> }

When a cache hit occurs we save one query to the HLR [a VLR query is involved in both *CacheFIND*( ) and *BasicFIND*( )], and we also save traffic along some of the signalling links; instead of four message transmissions, as in *BasicFIND*( ), only two are needed. In steady-state operation, the cached pointer for any given user is updated only upon a miss.

Note that the *BasicFIND*( ) procedure differs from that specified for roaming subscribers in the IS-41 standard EIA/TIA, [6]. In the IS-41 standard, the second line in the *BasicFIND*( ) procedure is omitted, i.e., every call results in a query of the called user's HLR. Thus, in fact, the procedure specified in the standard will result in an even higher network load than the *BasicFIND*( ) procedure specified here. To make a fair assessment of the benefits of *CacheFIND*( ), however, we have compared it against *BasicFIND*( ). Thus, the benefits of *CacheFIND*( ) investigated here depend specifically on the use of caching and not simply on the availability of user location information at the local VLR.

20.8 Caching Threshold Analysis

In this section we investigate the classes of users for which the caching strategy yields net reductions in signalling traffic and database loads. We characterize classes of users by their CMR. The CMR of a user is the average number of calls to a user per unit time, divided by the average number of times the user changes registration areas per unit time. We also define a LCMR, which is the average number of calls to a user from a given originating switch per unit time, divided by the average number of times the user changes registration areas per unit time.

For each user, the amount of savings due to caching is a function of the probability that the cached pointer correctly points to the user's location and increases with the user's LCMR. In this section we quantify the minimum value of LCMR for caching to be worthwhile. This caching threshold is parameterized with respect to costs of traversing signalling network elements and network databases and can be used as a guide to select the subset of users to whom caching should be applied. The analysis in this section shows that estimating user's LCMRs, preferably dynamically, is very important in order to apply the caching strategy. The next section will discuss methods for obtaining this estimate.

From the pseudocode for *BasicFIND*( ), the signalling network cost incurred in locating a PCS user in the event of an incoming call is the sum of the cost of querying the HLR (and receiving the response), and the cost of querying the VLR which the HLR points to (and receiving the response). Let

α = cost of querying the HLR and receiving a response

β = cost of querying the pointed VLR and receiving a response

Then, the cost of *BasicFIND*( ) operation is

$$C_B = \alpha + \beta \tag{20.2}$$

To quantify this further, assume costs for traversing various network elements as follows.

A_l = cost of transmitting a location request or response message on A link between SSP and LSTP
D = cost of transmitting a location request on response message or D link
A_r = cost of transmitting a location request or response message on A link between RSTP and SCP
L = cost of processing and routing a location request or response message by LSTP
R = cost of processing and routing a location request or response message by RSTP
H_Q = cost of a query to the HLR to obtain the current VLR location
V_Q = cost of a query to the VLR to obtain the routing address

Then, using the PCS reference network architecture (Fig. 80.2),

$$\alpha = 2\,(A_l + D + A_r + L + R) + H_Q \tag{20.3}$$
$$\beta = 2\,(A_l + D + A_r + L + R) + V_Q \tag{20.4}$$

From Eqs. (20.2)–(20.4)

$$C_B = 4\,(A_l + D + A_r + L + R) + H_Q + V_Q \tag{20.5}$$

We now calculate the cost of *CacheFIND*( ). We define the *hit ratio* as the relative frequency with which the cached pointer correctly points to the user's location when it is consulted. Let

p = cache hit ratio
C_H = cost of the *CacheFIND*( ) procedure when there is a hit
C_M = cost of the *CacheFIND*( ) procedure when there is a miss

Then the cost of *CacheFIND*( ) is

$$C_C = p\,C_H + (1-p)C_M \tag{20.6}$$

For *CacheFIND*( ), the signalling network costs incurred in locating a user in the event of an incoming call depend on the hit ratio as well as the cost of querying the VLR, which is stored in the cache; this VLR query may or may not involve traversing the RSTP. In the following, we say a VLR is a *local* VLR if it is served by the same LSTP as the originating switch, and a *remote* VLR otherwise. Let

q = Prob (VLR in originating switch's cache is a local VLR)
δ = cost of querying a local VLR
ϵ = cost of querying a remote VLR
η = cost of updating the cache upon a miss

Then,

$$\delta = 4A_l + 2L + V_Q \tag{20.7}$$
$$\epsilon = 4\,(A_l + D + L) + 2R + V_Q \tag{20.8}$$
$$C_H = q\delta + (1-q)\epsilon \tag{20.9}$$

Since updating the cache involves an operation to a fast local memory rather than a database operation, we shall assume in the following that $\eta = 0$. Then,

$$C_M = C_H + C_B = q\delta + (1-q)\epsilon + \alpha + \beta \tag{20.10}$$

From Eqs. (20.6), (20.9) and (20.10) we have

$$C_C = \alpha + \beta + \epsilon - p(\alpha + \beta) + q(\delta - \epsilon) \tag{20.11}$$

For net cost savings we require $C_C < C_B$, or that the hit ratio exceeds a *hit ratio threshold* p_T, derived using Eqs. (20.6), (20.9), and (20.2),

$$p > p_T = \frac{C_H}{C_B} = \frac{\epsilon + q(\delta - \epsilon)}{\alpha + \beta} \tag{20.12}$$

$$= \frac{4A_l + 4D + 4L + 2R + V_Q - q(4D + 2L + 2R)}{4A_l + 4D + 4A_r + 4L + 4R + H_Q + V_Q} \tag{20.13}$$

Equation (20.13) specifies the hit ratio threshold for a user, evaluated at a given switch, for which local maintenance of a cached location entry produces cost savings. As pointed out earlier, a given user's hit ratio may be location dependent, since the rates of calls destined for that user may vary widely across switches.

The hit ratio threshold in Eq. (20.13) is comprised of heterogeneous cost terms, i.e., transmission link utilization, packet switch processing, and database access costs. Therefore, numerical evaluation of the hit ratio threshold requires either detailed knowledge of these individual quantities or some form of simplifying assumptions. Based on the latter approach, the following two possible methods of evaluation may be employed.

1. Assume one or more cost terms dominate, and simplify Eq. (20.13) by setting the remaining terms to zero.
2. Establish a common unit of measure for all cost terms, for example, *time delay*. In this case, A_l, A_r, and D may represent transmission delays of fixed transmission speed (e.g., 56 kb/s) signalling links, L and R may constitute the sum of queueing and service delays of packet switches (i.e., STPs), and H_Q and V_Q the transaction delays for database queries.

In this section we adopt the first method and evaluate Eq. (20.13) assuming a single term dominates. (In Section 20.9 we present results using the second method). Table 20.3 shows the hit ratio threshold required to obtain net cost savings, for each case in which one of the cost terms is dominant.

In Table 20.3 we see that if the cost of querying a VLR or of traversing a local A link is the dominant cost, caching for users who may move is never worthwhile, regardless of users' call reception and mobility patterns. This is because the caching strategy essentially distributes the functionality of the HLR to the VLRs. Thus, the load on the VLR and the local A link is always increased, since any move by a user results in a cache miss. On the other hand, for a fixed user (or telephone), caching is always worthwhile. We also observe that if the remote A links or HLR querying are the bottlenecks, caching is worthwhile even for users with very low hit ratios.

As a simple average-case calculation, consider the net network benefit of caching when HLR access and update is the performance bottleneck. Consider a scenario where $u = 50\%$ of PCS users receive $c = 80\%$ of their calls from $s = 5$ RAs where their hit ratio $p > 0$, and $s' = 4$ of the SSPs at those RAs contain sufficiently large caches. Assume that caching

TABLE 20.3   Minimum Hit Ratios and LCMRs for Various Individual Dominant Signalling Network Cost Terms

Dominant Cost Term	Hit ratio Threshold, p_T	LCMR Threshold, $LCMR_T$	LCMR Threshold ($q = 0.043$)	LCMR Threshold ($q = 0.25$)
A_l	1	∞	∞	∞
A_r	0	0	0	0
D	$1 - q$	$1/q - 1$	22	3
L	$1 - q/2$	$2/q - 1$	45	7
R	$1 - q/2$	$2/q - 1$	45	7
H_Q	0	0	0	0
V_Q	1	∞	∞	∞

is applied only to this subset of users and to no other users. Suppose that the average hit ratio for these users is $p = 80\%$, so that 80% of the HLR accesses for calls to these users from these RA are avoided. Then the net saving in the accesses to the system's HLR is $H = (u\,c\,s'\,p)/s = 25\%$.

We discuss other quantities in Table 20.3 next. It is first useful to relate the cache hit ratio to users' calling and mobility patterns directly via the LCMR. Doing so requires making assumptions about the distribution of the user's calls and moves. We consider the steady state where the incoming call stream from an SSP to a user is a Poisson process with arrival rate λ, and the time that the user resides in an RA has a general distribution $F(t)$ with mean $1/\mu$. Thus,

$$LCMR = \frac{\lambda}{\mu} \tag{20.14}$$

Let t be the time interval between two consecutive calls from the SSP to the user and t_1 be the time interval between the first call and the time when the user moves to a new RA. From the random observer property of the arrival call stream [7], the hit ratio is

$$p = \ Pr[t < t_1] = \int_{t=0}^{\infty} \lambda e^{-\lambda t} \int_{t_1=t}^{\infty} \mu\left[1 - F\left(t_1\right)\right] \, dt_1 \, dt$$

If $F(t)$ is an exponential distribution, then

$$p = \frac{\lambda}{\lambda + \mu} \tag{20.15}$$

and we can derive the *LCMR threshold*, the minimum LCMR required for caching to be beneficial assuming incoming calls are a Poisson process and intermove times are exponentially distributed,

$$LCMR_T = \frac{p_T}{1 - p_T} \tag{20.16}$$

Equation (20.16) is used to derive LCMR thresholds assuming various dominant costs terms, as shown in Table 20.3.

Several values for $LCMR_T$ in Table 20.3 involve the term q, i.e., the probability that the pointed VLR is a local VLR. These values may be numerically evaluated by simplifying assumptions. For example, assume that all of the SSPs in the network are uniformly distributed amongst l LSTPs. Also, assume that all of the PCS subscribers are uniformly distributed in location across all SSPs and that each subscriber exhibits the same incoming call rate at every SSP. Under those conditions, q is simply $1/l$. Consider the case of the public switched telephone network. Given that there are a total of 160 local access transport area (LATA) across the 7 Regional Bell Operating Company (RBOC) regions [4], the average number of LATAs, or l, is 160/7 or 23. Table 20.3 shows the results with $q = 1/l$ in this case.

We observe that the assumption that all users receive calls uniformly from all switches in the network is extremely conservative. In practice, we expect that user call reception patterns would display significantly more locality, so that q would be larger and the LCMR thresholds required to make caching worthwhile would be smaller. It is also worthwhile to consider the case of a RBOC region with PCS deployed in a few LATA only, a likely initial scenario, say, 4 LATAs. In either case the value of q would be significantly higher; Table 20.3 shows the LCMR threshold when $q = 0.25$.

It is possible to quantify the net costs and benefits of caching in terms of signalling network impacts in this way and to determine the hit ratio and LCMR threshold above which users should have the caching strategy applied. Applying caching to users whose hit ratio and LCMR is below this threshold results in net increases in network impacts. It is, thus, important to estimate users' LCMRs accurately. The next section discusses how to do so.

20.9   Techniques for Estimating Users' LCMR

Here we sketch some methods of estimating users' LCMR. A simple and attractive policy is to not estimate these quantities on a per-user basis at all. For instance, if the average LCMR over all users in a PCS system is high enough (and from Table 20.3, it need not be high depending on which network elements are the dominant costs), then caching could be used at every SSP to yield net system-wide benefits. Alternatively, if it is known that at any given SSP the average LCMR over all users is high enough, a cache can be installed at that SSP. Other variations can be designed.

One possibility for deciding about caching on a per-user basis is to maintain information about a user's calling and mobility pattern at the HLR and to download it periodically to selected SSPs during off-peak hours. It is easy to envision numerous variations on this idea.

In this section we investigate two possible techniques for estimating LCMR on a per-user basis when caching is to be deployed. The first algorithm, called the *running average* algorithm, simply maintains a running average of the hit ratio for each user. The second algorithm, called the *reset-K* algorithm, attempts to obtain a measure of the hit ratio over the recent history of the user's movements. We describe the two algorithms next and evaluate their effectiveness using a stochastic analysis taking into account user calling and mobility patterns.

20.9.1   The Running Average Algorithm

The running average algorithm maintains, for every user that has a cache entry, the running average of the hit ratio. A running count is kept of the number of calls to a given user, and, regardless of the *FIND* procedure used to locate the user, a running count of the number of times that the user was at the same location for any two consecutive calls; the ratio of these numbers provides the measured running average of the hit ratio. We denote the measured running average of the hit ratio by p_M; in steady state, we expect that $p_M = p$. The user's previous location as stored in the cache entry is used only if the running average of the hit ratio p_M is greater than the cache hit threshold p_T. Recall that the cache scheme outperforms the basic scheme if $p > p_T = C_H/C_B$. Thus, in steady state, the running average algorithm will outperform the basic scheme when $p_M > p_T$.

We consider, as before, the steady state where the incoming call stream from an SSP to a user is a Poisson process with arrival rate λ, and the time that the user resides in an RA has an exponential distribution with mean $1/\mu$. Thus $LCMR = \lambda/\mu$ [Eq. (20.14)] and the location tracking cost at steady state is

$$C_C = \begin{cases} p_M C_H + (1 - p_M) C_B, & p_M > p_T \\ C_B, & \text{otherwise} \end{cases} \qquad (20.17)$$

Figure 20.6 plots the cost ratio C_C/C_B from Eq. (20.17) against $LCMR$. (This corresponds to assigning uniform units to all cost terms in Eq. (20.13), i.e., the second evaluation method as discussed in Section 20.8. Thus, the ratio C_C/C_B may represent the percentage reduction in user location time with the caching strategy compared to the basic strategy.) The figure indicates that in the steady state, the caching strategy with the running average algorithm for estimating LCMR can significantly outperform the basic scheme if $LCMR$ is sufficiently large. For instance with $LCMR \sim 5$, caching can lead to cost savings of 20–60% over the basic strategy.

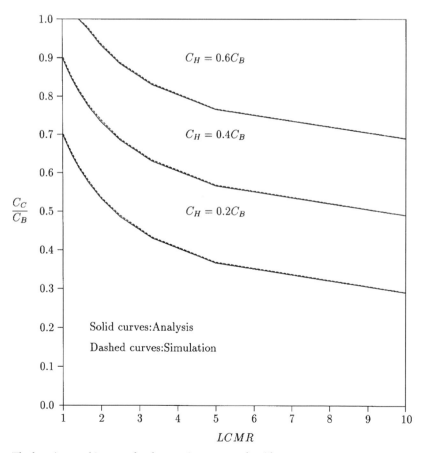

FIGURE 20.6   The location tracking cost for the running average algorithm.

Equation (20.17) (cf., solid curves in Fig. 20.6) is validated against a simple Monte Carlo simulation (cf., dashed curves in Fig. 20.6). In the simulation, the confidence interval for the 95% confidence level of the output measure C_C/C_B is within 3% of the mean value. This simulation model will later be used to study the running average algorithm when the mean of the movement distribution changes from time to time [which cannot be modeled by using Eq. (20.17)].

One problem with the running average algorithm is that the parameter p is measured from the entire past history of the user's movement, and the algorithm may not be sufficiently dynamic to adequately reflect the recent history of the user's mobility patterns.

20.9.2   The Reset-K Algorithm

We may modify the running average algorithm such that p is measured from the recent history. Define every K incoming calls as a *cycle*. The modified algorithm, which is referred to as the reset-K algorithm, counts the number of cache hits n in a cycle. If the measured hit ratio for a user, $p_M = n/K \geq p_T$, then the cache is enabled for that user, and the cached information is always used to locate the user in the next cycle. Otherwise, the cache is disabled for that user, and the basic scheme is used. At the beginning of a cycle, the cache hit count is reset, and a new p_M value is measured during the cycle.

To study the performance of the reset-K algorithm, we model the number of cache misses in a cycle by a Markov process. Assume as before that the call arrivals are a Poisson process with arrival rate λ and the time period the user resides in an RA has an exponential distribution with mean $1/\mu$. A pair (i, j), where $i > j$, represents the state that there are j cache misses before the first i incoming phone calls in a cycle. A pair $(i, j)^*$, where $i \geq j \geq 1$, represents the state that there are $j - 1$ cache misses before the first i incoming phone calls in a cycle, and the user moves between the ith and the $i + 1$ phone calls. The difference between (i, j) and $(i, j)^*$ is that if the Markov process is in the state (i, j) and the user moves, then the process moves into the state $(i, j + 1)^*$. On the other hand, if the process is in state $(i, j)^*$ when the user moves, the process remains in $(i, j)^*$ because at most one cache miss occurs between two consecutive phone calls.

Figure 20.7(a) illustrates the transitions for state $(i, 0)$ where $2 < i < K + 1$. The Markov process moves from $(i - 1, 0)$ to $(i, 0)$ if a phone call arrives before the user moves out. The rate is λ. The process moves from $(i, 0)$ to $(i, 1)^*$ if the user moves to another RA before the $i + 1$ call arrival. Let $\pi(i, j)$ denote the probability of the process being in state (i, j). Then the transition equation is

$$\pi(i, 0) = \frac{\lambda}{\lambda + \mu}\pi(i - 1, 0), \qquad 2 < i < K + 1 \tag{20.18}$$

Figure 20.7(b) illustrates the transitions for state $(i, i - 1)$ where $1 < i < K + 1$. The only transition into the state $(i, i - 1)$ is from $(i - 1, i - 1)^*$, which means that the user always moves to another RA after a phone call. [Note that there can be no state $(i - 1, i - 1)$ by definition and, hence, no transition from such a state.] The transition rate is λ. The process moves from $(i, i - 1)$ to $(i, i)^*$ with rate μ, and moves to $(i + 1, i - 1)$ with rate λ. Let $\pi^*(i, j)$ denote the probability of the process being in state $(i, j)^*$. Then the transition equation is

$$\pi(i, i - 1) = \frac{\lambda}{\lambda + \mu}\pi^*(i - 1, i - 1), \qquad 1 < i < K + 1 \tag{20.19}$$

Figure 20.7(c) illustrates the transitions for state (i, j) where $2 < i < K + 1, 0 < j < i - 1$. The process may move into state (i, j) from two states $(i - 1, j)$ and $(i - 1, j)^*$ with rate

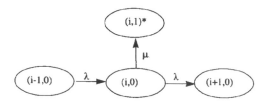

(a) Transitions for state (i,0) (2 < i < K+1)

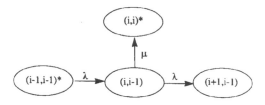

(b) Transitions for state (i,i-1)(1 < i < K+1)

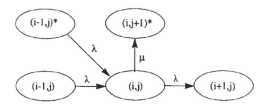

(c) Transitions for state (i,j) (2< i < K+1, 0 < j < i-1)

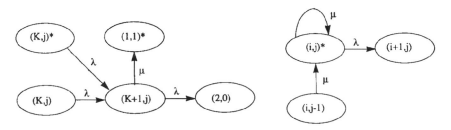

(d) Transitions for state (K+1,j) (0 < j < K+1)   (e) Transitions for state (i,j)* (0 < j ≤ i, 1 < i < K

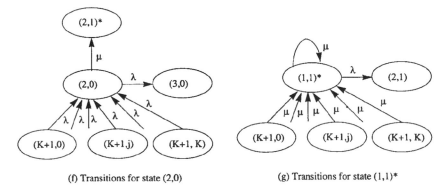

(f) Transitions for state (2,0)   (g) Transitions for state (1,1)*

FIGURE 20.7   State transitions.

λ, respectively. The process moves from (i, j) to $(i, j + 1)^*$ or $(i + 1, j)$ with rates μ and λ, respectively. The transition equation is

$$\pi(i, j) \quad = \quad \frac{\lambda}{\lambda + \mu} [\pi(i - 1, j) + \pi^*(i - 1, j)] ,$$
$$2 < i < K + 1, \qquad 0 < j < i - 1 \qquad (20.20)$$

Figure 20.7(d) illustrates the transitions for state $(K + 1, j)$ where $0 < j < K + 1$. Note that if a phone call arrives when the process is in (K, j) or $(K, j)^*$, the system enters a new cycle (with rate λ), and we could represent the new state as $(1, 0)$. In our model, we introduce the state $(K + 1, j)$ instead of $(1, 0)$, where

$$\sum_{0 \le j \le K} \pi(K + 1, j) = \pi(1, 0)$$

so that the hit ratio, and thus the location tracking cost, can be derived [see Eq. (20.25)]. The process moves from $(K + 1, j)$ [i.e., $(1, 0)$] to $(1, 1)^*$ with rate μ if the user moves before the next call arrives. Otherwise, the process moves to $(2, 0)$ with rate λ. The transition equation is

$$\pi(K + 1, j) = \frac{\lambda}{\lambda + \mu} [\pi(K, j) + \pi^*(K, j)], \qquad 0 < j < K + 1 \qquad (20.21)$$

For $j = 0$, the transition from $(K, j)^*$ to $(K + 1, 0)$ should be removed in Fig. 20.7(d) because the state $(K, 0)^*$ does not exist. The transition equation for $(K + 1, 0)$ is given in Eq. (20.18). Figure 20.7(e) illustrates the transitions for state $(i, j)^*$ where $0 < j < i$, $1 < i < K + 1$. The process can only move to $(i, j)^*$ from $(i, j - 1)$ (with rate μ). From the definition of $(i, j)^*$, if the user moves when the process is in $(i, j)^*$, the process remains in $(i, j)^*$ (with rate μ). Otherwise, the process moves to $(i + 1, j)$ with rate λ. The transition equation is

$$\pi^*(i, j) = \frac{\mu}{\lambda} \pi(i, j - 1), \qquad 0 < j \le i, \qquad 1 < i < K + 1, \qquad i \ge 2 \qquad (20.22)$$

The transitions for $(2, 0)$ are similar to the transitions for $(i, 0)$ except that the transition from $(1, 0)$ is replaced by $(K + 1, 0), \ldots, (K + 1, K)$ [cf., Fig. 20.7(f)]. The transition equation is

$$\pi(2, 0) = \frac{\lambda}{\lambda + \mu} \left[\sum_{0 \le j \le K} \pi(K + 1, j) \right] \qquad (20.23)$$

Finally, the transitions for $(1, 1)^*$ is similar to the transitions for $(i, j)^*$ except that the transition from $(1, 0)$ is replaced by $(K + 1, 0), \ldots, (K + 1, K)$ [cf., Fig. 20.7(g)]. The transition equation is

$$\pi^*(1, 1) = \frac{\mu}{\lambda} \left[\sum_{0 \le j \le K} \pi(K + 1, j) \right] \qquad (20.24)$$

Suppose that at the beginning of a cycle, the process is in state $(K + 1, j)$, then it implies that there are j cache misses in the previous cycle. The cache is enabled if and only if

$$p_M \ge p_T = \frac{C_H}{C_B} \Rightarrow 1 - \frac{j}{K} \ge \frac{C_H}{C_B} \Rightarrow 0 \le j \le \left\lceil K \left(1 - \frac{C_H}{C_B} \right) \right\rceil$$

Thus, the probability that the measured hit ratio $p_M < p_T$ in the previous cycle is

$$Pr\left[p_M < p_T\right] = \frac{\displaystyle\sum_{\lceil k[1-(C_H/C_B)]\rceil < j \le K} \pi(K+1, j)}{\displaystyle\sum_{0 \le j \le K} \pi(K+1, j)}$$

and the location tracking cost for the reset-K algorithm is

$$\begin{aligned}
C_C &= C_B Pr\left[p_M < p_T\right] + (1 - Pr\left[p_M < p_T\right]) \\
&\times \left\{ \sum_{0 \le j \le K} \left(\frac{(K-j)C_H}{K} + \frac{j(C_H + C_B)}{K} \right) \left[\frac{\pi(K+1, j)}{\displaystyle\sum_{0 \le i \le K} \pi(K+1, i)} \right] \right\}
\end{aligned} \tag{20.25}$$

The first term Eq. (20.25) represents the cost incurred when caching is disabled because the hit ratio threshold exceeds the hit ratio measured in the previous cycle. The second term is the cost when the cache is enabled and consists of two parts, corresponding to calls during which hits occur and calls during which misses occur. The ratio in square brackets is the conditional probability of being in state $\pi(K+1, j)$ during the current cycle.

The numerical computation of $\pi(K+1, j)$ can be done as follows. First, compute $a_{i,j}$ and $b_{i,j}$ where $\pi(i, j) = a_{i,j} \pi^*(1, 1)$ and $\pi^*(i, j) = b_{i,j} \pi^*(1, 1)$. Note that $a_{i,j} = 0(b_{i,j} = 0)$ if $\pi(i, j)[\pi^*(i, j)]$ is not defined in Eqs. (20.18)–(20.24). Since

$$\sum_{i,j} [\pi(i, j) + \pi^*(i, j)] = 1$$

we have

$$\pi^*(1, 1) = \frac{1}{\displaystyle\sum_{i,j} (a_{i,j} + b_{i,j})}$$

and $\pi(K+1, j)$ can be computed and the location tracking cost for the reset-K algorithm is obtained using Eq. (20.25).

The analysis is validated by a Monte Carlo simulation. In the simulation, the confidence interval for the 98% confidence level of the output measure C_C/C_B is within 3% of the mean value. Figure 20.8 plots curves for Eq. (20.25) (the solid curves) against the simulation experiments (the dashed curves) for $K = 20$ and $C_H = 0.5C_B$ and $0.3C_B$, respectively. The figure indicates that the analysis is consistent with the simulation model.

20.9.3   Comparison of the LCMR Estimation Algorithms

If the distributions for the incoming call process and the user movement process never change, then we would expect the running average algorithm to outperform the reset-K algorithm (especially when K is small) because the measured hit ratio p_M in the running average algorithm approaches the true hit ratio value p in the steady state. Surprisingly, the performance for the reset-K algorithm is roughly the same as the running average algorithm even if K is as small as 10. Figure 20.9 plots the location tracking costs for the running average algorithm and the reset-K algorithm with different K values.

The figure indicates that in steady state, when the distributions for the incoming call process and the user movement process never change, the running average algorithm outperforms reset K, and a large value of K outperforms a small K but the differences are insignificant.

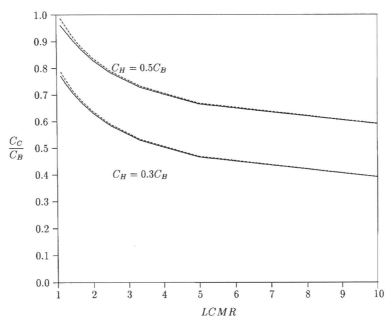

FIGURE 20.8   The location tracking costs for the reset-K algorithm; $K = 20$.

If the distributions for the incoming call process or the user movement process change from time to time, we expect that the reset-K algorithm outperforms the running average algorithm. We have examined this proposition experimentally. In the experiments, 4000 incoming calls are simulated. The call arrival rate changes from 0.1 to 1.0, 0.3, and then 5.0 for every 1000 calls (other sequences have been tested and similar results are observed). For every data point, the simulation is repeated 1000 times to ensure that the confidence interval for the 98% confidence level of the output measure C_C/C_B is within 3% of the mean value. Figure 20.10 plots the location tracking costs for the two algorithms for these experiments. By changing the distributions of the incoming call process, we observe that the reset-K algorithm is better than the running average algorithm for all C_H/C_B values.

20.10   Discussion

In this section we discuss aspects of the caching strategy presented here. Caching in PCS systems raises a number of issues not encountered in traditional computer systems, particularly with respect to architecture and locality in user call and mobility patterns. In addition, several variations in our reference assumptions are possible for investigating the implementation of the caching strategies. Here we sketch some of the issues involved.

20.10.1   Conditions When Caching Is Beneficial

We summarize the conditions for which the auxiliary strategies are worthwhile, under the assumptions of our analysis.

The caching strategy is very promising when the HLR update (or query load) or the remote A link is the performance bottleneck, since a low $LCMR (LCMR > 0)$ is required. For caching, the total database load and signalling network traffic is reduced whenever there is a cache hit. In addition, load and traffic is redistributed from the HLR and higher level SS7 network elements (RSTP, D links) to the VLRs and lower levels where excess network

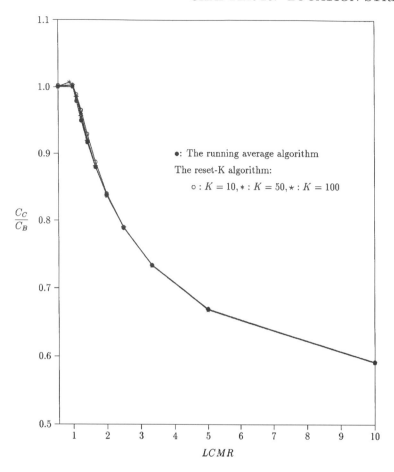

FIGURE 20.9   The location tracking costs for the running average algorithm and the reset-K algorithm; $C_H = 0.5C_B$.

capacity may be more likely to exist. If the VLR is the performance bottleneck, the caching strategy is not promising, unless the VLR capacity is upgraded.

The benefits of the caching strategy depend on user call and mobility patterns when the D link, RSTP, and LSTP are the performance bottlenecks. We have used a Poisson call arrival model and exponential intermove time to estimate this dependence. Under very conservative assumptions, for caching to be beneficial requires relatively high $LCMR$ (25–50); we expect that in practice this threshold could be lowered significantly (say, $LCMR > 7$). Further experimental study is required to estimate the amount of locality in user movements for different user populations to investigate this issue further. It is possible that for some classes of users data obtained from active badge location system studies (e.g., Fishman and Mazer, [8]) could be useful. In general, it appears that caching could also potentially provide benefits to some classes of users even when the D link, the RSTP, or the LSTP are the bottlenecks.

We observe that more accurate models of user calling and mobility patterns are required to help resolve the issues raised in this section. We are currently engaged in developing theoretical models for user mobility and estimating their effect on studies of various aspects of PCS performance [10].

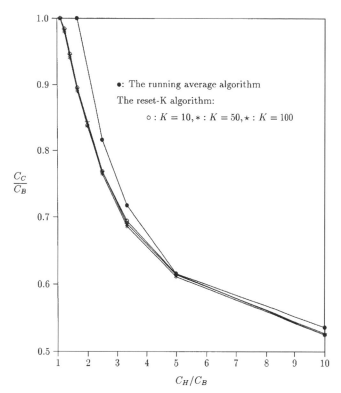

FIGURE 20.10 Comparing the running average algorithm and the reset-K algorithm under unstable call traffic.

20.10.2 Alternative Network Architectures

The reference architecture we have assumed (Fig. 20.2) is only one of several possible architectures. It is possible to consider variations in the placement of the HLR and VLR functionality, (e.g., placing the VLR at a local SCP associated with the LSTP instead of at the SSP), the number of SSPs served by an LSTP, the number of HLRs deployed, etc. It is quite conceivable that different regional PCS service providers and telecommunications companies will deploy different signalling network architectures, as well as placement of databases for supporting PCS within their serving regions [19]. It is also possible that the number and placement of databases in a network will change over time as the number of PCS users increases.

Rather than consider many possible variations of the architecture, we have selected a reference architecture to illustrate the new auxiliary strategies and our method of calculating their costs and benefits. Changes in the architecture may result in minor variations in our analysis but may not significantly affect our qualitative conclusions.

20.10.3 LCMR Estimation and Caching Policy

It is possible that for some user populations estimating the LCMR may not be necessary, since they display a relatively high-average LCMR. For some populations, as we have shown in Section 20.9, obtaining accurate estimates of user LCMR in order to decide whether or not to use caching can be important in determining the net benefits of caching.

In general, schemes for estimating the LCMR range from static to dynamic and from distributed to centralized. We have presented two simple distributed algorithms for estimating

LCMR, based on a long-range and short-range running calculation; the former is preferable if the call and mobility pattern of users is fairly Tuning the amount of history that is used to determine whether caching should be employed for a particular user is an obvious area for further study but is outside the scope of this chapter.

An alternative approach is to utilize some user-supplied information, by requesting profiles of user movements (e.g., see Tabbane, [22] and to integrate this with the caching strategy. A variation of this approach is to use some domain knowledge about user populations and their characteristics.

A related issue is that of cache size and management. In practice it is likely that the monetary cost of deploying a cache may limit its size. In that case, cache entries may not be maintained for some users; selecting these users carefully is important to maximize the benefits of caching. Note that the cache hit ratio threshold cannot necessarily be used to determine which users have cache entries, since it may be useful to maintain cache entries for some users even though their hit ratios have temporarily fallen below the threshold. A simple policy that has been found to be effective in computer systems in the least recently used (LRU) policy [20] in which cache entries that have been least recently used are discarded; LRU may offer some guidance in this context.

20.11   Conclusions

We began this chapter with an overview of the nuances of PCS, such as personal and terminal mobility, registration, deregistration, call delivery, etc. A tutorial was then provided on the two most common strategies for locating users in PCS, in North American interim standard IS-41 and the Pan-European standard GSM. A simplified analysis of the two standards was then provided to show the reader the extent to which database and signalling traffic is likely to be generated by PCS services. Suggestions were then made that are likely to result in reduced traffic.

Previous studies [12, 13, 14, 16] of PCS-related network signalling and data management functionalities suggest a high level of utilization of the signalling network in supporting call and mobility management activities for PCS systems. Motivated by the need to evolve location strategies to reduce signalling and database loads, we then presented an auxiliary strategy, called per-user caching, to augment the basic user location strategy proposed in standards [6, 18].

Using a reference system architecture for PCS, we quantified the criteria under which the caching strategy produces reductions in the network signalling and database loads in terms of users' LCMRs. We have shown that, if the HLR or the remote A link in an SS7 architecture is the performance bottleneck, caching is useful regardless of user call and mobility patterns. If the D link or STPs are the performance bottlenecks, caching is potentially beneficial for large classes of users, particularly if they display a degree of locality in their call reception patterns. Depending on the numbers of PCS users who meet these criteria, the system-wide impacts of these strategies could be significant. For instance, for users with $LCMR \sim 5$ and stable call and move patterns, caching can result in cost reduction of 20–60% over the basic user location strategy *BasicFIND*( ) under our analysis. Our results are conservative in that the *BasicFIND*( ) procedure we have used for comparison purposes already reduces the network impacts compared to the user location strategy specified in PCS standards such as IS-41.

We have also investigated in detail two simple on-line algorithms for estimating users' LCMRs and examined the call and mobility patterns for which each would be useful. The algorithms allow a system designer to tune the amount of history used to estimate a users'

LCMR and, hence, to attempt to optimize the benefits due to caching. The particular values of cache hit ratios and LCMR thresholds will change with variations in the way the PCS architecture and the caching strategy is implemented, but our general approach can still be applied. There are several issues deserving further study with respect to deployment of the caching strategy, such as the effect of alternative PCS architectures, integration with other auxiliary strategies such as the use of user profiles, and effective cache management policies.

Recently, we have augmented the work reported in this paper by a simulation study in which we have compared the caching and basic user location strategies [9]. The effect of using a time-based criterion for enabling use of the cache has also been considered [11]. We have proposed elsewhere, for users with low CMRs, an auxiliary strategy involving a system of forwarding pointers to reduce the signalling traffic and database loads [10], a description of which is beyond the scope of this chapter.

Acknowledgment

We acknowledge a number of our colleagues in Bellcore who have reviewed several previous papers by the authors and contributed to improving the clarity and readability of this work.

References

[1] Awerbuch, B. and Peleg, D., Concurrent online tracking of mobile users. In *Proc. SIGCOMM Symp. Comm. Arch. Prot.*, Oct. 1991.

[2] Bellcore., Advanced intelligent network release 1 network and operations plan, Issue 1. Tech. Rept. SR-NPL-001623. Bell Communications Research, Morristown, NJ, Jun. 1991.

[3] Bellcore., Personal communications services (PCS) network access services to PCS providers, Special Report SR-TSV-002459, Bell Communications Research, Morristown, NJ, Oct. 1993a.

[4] Bellcore., Switching system requirements for interexchange carrier interconnection using the integrated services digital network user part (ISDNUP). Tech. Ref. TR-NWT-000394. Bell Communications Research. Morristown, NJ, Dec. 1992c.

[5] Berman, R.K. and Brewster, J.H., Perspectives on the AIN architecture. *IEEE Comm. Mag.*, 1(2), 27–32, 1992.

[6] Electronic Industries Association/Telecommunications Industry Association., Cellular radio telecommunications intersystem operations. Tech. Rept. IS-41. Rev. B. Jul. 1991.

[7] Feller, W., *An Introduction to Probability Theory and Its Applications*. John Wiley & Sons, New York, 1966.

[8] Fishman, N. and Mazer, M., Experience in deploying an active badge system. In *Proc. Globecom Workshop on Networking for Pers. Comm. Appl.*, Dec. 1992.

[9] Harjono, H., Jain, R., and Mohan, S., Analysis and simulation of a cache-based auxiliary location strategy for PCS. In *Proc. IEEE Conf. Networks Pers. Comm.*, 1994.

[10] Jain, R. and Lin Y.-B., An auxiliary user location strategy employing forwarding pointers to reduce network impacts of PCS. *ACM Journal on Wireless Info. Networks (WINET)*, 1(2), 1995.

[11] Lin, Y.-B., Determining the user locations for personal communications networks. *IEEE Trans. Vehic. Tech.*, 466–473, Aug. 1994.

[12] Lo, C., Mohan, S., and Wolff, R., A comparison of data management alternatives for personal communications applications. Second Bellcore Symposium on Performance Modeling, SR-TSV-002424, Bell Communications Research, Morristown, NJ, Nov. 1993.

[13] Lo, C.N., Wolff, R.S., and Bernhardt, R.C., An estimate of network database transaction volume to support personal communications services. In *Proc. Intl. Conf. Univ. Pers. Comm.*, 1992.

[14] Lo, C. and Wolff, R., Estimated network database transaction volume to support wireless personal data communications applications. In *Proc. Intl. Conf. Comm.*, May 1993.

[15] Lycksell, E., GSM system overview. Tech. Rept. Swedish Telecom. Admin., Jan. 1991.

[16] Meier-Hellstern, K. and Alonso, E., The use of SS7 and GSM to support high density personal communications. In *Proc. Intl. Conf. Comm.*, 1992.

[17] Mohan, S. and Jain, R., Two user location strategies for PCS. *IEEE Pers. Comm. Mag.*, Premiere issue. 42–50, Feb. 1994.

[18] Mouly, M. and Pautet, M.B., *The GSM System for Mobile Communications*. M. Mouly, 49 rue Louise Bruneau, Palaiseau, France, 1992.

[19] Russo, P., Bechard, K., Brooks, E., Corn, R.L., Honig, W.L., Gove, R., and Young, J., In rollout in the United States. *IEEE Comm. Mag.*, 56–63, Mar. 1993.

[20] Silberschatz, A. and Peterson, J., *Operating Systems Concepts*. Addison-Wesley, Reading, MA, 1988.

[21] Tabbane, S., Comparison between the alternative location strategy (AS) and the classical location strategy (CS). Tech. Rept. Rutgers Univ. WINLAB. Rutgers, NJ, Jul. 1992.

[22] Tabbane, S., Evaluation of an alternative location strategy for future high density wireless communications systems. Tech. Rept. WINALAB-TR-51, Rutgers Univ. WINLAB. Rutgers, NJ, Jan. 1993.

[23] Thomas, R., Gilbert, H., and Mazziotto, G., Influence of the mobile station on the performance of a radio mobile cellular network. In *Proc. 3rd Nordic Seminar*. Sep. 1988.

21

Cell Design Principles

Michel Daoud Yacoub
University of Campinas

21.1 Introduction

Designing a cellular network is a challenging task that invites engineers to exercise all of their knowledge in telecommunications. Although it may not be necessary to work as an expert in all of the fields, the interrelationship among the areas involved impels the designer to naturally search for a deeper understanding of the main phenomena. In other words, the time for segregation, when radio engineers and traffic engineers would not talk to each other, at least through a common vocabulary, is probably gone.

A great many aspects must be considered in a cellular network planning. The main ones include the following.

Radio Propagation: Here the topography and the morphology of the terrain, the urbanization factor and the clutter factor of the city, and some other aspects of the target geographical region under investigation will constitute the input data for the radio coverage design.

Frequency Regulation and Planning: In most countries there is a centralized organization, usually performed by a government entity, regulating the assignment and use of the radio spectrum. The frequency planning within the assigned spectrum should then be made so that interferences are minimized and the traffic demand is satisfied.

Modulation: As far as analog systems are concerned, the narrowband FM is widely used due to its remarkable performance in the presence of fading. The North American

Digital Cellular Standard IS-54 proposes the $\pi/4$ differential quadrature phase-shift keying ($\pi/4$ DQPSK) modulation, whereas the Global Standard for Mobile Communications (GSM) establishes the use of the Gaussian minimum-shift keying (GMSK).

Antenna Design: To cover large areas and for low-traffic applications omnidirectional antennas are recommended. Some systems at their inception may have these characteristics, and the utilization of omnidirectional antennas certainly keeps the initial investment low. As the traffic demand increases, the use of some sort of capacity enhancement technique to meet the demand, such as replacing the omnidirectional by directional antennas, is mandatory.

Transmission Planning: The structure of the channels, both for signalling and voice, is one of the aspects to be considered in this topic. Other aspects include the performance of the transmission components (power capacity, noise, bandwidth, stability, etc.) and the design or specification of transmitters and receivers.

Switching Exchange: In most cases this consists of adapting the existing switching network for mobile radio communications purposes.

Teletraffic: For a given grade of service and number of channels available, how many subscribers can be accommodated into the system? What is the proportion of voice and signalling channels?

Software Design: With the use of microprocessors throughout the system there are software applications in the mobile unit, in the base station, and in the switching exchange.

Other aspects, such as human factors, economics, etc., will also influence the design.

This chapter outlines the aspects involving the basic design steps in cellular network planning. Topics, such as traffic engineering, cell coverage, and interference, will be covered, and application examples will be given throughout the section so as to illustrate the main ideas. We start by recalling the basic concepts including *cellular principles, performance measures and system requirements,* and *system expansion techniques.*

21.2   Cellular Principles

The basic idea of the cellular concept is *frequency reuse* in which the same set of channels can be reused in different geographical locations sufficiently apart from each other so that *cochannel interference* be within tolerable limits. The set of channels available in the system is assigned to a group of *cells* constituting the *cluster.* Cells are assumed to have a *regular hexagonal* shape and the number of cells per cluster determines the *repeat pattern.* Because of the hexagonal geometry only certain repeat patterns can tessellate. The number N of cells per cluster is given by

$$N = i^2 + ij + j^2 \tag{21.1}$$

where i and j are integers. From Eq. (21.1) we note that the clusters can accommodate only certain numbers of cells such as $1, 3, 4, 7, 9, 12, 13, 16, 19, 21, \ldots$, the most common being 4 and 7. The number of cells per cluster is intuitively related with system capacity as well as with transmission quality. The fewer cells per cluster, the larger the number of channels per cell (higher traffic carrying capacity) and the closer the cocells (potentially more cochannel interference). An important parameter of a cellular layout relating these entities is the D/R ratio, where D is the distance between cocells and R is the cell radius. In a hexagonal geometry it is found that

$$D/R = \sqrt{3N} \tag{21.2}$$

21.3 Performance Measures and System Requirements

Two parameters are intimately related with the grade of service of the cellular systems: carrier-to-cochannel interference ratio and blocking probability.

A high carrier-to-cochannel interference ratio in connection with a low-blocking probability is the desirable situation. This can be accomplished, for instance, in a large cluster with a low-traffic condition. In such a case the required grade of service can be achieved, although the resources may not be efficiently utilized. Therefore, a measure of efficiency is of interest. The **spectrum efficiency** η_s expressed in erlang per square meter per hertz, yields a measure of how efficiently space, frequency, and time are used, and it is given by

$$\eta_s = \frac{\text{number of reuses}}{\text{coverage area}} \times \frac{\text{number of channels}}{\text{bandwidth available}} \times \frac{\text{time the channel is busy}}{\text{total time of the channel}}$$

Another measure of interest is the **trunking efficiency** in which the number of subscribers per channel is obtained as a function of the number of channels per cell for different values of blocking probability. As an example, assume that a cell operates with 40 channels and that the mean blocking probability is required to be 5%. Using the erlang-B formula (refer to the Traffic Engineering section of this chapter), the traffic offered is calculated as 34.6 erlang. If the traffic per subscriber is assumed to be 0.02 erl, a total of $34.6/0.02 = 1730$ subscribers in the cell is found. In other words, the trunking efficiency is $1730/40 = 43.25$ subscribers per channel in a 40-channel cell. Simple calculations show that the trunking efficiency decreases rapidly when the number of channels per cell falls below 20.

The basic specifications require cellular services to be offered with a fixed telephone network quality. Blocking probability should be kept below 2%. As for the transmission aspect, the aim is to provide good quality service for 90% of the time. Transmission quality concerns the following parameters:

- Signal-to-cochannel interference (S/I_c) ratio
- Carrier-to-cochannel interference ratio (C/I_c)
- Signal plus noise plus distortion-to-noise plus distortion $(SINAD)$ ratio
- Signal-to-noise (S/N) ratio
- Adjacent channel interference selectivity (ACS)

The S/I_c is a subjective measure, usually taken to be around 17 dB. The corresponding C/I_c depends on the modulation scheme. For instance, this is around 8 dB for 25-kHz FM, 12 dB for 12.5-kHz FM, and 7 dB for GMSK, but the requirements may vary from system to system. A common figure for $SINAD$ is 12 dB for 25-kHz FM. The minimum S/N requirement is 18 dB, whereas ACS is specified to be no less than 70 dB.

21.4 System Expansion Techniques

The obvious and most common way of permitting more subscribers into the network is by allowing a system performance degradation but within acceptable levels. The question is how to objectively define what is acceptable. In general, the subscribers are more likely to tolerate a poor quality service rather than not having the service at all. Some alternative expansion techniques, however, do exist that can be applied to increase the system capacity. The most widely known are as follows.

Adding New Channels: In general, when the system is set up not all of the channels need be used, and growth and expansion can be planned in an orderly manner by utilizing the channels that are still available.

Frequency Borrowing: If some cells become more overloaded than others, it may be possible to reallocate channels by transferring frequencies so that the traffic demand can be accommodated.

Change of Cell Pattern: Smaller clusters can be used to allow more channels to attend a bigger traffic demand at the expense of a degradation of the transmission quality.

Cell Splitting: By reducing the size of the cells, more cells per area, and consequently more channels per area, are used with a consequent increase in traffic capacity. A radius reduction by a factor of f reduces the coverage area and increases the number of base stations by a factor of f^2. Cell splitting usually takes place at the midpoint of the congested areas and is so planned in order that the old base stations are kept.

Sectorization: A cell is divided into a number of sectors, three and six being the most common arrangements, each of which is served by a different set of channels and illuminated by a directional antenna. The sector, therefore, can be considered as a new cell. The base stations can be located either at the center or at the corner of the cell. The cells in the first case are referred to as center-excited cells and in the second as corner-excited cells. Directional antennas cut down the cochannel interference, allowing the cocells to be more closely spaced. Closer cell spacing implies smaller D/R ratio, corresponding to smaller clusters, i.e., higher capacity.

Channel Allocation Algorithms: The efficient use of channels determines the good performance of the system and can be obtained by different channel assignment techniques. The most widely used algorithm is based on fixed allocation. Dynamic allocation strategies may give better performance but are very dependent on the traffic profile and are usually difficult to implement.

21.5   Basic Design Steps

Engineering a cellular system to meet the required objectives is not a straightforward task. It demands a great deal of information, such as market demographics, area to be served, traffic offered, and other data not usually available in the earlier stages of system design. As the network evolves, additional statistics will help the system performance assessment and replanning. The main steps in a cellular system design are as follows.

Definition of the Service Area: In general, the responsibility for this step of the project lies on the operating companies and constitutes a tricky task, because it depends on the market demographics and, consequently, on how much the company is willing to invest.

Definition of the Traffic Profile: As before, this step depends on the market demographics and is estimated by taking into account the number of potential subscribers within the service area.

Choice of Reuse Pattern: Given the traffic distribution and the interference requirements a choice of the reuse pattern is carried out.

Location of the Base Stations: The location of the first base station constitutes an important step. A significant parameter to be taken into account in this is the relevance of the region to be served. The base station location is chosen so as to be at the center of or as close as possible to the target region. Data, such as available infrastructure and land, as well as local regulations are taken into consideration in this step. The cell radius is defined as a function of the traffic distribution. In urban areas, where the traffic is more heavily concentrated, smaller cells are chosen so as to attend the demand with the available channels. In suburban and in rural areas, the radius is chosen to be large because the traffic demand tends to be small. Once the placement of the first base station has been defined, the others will be accommodated in accordance with the repeat pattern chosen.

Radio Coverage Prediction: Given the topography and the morphology of the terrain, a radio prediction algorithm, implemented in the computer, can be used to predict the signal strength in the geographic region. An alternative to this relies on field measurements with the use of appropriate equipment. The first option is usually less costly and is widely used.

Design Checkup: At this point it is necessary to check whether or not the parameters with which the system has been designed satisfy the requirements. For instance, it may be necessary to re-evaluate the base station location, the antenna height, etc., so that better performance can be attained.

Field Measurements: For a better tuning of the parameters involved, field measurements (radio survey) should be included in the design. This can be carried out with transmitters and towers provisionally set up at the locations initially defined for the base station.

The cost assessment may require that a redesign of the system should be carried out.

21.6   Traffic Engineering

The starting point for engineering the traffic is the knowledge of the required grade of service. This is usually specified to be around 2% during the busy hour. The question lies on defining the busy hour. There are usually three possible definitions: (1) busy hour at the busiest cell, (2) system busy hour, and (3) system average over all hours.

The estimate of the subscriber usage rate is usually made on a demographic basis from which the traffic distribution can be worked out and the cell areas identified. Given the repeat pattern (cluster size), the cluster with the highest traffic is chosen for the initial design. The traffic A in each cell is estimated and, with the desired blocking probability $E(A, M)$, the erlang-B formula as given by Eq. (21.3) is used to determine the number of channels per cell, M

$$E(M, A) = \frac{A^M/M!}{\sum\limits_{i=0}^{M} A^i/i!} \tag{21.3}$$

In case the total number of available channels is not large enough to provide the required grade of service, the area covered by the cluster should be reduced in order to reduce the traffic per cell. In such a case, a new study on the interference problems must be carried out. The other clusters can reuse the same channels according to the reuse pattern. Not all channels need be provided by the base stations of those cells where the traffic is supposedly smaller than that of the heaviest loaded cluster. They will eventually be used as the system grows.

The traffic distribution varies in time and space, but it is commonly bell shaped. High concentrations are found in the city center during the rush hour, decreasing toward the outskirts. After the busy hour and toward the end of the day, this concentration changes as the users move from the town center to their homes. Note that because of the mobility of the users handoffs and roaming are always occurring, reducing the channel holding times in the cell where the calls are generated and increasing the traffic in the cell where the mobiles travel. Accordingly, the erlang-B formula is, in fact, a rough approximation used to model the traffic process in this ever-changing environment. A full investigation of the traffic performance in such a dynamic system requires all of the phenomena to be taken into account, making any traffic model intricate. Software simulation packages can be used so as to facilitate the understanding of the main phenomena as well as to help system planning. This is a useful alternative to the complex modeling, typically present in the analysis of cellular networks, where closed-form solutions are not usually available

On the other hand, conventional traffic theory, in particular, the erlang-B formula, is a handy tool widely used in cellular planning. At the inception of the system the calculations are carried out based on the best available traffic estimates, and the system capacity is obtained by grossly exaggerating the calculated figures. With the system in operation some adjustments must be made so that the requirements are met.

The approach just mentioned assumes the simplest channel assignment algorithm: the fixed allocation. It has the maximum spatial efficiency in channel reuse, since the channels are always assigned at the minimum reuse distance. Moreover, because each cell has a fixed set of channels, the channel assignment control for the calls can be distributed among the base stations.

The main problem of fixed allocation is its inability to deal with the alteration of the traffic pattern. Because of the mobility of the subscribers, some cells may experience a sudden growth in the traffic offered, with a consequent deterioration of the grade of service, whereas other cells may have free channels that cannot be used by the congested cells.

A possible solution for this is the use of dynamic channel allocation algorithms in which the channels are allocated on a demand basis There is an infinitude of strategies using the dynamic assignment principles, but they are usually complex to implement. An interim solution can be exercised if the change of the traffic pattern is predictable. For instance, if a region is likely to have an increase of the traffic on a given day (say, a football stadium on a match day), a mobile base station can be moved toward such a region in order to alleviate the local base.

Another specific solution uses the traffic available at the boundary between cells that may well communicate with more than one base station. In this case, a call that is blocked in its own cell can be directed to the neighboring cell to be served by its base station. This strategy, called *directed retry,* is known to substantially improve the traffic capacity. On the other hand, because channels with marginally acceptable transmission quality may be used, an increase in the interference levels, both for adjacent channel and cochannel, can be expected. Moreover, subscribers with radio access only to their own base will experience an increase in blocking probability.

21.7 Cell Coverage

The propagation of energy in a mobile radio environment is strongly influenced by several factors, including the natural and artificial relief, propagation frequency, antenna heights, and others. A precise characterization of the signal variability in this environment constitutes a hard task. Deterministic methods, such as those described by the *free space, plane earth,* and *knife-edge diffraction* propagation models, are restricted to very simple situations. They are useful, however, in providing the basic mechanisms of propagation. Empirical methods, such as those proposed by many researchers (e.g., [1, 4, 5, 8]; and others), use curves and/or formulas based on field measurements, some of them including deterministic solutions with various correction factors to account for the propagation frequency, antenna height, polarization, type of terrain, etc. Because of the random characteristics of the mobile radio signal, however, a single deterministic treatment of this signal will certainly lead the problem to a simplistic solution. Therefore, we may treat the signal on a statistical basis and interpret the results as random events occurring with a given probability. The cell coverage area is then determined as the proportion of locations where the received signal is greater than a certain threshold considered to be satisfactory.

Suppose that at a specified distance from the base station the *mean signal strength* is considered to be known. Given this we want to determine the cell radius such that the mobiles experience a received signal above a certain threshold with a stipulated probability. The mean signal strength can be determined either by any of the prediction models or by field measurements. As for the statistics of the mobile radio signal, five distributions are widely accepted today: lognormal, Rayleigh, Suzuki [11], Rice, and Nakagami. The lognormal distribution describes the variation of the mean signal level (large-scale variations) for points having the same transmitter–receiver antennas separation, whereas the other distributions characterize the instantaneous variations (small-scale variations) of the signal. In the calculations that follow we assume a lognormal environment. The other environments can be analyzed in a like manner; although this may not be of interest if some sort of diversity is implemented, because then the effects of the small-scale variations are minimized.

21.7.1 Propagation Model

Define m_w and k as the mean powers at distances x and x_0, respectively, such that

$$m_w = k \left(\frac{x}{x_0} \right)^{-\alpha} \tag{21.4}$$

where α is the path loss coefficient. Expressed in decibels, $M_w = 10 \log m_w$, $K = 10 \log k$ and

$$M_w = K - 10\alpha \log \left(\frac{x}{x_0} \right) \tag{21.5}$$

Define the received power as $w = v^2/2$, where v is the received envelope. Let $p(W)$ be the probability density function of the received power W, where $W = 10 \log w$. In a lognormal environment, v has a lognormal distribution and

$$p(W) = \frac{1}{\sqrt{2\pi}\sigma_w} \exp \left(-\frac{(W - M_w)^2}{2\sigma_w^2} \right) \tag{21.6}$$

where M_W is the mean and σ_w is the standard deviation, all given in decibels. Define w_T and $W_T = 10 \log w_T$ as the threshold above which the received signal is considered to be satisfactory. The probability that the received signal is below this threshold is its *probability distribution function* $P(W_T)$, such that

$$P(W_T) = \int_{-\infty}^{W_T} p(W) \, \mathrm{d}W = \frac{1}{2} + \frac{1}{2} \operatorname{erf} \left[\frac{(W_T - M_W)^2}{2\sigma_w^2} \right] \tag{21.7}$$

where erf( ) is the error function defined as

$$\operatorname{erf}(y) = \frac{2}{\sqrt{\pi}} \int_0^y \exp \left(-t^2 \right) \, \mathrm{d}t \tag{21.8}$$

21.7.2 Base Station Coverage

The problem of estimating the cell area can be approached in two different ways. In the first approach, we may wish to determine the proportion β of locations at x_0 where the received signal power w is above the threshold power w_T. In the second approach, we may estimate the proportion μ of the circular area defined by x_0 where the signal is above this threshold. In the first case, this proportion is averaged over the perimeter of the circumference (cell border); whereas in the second approach, the average is over the circular area (cell area).

The proportion β equals the probability that the signal at x_0 is greater than this threshold. Hence,

$$\beta = \text{prob}\,(W \geq W_T) = 1 - P\,(W_T) \tag{21.9}$$

Using Eqs. (21.5) and (21.7) in Eq. (21.9) we obtain

$$\beta = \frac{1}{2} - \frac{1}{2}\,\text{erf}\left[\frac{W_T - K + 10\alpha\log\,(x/x_0)}{\sqrt{2}\sigma_w}\right] \tag{21.10}$$

This probability is plotted in Fig. 21.1, for $x = x_0$ (cell border).

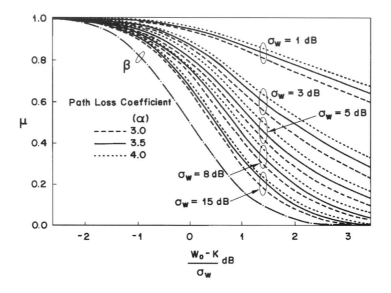

FIGURE 21.1   Proportion of locations where the received signal is above a given threshold; the dashdot line corresponds to the β approach and the other lines to the μ approach.

Let $\text{prob}(W \geq W_T)$ be the probability of the received power W being above the threshold W_T within an infinitesimal area dS. Accordingly, the proportion μ of locations within the circular area S experiencing such a condition is

$$\mu = \frac{1}{S}\int_S [1 - P\,(W_T)]\,dS \tag{21.11}$$

where $S = \pi r^2$ and $dS = x\,dx\,d\theta$. Note that $0 \leq x \leq x_0$ and $0 \leq \theta \leq 2\pi$. Therefore, solving for $d\theta$, we obtain

$$\mu = 2\int_0^1 u\beta\,du \tag{21.12}$$

where $u = x/x_0$ is the normalized distance.

Inserting Eq. (21.10) in Eq. (21.12) results in

$$\mu = 0.5\left\{1 + \text{erf}\,(a) + \exp\left(\frac{2ab+1}{b^2}\right)\left[1 - \text{erf}\left(\frac{ab+1}{b}\right)\right]\right\} \tag{21.13}$$

where $a = (K - W_T)/\sqrt{2}\sigma_w$ and $b = 10\alpha \log(e)/\sqrt{2}\sigma_w$.

These probabilities are plotted in Fig. 21.1 for different values of standard deviation and path loss coefficients.

21.7.3 Application Examples

From the theory that has been developed it can be seen that the parameters affecting the probabilities β and μ for cell coverage are the path loss coefficient α, the standard deviation σ_w, the required threshold W_T, and a certain power level K, measured or estimated at a given distance from the base station.

The applications that follow are illustrated for two different standard deviations: $\sigma_w = 5$ dB and $\sigma_w = 8$ dB. We assume the path loss coefficient to be $\alpha = 4$ (40 dB/decade), the mobile station receiver sensitivity to be -116 dB (1 mW), and the power level estimated at a given distance from the base station as being that at the cell border, $K = -102$ dB (1 mW). The receiver is considered to operate with a *SINAD* of 12 dB for the specified sensitivity. Assuming that cochannel interference levels are negligible and given that a signal-to-noise ratio S/N of 18 dB is required, the threshold W_T will be -116 dB (1 mW) $+ (18 - 12)$ dB (1 mW) $= -110$ dB (1 mW).

Three cases will be explored as follows.

Case 1: We want to estimate the probabilities β and μ that the received signal exceeds the given threshold 1) at the border of the cell, probability β and 2) within the area delimited by the cell radius, probability μ.

Case 2: It may be interesting to estimate the cell radius x_0 such that the received signal be above the given threshold with a given probability (say 90%) (1) at the perimeter of the cell and (2) within the cell area. This problem implies the calculation of the mean signal strength K at the distance x_0 (the new cell border) of the base station. Given K and given that at a distance x_0 (the former cell radius) the mean signal strength M_w is known [note that in this case $M_w = -102$ dB (1 mW)], the ratio x_0/x can be estimated.

Case 3: To fulfill the coverage requirement, rather than calculating the new cell radius, as in Case 2, a signal strength at a given distance can be estimated such that a proportion of the locations at this distance, proportion β, or within the area delimited by this distance, proportion μ, will experience a received signal above the required threshold. This corresponds to calculating the value of the parameter K already carried out in Case 2 for the various situations.

The calculation procedures are now detailed for $\sigma_w = 5$ dB. Results are also shown for $\sigma_w = 8$ dB.

Case 1: Using the given parameters we obtain $(W_T - K)/\sigma_w = -1.6$. With this value in Fig. 21.1, we obtain the probability that the received signal exceeds -116 dB (1 mW) for $S/N = 18$ dB given that at the cell border the mean signal power is -102 dB (1 mW) given in Table 21.1.

TABLE 21.1   Case 1 Coverage Probability

Standard Deviation, dB	β Approach (Border Coverage), %	μ Approach (Area Coverage), %
5	97	100
8	84	95

Note, from Table 21.1, that the signal at the cell border exceeds the receiver sensitivity with 97% probability for $\sigma_w = 5$ dB and with 84% probability for $\sigma_w = 8$ dB. If, on the other hand, we are interested in the area coverage rather than in the border coverage, then these figures change to 100% and 95%, respectively.

Case 2: From Fig. 21.1, with $\beta = 90\%$ we find $(W_T - K)/\sigma_w = -1.26$. Therefore, $K = -103.7$ dB (1 mW). Because $M_w - K = -10\alpha \log(x/x_0)$, then $x_0/x = 1.10$. Again, from Fig. 21.1, with $\mu = 90\%$ we find $(W_T - K)/\sigma_w = -0.48$, yielding $K = -107.6$ dB (1 mW). Because $M_w - K = -10\alpha \log(x/x_0)$, then $x_0/x = 1.38$. These results are summarized in Table 21.2, which shows the normalized radius of a cell where the received signal power is above -116 dB (1 mW) with 90% probability for $S/N = 18$ dB, given that at a reference distance from the base station (the cell border) the received mean signal power is -102 dB (1 mW).

TABLE 21.2 Case 2 Normalized Radius

Standard Deviation, dB	β Approach (Border Coverage)	μ Approach (Area Coverage)
5	1.10	1.38
8	0.88	1.27

Note, from Table 21.2, that in order to satisfy the 90% requirement at the cell border the cell radius can be increased by 10% for $\sigma_w = 5$ dB. If, on the other hand, for the same standard deviation the 90% requirement is to be satisfied within the cell area, rather than at the cell border, a substantial gain in power is achieved. In this case, the cell radius can be increased by a factor of 1.38. For $\sigma_w = 8$ dB and 90% coverage at the cell border, the cell radius should be reduced to 88% of the original radius. For area coverage, an increase of 27% of the cell radius is still possible.

Case 3: The values of the mean signal power K are taken from Case 2 and shown in Table 21.3, which shows the signal power at the cell border such that 90% of the locations will experience a received signal above -116 dB for $S/N = 18$ dB.

TABLE 21.3 Case 3 Signal Power

Standard Deviation dB	β Approach (Border Coverage), dB (1 mW)	μ Approach (Area Coverage), dB (1 mW)
5	-103.7	-107.6
8	-99.8	-106.2

21.8 Interference

Radio-frequency interference is one of the most important issues to be addressed in the design, operation, and maintenance of mobile communication systems. Although both intermodulation and intersymbol interferences also constitute problems to account for in system planning, a mobile radio system designer is mainly concerned about adjacent-channel and cochannel interferences.

21.8.1 Adjacent Channel Interference

Adjacent-channel interference occurs due to equipment limitations, such as frequency instability, receiver bandwidth, filtering, etc. Moreover, because channels are kept very close to each other for maximum spectrum efficiency, the random fluctuation of the signal, due to fading and near–far effect, aggravates this problem.

Some simple, but efficient, strategies are used to alleviate the effects of adjacent channel interference. In narrowband systems, the total frequency spectrum is split into two halves so that the reverse channels, composing the uplink (mobile to base station) and the forward channels, composing the downlink (base station to mobile), can be separated by half of the spectrum. If other services can be inserted between the two halves, then a greater frequency separation, with a consequent improvement in the interference levels, is accomplished. Adjacent channel interference can also be minimized by avoiding the use of adjacent channels within the same cell. In the same way, by preventing the use of adjacent channels in adjacent cells a better performance is achieved. This strategy, however, is dependent on the cellular pattern. For instance, if a seven-cell cluster is chosen, adjacent channels are inevitably assigned to adjacent cells.

21.8.2 Cochannel Interference

Undoubtedly the most critical of all interferences that can be engineered by the designer in cellular planning is cochannel interference. It arises in mobile radio systems using cellular architecture because of the frequency reuse philosophy.

A parameter of interest to assess the system performance in this case is the carrier-to-cochannel interference ratio C/I_c. The ultimate objective of estimating this ratio is to determine the reuse distance and, consequently, the repeat pattern. The C/I_c ratio is a random variable, affected by random phenomena such as (1) location of the mobile, (2) fading, (3) cell site location, (4) traffic distribution, and others. In this subsection we shall investigate the **outage probability,** i.e., the probability of failing to achieve adequate reception of the signal due to cochannel interference. This parameter will be indicated by $p(CI)$. As can be inferred, this is intrinsically related to the repeat pattern.

Cochannel interference will occur whenever the wanted signal does not simultaneously exceed the minimum required signal level s_0 and the n interfering signals, $i_1, i_2, \ldots, i_n$, by some protection ratio r. Consequently, the conditional outage probability, given n interferers, is

$$
\begin{aligned}
p\left(CI \,|\, n\right) &= 1 - \int_{s_0}^{\infty} p(s) \int_0^{s/r} p\left(i_1\right) \int_0^{(s/r)-i_1} p\left(i_2\right) \cdots \\
&\times \int_0^{(s/r)-i_1-\cdots-i_{n-1}} p\left(i_n\right) \, \mathrm{d}i_n \cdots \mathrm{d}i_2 \, \mathrm{d}i_1 \, \mathrm{d}s
\end{aligned}
\tag{21.14}
$$

The total outage probability can then be evaluated by

$$
p(CI) = \sum_n p\left(CI \,|\, n\right) p(n)
\tag{21.15}
$$

where $p(n)$ is the distribution of the number of active interferers.

In the calculations that follow we shall assume an interference-only environment, i.e., $s_0 = 0$, and the signals to be Rayleigh faded. In such a fading environment the probability density function of the signal-to-noise ratio x is given by

$$
p(x) = \frac{1}{x_m} \exp\left(-\frac{x}{x_m}\right)
\tag{21.16}
$$

where x_m is the mean signal-to-noise ratio. Note that $x = s$ and $x_m = s_m$ for the wanted signal, and $x = i_j$ and $x_m = i_{mj}$ for the interfering signal j, with s_m and i_{mj} being the mean of s and i_j, respectively.

By using the density of Eq. (21.16) in Eq. (21.14) we obtain

$$p\left(CI \mid n\right) = \sum_{j=1}^{n} \prod_{k=1}^{j} \frac{z_k}{1 + z_k} \tag{21.17}$$

where $z_k = rs_m/i_{mk}$

If the interferers are assumed to be equal, i.e., $z_k = z$ for $k = 1, 2, \ldots, n$, then

$$p\left(CI \mid n\right) = 1 - \left(\frac{z}{1 + z}\right)^n \tag{21.18}$$

Define $Z = 10 \log z$, $S_m = 10 \log s_m$, $I_m = 10 \log i_m$, and $R_r = 10 \log r$. Then, $Z = S_m - (I_m + R_r)$. Equation (21.18) is plotted in Fig. 21.2 as a function of Z for $n = 1$ and $n = 6$, for the situation in which the interferers are equal.

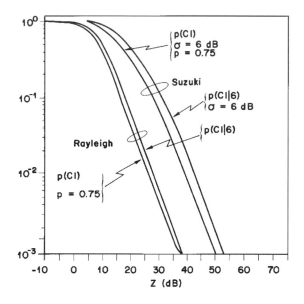

FIGURE 21.2 Conditional and unconditional outage probability for $n = 6$ interferes in a Rayleigh environment and in a Suzuki environment with $\sigma = 6$ dB.

If the probability of finding an interferer active is p, the distribution of active interferers is given by the binomial distribution. Considering the closest surrounding cochannels to be the most relevant interferers we then have six interferers. Thus

$$p(n) = \binom{6}{n} p^n (1 - p)^{6-n} \tag{21.19}$$

For equal capacity cells and an evenly traffic distribution system, the probability p is approximately given by

$$p = \sqrt[M]{B} \tag{21.20}$$

where B is the blocking probability and M is the number of channels in the cell.

Now Eqs. (21.20), (21.19), and (21.18) can be combined into Eq. (21.15) and the outage probability is estimated as a function of the parameter Z and the channel occupancy p. This is shown in Fig. 21.2 for $p = 75\%$ and $p = 100\%$.

A similar, but much more intricate, analysis can be carried out for the other fading environments. Note that in our calculations we have considered only the situation in which both the wanted signal and the interfering signals experience Rayleigh fading. For a more complete analysis we may assume the wanted signal to fade differently from the interfering signals, leading to a great number of possible combinations. A case of interest is the investigation of the influence of the standard deviation in the outage probability analysis. This is illustrated in Fig. 21.2 for the Suzuki (lognormal plus Rayleigh) environment with $\sigma = 6$ dB.

Note that by definition the parameter z is a function of the carrier-to-cochannel interference ratio, which, in turn, is a function of the reuse distance. Therefore, the outage probability can be obtained as a function of the cluster size, for a given protection ratio.

The ratio between the mean signal power s_m and the mean interfering power i_m equals the ratio between their respective distances d_s and d_i such that

$$\frac{s_m}{i_m} = \left(\frac{d_s}{d_i}\right)^{-\alpha} \tag{21.21}$$

where α is the path loss coefficient. Now, (1) let D be the distance between the wanted and interfering base stations, and (2) let R be the cell radius. The cochannel interference worst case occurs when the mobile is positioned at the boundary of the cell, i.e., $d_s = R$ and $d_i = D - R$. Then,

$$\frac{i_m}{s_m} = \left(\frac{D}{R} - 1\right)^{-\alpha} \tag{21.22a}$$

or, equivalently,

$$S_m - I_m = 10\alpha \log\left(\frac{D}{R} - 1\right) \tag{21.22b}$$

In fact, $S_m - I_m = Z + R_r$. Therefore,

$$Z + R_r = 10\alpha\log\left(\sqrt{3N} - 1\right) \tag{21.23}$$

With Eq. (21.23) and the curves of Fig. 21.2, we can compare some outage probabilities for different cluster sizes. The results are shown in Table 21.4 where we have assumed a protection ratio $R_r = 0$ dB. The protection ratio depends on the modulation scheme and varies typically from 8 dB (25-kHz FM) to 20 dB [single sideband (SSB) modulation].

Note, from Table 21.4, that the standard deviation has a great influence in the calculations of the outage probability.

21.9 Conclusions

The interrelationship among the areas involved in a cellular network planning is substantial. Vocabularies belonging to topics, such as radio propagation, frequency planning and regulation, modulation schemes, antenna design, transmission, teletraffic, and others, are common to all cellular engineers.

TABLE 21.4   Probability of Cochannel Interference in Different Cell Clusters

| | | Outage Probability, % | | | |
| | | Rayleigh | | Suzuki $\sigma = 6$ dB | |
N	$Z + R$, dB	$p = 75\%$	$p = 100\%$	$p = 75\%$	$p = 100\%$
1	−4.74	100	100	100	100
3	10.54	31	40	70	86
4	13.71	19	26	58	74
7	19.40	4.7	7	29	42
12	24.46	1	2.1	11	24
13	25.19	0.9	1.9	9	22

Designing a cellular network to meet system requirements is a challenging task which can only be partially and roughly accomplished at the design desk. Field measurements play an important role in the whole process and constitute an essential step used to tune the parameters involved.

Defining Terms

Outage probability: The probability of failing to achieve adequate reception of the signal due to, for instance, cochannel interference.

Spectrum efficiency: A measure of how efficiently space, frequency, and time are used. It is expressed in erlang per square meter per hertz.

Trunking efficiency: A function relating the number of subscribers per channel and the number of channels per cell for different values of blocking probability.

References

[1] Egli, J., Radio above 40 Mc over irregular terrain. *Proc. IRE.*, 45(10), 1383–1391, 1957.

[2] Hata, M., Empirical formula for propagation loss in land-mobile radio services. *IEEE Trans. Vehicular Tech.*, VT-29, 317–325, 1980.

[3] Ho, M.J. and Stüber, G.L., Co-channel interference of microcellular systems on shadowed Nakagami fading channels. *Proc. IEEE Vehicular Tech. Conf.*, 568–571, 1993.

[4] Ibrahim, M.F. and Parsons, J.D., Signal strength prediction in built-up areas, Part I: median signal strength. *Proc. IEEE.* Pt. F. (130), 377–384, 1983.

[5] Lee, W.C.Y., *Mobile Communications Design Fundamentals*, Howard W. Sams, Indianapolis, IN, 1986.

[6] Leonardo, E.J. and Yacoub, M.D., A statistical approach for cell coverage area in land mobile radio systems. Proceedings of the 7th IEE. Conf. on Mobile and Personal Comm., Brighton, UK, 16–20, Dec. 1993a.

[7] Leonardo, E.J. and Yacoub, M.D., (Micro) Cell coverage area using statistical methods. Proceedings of the IEEE Global Telecom. Conf. GLOBECOM'93, Houston, TX, 1227–1231, Dec. 1993b.

[8] Okumura, Y., Ohmori, E., Kawano, T., and Fukuda, K., Field strength and its variability in VHF and UHF land mobile service. *Rev. Elec. Comm. Lab.*, 16, 825–873, Sept.-Oct. 1968.

[9] Reudink, D.O., Large-scale variations of the average signal. In *Microwave Mobile Communications*, 79–131, John Wiley & Sons, New York, 1974.

[10] Sowerby, K.W. and Williamson, A.G., Outage probability calculations for multiple cochannel interferers in cellular mobile radio systems. *IEE Proc.,* Pt. F. 135(3), 208–215, 1988.

[11] Suzuki, H., A statistical model for urban radio propagation. *IEEE Trans. Comm.,* 25(7), 673–680, 1977.

Further Information

The fundamentals of mobile radio engineering in connection with many practical examples and applications as well as an overview of the main topics involved can be found in Yacoub, M.D., *Foundations of Mobile Radio Engineering,* CRC Press, Boca Raton, FL, 1993.

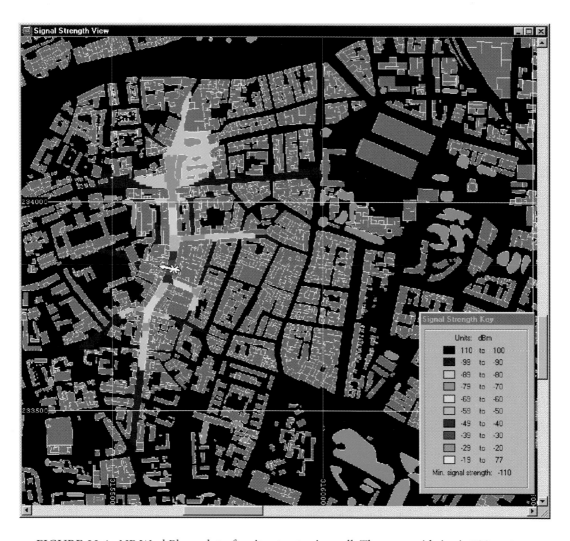

FIGURE 22.4 NP WorkPlace plot of a city street microcell. The map grid size is 500 meters.

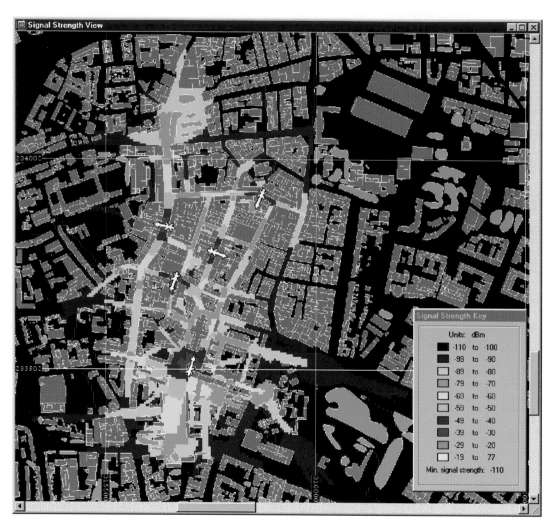

FIGURE 22.5 Cluster of microcells in a city center. The map grid size is 500 meters.

22

Microcellular Radio Communications

Raymond Steele
Southampton University and Multiple Access Communications

22.1 Introducing Microcells

In mobile radio communications an operator will be assigned a specific bandwidth W in which to operate a service. The operator will, in general, not design the mobile equipment, but purchase equipment that has been designed and standardized by others. The performance of this equipment will have a profound effect on the number of subscribers the network can support, as we will show later. Suppose the equipment requires a radio channel of bandwidth B. The operator can therefore fit $N_T = W/B$ channels into the allocated spectrum W.

Communications with mobiles are made from fixed sites, known as base stations (BSs). Clearly, if a mobile travels too far from its BS, the quality of the communications link becomes unacceptable. The perimeter around the BS where acceptable communications occur is called a cell and, hence, the term cellular radio. BSs are arranged so that their radio coverage areas, or cells, overlap, and each BS may be given $N = N_T/M$ channels. This implies that there are M BS and each BS uses a different set of channels.

The number N_T is relatively low, perhaps only 1000. As radio channels cannot operate with 100% utilization, the cluster of BSs or cells has fewer than 1000 simultaneous calls. In order to make the business viable, more users must be supported by the network. This is achieved by repeatedly reusing the channels. Clusters of BSs are tessellated with each cluster using the same N_T channels. This means that there are users in each cluster using the same frequency band at the same time, and inevitably there will be interference. This interference is known as cochannel interference. Cochannel cells, i.e., cells using the same channels, must be spaced sufficiently far apart for the interference levels to be acceptable. A mobile will therefore receive the wanted signal of power S and a total interference power of I, and the signal-to-interference ratio (SIR) is a key system design parameter.

Suppose we have large cells, a condition that occurs during the initial stages of deploying a network when coverage is important. For a given geographical area G_A we may have only one cluster of seven cells, and this may support some 800 simultaneous calls in our example. As the subscriber base grows, the number of clusters increases to, say, 100 with the area of each cluster being appropriately decreased. The network can now support some 80,000 simultaneous calls in the area G_A. As the number of subscribers continues to expand, we increase the number of clusters. The geographical area occupied by each cluster is now designed to match the number of potential users residing in that area. Consequently, the smallest clusters and, hence, the highest density of channels per area is found in the center of cities. As each cluster has the same number of channels, the smaller the clusters and, therefore, the smaller the cells, the greater the **spectral efficiency** measured in erlang per hertz per square meter. Achieving this higher spectral efficiency requires a concomitant increase in the infrastructure that connects the small cell BSs to their base station controller (BSC). The BSCs are part of the nonradio part of the mobile network that is interfaced with the public switched telephone network (PSTN) or the integrated service digital network (ISDN).

As we make the cells smaller, we change from locating the BS antennas on top of tall buildings or hills, where they produce large cells or macrocells, to the tops of small buildings or the sides of large buildings, where they form minicells, to lamp post elevations, where they form **street microcells.** Each decrease in cell size is accompanied by a reduction in the radiated power levels from the BSs and from the mobiles. As the BS antenna height is lowered, the neighboring buildings and streets increasingly control the radio propagation. This chapter is concerned with microcells and microcellular networks. We commence with the simplest type of microcells, namely, those used for highways.

22.2   Highway Microcells

Since their conception by Steele and Prabhu, [10], many papers have been published on **highway microcells,** ranging from propagation measurements to teletraffic issues [1]–[8], [11]. Figure 22.1 shows the basic concepts for a highway microcellular system having two cells per cluster. The highway is partitioned into contiguous cigar-shaped segments formed by directional antennas. Omnidirectional antennas can be used at junctions, roundabouts, cloverleaf, and other road intersections. The BS antennas are mounted on poles at elevations of some 6–12 m. Figure 22.2 shows received signal levels as a function of the distance d between BS and MS for different roads [1]. The average loss in received signal level, or path loss, is approximately inversely proportional to d^4. The path loss is associated with a slow fading component that is due to the variations in the terrain, the road curvature and cuttings, and the presence of other vehicles.

The curves in the figure are plotted for an 18-element yagi BS antenna having a gain of 15 dB and a front-to-back ratio of 25 dB. In Fig. 22.2 reference is made to junctions on different motorways, e.g., junction 5 on motorway M4. This is because the BS antennas are mounted at these road junctions with the yagi antenna pointing along the highway in order to create a cigar-shaped cell. The flat part of the curve near the BS is due to the MS receiver being saturated by high-signal levels. Notice that the curve related to M25, junction 11,

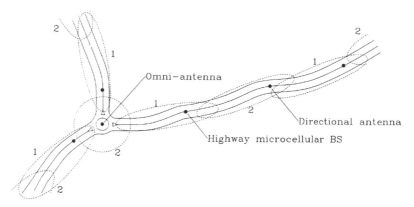

FIGURE 22.1   Highway microcellular clusters. Microcells with the same number use the same frequency set.

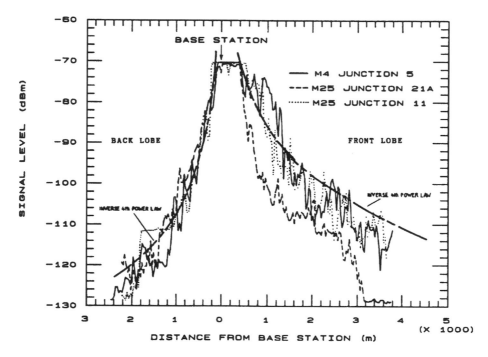

FIGURE 22.2   Overlayed received signal strength profiles of various highway cells including the inverse fourth power law curve for both the front and back antenna lobes. *Source:* Chia et al. 1987. Propagation and bit error ratio measurements for a microcellular system. *JIERE.* 57(6), 5255–5266. With permission.

decreases rapidly with distance when the MS leaves the immediate vicinity of the BS. This is due to the motorway making a sharp turn into a cutting, causing the MS to lose line of sight (LOS) with the BS. Later the path loss exponent is approximately 4. Experiments have shown that using the arrangement just described, with each BS transmitting 16 mW at 900 MHz, 16 kb/s noncoherent frequency shift keying, two-cell clusters could be formed where each cell has a length along the highway ranging from 1 to 2 km. For MSs traveling at 110 km/h the average handover rate is 1.2 per minute [1].

22.2.1 Spectral Efficiency of Highway Microcellular Network

Spectral efficiency is a key system parameter. The higher the efficiency, the greater will be the teletraffic carried by the network for the frequency band assigned by the regulating authorities per unit geographical area. We define the spectral efficiency in mobile radio communications in erlang per hertz per square meter as

$$\eta \triangleq A_{CT}/S_T W \tag{22.1}$$

although erlang per megahertz per square kilometer is often used. In this equation, A_{CT} is the total traffic carried by the microcellular network,

$$A_{CT} = CA_C \tag{22.2}$$

where C is the number of microcells in the network and A_C the carried traffic by each microcellular BS. The total area covered of the tessellated microcells is

$$S_T = CS \tag{22.3}$$

where S is the average area of a microcell, whereas the total bandwidth available is

$$W = MNB \tag{22.4}$$

whose terms M, N, and B were defined in Section 22.1. Substituting Eqs. (22.2)–(22.4) into Eq. (22.1), yields

$$\eta = \frac{\rho}{SMB} \tag{22.5}$$

where

$$\rho = A_C/N \tag{22.6}$$

is the utilization of each BS channel.

If the length of each microcell is L, there are n up lanes and n down lanes, and each vehicle occupies an effective lane length V, which is speed dependent, then the total number of vehicles in a cell is

$$K = 2nL/V \tag{22.7}$$

Given that all vehicles have a mobile terminal, the maximum number of mobiles in a cell is K. In a highway microcell we are not interested in the actual area $S = 2nL$ but in how many vehicles can occupy this area, namely, the effective area K. Notice that K is largest in a traffic jam when all vehicles are stationary and V only marginally exceeds the vehicle length. Given that N is sufficiently large, η is increased when the traffic flow is decreased.

Using fixed channel assignment (FCA) with frequency division multiple access (FDMA) or with time division multiple access (TDMA), the cluster size M can be two. Using dynamic channel assignment (DCA) with TDMA, or code division multiple access (CDMA), causes the spectral efficiency η to be very high because for a given traffic utilization ρ and channel bandwidth B, the S is small (as we are considering microcells), and M may be thought of as 1, or less, due to sectorization. The total traffic A_{CT}, given by Eq. (22.2), is also very high because by making L relatively short, C is accordingly high.

The traffic carried by a microcellular BS is

$$A_C = [\lambda_N (1 - P_{bn}) + \lambda_H (1 - P_{fhm})] \overline{T}_H \tag{22.8}$$

where P_{bn} is the probability of a new call being blocked, P_{fhm} is the probability of handover failure when mobiles enter the microcell while making a call and concurrently no channel is available, λ_N and λ_H are the new call and handover rates, respectively, and $\overline{T}_H$ is the mean channel holding time of all calls. For the simple case where no channels are exclusively reserved for handovers, $P_{bn} = P_{fhm}$, and

$$A_C = \lambda_T \overline{T}_H (1 - P_{bn}) = A (1 - P_{bn}) \tag{22.9}$$

where

$$\lambda_T = \lambda_N + \lambda_H \tag{22.10}$$

and A is the total offered traffic. The mathematical complexity resides in calculating A and P_{bn}, and the reader is advised to consult El-Dolil, Wong, and Steele, [2], and Steele and Nofal, [8].

Priority schemes have been proposed whereby N channels are available for handover, but only $N - N_h$ for new calls. Thus N_h channels are exclusively reserved for handover [2]. While P_{bn} marginally increases, P_{fhm} decreases by orders of magnitude for the same average number of new calls per sec per microcell. This is important as people prefer to be blocked while attempting to make a call compared to having a call in progress terminated due to no channel being available on handover. An important enhancement is to use an oversailing macrocellular cluster, where each macrocell supports a microcellular cluster. The role of the macrocell is to provide channels to support microcells that are overloaded and to provide communications to users who are in areas not adequately covered by the microcells [2]. When vehicles are in a solid traffic jam, there are no handovers and so N_h should be zero. When traffic is flowing fast, N_h should be high. Accordingly a useful strategy is to make N_h adaptive to the new call and handover rates [9].

22.3   City Street Microcells

We will define a city street microcell as one where the BS antenna is located below the lowest building. As a consequence, the diffraction over the buildings can be ignored, and the heights of the buildings are of no consequence. Roads and their attendant buildings form trenches or canyons through which the mobiles travel. If there is a direct line-of-sight path between the BS and a MS and a ground-reflected path, the received signal level vs BS–MS distance is as shown in Fig. 22.3. Should there be two additional paths from rays reflected from the buildings, then the profile for this four-ray situation is also shown in Fig. 22.3. These theoretical curves show that as the MS travels from the BS the average received signal level is relatively constant and then decreases relatively rapidly. This is a good characteristic as it offers a good signal level within the microcell, and the interference into adjacent microcells falls off rapidly with distance.

In practice there are many paths, but there is often a dominant one. As a consequence the fading is Rician [7]. The Rician distribution approximates to a Gaussian one when the received signal is from a dominant path with the power in the scattered paths being negligible, to a Rayleigh one when there is no dominant path. Macrocells usually have Rayleigh fading, whereas in microcells the fading only occasionally becomes Rayleigh and is more likely to be closer to Gaussian. This means that the depth of the fades in microcells are usually significantly smaller than in macrocells enabling microcellular communications to operate closer to the receiver noise floor without experiencing error bursts and to accommodate higher cochannel interference levels. Because of the small dimensions of the microcells, the delays between the first and last significant paths is relatively small compared to the corresponding delays in macrocells. Consequently, the impulse response is generally shorter in microcells and, therefore, the transmitted bit rate can be significantly higher before in-

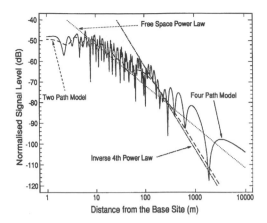

FIGURE 22.3   Signal level profiles for the two- and four-path models. Also shown are the free space and inverse fourth power laws. *Source*: Green. 1990. Radio link design for microcellular system, *British Telecom. Tech. J.*, 8(1), 85–96. With permission.

tersymbol interference is experienced compared to the situation in macrocells. Microcells are, therefore, more spectrally efficient with an enhanced propagation environment.

There are two types of these city street microcells, one for pedestrians and the other for vehicles. In general, there will be more portables carried by pedestrians than mobile stations in cars. Also, as cars travel more quickly than people, their microcells are accordingly larger than for pedestrians. The handover rates for portables and vehicular MS may be similar, and networks must be capable of handling the many handovers per call that may occur. In addition, the time available to complete a handover may be very short compared to those in macrocells.

City street microcells are irregular when the streets are irregular as demonstrated by the NP WorkPlace[1] plot of a BS in a city area displayed in Fig. 22.4. To achieve a contiguous coverage we site the BSs one at a time. Having sited the first BS and located the microcellular boundary along the streets, we locate adjacent BSs such that their boundaries butt with each other along the main streets. Unless many microcellular BSs are deployed, there will be some secondary streets where there will be insufficient signal levels. Those areas that are not covered by the microcellular BS will be accommodated by an oversailing macrocellular BS that services the complete cluster of microcellular BSs. Figure 22.5 shows a cluster of microcells; the oversailing macrocell could be sited outside the area of this figure. We emphasize that total coverage by microcells in a typical city center is difficult to achieve, and it is vital that oversailing macrocells are used to cover these dead spots. The macrocell also facilitates handovers and efficient microcellular channel utilization.

There are important differences between highway microcells and city microcells, which relate to their one- and two-dimensional characteristics. A similar comment applies to street microcells and hexagonal cells. Basically, the buildings have a profound effect on cochannel interference. The buildings shield much of the cochannel interference, and the double regression path loss law of microcells [3] also decreases interference if the break-distance constitutes the notional microcell boundary. City microcellular clusters may have

---

[1]NP WorkPlace is a propriety software outdoor planning tool developed by Multiple Access Communications Ltd.

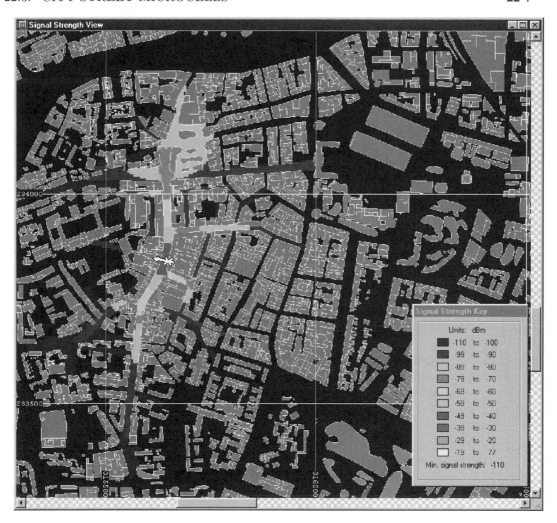

FIGURE 22.4 NP WorkPlace plot of a city street microcell. The map gridsize is 500 meters. The color version of this plot is presented elsewhere in this Handbook.

as few as two microcells, but four is more typical, and in some topologies six or more may be required. The irregularity of city streets means that some signals can find paths through building complexes to give cochannel interference where it is least expected.

22.3.1 Teletraffic Issues

Consider the arrangement where each microcellular cluster is overlaid by a macrocell. The macrocells are also clustered. The arrangement is shown in Fig. 22.6. The total traffic carried is

$$A_{CT} = C_m A_{cm} + C_M A_{CM} \qquad (22.11)$$

where C_m and C_M are number of microcells and macrocells in the network, respectively. Each microcellular BS has N channels and carries A_{cm} erlang. The corresponding values

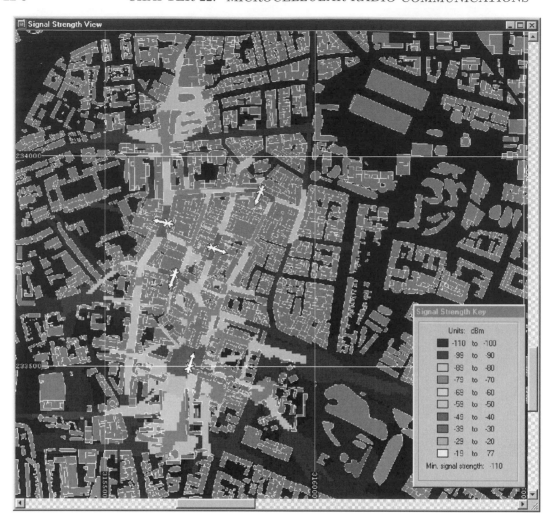

FIGURE 22.5   Cluster of microcells in a city centre. The map gridsize is 500 meters. The color version of this plot is presented elsewhere in this Handbook.

for the macrocellular BSs are N_0 and A_{CM}. The channel utilization for the network is

$$\rho_2 = \frac{C_m A_{cm} + C_M A_{CM}}{C_m N + C_M N_0} = \frac{M A_{cm} + A_{CM}}{MN + N_0} \qquad (22.12)$$

where M is the number of microcells per cluster.

The spectral efficiency is found by noting that the total bandwidth is

$$B_T = B_c \left(MN + M_0 N_0 \right) \qquad (22.13)$$

where B_c is the effective channel bandwidth, and M_0 is the number of macrocells per macrocellular cluster. The traffic carried by a macrocellular cluster and its cluster of microcells is

$$A_M = A_{CM} M_0 + M_0 \left(A_{cm} M \right) \qquad (22.14)$$

over an area of

$$S_M = \left(S_m M \right) M_0 \qquad (22.15)$$

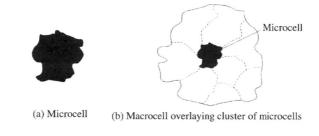

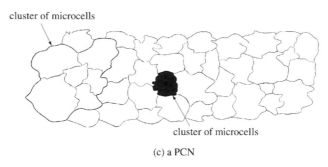

(c) a PCN

FIGURE 22.6 Microcellular clusters with oversailing macrocellular clusters. Each macrocell is associated with a particular microcellular cluster. *Source*: Steele and Williams. 1993. Third generation PCN and the intelligent multimode mobile portable. *IEE Elect. and Comm. Eng. J.*, 5(3), 147–156. With permission.

where S_m is the area of each microcell.

The spectral efficiency is, therefore,

$$
\begin{aligned}
\eta &= \frac{A_{CM} + A_{cm}M}{B_c\left(MN + M_0 N_0\right) S_m M} \\
&= \frac{\rho_2}{B_c S_m M}\left\{\frac{MN + N_0}{MN + M_0 N_0}\right\}
\end{aligned}
\tag{22.16}
$$

We note that by using oversailing macrocells to assist microcells experiencing overloading we are able to operate the microcells at high levels of channel utilization. However, the channel utilization of the macrocells must not be high if the probability of calls forced to terminate due to handover failure is to be minuscule.

22.4   Indoor Microcells

Microcellular BSs may be located within buildings to produce **indoor microcells** whose dimensions may extend from a small office, to part of larger offices, to a complete floor, or to a number of floors. The microcells are box-like for a single office microcell; or may contain many boxes, e.g., when a microcell contains contiguous offices. Furniture, such as bookcases, filing cabinets, and desks may represent large obstacles that may introduce shadowing effects. The signal attenuation through walls, floors, and ceilings may vary dramatically depending on the construction of the building. There is electromagnetic leakage down stairwells and through service ducting, and signals may leave the building and re-enter it after reflection and diffractions from other buildings.

Predicting the path loss in office microcells is, therefore, fraught with difficulties. At the outset it may not be easy to find out the relevant details of the building construction. Even if these are known, an estimation of the attenuation factors for walls, ceilings, and floors from the building construction is far from simple. Then there is a need to predict the effect of the furniture, effect of doors, the presence of people, and so on. Simple equations have been proposed. For example, the one by Keenan and Motley, [4], who represent the path loss in decibels by

$$PL = L(V) + 20 \log_{10} d + n_f a_f + n_w a_w \qquad (22.17)$$

where d is the straight line distance between the BS and the MS, a_f and a_w are the attenuation of a floor and a wall, respectively, n_f and n_w are the number of floors and walls along the line d, respectively, and $L(V)$ is a so-called clutter loss, which is frequency dependent. This equation should be used with caution. Researchers have made many measurements and found that even when they use computer ray tracing techniques the results can be considerably disparate. Errors having a standard deviation of 8–12 dB are not unusual at the time of writing. Given the wide range of path loss in mobile communications, however, and the expense of making measurements, particularly when many BS locations are examined, means that there is nevertheless an important role for planning tools to play, albeit their poorer accuracy compared to street microcellular planning tools.

As might be expected, the excess path delays in buildings is relatively small. The maximum delay spread within rooms and corridors may be <200 and 300 ns, respectively [6]. The digital European cordless telecommunication (DECT) indoor system operates at 1152 kb/s without either channel coding or equalization [7]. This means that the median rms values are relatively low, and 25 ns has been measured [6]. When the delay spread becomes too high resulting in bit errors, the DECT system hops the user to a better channel using DCA.

22.5 Microcellular Infrastructure

An important requirement of first and second generation mobile networks is to restrict the number of base stations to achieve sufficient network capacity with an acceptably low probability of blocking. This approach is wise given the cost of base stations and their associated equipment, plus the cost and difficulties in renting sites. It is somewhat reminiscent of the situation faced by early electronic circuit designers who needed to minimize the number of tubes and later the number of discrete transistors in their equipment. It was the introduction of microelectronics that freed the circuit designer. We are now in an analogous situation where we need to free the network designers of the third generation communication networks, allowing the design to have microcellular BSs in the position where they are required, without being concerned if they are rarely used, and knowing that the cost of the microcellular network is a minor one compared to the overall network cost. This approach is equivalent to installing electric lighting where we are not unduly concerned if not all the lights are switched on at any particular time, preferring to be able to provide illumination where and when it is needed.

To realize high-capacity mobile communications we need to design microcellular BSs of negligible costs, of coffee mug dimensions, and with the ability to connect them at the cost of, say, electrical wiring in streets and buildings. Cordless telecommunication (CT) BSs are already of shoe-box size, and companies are designing coffee mug-size versions. The cost of these BSs will be low in mass production, and many BSs will be equivalent in cost to one first generation analog cellular BS. Microcellular BSs could be miniaturized, fully functional BSs achieved by using microelectronic techniques and by exploiting the low-radiated power levels (<10 mW) required. At the other extreme, the microcells could be formed using distribution points (DPs) that only have optical-to-microwave converters, microwave-to-optical converters, and linear amplifiers, with the remainder of the BS at another location. In between the miniaturized, fully functional BSs and the DPs there is a range of options that depends on how much complexity is built into the microcellular BS and how the intelligence of the network is distributed.

22.5.1 Radio over Fiber

The method of using DPs to form microcells is often referred to as radio over fiber (ROF) [5]. Figure 22.7(a) shows a microcellular BS transmitting to a MS. When the DP concept is evoked, the microcellular BS contains electrical-to-optical (E/O) and optical-to-electrical (O/E) converters as shown in Fig. 22.7(b). The microwave signal that would have been radiated to the mobile is now applied, after suitable attenuation, to a laser transmitter. Essentially, the microwave signal amplitude modulates the laser, and the modulated signal is conveyed over a single-mode optical fiber to the distribution point. O/E conversion ensues followed by power amplification, and the resulting signal is transmitted to the MS. Signals from the MS are low-noise amplified and applied to the laser transmitter in the DP. Optical signals are sent from the DP to the BS where O/E conversion is performed followed by radio reception.

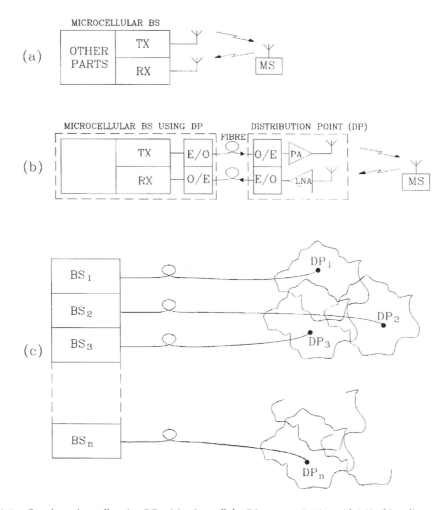

FIGURE 22.7 Creating microcells using DPs: (a) microcellular BS communicating with MS, (b) radio over fiber to distribution point, and (c) microcellular clusters using DPs.

In general, the BS transceiver will be handling multicarrier signals for many mobiles, and the DP will accordingly be transceiving signals with many mobiles whose power levels may be significantly different, even when power control is used. Care must be exercised to avoid serious intermodulation products arising in the optical components.

Figure 22.7(c) shows the cositing of n microcellular BSs for use with DPs. This cositing may be conveniently done at a mobile switching center (MSC). Shown in the figure are the DPs and their irregular shaped overlapping microcells. The DPs can be attached to lamp posts in city streets, using electrical power from the electric light supply and the same ducting as used by the electrical wiring, or local telephone ducting. The DPs can also be attached to the outside of buildings. DPs within buildings may be conveniently mounted on ceilings.

The DP concept allows small, lightweight equipment in the form of DPs to be geographically distributed to form microcells; however, there are problems. The N radio carriers cause intermodulation products (IMPs), which may be reduced by decreasing the depth of amplitude modulation for each radio channel. Unfortunately this also decreases the carrier-to-noise ratio (CNR) and the dynamic range of the link. With TDMA having many channels per carrier, we can decrease the number of radio carriers and make the IMPs more controllable. CDMA is particularly adept at coping with IMPs. The signals arriving from the MSs may have different power levels, in spite of power control. Because of the small size cells, the dynamic range of the signals arriving from MSs having power control should be <20 dB. If not, the power levels of the signals arriving at the DP from the MSs may need to be made approximately similar by individual amplification at the DP. We also must be careful to limit the length of the fiber as optical signals propagate along fibers much more slowly than radio signals propagate in free space. This should not be a problem in microcells, unless the fiber makes many detours before arriving at its DP.

The current cost of lasers is not sufficiently low for the ROF DP technique to be deployed. However, there is research into lasers, which are inherently simple, low cost, robust, and provide narrow line widths. There are also the developments in optoelectronic integrated circuits that may ultimately bring costs down. In addition, wavelength division multiplexing will bring benefits.

22.5.2   Miniaturized Microcellular BSs

The low-radiated power levels used by BSs and MSs in microcells have important ramifications on BS equipment design. Small fork combiners can be used along with linear amplifiers. Even FDMA BSs become simple when the high-radiated power levels are abandoned. It is the changes to the size of the RF components that enables the size of the microcellular BSs to be small.

The interesting question that next arises is, how much baseband signal processing complexity should the microcellular BS have? If the microcellular BSs are connected to nodes in an optical LAN, we can convey the final baseband signals to the BS, leaving the BS with the IF and front-end RF components. This means that processing of the baseband signals will be done at the group station (GS), which may be a MSC connected to the LAN. The GS will, therefore, transcode the signals from the ISDN into a suitable format. Using an optical LAN, however, and with powerful microelectronics, the transcoding and full BS operations could be done at each microcellular BS. Indeed, the microcellular BS may eventually execute many of the operations currently handled by the MSC.

22.6 Multiple Access Issues

There are three basic multiple access methods. Time division multiple access, frequency division multiple access, and spread spectrum multiple access (SSMA). SSMA comes in two versions; frequency-hopping SSMA and discrete-sequence SSMA. The latter is usually referred to as CDMA. There are also many hybrids of these systems. The principles of multiple access are described elsewhere in this book and will not be repeated here. Instead, we will comment on key factors that effect the choice of the multiple access method in microcellular environments.

As a preamble, if we observe the equations for spectral efficiency η, we see that η is inversely proportional to the number of microcells per cluster M. The smallest value of M is unity, where every microcell uses the same frequencies. Under these conditions the SIR will be low. Thus to achieve high η, we need a low value of M, and for an acceptable bit error rate (BER), we require the radio link to be able to operate with low values of SIRs. Because cellular radio operates in an intentional jamming environment, whereas CDMA was conceived to operate in a military environment where jamming by the enemy is expected, CDMA is a most appropriate multiple access method for cellular radio. The CDMA system, IS-95, will operate efficiently in single cell ($M = 1$) clusters where each cell is sectorized.

In highway microcells two-cell clusters can be used with TDMA and FDMA. Street microcells have complex shapes, see Figs. 22.4 and 22.5, and when FCA is used with TDMA and FDMA, there is a danger that high levels of interference will be ducted through streets and cause high-interference levels in a small segment of a microcell. To accommodate this phenomenon, the system must have a rapid handover (HO) capability, with HO to either a different channel at the same BS, to the interfering cell, or to an oversailing macrocell. CDMA is much more adept at handling this situation. The irregularity of street microcells, except in regularly shaped cities, such as midtown Manhattan, suggests that FCA should not be used. If it is, it requires $M \geq 4$. Instead, DCA should be employed. For example, when DCA is used with TDMA we abandon the notion of clusters of microcells. We may arrange for all microcells to have the same frequency set and design the system with accurate power control to contain the cochannel interference, and to encourage the MS to switch to another channel at the current BS or switch to a new BS directly when the SIR becomes below a threshold at either end of the duplex link. The application of DCA increases the capacity and can also contend with the situation where a MS suddenly experiences a rapid peak of cochannel interference during its travels.

The interference levels in CDMA in street microcells is mainly from users within its microcell, rather than from users in other microcells due to the shielding of the buildings. For CDMA to operate efficiently in street microcells, it should increase its chip rate to ensure it can exploit path diversity in its RAKE receiver. By increasing the chip rate, higher data rates can be accommodated and, hence, a greater variety of services. CDMA should be used in a similar way in office microcells. If the chip rate cannot be increased, however, the equipment installer can deploy a distributed antenna system where between each antenna a delay element is introduced. By this means path diversity gains are realized.

TDMA/DCA is appropriate for indoor microcells, where the complexity of the DCA is easier to implement compared to street microcells. FDMA should be considered for indoor microcells where it is well suited to provide high-bit-rate services since the transmitted rate is the same as the source rate. It also benefits from the natural shielding that exists within buildings to contain the cochannel interference and the low-power levels that simplify equipment design.

22.7   Discussion

At the time of writing, microcells are used in cordless telecommunications, where indoor microcells and outdoor telepoint microcells are used. There are very few microcells in cellular systems because there are no commercially available microcellular BSs. Nevertheless, operators have formed microcells using existing macrocellular BSs. Microcellular BSs, however, do exist in manufacturer's laboratories, and their entrance into the market is imminent. When microcells are deployed in large numbers, the vast increase in teletraffic will call for new network topologies and protocols.

In our deliberations we focused on highway microcells, city-street microcells, and indoor microcells. Minicells, where the BS antenna is below most of the buildings but above others, are currently being deployed. We may anticipate the fusion of the types of minicells and microcells. We will have microcells of strange shapes, like city street microcells but in three dimensions. Street microcells may serve the lower floors of buildings and vice versa. Microcells, located in minicell environments, may cover the streets as well as floors in neighboring buildings. We may also anticipate very small microcells, the so-called picocells. Indeed, we will have multicellular networks with multimode radio interfaces. This means that an intelligent multimode terminal with its supporting network will be required [11]. The role of microcells is to carry the high-bit-rate traffic and, hence, support a wide range of services. Our teletraffic equations tell us that microcellular personal communication networks will support orders more teletraffic than current conventional systems. Technology advancements will produce coffee cup size microcellular BSs and facilitate new network architectures that will eventually lead to the widespread concentration of intelligence at the BSs.

Defining Terms

Highway microcells: Segments of a highway having a base station and supporting mobile communications.

Indoor microcells: Small volumes of a building, e.g., an office, having a base station and supporting mobile communications.

Spectral efficiency: Has a special meaning in cellular radio. It is the traffic carried in erlang per hertz (or kilohertz) per area in square meters (or square kilometeres).

Street microcells: Small cells whose shape are determined by the street topology and their buildings. The base station antennas are below the urban skyline.

References

[1] Chia, S.T.S., Steele, R., Green, E., and Baran, A., Propagation and bit error ratio measurements for a microcellular system. *JIERE,* Supplement 57(6), 5255–5266, 1987.

[2] El-Dolil, S.A., Wong, W.C., and Steele, R., Teletraffic performance of highway microcells with overlay macrocell. *IEEE JSAC,* 7(1), 71–78, 1989.

[3] Green, E., Radio link design for microcellular systems. *British Telecom Tech. J.,* 8(1), 85–96, 1990.

[4] Keenan, J.M. and Motley, A.J., Radio coverage in buildings. *British Telecom. Tech. J.,* 8(1), 19–24, 1990.

[5] Merrett, R.P., Cooper, A.J., and Symington, I.C., A cordless access system using radio-over-fiber techniques. IEEE VT-91, 921–924.

[6] Saleh, A.A.M. and Valenzula, R.A., A statistical model for indoor multipath propagation. *IEEE JSAC*, 128–137, Feb. 1987.

[7] Steele, R., *Mobile Radio Communications*, Pentech Press, London, 1992.

[8] Steele, R. and Nofal, M., Teletraffic performance of microcellular personal communication networks. *IEE Proc-I*, 139(4), 448–461, 1992.

[9] Steele, R., Nofal, M., and El-Dolil, S., An adaptive algorithm for variable teletraffic demand in highway microcells. *Electronic Letters*, 26(14), 988–990, 1990.

[10] Steele R. and Prabhu, V.K., Mobile radio cellular structures for high user density and large data rates. In *Proc of the IEE*, Pt. F. (5), 396–404, 1985.

[11] Steele, R. and Williams, J.E.B., Third generation PCN and the intelligent multimode mobile portable, *IEE Elec. and Comm. Eng. J.*, 5(3), 147–156, 1990.

Further Information

The *IEEE Communications Magazine Special Issue* on an update on personal communications, Vol. 30, No. 12, Dec. 1992 provides a good introduction to microcells, particularly the paper by L.J. Greenstein, et al.

23

Fixed and Dynamic Channel Assignment

Bijan Jabbari
George Mason University

23.1 Introduction

One of the important aspects of frequency reuse-based cellular radio as compared to early land mobile telephone systems is the potential for dynamic allocation of channels to traffic demand. This fact had been recognized from the early days of research (e.g., see [18, Chapter 7], [7, 8]) in this field. With the emergence of wireless personal communications and use of microcell with nonuniform traffic, radio resource assignment becomes essential to network operation and largely determines the available spectrum efficiency. The primary reason for this lies in the use of microcell in dense urban areas where distinct differences exist as compared to large cell systems due to radio propagation and fading effects that affect the interference conditions.

In this chapter, we will first review the channel reuse constraint and then describe methods to accomplish the assignment. Subsequently, we will consider variations of fixed channel assignment and discuss dynamic resource assignment. Finally, we will briefly discuss the traffic modeling aspect.

23.2 The Resource Assignment Problem

The resources in a wireless cellular network are derived either in frequency division multiple access (FDMA), time division multiple access (TDMA), or joint frequency–time (MC-TDMA) [1]. In these channel derivation techniques, the frequency reuse concept is used throughout the service areas comprised of cells and microcells. The same channel is used by distinct terminals in different cells, with the only constraint of meeting a given interference threshold. In spread spectrum multiple access (SSMA) such as the implemented code division multiple access (CDMA) system (IS-95) [24] each subscriber spreads its transmitted signal over the same frequency band by using a pseudorandom sequence simultaneously. As any active channel is influenced by the others, a new channel can be set up only if the overall interference is below a given threshold. Thus, the problem of resource assignment in CDMA relates to transmission power control in forward (base station to mobile terminal) and reverse (mobile terminal to base station) channels. Of course, the problem of power control applies to TDMA and FDMA as well, but not to the extent that it impacts the capacity of CDMA. Here we first focus on time- and frequency-based access methods. We subsequently present some preliminary discussion on spread spectrum CDMA systems.

Fixed channel assignment (FCA) and dynamic channel assignment (DCA) techniques are the two extremes of allocating radio channels to mobile subscribers. For a specific grade of service and quality of transmission, the assignment scheme provides a tradeoff between spectrum utilization and implementation complexity. The performance parameters from a radio resource assignment point of view are interference constraints (quality of transmission link), probability of call blocking (grade of service), and the system capacity (spectrum utilization) described by busy hour erlang traffic that can be carried by the network. In a cellular system, however, there exist other functions, such as handoff and its execution or radio access control. These functions may be facilitated by the use of specific assignment schemes and, therefore, they should be considered in such a tradeoff [14].

The problem of channel assignment can be described as the following: Given a set of channels derived from the specified spectrum, assign the channels and their transmission power such that for every set of assigned channels to cell i, $(C/I)_i > (C/I)_0$. Here, $(C/I)_0$ represents the minimum allowed carrier to interference and $(C/I)_i$ represents carrier to interference at cell i.

23.3   Fixed Channel Assignment

In fixed channel assignment the interference constraints are ensured by a frequency plan independent of the number and location of active mobiles. Each cell is then assigned a fixed number of carriers, dependent on the traffic density and cell size. The corresponding frequency plan remains fixed on a long-term basis. In reconfigurable FCA (sometimes referred to as flexible FCA), however, it is possible to reconfigure the frequency plan periodically in response to near/medium term changes in predicted traffic demand.

In FCA, for a given set of communications system parameters, $(C/I)_0$ relates to a specific quality of transmission link (e.g., probability of bit error or voice quality). This parameter in turn relates to the number of channel sets [18] (or cluster size) given by $K = 1/3 (D/R)^2$. Thus, the ratio D/R is determined by $(C/I)_0$. Here D is the cochannel reuse distance and R is the cell radius. For example, in the North American cellular system advanced mobile phone service (AMPS), $(C/I)_0 = 18$ dB, which results in $K = 7$ or $D = 4.6R$. Here, we have used a propagation attenuation proportional to the fourth power of the distance. The radius of the cell is determined mainly by the projected traffic density. In Fig. 23.1 a seven cell cluster with frequency sets F1 through F7 (in cells designated A–G) has been illustrated. It is seen that the same set of frequencies is repeated two cells away.

The number of channels for each cell can be determined through the erlang-B formula (for example, see Cooper [6]) by knowing the busy hour traffic and the desired probability of blocking (grade of service). Probability of blocking P_B is related to offered traffic A, and

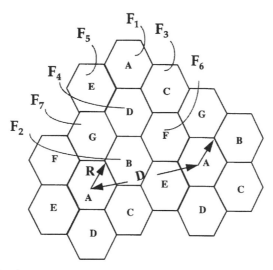

FIGURE 23.1 Fixed channel assignment.

the number of channels per cell N by

$$P_B = \frac{A^N/N!}{\displaystyle\sum_{i=0}^{N} A^i/i!}$$

This applies to the case of blocked calls cleared. If calls are delayed, the grade of service becomes the probability of calls being delayed P_Q and is given by the erlang-C formula [6],

$$P_Q = \frac{\dfrac{A^N}{N!(1-A/N)}}{\displaystyle\sum_{i=0}^{N-1} A^i/i! + \dfrac{A^N}{N!(1-A/N)}}$$

FCA is used in almost all existing cellular mobile networks employing FDMA or TDMA. To illustrate resource assignment in FCA, we describe the FCA schemes used in global system for mobile communications (GSM or DCS) [17]. Here, the mobile terminal continuously monitors the received signal strength and quality of a broadcast channel along with the identity of the transmitting base station. This mechanism allows both the mobile and the network to keep track of the subscriber movements throughout the service area. When the mobile terminal tries and eventually succeeds in accessing the network (see Fig. 23.2), a two way control channel is assigned that allows for both authentication and ciphering mode establishment. On completion of these two phases, the setup phase initiates and a traffic channel is eventually assigned to the mobile terminal. Here, the terminal only knows the identity of the serving cell; the (control and traffic) radio channel assignment is a responsibility of the network and fulfills the original carrier to cell association, while simplifying the control functions during normal system operation.

23.4 Enhanced Fixed Channel Assignment

FCA has the advantage of having simple realization. Since the frequency sets are preassigned to cells based on long-term traffic demand, however, it cannot adapt to traffic variation

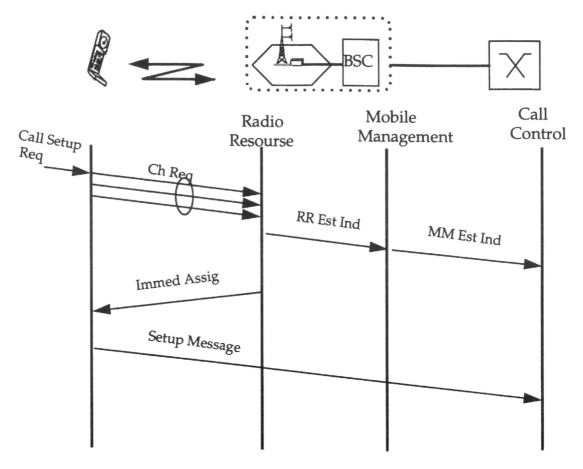

FIGURE 23.2   Radio resource assignment in GSM/DCS. *Source*: Jabbari, B., et al., Network issues for wireless personal communications. *IEEE Commun. Mag.,* 33(1), 1995.

across cells and, therefore, FCA will result in poor bandwidth utilization. To improve the utilization while maintaining the implementation simplicity, various strategies have been proposed as enhancements to FCA and deployed in existing networks. Two often used methods are *channel borrowing* and *directed retry*, which are described here.

In the channel borrowing strategy [11, 12, 18, 29], channels that are not in use in their cells may be borrowed by adjacent cells with high offered traffic on a call-by-call basis. Borrowing of channels allows the arriving calls to be served in their own cell. This implies that there will be further restrictions in using the borrowed channels in other cells. Various forms of borrowing have been surveyed in Tekinay and Jabbari [23].

In directed retry [10, 13, 14], a call to or from a mobile subscriber may try other cells with channels with sufficient signal strength meeting the C/I constraint if there are no channels available in its own cell to be served. In some cases it may be necessary to direct some of the calls in progress in a given congested cell to adjacent lightly loaded cells (those calls that can be served by adjacent cells) in order to accommodate the new calls in that given cell. This is referred to as directed handoff [13, 20]. The combination of these two capabilities provides a significant increase in bandwidth utilization.

23.5 Dynamic Channel Assignment

In dynamic channel assignment [2, 3, 4, 7, 8, 9, 12, 21], the assignment of channels to cells occurs based on the traffic demand in the cells. In other words, channels are pooled together and assignments are made and modified in real time. Therefore, this assignment scheme has the potential to achieve a significantly improved bandwidth utilization when there are temporal or spatial traffic variations.

In DCA, the interference constraints are ensured by a real-time evaluation of the most suitable (less interfered) channels that can be activated in a given cell (reassignment may be needed). That is, the system behaves as if the frequency plan was dynamically changing to meet the actual radio link quality and traffic loads, realizing an implicit sharing of the frequency band under interference constraints.

The implementation of DCA generally is more complex due to the requirement for system-wide state information where the state refers to which channel in which cell is being used. Obviously, it is impractical to update the system state in a large cellular network at any time, especially those based on microcells, as the controller will be overloaded or call set delay will be unacceptable. Therefore, methods have been devised based on a limited state space centralized control [15] or based on a distributed control to perform the necessary updating. A method referred to as maximum packing suggested in Everitt and Macfadyen [15], records only the number of channels in use. Distributed control schemes, however, where channel assignments are made at mobile stations or base stations may be attractive. These schemes are particularly suitable for a mobile controlled resource assignment where each mobile measures the actually perceived interference and decides to utilize a radio resource in a completely decentralized way.

Design of DCA algorithms is critical to achieve the potential advantages in efficiency and robustness to traffic heterogeneity throughout cells as compared to FCA and enhanced FCA. Poor DCA algorithms, however, might lead to an uncontrolled global situation, i.e., the locally selected channel might be very good for the specific mobile terminal and at the same time very poor for the interference level induced to other traffic sources.

Two classes of *regulated DCA* and *segregation DCA* are discussed here due to their importance. In a regulated DCA, appropriate thresholds tend to maintain the current channel, avoiding useless handoffs that can tie up channels; in a segregation DCA, channel acquisition obeys priorities assigned to each channel. In the latter, the channels successfully activated in a cell have a higher assignment probability than those found to have a high interference in the cell; of course, priorities change in time depending on changes in the system status. These types of assignment lead to a (frequency) plan still changing dynamically but more slowly and, in general, only because of substantial load imbalance. In steady-state conditions, the plan either tends to be confirmed or fluctuates around a basic configuration, proven to be suboptimal in terms of bandwidth efficiency.

Both digital European cordless telecommunications (DECT) and cordless technology known as CT2 are employing the DCA technique [25], and the next generation cellular systems are foreseen to deploy it. In CT2 the handset and the base station jointly determine a free suitable channel to serve the call. In DECT, the mobile terminal (portable handset) not only recognizes the visited base station (radio fixed part) through a pilot signal but continuously scans all of the system channels and holds a list of the less interfered ones. The potentially available channels are ordered by the mobile terminal with respect to the measured radio parameters, namely, the radio signal strength indicator, measuring and combining cochannel, adjacent-channel, and intermodulation interference. When a connection has to be established, the best channel is used to communicate on the radio interface.

It is possible to have a hybrid of DCA and FCA in a cellular network in which a fraction of channels are fixed assigned and the remainder are allocated based on FCA. This scheme has less system implementation complexity than the DCA scheme but provides performance improvement (lower probability of call blocking) depending on the DCA–FCA channel partitioning [19].

In general, DCA schemes cannot be considered independently of the adopted power control mechanism because the transmitted power from both mobile terminal and base station substantially affects the interference within the network. For a detailed discussion of DCA and power control, the readers are referred to [5]. Despite the availability of several power control algorithms, much work is needed to identify realizable power and channel control algorithms that maximize bandwidth efficiency. Nevertheless, realization of power control mechanisms would require distribution of the system status information and information exchange between mobile terminal and network entities. This in turn will involve overhead and terminal power usage.

The performance of DCA depends on the algorithm implementing this capability [13, 16] In general, due to interactions between different cells the performance of the system will involve modeling the system as a whole, as opposed to in FCA where cells are treated independently. Therefore, mathematical modeling and performance evaluation of DCA becomes quite complex. Simplifying assumptions may, therefore, be necessary to obtain approximate results [15, 22]. Simulation techniques have been widely used in evaluation of DCA performance. For a representative performance characteristics of DCA and a comparison to enhanced FCA schemes the readers are referred to [14].

23.6   CDMA Systems

Now we turn our attention to systems based on direct sequence spread spectrum CDMA. Examples include the second generation system based on IS-95 and the proposed third generation system based on Wideband CDMA (W-CDMA) [28]. Due to the simultaneous availability and reuse of the entire frequency spectrum in every cell, opportunity for dynamic channel assignment exists inherently. However, the problem of dynamic channel allocation turns into the problem of allocation of resources such as spreading codes, equipment limitations, intracell and intercell interference. Added to this is the practical implication of the soft capacity, which allows temporary degradation of the links. Furthermore, the use of voice activity compression technique allows increase in channel utilization while adding new performance measures such as speech freeze-out fraction.

The capacity of the CDMA system in terms of Erlang for both forward and reverse directions has been derived in [27]. The effect of imperfect power control, other cell interference has been taken into account for traffic sources with a given user activity. This capacity reflects the average traffic load the system can carry under a specified performance measure. This measure, although it is treated similar to the probability of blocking, in actuality is somewhat different from those commonly used in orthogonal time and frequency systems and indeed reflects the outage probability. The performance measure has been defined in [27] as the ratio of background noise density to total interference density, N_0/I_0, being greater than a specified value (for example 0.1). For the simple equally loaded cells case where the limiting direction is the reverse link, with a bandwidth of W, source data rate of R bits/second, required energy to interference ratio of E_b/I_0, and ratio of other cells to own cell (sector) interference of f, the condition to have an outage is given by [27]:

$$P_{\text{outage}} = \Pr \left[\sum_{i=1}^{K(1+f)} v_i \left(E_b/I_0 \right)_i > \left(W/R \right) \left(1 - N_0/I_0 \right) \right]$$

In the above formula, v_i represents the event the bi-state source is in the active state and K is the number of sources in a single cell. Note that $(E_b/I_0)_i$ is the energy per bit-to-interference ratio of source i as received at the observed base station. Additional scenarios and some approximations for special cases have been further carried out and discussed in [26]. Such analysis can be used to evaluate the system capacity for when traffic is nonuniformly distributed across the cells/sectors.

23.7 Conclusion

In this chapter we have classified and reviewed channel assignment techniques. We have emphasized the advantages of DCA schemes over FCA in terms of bandwidth utilization in a heterogeneous traffic environment at the cost of implementation complexity. The DCA schemes are expected to play an essential role in future cellular and microcellular networks.

CDMA inherently provides the DCA capability. However, the system capacity depends on factors including power control accuracy and other cell interference. We presented a brief discussion of capacity of CDMA.

References

[1] Abramson, N., Multiple access techniques for wireless networks. *Proc. of IEEE*, 82(9), 1994.

[2] Anderson, L.G., A simulation study of some dynamic channel assignment algorithms in a high capacity mobile telecommunications system. *IEEE Trans. on Comm.*, COM-21(11), 1973.

[3] Beck, R. and Panzer, H., Strategies for handover and dynamic channel allocation in micro-cellular mobile radio systems. *Proceedings of the IEEE Vehicular Technology Conference*, 1989.

[4] Chuang, J.C.-I., Performance issues and algorithms for dynamic channel assignment. *IEEE J. on Selected Areas in Comm.*, 11(6), 1993.

[5] Chuang, J.C.-I., Sollenberger, N.R., and Cox, D.C., A pilot-based dynamic channel assignment schemes for wireless access TDMA/FDMA systems. *Internat. J. of Wireless Inform. Networks*, Jan. 1994.

[6] Cooper, R.B.,*Introduction to Queueing Theory*, 3rd ed., CEEPress Books, 1990.

[7] Cox, D.C. and Reudnik, D.O., A comparison of some channel assignment strategies in large-scale mobile communications systems. *IEEE Trans. on Comm.*, COM-20(2), 1972.

[8] Cox, D.C. and Reudnik, D.O., Increasing channel occupancy in large-scale mobile radio environments: dynamic channel reassignment. *IEEE Trans. on Comm.*, COM-21(11), 1973.

[9] Dimitrijevic, D. and Vucetic, J.F., Design and performance analysis of algorithms for channel allocation in cellular networks. *IEEE Trans. on Vehicular Tech.*, 42(4), 1993.

[10] Eklundh, B., Channel utilization and blocking probability in a cellular mobile telephone system with directed retry. *IEEE Trans. on Comm.*, COM 34(4), 1986.

[11] Elnoubi, S.M., Singh, R., and Gupta, S.C., A new frequency channel assignment algorithm in high capacity mobile communications. *IEEE Trans. on Vehicular Techno.*, 31(3), 1982.

[12] Engel, J.S. and Peritsky, M.M., Statistically-optimum dynamic server assignment in systems with interfering servers. *IEEE Trans. on Comm.*, COM-21(11), 1973.

[13] Everitt, D., Traffic capacity of cellular mobile communications systems. *Computer Networks and ISDN Systems*, ITC Specialist Seminar, Sept. 25–29, 1989, 1990.

[14] Everitt, D., Traffic engineering of the radio interface for cellular mobile networks. *Proc. of IEEE,* 82(9), 1994.

[15] Everitt, D.E. and Macfadyen, N.W., Analysis of multicellular mobile radiotelephone systems with loss. *BT Tech. J.*, 2, 1983.

[16] Everitt, D. and Manfield, D., Performance analysis of cellular mobile communication systems with dynamic channel assignment. *IEEE J. on Selected Areas in Comm.*, 7(8), 1989.

[17] Jabbari, B., Colombo, G., Nakajima, A., and Kulkarni, J., Network issues for wireless personal communications. *IEEE Comm. Mag.*, 33(1), 1995.

[18] Jakes, W.C. Ed. *Microwave Mobile Communications*, Wiley, New York, 1974, reissued by IEEE Press, 1994.

[19] Kahwa, T.J. and Georganas, N.D., A hybrid channel assignment scheme in Large scale, cellular-structured mobile communications systems. *IEEE Trans. on Comm.*, COM-26(4), 1978.

[20] Karlsson, J. and Eklundh, B., A cellular mobile telephone system with load sharing—an enhancement of directed retry. *IEEE Trans. on Comm.*, COM 37(5), 1989.

[21] Panzer, H. and Beck, R., Adaptive resource allocation in metropolitan area cellular mobile radio systems. *Proceedings of the IEEE Vehicular Technology Conference*, 1990.

[22] Prabhu, V. and Rappaport, S.S., Approximate analysis for dynamic channel assignment in large systems with cellular structure. *IEEE Trans. on Comm.*, COM-22(10), 1974.

[23] Tekinay, S. and Jabbari, B., Handover and channel assignment in mobile cellular networks. *IEEE Comm. Mag.*, 29(11), 1991.

[24] Telecommunications Industry Association. TIA Interim Standard IS-95, CDMA Specifications, 1993.

[25] Tuttlebee, W.H.W., Cordless personal communications. *IEEE Comm. Mag.*, Dec. 1992.

[26] Viterbi, A.J., *CDMA—Principles of Spread Spectrum Communication*, Addison Wesley, Reading, MA, 1995.

[27] Viterbi, A.J. and Viterbi, A.J., Erlang capacity of a power controlled CDMA system. *IEEE JSAC*, 11(6), 892–900, 1993.

[28] W-CDMA. Feature topic. *IEEE Communications Magazine*, Sept. 1998.

[29] Zhang, M. and Yum, T.-S.P., Comparison of channel-assignment strategies in cellular mobile telephone systems. *IEEE Trans. on Vehicular Tech.*, 38(4), 1989.

24

Radiolocation Techniques

Gordon L. Stüber
Georgia Institute of Technology

James J. Caffery, Jr.
Georgia Institute of Technology

24.1 Introduction

Several location technologies have been developed and commercially deployed for locating wireless radios including Decca, Loran, Omega, the Global Positioning System (GPS), and the Global Navigation Satellite System (GLONASS). GPS, originally developed for military use, is perhaps the most popular commercial location system today, providing location accuracies to within 100 m.

All of the above systems use a location technique known as radiolocation. A radiolocation system operates by measuring radio signals traveling between a mobile station (MS) and a set of fixed stations (FSs). The measurements are then used to determine the length and/or direction of the radio paths, and the MS's position is derived from known geometrical relationships. In general, measurements from $n + 1$ FSs are necessary to locate an MS in n dimensions. To achieve high accuracy in a radiolocation system, it is necessary that a line-of-sight (LoS) exist between the MS and FSs. Otherwise, large errors are likely to be incurred.

With the above mentioned radiolocation technologies, the MS formulates its own position by using signals received from the FSs. This form of location is often referred to as *self-positioning*. In these systems, a special receiver is placed in the MS to calculate the MS's position. Alternatively, the position of the MS could be calculated at a remote location by using signals received at the FSs. This form of radiolocation is known as *remote-positioning* and requires a transmitter for the MS.

Over the last several years, there has been increased interest in developing location services for wireless communications systems. An array of applications for such technology exists including location sensitive billing, fraud detection, cellular system design and resource management, fleet management, and Intelligent Transportation Services (ITS) [8]. The greatest driving force behind location system development in wireless systems has been the FCC's Emergency-911 (E-911) requirements, where a wireless E-911 caller must be located within an *rms* accuracy of 125 m in 67% of the cases [4]. Adding GPS to the handset is not a universal solution because of the large pool of existing handsets. Remote-positioning radiolocation is a natural choice since it requires no modification of existing MS handsets and most, if not all, of the complexity could be incorporated into the network side.

In this chapter, techniques for locating wireless users using measurements from radio signals are described. A few algorithms for location estimation are developed along with a discussion of measures of accuracy. The radiolocation methods that are appropriate for wireless location and the major sources of error in various mobile cellular networks are discussed.

24.2   Description of Radiolocation Methods

Radiolocation systems can be implemented that are based on angle-of-arrival (AoA), signal strength, time-of-arrival (ToA), time-difference-of-arrival (TDoA), or their combinations. These are briefly discussed below.

24.2.1   Angle of Arrival

AoA techniques estimate the MS location by first measuring the arrival angles of a signal from a MS at several FSs (Fig. 24.1). The AoA can be determined through the use of directive antennas or antenna arrays. Simple geometric relationships are then used to determine the location by finding the intersections of the lines-of-position (see Fig. 24.1). To generate a position fix, the AoA method requires that the signal transmitted from the MS be received by a minimum of two FSs.

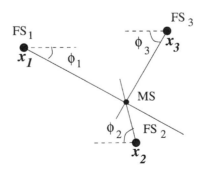

FIGURE 24.1   The measured angles, ϕ_i, determine the position of the MS for a given FS geometry.

24.2.2   Signal Strength

Radiolocation based on signal strength measurements uses a known mathematical model describing the **path loss** attenuation with distance. Since measurement of signal strength provides a distance estimate between a MS and FS, the MS must lie on a circle centered at the FS. Hence, for signal strength based radiolocation, the lines-of-position are defined by circles. By using measurements from multiple FSs, the location of the MS can be determined from the intersection of circles (Fig. 24.2).

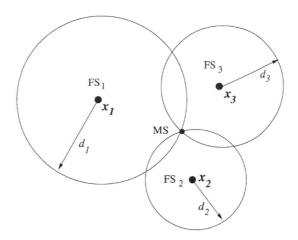

FIGURE 24.2 For signal strength and ToA based radiolocation systems the location of the MS is at the intersection of circles of radius d_i.

A second method makes use of premeasured signal strength contours around each FS. Received signal strength measured at multiple FSs can be mapped to a location by overlaying the contours for each FS. This technique can be used to combat **shadowing**, as discussed in Section 24.6.1.

24.2.3 Time-Based Location

The final class of radiolocation techniques are those based on estimating the ToAs of a signal transmitted by the MS and received at multiple FSs or the TDoAs of a signal received at multiple pairs of FSs. In the ToA approach, the distance between a MS and FS is measured by finding the one-way propagation time between the MS and FS. Geometrically, this provides a circle, centered at the FS, on which the MS must lie. Given a ToA at FS i, the equation for the circle is given by

$$\tau_i = D_i\left(\boldsymbol{x}_s\right)/c \tag{24.1}$$

where $D_i(\boldsymbol{x}_s) = \|\boldsymbol{x}_i - \boldsymbol{x}_s\|$, $\boldsymbol{x}_i$ is the position of ith FS, $\boldsymbol{x}_s$ is the position of the MS and c is the speed of light. By using at least three base stations to resolve ambiguities, the MS's position is given by the intersection of circles (Fig. 24.2). Since the ToA and path loss based signal strength methods are based on distance measurements between the MS and FSs, they are often referred to as *ranging* systems.

In the TDoA approach, time differences of arrival are used. Hence, the time of signal transmission need not be known. Since the hyperbola is a curve of constant time *difference* of arrival for two FSs, a TDoA measurement defines a line-of-position as a hyperbola, with the foci located at one of the two FSs. For FSs i and j, the equation of the hyperbola, $\rho_{i,j}$, for a given TDoA is

$$\rho_{i,j} = \frac{D_i\left(\boldsymbol{x}_s\right) - D_j\left(\boldsymbol{x}_s\right)}{c} \; . \tag{24.2}$$

The location of the MS is at the intersection of the hyperbolas (Fig. 24.3). In general, for N FSs receiving the signal from the MS, $N-1$ non-redundant TDoA measurements can be made. Thus, a MS can be located in $N-1$ dimensions.

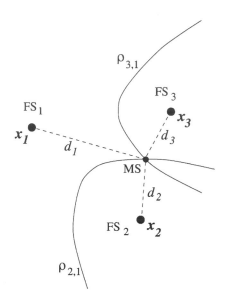

FIGURE 24.3   For TDoA based radiolocation the position of the MS is at the intersection of hyperbolas. The curves represent constant differences in distance to the MS with respect to the first FS, $\rho_{i,1} = d_i - d_1$.

24.3   Location Algorithms

When there are no measurement errors, the lines-of-position intersect at a point and a geometric solution can be obtained for the MS's location by finding the intersection of the lines of position. However, in practice, measurement errors occur and the lines of position do not intersect at a point. Consequently, other solution methods must be utilized. In the following, a solution approach is developed for two-dimensional location systems where the MS is located at $\boldsymbol{x}_s = [x_s, y_s]^T$ and the FSs are located at $\boldsymbol{x}_i = [x_i, y_i]^T$ for $i = 1, \ldots, N$.

24.3.1   Problem Formulation

In general, the $N \times 1$ vector of noisy measurements, $\boldsymbol{r}$, from a set of N FSs can be modeled by

$$\boldsymbol{r} = \boldsymbol{C}\left(\boldsymbol{x}_s\right) + \boldsymbol{n} \tag{24.3}$$

where $\boldsymbol{n}$ is an $N \times 1$ measurement noise vector, generally assumed to have zero mean and an $N \times N$ covariance matrix $\boldsymbol{\Sigma}$. The system measurement model $\boldsymbol{C}(\boldsymbol{x}_s)$ depends on the location method used:

$$C\left(\boldsymbol{x}_s\right) = \begin{cases} \boldsymbol{D}(\boldsymbol{x}_s) & \text{for ToA} \\ \boldsymbol{\mathcal{R}}\left(\boldsymbol{x}_s\right) & \text{for TDoA} \\ \boldsymbol{\Phi}\left(\boldsymbol{x}_s\right) & \text{for AoA} \end{cases} \tag{24.4}$$

where

$$\boldsymbol{D}\left(\boldsymbol{x}_s\right) = \left[\tau_1, \tau_2, \ldots, \tau_N\right]^T \tag{24.5}$$

$$\boldsymbol{\mathcal{R}}\left(\boldsymbol{x}_s\right) = \left[\rho_{2,1}, \rho_{3,1}, \ldots \rho_{N,1}\right]^T \tag{24.6}$$

$$\boldsymbol{\Phi}\left(\boldsymbol{x}_s\right) = \left[\phi_1, \phi_2, \ldots, \phi_N\right]^T . \tag{24.7}$$

The terms τ_i and $\rho_{i,1}$ are the ToAs and TDoAs defined in Eqs. (24.1) and (24.2), respectively, where without loss of generality, the TDoAs are referenced to the first FS. If the time of transmission τ_s is needed to form the ToA estimates, it can be incorporated into $\boldsymbol{x}_s$ as a parameter to be estimated along with x_s and y_s. The unknown parameter vector can then be modified to $\boldsymbol{x}_s = [x_s, y_s, \tau_s]^T$, while the system measurement model becomes $\boldsymbol{C}(\boldsymbol{x}_s) = \boldsymbol{D}(x_s, y_s) + \tau_s \boldsymbol{1}$.

The AoAs are defined by

$$\phi_i = \tan^{-1}\left(\frac{y_i - y_s}{x_i - x_s}\right) . \tag{24.8}$$

Although not explicitly shown in the above equations, τ_i, $\rho_{i,1}$ and ϕ_i are nonlinear functions of $\boldsymbol{x}_s$.

A well-known approach for determining an estimate from a noisy set of measurements is the method of **least squares** (LS) estimation. The weighted least squares (WLS) solution is formed as the vector $\hat{\boldsymbol{x}}_s$ that minimizes the cost function

$$\mathcal{E}\left(\hat{\boldsymbol{x}}_s\right) = \left[\boldsymbol{r} - \boldsymbol{C}\left(\hat{\boldsymbol{x}}_s\right)\right]^T \boldsymbol{W} \left[\boldsymbol{r} - \boldsymbol{C}\left(\hat{\boldsymbol{x}}_s\right)\right] . \tag{24.9}$$

LS methods can achieve the **maximum likelihood** (ML) estimate when the measurement noise vector is Gaussian with $\mathrm{E}[\boldsymbol{n}] = 0$ and equal variances, i.e., $\boldsymbol{\Sigma} = \sigma_n^2 \boldsymbol{I}$. For unequal variances, WLS with $\boldsymbol{W} = \boldsymbol{\Sigma}^{-1}$ gives the ML estimate. In the following, $\boldsymbol{W} = \boldsymbol{I}$ will be assumed.

24.3.2 Location Solutions

As Eq. (24.4) indicates, $\boldsymbol{C}(\boldsymbol{x}_s)$ is a nonlinear function of the unknown parameter vector $\boldsymbol{x}_s$ so that the LS problem is a nonlinear one. One straightforward approach is to iteratively search for the minimum of the function using a **gradient descent method**. With this approach, an initial guess is made of the MS location and successive estimates are updated according to

$$\hat{\boldsymbol{x}}_s^{(k+1)} = \hat{\boldsymbol{x}}_s^{(k)} - \boldsymbol{\nu}\, \nabla\mathcal{E}\left(\hat{\boldsymbol{x}}_s^{(k)}\right) \tag{24.10}$$

where the matrix $\boldsymbol{\nu} = \mathrm{diag}(\nu_x, \nu_y)$ is the step size, $\hat{\boldsymbol{x}}_s^{(k)}$ is the estimate at time k, and $\nabla = \partial/\partial\boldsymbol{x}$ denotes the gradient vector with respect to the vector $\boldsymbol{x}$.

In order to mold the problem into a linear LS problem, the nonlinear function $\boldsymbol{C}(\boldsymbol{x}_s)$ can be linearized by using a Taylor series expansion about some reference point $\boldsymbol{x}_0$ so that

$$C(\boldsymbol{x}_s) \approx C\left(\boldsymbol{x}_0\right) + \boldsymbol{H}\left(\boldsymbol{x}_s - \boldsymbol{x}_0\right) \tag{24.11}$$

where $\boldsymbol{H}$ is the Jacobian matrix of $\boldsymbol{C}(\boldsymbol{x}_s)$. Then the LS solution can be formed as

$$\hat{\boldsymbol{x}}_s = \boldsymbol{x}_0 + \left(\boldsymbol{H}^T\boldsymbol{H}\right)^{-1}\boldsymbol{H}^T\left[\boldsymbol{r} - \boldsymbol{C}\left(\boldsymbol{x}_0\right)\right] . \tag{24.12}$$

This approach can be performed iteratively, with each successive estimate being closer to the final estimate. A key drawback to this approach is that an initial guess, x_0, must be made of the MS's position.

The Taylor series approach introduces error when the linearized function $C(x_s)$ does not accurately approximate the nonlinear function. Other approaches have been developed for TDoA that avoid linearization by transforming the TDoA measurements into "pseudo-measurements." The pseudo-measurements are given by [5],

$$\varphi = \Delta x_s + D_1(x_s) r \tag{24.13}$$

where

$$\Delta = \begin{bmatrix} (x_2 - x_1)^T \\ \vdots \\ (x_N - x_1)^T \end{bmatrix} \qquad \varphi = \frac{1}{2} \begin{bmatrix} \|x_2\|^2 - \|x_1\|^2 - \rho_{2,1}^2 \\ \vdots \\ \|x_N\|^2 - \|x_1\|^2 - \rho_{N,1}^2 \end{bmatrix}. \tag{24.14}$$

The term $D_1(x_s)$ is nonlinear in the unknown vector x_s and can be removed by using a projection matrix that has r in its null space. A suggested projection is $P = (I - Z)[\text{diag}(r)]^{-1}$ where Z is a circular shift matrix [5]. Projecting (24.13) with P, the following linear equation results:

$$P\varphi = P\Delta x_s \tag{24.15}$$

which leads to the following linear LS solution for the location of the MS

$$\hat{x}_s = \left(\Delta^T P^T P \Delta \right)^{-1} \Delta^T P^T P\varphi . \tag{24.16}$$

24.4   Measures of Location Accuracy

To evaluate the performance of a location method, several benchmarks have been proposed. A common measure of accuracy is the comparison of the mean-squared-error (MSE) of the location estimate with the Cramér-Rao lower bound (CRLB) [10]. The concepts of circular error probability (CEP) [9] and geometric dilution of precision (GDOP) [6] have also been used as accuracy measures.

24.4.1   Cramér-Rao Lower Bound

For location in M dimensions, the MSE of the position estimate is given by

$$\text{MSE} = \sqrt{\text{E}\left[(x_s - \hat{x}_s)^T (x_s - \hat{x}_s) \right]} \tag{24.17}$$

where E[·] denotes expectation. The calculated MSE is often compared to the theoretical minimum MSE given by the CRLB which sets a lower bound on the variance of any unbiased estimator. The CRLB is the inverse of the information matrix J defined as [10]

$$J = \text{E}\left[\left(\frac{\partial p(r|x)}{\partial x} \right) \left(\frac{\partial p(r|x)}{\partial x} \right)^T \right]\Bigg|_{x=x_s} \tag{24.18}$$

where r is the vector of TDoA, ToA, or AoA estimates and $p(r|x)$ is the probability density function of r conditioned on the parameter vector x. Assuming Gaussian measurement noise, $p(\boldsymbol{r}|\boldsymbol{x})$ is Gaussian with mean $\boldsymbol{r}_0$ and covariance matrix $\boldsymbol{Q}$, and the CRLB reduces to

$$\text{CRLB} = \boldsymbol{J}^{-1} = c^2 \left(\frac{\partial \boldsymbol{r}_o^T}{\partial \boldsymbol{x}} \boldsymbol{Q}^{-1} \frac{\partial \boldsymbol{r}_0}{\partial \boldsymbol{x}^T} \right)^{-1} \Bigg|_{\boldsymbol{x} = \boldsymbol{x}_s} . \tag{24.19}$$

24.4.2 Circular Error Probability

A simple measure of accuracy is the CEP which is defined as the radius of the circle that has its center at the mean and contains half the realizations of a random vector. The CEP is a measure of the uncertainty in the location estimator $\hat{\boldsymbol{x}}_s$ relative to its mean $\text{E}[\,\hat{\boldsymbol{x}}_s\,]$. If the location estimator is unbiased, the CEP is a measure of the estimator uncertainty relative to the true MS position. If the magnitude of the bias vector is bounded by B, then with a probability of one-half, a particular estimate is within a distance of $B+$CEP from the true position.

Because it is difficult to derive an exact expression for the CEP, an approximation that is accurate to within 10% is often used. The approximation for CEP is given as [9]

$$\text{CEP} \approx 0.75 \sqrt{\text{E}\left[(\hat{\boldsymbol{x}}_s - \hat{\boldsymbol{\mu}})^T (\hat{\boldsymbol{x}}_s - \hat{\boldsymbol{\mu}}) \right]} \tag{24.20}$$

$$= 0.75 \sqrt{\sum_{i=1}^{M} \sigma_{\hat{x}_{s,i}}^2} \tag{24.21}$$

where $\hat{\boldsymbol{\mu}} = \text{E}[\hat{\boldsymbol{x}}_s]$ is the mean location estimate and $\sigma_{\hat{x}_{s,i}}^2$ is the variance of the ith estimated coordinate, $i = 1, \ldots, M$.

24.4.3 Geometric Dilution of Precision

The GDOP provides a measure of the effect of the geometric configuration of the FSs on the location estimate. It is defined as the ratio of the *rms* position error to the *rms* ranging error [9, 6]. Hence, for an unbiased estimator, the GDOP is given by

$$\text{GDOP} = \frac{\sqrt{\text{E}\left[(\hat{\boldsymbol{x}}_s - \hat{\boldsymbol{\mu}})^T (\hat{\boldsymbol{x}}_s - \hat{\boldsymbol{\mu}}) \right]}}{\sigma_r} \tag{24.22}$$

where σ_r denotes the fundamental ranging error for ToA and TDoA systems. For AoA, σ_r^2 is the average variance of the distance between each FS and a reference point near the true position of the MS.

The GDOP is an indicator of the extent to which the fundamental ranging error is magnified by the geometric relation between the MS and FSs. Furthermore, comparing (24.21) and (24.22), we find that the CEP and GDOP are related by

$$\text{CEP} \approx (0.75\sigma_r)\,\text{GDOP} . \tag{24.23}$$

The GDOP serves as a useful criterion for selecting the set of FSs from a large set to produce the minimum location error. In addition, it may aid cell site planning for cellular networks which plan to provide location services to their users.

24.5   Location in Cellular Systems

The location requirements set forth by the FCC [4] must be met not only by the new digital cellular systems, but the older analog system as well. In cellular networks, the BSs serve the role of the FSs in the algorithms of Section 24.3.1. With several different wireless systems on the market (**AMPS**, IS-54/136 **TDMA**, **GSM**, IS-95 **CDMA**), different methods may be necessary to implement location services in each of those systems. The signal strength method is often not implemented for cellular systems because of the large variability of received signal strength resulting from shadowing and **multipath fading** (see Section 24.6.1). AoA requires the placement of antenna arrays at the BSs which may be extremely costly. The AoA measurements can be obtained from array signal processing and is not dependent on the type of cellular system deployed. Unlike AoA, the ToA and TDoA methods require that timing information be obtained from the signals transmitted by a MS which may be implemented in different ways for each cellular system. The time-based methods may also require strict synchronization of the BSs, especially the TDoA approach. The remainder of this section discusses implementation strategies for the ToA and TDoA location methods in current and future generation cellular systems.

The most straightforward approach for obtaining timing information for ToA or TDoA location is the use of signal correlation methods. Specifically, maximizing cross-correlations between the signals received at pairs of BSs will provide an estimate of the TDoAs for each pair of BSs. Of course, this approach requires that the BSs be synchronized. These techniques are necessary for implementing a location system in AMPS since no system message parameters provide useful radiolocation information.

For CDMA, different methods can be used for the uplink (MS to BS) and downlink (BS to MS). On the uplink, the timing information for ToA or TDoA can be obtained using correlation techniques. Since the BSs in IS-95 are synchronized to a GPS time reference, the time of detection of the signal from the MS can serve as a ToA time stamp. Similarly, segments of the detected signal can be sent to a central processing office for cross-correlation in order to determine the set TDoAs for the BSs. The signals for the ToA/TDoA measurements can come from the reverse traffic channel or the access channel. The reverse traffic channel could be used for E-911 calls, for example, since a voice call must be initially made. For other location applications, the location may be desired when the MS is not actively transmitting. In these cases, the MS could be prompted to transmit messages on the access channel in response to commands from its serving BS on the paging channels. Unfortunately, it may be impossible to detect the MS transmissions at other BSs due to the **near–far effect** (see Section 24.6.3), although this problem can be alleviated by having the MS power-up to its maximum power for a short time. However, the use of the power up function must be limited to emergencies (such as E-911) in order to avoid excessive interference to other users.

An alternative for location in CDMA is to utilize pilot monitoring in the MS on the downlink. To assist in the handoff process, the MS monitors the strongest pilots from the surrounding BSs. The serving BS can send a pilot measurement request order (PMRO) causing the BS to respond with a message which includes the magnitudes of the pilots in the candidate set as well as the code phase of each pilot relative to its serving BS [2]. Hence, it is possible to construct TDoA estimates from these system messages. The accuracy of the TDoA estimates is dependent on the resolution of the code phase and the synchronization of the BSs. Fortunately, for IS-95, the BSs are synchronized to a GPS time reference. However, the code phase resolution is limited to a chip time, T_c, which implies a TDoA resolution of approximately 244 m. Finally, the **soft handoff**, during which the MS communicates with nearby BSs during a handoff, can be used for location in CDMA systems as long as at least three BSs are in a soft handoff with the MS.

The TDMA-based systems also provide timing information in their system messages that can be used for ToA or TDoA location. The time alignment parameter in IS-54/136 and timing advance in GSM (both abbreviated TA) are used by each of those networks to ensure that the transmissions of MSs arrive at their serving BSs in the appropriate time slots. Each BS sends the MSs a TA value which is the amount the MS must advance or retard the timing of its transmissions. Additionally, the TA serves as a measure of the propagation time between the MS and BS. By artificially forcing the MS to handoff to two or more BSs, the location of the MS could be found using the ToA method. A primary consideration is the accuracy of the TA. For IS-54, the timing of MS transmissions are advanced or retarded in units of $T_b/2$, where $T_b = 20.6\,\mu s$ is the bit duration [1]. Hence, the TAs are accurate to $T_b/4$, or 1543 m. For GSM, the TA messages are reported in units of bits, with $T_b = 3.7\,\mu s$, which gives a TA resolution of $T_b/2$, or 554 m, in GSM [3].

An alternative for GSM is to use the observed time difference (OTD) measurements which are made at the MS without forcing additional handoffs. The OTDs are used to facilitate handoffs by estimating the amount the timing of the MS would have to be advanced or retarded if it were to be handed over to another BS. With a synchronized network, the OTDs could be used to implement a TDoA location system. Unfortunately, the GSM standard does not require that the network be synchronized. Additionally, the OTD measurements are made to the same accuracy of the TA measurements, 554 m.

Because of the high chip rate and good correlation properties of the spreading code sequences used in CDMA systems, these systems have greater potential than the other systems for accurate location estimates. It is apparent that the resolution of the timing parameters in the system messages needs to be improved in order to provide more accurate estimates of location.

24.6   Sources of Location Error

In all cellular systems, several factors can introduce error in the location estimates. Sources of error that are common to all cellular systems include multipath propagation, non-line-of-sight (NLoS) propagation and multiple-access interference (MAI). However, MAI poses a more significant problem in CDMA systems because of **power control** and the near–far effect. These effects are described below.

24.6.1   Multipath

Multipath propagation can introduce error in signal strength, AoA, ToA, and TDoA measurements. For signal-strength-based location systems, multipath fading and shadowing cause variations in the signal strength that can be as great as 30–40 dB over distances in the order of a half wavelength. Signal strength averaging can help, but low mobility MSs may not be able to average out the effects of multipath fading and there will still be the variability due to shadowing. The errors due to shadowing can be combated by using pre-measured signal strength contours that are centered at the BSs. However, this approach assumes a constant physical topography since shadows will change with the tree foliage, construction/destruction of structures, etc.

For AoA-based systems, scattering near and around the MS and BS will affect the measured AoA. Multipath will interfere with the angle measurement even when a LoS component is present. For macrocells, scattering objects are primarily within a small distance of the MS and the BSs are usually elevated well above the local terrain. Consequently, the signals arrive with a relatively narrow AoA spread at the BSs. For microcells, the BSs may be placed below roof top level. Consequently, the BSs will often be surrounded by local scatterers such that the signals arrive at the BSs with a large AoA spread. Thus, the AoA method may be impractical for microcells.

In time-based location systems, the ToA or TDoA estimates can be in error even when there is a LoS path between the MS and BS. Conventional delay estimators, which are usually based on correlation techniques, are influenced by the presence of multipath fading. The result is a shift in the peak of the correlation away from the true value. Conventional delay estimators will detect a delay in the vicinity of these later arriving rays.

24.6.2 NLoS Propagation

With NLoS propagation, the signal transmitted from the MS (or BS) is reflected or diffracted and takes a path that is longer than the direct path or received at a different angle (Fig. 24.4). Obviously, the effect on an AoA system can be disastrous if the received AoAs are in a much different direction than the true AoAs. For the time-based systems, the measured distances can be considerably greater than true distances. For instance, for ToA location in the GSM system, the typical ranging error introduced by NLoS propagation can average 400–700 meters [7]. NLoS propagation will corrupt the ToA or TDoA measurements even when high resolution timing techniques are employed and even if there is no multipath fading.

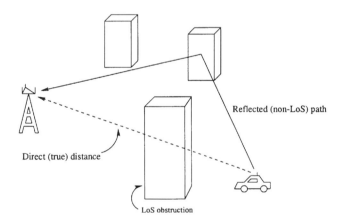

FIGURE 24.4   Propagation in a NLoS environment where signals are received from a reflection rather than a direct (LoS) path.

24.6.3 Multiple-Access Interference

All cellular systems suffer from cochannel interference. The transmissions of other users interfere with the signal of the desired user reducing the accuracy with which location measurements can be made. The problem is most evident in CDMA systems where the users share the same frequency band. As a result, signals from higher-powered users may mask the signals of the lower-powered users, a phenomenon known as the near–far effect. To combat the near–far effect, power control is used. However, for a location system where multiple BSs must receive the transmission from the MS, the near–far problem still exists because the MS is not power-controlled to the other BSs. Consequently, the signal from the MS may not be detected at enough BSs to form a location estimate. As mentioned in Section 24.5, it may be possible, for instance in E-911 situations, for the MS to power up to maximum level and, therefore, mitigate the near–far effect. A further possibility is to take advantage of soft handoffs. However, the MS must be in position for a three-way soft handoff to be located.

24.7 Summary and Conclusions

This chapter has provided a brief introduction to radiolocation techniques for wireless systems. Several algorithms were developed for locating a MS using AoA, ToA, and TDoA, and measures of locator accuracy were described. For location services in mobile cellular networks, many possibilities exist. However, it is apparent that none has the current capability of providing high accuracy location estimates to meet the FCC requirements.

Defining Terms

AMPS (advanced mobile phone service): Analog cellular system in North America.

CDMA (code-division multiple-access): A technique for spread-spectrum multiple-access digital communications that separates users through the use of unique code sequences.

Gradient descent method: A minimization technique which searches for the minimum of an error surface by taking steps along the direction of greatest slope.

GSM (global system for mobile communications): Pan-European digital cellular standard.

Least squares estimation: A method whose estimate is chosen as the value that minimizes the sum of squares of the measured error.

Maximum likelihood estimation: A method whose estimate is chosen as the parameter value from which the observed data was most likely to come.

Multipath fading: Rapid fluctuation of the complex envelope of the received signal caused by reception of multiple copies of the transmitted signal, each with different amplitude, phase, and delay.

Near–far effect: A phenomenon that arises from unequal received power levels from the MSs. Stronger signals mask the weaker signals.

Path loss: Description of the attenuation of signal power with distance from a transmitter.

Power control: System for controlling the transmission power of the MS. Used to reduce cochannel interference and mitigate the near–far effect on the uplink.

Shadowing: Slow variation in the mean envelope over a distance corresponding to tens of wavelengths.

Soft handoff: Reception and transmission of radio signals between an MS and two or more BSs to achieve a macrodiversity gain.

TDMA (time-division multiple access): A form of multiple access giving each user a different time slot for transmission and reception of signals.

References

[1] EIA/TIA Interim Standard, IS-54., Cellular System Dual Mode Mobile Station–Land Station Compatibility Specifications, May 1990.

[2] EIA/TIA Interim Standard IS-95., Mobile Station–Base Station Compatibility Standard for Dual-Mode Wideband Spread Spectrum Cellular System, May 1992.

[3] ETSI-SMG., GSM 05.10 Technical Specification, Version 4.0.1, Apr. 1992.

[4] FCC Docket No.96-254., Report and Order and Further Notice of Proposed Rulemaking in the Matter of Revision of the Commission's Rules to Ensure Compatibility with Enhanced 911 Emergency Calling Systems, Jun. 12, 1996.

[5] Friedlander, B., A passive location algorithm and its accuracy analysis, *IEEE Journal of Oceanic Engineering,* 234–244, Jan. 1987.

[6] Massatt, P. and Rudnick, K., Geometric formulas for dilution of precision calculations, *Journal of the Institute of Navigation,* 37, 379–391, Winter, 1991.

[7] Silventoinen, M. and Rantalainen, T., Mobile station emergency locating in GSM, in *Proceedings of the IEEE International Conference on Personal Wireless Communications,* 232–238, 1996.

[8] Stilp, L., Carrier and end-user applications for wireless location systems. *Proc. SPIE,* 119–126, 1996.

[9] Torrieri, D.J., Statistical theory of passive location systems, *IEEE Trans. on Aerospace and Electronic Systems,* AES-20, 183–197, Mar. 1984.

[10] Van Trees, H.L., *Detection, Estimation and Modulation Theory, Part I,* John Wiley & Sons, New York, Ch. 2, 1968.

Further Information

More information on radiolocation can be found in the April, 1998, issue of *IEEE Communications Magazine,* which provided several articles regarding location service issues for wireless communications networks.

Articles discussing location techniques and algorithms can be found in many IEEE journals including *Transactions on Vehicular Technology, Transactions on Aerospace and Electronic Systems,* and the *Journal of Oceanic Engineering.*

Proceedings of various IEEE conferences such as the *Vehicular Technology Conference* document some of the latest developments for location in cellular networks.

25

Power Control

Roman Pichna
Nokia Telecommunications

Qiang Wang
Nikko Research Center

25.1 Introduction

The growing demand for mobile communications is pushing the technological barriers of wireless communications. The available spectrum is becoming crowded and the old analog **FDMA** (frequency division multiple access) cellular systems no longer meet the growing demand for new services, higher quality, and spectral efficiency. A second generation of digital cellular mobile communication systems are being deployed all around the world. The second generation systems are represented by three major standards: the **GSM**, IS-136, and IS-95. The first two are TDMA-based digital cellular systems and offer a significant increase in spectral efficiency and quality of service as compared to the first generation systems, e.g., **AMPS, NMT,** and **TACS.** IS-95 is based on **DS/CDMA** technology. The standardization of the third generation systems, IMT-2000, (formerly known as FPLMTS) is being pursued at ITU. Similar efforts are being conducted at regional standardization bodies.

The channel capacity of any cellular system is significantly influenced by the cochannel interference. To minimize the cochannel interference, several techniques are proposed: frequency reuse patterns, which ensure that the same frequencies are not used in adjacent cells; efficient **power control,** which minimizes the transmitted power; cochannel interference cancellation techniques; and orthogonal signalling (time, frequency, or code). All of these are being intensively researched, and some have already been implemented.

This chapter provides a short overview of power control. Since power control is a very broad topic, it is not possible to exhaustively cover all facets associated with power control. The interested reader can find additional information in the recommended reading that is appended at the end of this chapter.

The following section (Section 25.2) provides a brief introduction into cellular networks and demonstrates the necessity of power control. The various types of power control are presented in this section. The next section (Section 25.3) illustrates some applications of power control employed in various systems such as analog AMPS, GSM, DS/CDMA cellular standard IS-95, and digital cordless telephone standard **CT2**. A glossary of definitions is provided at the end of the chapter.

25.2   Cellular Systems and Power Control

In cellular communication systems, the service area is divided into cells, each covered by a single base station. If, in the **forward link** (base station to mobile), all users served by all base stations share the same frequency, each communication between a base station and a particular user would also reach all other users in the form of cochannel interference. However, the greater the distance between the mobile and the interfering transmitter, the weaker the interference becomes due to the propagation loss. To ensure a good quality of service throughout the cell, the received signal in the fringe area of the cell must be strong. Once the signal has crossed the boundary of a cell, however, it becomes interference and is required to be as weak as possible. Since this is difficult, the channel frequency is usually not reused in adjacent cells in most of the cellular systems. If the frequency is reused, the cochannel interference impairs the signal reception in the adjacent cell, and the quality of service severely degrades unless other measures are taken to mitigate the interference. Therefore, a typical reuse pattern reuses the frequency in every seventh cell (frequency reuse factor = 1/7). The only exception is for CDMA-based systems where the users are separated by codes, and the allocated frequency may be shared by all users in all cells.

Even if the frequency is reused in every seventh cell, there is still some cochannel interference arriving at the receiver. It is, therefore, very important to maintain a minimal transmitted level at the base station to keep the cochannel interference low, frequency reuse factor high, and therefore the capacity of the system and quality of service high.

The same principle applies in the **reverse link** (mobile to base station)—the power control maintains the minimum necessary transmitted power for reliable communication. Several additional benefits can be gained from this strategy. The lower transmitted power conserves the battery energy allowing the mobile terminal (the portable) to be lighter and stay on the air longer. Furthermore, recent concerns about health hazards caused by the portable's electromagnetic emissions are also alleviated.

In the reverse link, the power control also serves to alleviate the near–far effect. If all mobiles transmitted at the same power level, the signal from a near mobile would be received as the strongest. The difference between the received signal strength from the nearest and the farthest mobile can be in the range of 100 dB, which would cause saturation of the weaker signals' receivers or an excessive amount of adjacent channel interference. To avoid this, the transmitted power at the mobile must be adjusted inversely proportional to the effective distance from the base station. The term effective distance is used since a closely located user in a propagation shadow or in a deep fade may have a weaker signal than a more distant user having excellent propagation conditions.

In a DS/CDMA system, power control is a vital necessity for system operation. The capacity of a DS/CDMA cellular system is interference limited since the channels are separated neither in frequency nor in time, and the cochannel interference is inherently strong. A single user exceeding the limit on transmitted power could inhibit the communication of all other users.

The power control systems have to compensate not only for signal strength variations due to the varying distance between base station and mobile but must also attempt to compensate for signal strength fluctuations typical of a wireless channel. These fluctuations are due to the changing propagation environment between the base station and the user as the user moves across the cell or as some elements in the cell move. There are two main groups of channel fluctuations: slow (i.e., **shadowing**) and fast **fading.**

As the user moves away from the base station, the received signal becomes weaker because of the growing propagation attenuation with the distance. As the mobile moves in uneven terrain, it often travels into a propagation shadow behind a building or a hill or other obstacle much larger than the wavelength of the frequency of the wireless channel. This phenomenon is called shadowing. Shadowing in a land-mobile channel is usually described as a stochastic process having log-normal distributed amplitude. For other types of channels other distributions are used, e.g., Nakagami.

Electromagnetic waves transmitted from the transmitter may follow multiple paths on the way from the transmitter to the receiver. The different paths have different delays and interfere at the antenna of the receiver. If two paths have the same propagation attenuation and their delay differs in an odd number of half-wavelengths (half-periods), the two waves may cancel each other at the antenna completely. If the delay is an even multiple of the half-wavelengths (half-periods), the two waves may constructively add, resulting in a signal of double amplitude. In all other cases (nonequal gains, delays not a multiple of half-wavelength), the resultant signal at the antenna of the receiver is between the two mentioned limiting cases. This fluctuation of the channel gain is called fading. Since the scattering and reflecting surfaces in the service area are randomly distributed (buildings, trees, furniture, walls, etc.), the amplitude of the resulting signal is also a random variable. The amplitude of fading is usually described by a Rayleigh, Rice, or Nakagami distributed random variable.

Since the mobile terminal may move at the velocity of a moving car or even of a fast train, the rate of channel fluctuations may be quite high and the power control has to react very quickly in order to compensate for it.

The performance of the **reverse link** of DS/CDMA systems is most affected by the near–far effect and, therefore, very sophisticated power control systems in the reverse link that attempt to alleviate the effects of channel fluctuations must be used. Together with other techniques, such as micro- and macrodiversity, interleaving and coding, interference cancellation, multiuser detection, and adaptive antennae, the DS/CDMA cellular system is able to cope with the wireless channel extremely well.

The effective use of the **power control** in DS/CDMA cellular system enables the frequency to be reused in every cell, which in turn enables features such as the soft hand-off and base station diversity. All together, these help enhance the capacity of the system.

In the **forward link** of a DS/CDMA system, power control may also be used. It may vary the transmitted power to the mobile, but the dynamic range is smaller due to the shared spectrum and, thus, shared interference.

We can distinguish between two kinds of power control, the open-loop power control and the closed-loop power control. The open-loop power control estimates the channel and adjusts the transmitted power accordingly but does not attempt to obtain feedback information on its effectiveness. Obviously, the open-loop power control is not very accurate, but since it does not have to wait for the feedback information it may be relatively fast. This can be advantageous in the case of a sudden channel fluctuation, such as a mobile driving from behind a big building or in case it should provide only the initial or rough transmitted power setting.

The principle operation of open-loop power control is shown in Fig. 25.1. The open-loop power control must base its action on the estimation of the channel state. In the reverse link it estimates the channel by measuring the received power level of the pilot from the base station in the forward link and sets the transmitted power level inversely proportional to it. Estimating the power of pilot is, in general, more reliable than estimating the power of the voice (or data) channel since the pilot is usually transmitted at higher power levels. Using the estimated value for setting the transmitted power ensures that the average power level received from the mobile at the base station remains constant irrespective of the channel variations. However, this approach assumes that the forward and the reverse link signal strengths are closely correlated. Although forward and reverse link may not share the same frequency and, therefore, the fading is significantly different, the long-term channel fluctuations due to shadowing and propagation loss are basically the same.

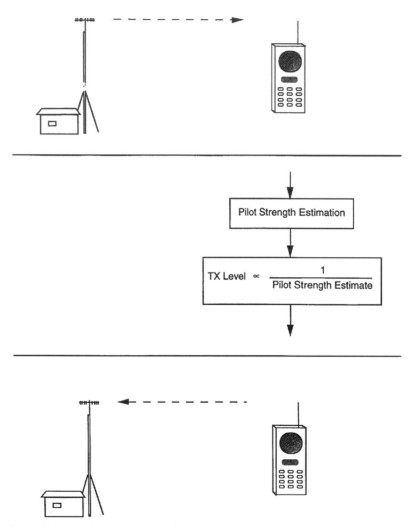

FIGURE 25.1   Reverse link open-loop power control.

The closed-loop power control system (Fig. 25.2a) may base its decision on an actual communication link performance metric, e.g., received signal power level, received signal-to-noise ratio, received bit-error rate, or received frame-error rate, or a combination of them. In the case of the reverse link power control, this metric may be forwarded to the mobile as a base for an autonomous power control decision, or the metric may be evaluated at the base station and only a power control adjustment command is transmitted to the mobile. If the reverse link power control decision is made at the base station, it may be based on the additional knowledge of the particular mobile's performance and/or a group of mobiles' performance (such as mobiles in a sector, cell, or even in a cluster of cells). If the power control decision for a particular mobile is made at the base station or at the switching office for all mobiles and is based on the knowledge of all other mobile's performance, it is called a centralized power control system. A centralized power control system may be more accurate than a distributed power control system, but it is much more complex in design, more costly, and technologically challenging.

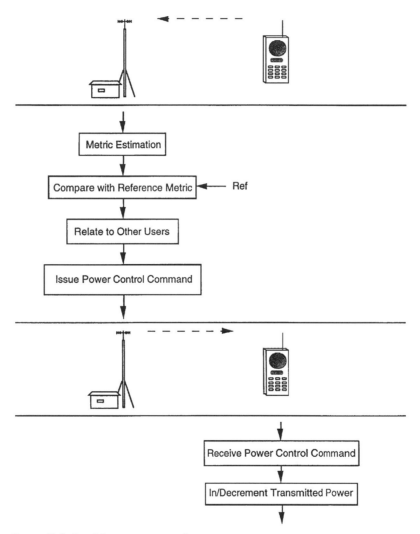

Figure 25.2a Reverse link closed-loop power control.

In principle, the same categorization may be used for the power control in the forward link (Fig. 25.2b) except that in the reverse link pilots from the mobiles are usually unavailable and only closed-loop power control is applied.

A special case for the design of power control are **TDD**-based systems [3]. In TDD systems, the forward and reverse link are highly correlated, and therefore a very good estimate of the channel gain in the forward link can be obtained from the estimate of

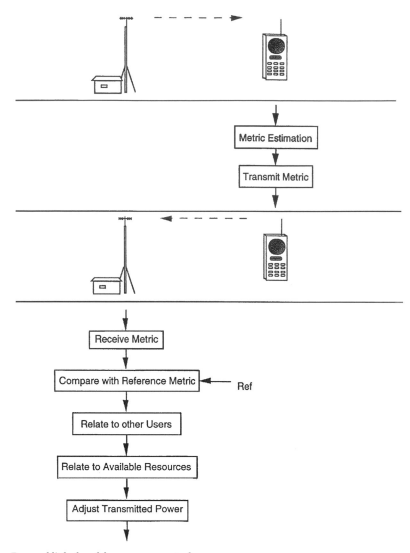

Figure 25.2b   Forward link closed-loop power control.

the reverse link gain and vice versa. An open-loop power control then performs with the precision of a closed-loop power control but much faster since no feedback information has to be transmitted.

In the ideal case, power control compensates for the propagation loss, shadowing, and fast fading. However, there are many effects that prevent the power control from becoming ideal. Fast fading rate, finite delays of the power control system, nonideal channel estimation, error in the power control command transmission, limited dynamic range, etc., all contribute to degrading the performance of the power control system. It is very important to examine the performance of power control under nonideal conditions since the research done has shown that the power control system is quite sensitive to some of these conditions [11]. Kudoh [5] simulated a nonideal closed-loop power control system. Errors in the system were represented by a log-normal distributed control error with standard deviation σ_E (dB). Some results on capacity reduction are presented in Table 25.1.

TABLE 25.1   Capacity Reduction versus Power Control Error

	$\sigma_E = 0.5$ dB, %	$\sigma_E = 1$ dB, %	$\sigma_E = 2$ dB, %	$\sigma_E = 3$ dB, %
Forward link	10	29	64	83
Reverse link	10	31	61	81

Source: Kudoh, E., On the capacity of DS/CDMA cellular mobile radios under imperfect transmitter power control. *IEICE Trans. Commun.*, E76-B, 886–893, Apr. 1993.

The authors have also studied the effects of Doppler and delay and feedback errors in power control loop on power control [8].

25.3   Power Control Examples

In the following section, several applications of power control of analog and digital cellular systems are presented.

In the analog networks we may see power control implemented in both the reverse link and forward link [6]. Power control in the reverse link:

- reduces the chance of receiver saturation by a closely located mobile
- reduces the cochannel interference and thus increases the frequency-reuse factor and capacity, and
- reduces the average transmitted power at the mobile thus conserving battery energy at the mobile.

The power control in the forward link:

- reduces cochannel interference and thus increases the frequency reuse factor and capacity and
- reduces adjacent-channel interference and improves the quality of service.

One example of a power control system shown by Lee [7] was of an air-to-ground communication system. The relevant airspace is divided into six zones based on the aircraft altitude. The transmitted power at the aircraft is then varied in six steps based on the zone in which the aircraft is located. The power control system exhibits a total of approximately 28 dB of dynamic range. This reduces the cochannel interference and, due to the excellent propagation conditions in the free air, has a significant effect on the capacity of the system.

Another example of a power control system in an analog wireless network is in the analog part of the TIA standard IS-95 [10]. IS-95 standardizes a dual-mode FDMA/CDMA cellular system compatible with the present day AMPS analog FDMA cellular system.

The analog part of IS-95 divides the mobiles into three classes according to nominal **ERP** (effective radiated power with respect to half-wave dipole) at the mobile. For each class, the standard specifies eight power levels. Based on the propagation conditions, the mobile station may receive a power control command that specifies at what power level the mobile should transmit. The maximum change is 4 dB per step. (See Table 25.2).

IS-95 supports further discontinuous transmission. This feature allows the mobile to vary its transmitted power between two states: low and high. These two states must be at least 8 dB apart.

As for the power control in a digital wireless system, three examples will be shown: GSM [1], CT2/CT2PLUS standard [2] for digital cordless telephones of second generation, and the IS-95 standard for digital cellular DS/CDMA system [10].

TABLE 25.2   Nominal ERP of the Mobile

Power level	Nominal ERP (dBW) of mobile		
	I	II	III
0	6	2	−2
1	2	2	−2
2	−2	−2	−2
3	−6	−6	−6
4	−10	−10	−10
5	−14	−14	−14
6	−18	−18	−18
7	−22	−22	−22

Source: Telecommunications Industry Association/Electronic Industries Association. Mobile Station-Base Station Compatibility Standard for Dual-Mode Wideband Spread Spectrum Cellular System, *TIA/EIA/IS-95 Interim Standard,* Jul. 1993.

GSM is a Pan-European digital cellular system that was introduced in many countries during the 1992–1993 period. GSM is a digital TDMA system with a frequency hopping feature. The power control in GSM ensures that the mobile station uses only the minimum power level necessary for reliable communication with the base station. GSM defines eight classes of base stations and five classes of mobiles according to their power output, as shown in Table 25.3.

TABLE 25.3   GSM Transmitter Classes

Power Class	Base Station Power (W)	Mobile Station power (W)
1	320	20
2	160	8
3	80	5
4	40	2
5	20	0.8
6	10	
7	5	
8	2.5	

Source: Balston, D.M. and Macario, R.C.V., *Cellular Radio Systems,* Artech House, Norwood, MA, 1993.

The transmitted power at the base station is controlled, nominally in 2-dB steps. The adjustment of the transmitted power reduces the intercell interference and, thus, increases the frequency reuse factor and capacity. The transmitted power at the base station may be decremented to a minimum of 13 dBm.

The power control of the mobile station is a closed-loop system controlled from the base station. The power control at the mobile sets the transmitted power to one of 15 transmission power levels spaced by 2 dB. Any change can be made only in steps of 2 dB during each time slot. Another task for the power control in GSM is to control graceful ramp-on and ramp-off of the TDMA bursts since too steep slopes would cause spurious frequency emissions.

The dynamic range of the received signal at the base station may be up to 116 dB [1] and, thus, the near–far problem may also by experienced, especially if the problem occurs in adjacent time slots. In addition to power control, a careful assignment of adjacent slots can also alleviate the near–far effect.

The CT2PLUS standard [2] is a Canadian enhancement of the **ETSI** CT2 standard. Both these standards allow power control in the forward and in the reverse link. Due to the expected small cell radius and relatively slow signal level fluctuation rate, given by the fact that the user of the portable is a pedestrian, the power control specifications are relatively simple. The transmission at the portable can have two levels: *normal* (full) and *low*. The low–normal difference is up to 20 dB.

The IS-95 standard represents a second generation digital wireless cellular system using direct-sequence code division multiple access (DS/CDMA). Since in a DS/CDMA system all users have the same frequency allocation, the cochannel interference is crucial for the performance of the system [4]. The near–far effect may cause the received signal level to change up to 100 dB [12]. This considerable dynamic range is disastrous for a DS/CDMA where the channels are separated by a finite correlation between spreading sequences. This is further aggravated by the shadowing and the fading. The fading may have a relatively high rate since the mobile terminal is expected to move at the speed of a car. Therefore, the power control system must be very sophisticated. Power control is employed in both the reverse link and in the forward link.

The reverse link power control serves to do the following:

- Equalizes the received power level from all mobiles at the base station. This function is vital for system operation. The better the power control performs, the more it reduces the cochannel interference and, thus, increases the capacity. The power control compensates for the near–far effect, shadowing, and partially for slow fading.

- Minimizes the necessary transmission power level to achieve good quality of service. This reduces the cochannel interference, which increases the system capacity and alleviates health concerns. In addition, it saves the battery power. Viterbi [12] has shown up to 20–30 dB average power reduction compared to the AMPS mobile user as measured in field trials.

The forward link power control serves to:

- Equalize the system performance over the service area (good quality signal coverage of the worst-case areas),

- Provide load shedding between unequally loaded cells in the service areas (e.g., along a busy highway) by controlling the intercell interference to the heavy loaded cells, and

- Minimize the necessary transmission power level to achieve good quality of service. This reduces the cochannel interference in other cells, which increases the system capacity and alleviates health concerns in the area around the base station.

The reverse link power control system is composed of two subsystems: the closed-loop and the open-loop. The system operates as follows. Prior to the application to access, closed-loop power control is inactive. The mobile estimates the mean received power of the received pilot from the base station and the open-loop power control estimates the mean output power at the access channel [10]. The system then sets the closed-loop probing and estimates the mean output power,

$$
\begin{aligned}
\text{mean output power (dBm)} \quad = \quad & - \text{ mean input power (dBm)} \\
& - 73 \\
& + \text{ NOM\_PWR (dB)} \\
& + \text{ INIT\_PWR (dB)}
\end{aligned}
\tag{25.1}
$$

where NOM_PWR and INIT_PWR are parameters obtained by the mobile prior to transmission. Subsequent probes are sent at increased power levels in steps until a response is obtained. The initial transmission on the reverse traffic channel is estimated as

$$
\begin{aligned}
\text{mean output power (dBm)} \quad = \quad & - \text{ mean input power (dBm)} \\
& - 73 \\
& + \text{ NOM\_PWR (dB)} \\
& + \text{ INIT\_PWR (dB)} \\
& + \text{ the sum of all access probe} \\
& \quad \text{corrections (dB)}
\end{aligned}
\tag{25.2}
$$

Once the first closed-loop power control bit is received the mean output power is estimated as

$$
\begin{aligned}
\text{mean output power (dBm)} \quad = \quad & - \text{ mean input power (dBm)} \\
& - 73 \\
& + \text{ NOM\_PWR (dB)} \\
& + \text{ INIT\_PWR (dB)} \\
& + \text{ the sum of all access probe corrections (dB)} \\
& + \text{ the sum of all closed-loop power control} \\
& \quad \text{corrections (dB)}
\end{aligned}
\tag{25.3}
$$

The ranges of the parameters NOM_PWR and INIT_PWR are shown in Table 25.4.

TABLE 25.4   NOM_PWR and INIT_PWR
Parameters

	Nominal value, dB	Range, dB
NOM_PWR	0	−8–7
INIT_PWR	0	−16–15

Source: Telecommunications Industry Association/Electronic Industries Association. Mobile Station-Base Station Compatibility Standard for Dual-Mode Wideband Spread Spectrum Cellular System. *TIA/EIA/IS-95 Interim Standard*, Jul. 1993.

The closed-loop power control command arrives at the mobile every 1.25 ms (i.e., 800 b/s). Therefore, the base station estimates the received power level for approximately 1.25 ms. A closed-loop power control command can have only two values: 0 to increase the power level and 1 to decrease the power level. The mobile must respond to the power control command by setting the required transmitted power level within 500 μs. The total range of the closed-loop power control system is ± 24 dB. The total supported range of power control (closed-loop and open-loop) must be at least ± 32 dB.

The behavior of the closed-loop power control system while the mobile receives base station diversity transmissions is straightforward. If all diversity transmitting base stations request the mobile to increase the transmitted power (all power control commands are 0), the mobile increases the power level. If at least one base station requests the mobile to decrease its power, the mobile decreases its power level.

The system also offers a feature of gated transmitted power for variable rate transmission mode. The gate-off state reduces the output power by at least 20 dB within 6 μs. This reduces the interference to the other users at the expense of transmitted bit rate. This feature may be used together with variable rate voice encoder or voice activated keying of the transmission.

The forward link power control works as follows. The mobile monitors the errors in the frames arriving from the base station. It reports the frame-error rate to the base station periodically. (Another mode of operation may report the error rate only if the error rate exceeds a preset threshold.) The base station evaluates the received frame-error rate reports and slightly adjusts its transmitting power. In this way, the base station may equalize the performance of the forward links in the cell or sector.

A system conforming with the standard has been field tested, and the results show that the power control is able to combat the channel fluctuation (together with other techniques such as RAKE reception) and achieve the bit energy to interference power density (E_b/I_0) necessary for a reliable service [12].

Power control together with soft handoff determines the feasibility of the DS/CDMA cellular system and is crucial to its performance. QUALCOMM, Inc. has shown on field trials that their system conforms with the theoretical predictions and surpasses the capacity of other currently proposed cellular systems [12].

25.4 Summary

We have shown the basic principles of power control in wireless cellular networks and have presented some examples of power control systems employed in some networks.

In a wireless channel the channel transmission or channel gain is a random variable. If all transmitters in the system transmitted at equal and constant power levels, the received powers would be random.

In the reverse link (mobile to base station) each user has its own wireless channel, generally uncorrelated with all other users. The received signals at the base station are independent and random. Furthermore, since the users are randomly distributed over the cell, the distance between the mobiles and the base station may vary and so does the propagation loss. The differences between the strongest and the weakest received signal level may approach the order of 100 dB. This power level difference may cause saturation of the receivers at the base station even if they are allocated a different frequency or time slot. This phenomenon is called the near–far effect.

The near–far effect is especially detrimental for a DS/CDMA system where the frequency band is shared by all users and, for any given user, all other users' transmissions form the cochannel interference. Therefore, for the DS/CDMA system it is vitally important to efficiently mitigate the near–far effect.

The most natural way to mitigate the near–far effect is to power control the transmission in such a way that the transmitted power counterfollows the channel fluctuations and compensates for them. Then the received signal at the base station arrives at a constant amplitude.

The use of power control is not limited to the reverse link, but is also employed in the forward link. The controlled transmission maintaining the transmitted level at the minimum acceptable level reduces the cochannel interference, which translates into an increased capacity of the system.

Since the DS/CDMA systems are most vulnerable to the near–far effect they have a very sophisticated power control system. In giving examples, we have concentrated on the DS/CDMA cellular system. We have also shown the power control used in other systems such as GSM, AMPS, and CT2.

Although there are more techniques available for mitigation of the near–far effect, power control is the most efficacious. As such, power control forms the core in the effort in combatting the near–far effect and channel fluctuations in general [12].

Defining Terms

AMPS: Advanced Mobile Phone Service. Analog cellular system in North America.

CT2: Cordless Telephone, Second Generation. A digital FDMA/TDD system.

DS/CDMA: Direct Sequence Code Division Multiple Access.

DOC: Department of Communications.

ERP: Effective Radiated Power.

ETSI: European Telecommunications Standard Institute.

Fading: Fast varying fluctuations of the wireless channel mainly due to the interference of time-delayed multipaths.

FDMA: Frequency Division Multiple Access.

Forward link: Link from the base (fixed) station to the mobile (user, portable).

GSM: Groupe Spéciale Mobile, recently referred to as the **Global System for Mobility**. An ETSI standard for digital cellular and microcellular systems.

NMT: Nordic Mobile Telephone. A cellular telephony standard used mainly in Northern Europe.

Power control: Control system for controlling the transmission power. Used to reduce the cochannel interference and mitigate the near–far effect in the reverse link.

Reverse link: Link from the mobile (user, portable) to the base (fixed) station.

Shadowing: Slowly varying fluctuations of the wireless channel due mainly to the shades in propagation of electromagnetic waves. Often described by log-normal probability density function.

TACS: Total Access Communication System. An analogue cellular system used mainly in UK.

TDD: Time Division Duplex.

TDMA: Time Division Multiple Access.

References

[1] Balston, D.M. and Macario, R.C.V., *Cellular Radio Systems*, Artech House, Norwood, MA, 1993.

[2] Department of Communications. ETI Interim Standard # I-ETS 300 131, Annex 1, Issue 2, Attachment 1. In *CT2PLUS Class 2: Specification for the Canadian Common Air Interface for Digital Cordless Telephony, Including Public Access Services, RS-130*. Communications Canada, Ottawa, ON, 1993.

[3] Esmailzadeh, R., Nakagawa, M., and Sourour, E.A., Time-division duplex CDMA communications, *IEEE Personal Comm.*, 4(3), 51–56, Apr. 1997.

[4] Gilhousen, K.S., Jacobs, I.S., Padovani, R, Viterbi, A.J., Weaver, L.A., and Wheatley III, C.E., On the capacity of cellular CDMA system. *IEEE Trans. Veh. Tech.*, 40, 303–312, May 1991.

[5] Kudoh, E., On the capacity of DS/CDMA cellular mobile radios under imperfect transmitter power control. *IEICE Trans. Commun.*, E76-B, 886–893, Apr. 1993.

[6] Lee, W.C.L., *Mobile Cellular Telecommunications Systems*, McGraw-Hill, New York, 1989.

[7] Lee, W.C.L., *Mobile Communications Design Fundamentals*, 2nd ed., John Wiley & Sons, New York, 1993.

[8] Pichna, R., Kerr, R., Wang, Q., Bhargava, V.K., and Blake, I.F., *CDMA Cellular Network Analysis Software*. Final Rep. Ref. No. 36-001-2-3560/01-ST, prepared for Department of Communications, Communications Research Centre, Ottawa, ON, Mar. 1993.

[9] Simon, M.K., Omura, J.K., Scholtz, R.A., and Levitt, B.K., *Spread Spectrum Communication Handbook*, McGraw-Hill, New York, 1994.

[10] Telecommunications Industry Association/Electronic Industries Association. Mobile Station-Base Station Compatibility Standard for Dual-Mode Wideband Spread Spectrum Cellular System. *TIA/EIA/IS-95 Interim Standard*, Jul. 1993.

[11] Viterbi, A.J. and Zehavi, E., Performance of power-controlled wideband terrestrial digital communication. *IEEE Trans. Comm.*, 41, 559–569, Apr. 1993.

[12] Viterbi, A.J., The orthogonal-random waveform dichotomy for digital mobile personal communication. *IEEE Personal Comm.*, 1(1st qtr.), 18–24, 1994.

Further Information

For general information see the following overview books:

[1] Balston, D.M. and Macario, R.C.V., *Cellular Radio Systems*, Artech House, Norwood, MA, 1993.

[2] Simon, M.K., Omura, J.K., Scholtz, R.A., and Levitt, B.K., *Spread Spectrum Communication Handbook*, McGraw-Hill, New York, 1994.

For more details on power control in DS/CDMA systems consult the following:

[3] Gilhousen, K.S., Jacobs, I.S., Padovani, R., Viterbi, A.J., Weaver, L.A., and Wheatley C.E., III., On the capacity of cellular CDMA system. *IEEE Trans. Veh. Tech.*, 40, 303–312, May 1991. or:

[4] Viterbi, A.J. and Zehavi, E., Performance of power-controlled wideband terrestrial digital communication. *IEEE Trans. Comm.*, 41, 559–569, Apr. 1993.

Readers deeply interested in power control are recommended the *IEEE Transactions on Communications, IEEE Transactions on Vehicular Technology*, and relevant issues of *IEEE Journal on Selected Areas in Communications*.

26

Enhancements in Second Generation Systems

Marc Delprat
Alcatel Mobile Communication Division

Vinod Kumar
Alcatel Mobile Communication Division

26.1 Introduction

Present digital **cellular** and **cordless** systems were optimized for voice services. At the very best, they can provide low and medium bit rate information services. The development of enhanced versions of these radio interfaces is motivated by the following:

- Provision of additional services, including high bit rate circuit-switched and packet-switched services that meet the short-term needs of mobile multimedia services.
- Improvements in the radio coverage of existing cellular systems. An extension to allow cordless coverage in the home is also being considered.
- Even more efficient utilization of the available frequency spectrum, which is a valuable but limited resource.

This chapter deals mainly with the air interface of second generation systems. The subject of enhancement in system performance is addressed from two directions. On the one hand, system features like adaptive multirate coders, packet transmission, which have been explicitly included in the standard, are discussed in detail. On the other hand, methods of equipment design or network design possible with the new or already existing features of the air interface are presented.

26.2   Overview of Second Generation Systems

Initially the need of a pan-European system to replace a large variety of disparate analog cellular systems was the major motivating factor behind the creation of the Global System for Mobile communications (GSM). In North America and Japan, where unique analog systems existed, the need to standardize respectively IS-54, IS-95, and Personal Digital Cellular (PDC) for digital cellular applications arose from the lack of spectrum to serve the high traffic density areas [3]. Additionally, some of the second generation systems like Digital European Cordless Telecommunications (DECT) and Personal Handy Phone Systems (PHS) are the result of a need to offer wireless services in residential and office environments with low cost subscriber equipment [16].

The physical layer characteristics of all these systems offer robust radio links paired with good spectral efficiency. The network related functionalities have been designed to offer secure communication to authenticated users even when roaming between various networks based on the same system. Table 26.1 provides the essential characteristics of the second generation systems as initially designed in the late eighties and early nineties. Since then several additions have been made to those standards. Enhancements of air interface as well as network subsystem functionalities have been incorporated.

26.3   Capacity Enhancement

The **capacity** of a mobile network can be defined as the Erlangs throughput by a cell, a cluster of cells, or by a portion of a network. For a given radio interface, the achievable capacity is a function of the robustness of the physical layer, the effectiveness of the medium access control (MAC) layer and the multiple access technique. Moreover, it is strongly dependent on the radio spectrum available for network planning.
If we define:

> BW   Available radio spectrum
> Wc   Spectrum occupied by a single radio carrier
> Nc   Number of circuits handled by a single carrier
>   (e.g., number of time slots per frame in a TDMA system)
> Cs   Permissible cluster size which guarantees a good
>   quality for a vast majority of active calls (like 90 to 95%)

The number of circuits per cell is given by (BW / Wc) × (Nc / Cs).

The Erlang capacity can be derived from this expression after consideration of the signalling overhead required by the pilot channel, signalling and traffic channel overheads for handover, and the specified call blocking rate.

This definition of capacity is applicable in case of TDMA/FDMA air interfaces and when the networks are designed using "deterministic frequency allocation." A slightly different approach to the problem is necessary in systems like DECT which use dynamic channel selection and for DS-CDMA systems. IS-95 networks pretend to use a cluster size (Cs) of one. However, the number of good quality circuits in a cell is a function of the available noninterfered spreading codes.

Capacity enhancement can be obtained through:

TABLE 26.1 Air Interface Characteristics of Second Generation Systems

Standard	CELLULAR				CORDLESS	
	GSM	IS-54	IS-95	PDC	DECT	PHS
Frequency band (MHz)	Europe	USA	USA	Japan	Europe	Japan
Uplink	890–915 (1710–1785)	824–849	824–849	940–956 (1429–1441)	1880–1900	1895–1907
Downlink	935–960 (1805–1880)	869–894	869–894	810–826 (1477–1489)		
Duplex spacing (MHz)	45 (95)	45	45	130 (48)	—	—
Carrier spacing (kHz)	200	30	1250	25	1728	300
Number of radio channels in the frequency band	124 (374)	832	20	640 (480)	10	77
Multiple access	TDMA	TDMA	CDMA	TDMA	TDMA	TDMA
Duplex mode	FDD	FDD	FDD	FDD	TDD	TDD
Number of channels per carrier	8 (half rate: 16)	3 (half rate: 6)	128	3 (half rate: 6)	12	4
Modulation	GMSK	Π/4 DQPSK	QPSK BPSK	Π/4 DQPSK	GFSK	Π/4 DQPSK
Carrier bit rate (kb/s)	270.8	48.6	1288	42	1152	384
Speech coder (full rate)	RPE-LTP	VSELP	QCELP	VSELP	ADPCM	ADPCM
Net bit rate (kb/s)	13	7.95	(var.rate: 8, 4, 2, 0.8)	6.7	32	32
Channel coder for speech channels	1/2 rate convolutional + CRC	1/2 rate convolutional + CRC	1/2 (downlink), 1/3 (uplink) convolutional + CRC	1/2 rate convolutional + CRC	no	no
Gross bit rate speech+channel coding (kb/s)	22.8	13	var. rate 19.2, 9.6, 4.8, 2.4	11.2	—	—
Frame size (ms)	4.6	40	20	20	10	5
MS transmission power (W)	Peak 8 2 (1) / Aver. 1 0.25 (0.125)	Peak 9 4.8 1.8 / Aver. 3 1.6 0.6	0.6	Peak 2 / Aver. 0.66	Peak 0.25 / Aver. 0.01	Peak 0.08 / Aver. 0.01
Power control MS BS	Y Y	Y Y	Y Y	Y Y	N N	Y Y
Operational C/I(dB)	9	16	6	17	21	26
Equalizer	needed	needed	Rake receiver	option	option	no
Handover	Y	Y	Soft handoff	Y	Y	Y

- An increase in number of radio carriers and/or the number of traffic channels (e.g., voice circuits)
- Improved interference management techniques
- Novel cellular configurations.

26.3.1   Capacity Enhancement Through Increase in Number of Carriers and/or Voice Circuits

This family of relatively simple solutions can be subdivided in three categories related to the availability of spectrum in the same band, or in a different band where the same radio interface can be utilized or the situation in which the number of voice circuits can be increased by the introduction of multiple speech codecs.

Increased Spectrum Availability

In such a fortunate situation, extra carriers can be added to the already installed base stations (BS). The existing Cs is maintained and the increase in capacity is according to the added carriers/circuits. An additional gain in Erlang capacity is available due to increased trunking efficiency.

Supplementary Spectrum Availability (in Different Frequency Bands)

Most of the second generation systems were implemented in the 800 and 900 MHz frequency bands. Their application has been extended to the 1800 and 1900 MHz bands too (e.g., GSM 1800 and PCS1900). Carriers from both the bands can be used at the same BSs—either according to a common or independent frequency reuse scheme. Increase in traffic throughput can be maximized by using dual band mobile stations (MSs) and by providing adequate handover mechanisms between the carriers of two bands. Also, due to difference in propagation for carriers from two widely separated bands, certain adjustments related to coverage planning might be required when the carriers are co-sited.

Multiple Speech Codecs for Increased Number of Voice Circuits

Originally, only full rate (FR) speech codecs were used in second generation systems. Hence, a one-to-one correspondence between the physical channels on the air interface and the available voice circuits was established. If the number of installed carriers in a cell is kept unchanged, the introduction of half rate (HR) codecs will double the number of voice circuits in the cell and a more than two-fold increase in Erlang capacity in the cell will be achievable. A similar possibility is offered by the adaptive multirate codec (AMR). The output bit rates of speech codec and channel codec can be adapted in order to minimize the carrier occupancy time necessary to offer a predetermined call quality. The statistical multiplexing gain thus obtained can be exploited to enhance the traffic throughput.

Such capacity enhancement methods can be implemented only if corresponding MSs or MSs with multiple codec capabilities are commercialized. Moreover, every cell site in a network will have to maintain a certain number of FR voice circuits necessary to offer downwards compatibility. Table 26.2 provides an applicability matrix related to the above mentioned capacity enhancement methods.

TABLE 26.2 Applicability Matrix of Methods for Capacity Enhancements Based on Increased Spectrum Availability

Second Generation System	Method for Capacity Enhancement		
	Additional Spectrum in Same Band	Additional Spectrum in Another Band	Multiple Speech Codecs
GSM Family	Available	Under trial in real networks	Under trial in real networks
IS-54/IS-136	Available	Applicable	Applicable
PDC	Available	Applicable	Applicable
IS-95 Family using DS-CDMA	Some of these solutions are applicable. However, their implementation shall adversely effect the functioning of soft hand over which is an essential feature of DS-CDMA networks.		
DECT	Available	Under Consideration	Under Consideration
PHS	Available	Not Applicable	Available

26.3.2 Improved Interference Management Techniques

Initially, the design of a cellular network is based on some simplifying assumptions like uniformly distributed subscriber density and traffic patterns or homogeneous propagation conditions. Usually, such design results in a worst case value of Cs. The situation can be improved through better managemnt of cochannel interference by implementing one or a combination of the following:

- Slow Frequency Hopping (SFH)
- Voice Activity Detection (VAD) and Discontinuous Transmission (DTx)
- Transmit Power Control (PC)
- Antenna beam forming

Slow Frequency Hopping

Every call is "spread" over all the carrier frequencies available in the cell. To avoid intracell interference, orthogonal (random or pseudo random) frequency hopping laws are used for different calls. Also, the worst-case intercell interference situation can last only one hop and an averaging out of cochannel interference in the network occurs. Statistically, the distribution of carrier-to-cochannel interference ratio in the network is more compact with frequency hopping than without it, and this can be exploited to reduce the cluster size Cs and increase capacity.

Voice Activity Detection (VAD) and Discontinuous Transmission (DTx)

Collection of statistics related to telephone conversations has demonstrated that the duty cycle of voice sources is around 40%. With DTx, the radio signal is transmitted according to the activity of voice source. An overall reduction in the averaged cochannel interference experienced by the calls can thus be observed. This offers the possibility of implementing smaller Cs. Also, saving energy with DTx in the up-link (UL) results in a prolonged autonomy for the MS.

Transmit Power Control

Usually, the full available transmit power is necessary for the initial access and for a short duration after call establishment. For the rest of the call, both DL and UL transmit powers can be reduced to a level necessary to maintain a good link quality. The overall improvement in radio interference in the network thus obtained is helpful for the reduction of cluster size. Like VAD/DTx, the DL transmit power control is not permitted for pilot carriers if the mobile assisted handover is implemented in the system.

Antenna Beamforming

Interference related to every call can be individually managed by "dynamic cell sectorization." Actually, the base station receiver captures the up-link signal in a narrow antenna beam dynamically "placed" around the MS. Similarly, the down-link signal is transmitted in a beam focused towards the MS. This sort of spatial filtering of cochannel interference is useful for implementing very compact frequency reuse schemes. Generally, both the down-link and up-link beamforming capabilities are placed at the BS where an antenna array and signal processing algorithms related to direction finding, signal source separation, and beam synthesis have to be implemented. Table 26.3 provides an applicability matrix for the second generation systems.

TABLE 26.3 Applicability Matrix of Methods for Capacity Enhancements Based on Improved Interference Management

Second Generation System	Method for Capacity Enhancement			
	Slow Frequency Hopping	VAD/DTx	Power Control	Antenna Beam Forming
GSM Family	AAP	AAP	AAP	APP
IS-95 (DS-CDMA)	Not Applicable	Essential requirements for satisfactory system operation and not capacity enhancement features		APP
IS-54/IS-136	ANP	ANP	AAP	APP
PDC	ANP	ANP	AAP	APP
DECT	ANP	ANP	ANP	APP
PHS	ANP	ANP	ANP	APP

Note: AAP Applicable and already provided by the standard.
ANP Applicable but not explicitly provided by the standard.
APP Applicable depending on BS equipment design.

Capacity enhancements of 200% or more have been reported [10, 13] through the implementation of combinations of SFH, VAD/DTx, and PC in GSM networks. With SFH and antenna beam forming, Cs of three is achievable for GSM networks [1]. Since VAD/DTx and PC cannot be applied to the pilot carriers, most of the operational networks deploy a dual cluster scheme where Cs for pilot carriers is slightly higher than Cs for traffic carriers. Moreover, some other issues related to pilot channels/carriers have somewhat impeded the introduction of antenna beam forming in GSM or IS-95 (DS-CDMA) networks for cellular applications. However, in wireless local loop applications, substantial capacity gains have been reported for IS-95 [12].

26.3.3 Novel Cellular Configurations

The traffic capacity throughput by a regular grid of homogeneous cells can be increased by cell splitting. Theoretically speaking, if the cell size is divided by two by adding base stations in the middle of existing ones and the current frequency reuse scheme is maintained the achievable traffic capacity is multiplied by four. However, a reduction in cell size beyond a limit leads to excessive overlap between cells due to increased difficulty of coverage prediction. Moreover, since the dwelling time of MS in a cell is reduced, the average number of handovers per call increases. Conventional cellular organizations can no longer meet the requirements for good quality of service due to the failure of handover mechanisms. For high capacity coverage, the following novel cellular configurations have been suggested/implemented:

- Microcells and the associated hierarchical network organization
- Concentric cells
- Frequency reuse of one through generalized slow frequency hopping.

Microcells and Associated Hierarchical Network Organization

Hot spot coverage in dense urban areas is realized in the form of isolated islands of microcells. This is implemented through micro-BS antennas placed below the roof tops of the surrounding buildings. Each antenna radiates very low power. Each island is covered by an umbrella cell which is a part of a continuous macrocellular network over a wider area. Traffic throughput is optimized by intelligent spectrum management performed either off-line or on-line. A set of carriers is assigned to the macrocellular layer organized in a conventional manner. The remaining carriers are repeatedly used in the islands of microcells. Cell selection parameters in the MS and the call admission control algorithms (e.g., Forced Directed Retry) in the base station controllers are adjusted such that a maximum of traffic in the hot spot is taken by the microcells. The umbrella cells are dimensioned to handle the spill-over traffic. This can be the fast moving MS which could generate too many handovers if kept with the microcells. An MS which experiences a sudden degradation of link budget with respect to its serving microcell and/or in the absence of a good target microcell can be temporarily handed over to the umbrella cell.

Despite its difficulty related to spectrum management, this technique has proved to be quite popular for the densification of parts of existing networks based on second generation systems using TDMA/FDMA. [6, 11] provide analysis and guidelines for spectrum management for optimized efficiency in hierarchical networks.

Concentric Cells

A concentric cell coverage is implemented by splitting the available traffic carriers in two groups. One group transmits at full power required to cover the complete cell and the other group transmits at a lower level thus providing the coverage of an inner zone concentric with the original cell. The pilot carrier is transmitted at full power. The localized transmission in the inner zone creates a lower level of interference for other cells. Hence, a smaller cluster size can be used for the frequencies of the inner zone leading to traffic capacity enhancement. Call admission control is designed to keep the MSs near the BS on the carriers of inner zone. Simple intracell handover mechanisms are used to ensure call continuity for the MSs moving across the boundary of the inner zone. Analysis and simulations have shown that optimized capacity enhancement is achieved if the inner zone is limited to 40% of the total cell area. Field trials of concentric cell networks have demonstrated 35% higher spectrum efficiency as compared to single cell networks.

Reuse of One Through Generalized Frequency Hopping and Fractional Loading

Micro- or picocellular networks with TDMA/FDMA systems can be deployed with frequency "reuse of one." Every base station has the capability of using all the available carriers by slow frequency hopping. The allocation of frequency hopping laws for MSs in clusters of adjacent cells is managed by a centralized control.

During the steady state of operation in a loaded network, only a fraction of the total available bandwidth is in active use in every cell (fractional loading). The level of interference for active calls, for unused circuits and the availability of noninterfered frequency hopping laws is constantly monitored by the network. New calls in a cell are accepted according to the availability of interference-free circuits. In case of unevenly distributed traffic load, an average circuit occupancy in the network is maintained at a level necessary to keep the interference level for active calls below a predetermined threshold. Extreme situations where the same circuit and/or same frequency hopping law is used in two adjacent cells are remedied by intracell handover of one of the two calls. Very high capacity enhancement has been demonstrated in operational GSM networks by using this technique.

Table 26.4 provides the applicability matrix for novel cellular configurations.

TABLE 26.4   Applicability Matrix of Methods for Capacity Enhancements Based on Novel Cellular Configurations

Second Generation System	Method for Capacity Enhancement		
	Micro-cells and Hierarchical N/W	Concentric Cells	Reuse of One with GSFH
GSM Family	AAP	AAP	AAP
IS-54/IS-136	AAP	AAP	ANP
PDC	AAP	AAP	ANP
IS-95 Family using DS-CDMA	Not applicable due to the resulting near–far problem and the power control complexity		Inherent due to DS-CDMA
DECT	AAP	AAP	ANP. Reuse of one possible with DCS
PHS	AAP	AAP	ANP

Note: AAP Applicable and already provided by the standard.
ANP Applicable but not explicitly provided by the standard.
APP Applicable depending on BS equipment design.

26.4   Quality Enhancement

26.4.1   Quality Aspects and Definitions

The quality of service in a telecommunications network can simply be defined as the average performance perceived by the end user in setting up and maintaining a communication. However, its assessment is complex since it is influenced by many parameters, especially in digital wireless systems. In these systems the quality is primarily based on the end-to-end bit error rate and on the continuity of radio links between the two ends. Interference-free radio coverage with sufficient desired signal strength needs to be provided to achieve the above. Moreover, communication continuity has to be ensured between coverage areas with high traffic density (microcells) and low/medium traffic density (macrocells). Three main quality aspects will be distinguished in the following, namely the call handling quality, the communication quality, and the coverage quality.

For speech transmission, the communication quality strongly depends on the intrinsic performance of the speech coder, and its evaluation normally requires intensive listening tests. When it is comparable to the quality achieved on modern wire-line telephone networks, it is called "toll quality." But the **speech quality** is also influenced by other parameters linked to the communication characteristics like radio channel impairments (bit error rate), transmission delay, echo, background noise and tandeming (i.e., when several coding/decoding operations exist in the link).

For data transmission, the communication quality can be more easily quantified based on bit error rate and transmission delay. In synchronous circuit mode, delay is fixed and bit error rate depends on radio channel quality. In packet mode, bit error rate can be kept low thanks to retransmission mechanisms, but average delay increases and throughput decreases as the radio channel degrades.

The **coverage quality** is the percentage of the served area where a communication can be established. It is determined by the acceptable path loss of the radio link and by the propagation characteristics in the area. The radio link budget generally includes some margin depending on the type of terrain (for shadowing effects) and on operator's requirements (for indoor penetration). A coverage quality of 90% is a typical value for cellular networks.

The call handling quality mainly depends on the capacity of the mobile network. When the user is under coverage of the network, the call set-up performance is measured by the blocking rate which depends on the network load. Since in cellular networks mobility of the users is high, capacity requirements are less predictible than in fixed networks which results in a higher blocking rate (typically 1%). Requirements are even more stringent in cordless systems.

A last call-handling quality attribute specific to mobile networks is the success rate in maintaining the communication for mobile users. The handover procedure triggered in cellular networks when the user moves from one cell to another implies the establishment of a new link with some risk to lose the call. The performance is here given by the handover success rate. It must be noted that even if successful, a handover generally results in a short transmission break which degrades the communication quality.

In fact, capacity and quality are dual parameters and a compromise is needed when designing a mobile network. The call handling quality is strongly linked to the correct dimensioning of the network capacity. On the other hand, offering a high capacity in a mobile network implies an intensive use of the available radio spectrum and hence a high average interference level, which may in turn degrade the communication quality. In the following, some major enhancements in speech quality and coverage quality standardized or implemented in second generation systems are reviewed.

26.4.2 Speech Quality Enhancements

In digital cellular systems the first generation of speech coders (full rate) were standardized in the late 1980s to provide good communication quality at medium bit rate (6.7 to 13 kb/s). Operation at a fixed bit rate matched to the intended communication channel was the most important requirement in creating these standards. The main characteristics of the full rate speech coders standardized for second generation cellular systems are listed in Table 26.1 together with other air interface parameters.

With the recent advances in speech coding techniques, low delay toll quality at 8 kb/s and near-toll quality below 6 kb/s are now available [2]. This has enabled the standardization of half rate codecs in cellular systems, though with an increased complexity of implementation. In GSM/DCS, IS-54, and PDC this evolution was anticipated at an early stage, so the need for a half rate speech channel was taken into account in the design of the TDMA frame structure. Half rate codecs provide a doubling of the systems capacity (in terms of number of voice circuits per cell site) while maintaining a speech quality comparable to that available from related full rate codecs. A 5.6 kb/s VSELP coder for GSM and a 3.45 kb/s PSI-CELP coder for PDC were standardized in 1993. However, they have not been widely introduced up to now because of their slightly lower quality in some conditions (e.g., background noise, tandeming) compared to their full rate counterpart.

On the contrary, operators have pushed towards speech quality enhancements, and a new generation of speech coders has emerged. In 1996 ETSI standardized the GSM EFR (enhanced full rate) coder, adopting without any competitive selection process the US1 coder defined for PCS1900 in the U.S. The EFR coder has the remarkable feature of keeping the same channel coding as for the full rate channel, hence simplifying its implementation in the infrastructure. The ACELP coder (using an algebraic codebook with some benefits in computational efficiency like in the the 8 kb/s G.729 ITU standard) has a net bit rate of 12.2 kb/s which leaves room for additional protection (CRC).

Similarly the IS-641 coder has been standardized for IS-54 cellular systems. Its structure is very similar to that of G.729 but with a frame size of 20 ms and a bit rate of 7.4 kb/s. Also a 13 kb/s QCELP coder has been standardized for IS-95. These new coders provide near toll quality in ideal transmission conditions.

Concerning cordless systems, both DECT and PHS use 32 kb/s ADPCM and, therefore, provide toll quality in normal conditions thanks to their relatively high operational C/I. For these systems the concern is rather to introduce lower rate coders with equivalent quality in some kind of half rate mode, allowing for capacity increase, but such coders have not been standardized yet.

With the emergence of variable bit rate (VBR) techniques, some further developments in speech coding are now taking place to satisfy the requirements of cellular operators for toll quality speech with better robustness to radio channel impairments, combined with the capacity increase achievable with half rate operation. One VBR approach is to adapt the bit rate according to the source requirements, taking advantage of silence and stationary segments in the speech signal. Another VBR approach is to adapt the bit rate according to the radio channel quality, either by varying the bit rate allocation between speech and channel coding within a constant gross bit rate (for better quality) or by reducing the gross bit rate in good transmission conditions (for higher capacity). Adaptation to the source characteristics is fast (on a frame-by-frame basis) and adaptation to the channel quality is slower (maximum, a few times per second). In all cases, the underlying idea is that current coders are dimensioned for worst-case operation so that the system quality and capacity can be increased by exploiting the large variations over time of the bit rate requirements (for a given quality).

VBR capabilities with source driven adaptation were introduced since the beginning in IS-95. In CDMA systems the resulting average bit rate reduction directly translates into increased capacity. The initial IS-96B QCELP coder supports four bit rates (8, 4, 2, and 0.8 kb/s) with a scalable CELP architecture, and bit rate adjustment is performed based on adaptive energy thresholds. In practice the two extreme rates are used most frequently. The achievable speech quality was estimated to be lower than that of the IS-54 VSELP coder. Subsequently an enhanced variable rate coder (EVR) was standardized as IS-127. It is based on the relaxed CELP (RCELP) coding technique and supports three rates (8.5, 4, and 0.8 kbit/s).

In Europe, ETSI has launched in 1997 the standardization of an Adaptive Multirate coder (AMR) for GSM systems. Its output bit rate is continuously adapted to radio channel conditions and traffic load. The objective is to find the best compromise between speech quality and capacity by selecting an optimum combination of channel mode and codec mode. The AMR codec can operate in two channel modes, full rate and half rate. For each channel mode there are several possible codec modes (typically three) with different bit rate allocations between speech and channel coding. An adaptation algorithm tracks the variations in speech quality using specific metrics and decides upon the changes in codec mode (up to several times per second). Changes in codec mode is detected by the receiver either via in-band signalling or by automatic mode identification. As shown in Fig. 26.1, the multiplicity of codec modes gives significant performance improvement over any of the corresponding fixed rate codecs.

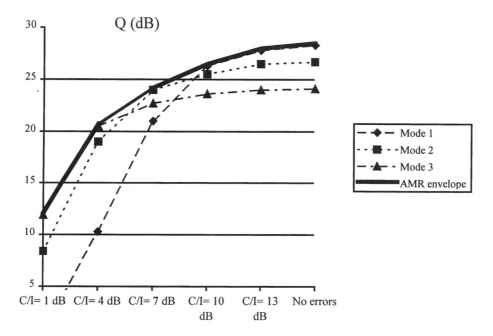

FIGURE 26.1 AMR codec quality as a function of C/I. (Typical results derived from the first AMR selection tests. Quality is given on the equivalent MNRU scale in dB).

AMR codec will also allow handovers between half rate and full rate channels using intracell handover mechanisms. The AMR codec will be selected by October 1998 and the full standard will be available in mid-1999. A potential extension is wideband coding, which could be added later to AMR as an option. The wideband option would extend the audio bandwidth from the current 300–3400 Hz to 50–5000 Hz or even 50–7000 Hz.

Speech quality enhancements in cellular networks do not concern only the speech coder itself. First, it is well known that tandeming (i.e., several cascaded coding/decoding operations) can be a source of significant degradation of speech quality. In the case of mobile-to-mobile calls there is no need to decode the speech signal in the network (assuming both mobiles use the same speech codec). Therefore, a tandem-free operation (TFO) will soon be introduced for GSM using in-band signalling between peer transcoders. The benefit of TFO, however, will largely depend on the percentage of intra-GSM mobile-to-mobile calls.

In CDMA systems like IS-95, the soft handoff feature enables a smooth transition from one cell to another. On the contrary in TDMA systems, handovers produce short breaks in the communication which locally degrade the speech quality in spite of the speech extrapolation mechanism used in the decoder. In GSM the duration of the transmission break can be as long as 160 ms due to the necessary time alignment of the mobile in the new cell. The speech interruption can be reduced by 40 ms in a synchronized network, but in practice local synchronization is only offered for co-sited cells. The same performance can be more easily achieved with the presynchronized handover where the mobile receives an indication of the distance to the base station together with the handover command. Some improvement can also be obtained on the uplink by switching at the right point in time and on the downlink by broadcasting the speech information to both the old and the new cell. As a result of all these improvements, the interruption time can be reduced down to 60 ms.

Some advances have also been made concerning the robustness to radio channel impairments. Unequal error protection (UEP) is now used in most codecs designed for cellular systems and it is often based on rate-compatible punctured convolutional codes. UEP enables the adjustment of the protection rate as a function of the sensitivity of bits output by the speech coder. Besides, most linear predictive coders use a speech extrapolation procedure, replacing potentially corrupted information in the current frame with more reliable information from a previously received frame. More sophisticated error concealment techniques, using both reliability information at the output of the channel decoder and a statistical source model (*a priori* knowledge), have been reported to provide up to 3 dB improvement in E_b/N_0 under adverse channel conditions [5].

Robustness to background noise is another topic of interest. Speech quality improvements in medium- to low-bit rate coders have been obtained with coding algorithms optimized for speech signals. Such algorithms may produce poor results in the presence of background noise. Therefore, optional noise cancellation processing has been introduced in recent standards. This can be performed either in the time domain (Kalman filtering) as in JDC half rate or in the frequency domain (spectral subtraction) as in IS-127 EVR.

26.4.3   Coverage Quality Enhancements

Various second generation systems provide the possibility of implementing mechanisms like slow frequency hopping (as in GSM), antenna diversity (also called microdiversity), macrodiversity or multisite transmission, and dynamic channel selection (as in DECT). Such mechanisms are useful to alleviate the effects of radio transmission phenomena like shadowing, fading, and cochannel interference. They are, therefore, particularly interesting to enhance the coverage quality in interference-limited or strong multipath environments (e.g., urban areas). Optional equalizers have been defined for cordless systems (DECT and PHS) with the introduction of a suitable training sequence ("prolonged preamble" in DECT), allowing for channel estimation in the presence of longer delay spread and thus providing increased coverage and/or quality.

Operators also have to face network planning issues linked to natural or artificial obstacles. Various solutions have been designed for the cases where the coverage cannot be efficiently ensured with regular base stations. Radiating cables (leaky feeders) are typically used for the coverage of tunnels. Radio repeaters have been standardized for GSM and for DECT (Wireless Relay Station, WRS) and are useful to fill coverage holes and to provide outdoor-to-indoor or outdoor-to-underground coverage extensions. Microcells (and microbase stations) may also be used as "gap fillers."

The use of smart antennas at cell sites helps to extend the cell range. In rural environments where traffic is low, the required number of cell sites is minimized by using antenna arrays and signal processing algorithms to ensure high sensitivity reception at the base station. Algorithms that implement either n-fold receive diversity or mobile station direction finding followed by beamforming for mobile station tracking have been shown to perform well in such radio channels. Adaptive beamforming techniques with an M-element antenna array generally provide a directivity gain of M, e.g., 9 dB with $M = 8$ (plus some diversity gain depending on channel type). It results in a significant increase of the uplink range (typically by a factor > 1.5 with $M = 8$), but an increased range in the downlink is also needed to get an effective reduction of the number of cell sites. An increased downlink range can be achieved using adaptive beamforming (but with a much higher complexity compared to the uplink-only implementation), a multibeam antenna (i.e., a phased array doing fixed beamforming), or an increased transmit power of the base station. However, the success of smart antenna techniques for range extension applications in second generation systems has been slowed down by their complexity of implementation and by operational constraints (multiple feeders, large antenna panels).

26.5 High Bit Rate Data Transmission

26.5.1 Circuit Mode Techniques

All second generation wireless systems support circuit mode data services with basic rates typically ranging from 9.6 kb/s (in cellular systems) to 32 kb/s (in cordless systems) for a single physical radio resource. With the growing needs for higher rates, new services have been developed based on multiple allocation or grouping of physical resource.

In GSM, HSCSD (High Speed Circuit Switched Data) enables multiple Full Rate Traffic Channels (TCH/F) to be allocated to a call so that a mobile subscriber can use n times the transmission capacity of a single TCH/F channel (Fig. 26.2). The n full rate channels over which the user data stream is split are handled completely independently in the physical layer and for layer 1 error control. The HSCSD channel resulting from the logical combination of n TCH/F channels is controlled as a single radio link during cellular operations such as handover. At the A interface, calls will be limited to a single 64 kb/s circuit. Thus HSCSD will support transparent (up to 64 kb/s) and nontransparent modes (up to $4 \times 9.6 = 38.4$ kb/s and, later, $4 \times 14.4 = 57.6$ kb/s). The initial allocation can be changed during a call if required by the user and authorized by the network. Initially the network allocates an appropriate HSCSD connection according to the requested user bit rate over the air interface. Both symmetric and asymmetric configurations for bidirectional HSCSD operation are authorized. The required TCH/F channels are allocated over consecutive or nonconsecutive timeslots.

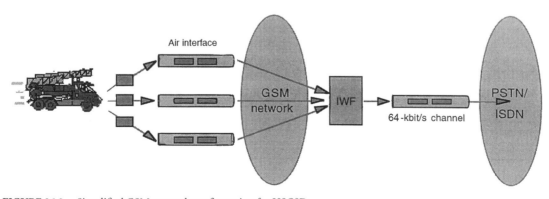

FIGURE 26.2 Simplified GSM network configuration for HSCSD.

Similar multislot schemes are envisaged or standardized for other TDMA systems. In IS-54 and PDC, where radio channels are relatively narrowband, no more than three time slots can be used per carrier and the achievable data rate is therefore limited to, say, 32 kb/s. On the contrary, in DECT up to 12 time slots can be used at 32 kb/s each, yielding a maximum data rate of 384 kb/s. Moreover, the TDD access mode of DECT allows asymmetric time slot allocation between uplink and downlink, thus enabling even higher data rates in one direction.

26.5.2   Packet Mode Techniques

There is a growing interest for packet data services in second generation wireless systems to support data applications with intermittent and bursty transmission requirements like the Internet, with a better usage of available radio resources, thanks to the multiplexing of data from several mobile users on the same physical channel. Cellular Digital Packet Data (CDPD) has been defined in the U.S. as a radio access overlay for AMPS or D-AMPS (IS-54) systems, allowing packet data transmission on available radio channels. However, CDPD is optimized for short data transmission and the bit rate is limited to 19.2 kb/s. A CDMA packet data standard has also been defined (IS-657) which supports CDPD and Internet protocols with a similar bit rate limitation but allowing use of the same backhaul as for voice traffic.

In Europe, ETSI has almost completed the standardization of GPRS (General Packet Radio Service) for GSM. A GPRS subscriber will be able to send and receive in an end-to-end packet transfer mode. Both point-to-point and point-to-multipoint modes are defined. A GPRS network coexists with a GSM PLMN as an autonomous network. In fact, the Serving GPRS Support Node (SGSN) interfaces with the GSM Base Station Controller (BSC), an MSC and a Gateway GPRS Service Node (GGSN). In turn, the GGSN interfaces with the GGSNs of other GPRS networks and with public Packet Data Networks (PDN). Typically, GPRS traffic can be set up through the common control channels of GSM, which are accessed in slotted ALOHA mode. The layer 2 protocol data units, which are about 2 kbytes in length, are segmented and transmitted over the air interface using one of the four possible channel coding schemes. The system is highly scaleable as it allows from one mobile using 8 radio time slots up to 16 mobiles per time slot, with separate allocation in up- and downlink. The resulting peak data rate per user ranges from 9 kb/s up to 170 kb/s. Time slot concatenation and variable channel coding to maximize the user information bit rate are envisaged for future implementations. This is indicated by the mobile station, which provides information concerning the desire to initiate in-call modifications and the channel coding schemes that can be used during the call set up phase. It is expected that use of the GPRS service will initially be limited and traffic growth will depend on the introduction of GPRS capable subscriber terminals. Easy scalability of the GPRS backbone (e.g., by introducing parallel GGSNs) is an essential feature of the system architecture (Fig. 26.3).

26.5.3   New Modulation Schemes

New modulation schemes are being studied as an option in several second generation wireless standards. The aim is to offer higher rate data services equivalent or close to the 2 Mb/s objective of the forthcoming third generation standards. Multilevel modulations (i.e., several bits per modulated symbol) represent a straightforward means to increase the carrier bit rate. However, it represents a significant change in the air interface characteristics, and the increased bit rate is achieved at the expense of a higher operational signal-to-noise plus interference ratio, which is not compatible with large cell dimensions. Therefore, the new

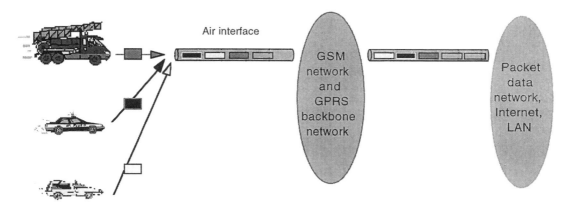

FIGURE 26.3   Simplified view of the GPRS architecture.

high bit rate data services are mainly targetting urban areas, and the effective bit rate allocated to data users will depend on the system load.

Such a new air interface option is being standardized for GSM under the name of EDGE (Enhanced Data rates for GSM Evolution). The selected modulation scheme is 8-PSK, suitable coding schemes are under study, whereas the other air interface parameters (carrier spacing, TDMA frame structure,...) are kept unchanged. Reusing HSCSD (for circuit data) and GPRS (for packet data) protocols and service capabilities, EDGE will provide similar ECSD and EGPRS services but with a three-fold increase of the user bit rate. The higher level modulation requires better radio link performances, typically a loss of 3 to 4 dB in sensitivity and a C/I increased by 6 to 7 dB. Operation will also be restricted to environments with limited time dispersion and limited mobile speed. Nevertheless, EGPRS will roughly double the mean throughput compared to GPRS (for the same average transmitted power). EDGE will also increase the maximum achievable data rate in a GSM system to 553.6 kb/s in multislot (unprotected) operation. Six different protection schemes are foreseen in EGPRS using convolutional coding with a rate ranging from 1/3 to 1 and corresponding to user rates between 22.8 and 69.2 kb/s per time slot. This is in addition to the four coding schemes already defined for GPRS. An intelligent link adaptation algorithm will dynamically select the most appropriate modulation and coding schemes, i.e., those yielding the highest throughput for a given channel quality. The first phase of EDGE standardization should be completed by end 1999. It should be noted that a similar EDGE option is being studied for IS-54/IS-136 (and their PCS derivatives). Initially, the 30 kHz channel spacing will be maintained and then extension to a 200 kHz channel will be provided in order to offer a convergence with its GSM counterpart.

A higher bit rate option is also under standardization for DECT. Here it is seen as an essential requirement to maintain backward compatibility with existing equipment so the new multilevel modulation will only affect the payload part of the bursts, keeping the control and signalling parts unchanged. This ensures that equipment with basic modulation and equipment with a higher rate option can efficiently share a common base station infrastructure. Only 4-level and 8-level modulations are considered and the symbol length, carrier spacing, and slot structure remain unchanged. The requirements on transmitter modulation accuracy need to be more stringent for 4- and 8-level modulation than for the current 2-level scheme. An increased accuracy can provide for coherent demodulation, whereby some (or most) of the sensitivity and C/I loss when using the multilevel mode can be regained. In combination with other new air interface features like forward error correction and double slots (with reduced overhead), the new modulation scheme will provide a wide range of data rates up to 2 Mb/s. For instance using (Π/4-DQPSK modulation (a possible/suitable choice), an unprotected connection with two double slots in each direction gives a data rate of 384 kb/s. Asymmetric connections with a maximum of 11 double slots in one direction will also be supported.

26.6   Conclusion

Since their introduction in the early 1990s, most of the second generation systems have been enjoying exponential growth. With more than 100 million subscribers acquired worldwide in less than ten years of lifetime, the systems based on the GSM family of standards have demonstrated the most spectacular development. Despite a more regional implementation of other second generation systems, each one of those can boast a multimillion subscriber base in mobile or fixed wireless networks.

A variety of service requirements of third generation mobile communication systems are being already met by the upcoming enhancements of second generation systems. Two important trends are reflected by this:

- The introduction of third generation systems like Universal Mobile Telecommunication System (UMTS) or International Mobile Telecommunication-2000 (IMT-2000) might be delayed to a point in time where the evolutionary capabilities of second generation systems have been exhausted.
- The deployment of networks based on third generation systems will be progressive. Any new radio interface will be imposed worldwide if and only if it provides substantial advantages as compared to the present systems. Another essential requirement is the capability of downward compatibility to second generation systems.

Defining Terms

Capacity: In a mobile network it can be defined as the Erlangs throughput by a cell, a cluster of cells, or by a portion of a network. For a given radio interface, the achievable capacity is a function of the robustness of the physical layer, the effectiveness of the medium access control (MAC) layer and the multiple access technique. Moreover, it is strongly dependent on the radio spectrum available for network planning.

Cellular: Refers to public land mobile radio networks for generally wide area (e.g., national) coverage, to be used with medium- or high-power vehicular mobiles or portable stations and for providing mobile access to the Public Switched Telephone Network (PSTN). The network implementation exhibits a cellular architecture which enables frequency reuse in nonadjacent cells.

Cordless: These are systems to be used with simple low power portable stations operating within a short range of a base station and providing access to fixed public or private networks. There are three main applications, namely, residential (at home, for Plain Old Telephone Service, POTS), public-access (in public places and crowded areas, also called Telepoint), and Wireless Private Automatic Branch eXchange (WPABX, providing cordless access in the office environment), plus emerging applications like radio access for local loop.

Coverage quality: It is the percentage of the served area where a communication can be established. It is determined by the acceptable path loss of the radio link and by the propagation characteristics in the area. The radio link budget generally includes some margin depending on the type of terrain (for shadowing effects) and on operator's requirements (for indoor penetration). A coverage quality of 90% is a typical value for cellular networks.

Speech quality: It strongly depends on the intrinsic performance of the speech coder and its evaluation normally requires intensive listening tests. When it is comparable to the quality achieved on modern wire-line telephone networks, it is called "toll quality." In wireless systems it is also influenced by other parameters linked to the communication characteristics like radio channel impairments (bit error rate), transmission delay, echo, background noise, and tandeming (i.e., when several coding/decoding operations are involved in the link).

References

[1] Anderson, S., Antenna Arrays in Mobile Communication Systems, *Proc. Second Workshop on Smart Antennas in Wireless Mobile Communications*, Stanford University, Jul. 1995.

[2] Budagavi, M. and Gibson, J.D., Speech coding in mobile radio communications, *Proceedings of the IEEE*, 86(7), 1402–1412, Jul. 1998.

[3] Cox, D.C., Wireless network access for personal communications, *IEEE Communications Magazine*, 96–115, Dec. 1992.

[4] DECT, *Digital European Cordless Telecommunications Common Interface*, ETS-300-175, ETSI, 1992.

[5] Fingscheidt, T. and Vary, P., Robust Speech Decoding: A Universal Approach to Bit Error Concealment, *Proc. IEEE ICASSP*, 1667–1670, Apr. 1997.

[6] Ganz, A., et al., On optimal design of multitier wireless cellular systems, *IEEE Communications Magazine*, 88–93, Feb. 1997.

[7] GSM, *GSM Recommendations Series 01-12*, ETSI, 1990.

[8] IS-54, *Cellular System, Dual-Mode Mobile Station-Base Station Compatibility Standard*, EIA/TIA Interim Standard, 1991.

[9] IS-95, *Mobile Station-Base Station Compatibility Standard for Dual-Mode Wideband Spread Spectrum Cellular System*, EIA/TIA Interim Standard, 1993.

[10] Kuhn, A., et al., Validation of the Feature Frequency Hopping in a Live GSM Network, *Proc. 46th IEEE Vehic. Tech. Conf.*, 321–325, Apr. 1996.

[11] Lagrange, X., Multitier cell design, *IEEE Communications Magazine*, 60–64, Aug. 1997.

[12] Lee, D. and Xu, C., The effect of narrowbeam antenna and multiple tiers on system capacity in CDMA wireless local loop, *IEEE Communications Magazine*, 110–114, Sep. 1997.

[13] Olofsson, H., et al., Interference Diversity as Means for Increased Capacity in GSM, *Proc. EPMCC'95*, 97–102, Nov. 1995.

[14] PDC, *Personal Digital Cellular System Common Air Interface*, RCR-STD27B, 1991.

[15] PHS, *Personal Handy Phone System: Second Generation Cordless Telephone System Standard*, RCR-STD28, 1993.

[16] Tuttlebee, W.H.W., Cordless personal communications, *IEEE Communications Magazine*, 42–53, Dec. 1992.

Further Information

European standards (GSM, CT2, DECT, TETRA) are published by ETSI Secretariat, 06921 Sophia Antipolis Cedex, France.

U.S. standards (IS-54, IS-95, APCO) are published by Electronic Industries Association, Engineering Department, 2001 Eye Street, N.W., Washington D.C. 20006, U.S.A.

Japanese standards (PDC, PHS) are published by RCR (Research and Development Center for Radio Systems), 1-5-16, Toranomon, Minato-ku, Tokyo 105, Japan.

27

The Pan-European Cellular System

Lajos Hanzo
University of Southampton

27.1 Introduction

Following the standardization and launch of the Pan-European digital mobile cellular radio system known as GSM, it is of practical merit to provide a rudimentary introduction to the system's main features for the communications practitioner. Since GSM operating licenses have been allocated to 126 service providers in 75 countries, it is justifiable that the GSM system is often referred to as the Global System of Mobile communications.

The GSM specifications were released as 13 sets of recommendations [1], which are summarized in Table 27.1, covering various aspects of the system [3].

After a brief system overview in Section 27.2 and the introduction of physical and logical channels in Section 27.3 we embark upon describing aspects of mapping logical channels onto physical resources for speech and control channels in Sections 27.4 and 27.5, respectively. These details can be found in recommendations R.05.02 and R.05.03. These recommendations and all subsequently enumerated ones are to be found in [1]. Synchronization issues are considered in Section 27.6. Modulation (R.05.04), transmission via the standardized wideband GSM channel models (R.05.05), as well as adaptive radio link control (R.05.06 and R.05.08), discontinuous transmission (**DTX**) (R.06.31), and voice activity detection (**VAD**) (R.06.32) are highlighted in Sections 27.7–27.10, whereas a summary of the fundamental GSM features is offered in Section 27.11.

TABLE 27.1   GSM Recommendations [R.01.01]

R.00	*Preamble* to the GSM recommendations
R.01	*General structure* of the recommendations, description of a GSM network, associated recommendations, vocabulary, etc.
R.02	*Service aspects:* bearer-, tele- and supplementary services, use of services, types and features of mobile stations (MS), licensing and subscription, as well as transferred and international accounting, etc.
R.03	*Network aspects,* including network functions and architecture, call routing to the MS, technical performance, availability and reliability objectives, handover and location registration procedures, as well as discontinuous reception and cryptological algorithms, etc.
R.04	*Mobile/base station (BS) interface and protocols,* including specifications for layer 1 and 3 aspects of the open systems interconnection (OSI) seven-layer structure.
R.05	*Physical layer on the radio path,* incorporating issues of multiplexing and multiple access, channel coding and modulation, transmission and reception, power control, frequency allocation and synchronization aspects, etc.
R.06	*Speech coding specifications,* such as functional, computational and verification procedures for the speech codec and its associated voice activity detector (VAD) and other optional features.
R.07	*Terminal adaptors for MSs,* including circuit and packet mode as well as voiceband data services.
R.08	*Base station and mobile switching center* (MSC) *interface,* and transcoder functions.
R.09	*Network interworking* with the public switched telephone network (PSTN), integrated services digital network (ISDN) and, packet data networks.
R.10	*Service interworking, short message service.*
R.11	*Equipment specification and type approval specification* as regards to MSs, BSs, MSCs, home (HLR) and visited location register (VLR), as well as system simulator.
R.12	*Operation and maintenance,* including subscriber, routing tariff and traffic administration, as well as BS, MSC, HLR and VLR maintenance issues.

27.2   Overview

The system elements of a GSM public land mobile network (**PLMN**) are portrayed in Fig. 27.1, where their interconnections via the standardized interfaces A and Um are indicated as well. The mobile station (**MS**) communicates with the serving and adjacent base stations (**BS**) via the radio interface Um, whereas the BSs are connected to the mobile switching center (**MSC**) through the network interface A. As seen in Fig. 27.1, the MS includes a mobile termination (MT) and a terminal equipment (TE). The TE may be constituted, for example, by a telephone set and fax machine. The MT performs functions needed to support the physical channel between the MS and the base station, such as radio transmissions, radio channel management, channel coding/decoding, speech encoding/decoding, and so forth.

The BS is divided functionally into a number of base transceiver stations (BTS) and a base station controller (BSC). The BS is responsible for channel allocation (R.05.09), link quality and power budget control (R.05.06 and R.05.08), signalling and broadcast traffic control, frequency hopping (**FH**) (R.05.02), handover (**HO**) initiation (R.03.09 and R.05.08), etc. The MSC represents the gateway to other networks, such as the public switched telephone network (**PSTN**), integrated services digital network (**ISDN**) and packet data networks using the interworking functions standardized in recommendation R.09. The MSC's further functions include paging, MS location updating (R.03.12), HO control (R.03.09), etc. The MS's mobility management is assisted by the home location register (**HLR**) (R.03.12), storing part of the MS's location information and routing incoming calls to the visitor location register (**VLR**) (R.03.12) in charge of the area, where the paged MS roams. Location update is asked for by the MS, whenever it detects from the received and decoded broadcast

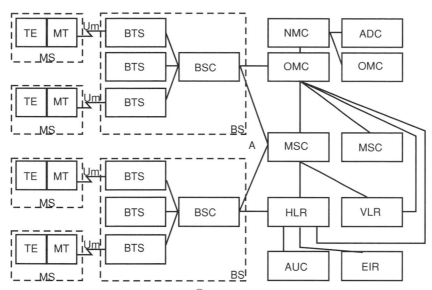

FIGURE 27.1 Simplified structure of GSM PLMN © ETT [4].

control channel (**BCCH**) messages that it entered a new location area. The HLR contains, amongst a number of other parameters, the international mobile subscriber identity (**IMSI**), which is used for the authentication (R.03.20) of the subscriber by his authentication center (**AUC**). This enables the system to confirm that the subscriber is allowed to access it. Every subscriber belongs to a home network and the specific services that the subscriber is allowed to use are entered into his HLR. The equipment identity register (**EIR**) allows for stolen, fraudulent, or faulty mobile stations to be identified by the network operators. The VLR is the functional unit that attends to a MS operating outside the area of its HLR. The visiting MS is automatically registered at the nearest MSC, and the VLR is informed of the MSs arrival. A roaming number is then assigned to the MS, and this enables calls to be routed to it. The operations and maintenance center (**OMC**), network management center (**NMC**) and administration center (**ADC**) are the functional entities through which the system is monitored, controlled, maintained and managed (R.12).

The MS initiates a call by searching for a BS with a sufficiently high received signal level on the BCCH carrier; it will await and recognize a frequency correction burst and synchronize to it (R.05.08). Now the BS allocates a bidirectional signalling channel and also sets up a link with the MSC via the network. How the control frame structure assists in this process will be highlighted in Section 27.5. The MSC uses the IMSI received from the MS to interrogate its HLR and sends the data obtained to the serving VLR. After authentication (R.03.20) the MS provides the destination number, the BS allocates a traffic channel, and the MSC routes the call to its destination. If the MS moves to another cell, it is reassigned to another BS, and a handover occurs. If both BSs in the handover process are controlled by the same BSC, the handover takes place under the control of the BSC, otherwise it is performed by the MSC. In case of incoming calls the MS must be paged by the BSC. A paging signal is transmitted on a paging channel (**PCH**) monitored continuously by all MSs, and which covers the location area in which the MS roams. In response to the paging signal, the MS performs an access procedure identical to that employed when the MS initiates a call.

27.3   Logical and Physical Channels

The GSM logical traffic and control channels are standardized in recommendation R.05.02, whereas their mapping onto physical channels is the subject of recommendations R.05.02 and R.05.03. The GSM system's prime objective is to transmit the logical traffic channel's (**TCH**) speech or data information. Their transmission via the network requires a variety of logical control channels. The set of logical traffic and control channels defined in the GSM system is summarized in Table 27.2. There are two general forms of speech and data traffic channels: the full-rate traffic channels (**TCH/F**), which carry information at a gross rate of 22.8 kb/s, and the half-rate traffic channels (**TCH/H**), which communicate at a gross rate of 11.4 kb/s. A physical channel carries either a full-rate traffic channel, or two half-rate traffic channels. In the former, the traffic channel occupies one timeslot, whereas in the latter the two half-rate traffic channels are mapped onto the same timeslot, but in alternate frames.

TABLE 27.2   GSM Logical Channels ©ETT [4]

Logical Channels					
Duplex BS ↔ MS Traffic Channels: TCH		Control Channels: CCH			
FEC-coded Speech	FEC-coded Data	Broadcast CCH BCCH BS → MS	Common CCH CCCH	Stand-alone Dedicated CCH SDCCH BS ↔ MS	Associated CCH ACCH BS ↔ MS
TCH/F 22.8 kb/s	TCH/F9.6 TCH/F4.8 TCH/F2.4 22.8 kb/s	Freq. Corr. Ch: FCCH	Paging Ch: PCH BS → MS	SDCCH/4	Fast ACCH: FACCH/F FACCH/H
TCH/H 11.4 kb/s	TCH/H4.8 TCH/H2.4 11.4 kb/s	Synchron. Ch: SCH	Random Access Ch: RACH MS → BS	SDCCH/8	Slow ACCH: SACCH/TF SACCH/TH SACCH/C4 SACCH/C8
		General Inf.	Access Grant Ch: AGCH BS → MS		

For a summary of the logical control channels carrying signalling or synchronisation data, see Table 27.2. There are four categories of logical control channels, known as the BCCH, the common control channel (**CCCH**), the stand-alone dedicated control channel (**SDCCH**), and the associated control channel (**ACCH**). The purpose and way of deployment of the logical traffic and control channels will be explained by highlighting how they are mapped onto physical channels in assisting high-integrity communications.

A physical channel in a time division multiple access (**TDMA**) system is defined as a timeslot with a timeslot number (TN) in a sequence of TDMA frames. The GSM system, however, deploys TDMA combined with frequency hopping (FH) and, hence, the physical channel is partitioned in both time and frequency. Frequency hopping (R.05.02) combined with interleaving is known to be very efficient in combatting channel fading, and it results in near-Gaussian performance even over hostile Rayleigh-fading channels. The principle of FH is that each TDMA burst is transmitted via a different RF channel (**RFCH**). If the present TDMA burst happened to be in a deep fade, then the next burst most probably will not be. Consequently, the physical channel is defined as a sequence of radio frequency channels and timeslots. Each carrier frequency supports eight physical channels mapped onto eight timeslots within a TDMA frame. A given physical channel always uses the same TN in every TDMA frame. Therefore, a timeslot sequence is defined by a TN and a TDMA frame number FN sequence.

27.4 Speech and Data Transmission

The speech coding standard is recommendation R.06.10, whereas issues of mapping the logical speech traffic channel's information onto the physical channel constituted by a timeslot of a certain carrier are specified in recommendation R.05.02. Since the error correction coding represents part of this mapping process, recommendation R.05.03 is also relevant to these discussions. The example of the full-rate speech traffic channel (**TCH/FS**) is used here to highlight how this logical channel is mapped onto the physical channel constituted by a so-called normal burst (**NB**) of the TDMA frame structure. This mapping is explained by referring to Figs. 27.2 and 27.3. Then this example will be extended to other physical bursts such as the frequency correction (**FCB**), synchronization (**SB**), access (**AB**), and dummy burst (**DB**) carrying logical control channels, as well as to their TDMA frame structures, as seen in Figs. 27.2 and 27.6.

The regular pulse excited (**RPE**) speech encoder is fully characterized in the following references: [3, 5, 7]. Because of its complexity, its description is beyond the scope of this chapter. Suffice to say that, as it can be seen in Fig. 27.3, it delivers 260 b/20 ms at a bit rate of 13 kb/s, which are divided into three significance classes: class 1a (50 b), class 1b (132 b) and class 2 (78 b). The class-1a bits are encoded by a systematic (53, 50) cyclic error detection code by adding three parity bits. Then the bits are reordered and four zero tailing bits are added to periodically reset the memory of the subsequent half-rate, constraint length five convolutional codec (**CC**) $CC(2, 1, 5)$, as portrayed in Fig. 27.3. Now the unprotected 78 class-2 bits are concatenated to yield a block of 456 b/20 ms, which implies an encoded bit rate of 22.8 kb/s. This frame is partitioned into eight 57-b subblocks that are block diagonally interleaved before undergoing intraburst interleaving. At this stage each 57-b subblock is combined with a similar subblock of the previous 456-b frame to construct a 116-b burst, where the flag bits hl and hu are included to classify whether the current burst is really a TCH/FS burst or it has been stolen by an urgent fast associated (**FACCH**) control channel message. Now the bits are encrypted and positioned in a NB, as depicted at the bottom of Fig. 27.2, where three tailing bits (**TB**) are added at both ends of the burst to reset the memory of the Viterbi channel equalizer (**VE**), which is responsible for removing both the channel-induced and the intentional controlled intersymbol interference [6].

The 8.25-b interval duration guard period (GP) at the bottom of Fig. 27.2 is provided to prevent burst overlapping due to propagation delay fluctuations. Finally, a 26-b equalizer training segment is included in the center of the normal traffic burst. This segment is constructed by a 16-b Viterbi channel equalizer training pattern surrounded by five quasiperiodically repeated bits on both sides. Since the MS has to be informed about which BS it communicates with, for neighboring BSs one of eight different training patterns is used, associated with the so-called BS color codes, which assist in identifying the BSs.

This 156.25-b duration TCH/FS NB constitutes the basic timeslot of the TDMA frame structure, which is input to the Gaussian minimum shift keying (**GMSK**) modulator to be highlighted in Section 27.7, at a bit rate of approximately 271 kb/s. Since the bit interval is $1/(271 \text{ kb/s}) = 3.69 \ \mu$s, the timeslot duration is $156.25 \cdot 3.69 \approx 0.577$ ms. Eight such normal bursts of eight appropriately staggered TDMA users are multiplexed onto one (**RF**) carrier giving, a TDMA frame of $8 \cdot 0.577 \approx 4.615$-ms duration, as shown in Fig. 27.2. The

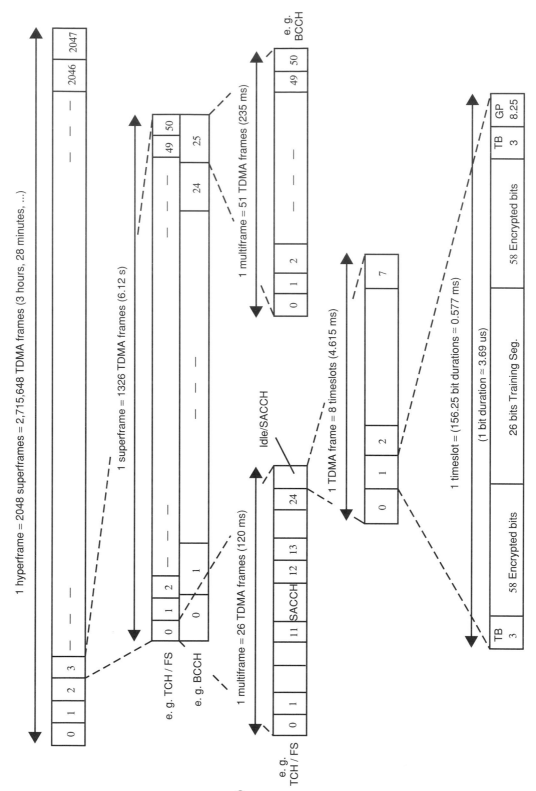

FIGURE 27.2 The GSM TDMA frame structure ©ETT [4].

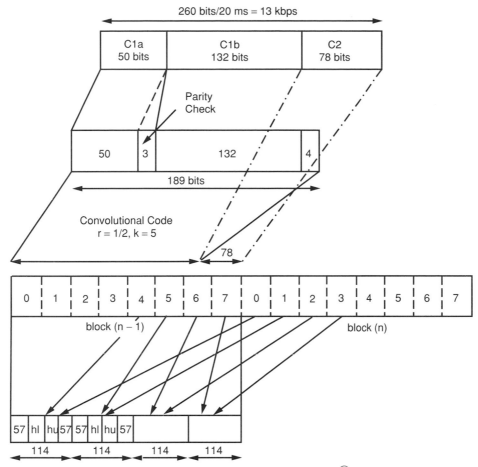

FIGURE 27.3 Mapping the TCH/FS logical channel onto a physical channel, ©ETT [4].

physical channel as characterized earlier provides a physical timeslot with a throughput of 114 b/4.615 ms = 24.7 kb/s, which is sufficiently high to transmit the 22.8 kb/s TCH/FS information. It even has a reserved capacity of 24.7 − 22.8 = 1.9 kb/s, which can be exploited to transmit slow control information associated with this specific traffic channel, i.e., to construct a so-called slow associated control channel (**SACCH**), constituted by the SACCH TDMA frames, interspersed with traffic frames at multiframe level of the hierarchy, as seen in Fig. 27.2.

Mapping logical data traffic channels onto a physical channel is essentially carried out by the channel codecs [8], as specified in recommendation R.05.03. The full- and half-rate data traffic channels standardized in the GSM system are: **TCH/F9.6, TCH/F4.8, TCH/F2.4,** as well as **TCH/H4.8, TCH/H2.4,** as was shown earlier in Table 27.2. Note that the numbers in these acronyms represent the data transmission rate in kilobits per second. Without considering the details of these mapping processes we now focus our attention on control signal transmission issues.

27.5   Transmission of Control Signals

The exact derivation, forward error correcting (**FEC**) coding and mapping of logical control channel information is beyond the scope of this chapter, and the interested reader is referred to ETSI, 1988 (R.05.02 and R.05.03) and Hanzo and Stefanov, 1992, for a detailed discussion. As an example, the mapping of the 184-b SACCH, FACCH, BCCH, SDCCH, PCH, and access grant control channel (**AGCH**) messages onto a 456-b block, i.e., onto four 114-b bursts is demonstrated in Fig. 27.4. A double-layer concatenated FIRE-code/convolutional code scheme generates 456 bits, using an overall coding rate of $R = 184/456$, which gives a stronger protection for control channels than the error protection of traffic channels.

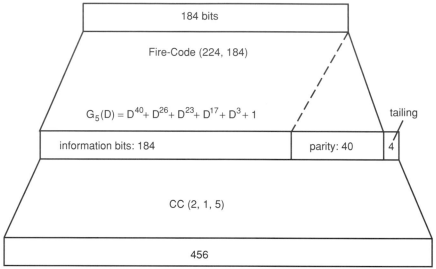

FIGURE 27.4   FEC in SACCH, FACCH, BCCH, SDCCH, PCH and AGCH, ©ETT [4].

Returning to Fig. 27.2 we will now show how the SACCH is accommodated by the TDMA frame structure. The TCH/FS TDMA frames of the eight users are multiplexed into multiframes of 24 TDMA frames, but the 13th frame will carry a SACCH message, rather than the 13th TCH/FS frame, whereas the 26th frame will be an idle or dummy frame, as seen at the left-hand side of Fig. 27.2 at the multiframe level of the traffic channel hierarchy. The general control channel frame structure shown at the right of Fig. 27.2 is discussed later. This way 24-TCH/FS frames are sent in a 26-frame multiframe during $26 \cdot 4.615 = 120$ ms. This reduces the traffic throughput to $(24/26) \cdot 24.7 = 22.8$ kb/s required by TCH/FS, allocates $(1/26) \cdot 24.7 = 950$ b/s to the SACCH and wastes 950 b/s in the idle frame. Observe that the SACCH frame has eight timeslots to transmit the eight 950-b/s SACCHs of the eight users on the same carrier. The 950-b/s idle capacity will be used in case of half-rate channels, where 16 users will be multiplexed onto alternate frames of the TDMA structure to increase system capacity. Then 16, 11.4-kb/s encoded half-rate speech TCHs will be transmitted in a 120-ms multiframe, where also 16 SACCHs are available.

The FACCH messages are transmitted via the physical channels provided by bits stolen from their own host traffic channels. The construction of the FACCH bursts from 184 control bits is identical to that of the SACCH, as also shown in Fig. 27.4 but its 456-b frame is mapped onto eight consecutive 114-b TDMA traffic bursts, exactly as specified for TCH/FS. This is carried out by stealing the even bits of the first four and the odd bits of the last four bursts, which is signalled by setting $hu = 1, hl = 0$ and $hu = 0, hl = 1$ in the first and last bursts, respectively. The unprotected FACCH information rate is 184 b/20 ms = 9.2 kb/s, which is transmitted after concatenated error protection at a rate of 22.8 kb/s. The repetition delay is 20 ms, and the interleaving delay is $8 \cdot 4.615 = 37$ ms, resulting in a total of 57-ms delay.

In Fig. 27.2 at the next hierarchical level, 51-TCH/FS multiframes are multiplexed into one superframe lasting $51 \cdot 120$ ms = 6.12 s, which contains $26 \cdot 51 = 1326$-TDMA frames. In the case of 1326-TDMA frames, however, the frame number would be limited to $0 \leq \textbf{\textit{FN}} \leq 1326$ and the encryption rule relying on such a limited range of FN values would not be sufficiently secure. Then 2048 superframes were amalgamated to form a hyperframe of $1326 \cdot 2048 = 2,715,648$-TDMA frames lasting $2048 \cdot 6.12$ s ≈ 3 h 28 min, allowing a sufficiently high FN value to be used in the encryption algorithm. The uplink and downlink traffic-frame structures are identical with a shift of three timeslots between them, which relieves the MS from having to transmit and receive simultaneously, preventing high-level transmitted power leakage back to the sensitive receiver. The received power of adjacent BSs can be monitored during unallocated timeslots.

In contrast to duplex traffic and associated control channels, the simplex BCCH and CCCH logical channels of all MSs roaming in a specific cell share the physical channel provided by timeslot zero of the so-called BCCH carriers available in the cell. Furthermore, as demonstrated by the right-hand side section of Fig. 27.2, 51 BCCH and CCCH TDMA frames are mapped onto a $51 \cdot 4.615 = 235$-ms duration multiframe, rather than on a 26-frame, 120-ms duration multiframe. In order to compensate for the extended multiframe length of 235 ms, 26 multiframes constitute a 1326-frame superframe of 6.12-s duration. Note in Fig. 27.5, that the allocation of the uplink and downlink frames is different, since these control channels exist only in one direction.

Specifically, the random access channel (**RACH**) is only used by the MSs in the uplink direction if they request, for example, a bidirectional SDCCH to be mapped onto an RF channel to register with the network and set up a call. The uplink RACH has a low capacity, carrying messages of 8-b/235-ms multiframe, which is equivalent to an unprotected control information rate of 34 b/s. These messages are concatenated FEC coded to a rate of 36 b/235 ms = 153 b/s. They are not transmitted by the NB derived for TCH/FS, SACCH, or FACCH logical channels, but by the AB, depicted in Fig. 27.6 in comparison to a NB and other types of bursts to be described later. The FEC coded, encrypted 36-b AB messages of Fig. 27.6 contain among other parameters, the encoded 6-b BS identifier code (**BSIC**) constituted by the 3-b PLMN color code and 3-b BS color code for unique BS identification. These 36 b are positioned after the 41-b synchronization sequence, which has a high wordlength in order to ensure reliable access burst recognition and a low probability of being emulated by interfering stray data. These messages have no interleaving delay, while they are transmitted with a repetition delay of one control multiframe length, i.e., 235 ms.

Adaptive time frame alignment is a technique designed to equalize propagation delay differences between MSs at different distances. The GSM system is designed to allow for cell sizes up to 35 km radius. The time a radio signal takes to travel the 70 km from the base station to the mobile station and back again is 233.3 μs. As signals from all the mobiles in the cell must reach the base station without overlapping each other, a long guard period of 68.25 b (252 μs) is provided in the access burst, which exceeds the maximum possible propagation delay of 233.3 μs. This long guard period in the access burst is needed when the mobile station attempts its first access to the base station or after a handover has occurred.

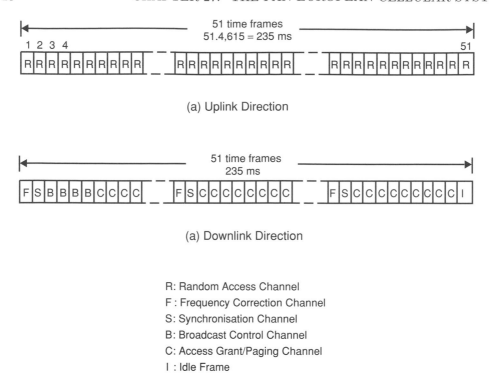

(a) Uplink Direction

(a) Downlink Direction

R: Random Access Channel
F : Frequency Correction Channel
S: Synchronisation Channel
B: Broadcast Control Channel
C: Access Grant/Paging Channel
I : Idle Frame

FIGURE 27.5   The control multiframe, ©ETT [4].

When the base station detects a 41-b random access synchronization sequence with a long guard period, it measures the received signal delay relative to the expected signal from a mobile station of zero range. This delay, called the timing advance, is signalled using a 6-b number to the mobile station, which advances its timebase over the range of 0–63 b, i.e., in units of 3.69 μs. By this process the TDMA bursts arrive at the BS in their correct timeslots and do not overlap with adjacent ones. This process allows the guard period in all other bursts to be reduced to $8.25 \cdot 3.69$ μs ≈ 30.46 μs (8.25 b) only. During normal operation, the BS continuously monitors the signal delay from the MS and, if necessary, it will instruct the MS to update its time advance parameter. In very large traffic cells there is an option to actively utilize every second timeslot only to cope with higher propagation delays, which is spectrally inefficient, but in these large, low-traffic rural cells it is admissible.

As demonstrated by Fig. 27.2, the downlink multiframe transmitted by the BS is shared amongst a number of BCCH and CCCH logical channels. In particular, the last frame is an idle frame (I), whereas the remaining 50 frames are divided in five blocks of ten frames, where each block starts with a frequency correction channel (**FCCH**) followed by a synchronization channel (**SCH**). In the first block of ten frames the FCCH and SCH frames are followed by four BCCH frames and by either four AGCH or four PCH. In the remaining four blocks of ten frames, the last eight frames are devoted to either PCHs or AGCHs, which are mutually exclusive for a specific MS being either paged or granted a control channel.

The FCCH, SCH, and RACH require special transmission bursts, tailored to their missions, as depicted in Fig. 27.6. The FCCH uses frequency correction bursts (FCB) hosting a specific 142-b pattern. In partial response GMSK it is possible to design a modulating data sequence, which results in a near-sinusoidal modulated signal imitating an unmodulated carrier exhibiting a fixed frequency offset from the RF carrier utilized. The synchronization

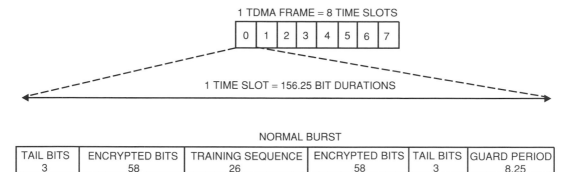

FIGURE 27.6 GSM burst structures, ©ETT [4].

channel transmits SB hosting a $16 \cdot 4 = 64$-b extended sequence exhibiting a high-correlation peak in order to allow frame alignment with a quarter-bit accuracy. Furthermore, the SB contains $2 \cdot 39 = 78$ encrypted FEC-coded synchronization bits, hosting the BS and PLMN color codes, each representing one of eight legitimate identifiers. Lastly, the AB contain an extended 41-b synchronization sequence, and they are invoked to facilitate initial access to the system. Their long guard space of 68.25-b duration prevents frame overlap, before the MS's distance, i.e., the propagation delay becomes known to the BS and could be compensated for by adjusting the MS's timing advance.

27.6 Synchronization Issues

Although some synchronization issues are standardized in recommendations R.05.02 and R.05.03, the GSM recommendations do not specify the exact BS-MS synchronization algorithms to be used, these are left to the equipment manufacturers. A unique set of timebase counters, however, is defined in order to ensure perfect BS-MS synchronism. The BS sends FCB and SB on specific timeslots of the BCCH carrier to the MS to ensure that the MS's frequency standard is perfectly aligned with that of the BS, as well as to inform the MS about the required initial state of its internal counters. The MS transmits its uniquely numbered traffic and control bursts staggered by three timeslots with respect to those of the BS to prevent simultaneous MS transmission and reception, and also takes into account the required timing advance (**TA**) to cater for different BS-MS-BS round-trip delays.

The timebase counters used to uniquely describe the internal timing states of BSs and MSs are the quarter-bit number ($\boldsymbol{QN} = 0$–624) counting the quarter-bit intervals in bursts, bit number ($\boldsymbol{BN} = 0$–156), timeslot number ($\boldsymbol{TN} = 0$–7) and TDMA Frame Number ($FN = 0$–$26 \cdot 51 \cdot 2048$), given in the order of increasing interval duration. The MS sets up its timebase counters after receiving a SB by determining QN from the 64-b extended training sequence in the center of the SB, setting $TN = 0$ and decoding the 78-encrypted, protected bits carrying the 25-SCH control bits.

The SCH carries frame synchronization information as well as BS identification information to the MS, as seen in Fig. 27.7, and it is provided solely to support the operation of the radio subsystem. The first 6 b of the 25-b segment consist of three PLMN color code bits

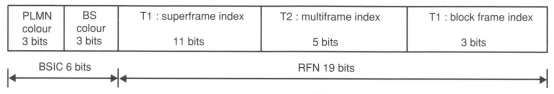

FIGURE 27.7   Synchronization channel (SCH) message format, ©ETT [4].

and three BS color code bits supplying a unique BS identifier code (BSIC) to inform the MS which BS it is communicating with. The second 19-bit segment is the so-called reduced TDMA frame number **RFN** derived from the full TDMA frame number FN, constrained to the range of $[0$–$(26 \cdot 51 \cdot 2048) - 1] = (0$–$2{,}715{,}647)$ in terms of three subsegments $T1$, $T2$, and $T3$. These subsegments are computed as follows: $T1(11\,\text{b}) = [FN \text{ div } (26 \cdot 51)]$, $T2(5\,\text{b}) = (FN \bmod 26)$ and $T3'(3\text{b}) = [(T3 - 1) \text{ div } 10]$, where $T3 = (FN \bmod 5)$, whereas div and mod represent the integer division and modulo operations, respectively. Explicitly, in Fig. 27.7 $T1$ determines the superframe index in a hyperframe, $T2$ the multiframe index in a superframe, $T3$ the frame index in a multiframe, whereas $T3'$ is the so-called signalling block index [1–5] of a frame in a specific 51-frame control multiframe, and their roles are best understood by referring to Fig. 27.2. Once the MS has received the SB, it readily computes the FN required in various control algorithms, such as encryption, handover, etc., as

$$FN = 51\left[(T3 - T2) \bmod 26\right] + T3 + 51 \cdot 26 \cdot T1, \qquad \text{where } T3 = 10 \cdot T3' + 1$$

27.7   Gaussian Minimum Shift Keying Modulation

The GSM system uses constant envelope partial response GMSK modulation [6] specified in recommendation R.05.04. Constant envelope, continuous-phase modulation schemes are robust against signal fading as well as interference and have good spectral efficiency. The slower and smoother are the phase changes, the better is the spectral efficiency, since the signal is allowed to change less abruptly, requiring lower frequency components. The effect of an input bit, however, is spread over several bit periods, leading to a so-called partial response system, which requires a channel equalizer in order to remove this controlled, intentional intersymbol interference (ISI) even in the absence of uncontrolled channel dispersion.

The widely employed partial response GMSK scheme is derived from the full response minimum shift keying (MSK) scheme. In MSK the phase changes between adjacent bit periods are piecewise linear, which results in discontinuous-phase derivative, i.e., instantaneous frequency at the signalling instants, and hence widens the spectrum. Smoothing these phase changes, however, by a filter having a Gaussian impulse response [6], which is known to have the lowest possible bandwidth, this problem is circumvented using the schematic of Fig. 27.8, where the GMSK signal is generated by modulating and adding two quadrature carriers. The key parameter of GMSK in controlling both bandwidth and interference resistance is the 3-dB down filter-bandwidth × bit interval product ($B \cdot T$), referred to as normalized bandwidth. It was found that as the $B \cdot T$ product is increased from 0.2 to 0.5, the interference resistance is improved by approximately 2 dB at the cost of increased bandwidth occupancy, and best compromise was achieved for $B \cdot T = 0.3$. This corresponds to spreading the effect of 1 b over approximately 3-b intervals. The spectral efficiency gain due to higher interference tolerance and, hence, more dense frequency reuse was found to be more significant than the spectral loss caused by wider GMSK spectral lobes.

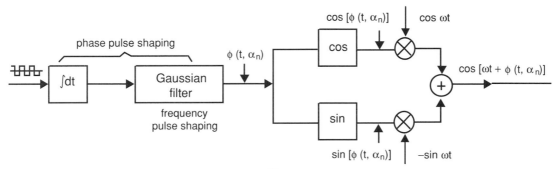

FIGURE 27.8   GMSK modulator schematic diagram, ©ETT [4].

The channel separation at the TDMA burst rate of 271 kb/s is 200 kHz, and the modulated spectrum must be 40 dB down at both adjacent carrier frequencies. When TDMA bursts are transmitted in an on-off keyed mode, further spectral spillage arises, which is mitigated by a smooth power ramp up and down envelope at the leading and trailing edges of the transmission bursts, attenuating the signal by 70 dB during a 28- and 18-μs interval, respectively.

27.8   Wideband Channel Models

The set of 6-tap GSM impulse responses [2] specified in recommendation R.05.05 is depicted in Fig. 27.9, where the individual propagation paths are independent Rayleigh fading paths, weighted by the appropriate coefficients h_i corresponding to their relative powers portrayed in the figure. In simple terms the wideband channel's impulse response is measured by transmitting an impulse and detecting the received echoes at the channel's output in every D-spaced so-called delay bin. In some bins no delayed and attenuated multipath component is received, whereas in others significant energy is detected, depending on the typical reflecting objects and their distance from the receiver. The path delay can be easily related to the distance of the reflecting objects, since radio waves are travelling at the speed of light. For example, at a speed of 300,000 km/s, a reflecting object situated at a distance of 0.15 km yields a multipath component at a round-trip delay of 1 μs.

The typical urban (**TU**) impulse response spreads over a delay interval of 5 μs, which is almost two 3.69-μs bit-intervals duration and, therefore, results in serious ISI. In simple terms, it can be treated as a two-path model, where the reflected path has a length of 0.75 km, corresponding to a reflector located at a distance of about 375 m. The hilly

TYPICAL URBAN (TU) IMPULSE RESPONSE

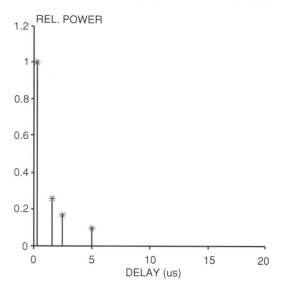

HILLY TERRAIN (HT) IMPULSE RESPONSE

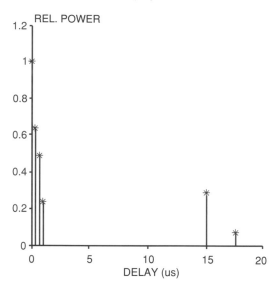

RURAL AREA (RA) IMPULSE RESPONSE

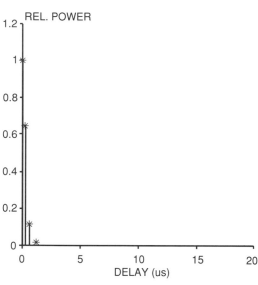

EQUALISER TEST (EQ) IMPULSE RESPONSE

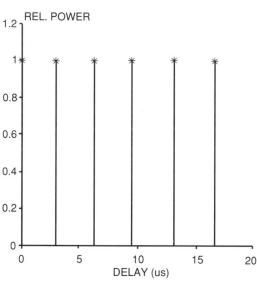

FIGURE 27.9   Typical GSM channel impulse responses, ©ETT [4].

terrain (HT) model has a sharply decaying short-delay section due to local reflections and a long-delay path around 15 μs due to distant reflections. Therefore, in practical terms it can be considered a two- or three-path model having reflections from a distance of about 2 km. The rural area (**RA**) response seems the least hostile amongst all standardized responses, decaying rapidly inside 1-b interval and, therefore, is expected to be easily combated by the channel equalizer. Although the type of the equalizer is not standardized, partial response systems typically use VEs. Since the RA channel effectively behaves as a single-path nondispersive channel, it would not require an equalizer. The fourth standardized impulse response is artificially contrived in order to test the equalizer's performance and is constituted by six equidistant unit-amplitude impulses representing six equal-powered independent Rayleigh-fading paths with a delay spread over 16 μs. With these impulse responses in mind, the required channel is simulated by summing the appropriately delayed and weighted received signal components. In all but one case the individual components are assumed to have Rayleigh amplitude distribution, whereas in the RA model the main tap at zero delay is supposed to have a Rician distribution with the presence of a dominant line-of-sight path.

27.9 Adaptive Link Control

The adaptive link control algorithm portrayed in Fig. 27.10 and specified in recommendation R.05.08 allows for the MS to favor that specific traffic cell which provides the highest probability of reliable communications associated with the lowest possible path loss. It also decreases interference with other cochannel users and, through dense frequency reuse, improves spectral efficiency, whilst maintaining an adequate communications quality, and facilitates a reduction in power consumption, which is particularly important in hand-held MSs. The handover process maintains a call in progress as the MS moves between cells, or when there is an unacceptable transmission quality degradation caused by interference, in which case an intracell handover to another carrier in the same cell is performed. A radio-link failure occurs when a call with an unacceptable voice or data quality cannot be improved either by RF power control or by handover. The reasons for the link failure may be loss of radio coverage or very high-interference levels. The link control procedures rely on measurements of the received RF signal strength (**RXLEV**), the received signal quality (**RXQUAL**), and the absolute distance between base and mobile stations (DISTANCE).

RXLEV is evaluated by measuring the received level of the BCCH carrier which is continuously transmitted by the BS on all time slots of the B frames in Fig. 27.5 and without variations of the RF level. A MS measures the received signal level from the serving cell and from the BSs in all adjacent cells by tuning and listening to their BCCH carriers. The root mean squared level of the received signal is measured over a dynamic range from -103 to -41 dBm for intervals of one SACCH multiframe (480 ms). The received signal level is averaged over at least 32 SACCH frames ($\approx$15 s) and mapped to give RXLEV values between 0 and 63 to cover the range from -103 to -41 dBm in steps of 1 dB. The RXLEV parameters are then coded into 6-b words for transmission to the serving BS via the SACCH.

RXQUAL is estimated by measuring the bit error ratio (**BER**) before channel decoding, using the Viterbi channel equalizer's metrics [6] and/or those of the Viterbi convolutional decoder [8]. Eight values of RXQUAL span the logarithmically scaled BER range of 0.2–12.8% before channel decoding.

The absolute DISTANCE between base and mobile stations is measured using the timing advance parameter. The timing advance is coded as a 6-b number corresponding to a propagation delay from 0 to $63 \cdot 3.69$ μs $= 232.6$ μs, characteristic of a cell radius of 35 km.

While roaming, the MS needs to identify which potential target BS it is measuring, and the BCCH carrier frequency may not be sufficient for this purpose, since in small cluster sizes the same BCCH frequency may be used in more than one surrounding cell. To avoid ambiguity a 6-b BSIC is transmitted on each BCCH carrier in the SB of Fig. 27.6. Two other parameters transmitted in the BCCH data provide additional information about the BS. The binary flag called **PLMN_PERMITTED** indicates whether the measured BCCH carrier belongs to a PLMN that the MS is permitted to access. The second Boolean flag, **CELL_BAR_ACCESS,** indicates whether the cell is barred for access by the MS, although

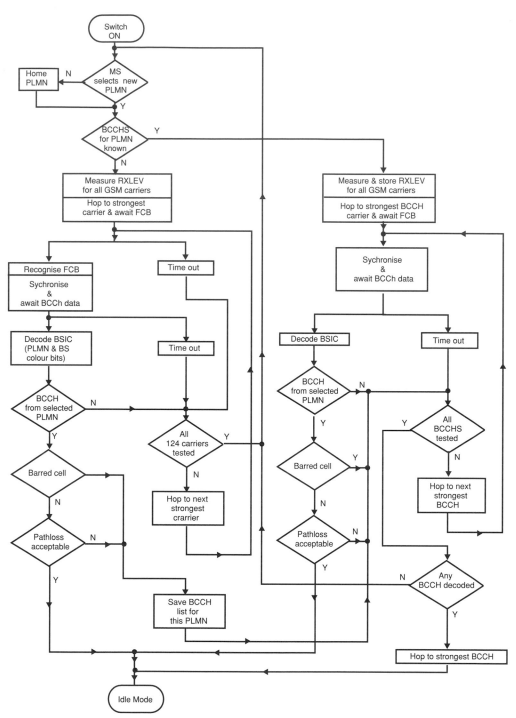

FIGURE 27.10   Initial cell selection by the MS, ©ETT [4].

it belongs to a permitted PLMN. A MS in idle mode, i.e., after it has just been switched on or after it has lost contact with the network, searches all 125 RF channels and takes readings of RXLEV on each of them. Then it tunes to the carrier with the highest RXLEV and searches for FCB in order to determine whether or not the carrier is a BCCH carrier. If it is not, then the MS tunes to the next highest carrier, and so on, until it finds a BCCH carrier, synchronizes to it and decodes the parameters BSIC, PLMN_PERMITTED and CELL_BAR_ ACCESS in order to decide whether to continue the search. The MS may store the BCCH carrier frequencies used in the network accessed, in which case the search time would be reduced. Again, the process described is summarized in the flowchart of Fig. 27.10.

The adaptive power control is based on RXLEV measurements. In every SACCH multi-frame the BS compares the RXLEV readings reported by the MS or obtained by the base station with a set of thresholds. The exact strategy for RF power control is determined by the network operator with the aim of providing an adequate quality of service for speech and data transmissions while keeping interferences low. Clearly, adequate quality must be achieved at the lowest possible transmitted power to keep cochannel interferences low, which implies contradictory requirements in terms of transmitted power. The criteria for reporting radio link failure are based on the measurements of RXLEV and RXQUAL performed by both the mobile and base stations, and the procedures for handling link failures result in the re-establishment or the release of the call, depending on the network operator's strategy.

The handover process involves the most complex set of procedures in the radio-link control. Handover decisions are based on results of measurements performed both by the base and mobile stations. The base station measures RXLEV, RXQUAL, DISTANCE, and also the interference level in unallocated time slots, whereas the MS measures and reports to the BS the values of RXLEV and RXQUAL for the serving cell and RXLEV for the adjacent cells. When the MS moves away from the BS, the RXLEV and RXQUAL parameters for the serving station become lower, whereas RXLEV for one of the adjacent cells increases.

27.10 Discontinuous Transmission

Discontinuous transmission (DTX) issues are standardized in recommendation R.06.31, whereas the associated problems of voice activity detection VAD are specified by R.06.32. Assuming an average speech activity of 50% and a high number of interferers combined with frequency hopping to randomize the interference load, significant spectral efficiency gains can be achieved when deploying discontinuous transmissions due to decreasing interferences, while reducing power dissipation as well. Because of the reduction in power consumption, full DTX operation is mandatory for MSs, but in BSs, only receiver DTX functions are compulsory.

The fundamental problem in voice activity detection is how to differentiate between speech and noise, while keeping false noise triggering and speech spurt clipping as low as possible. In vehicle-mounted MSs the severity of the speech/noise recognition problem is aggravated by the excessive vehicle background noise. This problem is resolved by deploying a combination of threshold comparisons and spectral domain techniques [1, 3]. Another important associated problem is the introduction of noiseless inactive segments, which is mitigated by comfort noise insertion (**CNI**) in these segments at the receiver.

27.11   Summary

Following the standardization and launch of the GSM system its salient features were summarized in this brief review. Time division multiple access (TDMA) with eight users per carrier is used at a multiuser rate of 271 kb/s, demanding a channel equalizer to combat dispersion in large cell environments. The error protected chip rate of the full-rate traffic channels is 22.8 kb/s, whereas in half-rate channels it is 11.4 kb/s. Apart from the full- and half-rate speech traffic channels, there are 5 different rate data traffic channels and 14 various control and signalling channels to support the system's operation. A moderately complex, 13 kb/s regular pulse excited speech codec with long term predictor (**LTP**) is used, combined with an embedded three-class error correction codec and multilayer interleaving to provide sensitivity-matched unequal error protection for the speech bits. An overall speech delay of 57.5 ms is maintained. Slow frequency hopping at 217 hops/s yields substantial performance gains for slowly moving pedestrians.

TABLE 27.3   Summary of GSM Features

System feature	Specification
Up-link bandwidth, MHz	890–915 = 25
Down-link bandwidth, MHz	935–960 = 25
Total GSM bandwidth, MHz	50
Carrier spacing, KHz	200
No. of RF carriers	125
Multiple access	TDMA
No. of users/carrier	8
Total No. of channels	1000
TDMA burst rate, kb/s	271
Modulation	GMSK with BT = 0.3
Bandwidth efficiency, b/s/Hz	1.35
Channel equalizer	yes
Speech coding rate, kb/s	13
FEC coded speech rate, kb/s	22.8
FEC coding	Embedded block/ convolutional
Frequency hopping, hop/s	217
DTX and VAD	yes
Maximum cell radius, km	35

Constant envelope partial response GMSK with a channel spacing of 200 kHz is deployed to support 125 duplex channels in the 890–915-MHz up-link and 935–960-MHz down-link bands, respectively. At a transmission rate of 271 kb/s a spectral efficiency of 1.35-bit/s/Hz is achieved. The controlled GMSK-induced and uncontrolled channel-induced intersymbol interferences are removed by the channel equalizer. The set of standardized wideband GSM channels was introduced in order to provide bench markers for performance comparisons. Efficient power budgeting and minimum cochannel interferences are ensured by the combination of adaptive power and handover control based on weighted averaging of up to eight up-link and down-link system parameters. Discontinuous transmissions assisted by reliable spectral-domain voice activity detection and comfort-noise insertion further reduce interferences and power consumption. Because of ciphering, no unprotected information is sent via the radio link. As a result, spectrally efficient, high-quality mobile communications with a variety of services and international roaming is possible in cells of up to 35 km radius for signal-to-noise and interference ratios in excess of 10–12 dBs. The key system features are summarized in Table 27.3.

Defining Terms

A3: Authentication algorithm

A5: Cyphering algorithm

A8: Confidential algorithm to compute the cyphering key

AB: Access burst

ACCH: Associated control channel

ADC: Administration center

AGCH: Access grant control channel

AUC: Authentication center

AWGN: Additive Gaussian noise

BCCH: Broadcast control channel

BER: Bit error ratio

BFI: Bad frame indicator flag

BN: Bit number

BS: Base station

BS-PBGT: BS powerbudget: to be evaluated for power budget motivated handovers

BSIC: Base station identifier code

CC: Convolutional codec

CCCH: Common control channel

CELL_BAR_ACCESS: Boolean flag to indicate, whether the MS is permitted to access the specific traffic cell

CNC: Comfort noise computation

CNI: Comfort noise insertion

CNU: Comfort noise update state in the DTX handler

DB: Dummy burst

DL: Down link

DSI: Digital speech interpolation to improve link efficiency

DTX: Discontinuous transmission for power consumption and interference reduction

EIR: Equipment identity register

EOS: End of speech flag in the DTX handler

FACCH: Fast associated control channel

FCB: Frequency correction burst

FCCH: Frequency correction channel

FEC: Forward error correction

FH: Frequency hopping

FN: TDMA frame number

GMSK: Gaussian minimum shift keying

GP: Guard space

HGO: Handover in the VAD

HLR: Home location register

HO: Handover

HOCT: Handover counter in the VAD

HO_MARGIN: Handover margin to facilitate hysteresis

HSN: Hopping sequence number: frequency hopping algorithm's input variable

IMSI: International mobile subscriber identity

ISDN: Integrated services digital network

LAI: Location area identifier

LAR: Logarithmic area ratio

LTP: Long term predictor

MA: Mobile allocation: set of legitimate RF channels, input variable in the frequency hopping algorithm

MAI: Mobile allocation index: output variable of the FH algorithm

MAIO: Mobile allocation index offset: initial RF channel offset, input variable of the FH algorithm

MS: Mobile station

MSC: Mobile switching center

MSRN: Mobile station roaming number

MS_TXPWR_MAX: Maximum permitted MS transmitted power on a specific traffic channel in a specific traffic cell

MS_TXPWR_MAX(n): Maximum permitted MS transmitted power on a specific traffic channel in the nth adjacent traffic cell

NB: Normal burst

NMC: Network management center

NUFR: Receiver noise update flag

NUFT: Noise update flag to ask for SID frame transmission

OMC: Operation and maintenance center

PARCOR: Partial correlation

PCH: Paging channel

PCM: Pulse code modulation

PIN: Personal identity number for MSs

PLMN: Public land mobile network

PLMN_ PERMITTED: Boolean flag to indicate whether the MS is permitted to access the specific PLMN

PSTN: Public switched telephone network

QN: Quarter bit number

R: Random number in the authentication process

RA: Rural area channel impulse response

RACH: Random access channel

RF: Radio frequency

RFCH: Radio frequency channel

RFN: Reduced TDMA frame number: equivalent representation of the TDMA frame number that is used in the synchronization channel

RNTABLE: Random number table utilized in the frequency hopping algorithm

RPE: Regular pulse excited

RPE-LTP: Regular pulse excited codec with long term predictor

RS-232: Serial data transmission standard equivalent to CCITT V24. interface

RXLEV: Received signal level: parameter used in handovers

RXQUAL: Received signal quality: parameter used in handovers

S: Signed response in the authentication process

SACCH: Slow associated control channel

SB: Synchronization burst

SCH: Synchronization channel

SCPC: Single channel per carrier

SDCCH: Stand-alone dedicated control channel

SE: Speech extrapolation

SID: Silence identifier

SIM: Subscriber identity module in MSs

SPRX: Speech received flag

SPTX: Speech transmit flag in the DTX handler

STP: Short term predictor

TA: Timing advance

TB: Tailing bits

TCH: Traffic channel

TCH/F: Full-rate traffic channel

TCH/F2.4: Full-rate 2.4-kb/s data traffic channel

TCH/F4.8: Full-rate 4.8-kb/s data traffic channel

TCH/F9.6: Full-rate 9.6-kb/s data traffic channel

TCH/FS: Full-rate speech traffic channel

TCH/H: Half-rate traffic channel

TCH/H2.4: Half-rate 2.4-kb/s data traffic channel

TCH/H4.8: Half-rate 4.8-kb/s data traffic channel

TDMA: Time division multiple access

TMSI: Temporary mobile subscriber identifier

TN: Time slot number

TU: Typical urban channel impulse response

TXFL: Transmit flag in the DTX handler

UL: Up link

VAD: Voice activity detection

VE: Viterbi equalizer

VLR: Visiting location register

References

[1] European Telecommunications Standardization Institute. Group Speciale Mobile or Global System of Mobile Communication (GSM) Recommendation, ETSI Secretariat, Sophia Antipolis Cedex, France, 1988.

[2] Greenwood, D. and Hanzo, L., Characterisation of mobile radio channels, In *Mobile Radio Communications*. Steele, R., Ed., Chap. 2, 92–185. IEEE Press–Pentech Press, London, 1992.

[3] Hanzo, L. and Stefanov, J., The Pan-European digital cellular mobile radio system—known as GSM. In *Mobile Radio Communications*, Steele, R., Ed., Chap. 8, 677–773, IEEE Press–Pentech Press, London, 1992.

[4] Hanzo, L. and Steele, R., The Pan-European mobile radio system, Pts. 1 and 2, *European Trans. on Telecomm.*, 5(2), 245–276, 1994.

[5] Salami, R.A., Hanzo, L., et al., Speech coding. In *Mobile Radio Communications*, Steele, R., Ed., Chap. 3, 186–346. IEEE Press–Pentech Press, London, 1992.

[6] Steele, R. Ed., *Mobile Radio Communications*, IEEE Press–Pentech Press, London, 1992.

[7] Vary, P. and Sluyter, R.J., MATS-D speech codec: Regular-pulse excitation LPC, *Proceedings of Nordic Conference on Mobile Radio Communications*. 257–261, 1986.

[8] Wong, K.H.H. and Hanzo, L., Channel coding. In *Mobile Radio Communications*. Steele, R., Ed., Chap. 4, 347–488. IEEE Press–Pentech Press, London, 1992.

28

Speech and Channel Coding for North American TDMA Cellular Systems

Paul Mermelstein
INRS-Télécommunications
University of Québec

28.1 Introduction

The goals of this chapter are to give the reader a tutorial introduction and high-level understanding of the techniques employed for speech transmission by the IS-54 digital cellular standard. It builds on the information provided in the standards document but is not meant to be a replacement for it. Separate standards cover the control channel used for the setup of calls and their handoff to neighboring cells, as well as the encoding of data signals for transmission. For detailed implementation information the reader should consult the most recent standards document [9].

IS-54 provides for encoding bidirectional speech signals digitally and transmitting them over cellular and microcellular mobile radio systems. It retains the 30-kHz channel spacing of the earlier advanced mobile telephone service (AMPS), which uses analog frequency modulation for speech transmission and frequency shift keying for signalling. The two directions of transmission use frequencies some 45 MHz apart in the band between 824 and 894 MHz. AMPS employs one channel per conversation in each direction, a technique known as frequency division multiple access (FDMA). IS-54 employs time division multiple access (TDMA) by allowing three, and in the future six, simultaneous transmissions to share each frequency band. Because the overall 30-kHz channelization of the allocated 25 MHz of spectrum in each direction is retained, it is also known as a FDMA-TDMA system. In contrast, the later IS-95 standard employs code division multiple access (CDMA) over bands of 1.23 MHz by combining several 30-kHz frequency channels.

Each frequency channel provides for transmission at a digital bit rate of 48.6 kb/s through use of differential quadrature-phase shift key (DQPSK) modulation at a 24.3-kBd channel rate. The channel is divided into six time slots every 40 ms. The full-rate voice coder employs every third time slot and utilizes 13 kb/s for combined speech and channel coding. The six slots provide for an eventual half-rate channel occupying one slot per 40 ms frame and utilizing only about 6.5 kb/s for each call. Thus, the simultaneous call carrying capacity with IS-54 is increased by a factor 3(factor 6 in the future) above that of AMPS. All digital transmission is expected to result in a reduction in transmitted power. The resulting reduction in intercell interference may allow more frequent reuse of the same frequency channels than the reuse pattern of seven cells for AMPS. Additional increases in erlang capacity (the total call-carrying capacity at a given blocking rate) may be available from the increased trunking efficiency achieved by the larger number of simultaneously available channels. The first systems employing dual-mode AMPS and TDMA service were put into operation in 1993.

In 1996 the TIA introduced the IS-641 enhanced full rate codec. This codec consists of 7.4 kb/s speech coding following the algebraic code-excited linear prediction (ACELP) technique [7], and 5.6 kb/s channel coding. The 13 kb/s coded information replaces the combined 13 kb/s for speech and channel coding introduced by the IS-54 standard. The new codec provides significant enhancements in terms of speech quality and robustness to transmission errors. The quality enhancement for clear channels results from the improved modeling of the stochastic excitation by means of an algebraic **codebook** instead of the two trained VSELP codebooks. Improved robustness to transmission errors is achieved by employing predictive quantization techniques for the linear-prediction filter and gain parameters, and increasing the number of bits protected by forward error correction.

28.2   Modulation of Digital Voice and Data Signals

The modulation method used in IS-54 is $\pi/4$ shifted differentially encoded quadrature phase-shift keying (DPSK). Symbols are transmitted as changes in phase rather than their absolute values. The binary data stream is converted to two binary streams X_k and Y_k formed from the odd- and even-numbered bits, respectively. The quadrature streams I_k and Q_k are formed according to

$$I_k = I_{k-1}\cos\left[\Delta\phi\left(X_k, Y_k\right)\right] - Q_{k-1}\sin\left[\Delta\phi\left(X_k, Y_k\right)\right]$$
$$Q_k = I_{k-1}\sin\left[\Delta\phi\left(X_k, Y_k\right)\right] + Q_{k-1}\cos\left[\Delta\phi\left(X_k, Y_k\right)\right]$$

where I_{k-1} and Q_{k-1} are the amplitudes at the previous pulse time. The phase change $\Delta\phi$ takes the values $\pi/4, 3\pi/4, -\pi/4$, and $-3\pi/4$ for the dibit (X_k, Y_k) symbols (0,0), (0,1),

(1,0) and (1,1), respectively. This results in a rotation by $\pi/4$ between the constellations for odd and even symbols. The differential encoding avoids the problem of 180° phase ambiguity that may otherwise result in estimation of the carrier phase.

The signals I_k and Q_k at the output of the differential phase encoder can take one of five values, $0, \pm 1, \pm 1/\sqrt{2}$ as indicated in the constellation of Fig. 28.1. The corresponding impulses are applied to the inputs of the I and Q baseband filters, which have linear phase and square root raised cosine frequency responses. The generic modulator circuit is shown in Fig. 28.2. The rolloff factor α determines the width of the transition band and its value is 0.35,

$$|H(f)| = \begin{cases} 1, & 0 \le f \le (1-\alpha)/2T \\ \sqrt{1/2\{1 - \sin[\pi(2fT - 1)/2\alpha]\}}, & (1-\alpha)/2T \le f \le (1+\alpha)/2T \\ 0, & f > (1+\alpha)/2T \end{cases}$$

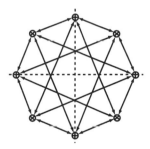

FIGURE 28.1   Constellation for $\pi/4$ shifted QPSK modulation. *Source*: TIA, 1992. Cellular System Dual-mode Mobile Station–Base Station Compatibility Standard TIA/EIA IS-54. With permission.

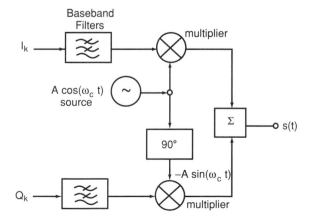

FIGURE 28.2   Generic modulation circuit for digital voice and data signals. *Source*: TIA, 1992. Cellular System Dual-mode Mobile Station–Base Station Compatibility Standard TIA/EIA IS-54.

28.3   Speech Coding Fundamentals

The IS-54 standard employs a vector-sum excited linear prediction (VSELP) coding technique. It represents a specific formulation of the much larger class of code-excited linear prediction (CELP) coders [2] that have proved effective in recent years for the coding of speech at moderate rates in the range 4–16 kb/s. VSELP provides reconstructed speech with a quality that is comparable to that available with frequency modulation and analog transmission over the AMPS system. The coding rate employed is 7.95 kb/s. Each of the six slots per frame carry 260 b of speech and channel coding information for a gross information rate of 13 kb/s. The 260 b correspond to 20 ms of real time speech, transmitted as a single burst.

For an excellent recent review of speech coding techniques for transmission, the reader is referred to Gersho, 1994 [3]. Most modern speech coders use a form of analysis by synthesis coding where the encoder determines the coded signal one segment at a time by feeding candidate excitation segments into a replica of a synthesis filter and selecting the segment that minimizes the distortion between the original and reproduced signals. Linear prediction coding (LPC) techniques [1] encode the speech signal by first finding an optimum linear filter to remove the short-time correlation, passing the signal through that LPC filter to obtain a residual signal, and encoding this residual using much fewer bits than would have been required to code the original signal with the same fidelity. In most cases the coding of the residual is divided into two steps. First, the long-time correlation due to the periodic pitch excitation is removed by means of an optimum one-tap filter with adjustable gain and lag. Next, the remaining residual signal, which now closely resembles a white-noise signal, is encoded. Code-excited linear predictors use one or more codebooks from which they select replicas of the residual of the input signal by means of a closed-loop error-minimization technique. The index of the codebook entry as well as the parameters of all the filters are transmitted to allow the speech signal to be reconstructed at the receiver. Most code-excited coders use trained codebooks. Starting with a codebook containing Guassian signal segments, entries that are found to be used rarely in coding a large body of speech data are iteratively eliminated to result in a smaller codebook that is considered more effective.

The speech signal can be considered quasistationary or stationary for the duration of the speech frame, of the order of 20 ms. The parameters of the short-term filter, the LPC coefficients, are determined by analysis of the autocorrelation function of a suitably windowed segment of the input signal. To allow accurate determination of the time-varying pitch lag as well as simplify the computations, each speech frame is divided into four 5-ms subframes. Independent pitch filter computations and residual coding operations are carried out for each subframe.

The speech decoder attempts to reconstruct the speech signal from the received information as best possible. It employs a codebook identical to that of the encoder for excitation generation and, in the absence of transmission errors, would produce an exact replica of the signal that produced the minimized error at the encoder. Transmission errors do occur, however, due, to signal fading and excessive interference. Since any attempt at retransmission would incur unacceptable signal delays, sufficient error protection is provided to allow correction of most transmission errors.

28.4   Channel Coding Considerations

The sharp limitations on available bandwidth for error protection argue for careful consideration of the sensitivity of the speech coding parameters to transmission errors. Pairwise interleaving of coded blocks and convolutional coding of a subset of the parameters permit correction of a limited number of transmission errors. In addition, a cyclic redundancy check (CRC) is used to determine whether the error correction was successful. The coded information is divided into three blocks of varying sensitivity to errors. Group 1 contains the most sensitive bits, mainly the parameters of the LPC filter and frame energy, and is protected by both error detection and correction bits. Group 2 is provided with error correction only. The third group, comprising mostly the fixed codebook indices, is not protected at all.

The speech signal contains significant temporal redundancy. Thus, speech frames within which errors have been detected may be reconstructed with the aid of previously correctly received information. A bad-frame masking procedure attempts to hide the effects of short fades by extrapolating the previously received parameters. Of course, if the errors persist, the decoded signal must be muted while an attempt is made to hand off the connection to a base station to/from which the mobile may experience better reception.

28.5 VSELP Encoder

A block diagram of the VSELP speech encoder [4] is shown in Fig. 28.3. The excitation signal is generated from three components, the output of a long term or pitch filter, as well as entries from two codebooks. A weighted synthesis filter generates a synthesized approximation to the frequency-weighted input signal. The weighted mean square error between these two signals is used to drive the error minimization process. This weighted error is considered to be a better approximation to the perceptually important noise components than the unweighted mean square error. The total weighted square error is minimized by adjusting the pitch lag and the codebook indices as well as their gains. The decoder follows the encoder closely and generates the excitation signal identically to the encoder but uses an unweighted linear-prediction synthesis filter to generate the decoded signal. A spectral postfilter is added after the synthesis filter to enhance the quality of the reconstructed speech.

The precise data rate of the speech coder is 7950 b/s or 159 b per time slot, each corresponding to 20 ms of signal in real time. These 159 b are allocated as follows: 1) short-term filter coefficients, 38 bits; 2) frame energy, 5 bits; 3) pitch lag, 28 bits; 4) codewords, 56 bits; and 5) gain values, 32 bits.

28.6 Linear Prediction Analysis and Quantization

The purpose of the LPC analysis filter is to whiten the spectrum of the input signal so that it can be better matched by the codebook outputs. The corresponding LPC synthesis filter $A(z)$ restores the short-time speech spectrum characteristics to the output signal. The transfer function of the tenth-order synthesis filter is given by

$$A(z) = \frac{1}{1 - \sum_{i=1}^{N_p} \alpha_i z^{-i}}$$

The filter predictor parameters $\alpha_1, \ldots, \alpha_{N_p}$ are not transmitted directly. Instead, a set of **reflection coefficients** $r_1, \ldots, r_{N_p}$ are computed and quantized. The predictor parameters are determined from the reflection coefficients using a well-known backward recursion algorithm [6].

A variety of algorithms are known that determine a set of reflection coefficients from a windowed input signal. One such algorithm is the fixed point **covariance lattice**, FLAT,

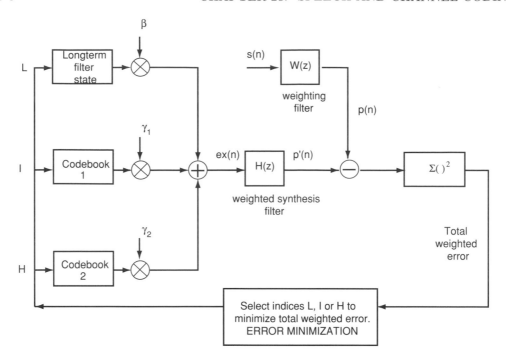

FIGURE 28.3   Black diagram of the speech encoder in VSELP. TIA. 1992. Cellular system Dual-mode Mobile Station–Base Station Compatibility Standard. TIA/EIA IS-54.

which builds an optimum inverse lattice stage by stage. At each stage j, the sum of the mean-squared forward and backward residuals is minimized by selection of the best reflection coefficient r_j. The analysis window used is 170 samples long, centered with respect to the middle of the fourth 5-ms subframe of the 20-ms frame. Since this centerpoint is 20 samples from the end of the frame, 65 samples from the next frame to be coded are used in computing the reflection coefficient of the current frame. This introduces a lookahead delay of 8.125 ms.

The FLAT algorithm first computes the covariance matrix of the input speech for $N_A = 170$ and $N_p = 10$,

$$\phi(i,k) = \sum_{n=N_p}^{N_A-1} s(n-i)s(n-k), \qquad 0 \le i, \quad k \le N_p \,,$$

Define the forward residual out of stage j as $f_j(n)$ and the backward residual as $b_j(n)$. Then the autocorrelation of the initial forward residual $F_0(i,k)$ is given by $\phi(i,k)$. The autocorrelation of the initial backward residual $B_0(i,k)$ is given by $\phi(i+1,k+1)$ and the initial cross correlation of the two residuals is given by $C_0(i,k) = \phi(i,k+1)$ for $0 \le i, k \le N_{p-1}$. Initially j is set to 1. The **reflection coefficient** at each stage is determined as the ratio of the cross correlation to the mean of the autocorrelations. A block diagram of the computations is shown in Fig. 28.4. By quantizing the reflection coefficients within the computation loops, reflection coefficients at subsequent stages are computed taking into account the quantization errors of the previous stages. Specifically,

$$
\begin{aligned}
C'_{j-1} &= C_{j-1}(0,0) + C_{j-1}(N_p - j, N_p - j) \\
F'_{j-1} &= F_{j-1}(0,0) + F_{j-1}(N_p - j, N_p - j) \\
B'_{j-1} &= B_{j-1}(0,0) + B_{j-1}(N_p - j, N_p - j)
\end{aligned}
$$

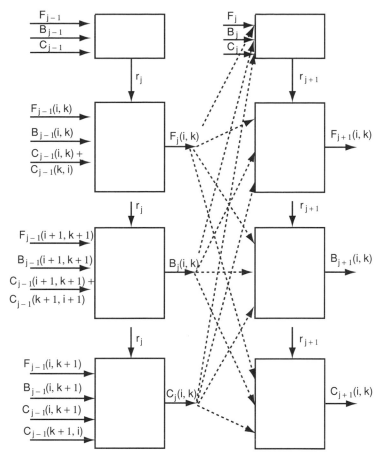

FIGURE 28.4 Block diagram for lattice covariance computations.

and

$$r_j = \frac{-2C'_{j-1}}{F'_{j-1} + B'_{j-1}}$$

Use of two sets of correlation values separated by $N_p - j$ samples provides additional stability to the computed reflection coefficients in case the input signal changes form rapidly.

Once a quantized reflection coefficient r_j has been determined, the resulting auto- and cross correlations can be determined iteratively as

$$
\begin{aligned}
F_j(i,k) &= F_{j-1}(i,k) + r_j[C_{j-1}(i,k) + C_{j-1}(k,i)] + r_j^2 B_{j-1}(i,k) \\
B_j(i,k) &= B_{j-1}(i+1,k+1) + r_j[C_{j-1}(i+1,k+1) + C_{j-1}(k+1,i+1)] \\
&\quad + r_j^2 F_{j-1}(i+1,k+1)
\end{aligned}
$$

and

$$
\begin{aligned}
C_j(i,k) &= C_{j-1}(i,k+1) + r_j[B_{j-1}(i,k+1) + F_{j-1}(i,k+1)] \\
&\quad + r_j^2 C_{j-1}(k+1,i)
\end{aligned}
$$

These computations are carried out iteratively for $r_j, j = 1, \ldots, N_p$.

28.7　Bandwidth Expansion

Poles with very narrow bandwidths may introduce undesirable distortions into the synthesized signal. Use of a binomial window with effective bandwidth of 80 Hz suffices to limit the ringing of the LPC filter and reduce the effect of the LPC filter selected for one frame on the signal reconstructed for subsequent frames. To achieve this, prior to searching for the reflection coefficients, the $\phi(i, k)$ is modified by use of a window function $w(j), j = 1, \ldots, 10$, as follows:

$$\phi'(i, k) = \phi(i, k)w(|i - k|)$$

28.8　Quantizing and Encoding the Reflection Coefficients

The distortion introduced into the overall spectrum by quantizing the reflection coefficients diminishes as we move to higher orders in the reflection coefficients. Accordingly, more bits are assigned to the lower order coefficients. Specifically, 6, 5, 5, 4, 4, 3, 3, 3, 3, and 2 b are assigned to $r_1, \ldots, r_{10}$, respectively. Scalar quantization of the reflection coefficients is used in IS-54 because it is particularly simple. **Vector quantization** achieves additional quantizing efficiencies at the cost of significant added complexity.

It is important to preserve the smooth time evolution of the linear prediction filter. Both the encoder and decoder linearly interpolate the coefficients α_i for the first, second and third subframes of each frame using the coefficients determined for the previous and current frames. The fourth subframe uses the values computed for that frame.

28.9　VSELP Codebook Search

The codebook search operation selects indices for the long-term filter (pitch lag L) and the two codebooks I and H so as to minimize the total weighted error. This closed-loop search is the most computationally complex part of the encoding operation, and significant effort has been invested to minimize the complexity of these operations without degrading performance. To reduce complexity, simultaneous optimization of the codebook selections is replaced by a sequential optimization procedure, which considers the long-term filter search as the most significant and therefore executes it first. The two vector-sum codebooks are considered to contribute less and less to the minimization of the error, and their search follows in sequence. Subdivision of the total codebook into two vector sums simplifies the processing and makes the result less sensitive to errors in decoding the individual bits arising from transmission errors.

Entries from each of the two vector-sum codebooks can be expressed as the sum of basis vectors. By orthogonalizing these basis vectors to the previously selected codebook component(s), one ensures that the newly introduced components reduce the remaining errors. The subframes over which the codebook search is carried out are 5 ms or 40 samples long. An optimal search would need exploration of a 40-dimensional space. The vector-sum approximation limits the search to 14 dimensions after the optimal pitch lag has been selected. The search is further divided into two stages of 7 dimensions each. The two codebooks are specified in terms of the fourteen, 40-dimensional basis vectors stored at the encoder and decoder. The two 7-b indices indicate the required weights on the basic vectors to arrive at the two optimum codewords.

The codebook search can be viewed as selecting the three best directions in 40-dimensional space, which when summed result in the best approximation to the weighted input signal.

The gains of the three components are determined through a separate error minimization process.

28.10 Long-Term Filter Search

The long-term filter is optimized by selection of a lag value that minimizes the error between the weighted input signal $p(n)$ and the past excitation signal filtered by the current weighted synthesis filter $H(z)$. There are 127 possible coded lag values provided corresponding to lags of 20–146 samples. One value is reserved for the case when all correlations between the input and the lagged residuals are negative and use of no long term filter output would be best. To simplify the convolution operation between the impulse response of the weighted synthesis filter and the past excitation, the impulse response is truncated to 21 samples or 2.5 ms. Once the lag is determined, the untruncated impulse response is used to compute the weighted long-term lag vector.

28.11 Orthogonalization of the Codebooks

Prior to the search of the first codebook, each filtered basis vector may be made orthogonal to the long-term filter output, the zero-state response of the weighted synthesis filter $H(z)$ to the long-term prediction vector. Each orthogonalized filtered basis vector is computed by subtracting its projection onto the long-term filter output from itself.

Similarly, the basis vectors of the second codebook can be orthogonalized with respect to both the long-term filter output and the first codebook output, the zero-state response of $H(z)$ to the previously selected summation of first-codebook basis vectors. In each case the codebook excitation can be reconstituted as

$$u_{k,i}(n) = \sum_{m=1}^{M} \theta_{im} v_{k,m}(n)$$

where $k = 1, 2$ for the two codebooks, $i = I$ or H the 7-b code vector received, $v_{k,m}$ are the two sets of basis vectors, and $\theta_{im} = +1$ if bit m of codeword $i = 1$ and -1 if bit m of codeword $i = 0$. Orthogonalization is not required at the decoder since the gains of the codebooks outputs are determined with respect to the weighted nonorthogonalized code vectors.

28.12 Quantizing the Excitation and Signal Gains

The three codebook gain values β, γ_1, and γ_2 are transformed to three new parameters $GS, P0$ and $P1$ for quantization purposes. GS is an energy offset parameter that equalizes the input and output signal energies. It adjusts the energy of the output of the LPC synthesis filter to equal the energy computed for the same subframe at the encoder input. $P0$ is the energy contribution of the long-term prediction vector as a fraction of the total excitation energy within the subframe. Similarly, $P1$ is the energy contribution of the code vector selected from the first codebook as a fraction of the total excitation energy of the subframe. The transformation reduces the dynamic range of the parameters to be encoded. An 8-b **vector quantizer** efficiently encodes the appropriate $(GS, P0, P1)$ vectors by selecting the vector which minimizes the weighted error. The received and decoded values β, γ_1, and γ_2 are computed from the received $(GS, P0, P1)$ vector and applied to reconstitute the decoded signal.

28.13   Channel Coding and Interleaving

The goals of channel coding are to reduce the impairments in the reconstructed speech due to transmission errors. The 159 b characterizing each 20-ms block of speech are divided into two classes, 77 in class 1 and 82 in class 2. Class 1 includes the bits in which errors result in a more significant impairment, whereas the speech quality is considered less sensitive to the class- 2 bits. Class 1 generally includes the gain, pitch lag, and more significant reflection coefficient bits. In addition, a 7-b cyclic redundancy check is applied to the 12 most perceptually significant bits of class 1 to indicate whether the error correction was successful. Failure of the CRC check at the receiver suggests that the received information is so erroneous that it would be better to discard it than use it. The error correction coding is illustrated in Fig. 28.5.

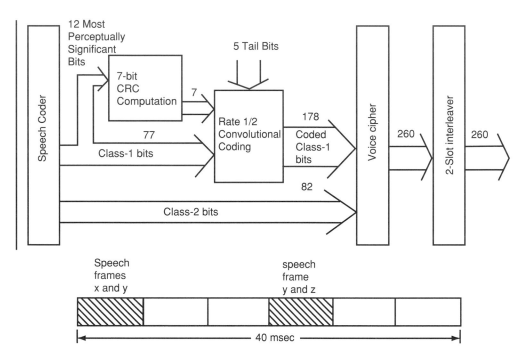

FIGURE 28.5   Error correction insertion for speech coder. Source TIA, 1992. Cellular Systems Dual-Mode Mobile Station–Base Station Compatibility Standards. TIA/EIA IS-54. With permission.

The error correction technique used is rate 1/2 convolutional coding with a constraint length of 5 [5]. A tail of 5 b is appended to the 84 b to be convolutionally encoded to result in a 178-b output. Inclusion of the tail bits ensures independent decoding of successive time slots and no propagation of errors between slots.

Interleaving the bits to be transmitted over two time slots is introduced to diminish the effects of short deep fades and to improve the error-correction capabilities of the channel coding technique. Two speech frames, the previous and the present, are interleaved so that the bits from each speech block span two transmission time slots separated by 20 ms. The interleaving attempts to separate the convolutionally coded class-1 bits from one frame as much as possible in time by inserting noncoded class-2 bits between them.

28.14 Bad Frame Masking

A CRC failure indicates that the received data is unusable, either due to transmission errors resulting from a fade, or from pre-emption of the time slot by a control message (fast associated control channel, FACCH). To mask the effects that may result from leaving a gap in the speech signal, a masking operation based on the temporal redundancy between adjacent speech blocks has been proposed. Such masking can at best bridge over short gaps but cannot recover loss of signal of longer duration. The bad frame masking operation may follow a finite state machine where each state indicates an operation appropriate to the elapsed duration of the fade to which it corresponds. The masking operation consists of copying the previous LPC information and attenuating the gain of the signal. State 6 corresponds to error sequences exceeding 100 ms, for which the output signal is muted. The result of such a masking operation is generation of an extrapolation in the gap to the previously received signal, significantly reducing the perceptual effects of short fades. No additional delay is introduced in the reconstructed signal. At the same time, the receiver will report a high frequency of bad frames leading the system to explore handoff possibilities immediately. A quick successful handoff will result in rapid signal recovery.

28.15 ACELP Encoder

The ACELP encoder employs linear prediction analysis and quantization techniques similar to those used in VSELP and discussed in Section 28.6. The frame structure of 20 ms frames and 5 ms subframes is preserved. Linear prediction analysis is carried out for every frame. The ACELP encoder uses a long-term filter similar to the one discussed in Section 28.10 and represented as an adaptive codebook. The nonpredictable part of the LPC residual is represented in terms of ACELP codebooks, which replace the two VSELP codebooks shown in Fig. 28.3.

Instead of encoding the reflection coefficients as in VSELP, the information is transformed into line-spectral frequency pairs (LSP) [8]. The LSPs can be derived from linear prediction coefficients, a 10th order analysis generating 10 line-spectral frequencies (LSF), 5 poles, and 5 zeroes. The LSFs can be vector quantized and the LPC coefficients recalculated from the quantized LSFs. As long as the interleaved order of the poles and zeroes is preserved, quantization of the LSPs preserves the stability of the LPC synthesis filters. The LSPs of any frame can be better predicted from the values calculated and transmitted corresponding to previous frames, resulting in additional advantages. The long-term means of the LSPs are calculated for a large body of speech data and stored at both the encoder and decoder. First-order moving-average prediction is then used for the mean-removed LSPs. The time-prediction technique also permits use of predicted values for the LSPs in case uncorrectable transmissions errors are encountered, resulting in reduced speech degradation. To simplify the vector quantization operations, each LSP vector is split into 3 subvectors of dimensions 3, 3, and 4. The three subvectors are quantized with 8, 9, and 9 bits respectively, corresponding to a total bit assignment of 26 bits per frame for LPC information.

28.16 Algebraic Codebook Structure and Search

Algebraic codebooks contain relatively few pulses having nonzero values leading to rapid search of the possible innovation vectors, the vectors which together with the ACB output form the excitation of the LPC filter for the current subframe. In this implementation the 40-position innovation vector contains only four nonzero pulses and each can take on only values $+1$ and -1. The 40 positions are divided into four tracks and one pulse is selected from each track. The tracks are generally equally spaced but differ in their starting value, thus the first pulse can take on positions 0, 5, 10, 15, 20, 25, 30, or 35 and the second has possible positions 1, 6, 11, 16, 21, 26, 31, or 36. The first three pulse positions are coded with 3 bits and the fourth pulse position (starting positions 3 or 4) with 4 bits, resulting in a 17-bit sequence for the algebraic code of each subframe.

The algebraic codebook is searched by minimizing the mean square error between the weighted input speech and the weighted synthesized speech over the time span of each subframe. In each case the weighting is that produced by a perceptual weighting filter that has the effect of shaping the spectrum of the synthesis error signal so that it is better masked by spectrum of the current speech signal.

28.17   Quantization of the Gains for ACELP Encoding

The adaptive codebook gain and the fixed (algebraic) codebook gains are vector quantized using a 7-bit codebook. The gain codebook search is performed by minimizing the mean-square of the weighted error between the original and the reconstructed speech, expressed as a function of the adaptive codebook gain and a fixed codebook correction factor. This correction factor represents the log energy difference between a predicted gain and an esti-mated gain. The predicted gain is computed using fourth-order moving-average prediction with fixed coefficients on the innovation energy of each subframe. The result is a smoothed energy profile even in the presence of modest quantization errors. As discussed above in case of the LSP quantization, the moving-average prediction serves to provide predicted values even when the current frame information is lost due to transmission errors. Degradations resulting from loss of one or two frames of information are thereby mitigated.

28.18   Channel Coding for ACELP Encoding

The channel coding and interleaving operations for ACELP speech coding are similar to those discussed in Section 28.13 for VSELP coding. The number of bits protected by both error-detection (parity) and error-correction convolutional coding is increased to 48 from 12. Rate 1/2 convolutional coding is used on the 108 more significant bits, 96 class-1 bits, 7 CRC bits and the 5 tail bits of the convolutional coder, resulting in 216 coded class-1 bits. Eight of the 216 bits are dropped by puncturing, yielding 208 coded class-1 bits which are then combined with 52 nonprotected class-2 bits. As compared to the channel coding of the VSELP encoder, the numbers of protected bits is increased and the number of unprotected bits is reduced while keeping the overall coding structure unchanged.

28.19   Conclusions

The IS-54 digital cellular standard specifies modulation and speech coding techniques for mobile cellular systems that allow the interoperation of terminals built by a variety of manu-facturers and systems operated across the country by a number of different service providers. It permits speech communication with good quality in a transmission environment charac-terized by frequent multipath fading and significant intercell interference. Generally, the quality of the IS-54 decoded speech is better at the edges of a cell than the corresponding AMPS transmission due to the error mitigation resulting from channel coding. Near a base station or in the absence of significant fading and interference, the IS-54 speech quality is reported to be somewhat worse than AMPS due to the inherent limitations of the analysis–synthesis model in reconstructing arbitrary speech signals with limited bits. The IS-641 standard coder achieves higher speech quality, particularly at the edges of heavily occupied cells where transmission errors may be more numerous. At this time no new systems following the IS-54 standard are being introduced. Most base-stations have been converted to transmit and receive on the IS-641 standard as well and use of IS-54 transmissions is dropping rapidly. At the time of its introduction in 1996 the IS-641 coder represented the state of the art in terms of toll quality speech coding near 8 kb/s, a significant improvement over the IS-54 coder introduced in 1990. These standards represent reasonable engineering compromises between high performance and complexity sufficiently low to permit single-chip implementations in mobile terminals.

Both IS-54 and IS-641 are considered second generation cellular standards. Third generation cellular systems promise higher call capacities through better exploitation of the time-varying transmission requirements of speech conversations, as well as improved modulation and coding in wider spectrum bandwidths that achieve similar bit-error ratios but reduce the required transmitted power. Until such systems are introduced, the second generation TDMA systems can be expected to provide many years of successful cellular and personal communications services.

Defining Terms

Codebook: A set of signal vectors available to both the encoder and decoder.

Covariance lattice algorithm: An algorithm for reduction of the covariance matrix of the signal consisting of several lattice stages, each stage implementing an optimal first-order filter with a single coefficient.

Reflection coefficient: A parameter of each stage of the lattice linear prediction filter that determines 1) a forward residual signal at the output of the filter-stage by subtracting from the forward residual at the input a linear function of the backward residual, also 2) a backward residual at the output of the filter stage by subtracting a linear function of the forward residual from the backward residual at the input.

Vector quantizer: A quantizer that assigns quantized vectors to a vector of parameters based on their current values by minimizing some error criterion.

References

[1] Atal, B.S. and Hanauer, S.L., Speech analysis and synthesis by linear prediction of the speech wave. *J. Acoust. Soc. Am.*, 50, 637–655, 1971.

[2] Atal, B.S. and Schroeder, M., Stochastic coding of speech signals at very low bit rates. *Proc. Int. Conf. Comm.*, 1610–1613, 1984.

[3] Gersho, A., Advances in speech and audio compression. *Proc. IEEE*, 82, 900–918, 1994.

[4] Gerson, I.A. and Jasiuk, M.A., Vector sum excited linear prediction (VSELP) speech coding at 8 kbps. *Int. Conf. Acoust. Speech and Sig. Proc.*, ICASSP90, 461–464, 1990.

[5] Lin S. and Costello, D., *Error Control Coding: Fundamentals and Application,* Prentice Hall, Englewood Cliffs, NJ, 1983.

[6] Makhoul, J., Linear prediction, a tutorial review. *Proc. IEEE*, 63, 561–580, 1975.

[7] Salami, R., Laflamme, C., Adoul, J.P., and Massaloux, D., A toll quality 8 kb/s speech codec for the personal communication system (PCS). *IEEE Trans. Vehic. Tech.*, 43, 808–816, 1994.

[8] Soong, F.K. and Juang, B.H., Line spectrum pair (LSP) and speech data compression. *Proc. ICASSP'84*, 1.10.1–1.10.4, 1984.

[9] Telecommunications Industry Association, EIA/TIA Interim Standard, Cellular System Dual-mode Mobile Station–Base Station Compatibility Standard IS-54B, TIA/EIA, Washington, D.C., 1992.

Further Information

For a general treatment of speech coding for telecommunications, see N.S. Jayant and P. Noll, *Digital Coding of Waveforms*, Prentice Hall, Englewood, NJ, 1984. For a more detailed treatment of linear prediction techniques, see J. Markel and A. Gray, *Linear Prediction of Speech*, Springer–Verlag, NY, 1976.

29

The British Cordless Telephone Standard: CT-2

Lajos Hanzo
University of Southampton

29.1 History and Background

Following a decade of world-wide research and development (**R&D**), cordless telephones (**CT**) are now becoming widespread consumer products, and they are paving the way towards ubiquitous, low-cost personal communications networks (**PCN**) [7, 8]. The two most well-known European representatives of CTs are the digital European cordless telecommunications (**DECT**) system [1, 5] and the CT-2 system [2, 6]. Three potential application areas have been identified, namely, domestic, business, and public access, which is also often referred to as telepoint (**TP**).

In addition to conventional voice communications, CTs have been conceived with additional data services and local area network (**LAN**) applications in mind. The fundamental difference between conventional mobile radio systems and CT systems is that CTs have been designed for small to very small cells, where typically benign low-dispersion, dominant line-of-sight (**LOS**) propagation conditions prevail. Therefore, CTs can usually dispense with channel equalizers and complex low-rate speech codecs, since the low-signal dispersion allows for the employment of higher bit rates before the effect of channel dispersion becomes a limiting factor. On the same note, the LOS propagation scenario is associated with mild fading or near-constant received signal level, and when combined with appropriate small-cell power-budget design, it ensures a high average signal-to-noise ratio (**SNR**).

0-8493-8597-0/99/$0.00+$.50

These prerequisites facilitate the employment of high-rate, low-complexity speech codecs, which maintain a low battery drain. Furthermore, the deployment of forward error correction codecs can often also be avoided, which reduces both the bandwidth requirement and the power consumption of the portable station (PS).

A further difference between public land mobile radio (**PLMR**) systems [3] and CTs is that whereas the former endeavor to standardize virtually all system features, the latter seek to offer a so-called access technology, specifying the common air interface (**CAI**), access and signalling protocols, and some network architecture features, but leaving many other characteristics unspecified. By the same token, whereas PLMR systems typically have a rigid frequency allocation scheme and fixed cell structure, CTs use dynamic channel allocation (**DCA**) [4]. The DCA principle allows for a more intelligent and judicious channel assignment, where the base station (BS) and PS select an appropriate traffic channel on the basis of the prevailing traffic and channel quality conditions, thus minimizing, for example, the effect of cochannel interference or channel blocking probability.

In contrast to PLMR schemes, such as the Pan-European global system of mobile communications (GSM) system [3], CT systems typically dispense with sophisticated mobility management, which accounts for the bulk of the cost of PLMR call charges, although they may facilitate limited hand-over capabilities. Whereas in residential applications CTs are the extension of the public switched telephone network (PSTN), the concept of omitting mobility management functions, such as location update, etc., leads to telepoint CT applications where users are able to initiate but not to receive calls. This fact drastically reduces the network operating costs and, ultimately, the call charge at a concomittant reduction of the services rendered.

Having considered some of the fundamental differences between PLMR and CT systems let us now review the basic features of the CT-2 system.

29.2   The CT-2 Standard

The European CT-2 recommendation has evolved from the British standard **MPT-1375** with the aim of ensuring the compatibility of various manufacturers' systems as well as setting performance requirements, which would encourage the development of cost-efficient implementations. Further standardization objectives were to enable future evolution of the system, for example, by reserving signalling messages for future applications and to maintain a low PS complexity even at the expense of higher BS costs. The CT-2 or MPT 1375 CAI recommendation is constituted by the four following parts.

1. *Radio interface:* Standardizes the radio frequency (RF) parameters, such as legitimate channel frequencies, the modulation method, the transmitter power control, and the required receiver sensitivity as well as the carrier-to-interference ratio (**CIR**) and the time division duplex (**TDD**) multiple access scheme. Furthermore, the transmission burst and master/slave timing structures to be used are also laid down, along with the scrambling procedures to be applied.

2. *Signalling layers one and two:* Defines how the bandwidth is divided among signalling, traffic data, and synchronization information. The description of the first signalling layer includes the dynamic channel allocation strategy, calling channel detection, as well as link setup and establishment algorithms. The second layer is concerned with issues of various signalling message formats, as well as link establishment and re-establishment procedures.

3. *Signalling layer three:* The third signalling layer description includes a range of message sequence diagrams as regards to call setup to telepoint BSs, private BSs, as well as the call clear down procedures.

4. *Speech coding and transmission:* The last part of the standard is concerned with the algorithmic and performance features of the audio path, including frequency responses, clipping, distortion, noise, and delay characteristics.

Having briefly reviewed the structure of the CT-2 recommendations let us now turn our attention to its main constituent parts and consider specific issues of the system's operation.

29.3 The Radio Interface

29.3.1 Transmission Issues

In our description of the system we will adopt the terminology used in the recommendation, where the PS is called cordless portable part (**CPP**), whereas the BS is referred to as cordless fixed part (**CFP**). The channel bandwidth and the channel spacing are 100 kHz, and the allocated system bandwidth is 40 MHz, which is hosted in the range of 864.15–868.15 MHz. Accordingly, a total of 40 RF channels can be utilized by the system.

The accuracy of the radio frequency must be maintained within ± 10 kHz of its nominal value for both the CFP and CPP over the entire specified supply voltage and ambient temperature range. To counteract the maximum possible frequency drift of 20 kHz, automatic frequency correction (**AFC**) may be used in both the CFP and CPP receivers. The AFC may be allowed to control the transmission frequency of only the CPP, however, in order to prevent the misalignment of both transmission frequencies.

Binary frequency shift keying (FSK) is proposed, and the signal must be shaped by an approximately Gaussian filter in order to maintain the lowest possible frequency occupancy. The resulting scheme is referred to as Gaussian frequency shift keying (**GFSK**), which is closely related to Gaussian minimum shift keying (GMSK) [7] used in the DECT [1] and GSM [3] systems.

Suffice to say that in M-arry FSK modems the carrier's frequency is modulated in accordance with the information to be transmitted, where the modulated signal is given by

$$S_i(t) = \sqrt{\frac{2E}{T}} \cos\left[\omega_i t + \Phi\right] \qquad i = 1, \ldots, M$$

and E represents the bit energy, T the signalling interval length, ω_i has M discrete values, whereas the phase Φ is constant.

29.3.2 Multiple Access and Burst Structure

The so-called TDD multiple access scheme is used, which is demonstrated in Fig. 29.1. The simple principle is to use the same radio frequency for both uplink and downlink transmissions between the CPP and the CFP, respectively, but with a certain staggering in time. This figure reveals further details of the burst structure, indicating that 66 or 68 b per TDD frame are transmitted in both directions.

There is a 3.5- or 5.5-b duration guard period (**GP**) between the uplink and downlink transmissions, and half of the time the CPP (the other half of the time the CFP) is transmitting with the other part listening, accordingly. Although the guard period wastes some channel capacity, it allows a finite time for both the CPP and CFP for switching from transmission to reception and vice versa. The burst structure of Fig. 29.1 is used during normal operation across an established link for the transmission of adaptive differential pulse code modulated (ADPCM) speech at 32 kb/s according to the CCITT G721 standard in a so-called B channel or bearer channel. The D channel, or signalling channel, is used for the transmission of link control signals. This specific burst structure is referred to as a multiplex one (**M1**) frame.

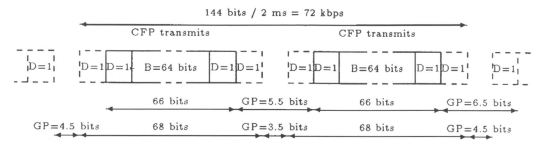

FIGURE 29.1   M1 burst and TDD frame structure.

Since the speech signal is encoded according to the CCITT G721 recommendation at 32 kb/s the TDD bit rate must be in excess of 64 kb/s in order to be able to provide the idle guard space of 3.5- or 5.5-b interval duration plus some signalling capacity. This is how channel capacity is sacrificed to provide the GP. Therefore, the transmission bit rate is stipulated to be 72 kb/s and the transmission burst length is 2 ms, during which 144-b intervals can be accommodated. As it was demonstrated in Fig. 29.1, 66 or 68 b are transmitted in both the uplink and downlink burst, and taking into account the guard spaces, the total transmission frame is constituted by $(2 \cdot 68) + 3.5 + 4.5 = 144$ b or equivalently, by $(2 \cdot 66) + 5.5 + 4.5 = 144$ b. The 66-b transmission format is compulsory, whereas the 68-b format is optional. In the 66-b burst there is one D bit dedicated to signalling at both ends of the burst, whereas in the 68-b burst the two additional bits are also assigned to signalling. Accordingly, the signalling rate becomes 2 b/2 ms or 4 b/2 ms, corresponding to 1 kb/s or 2 kb/s signalling rates.

29.3.3   Power Ramping, Guard Period, and Propagation Delay

As mentioned before and suggested Fig. 29.1, there is a 3.5- or 5.5-b interval duration GP between transmitted and received bursts. Since the signalling rate is 72 kb/s, the bit interval becomes about $1/(72 \text{ kb/s}) \approx 13.9$ μs and, hence, the GP duration is about 49 μs or 76 μs. This GP serves a number of purposes. Primarily, the GP allows the transmitter to ramp up and ramp down the transmitted signal level smoothly over a finite time interval at the beginning and end of the transmitted burst. This is necessary, because if the transmitted signal is toggled instantaneously, that is equivalent to multiplying the transmitted signal by a rectangular time-domain window function, which corresponds in the frequency domain to convolving the transmitted spectrum with a sinc function. This convolution would result in spectral side-lobes over a very wide frequency range, which would interfere with adjacent channels. Furthermore, due to the introduction of the guard period, both the CFP and CPP can tolerate a limited propagation delay, but the entire transmitted burst must arrive within the receivers' window, otherwise the last transmitted bits cannot be decoded.

29.3.4 Power Control

In order to minimize the battery drain and the cochannel interference load imposed upon cochannel users, the CT-2 system provides a power control option. The CPPs must be able to transmit at two different power levels, namely, either between 1 and 10 mW or at a level between 12 and 20 dB lower. The mechanism for invoking the lower CPP transmission level is based on the received signal level at the CFP. If the CFP detects a received signal strength more than 90 dB relative to 1 μV/m, it may instruct the CPP to drop its transmitted level by the specified 12–20 dB. Since the 90-dB gain factor corresponds to about a ratio of 31,623, this received signal strength would be equivalent for a 10-cm antenna length to an antenna output voltage of about 3.16 mV. A further beneficial ramification of using power control is that by powering down CPPs that are in the vicinity of a telepoint-type multiple-transceiver CFP, the CFP's receiver will not be so prone to being desensitised by the high-powered close-in CPPs, which would severely degrade the reception quality of more distant CPPs.

29.4 Burst Formats

As already mentioned in the previous section on the radio interface, there are three different subchannels assisting the operation of the CT-2 system, namely, the *voice/data channel* or *B channel,* the *signalling channel* or *D channel,* and the *burst synchronization channel* or **SYN** *channel.* According to the momentary system requirements, a variable fraction of the overall channel capacity or, equivalently, a variable fraction of the bandwidth can be allocated to any of these channels. Each different channel capacity or bandwidth allocation mode is associated with a different burst structure and accordingly bears a different name. The corresponding burst structures are termed as multiplex one (M1), multiplex two (**M2**), and multiplex three (**M3**), of which multiplex one used during the normal operation of established links has already been described in the previous section. Multiplex two and three will be extensively used during link setup and establishment in subsequent sections, as further details of the system's operation are unravelled.

Signalling layer one (L1) defines the burst formats multiplex one–three just mentioned, outlines the calling channel detection procedures, as well as link setup and establishment techniques. *Layer two* (**L2**) deals with issues of acknowledged and unacknowledged information transfer over the radio link, error detection and correction by retransmission, correct ordering of messages, and link maintenance aspects.

The burst structure multiplex two is shown in Fig. 29.2. It is constituted by two 16-b D-channel segments at both sides of the 10-b *preamble (P)* and the 24-b frame synchronization pattern (SYN), and its signalling capacity is 32 b/2 ms = 16 kb/s. Note that the M2 burst does not carry any B-channel information, it is dedicated to synchronization purposes. The 32-b D-channel message is split in two 16-b segments in order to prevent that any 24-b fraction of the 32-b word emulates the 24-b SYN segment, which would result in synchronization misalignment.

Since the CFP plays the role of the master in a telepoint scenario communicating with many CPPs, all of the CPP's actions must be synchronized to those of the CFP. Therefore, if the CPP attempts to initiate a call, the CFP will reinitiate it using the M2 burst, while imposing its own timing structure. The 10-b preamble consists of an alternate zero/one sequence and assists in the operation of the clock recovery circuitry, which has to be able to recover the clock frequency before the arrival of the SYN sequence, in order to be able to detect it. The SYN sequence is a unique word determined by computer search, which has a sharp autocorrelation peak, and its function is discussed later. The way the M2

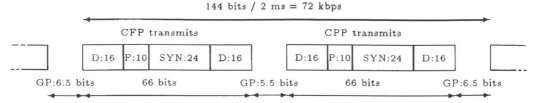

FIGURE 29.2   CT2 multiplex two burst structure.

and M3 burst formats are used for signalling purposes will be made explicit in our further discussions when considering the link setup procedures.

The specific SYN sequences used by the CFP and the CPP are shown in Table 29.1 along with the so-called *channel marker* (**CHM**) sequences used for synchronization purposes by the M3 burst format. Their differences will be made explicit during our further discourse. Observe from the table that the sequences used by the CFP and CPP, namely, SYNF, **CHMF** and SYNP, **CHMP,** respectively, are each other's bit-wise inverses. This was introduced in order to prevent CPPs and CFPs from calling each other directly. The CHM sequences are used, for instance, in residential applications, where the CFP can issue an M2 burst containing a 24-b CHMF sequence and a so-called poll message mapped on to the D-channel bits in order to wake up the specific CPP called. When the called CPP responds, the CFP changes the CHMF to SYNF in order to prevent waking up further CPPs unnecessarily.

TABLE 29.1   CT-2 Synchronization Patterns

	MSB (sent last)					LSB (sent first)
CHMF	1011	1110	0100	1110	0101	0000
CHMP	0100	0001	1011	0001	1010	1111
SYNCF	1110	1011	0001	1011	0000	0101
SYNCP	0001	0100	1110	0100	1111	1010

Since the CT-2 system does not entail mobility functions, such as registration of visiting CPPs in other than their own home cells, in telepoint applications all calls must be initiated by the CPPs. Hence, in this scenario when the CPP attempts to set up a link, it uses the so-called multiplex three burst format displayed in Fig. 29.3. The design of the M3 burst reflects that the CPP initiating the call is oblivious of the timing structure of the potentially suitable target CFP, which can detect access attempts only during its receive window, but not while the CFP is transmitting. Therefore, the M3 format is rather complex at first sight, but it is well structured, as we will show in our further discussions. Observe in the figure that in the M3 format there are five consecutive 2-ms long 144-b transmitted bursts, followed by two idle frames, during which the CPP listens in order to determine whether its 24-b CHMP sequence has been detected and acknowledged by the CFP. This process can be followed by consulting Fig. 29.6, which will be described in depth after considering the detailed construction of the M3 burst.

The first four of the five 2-ms bursts are identical D-channel bursts, whereas the fifth one serves as a synchronization message and has a different construction. Observe, furthermore, that both the first four 144-b bursts as well as the fifth one contain four so-called submultiplex segments, each of which hosts a total of $(6 + 10 + 8 + 10 + 2) = 36$ b. In

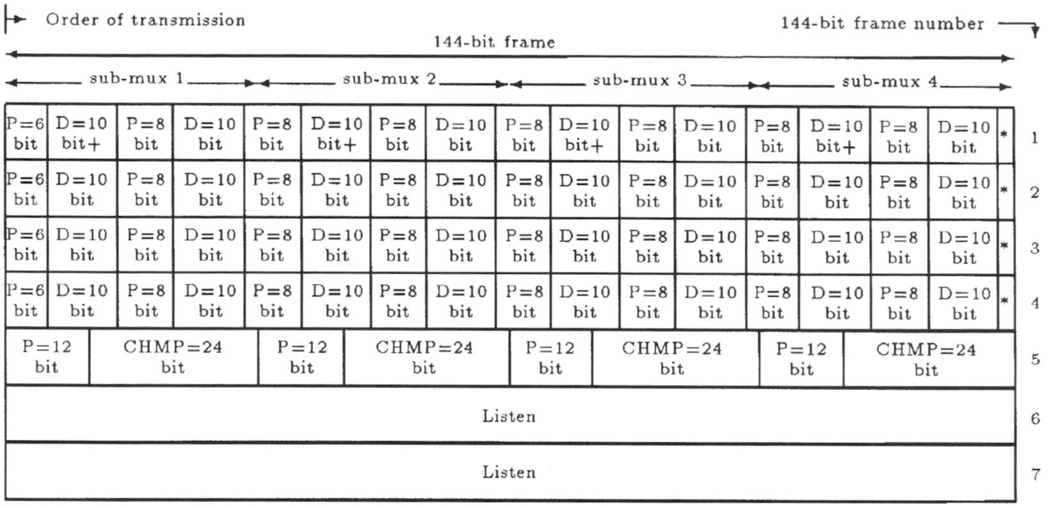

Order of transmission		144-bit frame														144-bit frame number ⟶
⟵ sub-mux 1 ⟶				⟵ sub-mux 2 ⟶				⟵ sub-mux 3 ⟶				⟵ sub-mux 4 ⟶				
P=6 bit	D=10 bit+	P=8 bit	D=10 bit	P=8 bit	D=10 bit+	P=8 bit	D=10 bit	P=8 bit	D=10 bit	P=8 bit	D=10 bit+	P=8 bit	D=10 bit	P=8 bit	D=10 bit	* 1
P=6 bit	D=10 bit	P=8 bit	D=10 bit	P=8 bit	D=10 bit	P=8 bit	D=10 bit	P=8 bit	D=10 bit	P=8 bit	D=10 bit	P=8 bit	D=10 bit	P=8 bit	D=10 bit	* 2
P=6 bit	D=10 bit	P=8 bit	D=10 bit	P=8 bit	D=10 bit	P=8 bit	D=10 bit	P=8 bit	D=10 bit	P=8 bit	D=10 bit	P=8 bit	D=10 bit	P=8 bit	D=10 bit	* 3
P=6 bit	D=10 bit	P=8 bit	D=10 bit	P=8 bit	D=10 bit	P=8 bit	D=10 bit	P=8 bit	D=10 bit	P=8 bit	D=10 bit	P=8 bit	D=10 bit	P=8 bit	D=10 bit	* 4
P=12 bit	CHMP=24 bit			P=12 bit	CHMP=24 bit			P=12 bit	CHMP=24 bit			P=12 bit	CHMP=24 bit			5
Listen																6
Listen																7

Notes: * 2 bit P

1/ Transmission is continuous for five bursts or 10 ms, then off for two burst periods or 4 ms

2/ The 20 bits of D chan are repeated in each of the 4 sub-mux's before D changes

3/ The D chan sync. word (SYNCD) always begins at the start of the slots marked +

FIGURE 29.3 CT2 multiplex three burst structure.

the first four 144-b bursts there are $(6 + 8 + 2) = 16$ one/zero clock-synchronizing P bits and $(10 + 10) = 20$ D bits or signalling bits. Since the D-channel message is constituted by two 10-b half-messages, the first half of the D-message is marked by the + sign in the figure. As mentioned in the context of M2, the D-channel bits are split in two halves and interspersed with the preamble segments in order to ensure that these bits do not emulate valid CHM sequences. Without splitting the D bits this could happen upon concatenating the one/zero P bits with the D bits, since the tail of the SYNF and SYNP sequences is also a one/zero segment. In the fifth 144-b M3 burst, each of the four submultiplex segments is constituted by 12 preamble bits and 24 CPP channel marker (CHMP) bits.

The four-fold submultiplex M3 structure ensures that irrespective of how the CFP's receive window is aligned with the CPP's transmission window, the CFP will be able to capture one of the four submultiplex segments of the fifth M3 burst, establish clock synchronization during the preamble, and lock on to the CHMP sequence. Once the CFP has successfully locked on to one of the CHMP words, the corresponding D-channel messages comprising the CPP identifier can be decoded. If the CPP identifier has been recognized, the CFP can attempt to reinitialize the link using its own master synchronization.

29.5 Signalling Layer Two (L2)

29.5.1 General Message Format

The signalling L2 is responsible for acknowledged and un-acknowledged information transfer over the air interface, error detection and correction by retransmission, as well as for the correct ordering of messages in the acknowledged mode. Its further functions are the link end-point identification and link maintenance for both CPP and CFP, as well as the definition of the L2 and **L3** interface.

Compliance with the L2 specifications will ensure the adequate transport of messages between the terminals of an established link. The L2 recommendations, however, do not define the meaning of messages, this is specified by L3 messages, albeit some of the messages are undefined in order to accommodate future system improvements.

The L3 messages are broken down to a number of standard packets, each constituted by one or more codewords (CW), as shown in Fig. 29.4. The codewords have a standard length of eight octets, and each packet contains up to six codewords. The first codeword in a packet is the so-called address codeword (ACW) and the subsequent ones, if present, are data codewords (**DCW**). The first octet of the ACW of each packet contains a variety of parameters, of which the binary flag **L3_END** is indicated in Fig. 29.4, and it is set to zero in the last packet. If the L3 message transmitted is mapped onto more than one packet, the packets must be numbered up to N. The address codeword is always preceded by a 16-b D-channel frame synchronization word **SYNCD**. Furthermore, each eight-octet CW is protected by a 16-b parity-check word occupying its last two octets. The binary Bose–Chaudhuri–Hocquenghem BCH(63,48) code is used to encode the first six octets or 48 b by adding 15 parity b to yield 63 b. Then bit 7 of octet 8 is inverted and bit 8 of octet 8 added such that the 64-b codeword has an even parity. If there are no D-channel packets to send, a 3-octet idle message **IDLE_D** constituted by zero/one reversals is transmitted. The 8-octet format of the ACWs and DCWs is made explicit in Fig. 29.5, where the two parity check octets occupy octets 7 and 8. The first octet hosts a number of control bits. Specifically, bit 1 is set to logical one for an ACW and to zero for a DCW, whereas bit 2 represents the so-called format type **FT** bit. $FT = 1$ indicates that variable length packet format is used for the transfer of L3 messages, whereas $FT = 0$ implies that a fixed length link setup is used for link end point addressing end service requests. FT is only relevant to ACWs, and in DCWs it has to be set to one.

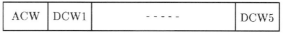

a/ L2 packet: Contains at least an ACW and 0...5 DCWs

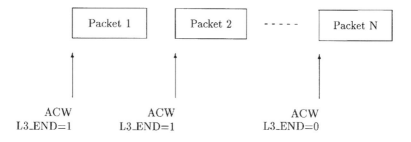

b/ L3 message: N packets

FIGURE 29.4   General L2 and L3 message format.

8	7	6	5	4	3	2	1	

HIC			SR	LS1	LS0	FT	CW	Octet 1
Handset Identifier Code (HIC)								Octet 2
HIC								Octet 3
Manufacturer Identifier Code (MIC)								Octet 4
Link Identifier Code (LID)								Octet 5
LID								Octet 6
Parity								Octet 7
Parity								Octet 8

FIGURE 29.5 Fixed format packets mapped on M1, M2, and M3 during link initialization and on M1 and M2 during handshake.

29.5.2 Fixed Format Packet

As an example, let us focus our attention on the fixed format scenario associated with $FT = 0$. The corresponding codeword format defined for use in M1, M2, and M3 for link initiation and in M1 and M2 for handshaking is displayed in Fig. 29.5. Bits 1 and 2 have already been discussed, whereas the 2-bit link status (**LS**) field is used during call setup and handshaking. The encoding of the four possible LS messages is given in Table 29.2. The aim of these LS messages will become more explicit during our further discussions with reference to Fig. 29.6 and Fig. 29.7. Specifically, **link_request** is transmitted from the CPP to the CFP either in an M3 burst as the first packet during CPP-initiated call setup and link re-establishment, or returned as a poll response in an M2 burst from the CPP to the CFP, when the CPP is responding to a call. **Link_grant** is sent by the CFP in response to a link_request originating from the CPP. In octets 5 and 6 it hosts the so-called link identification (**LID**) code, which is used by the CPP, for example, to address a specific CFP or a requested service. The LID is also used to maintain link reference during handshake exchanges and link re-establishment. The two remaining link status handshake messages, namely, ID_OK and ID_lost, are used to report to the far end whether a positive confirmation of adequate link quality has been received within the required time-out period. These issues will be revisited during our further elaborations. Returning to Fig. 29.5, we note that the fixed packet format ($FT = 0$) also contains a 19-b handset identification code (**HIC**) and an 8-b manufacturer identification code (**MIC**). The concatenated HIC and MIC fields jointly from the unique 27-b portable identity code (**PIC**), serving as a link end-point identifier. Lastly, we have to note that bit 5 of octet 1 represents the signalling rate (**SR**) request/response bit, which is used by the calling party to specify the choice of the 66- or 68-b M1 format. Specifically, $SR = 1$ represents the four bit/burst M1 signalling format. The first 6 octets are then protected by the parity check information contained in octets 7 and 8.

TABLE 29.2   Encoding of
Link Status Messages

LS1	LS0	Message
0	0	Link_request
0	1	Link_grant
1	0	ID_OK
1	1	ID_lost

29.6   CPP-Initiated Link Setup Procedures

Calls can be initiated at both the CPP and CFP, and the call initiation and detection procedures invoked depend on which party initiated the call. Let us first consider calling channel detection at the CFP, which ensues as follows. Under the instruction of the CFP control scheme, the RF synthesizer tunes to a legitimate RF channel and after a certain settling time commences reception. Upon receiving the M3 bursts from the CPP, the automatic gain control (AGC) circuitry adjusts its gain factor, and during the 12-b preamble in the fifth M3 burst, bit synchronization is established. This specific 144-b M3 burst, is transmitted every 14 ms, corresponding to every seventh 144-b burst. Now the CFP is ready to bit-synchronously correlate the received sequences with its locally stored CHMP word in order to identify any CHMP word arriving from the CPP. If no valid CHMP word is detected, the CFP may retune itself to the next legitimate RF channel, etc.

As mentioned, the call identification and link initialization process is shown in the flowchart of Fig. 29.6. If a valid 24-b CHMP word is identified, D-channel frame synchronization can take place using the 16-b SYNCD sequence and the next 8-octet L2 D-channel message delivering the link_request handshake portrayed earlier in Fig. 29.5 and Table 29.2 is decoded by the CFP. The required $16 + 64 = 80$ D bits are accommodated in this scenario by the $4 \cdot 20 = 80$ D bits of the next four 144-b bursts of the M3 structure, where the 20 D bits of the four submultiplex segments are transmitted four times within the same burst before the D message changes. If the decoded LID code of Fig. 29.5 is recognized by the CFP, the link may be reinitialized based on the master's timing information using the M2 burst associated with SYNF and containing the link_grant message addressed to the specific CPP identified by its PID.

Otherwise the CFP returns to its scanning mode and attempts to detect the next CHMP message. The reception of the CFP's 24-b SYNF segment embedded in the M2 message shown previously in Fig. 29.2 allows the CPP to identify the position of the CFP's transmit and receive windows and, hence, the CPP now can respond with another M2 burst within the receive window of the CFP. Following a number of M2 message exchanges, the CFP then sends a L3 message to instruct the CPP to switch to M1 bursts, which marks the commencement of normal voice communications and the end of the link setup session.

29.7   CFP-Initiated Link Setup Procedures

Similar procedures are followed when the CPP is being polled. The CFP transmits the 24-b CHMF words hosted by the 24-b SYN segment of the M2 burst shown in Fig. 29.2 in order to indicate that one or more CPPs are being paged. This process is displayed in the flowchart of Fig. 29.7, as well as in the timing diagram displayed in Fig. 29.8. The M2 D-channel messages convey the identifiers of the polled CPPs.

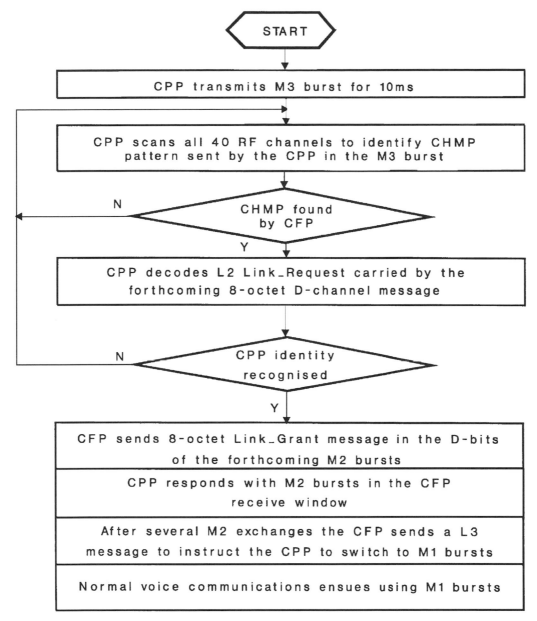

FIGURE 29.6   Flowchart of the CT-2 link initialization by the CPP.

The CPPs keep scanning all 40 legitimate RF channels in order to pinpoint any 24-b CHMF words. Explicitly, the CPP control scheme notifies the RF synthesizer to retune to the next legitimate RF channel if no CHMF words have been found on the current one. The synthesizer needs a finite time to settle on the new center frequency and then starts receiving again. Observe in Fig. 29.8 that at this stage only the CFP is transmitting the M2 bursts; hence, the uplink-half of the 2-ms TDD frame is unused.

Since the M2 burst commences with the D-channel bits arriving from the CFP, the CPP receiver's AGC will have to settle during this 16-b interval, which corresponds to about $16 \cdot 1/[72 \text{ kb/s}] \approx 0.22$ ms. Upon the arrival of the 10 alternating one–zero preamble bits,

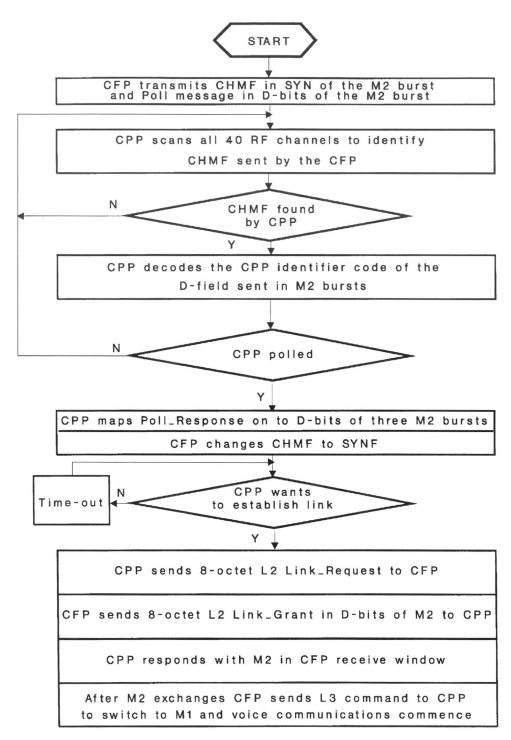

FIGURE 29.7   Flowchart of the CT-2 link initialization by the CFP.

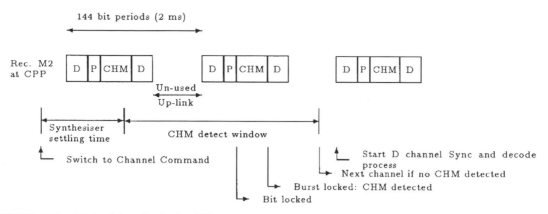

FIGURE 29.8 CT-2 call detection by the CPP.

bit synchronization is established. Now the CPP is ready to detect the CHMF word using a simple correlator circuitry, which establishes the appropriate frame synchronization. If, however, no CHMF word is detected within the receive window, the synthesizer will be retuned to the next RF channel, and the same procedure is repeated, until a CHMF word is detected.

When a CHMF word is correctly decoded by the CPP, the CPP is now capable of frame and bit synchronously decoding the D-channel bits. Upon decoding the D-channel message of the M2 burst, the CPP identifier (**ID**) constituted by the LID and PID segments of Fig. 29.5 is detected and compared to the CPP's own ID in order to decide as to whether the call is for this specific CPP. If so, the CPP ID is reflected back to the CFP along with a SYNP word, which is included in the SYN segment of an uplink M2 burst. This channel scanning and retuning process continues until a legitimate incoming call is detected or the CPP intends to initiate a call.

More precisely, if the specific CPP in question is polled and its own ID is recognized, the CPP sends its poll_response message in three consecutive M2 bursts, since the capacity of a single M2 burst is 32 D bits only, while the handshake messages of Fig. 29.5 and Table 29.2 require 8 octets preceded by a 16-b SYNCD segment. If by this time all paged CPPs have responded, the CFP changes the CHMF word to a SYNF word, in order to prevent activating dormant CPPs who are not being paged. If any of the paged CPPs intends to set up the link, then it will change its poll_response to a L2 link_request message, in response to which the CFP will issue an M2 link_grant message, as seen in Fig. 29.7, and from now on the procedure is identical to that of the CPP-initiated link setup portrayed in Fig. 29.6.

29.8 Handshaking

Having established the link, voice communications is maintained using M1 bursts, and the link quality is monitored by sending handshaking (**HS**) signalling messages using the D-channel bits. The required frequency of the handshaking messages must be between once every 400 ms and 1000 ms. The CT-2 codewords ID_OK, ID_lost, link_request and link_grant of Table 29.2 all represent valid handshakes. When using M1 bursts, however, the transmission of these 8-octet messages using the 2- or 4-b/2ms D-channel segment must be spread over 16 or 32 M1 bursts, corresponding to 32 or 64 ms.

Let us now focus our attention on the *handshake protocol* shown in Fig. 29.9. Suppose that the CPP's handshake interval of Thtx_p = 0.4 s since the start of the last transmitted handshake has expired, and hence the CPP prepares to send a handshake message HS_p.

If the CPP has received a valid HS_f message from the CFP within the last Thrx_p = 1s, the CPP sends an HS_p = ID_OK message to the CFP, otherwise an ID_Lost HS_p. Furthermore, if the valid handshake was HS_f = ID_OK, the CPP will reset its HS_f lost timer Thlost_p to 10 s. The CFP will maintain a 1-s timer referred to as Thrx_f, which is reset to its initial value upon the reception of a valid HS_p from the CPP.

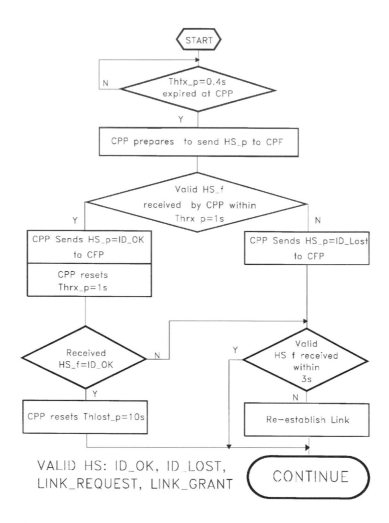

FIGURE 29.9   CT-2 handshake algorithms.

The CFP's actions also follow the structure of Fig. 29.9 upon simply interchanging CPP with CFP and the descriptor _p with _f. If the Thrx_f = 1 s timer expires without the reception of a valid HS_p from the CPP, then the CFP will send its ID_Lost HS_f message to the CPP instead of the ID_OK message and will not reset the Thlost_f = 10 s timer. If, however, the CFP happens to detect a valid HS_p, which can be any of the ID_OK, ID_Lost, link_request and link_grant messages of Table 29.2, arriving from the CPP, the CFP will reset its Thrx_f = 1 s timer and resumes transmitting the ID_OK HS_f message instead of the ID_Lost. Should any of the HS messages go astray for more than 3 s, the CPP or the CFP may try and re-establish the link on the current or another RF channel.

Again, although any of the ID_OK, ID_Lost, link_request and link_grant represent valid handshakes, only the reception of the ID_OK HS message is allowed to reset the Thlost = 10 s timer at both the CPP and CFP. If this timer expires, the link will be relinquished and the call dropped.

The handshake mechanism is further augmented by referring to Fig. 29.10, where two different scenarios are exampified, portraying the situation when the HS message sent by the CPP to the CFP is lost or, conversely, that transmitted by the CFP is corrupted.

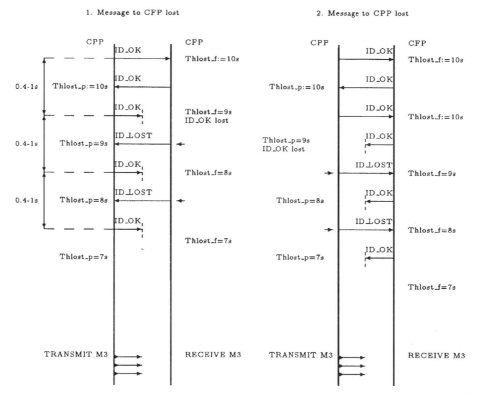

FIGURE 29.10   Handshake loss scenarios.

Considering the first scenario, during error-free communications the CPP sends HS_p = ID_OK, and upon receiving it the CFP resets its Thlost_f timer to 10 s. In due course it sends an HS_f = ID_OK acknowledgement, which also arrives free from errors. The CPP resets the Thlost_f timer to 10 s and, after the elapse of the 0.4–1 s handshake interval, issues an HS_p = ID_OK message, which does not reach the CFP. Hence, the Thlost_f timer is now reduced to 9 s and an HS_f = ID_Lost message is sent to the CPP. Upon reception of this, the CPP now cannot reset its Thlost_p timer to 10 s but can respond with an HS_p = ID_OK message, which again goes astray, forcing the CFP to further reduce its Thlost_f timer to 8 s. The CFP issues the valid handshake HS_f = ID_Lost, which arrives at the CPP, where the lack of HS_f = ID_OK reduces Thlost_p to 8 s. Now the corruption of the issued HS_p = ID_OK reduces Thlost_f to 7 s, in which event the link may be reinitialized using the M3 burst. The portrayed second example of Fig. 29.10 can be easily followed in case of the scenario when the HS_f message is corrupted.

29.9   Main Features of the CT-2 System

In our previous discourse we have given an insight in the algorithmic procedures of the CT-2 MPT 1375 recommendation. We have briefly highlighted the four-part structure of the standard dealing with the radio interface, signalling layers 1 and 2, signalling layer 3, and the speech coding issues, respectively. There are forty 100-kHz wide RF channels in the band 864.15–868.15 MHz, and the 72 kb/s bit stream modulates a Gaussian filtered FSK modem. The multiple access technique is TDD, transmitting 2-ms duration, 144-b M1 bursts during normal voice communications, which deliver the 32-kb/s ADPCM-coded speech signal. During link establishment the M2 and M3 bursts are used, which were also portrayed in this treatise, along with a range of handshaking messages and scenarios.

Defining Terms

AFC: Automatic frequency correction

CAI: Common air interface

CFP: Cordless fixed part

CHM: Channel marker sequence

CHMF: CFP channel marker

CHMP: CPP channel marker

CPP: Cordless portable part

CT: Cordless telephone

DCA: Dynamic channel allocation

DCW: Data code word

DECT: Digital European cordless telecommunications system

FT: Frame format type bit

GFSK: Gaussian frequency shift keying

GP: Guard period

HIC: Handset identification code

HS: Handshaking

ID: Identifier

L2: Signalling layer 2

L3: Signalling layer 3

LAN: Local area network

LID: Link identification

LOS: Line of sight

LS: Link status

M1: Multiplex one burst format

M2: Multiplex two burst format

M3: Multiplex three burst format

MIC: Manufacturer identification code

MPT-1375: British CT2 standard

PCN: Personal communications network

PIC: Portable identification code

PLMR: Public land mobile radio

SNR: Signal-to-noise ratio

SR: Signalling rate bit

SYN: Synchronization sequence

SYNCD: 16-b D-channel frame synchronization word

TDD: Time division duplex multiple access scheme

TP: Telepoint

References

[1] Asghar, S., Digital European cordless telephone (DECT), In *The Mobile Communications Handbook,* Chap. 30, CRC Press, Boca Raton, FL, 1995.

[2] Gardiner, J.G., Second generation cordless (CT-2) telephony in the UK: telepoint services and the common air-interface, *Elec. & Comm. Eng. J.,* 71–78, Apr. 1990.

[3] Hanzo, L., The Pan-European mobile radio system, In *The Mobile Communications Handbook,* Chap. 25, CRC Press, Boca Raton, FL, 1995.

[4] Jabbari, B., Dynamic channel assignment, In *The Mobile Communications Handbook,* Chap. 21, CRC Press, Boca Raton, FL, 1995.

[5] Ochsner, H., The digital European cordless telecommunications specification DECT. In *Cordless telecommunication in Europe.* Tuttlebee, W.H.M., Ed., 273–285. Springer-Verlag, 1990.

[6] Steedman, R.A.J., The Common Air Interface MPT 1375. In *Cordless Telecommunication in Europe.* Tuttlebee, W.H.W. Ed., 261–272, Springer-Verlag, 1990.

[7] Steele, R., Ed., *Mobile Radio Communications,* Pentech Press, London, 1992.

[8] Tuttlebee, W.H.W., Ed., *Cordless Telecommunication in Europe,* Springer-Verlag, 1990.

30

Half-Rate Standards

Wai-Yip Chan
Illinois Institute of Technology

Ira Gerson
*Motorola Corporate Systems
Research Laboratories*

Toshio Miki
*NTT Mobile Communication
Network, Inc.*

30.1 Introduction

A half-rate speech coding standard specifies a procedure for digital transmission of speech signals in a digital cellular radio system. The speech processing functions that are specified by a half-rate standard are depicted in Fig. 30.1. An input speech signal is processed by a

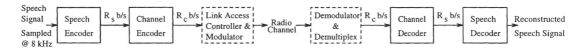

FIGURE 30.1 Digital speech transmission for digital cellular radio. Boxes with solid outlines represent processing modules that are specified by the half-rate standards.

speech encoder to generate a digital representation at a *net bit rate* of R_s bits per second. The encoded bit stream representing the input speech signal is processed by a *channel encoder* to generate another bit stream at a *gross bit rate* of R_c bits per second, where $R_c > R_s$. The channel encoded bit stream is organized into data frames, and each frame is transmitted as payload data by a radio-link access controller and modulator. The net bit rate R_s counts the number of bits used to describe the speech signal, and the difference between the gross and net bit rates ($R_c - R_s$) counts the number of error protection bits needed by the *channel decoder* to correct and detect transmission errors. The output of the channel decoder is given to the *speech decoder* to generate a *quantized* version of the speech encoder's input signal. In current digital cellular radio systems that use time-division multiple access (TDMA), a voice connection is allocated a fixed transmission rate (i.e., R_c is a constant). The operations performed by the speech and channel encoders and decoders and their input and output data formats are governed by the half-rate standards.

Globally, three major TDMA cellular radio systems have been developed and deployed. The initial digital speech services offered by these cellular systems were governed by *full-rate standards.* Because of the rapid growth in demand for cellular services, the available transmission capacity in some areas is frequently saturated, eroding customer satisfaction. By providing essentially the same voice quality but at half the gross bit rates of the full-rate standards, half-rate standards can readily double the number of callers that can be serviced by the cellular systems. The gross bit rates of the full-rate and half-rate standards for the European Groupe Speciale Mobile (GSM), Japanese Personal Digital Cellular[1] (PDC), and North American cellular (IS-54) systems are listed in Table 30.1. The three systems were developed and deployed under different time tables. Their disparate full- and half-bit rates partly reflect this difference. At the time of writing (January, 1995), the European and the Japanese systems have each selected an algorithm for their respective half-rate **codec.** Standardization of the North American half-rate codec has not reached a conclusion as none of the candidate algorithms has fully satisfied the standard's requirements. Thus, we focus here on the Japanese and European half-rate standards and will only touch upon the requirements of the North American standard.

TABLE 30.1   Gross Bit Rates Used for Digital Speech Transmission in Three TDMA Cellular Radio Systems

	Gross Bit Rate, b/s	
Standard Organization and Digital Cellular System	Full Rate	Half Rate
European Telecommunications Standards Institute (ETSI), GSM	22,800	11,400
Research & Development Center for Radio Systems (RCR), PDC	11,200	5,600
Telecommunication Industries Association (TIA), IS-54	13,000	6,500

30.2   Speech Coding for Cellular Mobile Radio Communications

Unlike the relatively benign transmission media commonly used in the public-switched telephone network (PSTN) for analog and digital transmission of speech signals, mobile radio channels are impaired by various forms of fading and interference effects. Whereas proper engineering of the radio link elements (modulation, power control, diversity, equalization, frequency allocation, etc.) ameliorates fading effects, burst and isolated bit errors still occur frequently. The net effect is such that speech communication may be required to be operational even for bit-error rates greater than 1%. In order to furnish reliable voice communication, typically half of the transmitted payload bits are devoted to error correction and detection.

---

[1]Personal Digital Cellular was formerly Japanese Digital Cellular (JDC).

It is common for low-bit-rate speech codecs to process samples of the input speech signal one frame at a time, e.g., 160 samples processed once every 20 ms. Thus, a certain amount of time is required to gather a block of speech samples, encode them, perform channel encoding, transport the encoded data over the radio channel, and perform channel decoding and speech synthesis. These processing steps of the speech codec add to the overall end-to-end transmission delay. Long transmission delay hampers conversational interaction. Moreover, if the cellular system is interconnected with the PSTN and a four-wire to two-wire (analog) circuit conversion is performed in the network, feedbacks called *echoes* may be generated across the conversion circuit. The echoes can be heard by the originating talker as a delayed and distorted version of his/her speech and can be quite annoying. The annoyance level increases with the transmission delay and may necessitate (at additional costs) the deployment of **echo cancellers**.

A consequence of user mobility is that the level and other characteristics of the acoustic background noise can be highly variable. Though acoustic noise can be minimized through suitable acoustic transduction design and the use of adaptive filtering/cancellation techniques [9, 13, 15], the speech encoding algorithm still needs to be robust against background noise of various levels and kinds (e.g., babble, music, noise bursts, and colored noise).

Processing complexity directly impacts the viability of achieving a circuit realization that is compact and has low-power consumption, two key enabling factors of equipment portability for the end user. Factors that tend to result in low complexity are fixed-point instead of floating-point computation, lack of complicated arithmetic operations (division, square roots, transcendental functions), regular algorithm structure, small data memory, and small program memory. Since, in general, better speech quality can be achieved with increasing speech and channel coding delay and complexity, the digital cellular mobile-radio environment imposes conflicting and challenging requirements on the speech codec.

30.3   Codec Selection and Performance Requirements

The half-rate speech coding standards are drawn up through competitive testing and selection. From a set of candidate codec algorithms submitted by contending organizations, the one algorithm that meets basic selection criteria and offers the best performance is selected to form the standard. The codec performance measures and codec testing and selection procedures are set out in a test plan under the auspices of the organization (Table 30.1) responsible for the standardization process (see, e.g., [16]). Major codec characteristics evaluated are speech quality, delay, and complexity. The full-rate codec is also evaluated as a *reference codec,* and its evaluation scores form part of the selection criteria for the codec candidates.

The speech quality of each candidate codec is evaluated through listening tests. To conduct the tests, each candidate codec is required to process speech signals and/or encoded bit streams that have been preprocessed to simulate a range of operating conditions: variations in speaker voice and level, acoustic background noise type and level, channel error rate, and stages of **tandem coding**. During the tests, subjects listen to processed speech signals and judge their quality levels or annoyance levels on a five-point opinion scale. The opinion scores collected from the tests are suitably averaged over all trials and subjects for each test condition (see [11], for mean opinion score (MOS) and degradation mean opinion score). The categorical opinion scales of the subjects are also calibrated using *modulated noise reference units (MNRUs)* [3]. Modulated noise better resembles the distortions created by speech codecs than noise that is uncorrelated with the speech signal. Modulated noise is generated by multiplying the speech signal with a noise signal. The resultant modulated noise is scaled to a desired power level and then added to the uncoded (clean) speech signal. The ratio between the power level of the speech signal and that of the modulated noise is expressed in decibels and given the notation *dBQ*. Under each test condition, subjects are presented with speech signals processed by the codecs as well as speech signals corrupted by modulated noise. Through presenting a range of modulated-noise levels, the subjects' opinions are calibrated on the dBQ scale. Thereafter, the mean opinion scores obtained for the codecs can also be expressed on that scale.

For each codec candidate, a profile of scores is compiled, consisting of speech quality scores, delay measurements, and complexity estimates. Each candidate's score profile is compared with that of the reference codec, ensuring that basic requirements are satisfied (see, e.g., [12]). An overall figure of merit for each candidate is also computed from the profile. The candidates, if any, that meet the basic requirements then compete on the basis of maximizing the figure of merit.

Basic performance requirements for each of the three half-rate standards are summarized in Table 30.2. In terms of speech quality, the GSM and PDC half-rate codecs

are permitted to underperform their respective full-rate codecs by no more than 1 dBQ averaging over all test conditions and no more than 3 dBQ within each test condition. More stringently, the North American half-rate codec is required to furnish a speech-quality profile that is statistically equivalent to that of the North American full-rate codec as determined by a specific statistical procedure for multiple comparisons [16]. Since various requirements on the half-rate standards are set relative to their full-rate counterparts, an indication of the *relative* speech quality between the three half-rate standards can be deduced from the test results of De Martino [2] comparing the three full-rate codecs. The maximum delays in Table 30.2 apply to the total of the delays through the speech and channel encoders and decoders (Fig. 30.1). Codec complexity is computed using a formula that counts the computational operations and memory usage of the codec algorithm. The complexity of the half-rate codecs is limited to 3 or 4 times that of their full-rate counterparts.

TABLE 30.2　　Basic Performance Requirements for the Three Half-Rate Standards

Digital Cellular Systems	Basic performance requirements		
	Min. Speech Quality, dBQ Rel. to Full Rate	Max. Delay, ms	Max. Complexity Rel. to Full Rate
Japanese (PDC)	−1 average, −3 maximum	94.8	3×
European (GSM)	−1 average, −3 maximum	90	4×
North American (IS-54)	Statistically equivalent	100	4×

30.4　Speech Coding Techniques in the Half-Rate Standards

Existing half-rate and full-rate standard coders can be characterized as *linear-prediction based analysis-by-synthesis* (LPAS) speech coders [4]. LPAS coding entails using a time-varying all-pole filter in the decoder to synthesize the quantized speech signal. A short segment of the signal is synthesized by driving the filter with an *excitation* signal that is either *quasiperiodic* (for *voiced* speech) or *random* (for *unvoiced* speech). In either case, the excitation signal has a *spectral envelope* that is relatively flat. The synthesis filter serves to shape the spectrum of the excitation input so that the spectral envelope of the synthesized output resembles the filter's magnitude frequency response. The magnitude response often has prominent peaks; they render the *formants* that give a speech signal its phonetic character. The synthesis filter has to be adapted to the current frame of input speech signal. This is accomplished with the encoder performing a linear prediction (LP) analysis of the frame: the inverse of the all-pole synthesis filter is applied as an LP *error filter* to the frame, and the values of the filter parameters are computed to minimize the energy of the filter's output error signal. The resultant filter parameters are quantized and conveyed to the decoder for it to update the synthesis filter.

Having executed an LP analysis and quantized the synthesis filter parameters, the LPAS encoder performs analysis-by-synthesis (ABS) on the input signal to find a suitable excitation signal. An ABS encoder maintains a *copy* of the decoder. The encoder examines the possible outputs that can be produced by the decoder copy in order to determine how best to instruct (using transmitted information) the actual decoder so that it would output (synthesize) a good approximation of the input speech signal. The decoder copy tracks the state of the actual decoder, since the latter evolves (under ideal channel conditions) according to information received from the encoder. The details of the ABS procedure vary with the particular excitation model employed in a specific coding scheme. One of the earliest seminal LPAS schemes is *code excited linear prediction (CELP)* [4]. In CELP, the excitation signal is obtained from a **codebook** of *code vectors,* each of which is a candidate for the excitation signal. The encoder searches the codebook to find the one code vector that would result in a best match between the resultant synthesis output signal and the encoder's input speech signal. The matching is considered best when the energy of the difference between the two signals being matched is minimized. A *perceptual weighting filter* is usually applied to the difference signal (prior to energy integration) to make the minimization more relevant to human perception of speech fidelity. Regions in the frequency spectrum where human listeners are more sensitive to distortions are given relatively stronger weighting by the filter and vice versa. For instance, the concentration of spectral energy around the formant frequencies gives rise to stronger *masking* of coder noise (i.e., rendering the noise less audible) and, therefore, weaker weighting can be applied to the formant frequency regions. For masking to be effective, the weighting filter has to be adapted to the time-varying speech spectrum. Adaptation is achieved usually by basing the weighting filter parameters on the synthesis filter parameters.

The CELP framework has evolved to form the basis of a great variety of speech coding algorithms, including all existing full- and half-rate standard algorithms for digital cellular systems. We outline next the basic CELP encoder-processing steps, in a form suited to our subsequent detailed descriptions of the PDC and GSM half-rate coders. These steps have accounted for various computational efficiency considerations and may, therefore, deviate from a conceptual functional description of the encoder constituents.

1. LP analysis on the current frame of input speech to determine the coefficients of the all-pole synthesis filter;
2. quantization of the LP filter parameters;
3. determination of the open-loop **pitch period** or lag;
4. adapting the perceptual weighting filter to the current LP information (and also pitch information when appropriate) and applying the adapted filter to the input speech signal;
5. formation of a filter cascade (which we shall refer to as *perceptually weighted synthesis filter*) consisting of the LP synthesis filter, as specified by the quantized parameters in step 2, followed by the perceptual weighting filter;
6. subtraction of the *zero-input response* of the perceptually weighted synthesis filter (the filter's decaying response due to past input) from the perceptually weighted

input speech signal obtained in step 4;

7. an *adaptive codebook* is searched to find the most suitable periodic excitation, i.e., when the perceptually weighted synthesis filter is driven by the best code vector from the adaptive codebook, the output of the filter cascade should best match the difference signal obtained in step 6;

8. one or more nonadaptive excitation codebooks are searched to find the most suitable random excitation vectors that, when added to the best periodic excitation as determined in step 7 and with the resultant sum signal driving the filter cascade, would result in an output signal best matching the difference signal obtained in step 6.

Steps 1–6 are executed once per frame. Steps 7 and 8 are executed once for each of the *subframes* that together constitute a frame. Step 7 may be skipped depending on the pitch information from step 3, or if step 7 were always executed, a *nonperiodic excitation* decision would be one of the possible outcomes of the search process in step 7. Integral to steps 7 and 8 is the determination of gain (scaling) parameters for the excitation vectors. For each frame of input speech, the filter and excitation and gain parameters determined as outlined are conveyed as encoded bits to the speech decoder.

In a properly designed system, the data conveyed by the channel decoder to the speech decoder should be free of errors most of the time, and the speech signal synthesized by the speech decoder would be identical to that as determined in the speech encoder's ABS operation. It is common to enhance the quality of the synthesized speech by using an adaptive *postfilter* to attenuate coder noise in the perceptually sensitive regions of the spectrum. The postfilter of the decoder and the perceptual weighting filter of the encoder may seem to be functionally identical. The weighting filter, however, influences the selection of the best excitation among available choices, whereas the postfilter actually shapes the spectrum of the synthesized signal. Since postfiltering introduces its own distortion, its advantage may be diminished if tandem coding occurs along the end-to-end communication path. Nevertheless, proper design can ensure that the net effect of postfiltering is a reduction in the amount of audible codec noise [1]. Excepting postfiltering, all other speech synthesis operations of an LPAS decoder are (effectively) duplicated in the encoder (though the converse is not true). Using this fact, we shall illustrate each coder in the sequel by exhibiting only a block diagram of its encoder or decoder but not both.

30.5   Channel Coding Techniques in the Half-Rate Standards

Crucial to the maintenance of quality speech communication is the ability to transport coded speech data across the radio channel with minimal errors. Low-bit-rate LPAS coders are particularly sensitive to channel errors; errors in the bits representing the LP parameters in one frame, for instance, could result in the synthesis of nonsensical sounds for longer than a frame duration. The error rate of a digital cellular radio channel with no channel coding can be catastrophically high for LPAS coders. The amount of tolerable transmission delay is limited by the requirement of interactive communication and, consequently, *forward error control* is used to remedy transmission errors. "Forward" means that channel errors are remedied in the receiver, with no additional information from the transmitter and, hence, no additional transmission delay. To enable the channel decoder to correct channel errors, the channel encoder conveys more bits than the amount generated by the speech encoder. The additional bits are for error *protection*, as errors may or may not occur in any particular transmission epoch. The ratio of the number of encoder input (information) bits to the number of encoder output (code) bits is called the (channel) *coding rate*. This is a number no more than one and generally decreases as the error protection power increases. Though a lower channel coding rate gives more error protection, fewer bits will be available for speech coding. When the channel is in good condition and, hence, less error protection is needed, the received speech quality could be better if bits devoted to channel coding were used for speech coding. On the other hand, if a high channel coding rate were used, there would be uncorrected errors under poor channel conditions and speech quality would suffer. Thus, when nonadaptive forward error protection is used over channels with nonstationary statistics, there is an inevitable tradeoff between quality degradation due to uncorrected errors and that due to expending bits on error protection (instead of on speech encoding).

Both the GSM and PDC half-rate coders use *convolutional coding* [14] for error correction. Convolutional codes are sliding or sequential codes. The encoder of a rate $m/n, m < n$ convolutional code can be realized using m shift registers. For every m information bits input to the encoder (one bit to each of the m shift registers), n code bits are output to the channel. Each code bit is computed as a modulo-2 sum of a subset of the bits in the shift registers. Error protection overhead can be reduced by exploiting the unequal sensitivity of speech quality to errors in different positions of the encoded bit stream. A family of *rate-compatible punctured convolutional codes* (RCPCCs) [10] is a collection of related convolutional codes; all of the codes in the collection except the one with the lowest rate are derived by *puncturing* (dropping) code bits from the convolutional code with the lowest rate. With an RCPCC, the channel coding rate can be varied on the fly (i.e., variable-rate coding) while a sequence of information bits is being encoded through the shift registers, thereby imparting on different segments in the sequence different degrees of error protection.

For decoding a convolutional coded bit stream, the *Viterbi algorithm* [14] is a computationally efficient procedure. Given the output of the demodulator, the algorithm determines the most likely sequence of data bits sent by the channel encoder. To fully utilize the error correction power of the convolutional code, the amplitude of the demodulated *channel symbol* can be quantized to more bits than the minimum number required, i.e., for subsequent *soft decision decoding*. The minimum number of bits is given by the number of channel-coded bits mapped by the modulator onto each channel symbol; decoding based on the minimum-rate bit stream is called *hard decision* decoding. Although soft decoding gives better error protection, decoding complexity is also increased.

Whereas convolutional codes are most effective against randomly scattered bit errors, errors on cellular radio channels often occur in bursts of bits. These bursts can be broken up if the bits put into the channel are rearranged after demodulation. Thus, in *block interleaving*, encoded bits are read into a matrix by row and then read out of the matrix by column (or vice versa) and then passed on to the modulator; the reverse operation is performed by a *deinterleaver* following demodulation. Interleaving increases the transmission delay to the extent that enough bits need to be collected in order to fill up the matrix.

Owing to the severe nature of the cellular radio channel and limited available transmission capacity, uncorrected errors often remain in the decoded data. A common countermeasure is to append an error detection code to the speech data stream prior to channel coding. When residual channel errors are detected, the speech decoder can take various remedial measures to minimize the negative impact on speech quality. Common measures are repetition of speech parameters from the most recent good frames and gradual muting of the possibly corrupted synthesized speech.

The PDC and GSM half-rate standard algorithms together embody some of the latest advances in speech coding techniques, including: *multimodal coding* where the coder configuration and bit allocation change with the type of speech input; *vector quantization* (*VQ*) [5] of the LP filter parameters; higher precision and improved coding efficiency for pitch-periodic excitation; and postfiltering with improved tandeming performance. We next explore the more distinctive features of the PDC and GSM speech coders.

30.6   The Japanese Half-Rate Standard

An algorithm was selected for the Japanese half-rate standard in April 1993, following the evaluation of 12 submissions in a first round, and four final candidates in a second round [12]. The selected algorithm, called pitch synchronous innovation CELP[2] (PSI-CELP), met all of the basic selection criteria and scored the highest among all candidates evaluated. A block diagram of the PSI-CELP encoder is shown in Fig. 30.2, and bit allocations are summarized in Table 30.3. The complexity of the coder is estimated to be approximately 2.4 times that of the PDC full-rate coder. The frame size of the coder is 40 ms, and its subframe size is 10 ms. These sizes are longer than those used in most existing CELP-type standard coders. However, LP analysis is performed twice per frame in the PSI-CELP coder.

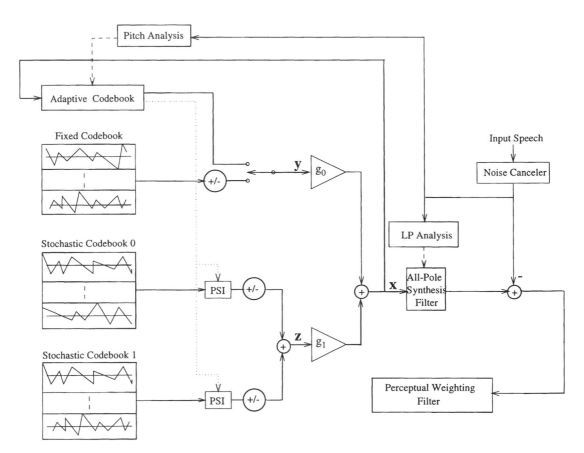

FIGURE 30.2   Basic structure of the PSI-CELP encoder.

---

[2]There were two candidate algorithms named PSI-CELP in the PDC half-rate competition. The algorithm described here was contributed by NTT Mobile Communications Network, Inc. (NTT DoCoMo).

TABLE 30.3 Bit Allocations for the PSI-CELP Half- Rate PDC Speech Coder

Parameter	Bits	Error Protected Bits
LP synthesis filter	31	15
Frame energy	7	7
Periodic excitation	8×4	8×4
Stochastic excitation	10×4	0
Gain	7×4	3×4
Total	138	66

A distinctive feature of the PSI-CELP coder is the use of an adaptive noise canceller [13, 15] to suppress noise in the input signal prior to coding. The input signal is classified into various modes, depending on the presence or absence of background noise and speech and their relative power levels. The current active mode determines whether *Kalman filtering* [9] is applied to the input signal and whether the parameters of the Kalman filter are adapted. Kalman filtering is applied when a significant amount of background noise is present or when both background noise and speech are strongly present. The filter parameters are adapted to the statistics of the speech and noise signals in accordance with whether they are both present or only noise is present.

The LP filter parameters in the PSI-CELP coder are encoded using VQ. A tenth-order LP analysis is performed every 20 ms. The resultant filter parameters are converted to 10 *line spectral frequencies* (LSFs).[3] The LSF parameters have a naturally increasing order, and together are treated as the ordered components of a vector. Since the speech spectral envelope tends to evolve slowly with time, there is intervector dependency between adjacent LSF vectors that can be exploited. Thus, the two LSF vectors for each 40-ms frame are paired together and jointly encoded. Each LSF vector in the pair is split into three subvectors. The pair of subvectors that cover the same vector component indexes are combined into one composite vector and vector quantized. Altogether, 31 b are used to encode a pair of LSF vectors. This three-way *split VQ*[4] scheme embodies a compromise between the prohibitively high complexity of using a large vector dimension and the performance gain from exploiting intra- and intervector dependency.

The PSI-CELP encoder uses a perceptual weighting filter consisting of a cascade of two filter sections. The sections exploit the pitch-harmonic structure and the LP spectral-envelope structure of the speech signal, respectively. The pitch-harmonic section has four parameters, a pitch lag and three coefficients, whose values are determined from an analysis of the periodic structure of the input speech signal. Pitch-harmonic weighting reduces the amount of noise in between the pitch harmonics by aggregating coder noise to be closer to the harmonic frequencies of the speech signal. In high-pitched voice, the harmonics are spaced relatively farther apart, and pitch-harmonic weighting becomes correspondingly more important.

The excitation vector x (Fig. 30.2) is updated once every subframe interval (10 ms) and is constructed as a *linear combination* of two vectors

$$x = g_0 y + g_1 z \tag{30.1}$$

where g_0 and g_1 are scalar gains, y is labeled as the *periodic* component of the excitation and z as the *stochastic* or *random* component. When the input speech is voiced, the ABS op-

---

[3]Also known as line spectrum pairs (LSPs).

[4]Matrix quantization is another possible description.

eration would find a value for y from the *adaptive codebook* (Fig. 30.2). The codebook is constructed out of past samples of the excitation signal x; hence, there is a feedback path into the adaptive codebook in Fig. 30.2. Each code vector in the adaptive codebook corresponds to one of the 192 possible pitch lag L values available for encoding; the code vector is populated with samples of x beginning with the Lth sample backward in time. L is not restricted to be an integer, i.e., *fractional pitch period* is permitted. Successive values of L are more closely spaced for smaller values of L; short, medium, and long lags are quantized to one-quarter, one-half, and one sampling-period resolution, respectively. As a result, the *relative* quantization error in the encoded pitch frequency (which is the reciprocal of the encoded pitch lag) remains roughly constant with increasing pitch frequency. When the input speech is unvoiced, y would be obtained from the fixed codebook (Fig. 30.2). To find the best value for y, the encoder searches through the aggregate of 256 code vectors from both the adaptive and fixed codebooks. The code vector that results in a synthesis output most resembling the input speech is selected. The best code vector thus chosen also implicitly determines the voicing condition (voiced/unvoiced) and the pitch lag value L^* most appropriate to the current subframe of input speech. These parameters are said to be determined in a *closed-loop* search.

The stochastic excitation z is formed as a sum of two code vectors, each selected from a *conjugate codebook* (Fig. 30.2) [13]. Using a pair of conjugate codebooks each of size 16 code vectors (4 b) has been found to improve robustness against channel errors, in comparison with using one single codebook of size 256 code vectors (8 b). The synthesis output due to z can be decomposed into a sum of two orthogonal components, one of which points in the same direction as the synthesis output due to the periodic excitation y and the other component points in a direction orthogonal to the synthesis output due to y. The latter synthesis output component of z is kept, whereas the former component is discarded. Such decomposition enables the two gain factors g_0 and g_1 to be separately quantized. For voiced speech, the conjugate code vectors are preprocessed to produce a set of *pitch synchronous innovation* (PSI) vectors. The first L^* samples of each code vector are treated as a fundamental period of samples. The fundamental period is replicated until there are enough samples to populate a subframe. If L^* is not an integer, interpolated samples of the code vectors are used (upsampled versions of the code vectors can be precomputed). PSI has been found to reinforce the periodicity and substantially improve the quality of synthesized voiced speech.

The postfilter in the PSI-CELP decoder has three sections, for enhancing the formants, the pitch harmonics, and the high frequencies of the synthesized speech, respectively. Pitch-harmonic enhancement is applied only when the adaptive codebook has been used. Formant enhancement makes use of the decoded LP synthesis filter parameters, whereas a refined pitch analysis is performed on the synthesized speech to obtain the values for the parameters of the pitch-harmonic section of the postfilter. A first-order high-pass filter section compensates for the low-pass spectral tilt [1] of the formant enhancement section.

Of the 138 speech data bits generated by the speech encoder every 40-ms frame, 66 b (Table 30.3) receive error protection and the remaining 72 speech data bits of the frame are not error protected. An error detection code of 9 *cyclic redundancy check* (*CRC*) bits is appended to the 66 b and then submitted to a rate 1/2, punctured convolutional encoder to generate a sequence of 152 channel coded bits. Of the unprotected 72 b, the 40 b that index the excitation codebooks (Table 30.3) are remapped or *pseudo-Gray coded* [17] so as to equalize their channel error sensitivity. As a result, a bit error occurring in an index word is likely to cause about the same amount of degradation regardless of the bit error position in the index word. For each speech frame, the channel encoder emits 224 b of payload data.

The payload data from two adjacent frames are interleaved before transmission over the radio link.

Uncorrected errors in the most critical 66 b are detected with high probability as a CRC error. A finite state machine keeps track of the recent history of CRC errors. When a sequence of CRC errors is encountered, the power level of the synthesized speech is progressively suppressed, so that muting is reached after four consecutive CRC errors. Conversely, following the cessation of a sequence of CRC errors, the power level of the synthesized speech is ramped up gradually.

30.7 The European GSM Half-Rate Standard

A *vector sum excited linear prediction* (VSELP) coder, contributed by Motorola, Inc., was selected in January 1994 by the main GSM technical committee as a basis for the GSM half-rate standard. The standard was finally approved in January 1995. VSELP is a generic name for a family of algorithms from Motorola; the North American full-rate and the Japanese full-rate standards are also based on VSELP. All VSELP coders make use of the basic idea of representing the excitation signal by a linear combination of *basis vectors* [6]. This representation renders the excitation codebook search procedure very computationally efficient. A block diagram of the GSM half-rate decoder is depicted in Fig. 30.3 and bit allocations are tabulated in Table 30.4. The coder's frame size is 20 ms, and each frame comprises four subframes of 5 ms each. The coder has been optimized for execution on a processor with 16-b word length and 32-b accumulator. The GSM standard is a *bit exact* specification: in addition to specifying the codec's processing steps, the numerical formats and precisions of the codec's variables are also specified.

TABLE 30.4 Bit Allocations for the VSELP Half-Rate GSM Coder

Parameter	Bits/subframe	Bits/frame
LP synthesis filter		28
Soft interpolation		1
Frame energy		5
Mode selection		2
Mode 0		
Excitation code I	7	28
Excitation code H	7	28
Gain code G_s, P_0	5	20
Mode 1, 2, and 3		
Pitch lag L (first subframe)		8
Difference lag (subframes 2, 3, 4)	4	12
Excitation code J	9	36
Gain code G_s, P_0	5	20
Total		112

The synthesis filter coefficients in GSM VSELP are encoded using the *fixed point lattice technique* (FLAT) [8] and vector quantization. FLAT is based on the *lattice filter* representation of the linear prediction error filter. The tenth-order lattice filter has 10 stages, with the ith stage, $i \in \{1, \ldots, 10\}$, containing a *reflection coefficient* parameter r_i. The lattice filter has an *order-recursion* property such that the best prediction error filters of all orders less than ten are all embedded in the best tenth-order lattice filter. This means that once the values of the lower order reflection coefficients have been optimized, they do not have to be reoptimized when a higher order predictor is desired; in other words, the coefficients can be optimized sequentially from low to high orders. On the other hand, if the lower order coefficients were suboptimal (as in the case when the coefficients are quantized), the higher

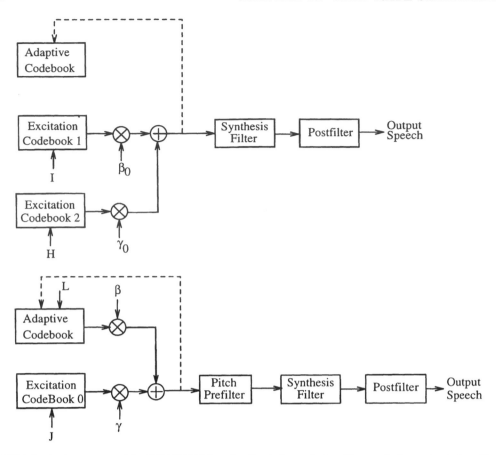

FIGURE 30.3 Basic structure of the GSM VSELP decoder. Top is for mode 0 and bottom is for modes 1, 2, and 3.

order coefficients could still be selected to minimize the prediction *residual* (or error) energy at the output of the higher order stages; in effect, the higher order stages can compensate for the suboptimality of lower order stages.

In the GSM VSELP coder, the ten reflection coefficients $\{r_1, \ldots, r_{10}\}$ that have to be encoded for each frame are grouped into three coefficient vectors $\boldsymbol{v}_1 = [r_1 r_2 r_3], \boldsymbol{v}_2 = [r_4 r_5 r_6], \boldsymbol{v}_3 = [r_7 r_8 r_9 r_{10}]$. The vectors are quantized sequentially, from $\boldsymbol{v}_1$ to $\boldsymbol{v}_3$, using a b_i-bit VQ codebook C_i for $\boldsymbol{v}_i$, where b_i, $i = 1, 2, 3$ are 11, 9, and 8 b, respectively. The vector $\boldsymbol{v}_i$ is quantized to minimize the prediction error at the energy output of the jth stage of the lattice filter where r_j is the highest order coefficient in the vector $\boldsymbol{v}_i$. The computational complexity associated with quantizing $\boldsymbol{v}_i$ is reduced by searching only a small subset of the code vectors in C_i. The subset is determined by first searching a *prequantizer* codebook of size c_i bits, where c_i, $i = 1, \ldots, 3$ are 6, 5, and 4 b, respectively. Each code vector in the prequantizer codebook is associated with $2^{b_i - c_i}$ code vectors in the target codebook. The subset is obtained by pooling together all of the code vectors in C_i that are associated with the top few best matching prequantizer code vectors. In this way, a factor of reduction in computational complexity of nearly $2^{b_i - c_i}$ is obtained for the quantization of $\boldsymbol{v}_i$.

The half-rate GSM coder changes its configuration of excitation generation (Fig. 30.3) in accordance with a *voicing mode* [7]. For each frame, the coder selects one of four possible voicing modes depending on the values of the *open-loop* pitch-prediction gains computed for the frame and its four subframes. Open loop refers to determining the pitch lag and the pitch-predictor coefficient(s) via a direct analysis of the input speech signal or, in the case of the half-rate GSM coder, the perceptually weighted (LP-weighting only) input signal. Open-loop analysis can be regarded as the opposite of closed-loop analysis, which in our context is synonymous with ABS. When the pitch-prediction gain for the frame is weak, the input speech signal is deemed to be unvoiced and mode 0 is used. In this mode, two 7-b *trained* codebooks (excitation codebooks 1 and 2 in Fig. 30.3) are used, and the excitation signal for each subframe is formed as a linear combination of two code vectors, one from each of the codebooks. A trained codebook is one designed by applying the coder to a representative set of speech signals while optimizing the codebook to suit the set. Mode 1, 2, or 3 is chosen depending on the strength of the pitch-prediction gains for the frame and its subframes. In these modes, the excitation signal is formed as a linear combination of a code vector from an 8-b adaptive codebook and a code vector from a 9-b trained codebook (Fig. 30.3). The code vectors that are summed together to form the excitation signal for a subframe are each scaled by a gain factor (β and γ in Fig. 30.3). Each mode uses a gain VQ codebook specific to that mode.

As depicted in Fig. 30.3, the decoder contains an adaptive pitch prefilter for the voiced modes and an adaptive postfilter for all modes. The filters enhance the perceptual quality of the decoded speech and are not present in the encoder. It is more conventional to locate the pitch prefilter as a section of the postfilter; the distinctive placement of the pitch prefilter in VSELP was chosen to reduce artifacts caused by the time-varying nature of the filter. In mode 0, the encoder uses an LP spectral weighting filter in its ABS search of the two excitation codebooks. In the other modes, the encoder uses a pitch-harmonic weighting filter in cascade with an LP spectral weighting filter for searching excitation codebook 0, whereas only LP spectral weighting is used for searching the adaptive codebook. The pitch-harmonic weighting filter has two parameters, a pitch lag and a coefficient, whose values are determined in the aforementioned open-loop pitch analysis.

A code vector in the 8-b adaptive codebook has a dimension of 40 (the duration of a subframe) and is populated with past samples of the excitation signal beginning with the Lth sample back from the present time. L can take on one of 256 different integer and fractional values. The best adaptive code vector for each subframe can be selected via a complete ABS; the required exhaustive search of the adaptive codebook is, however, computationally expensive. To reduce computation, the GSM VSELP coder makes use of the aforementioned open-loop pitch analysis to produce a list of *candidate lag values*. The open-loop pitch-prediction gains are ranked in decreasing order, and only the lags corresponding to top-ranked gains are kept as candidates. The final decisions for the four L values of the four subframes in a frame are made jointly. By assuming that the four L values can not vary over the entire range of all possible 256 values in the short duration of a frame, the L of the first subframe is coded using 8 b, and the L of each of the other three subframes is coded *differentially* using 4 b. The 4 b represent 16 possible values of deviation relative to the lag of the previous subframe. The four lags in a frame trace out a *trajectory* where the change from one time point to the next is restricted; consequently, only 20 b are needed instead of 32 b for encoding the four lags. Candidate trajectories are constructed by linking top ranked lags that are commensurate with differential encoding. The best trajectory among the candidates is then selected via ABS.

The trained excitation codebooks of VSELP have a special vector sum structure that facilitates fast searching [6]. Each of the 2^b code vectors in a b-bit trained codebook is formed as a linear combination of b *basis vectors*. Each of the b scalar weights in the linear combination is restricted to have a binary value of either 1 or -1. The 2^b code vectors in the codebook are obtained by taking all 2^b possible combinations of values of the weights. A substantial storage saving is incurred by storing only b basis vectors instead of 2^b code vectors. Computational saving is another advantage of the vector-sum structure. Since filtering is a linear operation, the synthesis output due to each code vector is a linear combination of the synthesis outputs due to the individual basis vectors, where the same weight values are used in the output linear combination as in forming the code vector. A vector sum codebook can be searched by first performing synthesis filtering on its b basis vectors. If, for the present subframe, another trained codebook (mode 0) or an adaptive codebook (mode 1, 2, 3) had been searched, the filtered basis vectors are further orthogonalized with respect to the signal synthesized from that codebook, i.e., each filtered basis vector is replaced by its own component that is orthogonal to the synthesized signal. Further complexity reduction is obtained by examining the code vectors in a sequence such that two successive code vectors differ in only one of the b scalar weight values; that is, the entire set of 2^b code vectors is searched in a *Gray coded* sequence. With successive code vectors differing in only one term in the linear combination, it is only necessary in the codebook search computation to progressively track the difference [6].

The total energy of a speech frame is encoded with 5 b (Table 30.4). The two gain factors (β and γ in Fig. 30.3) for each subframe are computed after the excitation codebooks have been searched and are then transformed to parameters G_s and P_0 to be vector quantized. Each mode has its own 5-b gain VQ codebook. G_s represents the energy of the subframe relative to the total frame energy, and P_0 represents the fraction of the subframe energy due to the first excitation source (excitation codebook 1 in mode 0, or the adaptive codebook in the other modes).

An *interpolation bit* (Table 30.4) transmitted for each frame specifies to the decoder whether the LP synthesis filter parameters for each subframe should be obtained from interpolating between the decoded filter parameters for the current and the previous frames. The encoder determines the value of this bit according to whether interpolation or no interpolation results in a lower prediction residual energy for the frame. The postfilter in the decoder operates in concordance with the actual LP parameters used for synthesis.

The speech encoder generates 112 b of encoded data (Table 30.4) for every 20-ms frame of the speech signal. These bits are processed by the channel encoder to improve, after channel decoding at the receiver, the uncoded bit-error rate and the detectability of uncorrected errors. Error detection coding in the form of 3 CRC bits is applied to the most critical 22 data bits. The combined 25 b plus an additional 73 speech data bits and 6 *tail bits* are input to an RCPCC encoder (the tail bits serve to bring the channel encoder and decoder to a fixed terminal state at the end of the payload data stream). The 3 CRC bits are encoded at rate 1/3 and the other 101 b are encoded at rate 1/2, generating a total of 211 channel coded bits. These are finally combined with the remaining 17 (uncoded) speech data bits to form a total of 228 b for the payload data of a speech frame. The payload data from two speech frames are interleaved for transmission over four timeslots of the GSM TDMA channel.

With the Viterbi algorithm, the channel decoder performs soft decision decoding on the demodulated and deinterleaved channel data. Uncorrected channel errors may still be present in the decoded speech data after Viterbi decoding. Thus, the channel decoder classifies each frame into three integrity categories: bad, unreliable, and reliable, in order to assist the speech decoder in undertaking error concealment measures. A frame is considered bad if the CRC check fails or if the received channel data is close to more than one candidate sequence. The latter evaluation is based on applying an adaptive threshold to the metric values produced by the Viterbi algorithm over the course of decoding the most critical 22 speech data bits and their 3 CRC bits. Frames that are not bad may be classified as unreliable, depending on the metric values produced by the Viterbi algorithm and on channel reliability information supplied by the demodulator.

Depending on the recent history of decoded data integrity, the speech decoder can take various error concealment measures. The onset of bad frames is concealed by repetition of parameters from previous reliable frames, whereas the persistence of bad frames results in power attenuation and ultimately muting of the synthesized speech. Unreliable frames are decoded with normality constraints applied to the energy of the synthesized speech.

30.8 Conclusions

The half-rate standards employ some of the latest techniques in speech and channel coding to meet the challenges posed by the severe transmission environment of digital cellular radio systems. By halving the bit rate, the voice transmission capacity of existing full-rate digital cellular systems can be doubled. Although advances are still being made that can address the needs of quarter-rate speech transmission, much effort is currently devoted to enhancing the speech quality and robustness of full-rate (GSM and IS-54) systems, aiming to be closer to *toll quality*. On the other hand, the imminent introduction of competing wireless systems that use different modulation schemes [e.g., coded division multiple access (CDMA)] and/or different radio frequencies [e.g., personal communications systems (PCS)] is poised to alleviate congestion in high-user-density areas.

Defining Terms

Codebook: An ordered collection of all possible values that can be assigned to a scalar or vector variable. Each unique scalar or vector value in a codebook is called a *codeword*, or *code vector* where appropriate.

Codec: A contraction of *(en)coder–decoder*, used synonymously with the word *coder*. The encoder and decoder are often designed and deployed as a pair. A half-rate standard codec performs speech as well as channel coding.

Echo canceller: A signal processing device that, given the source signal causing the echo signal, generates an estimate of the echo signal and subtracts the estimate from the signal being interfered with by the echo signal. The device is usually based on a discrete-time adaptive filter.

Pitch period: The fundamental period of a voiced speech waveform that can be regarded as periodic over a short-time interval (quasiperiodic). The reciprocal of pitch period is *pitch frequency* or simply, *pitch*.

Tandem coding: Having more than one encoder–decoder pair in an end-to-end transmission path. In cellular radio communications, having a radio link at each end of the communication path could subject the speech signal to two passes of speech encoding–decoding. In general, repeated encoding and decoding increases the distortion.

Acknowledgment

The authors would like to thank Erdal Paksoy and Mark A. Jasiuk for their valuable comments.

References

[1] Chen, J.-H. and Gersho, A., Adaptive postfiltering for quality enhancement of coded speech. *IEEE Trans. Speech & Audio Proc.*, 3(1), 59–71, 1995.

[2] De Martino, E., Speech quality evaluation of the European, North-American and Japanese speech codec standards for digital cellular systems. In *Speech and Audio Coding for Wireless and Network Applications*, Atal, B.S., Cuperman, V., and Gersho, A., Eds., 55–58, Kluwer Academic Publishers, Norwell, MA, 1993.

[3] Dimolitsas, S., Corcoran, F.L., and Baraniecki, M.R., Transmission quality of North American cellular, personal communications, and public switched telephone networks. *IEEE Trans. Veh. Tech.*, 43(2), 245–251, 1994.

[4] Gersho, A., Advances in speech and audio compression. *Proc. IEEE*, 82(6), 900–918, 1994.

[5] Gersho, A. and Gray, R.M., *Vector Quantization and Signal Compression*, Kluwer Academic Publishers, Norwell, MA, 1991.

[6] Gerson, I.A. and Jasiuk, M.A., Vector sum excited linear prediction (VSELP) speech coding at 8 kbps. In *Proceedings, IEEE Intl. Conf. Acoustics, Speech, & Sig. Proc.*, 461–464, April, 1990.

[7] Gerson, I.A. and Jasiuk, M.A., Techniques for improving the performance of CELP—type speech coders. *IEEE J. Sel. Areas Comm.*, 10(5), 858–865, 1992.

[8] Gerson, I.A., Jasiuk, M.A., Nowack, J.M., Winter, E.H., and Müller, J.-M., Speech and channel coding for the half-rate GSM channel. In *Proceedings, ITG-Report 130 on Source and Channel Coding*, 225–232. Munich, Germany, Oct., 1994.

[9] Gibson, J.D., Koo, B., and Gray, S.D., Filtering of colored noise for speech enhancement and coding. *IEEE Trans. Sig. Proc.*, 39(8), 1732–1742, 1991.

[10] Hagenauer, J., Rate-compatible punctured convolutional codes (RCPC codes) and their applications. *IEEE Trans. Comm.*, 36(4), 389–400, 1988.

[11] Jayant, N.S. and Noll, P., *Digital Coding of Waveforms*, Prentice-Hall, Englewood Cliffs, NJ, 1984.

[12] Masui, F. and Oguchi, M.,. Activity of the half rate speech codec algorithm selection for the personal digital cellular system. *Tech. Rept. of IEICE*, RCS93-77(11), 55–62 (in Japanese), 1993.

[13] Ohya, T., Suda, H., and Miki, T., 5.6 kbits/s PSI-CELP of the half-rate PDC speech coding standard. In *Proceedings, IEEE Veh. Tech. Conf.*, 1680–1684, June, 1994.

[14] Proakis, J.G., *Digital Communications*, 3rd ed., McGraw-Hill, New York, 1995.

[15] Suda, H., Ikeda, K., and Ikedo, J., Error protection and speech enhancement schemes of PSI-CELP, *NTT R & D*. (Special issue on PSI-CELP speech coding system for mobile communications), 43(4), 373–380, (in Japanese), 1994.

[16] Telecommunication Industries Association (TIA). Half-rate speech codec test plan V6.0. TR45.3.5/93.05.19.01, 1993.

[17] Zeger, K. and Gersho, A., Pseudo-Gray coding. *IEEE Trans. Comm.*, 38(12), 2147–2158, 1990.

Further Information

Additional technical information on speech coding can be found in the books, periodicals, and conference proceedings that appear in the list of references. Other relevant publications not represented in the list are *Speech Communication*, Elsevier Science Publishers; *Advances in Speech Coding*, B. S. Atal, V. Cuperman, and A, Gersho, eds., Kluwer Academic Publishers; and *Proceedings of the IEEE Workshop on Speech Coding*.

31

Wireless Video Communications

Madhukar Budagavi
Texas Instruments

Raj Talluri
Texas Instruments

31.1 Introduction

Recent advances in technology have resulted in a rapid growth in mobile communications. With this explosive growth, the need for reliable transmission of mixed media information—audio, video, text, graphics, and speech data—over wireless links is becoming an increasingly important application requirement. The bandwidth requirements of raw video data are very high (a 176 × 144 pixels, color video sequence requires over 8 Mb/s). Since the amount of bandwidth available on current wireless channels is limited, the video data has to be compressed before it can be transmitted on the wireless channel. The techniques used for video compression typically utilize predictive coding schemes to remove redundancy in the video signal. They also employ variable length coding schemes, such as Huffman codes, to achieve further compression.

The wireless channel is a noisy fading channel characterized by long bursts of errors [8]. When compressed video data is transmitted over wireless channels, the effect of channel errors on the video can be severe. The variable length coding schemes make the compressed bitstream sensitive to channel errors. As a result, the video decoder that is decoding the corrupted video bitstream can easily lose synchronization with the encoder. Predictive coding techniques, such as **block motion compensation,** which are used in current video compression standards, make the matter worse by quickly propagating the effects of channel errors across the video sequence and rapidly degrading the video quality. This may render the video sequence totally unusable.

Error control coding [5], in the form of **Forward Error Correction (FEC)** and/or **Automatic Repeat reQuest (ARQ),** is usually employed on wireless channels to improve the channel conditions. FEC techniques prove to be quite effective against random bit errors, but their performance is usually not adequate against longer duration burst errors. FEC techniques also come with an increased overhead in terms of the overall bitstream size; hence, some of the coding efficiency gains achieved by video compression are lost. ARQ techniques typically increase the delay and, therefore, might not be suitable for real-time videoconferencing. Thus, in practical video communication schemes, error control coding is typically used only to provide a certain level of error protection to the compressed video bitstream, and it becomes necessary for the video coder to accept some level of errors in the video bitstream. Error-resilience tools are introduced in the video codec to handle these residual errors that remain after error correction.

The emphasis in this chapter is on discussing relevant international standards that are making wireless video communications possible. We will concentrate on both the error control and source coding aspects of the problem. In the next section, we give an overview of a wireless video communication system that is a part of a complete wireless multimedia communication system. The International Telecommunication Union—Telecommunications Standardization Sector (ITU-T) H.223 [1] standard that describes a method of providing error protection to the video data before it is transmitted is also described. It should be noted that the main function of H.223 is to multiplex/demultiplex the audio, video, text, graphics, etc., which are typically communicated together in a videoconferencing application—error protection of the transmitted data becomes a requirement to support this functionality on error-prone channels. In Section 31.3, an overview of error-resilient video coding is given. The specific tools adopted into the International Standards Organization (ISO)/International Electrotechnical Commission (IEC) Motion Picture Experts Group (MPEG) v.4 (i.e., MPEG-4) [7] and the ITU-T H.263 [3] video coding standards to improve the error robustness of the video coder are described in Sections 31.4 and 31.5, respectively.

Table 31.1 provides a listing of some of the standards that are described or referred to in this chapter.

31.2   Wireless Video Communications

Figure 31.1 shows the basic block diagram of a wireless video communication system [10]. Input video is compressed by the video encoder to generate a compressed bitstream. The transport coder converts the compressed video bitstream into data units suitable for transmission over wireless channels. Typical operations carried out in the transport coder include channel coding, framing of data, modulation, and control operations required for accessing the wireless channel. At the receiver side, the inverse operations are performed to reconstruct the video signal for display.

In practice, the video communication system is part of a complete multimedia communication system and needs to interact with other system components to achieve the desired functionality. Hence, it becomes necessary to understand the other components of a multimedia communication system in order to design a good video communication system. Figure 31.2 shows the block diagram of a wireless multimedia terminal based on the ITU-T H.324 set of standards [4]. We use the H.324 standard as an example because it is the first videoconferencing standard for which mobile extensions were added to facilitate use on

TABLE 31.1   List of Relevant Standards

ISO/IEC 14496-2 (MPEG-4)	Information Technology—Coding of Audio-Visual Objects: Visual
H.263 (Version 1 and Version 2)	Video coding for low bitrate communication
H.261	Video codec for audiovisual services at p X 64 kbit/s
H.223	Multiplexing protocol for low bitrate multimedia communication
H.324	Terminal for low bitrate multimedia communication
H.245	Control protocol for multimedia communication
G.723.1	Dual rate speech coder for multimedia communication transmitting at 5.3 and 6.3 kbit/s

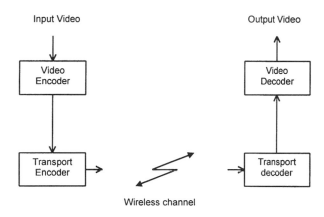

FIGURE 31.1   A wireless video communication system.

wireless channels. The system components of a multimedia terminal can be grouped into three processing blocks: (1) audio, video, and data (the word *data* is used here to mean still images/slides, shared files, documents etc.), (2) control, and (3) multiplex-demultiplex blocks.

1. Audio, video, and data processing blocks—These blocks basically produce/consume the multimedia information that is communicated. The aggregate bitrate generated by these blocks is restricted due to limitations of the wireless channel and, therefore, the total rate allowed has to be judiciously allocated among these blocks. Typically, the video blocks use up the highest percentage of the aggregate rate, followed by audio and then data. H.324 specifies the use of H.261/H.263 for video coding and G.723.1 for audio coding.

2. Control block—This block has a wide variety of responsibilities all aimed at setting up and maintaining a multimedia call. The control block facilitates the set-up of compression methods and preferred bitrates for audio, video, and data to be used in the multimedia call. It is also responsible for end-to-network signalling for accessing the network and end-to-end signalling for reliable operation of the multimedia call. H.245 is the control protocol in the H.324 suite of standards that specifies the control messages to achieve the above functionality.

3. Multiplex-Demultiplex (MUX) block—This block multiplexes the resulting audio, video, data, and control signals into a single stream before transmission on the network. Similarly, the received bitstream is demultiplexed to obtain the audio, video, data, and control signals, which are then passed to their respec-

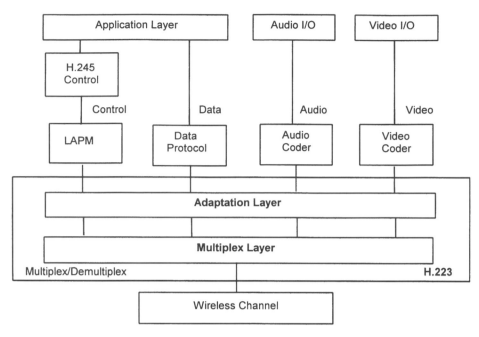

FIGURE 31.2   Configuration of a wireless multimedia terminal.

tive processing blocks. The MUX block accesses the network through a suitable network interface. The H.223 standard is the multiplexing scheme used in H.324.

Proper functioning of the MUX is crucial to the operation of the video communication system, as all the multimedia data/signals flow through it. On wireless channels, transmission errors can lead to a breakdown of the MUX resulting in, for example, nonvideo data being channeled to the video decoder or corrupted video data being passed on to the video decoder. Three annexes were specifically added to H.223 to enable its operation in error-prone environments. Below, we give a more detailed overview of H.223 and point out the levels of error protection provided by H.223 and its three annexes. It should also be noted that MPEG-4 does not specify a lower-level MUX like H.223, and thus H.223 can also be used to transmit MPEG-4 video data.

31.2.1 Recommendation H.223

Video, audio, data, and control information is transmitted in H.324 on distinct logical channels. H.223 determines the way in which the logical channels are mixed into a single bitstream before transmission over the physical channel (e.g., the wireless channel). The H.223 multiplex consists of two layers—the multiplex layer and the adaptation layer, as shown in Fig. 31.2. The multiplex layer is responsible for multiplexing the various logical channels. It transmits the multiplexed stream in the form of packets. The adaptation layer adapts the information stream provided by the applications above it to the multiplex layer below it by adding, where appropriate, additional octets for the purposes of error control and sequence numbering. The type of error control used depends on the type of information (audio/video/data/control) being conveyed in the stream. The adaptation layer provides error control support in the form of both FEC and ARQ.

H.223 was initially targeted for use on the benign general switched telephone network (GSTN). Later on, to enable its use on wireless channels, three annexes (referred to as

Levels 1–3, respectively), were defined to provide improved levels of error protection. The initial specification of H.223 is referred to as Level 0. Together, Levels 0–3 provide for a trade-off of error robustness against the overhead required, with Level 0 being the least robust and using the least amount of overhead and Level 3 being the most robust and also using the most amount of overhead.

1. H.223 Level 0—Default mode. In this mode the transmitted packet sizes are of variable length and are delimited by an 8-bit HDLC (High-level Data Link Control) flag (**01111110**). Each packet consists of a 1-octet header followed by the payload, which consists of a variable number of information octets. The header octet includes a Multiplex Code (MC) which specifies, by indexing to a multiplex table, the logical channels to which each octet in the information field belongs. To prevent emulation of the HDLC flag in the payload, bitstuffing is adopted.

2. H.223 Level 1 (Annex A)—Communication over low error-prone channels. The use of bitstuffing leads to poor performance in the presence of errors; therefore in Level 1, bitstuffing is not performed. The other improvement incorporated in Level 1 is the use of a longer 16-bit pseudo-noise synchronization flag to allow for more reliable detection of packet boundaries. The input bitstream is correlated with the synchronization flag and the output of the correlator is compared with a correlation threshold. Whenever the correlator output is equal to or greater than the threshold, a flag is detected. Since, bitstuffing is not performed, it is possible to have this flag emulated in the payload. However, the probability of such an emulation is low and is outweighed by the improvement gained by not using bitstuffing over error-prone channels.

3. H.223 Level 2 (Annex B)—Communication over moderately error-prone channels. When compared to the Level 1 operation, Level 2 increases the protection on the packet header. A Multiplex Payload Length (MPL) field, which gives the length of the payload in bytes, is introduced into the header to provide additional redundancy for detecting the length of the video packet. A (24,12,8) extended Golay code is used to protect the MC and the MPL fields. Use of error protection in the header enables robust delineation of packet boundaries. Note that the payload data is not protected in Level 2.

4. H.223 Level 3 (Annex C)—Communication over highly error-prone channels. Level 3 goes one step above Level 2 and provides for protection of the payload data. Rate Compatible Punctured Convolutional (RCPC) codes, various CRC polynomials, and ARQ techniques are used for protection of the payload data. Level 3 allows for the payload error protection overhead to vary depending on the channel conditions. RCPC codes are used for achieving this adaptive level of error protection because RCPC codes use the same channel decoder architecture for all the allowed levels of error protection, thereby reducing the complexity of the MUX.

31.3 Error Resilient Video Coding

Even after error control and correction, some amount of residual errors still exist in the compressed bitstream fed to the video decoder in the receiver. Therefore, the video decoder should be robust to these errors and should provide acceptable video quality even in the presence of some residual errors. In this section, we first describe a standard video coder configuration that is the basis of many international standards and also highlight the potential problems that are encountered when compressed video from these systems is transmitted over wireless channels. We then give an overview of the strategies that can be adopted to overcome these problems. Most of these strategies are incorporated in the MPEG-4 video coding standard and the H.263 (Version 2) video coding standard [3]. The original H.263 standard [2] which was standardized in 1996 for use in H.324 terminals connected to GSTN is referred to as Version 1. Version 2 of the H.263 standard provides additional improvements and functionalities (which include error-resilience tools) over the Version 1 standard. We will use H.263 to refer to both Version 1 and Version 2 standards and a distinction will be made only when required.

31.3.1 A Standard Video Coder

Redundancy exists in video signals in both spatial and temporal dimensions. Video coding techniques exploit this redundancy to achieve compression. A plethora of video compression techniques have been proposed in the literature, but a hybrid coding technique consisting of block motion compensation (BMC) and discrete cosine transforms (DCT) has been found to be very effective in practice. In fact, most of the current video coding standards such as H.263 and MPEG-4, which provide state-of-the-art compression performance, are all based on this hybrid coding technique. In this hybrid BMC/DCT coding technique, BMC is used to exploit temporal redundancy and the DCT is used to reduce spatial redundancy.

Figure 31.3 illustrates a standard hybrid BMC/DCT video coder configuration. Pictures are coded in either of two modes—interframe (INTER) or intraframe (INTRA) mode. In intraframe coding, the video image is encoded without any relation to the previous image, whereas in interframe coding, the current image is predicted from the previous image using BMC, and the difference between the current image and the predicted image, called the residual image, is encoded. The basic unit of data which is operated on is called a macroblock (MB) and is the data (both **luminance and chrominance** components) corresponding to a block of 16 × 16 pixels. The input image is split into disjoint macroblocks and the processing is done on a macroblock basis. Motion information, in the form of **motion vectors,** is calculated for each macroblock. The motion compensated prediction residual error is then obtained by subtracting each pixel in the macroblock with its motion shifted counterpart in the previous frame. Depending on the mode of coding used for the macroblock, either the image macroblock or the corresponding residual image macroblock is split into blocks of size 8 × 8 and an 8 × 8 DCT is applied to each of these 8 × 8 blocks. The resulting DCT coefficients are then quantized. Depending on the quantization step-size, this will result in a significant number of zero-valued coefficients. To efficiently encode the DCT coefficients that remain nonzero after quantization, the DCT coefficients are zig-zag scanned, and run-length encoded and the run-lengths are variable length encoded before transmission. Since a significant amount of correlation exists between the neighboring macroblocks' motion vectors, the motion vectors are themselves predicted from already transmitted motion vectors and the motion vector prediction error is encoded. The motion vector prediction error and the mode information are also variable length coded before transmission to achieve efficient compression.

The decoder uses a reverse process to reconstruct the macroblock at the receiver. The variable length codewords present in the received video bitstream are decoded first. For INTER macroblocks, the pixel values of the prediction error are reconstructed by inverse quantization and inverse DCT and are then added to the motion compensated pixels from the previous frame to reconstruct the transmitted macroblock. For INTRA macroblocks,

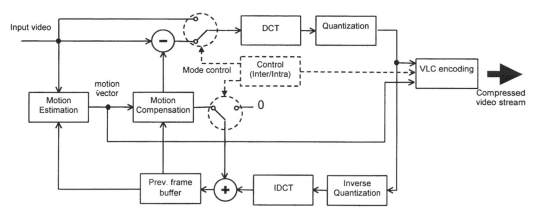

FIGURE 31.3   A standard video coder.

inverse quantization and inverse DCT directly result in the transmitted macroblock. All macroblocks of a given picture are decoded to reconstruct the whole picture.

31.3.2   Error Resilient Video Decoding

The use of predictive coding and variable length coding (VLC), though very effective from a compression point of view, makes the video decoding process susceptible to transmission errors. In VLC, the boundary between codewords is implicit. The compressed bitstream has to be read until a full codeword is encountered; the codeword is then decoded to obtain the information encoded in the codeword. When there are transmission errors, the implicit nature of the boundary between codewords typically leads to an incorrect number of bits being used in VLC decoding and, thus, subsequently results in a loss of synchronization with the encoder. In addition, the use of predictive coding leads to the propagation of these transmission errors to neighboring spatial blocks and to subsequently decoded frames, which leads to a rapid degradation in the reconstructed video quality.

To minimize the disastrous impact that transmission errors can have on the video decoding process, the following stages are incorporated in the video decoder to make it more robust:

- Error detection and localization
- Resynchronization
- Data recovery
- Error concealment

Figure 31.4 shows an error resilient video decoder configuration. The first step involved in robust video coding is the detection of errors in the bitstream. The presence of errors in the bitstream can be signaled by the FEC used in the multiplex layer. The video coder can also detect errors whenever illegal VLC codewords are encountered in the bitstream or when the decoding of VLC codewords leads to an illegal value of the decoded information (e.g., occurrence of more than 64 DCT coefficients for an 8 × 8 DCT block). Accurate detection of errors in the bitstream is a very important step, since most of the other error resilience techniques can only be invoked if an error is detected.

Due to the use of VLC, the location in the bitstream where the decoder detects an error is not the same location where the error has actually occurred but some undetermined distance away from it. This is shown in Fig. 31.5. Once an error is detected, it also implies that the

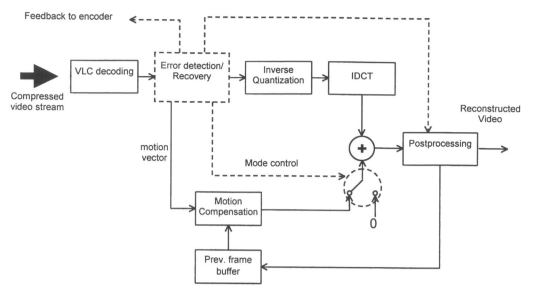

FIGURE 31.4   Error resilient video decoder.

decoder is not in synchronization with the encoder. Resynchronization schemes are then employed for the decoder to fall back into lock step with the encoder. While constructing the bitstream, the encoder inserts unique resynchronization words into the bitstream at approximately equally spaced intervals. These resynchronization words are chosen such that they are unique from the valid video bitstream. That is, no valid combination of the video algorithm's VLC tables can produce these words. The decoder, upon detection of an error, seeks forward in the bitstream looking for this known resynchronization word. Once this word is found, the decoder then falls back in synchronization with the encoder. At this point, the decoder has detected an error, regained synchronization with the encoder, and isolated the error to be between the two resynchronization points. Since the decoder can only isolate the error to be somewhere between the resynchronization points but not pinpoint its exact location, all of the data that corresponds to the macroblocks between these two resynchronization points needs to be discarded. Otherwise, the effects of displaying an image reconstructed from erroneous data can cause highly annoying visual artifacts.

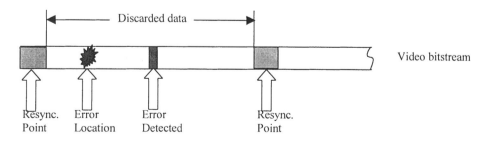

FIGURE 31.5   At the decoder, it is usually not possible to detect the error at the actual error occurrence location; hence, all the data between the two resynchronization points may need to be discarded.

Some data recovery techniques, such as "reversible decoding," enable the decoder to salvage some of the data between the two resynchronization points. These techniques advocate the use of a special kind of VLC table at the encoder in coding the DCTs and motion vector information. These special VLCs have the property that they can be decoded both in the forward and reverse directions. By comparing the forward and reverse decoded data, the exact location of the error in the bit stream can be localized more precisely and some of the data between the two resynchronization points can be salvaged. The use of these reversible VLCs (RVLCs) is part of the MPEG-4 standard and will be described in greater detail in the following sections.

After data recovery, the impact of the data that is deemed to be in error needs to be minimized. This is the error concealment stage. One simple error concealment strategy is to simply replace the luminance and chrominance components of the erroneous macroblocks with the luminance and chrominance of the corresponding macroblocks in the previous frame of the video sequence. While this technique works fairly well and is simple to implement, more complex techniques use some type of estimation strategies to exploit the local correlation that exists within a frame of video data to come up with a better estimate of the missing or erroneous data. These error concealment strategies are essentially postprocessing algorithms and are not mandated by the video coding standards. Different implementations of the wireless video systems utilize different kinds of error concealment strategies based on the available computational power and the quality of the channel.

If there is support for a decoder feedback path to the encoder as shown in Fig. 31.3, this path can be used to signal detected errors. The feedback information from the decoder can be used to retransmit data or to influence future encoder action so as to stop the propagation of detected errors in the decoder. Note that for the feedback to take place, the network must support a back channel.

31.3.3 Classification of Error-Resilience Techniques

In general, techniques to improve the robustness of the video coder can be classified into three categories based on whether the encoder or the decoder plays a primary part in improving the error robustness [10]. *Forward error resilience* techniques refer to those techniques where the encoder plays the primary part in improving the error robustness, typically by introducing redundancy in the transmitted information. In *postprocessing* techniques, the decoder plays the primary part and does concealment of errors by estimation and interpolation (e.g., spatial-temporal filtering) using information it has already received. In *interactive error* resilience techniques, the decoder and the encoder interact to improve the error resilience of the video coder. Techniques that use decoder feedback come under this category.

31.4 MPEG-4 Error Resilience Tools

MPEG-4 is an ISO/IEC standard being developed by the Motion Pictures Expert Group. Initially MPEG was aimed primarily at low-bit-rate communications; however, its scope was later expanded to be much more of a multimedia coding standard [7]. The MPEG-4 video coding standard is the first video coding standard to address the problem of efficient representation of visual objects of arbitrary shape. MPEG-4 was also designed to provide "universal accessibility," i.e., the ability to access audio-visual information over a wide range of storage and transmission media. In particular, because of the proliferation of wireless communications, this implied development of specific tools to enable error-resilient transmission of compressed data over noisy communication channels.

A number of tools have been incorporated into the MPEG-4 video coder to make it more error resilient. All these tools are basically forward error resilience tools. We describe below each of these tools and its advantages.

31.4.1 Resynchronization

As mentioned earlier, a video decoder that is decoding a corrupted bitstream may lose synchronization with the encoder (i.e., it is unable to identify the precise location in the image where the current data belongs). If remedial measures are not taken, the quality of the decoded video rapidly degrades and becomes unusable. One approach is for the encoder to introduce resynchronization markers in the bitstream at various locations. When the decoder detects an error, it can then look for this resynchronization marker and regain synchronization.

Previous video coding standards such as H.261 and H.263 (Version 1) logically partition each of the images to be encoded into rows of macroblocks called Group Of Blocks (GOBs). These GOBs correspond to a horizontal row of macroblocks for **QCIF** images. Figure 31.6 shows the GOB numbering scheme for H.263 (Version 1) for QCIF resolution. For error resilience purposes, H.263 (Version 1) provides the encoder an option of inserting resynchronization markers at the beginning of each of the GOBs. Hence, for QCIF images these resynchronization markers are allowed to occur only at the left edge of the images. The smallest region that the error can be isolated to and concealed in this case is thus one row of macroblocks.

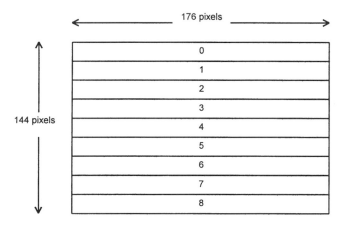

FIGURE 31.6 H.263 GOB numbering for a QCIF image.

In contrast, the MPEG-4 encoder is not restricted to inserting the resynchronization markers only at the beginning of each row of macroblocks. The encoder has the option of dividing the image into video packets. Each video packet is made up of an integer number of consecutive macroblocks in raster scan order. These macroblocks can span several rows of macroblocks in the image and can even include partial rows of macroblocks. One suggested mode of operation for the MPEG-4 encoder is for it to insert a resynchronization marker periodically at approximately every K bits. Note that resynchronization markers can only be placed at a macroblock boundary and, hence, the video packet length cannot be constrained to be exactly equal to K bits. When there is a significant activity in one part of the image, the macroblocks corresponding to these areas generate more bits than other parts of the image. If the MPEG-4 encoder inserts the resynchronization markers at uniformly spaced bit intervals, the macroblock interval between the resynchronization markers is a lot closer in the high activity areas and a lot farther apart in the low activity areas. Thus, in the presence of a short burst of errors, the decoder can quickly localize the error to within a few macroblocks in the important high activity areas of the image and preserve the image quality in these important areas. In the case of H.263 (Version 1), where the resynchronization markers are restricted to be at the beginning of the GOBs, it is only possible for the decoder to isolate the errors to a fixed GOB independent of the image content. Hence, effective coverage of the resynchronization marker is reduced when compared to the MPEG-4 scheme. The recommended spacing of the resynchronization markers in MPEG-4 is based on the bitrates. For 24 Kb/s, it is recommended to insert them at intervals of 480 bits and for bitrates between 25 Kb/s to 48 Kb/s, it is recommended to place them at every 736 bits. Figures 31.7(a) and (b) illustrate the placement of resynchronization markers for H.263 (Version 1) and MPEG-4.

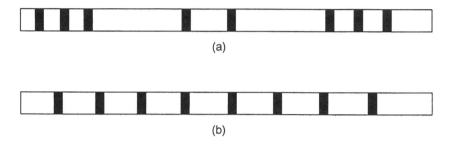

(a)

(b)

FIGURE 31.7   Position of resynchronization markers in the bitstream for (a) H.263 (Version 1) encoder with GOB headers and for (b) an MPEG-4 encoder with video packets.

Note that in addition to inserting the resynchronization markers at the beginning of each video packet, the encoder also needs to remove all data dependencies that exist between the data belonging to two different video packets within the same image. This is required so that even if one of the video packets in the current image is corrupted due to errors, the other packets can be decoded and utilized by the decoder. In order to remove these data dependencies, the encoder inserts two additional fields in addition to the resynchronization marker at the beginning of each video packet, as shown in Fig. 31.8. These are, (1) the absolute macroblock number of the first macroblock in the video packet, *Mb. No.,* (which indicates the spatial location of the macroblock in the current image), (2) the quantization parameter, QP, which denotes the initial quantization parameter used to quantize the DCT coefficients in the video packet. The encoder also modifies the predictive encoding method used for coding the motion vectors such that there are no predictions across the video packet boundaries. Also shown in Fig. 31.8 is a third field, labeled *HEC*. Its use is discussed in a later section.

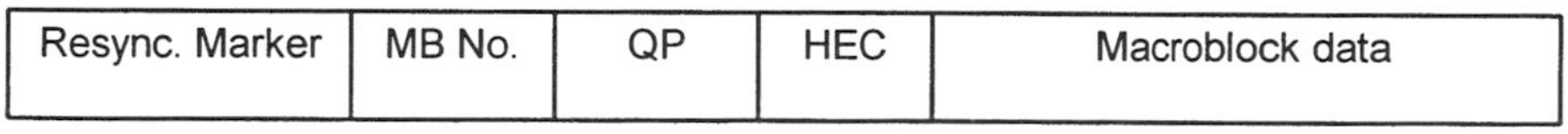

Resync. Marker	MB No.	QP	HEC	Macroblock data

FIGURE 31.8   An MPEG-4 video packet.

31.4.2 Data Partitioning

Data partitioning in MPEG-4 provides enhanced error localization and error concealment capabilities. The data partitioning mode partitions the data within a video packet into a motion part and a texture part (DCT coefficients) separated by a unique Motion Marker (MM), as shown in Fig. 31.9. All the syntactic elements of the video packet that have motion-related information are placed in the motion partition and all the remaining syntactic elements that relate to the DCT data are placed in the texture partition. If the texture information is lost, data partitioning enables the salvation of motion information, which can then be used to conceal the errors in a more effective manner.

Resync. Marker	MB No.	QP	HEC	Motion Data	MM	Texture data

FIGURE 31.9   A data partitioned MPEG-4 video packet.

The motion marker is computed from the motion VLC tables using a search program such that it is Hamming distance 1 from any possible valid combination of the motion VLC tables [9]. The motion marker is uniquely decodable from the motion VLC tables, and it indicates to the decoder the end of the motion information and the beginning of the DCT information. The number of macroblocks in the video packet is implicitly known after encountering the motion marker. Note that the motion marker is only computed once based on the VLC tables and is fixed in the standard. Based on the VLC tables in MEPG-4, the motion marker is a 17-bit word whose value is **1 1111 0000 0000 0001**.

31.4.3 Reversible Variable Length Codes (RVLCs)

As was shown in Fig. 31.5, if the decoder detects an error during the decoding of VLC codewords, it loses synchronization and hence typically has to discard all the data up to the next resynchronization point. RVLCs are designed such that they can be instantaneously decoded both in the forward and the backward direction. When the decoder detects an error while decoding the bitstream in the forward direction, it jumps to the next resynchronization marker and decodes the bitstream in the backward direction until it encounters an error. Based on the two error locations, the decoder can recover some of the data that would have otherwise been discarded. This is shown in Fig. 31.10, which shows only the texture part of the video packet—only data in the shaded area is discarded. Note that if RVLCs were not used, all the data in the texture part of the video packet would have to be discarded. RVLCs thus enable the decoder to better isolate the error location in the bitstream.

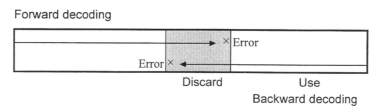

FIGURE 31.10   Use of reversible variable length codes.

Figure 31.11 shows the comparison of performance of resynchronization, data partitioning, and RVLC techniques for 24 Kb/s QCIF video data. The experiments involved transmission of three video sequences, each of duration 10s, over a bursty channel simulated by a 2-state Gilbert model [6]. The burst duration on the channel is 1 ms and the burst occurrence probability is 10^{-2}. Figure 31.11, which plots the average peak signal-to-noise ratios of the received video frames, shows that data partitioning and RVLC provide improved performance when compared to using only resynchronization markers.

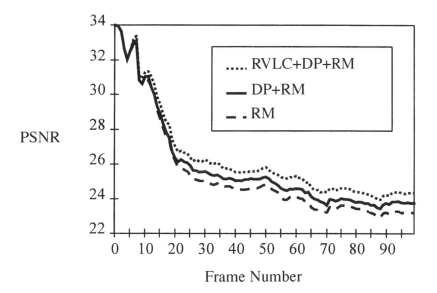

FIGURE 31.11 Performance comparison of resynchronization, data partitioning, and RVLC over a bursty channel simulated by a 2-state Gilbert model. Burst durations are 1ms long and the probability of occurrence of a burst is 10^{-2}. Legend: RM—resynchronization marker; DP—data partitioning; RVLC—reversible variable length codes.

31.4.4 Header Extension Code (HEC)

Some of the most important information that the decoder needs in order to decode the video bitstream is in the video frame header data. This data includes information about the spatial dimensions of the video data, the time stamps associated with the decoding and the presentation of this video data, and the type of the current frame (INTER/INTRA). If some of this information is corrupted due to channel errors, the decoder has no other recourse but to discard all the information belonging to the current video frame. In order to reduce the sensitivity of this data, a technique called Header Extension Code (HEC) was introduced into the MPEG-4 standard. In each video packet, a 1-bit field called HEC is present. The location of HEC in the video packet is shown in Fig. 31.8. For each video packet, when HEC is set, the important header information that describes the video frame is repeated in the bits following the HEC. This information can be used to verify and correct the header information of the video frame. The use of HEC significantly reduces the number of discarded video frames and helps achieve a higher overall decoded video quality.

31.4.5   Adaptive Intra Refresh (AIR)

Whenever an INTRA macroblock is received, it basically stops the temporal propagation of errors at its corresponding spatial location. The procedure of forcefully encoding some macroblocks in a frame in INTRA mode to flush out possible errors is called INTRA refreshing. INTRA refresh is very effective in stopping the propagation of errors, but it comes at the cost of a large overhead. Coding a macroblock in INTRA mode typically requires many more bits when compared to coding the macroblock in INTER mode. Hence, the INTRA refresh technique has to be used judiciously.

For areas with low motion, simple error concealment by just copying the previous frame's macroblocks works quite effectively. For macroblocks with high motion, error concealment becomes very difficult. Since the high motion areas are perceptually the most significant, any persistent error in the high motion area becomes very noticeable. The AIR technique of MPEG-4 makes use of the above facts and INTRA refreshes the motion areas more frequently, thereby allowing the corrupted high motion areas to recover quickly from errors.

Depending on the bitrate, the AIR approach only encodes a fixed and predetermined number of macroblocks in a frame in INTRA mode (the exact number is not standardized by MPEG-4). This fixed number might not be enough to cover all the macroblocks in the motion area; hence, the AIR technique keeps track of the macroblocks that have been refreshed (using a "refresh map") and in subsequent frames refreshes any macroblocks in the motion areas that might have been left out.

31.5   H.263 Error Resilience Tools

In this section, we discuss four error resilience techniques which are part of the H.263 standard—*slice structure mode* and *independent segment decoding,* which are forward error resilience features, and *error tracking* and *reference picture selection,* which are interactive error resilience techniques. Error tracking was introduced in H.263 (Version 1) as an appendix, whereas the remaining three techniques were introduced in H.263 (Version 2) as annexes.

31.5.1   Slice Structure Mode (Annex K)

The slice structured mode of H.263 is similar to the video packet approach of MPEG-4 with a slice denoting a video packet. The basic functionality of a slice is the same as that of a video packet—providing periodic resynchronization points throughout the bistream. The structure of a slice is shown in Fig. 31.12. Like an MPEG-4 video packet, the slice consists of a header followed by the macroblock data. The SSC is the slice start code and is identical to the resynchronization marker of MPEG-4. The MBA field, which denotes the starting macroblock number in the slice, and the SQUANT field, which is the quantizer scale coded nonpredictively, allow for the slice to be coded independently.

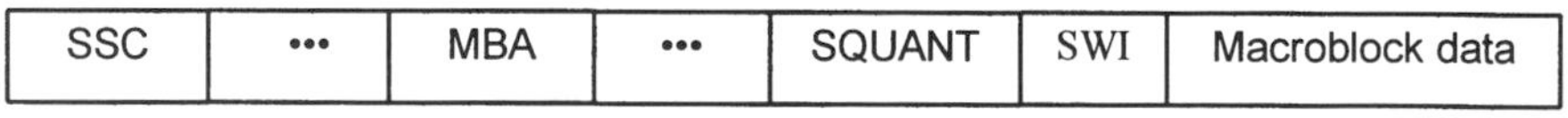

SSC	•••	MBA	•••	SQUANT	SWI	Macroblock data

FIGURE 31.12   Structure of a slice in H.263/Annex K.

The slice structured mode also contains two submodes which can be used to provide additional functionality. The submodes are

- Rectangular slice submode (RSS)—This allows for rectangular shaped slices. The rectangular region contained in the slice is specified by SWI+1 (See Fig. 31.12 for the location of the SWI field in the slice header), which gives the width of the rectangular region, and MBA, which specifies the upper left macroblock of the slice. Note that the height of the rectangular region gets specified by the number of macroblocks contained in the slice. This mode can be used, for example, to subdivide images into rectangular regions of interest for region-based coding.

- Arbitrary slice submode (ASO)—The default order of transmission of slices is such that the MBA field is strictly increasing from one slice to the next transmitted slice. When ASO is used, the slices may appear in any order within the bitstream. This mode is useful when the wireless network supports prioritization of slices which might result in out-of-order arrival of video slices at the decoder.

31.5.2   Independent Segment Decoding (ISD) (Annex R)

Even though the slice structured mode eliminates decoding dependency between neighboring slices, errors in slices can spatially propagate to neighboring slices in subsequent frames due to motion compensation. This happens because motion vectors in a slice can point to macroblocks of neighboring slices in the reference picture. Independent segment decoding eliminates this from happening by restricting the motion vectors within a predefined segment of the picture from pointing to other segments in the picture, thereby helping to contain the error to be within the erroneous segment. This improvement in the localization of errors, however, comes at a cost of a loss of coding efficiency. Because of this restriction on the motion vectors, the motion compensation is not as effective, and the residual error images use more bits.

For ease of implementation, the ISD mode puts restrictions on segment shapes and on the changes of segment shapes from picture to picture. The ISD mode cannot be used with the slice structured mode (Annex K) unless the rectangular slice submode of Annex K is active. This prevents the need for treating awkward shapes of slices that can otherwise arise when Annex K is not used with rectangular slice submode. The segment shapes are not allowed to change from picture to picture unless an INTRA frame is being coded.

31.5.3   Error Tracking (Appendix I)

The error tracking approach is an INTRA refresh technique but uses decoder feedback of errors to decide which macroblocks in the current image to code in INTRA mode to prevent the propagation of these errors. When there are no errors on the channel, normal coding (which usually results in the bit-efficient INTER mode being selected most of the time) is used. The use of decoder feedback allows the system to adapt to varying channel conditions and minimizes the use of forced INTRA updates to situations when there are channel errors.

Because of the time delay involved in the decoder feedback, the encoder has to track the propagation of an error from its original occurrence to the current frame to decide which macroblocks should be INTRA coded in the current frame. A low complexity algorithm was proposed in Appendix I of H.263 to track the propagation of errors. However, it should be noted that the use of this technique is not mandated by H.263. Also, H.263 itself does not standardize the mechanism by which the decoder feedback of error can be sent. Typically, H.245 control messages are used to signal the decoder feedback for error tracking purposes.

31.5.4   Reference Picture Selection (Annex N)

The Reference Picture Selection (RPS) mode of H.263 also relies on decoder feedback to efficiently stop the propagation of errors. The back channel used in RPS mode can be a separate logical channel (e.g., by using H.245), or if two-way communication is taking place, the back channel messages can be sent multiplexed with the encoded video data. In the presence of errors, the RPS mode allows the encoder to be instructed to select one of the several previously correctly received and decoded frames as the reference picture for motion compensation of the current frame being encoded. This effectively stops the propagation of error. Note that the use of RPS requires the use of multiple frame buffers at both the encoder and the decoder to store previously decoded frames. Hence, the improvement in performance in the RPS mode has come at the cost of increased memory requirements.

31.6   Discussion

In this chapter we presented a broad overview of the various techniques that enable wireless video transmission. Due to the enormous amount of bandwidth required, video data is typically compressed before being transmitted, but the errors introduced by the wireless channels have a severe impact on the compressed video information. Hence, special techniques need to be employed to enable robust video transmission. International standards play a very important role in communications applications. The two current standards that are most relevant to video applications are ISO MPEG-4 and ITU H.263. In this chapter, we detailed these two standards and explained the error resilient tools that are part of these standards to enable robust video communication over wireless channels. A tutorial overview of these tools has been presented and the performance of these tools has been described.

There are, however, a number of other methods that further improve the performance of a wireless video codec that the standards do not specify. If the encoder and decoder are aware of the limitations imposed by the communication channel, they can further improve the video quality by using these methods. These methods include encoding techniques such as rate control to optimize the allocation of the effective channel bit rate between various parts of video to be transmitted and intelligent decisions on when and where to place INTRA refresh macroblocks to limit the error propagation. Decoding methods such as superior error concealment strategies that further conceal the effects of erroneous macroblocks by estimating them from correctly decoded macroblocks in the spatiotemporal neighborhood can also significantly improve the effective video quality.

This chapter has mainly focused on the error resilience aspects of the video layer. There are a number of error detection and correction strategies, such as Forward Error Correction (FEC), that can further improve the reliability of the transmitted video data. These FEC codes are typically provided in the systems layer and the underlying network layer. If the video transmission system has the ability to monitor the dynamic error characteristics of the communication channel, joint source-channel coding techniques can also be effectively employed. These techniques enable the wireless communication system to perform optimal trade-offs in allocating the available bits between the source coder (video) and the channel coder (FEC) to achieve superior performance.

Current video compression standards also support *layered* coding methods. In this approach, the compressed video information can be separated into multiple layers. The *base* layer, when decoded, provides a certain degree of video quality and the *enhancement* layer, when received and decoded, then adds to the base layer to further improve the video quality. In wireless channels, these base and enhancement layers give a natural method of partitioning the video data into more important and less important layers. The base layer can be protected by a stronger level of error protection (higher overhead channel coder) and the enhancement layer by a lesser strength coder. Using this Unequal Error Protection (UEP) scheme, the communication system is assured of a certain degree of performance most of the time through the base layer, and when the channel is not as error prone and the decoder receives the enhancement layer, this scheme provides improved quality.

Given all these advances in video coding technology, coupled with the technological advances in processor technology, memory devices, and communication systems, wireless video communications is fast becoming a very compelling application. With the advent of higher bandwidth third generation wireless communication systems, it will be possible to transmit compressed video in many wireless applications, including mobile videophones, videoconferencing systems, PDAs, security and surveillance applications, mobile Internet terminals, and other multimedia devices.

Defining Terms

Automatic Repeat reQuest (ARQ): An error control system in which notification of erroneously received messages is sent to the transmitter which then simply retransmits the message. The use of ARQ requires a feedback channel and the receiver must perform error detection on received messages. Redundancy is added to the message before transmission to enable error detection at the receiver.

Block motion compensation (BMC): Motion compensated prediction that is done on a block basis; that is, blocks of pixels are assumed to be displaced spatially in a uniform manner from one frame to another.

Forward Error Correction: Introduction of redundancy in data to allow for correction of errors without retransmission.

Luminance and chrominance: Luminance is the brightness information in a video image, whereas chrominance is the corresponding color information.

Motion vectors: Specifies the spatial displacement of a block of pixels from one frame to another.

QCIF: Quarter Common Intermediate Format (QCIF) is a standard picture format that defines the image dimensions to be 176×144 (pixels per line $\times$ lines per picture) for luminance and 88×72 for chrominance.

References

[1] International Telecommunications Union—Telecommunications Standardization Sector, Recommendation H.223: Multiplexing protocol for low bitrate multimedia communications, Geneva, 1996.

[2] International Telecommunications Union—Telecommunications Standardization Sector, Recommendation H.263: Video coding for low bitrate communication, Geneva, 1996.

[3] International Telecommunications Union—Telecommunications Standardization Sector, Draft Recommendation H.263 (Version 2): video coding for low bitrate communication, Geneva, 1998.

[4] International Telecommunications Union—Telecommunications Standardization Sector, Recommendation H.324: terminal for low bit rate multimedia communications, Geneva, 1996.

[5] Lin, S. and Costello, D.J., Jr., *Error Control Coding: Fundamentals and Applications*, Prentice-Hall, Englewood Cliffs, NJ, 1983.

[6] Miki, T., et al., Revised error pattern generation programs for core experiments on error resilience, *ISO/IEC JTC1/SC29/WG11 MPEG96/1492*, Maceio, Brazil, Nov. 1996.

[7] International Organization for Standardization, Committee draft of Tokyo (N2202): information technology—coding of audio-visual objects: visual, *ISO/IEC 14496-2*, Mar. 1998.

[8] Sklar, B., Rayleigh fading channels in mobile digital communication systems, Pt. I: Characterization, *IEEE Commun. Mag.*, 35, 90–100, 1997.

[9] Talluri, R., et al., Error concealment by data partitioning, to appear in *Signal Processing: Image Commun.*, 1998.

[10] Wang, Y. and Zhu, Q., Error control and concealment for video communication: a review, *IEEE Trans. Circuits Syst. Video Technol.*, 86(5), 974–997, May 1998.

Further Information

A broader overview of wireless video can be found in the special issue of *IEEE Communications Magazine*, June 1998. Wang and Zhu [10] provide an exhaustive review of error concealment techniques for video communications. More details on MPEG-4 and ongoing Version 2 activities in MPEG-4 can be found on the web page `http://drogo.cselt.it/mpeg/standards/mpeg-4/mpeg-4.htm`. H.263 (Version 2) activities are tracked on the web page `http://www.ece.ubc.ca/spmg/research/motion/h263plus/`. Most of the ITU-T recommendations can be obtained from the web site `http://www.itu.org`. The special issue of *IEEE Communications Magazine*, December 1996, includes articles on H.324 and H.263.

Current research relevant to wireless video communications is reported in a number of journals including *IEEE Transactions on Circuits and Systems for Video Technology*, *IEEE Transactions on Image Processing*, *IEEE Transactions on Vehicular Technology*, *Signal Processing: Image Communication*. The *IEEE Communications Magazine* regularly reports review articles relevant to wireless video communications. Conferences of interest include the IEEE International Conference on Image Processing (ICIP), IEEE Vehicular Technology Conference (VTC), and IEEE International Conference on Communications (ICC).

32

Wireless LANs

Suresh Singh
Oregon State University

32.1 Introduction

A proliferation of high-performance portable computers combined with end-user need for communication is fueling a dramatic growth in wireless **local area network** (LAN) technology. Users expect to have the ability to operate their portable computer globally while remaining connected to communications networks and service providers. Wireless LANs and cellular networks, connected to high-speed networks, are being developed to provide this functionality.

Before delving deeper into issues relating to the design of wireless LANs, it is instructive to consider some scenarios of user mobility.

1. A simple model of user mobility is one where a computer is physically moved while retaining network connectivity at either end. For example, a move from one room to another as in a hospital where the computer is a hand-held device displaying patient charts and the nurse using the computer moves between wards or floors while accessing patient information.

2. Another model situation is where a group of people (at a conference, for instance) set up an ad-hoc LAN to share information as in Fig. 32.1.

3. A more complex model is one where several computers in constant communication are in motion and continue to be networked. For example, consider the problem of having robots in space collaborating to retrieve a satellite.

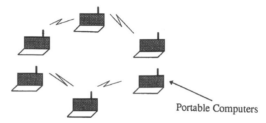

FIGURE 32.1   Ad-hoc wireless LAN.

A great deal of research has focused on dealing with physical layer and **medium access control (MAC)** layer protocols. In this chapter we first summarize standardization efforts in these areas. The remainder of the chapter is then devoted to a discussion of networking issues involved in wireless LAN design. Some of the issues discussed include routing in wireless LANs (i.e., how does data find its destination when the destination is mobile?) and the problem of providing service guarantees to end users (e.g., error-free data transmission or bounded delay and bounded bandwidth service, etc.).

32.2   Physical Layer Design

Two media are used for transmission over wireless LANs, infrared and radio frequency. RF LANs are typically implemented in the industrial, scientific, and medical (ISM) frequency bands 902–928 MHz, 2400–2483.5 MHz and 5725–5850 MHz. These frequencies do not require a license allowing the LAN product to be portable, i.e., a LAN can be moved without having to worry about licensing.

IR and RF technologies have different design constraints. IR receiver design is simple (and thus inexpensive) in comparison to RF receiver design because IR receivers only detect the amplitude of the signal not the frequency or phase. Thus, a minimal of filtering is required to reject interference. Unfortunately, however, IR shares the electromagnetic spectrum with the sun and incandescent or fluorescent light. These sources of modulated infrared energy reduce the signal-to-noise ratio of IR signals and, if present in extreme intensity, can make the IR LANs inoperable. There are two approaches to building IR LANs.

1. The transmitted signal can be focused and aimed. In this case the IR system can be used outdoors and has an area of coverage of a few kilometers.
2. The transmitted signal can be bounced off the ceiling or radiated omnidirectionally. In either case, the range of the IR source is 10–20 m (i.e., the size of one medium-sized room).

RF systems face harsher design constraints in comparison to IR systems for several reasons. The increased demand for RF products has resulted in tight regulatory constraints on the allocation and use of allocated bands. In the U.S., for example, it is necessary to implement spectrum spreading for operation in the ISM bands. Another design constraint is the requirement to confine the emitted spectrum to a band, necessitating amplification at higher carrier frequencies, frequency conversion using precision local oscillators, and selective components. RF systems must also cope with environmental noise that is either naturally occurring, for example, atmospheric noise or man made, for example, microwave ovens, copiers, laser printers, or other heavy electrical machinery. RF LANs operating in the ISM frequency ranges also suffer interference from amateur radio operators.

Operating LANs indoors introduces additional problems caused by multipath propagation, Rayleigh fading, and absorption. Many materials used in building construction are opaque to IR radiation resulting in incomplete coverage within rooms (the coverage depends on obstacles within the room that block IR) and almost no coverage outside closed rooms. Some materials, such as white plasterboard, can also cause reflection of IR signals. RF is relatively immune to absorption and reflection problems. Multipath propagation affects both IR and RF signals. The technique to alleviate the effects of multipath propagation in both types of systems is the same use of aimed (directional) systems for transmission enabling the receiver to reject signals based on their angle of incidence. Another technique that may be used in RF systems is to use multiple antennas. The phase difference between different paths can be used to discriminate between them.

Rayleigh fading is a problem in RF systems. Recall that Rayleigh fading occurs when the difference in path length of the same signal arriving along different paths is a multiple of half a wavelength. This causes the signal to be almost completely canceled out at the receiver. Because the wavelengths used in IR are so small, the effect of Rayleigh fading is not noticeable in those systems. RF systems, on the other hand, use wavelengths of the order of the dimension of a laptop. Thus, moving the computer a small distance could increase/decrease the fade significantly.

Spread spectrum transmission technology is used for RF-based LANs and it comes in two varieties: direct-sequence spread spectrum (DSSS) and frequency-hopping spread spectrum (FHSS). In a FHSS system, the available band is split into several channels. The transmitter transmits on one channel for a fixed time and then hops to another channel. The receiver is synchronized with the transmitter and hops in the same sequence; see Fig. 32.2(a). In DSSS systems, a random binary string is used to modulate the transmitted signal. The relative rate between this sequence and user data is typically between 10 and 100; see Fig. 32.2(b).

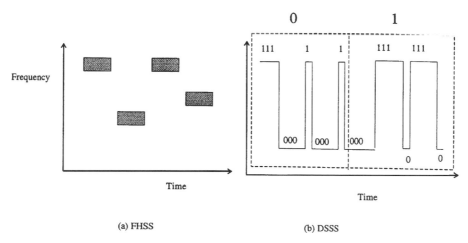

FIGURE 32.2   Spread spectrum.

The key requirements of any transmission technology is its robustness to noise. In this respect DSSS and FHSS show some differences. There are two possible sources of interference for wireless LANs: the presence of other wireless LANs in the same geographical area (i.e., in the same building, etc.) and interference due to other users of the ISM frequencies. In the latter case, FHSS systems have a greater ability to avoid interference because the hopping sequence could be designed to prevent potential interference. DSSS systems, on the other hand, do exhibit an ability to recover from interference because of the use of the spreading factor [Fig. 32.2(b)].

It is likely that in many situations several wireless LANs may be collocated. Since all wireless LANs use the same ISM frequencies, there is a potential for a great deal of interference. To avoid interference in FHSS systems, it is necessary to ensure that the hopping sequences are orthogonal. To avoid interference in DSSS systems, on the other hand, it is necessary to allocate different channels to each wireless LAN. The ability to avoid interference in DSSS systems is, thus, more limited in comparison to FHSS systems because FHSS systems use very narrow subchannels (1 MHz) in comparison to DSSS systems that use wider subchannels (for example, 25 MHz), thus, limiting the number of wireless LANs that can be collocated. A summary of design issues can be found in [1].

32.3   MAC Layer Protocols

MAC protocol design for wireless LANs poses new challenges because of the in-building operating environment for these systems. Unlike wired LANs (such as the ethernet or token ring), wireless LANs operate in strong multipath fading channels where channel characteristics can change in very short distances resulting in unreliable communication and unfair channel access due to capture. Another feature of the wireless LAN environment is that carrier sensing takes a long time in comparison to wired LANs; it typically takes between 30 and 50 μs (see [4]), which is a significant portion of the packet transmission time. This results in inefficiencies if the CSMA family of protocols is used without any modifications.

Other differences arise because of the mobility of users in wireless LAN environments. To provide a building (or any other region) with wireless LAN coverage, the region to be covered is divided into cells as shown in Fig. 32.3. Each cell is one wireless LAN, and adjacent cells use different frequencies to minimize interference. Within each cell there is an access point called a **mobile support station** (MSS) or base station that is connected to some wired network. The mobile users are called **mobile hosts (MH)**. The MSS performs the functions of channel allocation and providing connectivity to existing wired networks; see Fig. 32.4. Two problems arise in this type of an architecture that are not present in wired LANs.

1. The number of nodes within a cell changes dynamically as users move between cells. How can the channel access protocol dynamically adapt to such changes efficiently?

2. When a user moves between cells, the user has to make its presence known to the other nodes in the cell. How can this be done without using up too much bandwidth? The protocol used to solve this problem is called a handoff protocol and works along the following lines: A switching station (or the MSS nodes working together, in concert) collects signal strength information for each mobile host within each cell. Note that if a mobile host is near a cell boundary, the MSS node in its current cell as well as in the neighboring cell can hear its transmissions and determine signal strengths. If the mobile host is currently under the coverage of MSS M1 but its signal strength at MSS M2 becomes larger, the switching station initiates a handoff whereby the MH is considered as part of M2's cell (or network).

The mode of communication in wireless LANs can be broken in two: communication from the mobile to the MSS (called *uplink* communication) and communication in the reverse direction (called *downlink* communication). It is estimated that downlink communication accounts for about 70–80% of the total consumed bandwidth. This is easy to see because most of the time users request files or data in other forms (image data, etc.) that consume

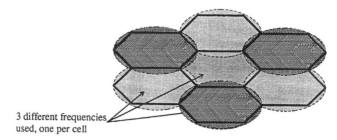

FIGURE 32.3 Cellular structure for wireless LANs (note frequency reuse).

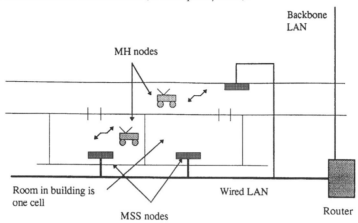

FIGURE 32.4 In-building LAN (made up of several wireless LANs).

much more transmission bandwidth than the requests themselves. In order to make efficient usc of bandwidth (and, in addition, guarantee service requirements for real-time data), most researchers have proposed that the downlink channel be controlled entirely by the MSS nodes. These nodes allocate the channel to different mobile users based on their current requirements using a protocol such as **time division multiple acccss** (TDMA). What about uplink traffic? This is a more complicated problem because the set of users within a cell is dynamic, thus making it infeasible to have a static channel allocation for the uplink. This problem is the main focus of MAC protocol design.

What are some of the design requirements of an appropriate MAC protocol? The IEEE 802.11 recommended standard for wireless LANs has identified almost 20 such requirements, some of which are discussed here (the reader is referred to [3], for further details). Clearly any protocol must maximize throughput while minimizing delays and providing fair access to all users. In addition to these requirements, however, mobility introduces several new requirements.

1. The MAC protocol must be independent of the underlying physical layer transmission technology adopted (be it DSSS, FHSS or IR).
2. The maximum number of users can be as high as a few hundred in a wireless LAN. The MAC protocol must be able to handle many users without exhibiting catastrophic degradation of service.
3. The MAC protocols must provide secure transmissions because the wireless medium is easy to tap.
4. The MAC protocol needs to work correctly in the presence of collocated networks.
5. It must have the ability to support ad-hoc networking (as in Fig. 32.1).

6. Other requirements include the need to support priority traffic, preservation of packet order, and an ability to support multicast.

Several contention-based protocols currently exist that could be adapted for use in wireless LANs. The protocols currently being looked by IEEE 802.11 include protocols based on **carrier sense multiple access (CSMA)**, polling, and TDMA. Protocols based on **code division multiple access** (CDMA) and **frequency division multiple access** (FDMA) are not considered because the processing gains obtained using these protocols are minimal while, simultaneously, resulting in a loss of flexibility for wireless LANs.

It is important to highlight an important difference between networking requirements of ad-hoc networks (as in Fig. 32.1) and networks based on cellular structure. In cellular networks, all communication occurs between the mobile hosts and the MSS (or base station) within that cell. Thus, the MSS can allocate channel bandwidth according to requirements of different nodes, i.e., we can use centralized channel scheduling for efficient use of bandwidth. In ad-hoc networks there is no such central scheduler available. Thus, any multiaccess protocol will be contention based with little explicit scheduling. In the remainder of this section we focus on protocols for cell-based wireless LANs only.

All multiaccess protocols for cell-based wireless LANs have a similar structure; see [3].

1. The MSS announces (explicitly or implicitly) that nodes with data to send may contend for the channel.
2. Nodes interested in sending data contend for the channel using protocols such as CSMA.
3. The MSS allocates the channel to successful nodes.
4. Nodes transmit packets (contention-free transmission).
5. MSS sends an explicit acknowledgment (ACK) for packets received.

Based on this model we present three MAC protocols.

32.3.1   Reservation-TDMA (R-TDMA)

This approach is a combination of TDMA and some contention protocol (see PRMA in [7]). The MSS divides the channel into slots (as in TDMA), which are grouped into frames. When a node wants to transmit it needs to reserve a slot that it can use in every consecutive frame as long as it has data to transmit. When it has completed transmission, other nodes with data to transmit may contend for that free slot. There are four steps to the functioning of this protocol.

a. At the end of each frame the MSS transmits a feedback packet that informs nodes of the current reservation of slots (and also which slots are free). This corresponds to steps 1 and 3 from the preceding list.
b. During a frame, all nodes wishing to acquire a slot transmit with a probability ρ during a free slot. If a node is successful it is so informed by the next feedback packet. If more than one node transmits during a free slot, there is a collision and the nodes try again during the next frame. This corresponds to step 2.
c. A node with a reserved slot transmits data during its slot. This is the contention-free transmission (step 4).
d. The MSS sends ACKs for all data packets received correctly. This is step 5.

The R-TDMA protocol exhibits several nice properties. First and foremost, it makes very efficient use of the bandwidth, and average latency is half the frame size. Another big benefit is the ability to implement power conserving measures in the portable computer. Since each node knows when to transmit (nodes transmit during their reserved slot only) it can move into a power-saving mode for a fixed amount of time, thus increasing battery life. This feature is generally not available in CSMA-based protocols. Furthermore, it is easy to implement priorities because of the centralized control of scheduling. One significant drawback of this protocol is that it is expensive to implement (see [2]).

32.3.2 Distributed Foundation Wireless MAC (DFWMAC)

The CSMA/CD protocol has been used with great success in the ethernet. Unfortunately, the same protocol is not very efficient in a wireless domain because of the problems associated with cell interference (i.e., interference from neighboring cells), the relatively large amount of time taken to sense the channel (see [6]) and the hidden terminal problem (see [12, 13]). The current proposal is based on a CSMA/collision avoidance (CA) protocol with a four-way handshake; see Fig. 32.5.

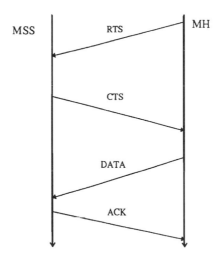

FIGURE 32.5   CSMA/CA and four-way handshaking protocol.

The basic operation of the protocol is simple. All MH nodes that have packets to transmit compete for the channel by sending ready to transmit (RTS) messages using nonpersistent CSMA. After a station succeeds in transmitting a RTS, the MSS sends a clear to transmit (CTS) to the MH. The MH transmits its data and then receives an ACK. The only possibility of collision that exists is in the RTS phase of the protocol and inefficiencies occur in the protocol, because of the RTS and CTS stages. Note that unlike R-TDMA it is harder to implement power saving functions. Furthermore, latency is dependent on system load making it harder to implement real-time guarantees. Priorities are also not implemented. On the positive side, the hardware for this protocol is very inexpensive.

32.3.3 Randomly Addressed Polling (RAP)

In this scheme, when a MSS is ready to collect uplink packets it transmits a READY message. At this point all nodes with packets to send attempt to grab the channel as follows.

a. Each MH with a packet to transmit generates a random number between 0 and P.

b. All active MH nodes simultaneously and orthogonally transmit their random numbers (using CDMA or FDMA). We assume that all of these numbers are received correctly by the MSS. Remember that more than one MH node may have selected the same random number.

c. Steps a and b are repeated L times.

d. At the end of L stages, the MSS determines a stage (say, k) where the total number of distinct random numbers was the largest. The MSS polls each distinct each random number in this stage in increasing order. All nodes that had generated the polled random number transmit packets to the MSS.

e. Since more than one node may have generated the same random number, collisions are possible. The MSS sends a ACK or NACK after each such transmission. Unsuccessful nodes try again during the next iteration of the protocol.

The protocol is discussed in detail in [4] and a modified protocol called GRAP (for group RAP) is discussed in [3]. The authors propose that GRAP can also be used in the contention stage (step 2) for TDMA- and CSMA-based protocols.

32.4 Network Layer Issues

An important goal of wireless LANs is to allow users to move about freely while still maintaining all of their connections (network resources permitting). This means that the network must route all packets destined for the mobile user to the MSS of its current cell in a transparent manner. Two issues need to be addressed in this context.

- How can users be addressed?
- How can active connections for these mobile users be maintained?

Ioanidis, Duchamp, and Maguire [8] propose a solution called the IPIP (IP-within-IP) protocol. Here each MH has a unique **internet protocol** (IP) address called its home address. To deliver a packet to a remote MH, the source MSS first broadcasts an address resolution protocol (ARP) request to all other MSS nodes to locate the MH. Eventually some MSS responds. The source MSS then encapsulates each packet from the source MH within another packet containing the IP address of the MSS in whose cell the MH is located. The destination MSS extracts the packet and delivers it to the MH. If the MH has moved away in the interim, the new MSS locates the new location of the MH and performs the same operation. This approach suffers from several problems as discussed in [11]. Specifically, the method is not scaleable to a network spanning areas larger than a campus for the following reasons.

1. IP addresses have a prefix identifying the campus subnetwork where the node lives; when the MH moves out of the campus, its IP address no longer represents this information.

2. The MSS nodes serve the function of routers in the mobile network and, therefore, have the responsibility of tracking all of the MH nodes globally causing a lot of overhead in terms of message passing and packet forwarding; see [5].

Teraoka and Tokoro [11], have proposed a much more flexible solution to the problem called virtual IP (VIP). Here every mobile host has a virtual IP address that is unchanging regardless of the location of the MH. In addition, hosts have physical network addresses (traditional IP addresses) that may change as the host moves about. At the transport layer, the target node is always specified by its VIP address only. The address resolution from the VIP address to the current IP address takes place either at the network layer of the same machine or at a gateway. Both the host machines and the gateways maintain a cache of VIP to IP mappings with associated timestamps. This information is in the form of a table called *address mapping table* (AMT). Every MH has an associated *home gateway*. When a MH moves into a new subnetwork, it is assigned a new IP address. It sends this new IP address and its VIP address to its home gateway via a *VipConn* control message. All intermediate gateways that relay this message update their AMT tables as well. During this process of updating the AMT tables, all packets destined to the MH continue to be sent to the old location. These packets are returned to the sender, who then sends them to the home gateway of the MH. It is easy to see that this approach is easily scaleable to large networks, unlike the IPIP approach.

32.4.1 Alternative View of Mobile Networks

The approaches just described are based on the belief that mobile networks are merely an extension of wired networks. Other authors [10] disagree with this assumption because there are fundamental differences between the mobile domain and the fixed wired network domain. Two examples follow.

1. The available bandwidth at the wireless link is small; thus, end-to-end packet retransmission for transmission control protocol (TCP)-like protocols (implemented over datagram networks) is a bad idea. This leads to the conclusion that transmission within the mobile network must be connection oriented. Such a solution, using virtual circuits (VC), is proposed in [5].

2. The bandwidth available for a MH with open connections changes dynamically since the number of other users present in each cell varies randomly. This is a feature not present in fixed high-speed networks where, once a connection is set up, its bandwidth does not vary much. Since bandwidth changes are an artifact of mobility and are dynamic, it is necessary to deal with the consequences (e.g., buffer overflow, large delays, etc.) locally to both, i.e., shield fixed network hosts from the idiosyncrasies of mobility as well as to respond to changing bandwidth quickly (without having to rely on end-to-end control). Some other differences are discussed in [10].

32.4.2 A Proposed Architecture

Keeping these issues in mind, a more appropriate architecture has been proposed in Ghai and Singh [5], and Singh [10]. Mobile networks are considered to be different and separate from wired networks. Within a mobile network is a three-layer hierarchy; see Fig. 32.6. At the bottom layer are the MHs. At the next level are the MSS nodes (one per cell). Finally, several MSS nodes are controlled by a **supervisor host (SH)** node (there may be one SH node per small building). The SH nodes are responsible for flow control for all MH connections within their domain; they are also responsible for tracking MH nodes and forwarding packets as MH nodes roam. In addition, the SH nodes serve as a *gateway* to the wired networks. Thus, any connection setup from a MH to a fixed host is broken in two, one from the MH to the SH and another from the SH to the fixed host. The MSS nodes in this design are simply connection endpoints for MH nodes. Thus, they are simple devices that implement the MAC protocols and little else. Some of the benefits of this design are

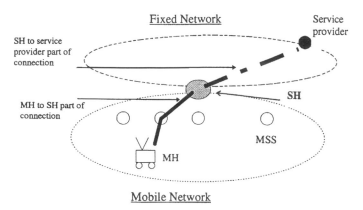

FIGURE 32.6   Proposed architecture for wireless networks.

as follows.

1. Because of the large coverage of the SH (i.e., a SH controls many cells) the MH remains in the domain of one SH much longer. This makes it easy to handle the consequences of dynamic bandwidth changes locally. For instance, when a MH moves into a crowded cell, the bandwidth available to it is reduced. If it had an open ftp connection, the SH simply buffers undelivered packets until they can be delivered. There is no need to inform the other endpoint of this connection of the reduced bandwidth.

2. When a MH node sets up a connection with a service provider in the fixed network, it negotiates some quality of service (QOS) parameters such as bandwidth, delay bounds, etc. When the MH roams into a crowded cell, these QOS parameters can no longer be met because the available bandwidth is smaller. If the traditional view is adopted (i.e., the mobile networks are extensions of fixed networks) then these QOS parameters will have to be renegotiated each time the bandwidth changes (due to roaming). This is a very expensive proposition because of the large number of control messages that will have to be exchanged. In the approach of Singh [10], the service provider will never know about the bandwidth changes since it deals only with the SH that is accessed via the wired network. The SH bears the responsibility of handling bandwidth changes by either buffering packets until the bandwidth available to the MH increases (as in the case of the ftp example) or it could discard a fraction of real-time packets (e.g., a voice connection) to ensure delivery of most of the packets within their deadlines. The SH could also instruct the MSS to allocate a larger amount of bandwidth to the MH when the number of buffered packets becomes large. Thus, the service provider in the fixed network is shielded from the mobility of the user.

32.4.3   Networking Issues

It is important for the network to provide connection-oriented service in the mobile environment (as opposed to connectionless service as in the internet) because bandwidth is at a premium in wireless networks, and it is, therefore, inadvisable to have end-to-end retransmission of packets (as in TCP). The proposed architecture is well suited to providing connection-oriented service by using VCs.

In the remainder of this section we look at how virtual circuits are used within the mobile network and how routing is performed for connections to mobile hosts. Every connection set up with one or more MH nodes as a connection endpoint is routed through the SH nodes and each connection is given a unique VC number. The SH node keeps track of all MH nodes that lie within its domain. When a packet needs to be delivered to a MH node, the SH first buffers the packet and then sends it to the MSS at the current location of the MH or to the predicted location if the MH is currently between cells. The MSS buffers all of these packets for the MH and transmits them to the MH if it is in its cell. The MSS discards packets after transmission or if the SH asks it to discard the packets. Packets are delivered in the correct order to the MH (without duplicates) by having the MH transmit the expected sequence number (for each VC) during the initial handshake (i.e., when the MH first enters the cell). The MH sends ACKs to the SH for packets received. The SH discards all packets that have been acknowledged. When a MH moves from the domain of SH1 into the domain of SH2 while having open connections, SH1 continues to forward packets to SH2 until either the connections are closed or until SH2 sets up its own connections with the other endpoints for each of MH's open connections (it also gives new identifiers to all these open connections). The detailed protocol is presented in [5].

The SH nodes are all connected over the fixed (wired) network. Therefore, it is necessary to route packets between SH nodes using the protocol provided over the fixed networks. The VIP protocol appears to be best suited to this purpose. Let us assume that every MH has a globally unique VIP address. The SHs have both a VIP as well as a fixed IP address. When a MH moves into the domain of a SH, the IP address affixed to this MH is the IP address of the SH. This ensures that all packets sent to the MH are routed through the correct SH node. The SH keeps a list of all VIP addresses of MH nodes within its domain and a list of open VCs for each MH. It uses this information to route the arriving packets along the appropriate VC to the MH.

32.5 Transport Layer Design

The transport layer provides services to higher layers (including the application layer), which include connectionless services like UDP or connection-oriented services like TCP. A wide variety of new services will be made available in the high-speed networks, such as continuous media service for real-time data applications such as voice and video. These services will provide bounds on delay and loss while guaranteeing some minimum bandwidth.

Recently variations of the TCP protocol have been proposed that work well in the wireless domain. These proposals are based on the traditional view that wireless networks are merely extensions of fixed networks. One such proposal is called I-TCP [2] for indirect TCP. The motivation behind this work stems from the following observation. In TCP the sender times out and begins retransmission after a timeout period of several hundred milliseconds. If the other endpoint of the connection is a mobile host, it is possible that the MH is disconnected for a period of several seconds (while it moves between cells and performs the initial greeting). This results in the TCP sender timing out and transmitting the same data several times over, causing the effective throughput of the connection to degrade rapidly. To alleviate this problem, the implementation of I-TCP separates a TCP connection into two pieces—one from the fixed host to another fixed host that is near the MH and another from this host to the MH (note the similarity of this approach with the approach in Fig. 32.6). The host closer to the MH is aware of mobility and has a larger timeout period. It serves as a type of gateway for the TCP connection because it sends ACKs back to the sender before receiving ACKs from the MH. The performance of I-TCP is far superior to traditional TCP for the mobile networks studied.

In the architecture proposed in Fig. 32.6, a TCP connection from a fixed host to a mobile host would terminate at the SH. The SH would set up another connection to the MH and would have the responsibility of transmitting all packets correctly. In a sense this is a similar idea to I-TCP except that in the wireless network VCs are used rather than datagrams. Therefore, the implementation of TCP service is made much easier.

A problem that is unique to the mobile domain occurs because of the unpredictable movement of MH nodes (i.e., a MH may roam between cells resulting in a large variation of available bandwidth in each cell). Consider the following example. Say nine MH nodes have opened 11-kb/s connections in a cell where the available bandwidth is 100 kb/s. Let us say that a tenth mobile host M10, also with an open 11-kb/s connection, wanders in. The total requested bandwidth is now 110 kb/s while the available bandwidth is only 100 kb/s. What is to be done? One approach would be to deny service to M10. However, this seems an unfair policy. A different approach is to penalize all connections equally so that each connection has 10-kb/s bandwidth allocated.

To reduce the bandwidth for each connection from 11 kb/s to 10 kb/s, two approaches may be adopted:

1. Throttle back the sender for each connection by sending control messages.
2. Discard 1-kb/s data for each connection at the SH. This approach is only feasible for applications that are tolerant of data loss (e.g., real-time video or audio).

The first approach encounters a high overhead in terms of control messages and requires the sender to be capable of changing the data rate dynamically. This may not always be possible; for instance, consider a teleconference consisting of several participants where each mobile participant is subject to dynamically changing bandwidth. In order to implement this approach, the data (video or audio or both) will have to be compressed at different ratios for each participant, and this compression ratio may have to be changed dynamically as each participant roams. This is clearly an unreasonable solution to the problem. The second approach requires the SH to discard 1-kb/s of data for each connection. The question is, how should this data be discarded? That is, should the 1 kb of discarded data be consecutive (or clustered) or uniformly spread out over the data stream every 1 s? The way in which the data is discarded has an effect on the final perception of the service by the mobile user. If the service is audio, for example, a random uniform loss is preferred to a clustered loss (where several consecutive words are lost). If the data is compressed video, the problem is even more serious because most random losses will cause the encoded stream to become unreadable resulting in almost a 100% loss of video at the user.

A solution to this problem is proposed in Seal and Singh [9], where a new sublayer is added to the transport layer called the *loss profile transport sublayer* (*LPTSL*). This layer determines how data is to be discarded based on special transport layer markers put by application calls at the sender and based on negotiated loss functions that are part of the QOS negotiations between the SH and service provider. Figure 32.7 illustrates the functioning of this layer at the service provider, the SH, and the MH. The original data stream is broken into *logical segments* that are separated by markers (or flags). When this stream arrives at the SH, the SH discards entire logical segments (in the case of compressed video, one logical segment may represent one frame) depending on the bandwidth available to the MH. The purpose of discarding entire logical segments is that discarding a part of such a segment of data makes the rest of the data within that segment useless—so we might as well discard the entire segment. Observe also that the flags (to identify logical segments) are inserted by the LPTSL via calls made by the application layer. Thus, the transport layer or the LPTSL does not need to know encoding details of the data stream. This scheme is currently being implemented at the University of South Carolina by the author and his research group.

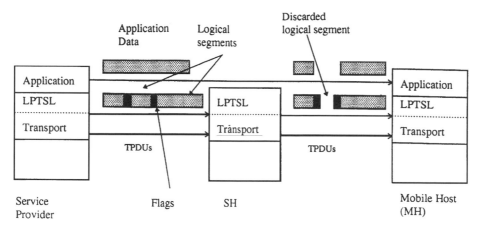

FIGURE 32.7   LPTSL, an approach to handle dynamic bandwidth variations.

32.6   Conclusions

The need for wireless LANs is driving rapid development in this area. The IEEE has proposed standards (802.11) for the physical layer and MAC layer protocols. A great deal of work, however, remains to be done at the network and transport layers. There does not appear to be a consensus regarding subnet design for wireless LANs. Our work has indicated a need for treating wireless LAN subnetworks as being fundamentally different from fixed networks, thus resulting in a different subnetwork and transport layer designs. Current efforts are underway to validate these claims.

Defining Terms

Carrier-sense multiple access (CSMA): Protocols such as those used over the ethernet.

Medium access control (MAC): Protocols arbitrate channel access between all nodes on a wireless LAN.

Mobile host (MH) nodes: The nodes of wireless LAN.

Supervisor host (SH): The node that takes care of flow-control and other protocol processing for all connections.

References

[1] Bantz, D.F. and Bauchot, F.J., Wireless LAN design alternatives. *IEEE Network,* 8(2), 43–53, 1994.

[2] Barke, A. and Badrinath, B.R., I-TCP: indirect TCP for mobile hosts. Tech. Rept. DCS-TR-314, Dept. Computer Science, Rutgers University, Piscataway, NJ, 1994.

[3] Chen, K.-C., Medium access control of wireless LANs for mobile computing. *IEEE Network,* 8(5), 50–63, 1994.

[4] Chen, K.-C. and Lee, C.H., RAP: a novel medium access control protocol for wireless data networks. *Proc. IEEE GLOBECOM'93,* IEEE Press, Piscataway, NJ, 08854. 1713–1717, 1993.

[5] Ghai, R. and Singh, S., An architecture and communication protocol for picocellular networks. *IEEE Personal Comm. Mag.,* 1(3), 36–46, 1994.

[6] Glisic, S.G., 1-Persistent carrier sense multiple access in radio channel with imperfect carrier sensing. *IEEE Trans. on Comm.,* 39(3), 458–464, 1991.

[7] Goodman, D.J., Cellular packet communications. *IEEE Trans. on Comm.,* 38(8), 1272–1280, 1990.

[8] Ioanidis, J., Duchamp, D., and Maguire, G.Q., IP-based protocols for mobile internetworking. *Proc. of ACM SIGCOMM'91,* ACM Press, New York, NY, 10036 (Sept.), 235–245, 1991.

[9] Seal, K. and Singh, S., Loss profiles: a quality of service measure in mobile computing. *J. Wireless Networks,* 2, 45–61, 1996.

[10] Singh, S., Quality of service guarantees in mobile computing. *J. of Computer Comm.,* 19, 359–371, 1996.

[11] Teraoka, F. and Tokoro, M., Host migration transparency in IP networks: the VIP approach. *Proc. of ACM SIGCOMM,* ACM Press, New York, NY, 10036 (Jan.), 45–65, 1993.

[12] Tobagi, F. and Kleinrock, L., Packet switching in radio channels: Part I carrier sense multiple access models and their throughput delay characteristic. *IEEE Trans. on Comm.,* 23(12), 1400–1416, 1975a.

[13] Tobagi, F. and Kleinrock, L., Packet switching in radio channels: Part II the hidden terminal problem in CSMA and busy-one solution. *IEEE Trans. on Comm.,* 23(12), 1417–1433, 1975b.

Further Information

A good introduction to physical layer issues is presented in Bantz [1] and MAC layer issues are discussed in Chen [3]. For a discussion of network and transport layer issues, see Singh [10] and Ghai and Singh [5].

33

Wireless Data

Allen H. Levesque
GTE Laboratories, Inc.

Kaveh Pahlavan
Worcester Polytechnic Institute

33.1 Introduction

Wireless data services and systems represent a steadily growing and increasingly important segment of the communications industry. While the wireless data industry is becoming increasingly diverse, one can identify two mainstreams that relate directly to users' requirement for data services. On one hand, there are requirements for relatively low-speed data services provided to mobile users over wide geographical areas, as provided by private mobile data networks and by data services implemented on common-carrier cellular telephone networks. On the other hand, there are requirements for high-speed data services in local areas, as provided by cordless private branch exchange (PBX) systems and wireless local area networks (LANs), as well as by the emerging personal communications services (PCS). Wireless LANs are treated in Chapter 32. In this chapter we mainly address wide-area wireless data systems, commonly called *mobile data systems*, and briefly touch upon data services to be incorporated into the emerging digital cellular systems.

Mobile data systems provide a wide variety of services for both business users and public safety organizations. Basic services supporting most businesses include electronic mail, enhanced paging, modem and facsimile transmission, remote access to host computers and office LANs, information broadcast services and, increasingly, Internet access. Public safety organizations, particularly law-enforcement agencies, are making increasing use of wireless data communications over traditional VHF and UHF radio dispatch networks, over commercial mobile data networks, and over public cellular telephone networks. In addition, there are wireless services supporting vertical applications that are more or less tailored to the needs of specific companies or industries, such as transaction processing, computer-aided delivery dispatch, customer service, fleet management, and emergency medical services. Work currently in progress to develop the national Intelligent Transportation System (ITS) includes the definition of a wide array of new traveler services, many of which will be supported by standardized mobile data networks.

Much of the growth in use of wireless data services has been spurred by the rapid growth of the paging service industry and increasing customer demand for more advanced paging services, as well as the desire to increase work productivity by extending to the mobile environment the suite of digital communications services readily available in the office environment. There is also a desire to make more cost-efficient use of the mobile radio and cellular networks already in common use for mobile voice communications by incorporating efficient data transmission services into these networks. The services and networks that have evolved to date represent a variety of specialized solutions and, in general, they are not interoperable with each other. As the wireless data industry expands, there is an increasing demand for an array of attractively priced standardized services and equipment accessible to mobile users over wide geographic areas. Thus, we see the growth of nationwide privately operated service networks as well as new data services built upon the first and second generation cellular telephone networks. The implementation of PCS networks in the 2-GHz bands as well as the eventual implementation of third generation (3G) wireless networks will further extend this evolution.

In this chapter we describe the principal existing and evolving wireless data networks and the related standards activities now in progress. We begin with a discussion of the technical characteristics of wireless data networks.

33.2   Characteristics of Wireless Data Networks

From the perspective of the data user, the basic requirement for wireless data service is convenient, reliable, low-speed access to data services over a geographical area appropriate to the user's pattern of daily business operation. By low speed we mean data rates comparable to those provided by standard data modems operating over the public switched telephone network (PSTN). This form of service will support a wide variety of short-message applications, such as notice of electronic mail or voice mail, as well as short file transfers or even facsimile transmissions that are not overly lengthy. The user's requirements and expectations for these types of services are different in several ways from the requirements placed on voice communication over wireless networks. In a wireless voice service, the user usually understands the general characteristics and limitations of radio transmission and is tolerant of occasional *signal fades* and brief dropouts. An overall level of acceptable voice quality is what the user expects. In a data service, the user is instead concerned with the accuracy of delivered messages and data, the time-delay characteristics of the service network, the ability to maintain service while traveling about, and, of course, the cost of the service. All of these factors are dependent on the technical characteristics of wireless data networks, which we discuss next.

33.2.1   Radio Propagation Characteristics

The chief factor affecting the design and performance of wireless data networks is the nature of radio propagation over wide geographic areas. The most important mobile data systems operate in various land–mobile radio bands from roughly 100 to 200 MHz, the specialized mobile radio (SMR) band around 800 MHz, and the cellular telephone bands at 824–894 MHz. In these frequency bands, radio transmission is characterized by distance-dependent field strength, as well as the well-known effects of *multipath fading*, signal shadowing, and signal blockage. The signal coverage provided by a radio transmitter, which in turn determines the area over which a mobile data receiving terminal can receive a usable signal, is governed primarily by the *power–distance relationship*, which gives signal power as a function of distance between transmitter and receiver. For the ideal case of single-path transmission in free space, the relationship between transmitted power P_t and received power P_r is given by

$$P_r/P_t = G_t G_r (\lambda/4\pi d)^2 \tag{33.1}$$

where G_t and G_r are the transmitter and receiver antenna gains, respectively, d is the distance between the transmitter and the receiver, and λ is the wavelength of the transmitted signal. In the mobile radio environment, the power-distance relationship is in general different from the free-space case just given. For propagation over an Earth plane at distances much greater than either the signal wavelength or the antenna heights, the relationship between P_t and P_r is given by

$$P_r/P_t = G_t G_r \left(h_1^2 h_2^2 / d^4 \right) \tag{33.2}$$

where h_1 and h_2 are the transmitting and receiving antenna heights. Note here that the received power decreases as the fourth power of the distance rather than the square of distance seen in the ideal free-space case. This relationship comes from a propagation model in which there is a single signal reflection with phase reversal at the Earth's surface, and the resulting received signal is the vector sum of the direct line-of-sight signal and the reflected signal. When user terminals are deployed in mobile situations, the received signal is generally characterized by rapid fading of the signal strength, caused by the vector summation of reflected signal components, the vector summation changing constantly as the mobile terminal moves from one place to another in the service area. Measurements made by many researchers show that when the fast fading is averaged out, the signal strength is described by a Rayleigh distribution having a log-normal mean. In general, the power-distance relationship for mobile radio systems is a more complicated relationship that depends on the nature of the terrain between transmitter and receiver.

Various propagation models are used in the mobile radio industry for network planning purposes, and a number of these models are described in [1]. Propagation models for mobile communications networks must take account of the terrain irregularities existing over the intended service area. Most of the models used in the industry have been developed from measurement data collected over various geographic areas. A very popular model is the *Longley–Rice model* [8, 14]. Many wireless networks are concentrated in urban areas. A widely used model for propagation prediction in urban areas is one usually referred to as the *Okumura–Hata model* [4, 9].

By using appropriate propagation prediction models, one can determine the range of signal coverage for a base station of given transmitted power. In a wireless data system, if one knows the level of received signal needed for satisfactory performance, the area of acceptable performance can, in turn, be determined. Cellular telephone networks utilize base stations that are typically spaced 1–5 mi apart, though in some mid-town areas, spacings of 1/2 mi or less are now being used. In packet-switched data networks, higher power transmitters are used, spaced about 5–15 mi apart.

An important additional factor that must be considered in planning a wireless data system is the in-building penetration of signals. Many applications for wireless data services involve the use of mobile data terminals inside buildings, for example, for trouble-shooting and servicing computers on customers' premises. Another example is wireless communications into hospital buildings in support of emergency medical services. It is usually estimated that in-building signal penetration losses will be in the range of 15–30 dB. Thus, received signal strengths can be satisfactory in the outside areas around a building but totally unusable inside the building. This becomes an important issue when a service provider intends to support customers using mobile terminals inside buildings.

One important consequence of the rapid fading experienced on mobile channels is that errors tend to occur in bursts, causing the transmission to be very unreliable for short intervals of time. Another problem is signal dropouts that occur, for example, when a data call is handed over from one base station to another, or when the mobile user moves into a location where the signal is severely attenuated. Because of this, mobile data systems employ various error-correction and error-recovery techniques to insure accurate and reliable delivery of data messages.

33.3   Market Issues

There are two important trends that are tending to propel growth in the use of wireless data services. The first is the rapidly increasing use of portable devices such as laptop computers, pen-pads, notebook computers, and other similar devices. Increasingly, the laptop or notebook computer is becoming a standard item of equipment for traveling professional or business person, along with the cellular telephone and pager. This trend has been aided by the steady decrease in prices, increases in reliability, and improvements in capability and design for such devices. The second important trend tending to drive growth in wireless data services is the explosive growth in the use of the Internet. As organizations become increasingly reliant upon the Internet for their everyday operations, they will correspondingly want their employees to have convenient access to the Internet while travelling, just as they do in the office environment. Wireless data services can provide the traveler with the required network access in many situations where wired access to the public network is impractical or inconvenient. Mobile data communication services discussed here provide a solution for wireless access over wide areas. Recent estimates of traffic composition indicate that data traffic now accounts for less than 1% of the traffic on wireless networks, compared to 50% for wireline networks. Therefore, the potential for growth in the wireless data market is seen to be enormous.

33.4   Modem Services Over Cellular Networks

A simple form of wireless data communication now in common use is data transmission using modems or facsimile terminals over analog cellular telephone links. In this form of communication, the mobile user simply accesses a cellular channel just as he would in making a standard voice call over the cellular network. The user then operates the modem or facsimile terminal just as would be done from office to office over the PSTN. A typical connection is shown in Fig. 33.1, where the mobile user has a laptop computer and portable modem in the vehicle, communicating with another modem and computer in the office. Typical users of this mode of communication include service technicians, real estate agents, and traveling sales people. In this form of communication, the network is not actually providing a data service but simply a voice link over which the data modem or fax terminal can interoperate with a corresponding data modem or fax terminal in the office or service center. The connection from the mobile telephone switching office (MTSO) is a standard landline connection, exactly the same as is provided for an ordinary cellular telephone call.

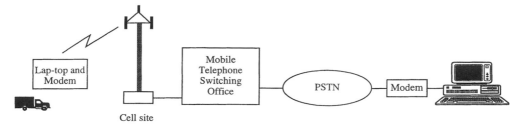

FIGURE 33.1   Modem operation over an analog cellular voice connection.

Many portable modems and fax devices are now available in the market and are sold as elements of the so-called "mobile office" for the traveling business person. Law enforcement personnel are also making increasing use of data communication over cellular telephone and dispatch radio networks to gain rapid access to databases for verification of automobile registrations and drivers' licenses. Portable devices are currently available that operate at transmission rates up to 9.6 or 14.4 kb/s. Error-correction modem protocols such as MNP-10, V.34, and V.42 are used to provide reliable delivery of data in the error-prone wireless transmission environment.

In another form of mobile data service, the mobile subscriber uses a portable modem or fax terminal as already described but now accesses a modem provided by the cellular service operator as part of a *modem pool*, which is connected to the MTSO. This form of service is shown in Fig. 33.2. The modem pool might provide the user with a choice of

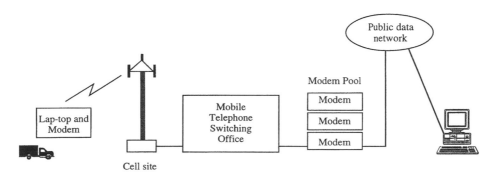

FIGURE 33.2   Cellular data service supported by modem pools in the network.

several standard modem types. The call connection from the modem pool to the office is a digital data connection, which might be supported by any of a number of public packet data networks, such as those providing X.25 service. Here, the cellular operator is providing a special service in the form of modem pool access, and this service in general carries a higher tariff than does standard cellular telephone service, due to the operator's added investment in the modem pools. In this form of service, however, the user in the office or service center does not require a modem but instead has a direct digital data connection to the desk-top or host computer.

Each of the types of wireless data transmission just described is in effect an appliqué onto an underlying cellular telephone service and, therefore, has limitations imposed by the characteristics of the underlying voice-circuit connection. That is, the cellular segment of the call connection is a circuit-mode service, which might be cost effective if the user needs to send long file transfers or fax transmissions but might be relatively costly if only short messages are to be transmitted and received. This is because the subscriber is being charged for a circuit-mode connection, which stays in place throughout the duration of the communication session, even if only intermittent short message exchanges are needed. The need for systems capable of providing cost-effective communication of relatively short message exchanges led to the development of wireless packet data networks, which we describe next.

33.5 Private Data Networks

Here we describe three packet data networks that provide mobile data services to users in major metropolitan areas of the United States.

33.5.1 ARDIS

ARDIS is a two-way radio service developed as a joint venture between IBM and Motorola and first implemented in 1983. In mid-1994, IBM sold its interest in ARDIS to Motorola and early in 1998 ARDIS was acquired by the American Mobile Satellite Corporation. The ARDIS network consists of four network control centers with 32 network controllers distributed through 1250 base station in 400 cities in the U.S. The service is suitable for two-way transfers of data files of size less than 10 kilobytes, and much of its use is in support of computer-aided dispatching, such as is used by field service personnel, often while they are on customers' premises. Remote users access the system from laptop radio terminals, which communicate with the base stations. Each of the ARDIS base stations is tied to one of the 32 radio network controllers, as shown in Fig. 33.3. The backbone of the network is implemented with leased telephone lines. The four ARDIS hosts, located in Chicago, New York, Los Angeles, and Lexington, KY, serve as access points for a customer's mainframe computer, which can be linked to an ARDIS host using async, bisync, SNA, or X.25 dedicated circuits.

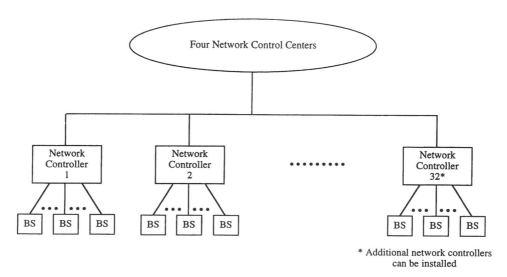

FIGURE 33.3   ARDIS network architecture.

The operating frequency band is 800 MHz, and the RF links use separate transmit and receive frequencies spaced by 45 MHz. The system was initially implemented with an RF channel data rate 4800 b/s per 25-kHz channel, using the MDC-4800 protocol. In 1993 the access data rate was upgraded to 19.2 kb/s, using the RD-LAP protocol, which provides a user data rate of about 8000 b/s. In the same year, ARDIS implemented a nationwide roaming capability, allowing users to travel between widely separated regions without having to preregister their portable terminals in each new region. The ARDIS system architecture is cellular, with cells overlapped to increase the probability that the signal transmission from a portable transmitter will reach at least one base station. The base station power is 40 W, which provides line-of-sight coverage up to a radius of 10–15 miles. The portable units operate with 4 W of radiated power. The overlapping coverage, combined with designed power levels, and error-correction coding in the transmission format, insures that the ARDIS can support portable communications from inside buildings, as well as on the street. This capability for in-building coverage is an important characteristic of the ARDIS service. The modulation technique is frequency-shift keying (FSK), the access method is frequency division multiple access (FDMA), and the transmission packet length is 256 bytes.

Although the use of overlapping coverage, almost always on the same frequency, provides reliable radio connectivity, it poses the problem of interference when signals are transmitted simultaneously from two adjacent base stations. The ARDIS network deals with this by turning off neighboring transmitters, for 0.5–1 s, when an outbound transmission occurs. This scheme has the effect of constraining overall network capacity.

The laptop portable terminals access the network using a random access method called data sense multiple access (DSMA) [11]. A remote terminal listens to the base station transmitter to determine if a "busy bit" is on or off. When the busy bit is off, the remote terminal is allowed to transmit. If two remote terminals begin to transmit at the same time, however, the signal packets may collide, and retransmission will be attempted, as in other contention-based multiple access protocols. The busy bit lets a remote user know when other terminals are transmitting and, thus, reduces the probability of packet collision.

33.5.2 MOBITEX

The MOBITEX system is a nationwide, interconnected trunked radio network developed by Ericsson and Swedish Telecom. The first MOBITEX network went into operation in Sweden in 1986, and networks have either been implemented or are being deployed in 22 countries. A MOBITEX operations association oversees the open technical specifications and coordinates software and hardware developments [6]. In the U.S., MOBITEX service was introduced by RAM Mobile Data in 1991. In 1992 Bell South Enterprises became a partner with RAM. Currently, RAM Mobile Data is a subsidiary of Bell South Corporation and operates under the name Bell South Wireless Data, LP. The Bell South MOBITEX service now covers over 90% of the U.S. urban business population with about 2000 base stations, and it provides automatic "roaming" across all service areas. By locating its base stations close to major business centers, the system provides a degree of in-building signal coverage. Although the MOBITEX system was designed to carry both voice and data service, the U.S. and Canadian networks are used to provide data service only. MOBITEX is an intelligent network with an open architecture that allows establishing virtual networks. This feature facilitates the mobility and expandability of the network [7, 12].

The MOBITEX network architecture is hierarchical, as shown in Fig. 33.4. At the top of the hierarchy is the network control center (NCC), from which the entire network is managed. The top level of switching is a national switch (MHX1) that routes traffic between service regions. The next level comprises regional switches (MHX2s), and below that are

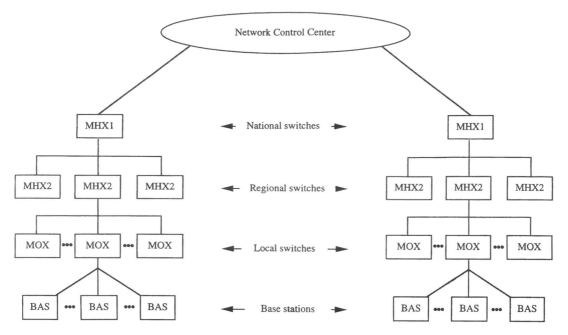

FIGURE 33.4   MOBITEX network architecture.

local switches (MOXs), each of which handles traffic within a given service area. At the lowest level in the network, multichannel trunked-radio base stations communicate with the mobile and portable data sets. MOBITEX uses packet-switching techniques, as does ARDIS, to allow multiple users to access the same channel at the same time. Message packets are switched at the lowest possible network level. If two mobile users in the same service area need to communicate with each other, their messages are relayed through the local base station, and only billing information is sent up to the network control center.

The base stations are laid out in a grid pattern using the same frequency reuse rules as are used for cellular telephone networks. In fact, the MOBITEX system operates in much the same way as a cellular telephone system, except that handoffs are not managed by the network. That is, when a radio connection is to be changed from one base station to another, the decision is made by the mobile terminal, not by a network computer as in cellular telephone systems.

To access the network, a mobile terminal finds the base station with the strongest signal and then registers with that base station. When the mobile terminal enters an adjacent service area, it automatically re-registers with a new base station, and the user's whereabouts are relayed to the higher level network nodes. This provides automatic routing of messages bound for the mobile user, a capability known as *roaming*. The MOBITEX network also has a store-and-forward capability.

The mobile units transmit at 896 to 901 MHz and the base stations at 935 to 940 MHz. The base stations use a trunked radio design employing 2 to 30 radio channels in each service area. The system uses dynamic power setting, in the range of 100 mW–10 W for mobile units and 100 mW–4 W for portable units. The Gaussian minimum shift keying (GMSK) modulation technique is used, with $BT = 0.3$ and noncoherent demodulation. The transmission rate is 8000 b/s half-duplex in 12.5-kHz channels, and the service is suitable for file transfers up to 20 kilobytes. The MOBITEX system uses a proprietary network-layer protocol called MPAK, which provides a maximum packet size of 512 bytes and a

24-b address field. Forward-error correction, as well as retransmissions, are used to ensure the bit-error-rate quality of delivered data packets. Fig. 33.5 shows the packet structure at various layers of the MOBITEX protocol stack. The system uses the reservation-slotted ALOHA (R-S-ALOHA) random access method.

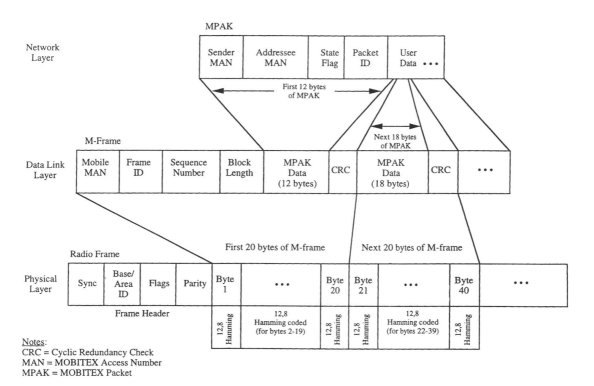

Notes:
CRC = Cyclic Redundancy Check
MAN = MOBITEX Access Number
MPAK = MOBITEX Packet

FIGURE 33.5   MOBITEX packet and frame structure at three layers of the protocol stack.

33.5.3 Ricochet Network

The Ricochet network is a wide-area wireless data network developed and operated by Metricom, Inc., headquartered in Los Gatos, California. Ricochet Network service is currently available in the greater San Francisco Bay Area, Seattle, and Washington, D.C. as well a number of university campuses, schools, and corporate campuses. Metricom plans to expand the service to other metropolitan areas as well. Ricochet is a spread-spectrum, frequency-hopping design, operating in an unlicensed band, the 902–928 MHz industrial, scientific, and medical (ISM) band. The Ricochet network employs shoebox-sized radio transceivers, called microcell radios, which are typically mounted on street light poles or utility poles. Each microcell unit accesses electric power from the street light itself, or from adjacent power lines, and is otherwise a self-contained unit. Each microcell radio employs 162 frequency-hopping channels and uses a randomly selected hopping sequence. This allows many subscribers to use the network simultaneously.

The microcell radios are typically installed every 1/4 to 1/2 mi in a regular geometric pattern. A mobile subscriber using a Ricochet modem accesses any of the microcell radios with data packets, and the packets are routed from one microcell radio to another by means of an efficient routing protocol. Within a 20 sq mi radius containing approximately 100 microcell radios, the network has a wired access point (WAP) that collects the radio packets and formats them for transmission over a wired IP network backbone through a T1 frame relay connection. Each WAP and the microcell radios that communicate with it can support thousands of subscribers.

Data packets transmitted by a Ricochet modem can be routed to another Ricochet modem or to one of a number of gateways that allow subscribers to access other services. The current network design includes gateways to the Internet, to the PSTN, to an X.25 network, and to corporate Intranets and LANs. The current Ricochet network design provides user data rates up to about 128 kb/s, and higher data rates are being planned.

33.6 Cellular Data Networks and Services

33.6.1 Cellular Digital Packet Data (CDPD)

The cellular digital packet data (CDPD) system was designed to provide packet data services as an overlay onto the existing analog cellular telephone network, which is called advanced mobile phone service (AMPS). CDPD was developed by IBM in collaboration with the major cellular carriers. Any cellular carrier owning a license for AMPS service is free to offer its customers CDPD service without any need for further licensing. A basic concept of the CDPD system is to provide data services on a noninterfering basis with the existing analog cellular telephone services using the same 30-kHz channels. This is accomplished in either of two ways. First, one or a few AMPS channels in each cell site can be devoted to CDPD service. Second, CDPD is designed to make use of a cellular channel that is temporarily not being used for voice traffic and to move to another channel when the current channel is assigned to voice service. In most of the CDPD networks deployed to date, the fixed-channel implementation is being used.

The compatibility of CDPD with the existing cellular telephone system allows it to be installed in any AMPS cellular system in North America, providing data services that are not dependent on support of a digital cellular standard in the service area. The participating companies issued release 1.0 of the CDPD specification in July 1993, and release 1.1 was issued in late 1994 [2]. At this writing (mid-1998), CDPD service is implemented in many of the major market areas in the U.S. Typical applications for CDPD service include: electronic mail, field support servicing, package delivery tracking, inventory control, credit card verification, security reporting, vehicle theft recovery, traffic and weather advisory services, and a wide range of information retrieval services.

Although CDPD cannot increase the number of channels usable in a cell, it can provide an overall increase in user capacity if data customers use CDPD instead of voice channels. This capacity increase results from the inherently greater efficiency of a connectionless packet data service relative to a connection-oriented service, given bursty data traffic. That is, a packet data service does not require the overhead associated with setup of a voice traffic channel in order to send one or a few data packets. In the following paragraphs we briefly describe the CDPD network architecture and the principles of operation of the system. Our discussion follows [13], closely.

The basic structure of a CDPD network (Fig. 33.6) is similar to that of the cellular network with which it shares transmission channels. Each mobile end system (M-ES) communicates with a mobile data base station (MDBS) using the protocols defined by the air-interface specification, to be described subsequently. The MDBSs are typically collocated with the cell equipment providing cellular telephone service to facilitate the channel-sharing procedures. All of the MDBSs in a service area are linked to a mobile data intermediate system

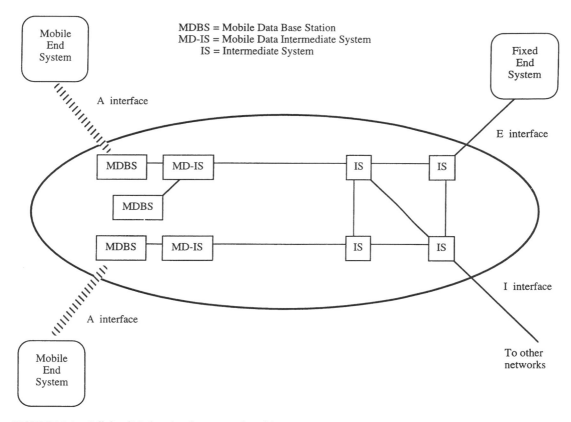

FIGURE 33.6 Cellular digital packet data network architecture.

(MD-IS) by microwave or wireline links. The MD-IS provides a function analogous to that of the mobile switching center (MSC) in a cellular telephone system. The MD-IS may be linked to other MD-ISs and to various services provided by end systems outside the CDPD network. The MD-IS also provides a connection to a network management system and supports protocols for network management access to the MDBSs and M-ESs in the network.

Service endpoints can be local to the MD-IS or remote, connected through external networks. A MD-IS can be connected to any external network supporting standard routing and data exchange protocols. A MD-IS can also provide connections to standard modems in the PSTN by way of appropriate modem interworking functions (modem emulators). Connections between MD-ISs allow routing of data to and from M-ESs that are roaming, that is, operating in areas outside their home service areas. These connections also allow MD-ISs to exchange information required for mobile terminal authentication, service authorization, and billing.

CDPD employs the same 30-kHz channelization as used in existing AMPS cellular systems throughout North America. Each 30-kHz CDPD channel supports channel transmission rates up to 19.2 kb/s. Degraded radio channel conditions, however, will limit the actual information payload throughput rate to lower levels, and will introduce additional time delay due to the error-detection and retransmission protocols.

The CDPD radio link physical layer uses GMSK modulation at the standard cellular carrier frequencies, on both forward (base-to-mobile) and reverse (mobile-to-base) links. The Gaussian pulse-shaping filter is specified to have bandwidth-time product $B_bT = 0.5$.

The specified B_bT product assures a transmitted waveform with bandwidth narrow enough to meet adjacent-channel interference requirements, while keeping the intersymbol interference small enough to allow simple demodulation techniques. The choice of 19.2 kb/s as the channel bit rate yields an average power spectrum that satisfies the emission requirements for analog cellular systems and for dual-mode digital cellular systems.

The forward channel carries data packets transmitted by the MDBS, whereas the reverse channel carries packets transmitted by the M-ESs. In the forward channel, the MDBS forms data frames by adding standard high level data link control (HDLC) terminating flags and inserted zero bits, and then segments each frame into blocks of 274 b. These 274 b, together with an 8-b *color code* for MDBS and MD-IS identification, are encoded into a 378-b coded block using a (63, 47) Reed–Solomon code over a 64-ary alphabet. A 6-b synchronization and flag word is inserted after every 9 code symbols. The flag words are used for reverse link access control. The forward link block structure is shown in Fig. 33.7.

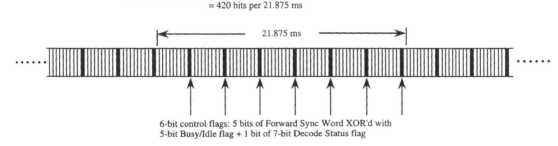

FIGURE 33.7   Cellular digital packet data forward link block structure.

In the reverse channel, when an M-ES has data frames to send, it formats the data with flags and inserted zeros in the same manner as in the forward link. That is, the reverse link frames are segmented and encoded into 378-b blocks using the same Reed–Solomon code as in the forward channel. The M-ES may form up to 64 encoded blocks for transmission in a single reverse channel transmission burst. During the transmission, a 7-b transmit continuity indicator is interleaved into each coded block and is set to all ones to indicate that more blocks follow, or all zeros to indicate that this is the last block of the burst. The reverse channel block structure is shown in Fig. 33.8.

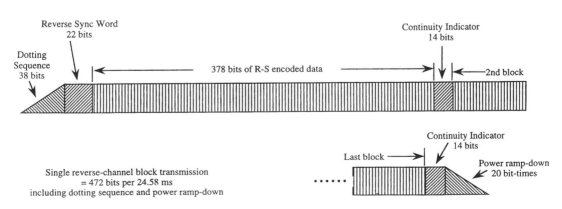

FIGURE 33.8   Cellular digital packet data reverse link block structure.

The media access control (MAC) layer in the forward channel is relatively simple. The receiving M-ES removes the inserted zeros and HDLC flags and reassembles data frames that were segmented into multiple blocks. Frames are discarded if any of their constituent blocks are received with uncorrectable errors.

On the reverse channel (M-ES to MDBS), access control is more complex, since several M-ESs must share the channel. CDPD uses a multiple access technique called digital sense multiple access (DSMA), which is closely related to the carrier sense multiple access/collision detection (CSMA/CD) access technique.

The network layer and higher layers of the CDPD protocol stack are based on standard ISO and Internet protocols. The CDPD specification stipulates that there be no changes to protocols above the network layer of the seven-layer ISO model, thus insuring the compatibility of applications software used by CDPD subscribers.

The selection of a channel for CDPD service is accomplished by the radio resource management entity in the MDBS. Through the network management system, the MDBS is informed of the channels in its cell or sector that are available either as dedicated data channels or as potential CDPD channels when they are not being used for analog cellular service, depending on which channel allocation method is implemented. For the implementation in which CDPD service is to use "channels of opportunity," there are two ways in which the MDBS can determine whether the channels are in use. If a communication link is provided between the analog system and the CDPD system, the analog system can inform the CDPD system directly about channel usage. If such a link is not available, the CDPD system can use a forward power monitor ("sniffer" antenna) to detect channel usage on the analog system. Circuitry to implement this function can be built into the cell sector interface.

Another version of CDPD called circuit-switched CDPD (C-SCDPD) is designed to provide service to subscribers traveling in areas where the local cellular service provider has not implemented the CDPD service. With C-SCDPD, the subscriber establishes a standard analog cellular circuit connection to a prescribed number, and then transmits and receives CDPD data packets over that circuit connection. The called number is a gateway that provides connection to the CDPD backbone packet network.

33.6.2 Digital Cellular Data Services

In response to the rapid growth in demand for cellular telephone service throughout the U.S and Canada, the Cellular Telecommunications Industry Association (CTIA) and the Telecommunications Industry Association (TIA) have been developing standards for new digital cellular systems to replace the existing analog cellular system (advanced mobile phone system or AMPS). Two air-interface standards have now been published. The IS-54 standard specifies a three-slot TDMA system, and the I-95 standard specifies a CDMA spread spectrum system. In both systems, a variety of data services are being planned.

Following the development of the IS-95 standard for CDMA voice service, the cellular industry has worked on defining various data services to operate in the same networks. The general approach taken in the definition of IS-95 data services has been to base the services on standard data protocols, to the greatest extent possible [17]. The previously-specified physical layer of the IS-95 protocol stack was adopted for the physical layer of the data services, with an appropriate radio link protocol (RLP) overlaid. The first CDMA data service to be defined was asynchronous ("start-stop" interface) data and Group-3 facsimile. This service provides for interoperability with many standard PSTN data modems as well as standard office fax machines. This service is in the category of *circuit-mode service,* since a circuit connection is first established, just as with a voice call, and the circuit connection is maintained until the user disconnects.

Following the standardization of the asynchronous data service, the industry defined a service that carries packet-formatted data over a CDMA circuit connection. It is important to note that this is not a true packet-data service over the radio link, since the full circuit connection is maintained regardless of how little packet data is transmitted. One potential application for this type of service is to provide subscribers with CDPD access from a CDMA network.

It is recognized that in order to make use more efficient, it will be highly desirable to provide a contention-based packet data service in CDMA cellular networks. This is currently a subject of study in CDMA data services standardization groups.

In parallel with the CDMA data services efforts, another TIA task group, TR45.3.2.5, has defined standards for digital data services for the TDMA digital cellular standard IS-54 [15, 18]. As with the IS-95 data services effort, initial priority was given to standardizing circuit-mode asynchronous data and Group-3 facsimile services [16].

33.7 Other Developing Standards

33.7.1 TErrestrial Trunked Radio (TETRA)

As has been the case in North America, there is interest in Europe in establishing fixed wide-area standards for mobile data communications. Whereas the Pan-European standard for digital cellular, termed Global Systems for Mobile communications (GSM), will provide an array of data services, data will be handled as a circuit-switched service, consistent with the primary purpose of GSM as a voice service system. Therefore, the European Telecommunications Standards Institute (ETSI) began developing a public standard in 1988 for trunked radio and mobile data systems, and this standardization process continues today. The standards, which are now known generically as TErrestrial Trunked Radio (TETRA) (formerly Trans-European Trunked Radio), were made the responsibility of the ETSI RES 6 subtechnical committee [3]. In 1996, the TETRA standardization activity was elevated within RES-6 with the creation of project TETRA.

TETRA is being developed as a family of standards. One branch of the family is a set of radio and network interface standards for trunked voice (and data) services. The other branch is an air-interface standard optimized for wide-area packet data services for both fixed and mobile subscribers and supporting standard network access protocols. Both versions of the standard will use a common physical layer, based on $\pi/4$ differential quadrature phase shift keying ($\pi/4$-DQPSK) modulation operating at a channel rate of 36 kb/s in each 25-kHz channel. The composite data rate of 36 kb/s comprises four 9 kb/s user channels multiplexed in a TDMA format. The TETRA standard provides both connection-oriented and connectionless data services, as well as mixed voice and data services.

TETRA has been designed to operate in the frequency range from VHF (150 MHz) to UHF (900 MHz). The RF carrier spacing in TETRA is 25 kHz. In Europe, harmonized bands have been designated in the frequency range 380–400 MHz for public safety users. It is expected that commercial users will adopt the 410–430 MHz band. The Conference of European Posts and Telecommunications Administrations (CEPT) has made additional recommendations for use in the 450–470 MHz and 870–876/915–921 MHz frequency bands.

Table 33.1 compares the chief characteristics and parameters of the wireless data services described.

TABLE 33.1 Characteristics and Parameters of Five Mobile Data Services

System:	ARDIS	MOBITEX	CDPD	IS-95[b]	TETRA[b]
Frequency band Base to mobile, (MHz). Mobile to base, (MHz).	(800 band, 45-kHz sep.)	935–940[a] 896–901	869–894 824–849	869–894 824–849	(400 and 900 Bands)
RF channel spacing	25 kHz (U.S.)	12.5 kHz	30 kHz	1.25 MHz	25 kHz
Channel access/ multiuser access	FDMA/ DSMA	FDMA/ dynamic- R-S-ALOHA	FDMA/ DSMA	FDMA/ CDMA-SS	FDMA/ DSMA & SAPR[c]
Modulation method	FSK, 4-FSK	GMSK	GMSK	4-PSK/DSSS	π/4-QDPSK
Channel bit rate, kb/s	19.2	8.0	19.2	9.6	36
Packet length	Up to 256 bytes (HDLC)	Up to 512 bytes	24–928 b	(Packet service-TBD)	192 b (short) 384 b (long)
Open architecture	No	Yes	Yes	Yes	Yes
Private or Public Carrier	Private	Private	Public	Public	Public
Service Coverage	Major metro. areas in U.S.	Major metro. areas in U.S.	All AMPS areas	All CDMA cellular areas	European trunked radio
Type of coverage	In-building & mobile	In-building & mobile	Mobile	Mobile	Mobile

[a] Frequency allocation in the U.S. in the U.K., 380–450 MHz band is used.
[b] IS-95 and TETRA data services standardization in progress.
[c] Slotted-ALOHA packet reservation.

33.8 Conclusions

Mobile data radio systems have grown out of the success of the paging-service industry and the increasing customer demand for more advanced services. The growing use of portable, laptop, and palmtop computers and other data services will propel a steadily increasing demand for wireless data services. Today, mobile data services provide length-limited wireless connections with in-building penetration to portable users in metropolitan areas. The future direction is toward wider coverage, higher data rates, and capability for wireless Internet access.

References

[1] Bodson, D., McClure, G.F., and McConoughey, S.R., Eds., *Land-Mobile Communications Engineering*, Selected Reprint Ser., IEEE Press, New York, 1984.

[2] CDPD Industry Coordinator. Cellular Digital Packet Data Specification, Release 1.1, November 1994, Pub. CDPD, Kirkland, WA, 1994.

[3] Haine, J.L., Martin, P.M., and Goodings, R.L.A., A European standard for packet-mode mobile data, *Proceedings of Personal, Indoor, and Mobile Radio Conference (PIMRC'92)*, Boston, MA. Pub. IEEE, New York, 1992.

[4] Hata, M., Empirical formula for propagation loss in land-mobile radio services. *IEEE Trans. on Vehicular Tech.*, 29(3), 317–325, 1980.

[5] International Standards Organization (ISO), Protocol for providing the connectionless-mode network service. Pub. ISO 8473, 1987.

[6] Khan, M. and Kilpatrick, J., MOBITEX and mobile data standards. *IEEE Comm. Maga.*, 33(3), 96–101, 1995.

[7] Kilpatrick, J.A., Update of RAM Mobile Data's packet data radio service. *Proceedings of the 42nd IEEE Vehicular Technology Conference (VTC'92)*, Denver, CO, 898–901, Pub. IEEE, New York, 1992.

[8] Longley, A.G. and Rice, P.L., Prediction of tropospheric radio transmission over irregular terrain. A computer method—1968, Environmental Sciences and Services Administration Tech. Rep. ERL 79-ITS 67, U.S. Government Printing Office, Washington, D.C., 1968.

[9] Okumura, Y., Ohmori, E., Kawano, T., and Fukuda, K., Field strength and its variability in VHF and UHF land-mobile service. *Review of the Electronic Communication Laboratory*, 16, 825–873, 1968.

[10] Pahlavan, K. and Levesque, A.H., Wireless data communications. *Proceedings of the IEEE*, 82(9), 1398–1430, 1994.

[11] Pahlavan, K. and Levesque, A.H., *Wireless Information Networks*, J. Wiley & Sons, New York, 1995.

[12] Parsa, K., The MOBITEX packet-switched radio data system. *Proceedings of the Personal, Indoor and Mobile Radio Conference, (PIMRC'92)*, Boston, MA, 534–538. Pub. IEEE, New York, 1992.

[13] Quick, R.R., Jr. and Balachandran, K., Overview of the cellular packet data (CDPD) system. *Proceedings of the Personal, Indoor and Mobile Radio Conference, (PIMRC'93)*, Yokohama, Japan, 338–343. Pub. IEEE, New York, 1993.

[14] Rice, P.L., Longley, A.G., Norton, K.A., and Barsis, A.P., Transmission loss predictions for tropospheric communication circuits. National Bureau of Standards, Tech. Note 101, Boulder, CO, 1967.

[15] Sacuta, A., Data standards for cellular telecommunications—a service-based approach. *Proceedings of the 42nd IEEE Vehicular Technology Conference*, Denver CO, 263–266. Pub. IEEE, New York, 1992.

[16] Telecommunications Industry Association. Async data and fax. Project No. PN-3123, and Radio link protocol 1. Project No. PN-3306, Nov. 14. Issued by TIA, Washington, D.C., 1994.

[17] Tiedemann, E., Data services for the IS-95 CDMA standard. Presented at Personal, Indoor and Mobile Radio Conf. PIMRC'93. Yokohama, Japan, 1993.

[18] Weissman, D., Levesque, A.H., and Dean, R.A., Interoperable wireless data. *IEEE Comm. Mag.*, 31(2), 68–77, 1993.

Further Information

Reference [11] provides a comprehensive survey of the wireless data field as of mid-1994. The monthly journals *IEEE Communications Magazine* and *IEEE Personal Communications Magazine,* and the bimonthly journal *IEEE Transactions on Vehicular Technology* report advances in many areas of mobile communications, including wireless data. For subscription information contact: IEEE Service Center, 445 Hoes Lane, P. O. Box 1331, Piscataway, NJ, 08855-1131. Phone (800)678-IEEE.

34

Wireless ATM: Interworking Aspects

Melbourne Barton
Bellcore

Matthew Cheng
Bellcore

Li Fung Chang
Bellcore

34.1 Introduction

The ATM Forum's **wireless asynchronous transfer mode (WATM)** Working Group (WG) is developing specifications intended to facilitate the use of ATM technology for a broad range of wireless network access and interworking scenarios, both public and private. These specifications are intended to cover the following two broad WATM application scenarios:

- *End-to-End WATM*—This provides seamless extension of ATM capabilities to mobile terminals, thus providing ATM virtual channel connections (VCCs) to wireless hosts. For this application, high data rates are envisaged, with limited coverage, and transmission of one or more ATM cells over the air.
- *WATM Interworking*—Here the fixed ATM network is used primarily for high-speed transport by adding mobility control in the ATM infrastructure network, without changing the non-ATM air interface protocol. This application will facilitate the use of ATM as an efficient and cost-effective infrastructure network for next generation non-ATM wireless access systems, while providing a smooth migration path to seamless end-to-end WATM.

This chapter focuses on the ATM interworking application scenario. It describes various interworking and non-ATM wireless access options and their requirements. A generic

personal communications services (PCS)[1]-to-ATM interworking scenario is described which enumerates the architectural features, protocol reference models, and signalling issues that are being addressed for mobility support in the ATM infrastructure network. Evolution strategies intended to eventually provide end-to-end WATM capabilities and a methodology to consistently support a range of quality of service (QoS) levels on the radio link are also described.

34.2   Background and Issues

ATM is the switching and multiplexing standard for **broadband integrated services digital network (BISDN),** which will ultimately be capable of supporting a broad range of applications over a set of high capacity multiservice interfaces. ATM holds out the promise of a single network platform that can simultaneously support multiple bandwidths and latency requirements for fixed access and wireless access services without being dedicated to any one of them. In today's wireline ATM network environment, the **user network interface (UNI)** is fixed and remains stationary throughout the connection lifetime of a call. The technology to provide fixed access to ATM networks has matured. Integration of fixed and wireless access to ATM will present a cost-effective and efficient way to provide future tetherless multimedia services, with common features and capabilities across both wireline and wireless network environments. Early technical results [19, 20, 25] have shown that standard ATM protocols can be used to support such integration and extend mobility control to the subscriber terminal by incorporating wireless specific layers into the ATM user and control planes.

Integration of wireless access features into wireline ATM networks will place additional demands on the fixed network infrastructure due primarily to the additional user data and signalling traffic that will be generated to meet future demands for wireless multimedia services. This additional traffic will allow for new signalling features including registration, call delivery, and handoff during the connection lifetime of a call. Registration keeps track of a wireless user's location, even though the user's communication link might not be active. Call delivery, establishes a connection link to/from a wireless user with the help of location information obtained from registration. The registration and call delivery functions are referred to as **location management.** Handoff is the process of switching (rerouting) the communication link from the old coverage area to the new coverage area when a wireless user moves during active communication. This function is also referred to as **mobility management.**

In June, 1996 the ATM Forum established a WATM WG to develop requirements and specifications for WATM. The WATM standards are to be compatible with ATM equipment adhering to the (then) current ATM Forum specifications. The technical scope of the WATM WG includes development of: (1) **radio access layer (RAL)** protocols for the physical (PHY), medium access control (MAC), and data link control (DLC) layers; (2) wireless control protocols for radio resource management; (3) mobile protocol extensions for ATM (mobile ATM) including handoff control, routing considerations, location management, traffic and QoS control, and wireless network management; and (4) wireless

---

[1]The term PCS is being used in a generic sense to mean emerging digital wireless systems, which support mobility in microcellular and other environments. It is currently defined in ANSI T1.702-1995 as "A set of capabilities that allows some combination of terminal mobility, personal mobility, and service profile management."

interworking functions for mapping between non-ATM wireless access and ATM signalling and control entities. Phase-1 WATM specifications are being developed for short-range, high-speed, end-to-end WATM devices using wireless terminals that operate in the 5 GHz frequency band. Operating speeds will be up to 25 Mb/s, with a range of 30 m–50 m indoor and 200 m–300 m outdoor. The European Telecommunications Standards Institute (ETSI) Broadband Radio Access Networks (BRAN) project is developing the RAL for Phase-1 WATM specifications using the **HIgh PERformance Radio LAN (HIPERLAN)** functional requirements. The ATM Forum plans to release the Phase-1 WATM specifications by the second quarter of 1999.

There are a number of emerging wireless access systems (including digital cellular, PCS, legacy LANs based on the IEEE 802.11 standards, satellite, and IMT-2000 systems), that could benefit from access to, and interworking with, the fixed ATM network. These wireless systems are based on different access technologies and require development of different **interworking functions (IWFs)** at the wireless access network and fixed ATM network boundaries to support WATM interworking. For example, a set of network interfaces have already been identified to support PCS access to the fixed ATM network infrastructure, without necessarily modifying the PCS air interface protocol to provide end-to-end ATM capabilities [4, 6, 28]. The WATM WG might consider forming sub-working groups which could work in parallel to identify other network interfaces and develop IWF specifications for each of (or each subset of) the wireless access options that are identified. These WATM interworking specifications would be available in the Phase-2 and later releases of the WATM standards.

Some service providers and network operators see end-to-end WATM as a somewhat limited service option at this time because it is being targeted to small enterprise networks requiring high-speed data applications, with limited coverage and low mobility. On the other hand, WATM interworking can potentially support a wider range of services and applications, including low-speed voice and data access, without mandatory requirements to provide over-the-air transmission of ATM cells. It will allow for wider coverage, possibly extending to macrocells, while supporting higher mobility. WATM interworking will provide potential business opportunities, especially for public switched telephone network (PSTN) operators and service providers, who are deploying emerging digital wireless technologies such as PCS. Existing wireless service providers (WSPs) with core network infrastructures in place can continue to use them while upgrading specific network elements to provide ATM transport. On the other hand, a new WSP entrant without such a network infrastructure can utilize the public (or private) ATM transport network to quickly deploy the WATM interworking service, and not be burdened with the cost of developing an overlay network. If the final goal is to provide end-to-end WATM services and applications, then WATM interworking can provide an incremental development path.

34.3   Wireless Interworking With Transit ATM Networks

Figure 34.1 shows one view of the architectural interworking that will be required between public/private wireless access networks and the fixed ATM network infrastructure. It identifies the network interfaces where modifications will be required to allow interworking between both systems. A desirable objective in formulating WATM specifications for this type of wireless access scenario should be to minimize modifications to the transit ATM network and existing/emerging wireless access system specifications. This objective can be largely met by limiting major modifications to the network interfaces between the boundaries of the transit ATM network and public/private wireless networks, and where possible, adopting existing network standard processes (i.e., SS7, IS-41, MAP, AIN.3, Q.931, Q.932, Q.2931, etc.) to minimize development costs and upgrades to existing service providers' network infrastructure. Development of standard network interfaces that allow interworking of a reasonable subset of non-ATM digital wireless access systems with the fixed ATM network infrastructure insure that:

- Large-scale revisions and modifications are not necessary to comply with later versions of the WATM specifications to accommodate other emerging digital wireless access systems that do not require end-to-end ATM connectivity
- WATM systems are supported by open interfaces with a rich set of functionality to provide access to both ATM and non-ATM wireless access terminal devices
- WATM services can reach a much larger potential market including those markets providing traditional large-scale support for existing voice services and vertical voice features.

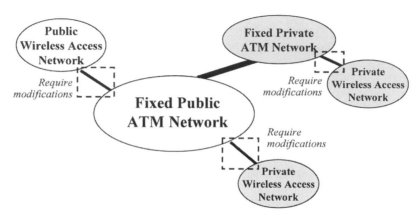

FIGURE 34.1   Wireless ATM interworking architecture.

34.3.1   Integrated Wireless–Wireline ATM Network Architecture

Figure 34.2 shows one example of a mature, multifunctional ATM transport network platform, which provides access to fixed and mobile terminals for wide-area coverage. Four distinct network interfaces are shown supporting: (1) fixed access with non-ATM terminal, (2) fixed access with ATM terminal, (3) wireless access with non-ATM terminal (WATM interworking), and (4) wireless access with ATM terminal (end-to-end WATM).

International Telecommunications Union (ITU) and ATM Forum standard interfaces either exist today or are being developed to support fixed access to ATM networks through various network interfaces. These include frame relay service (FRS) UNI, cell relay service (CRS) UNI, circuit emulation service (CES) UNI, and switched multimegabit data service (SMDS) subscriber NI (SNI). The **BISDN intercarrier interface (B-ICI)** specification [1] provides examples of wired IWFs that have been developed for implementation above the ATM layer to support intercarrier service-specific functions developed at the network nodes, and distributed in the public ATM/BISDN network. These distributed, service-specific functions are defined by B-ICI for FRS, CRS, CES, and SMDS. Examples of such functions include ATM cell conversion, clock recovery, loss of signal and alarm indi-

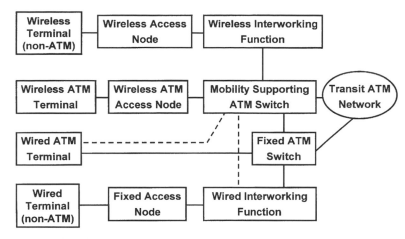

FIGURE 34.2   Wireless and wireline system integration with transit ATM Network.

cation detection, virtual channel identifier (VCI) mapping, access class selection, encapsulation/mapping, and QoS selection. In addition to the B-ICI, the **private network-network interface (PNNI)** specification [2] defines the basic call control signalling procedures (e.g., connection setup and release) in private ATM networks. It also has capabilities for autoconfiguration, scalable network hierarchy formation, topology information exchange, and dynamic routing.

On the wireless access side, existing ITU specifications provide for the transport of wireless services on public ATM networks (see, e.g., [10, 11]). For example, if the ATM UNI is at the mobile switching center (MSC), then message transfer part (MTP) 1 and 2 would be replaced by the PHY and ATM layers, respectively, the broadband ISDN user part (BISUP) replaced by MTP 3, and a common channel signalling (CCS) interface deployed in the ATM node. BISUP is used for ATM connection setup and any required feature control. If the ATM UNI is at the base station controller (BSC), then significant modifications are likely to be required. Equipment manufacturers have not implemented, to any large degree, the features that are available with the ITU specifications. In any case, these features are not sufficient to support the WATM scenarios postulated in this chapter.

The two sets of WATM scenarios postulated in this chapter are shown logically interfaced to the ATM network through a **mobility-enabled ATM (ME-ATM)** switch. This enhanced ATM access switch will have capabilities to support mobility management and location management. In addition to supporting handoff and rerouting of ATM connections, it will be capable of locating a mobile user anywhere in the network.

It might be desirable that functions related to mobility not be implemented in standard ATM switches so that network operators and service providers are not required to modify their ATM switches in order to accommodate WATM and related services. A feasible strategy that has been proposed is to implement mobility functions in servers (i.e., service control modules or SCMs) that are logically separated from the ATM switch. In these servers, all mobility features, service creation logic, and service management functions will be implemented to allow logical separation of ATM switching and service control from mobility support, service creation, and management. The open network interface between the ATM access switch and SCM will be standardized to enable multivendor operation. It would be left up to the switch manufacturer to physically integrate the SCM into a new ATM switching fabric, or implement the SCM as a separate entity.

34.3.2   Wireless Access Technology Options

Figure 34.3 identifies various digital wireless systems that could be deployed to connect mobile terminals to the transit ATM network through IWFs that have to be specified. The main emphasis is on developing mobility support in the transit ATM infrastructure network to support a range of wireless access technologies. Significant interests have been expressed, through technical contributions and related activities in the WATM WG, in developing specifications for the IWFs shown in Fig. 34.3 to allow access to the fixed ATM network through standard network interfaces. The main standardization issues that need to be addressed for each wireless access system are described below.

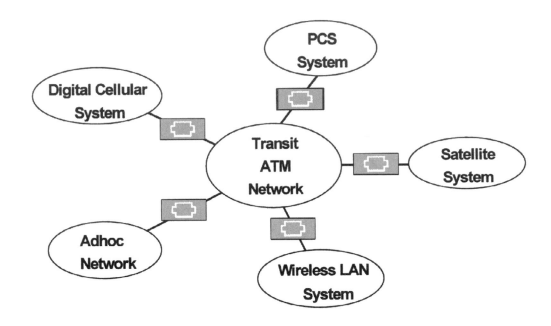

FIGURE 34.3   Various wireless access technologies supported by wireless ATM interworking.

PCS

This class of digital wireless access technologies include the low-tier PCS systems as described in Cox [8]. Digital cellular systems which are becoming known in the U.S. as high-tier PCS, especially when implemented in the 1.8-1.9 GHz PCS frequency bands, are addressed in the Digital Cellular section below. The PCS market has been projected to capture a significant share of the huge potential revenues to be generated by business and residential customers. In order to provide more flexible, widespread, tetherless portable communications than can be provided by today's limited portable communications approaches, low-power exchange access radio needs to be integrated with network intelligence provided by the wireline network.

Today, the network for supporting PCS is narrowband ISDN, along with network intelligence based on advanced intelligent network (AIN) concepts, and a signalling system for mobility/location management and call control based on the signalling system 7 (SS7) network architecture. It is expected that the core network will evolve to BISDN/ATM over time with the capability to potentially integrate a wide range of network services, both wired and wireless, onto a single network platform. Furthermore, there will be no need for an overlay network for PCS. In anticipation of these developments, the WATM WG included the development of specifications and requirements for PCS access to, and interworking with, the ATM transit network in its charter and work plan. To date, several contributions have been presented at WATM WG meetings that have identified some of the key technical issues relating to PCS-to-ATM interworking. These include (1) architectures, (2) mobility and location management signalling, (3) network evolution strategies, and (4) PCS service scenarios.

Wireless LAN

Today, wireless LAN (WLAN) is a mature technology. WLAN products are frequently used as LAN extensions to access areas of buildings with wiring difficulties and for cross-building interconnect and nomadic access. Coverage ranges from tens to a few hundreds of meters, with data rates ranging from hundreds of kb/s to more than 10 Mb/s. Several products provide 1 or 2 Mb/s. ATM LAN products provide LAN emulation services over the connection-oriented (CO) ATM network using various architectural alternatives [24]. In this case, the ATM network provides services that permit reuse of existing LAN applications by stations attached directly to an ATM switch, and allow interworking with legacy LANs. Furthermore, the increasing importance of mobility in data access networks and the availability of more usable spectrum are expected to speed up the evolution and adoption of mobile access to WLANs. Hence, it is of interest to develop wireless LAN products that have LAN emulation capabilities similar to wireline ATM LANs.

The ETSI BRAN project is developing the HIPERLAN RAL technology for wireless ATM access and interconnection. It will provide short-range wireless access to ATM networks at approximately 25 Mb/s in the 5 GHz frequency band. HIPERLAN is an ATM-based wireless LAN technology that will have end-to-end ATM capabilities. It does not require the development of an IWF to provide access to ATM. A previous version of HIPERLAN (called HIPERLAN I), which has been standardized, supports data rates from 1–23 Mb/s in the 5 GHz band using a non-ATM RAL [27]. This (and other non-ATM HIPERLAN standards being developed) could benefit from interworking with the backbone ATM as a means of extending the marketability of these products in the public domain such as areas of mass transit and commuter terminals.

Several proposals have been submitted to IEEE 802.11 to provide higher speed extensions of current IEEE 802.11 systems operating in the 2.4 GHz region and the development of specifications for new systems operating in the 5 GHz frequency band. The proposed 2.4 GHz extensions support different modulation schemes, but are interoperable with the current IEEE 802.11 low rate PHY and are fully compliant with the IEEE 802.11 defined MAC. The 5 GHz proposals are not interoperable with the current 2.4 GHz IEEE 802.11 systems. One of the final three 5 GHz proposals being considered is based on orthogonal frequency division multiplexing (OFDM), or multicarrier modulation. The other two are single carrier proposals using offset QPSK (OQPSK)/offset QAM (OQAM), and differential pulse position modulation (DPPM). The OFDM proposal has been selected. With 16-QAM modulation on each subcarrier and rate-3/4 convolutional coding, the OFDM system has a peak data rate capability of 30 Mb/s.

It is clear that a whole range of WLAN systems either exist today, or are emerging, that are not based on ATM technology, and therefore cannot provide seamless access to the ATM infrastructure network. The development of IWF specifications that allow these

WLANs to provide such access through standard network interfaces will extend the range of applications and service features provided by WLANs. In Pahlavan [17], a number of architectural alternatives to interconnect WLAN and WATM to the ATM and/or LAN backbone are discussed, along with service scenarios and market and product issues.

Digital Cellular

Digital cellular mobile radio systems include the 1.8–1.9 GHz (high-tier PCS) and the 800–900 MHz systems that provide high-mobility, wide-area coverage over macrocells. Cellular radio systems at 800–900 MHz have evolved to digital in the form of Global System for Mobile Communications (GSM) in Europe, Personal Digital Cellular (PDC) in Japan, and IS-54 Time Division Multiple Access (TDMA) and IS-95 Code Division Multiple Access (CDMA) in the U.S. The capabilities in place today for roaming between cellular networks provide for even wider coverage. Cellular networks have become widespread, with coverage extending beyond some national boundaries. These systems integrate wireless access with large-scale mobile networks having sophisticated intelligence to manage mobility of users.

Cellular networks (e.g., GSM) and ATM networks are evolving somewhat independently. The development of IWFs that allow digital cellular systems to utilize ATM transport will help to bridge the gap. Cellular networks have sophisticated mechanisms for authentication and handoff, and support for rerouting through the home network. In order to facilitate the migration of cellular systems to ATM transport, one of the first issues that should be addressed is that of finding ways to enhance the basic mobility functions already performed by them for implementation in WATM. Contributions presented at WATM WG meetings have identified some basic mobility functions of cellular mobile networks (and cordless terminal mobility) which might be adopted and enhanced for WATM interworking. These basic mobility functions include rerouting scenarios (including path extension), location update, call control, and authentication.

Satellite

Satellite systems are among the primary means of establishing connectivity to untethered nodes for long-haul radio links. This class of applications has been recognized as an essential component of the National Information Infrastructure (NII) [16]. Several compelling reasons have been presented in the ATM Forum's WATM WG for developing standard network interfaces for **satellite ATM (SATATM)** networks. These include (1) ubiquitous wide area coverage, (2) topology flexibility, (3) inherent point-to-multipoint and broadcast capability, and (4) heavy reliance by the military on this mode of communications. Although the geostationary satellite link represents only a fraction of satellite systems today, WATM WG contributions that have addressed this interworking option have focused primarily on geostationary satellites. Some of these contributions have also proposed the development of WATM specifications for SATATM systems having end-to-end ATM capabilities.

Interoperability problems between satellite systems and ATM networks could manifest themselves in at least four ways.

1. Satellite links operate at much higher bit error rates (BERs) with variable error rates and bursty errors.
2. The approximately 540 ms round trip delay for geosynchronous satellite communications can potentially have adverse impacts on ATM traffic and congestion control procedures.

3. Satellite communications bandwidth is a limited resource, and might be incompatible with less bandwidth efficient ATM protocols.

4. The high availability rates (at required BERs) for delivery of ATM (e.g., 99.95%) is costly, hence the need to compromise between performance relating to availability levels and cost.

A number of experiments have been performed to gain insights into these challenges [21]. The results can be used to guide the development of WATM specifications for the satellite interworking scenario. Among the work items that have been proposed for SATATM access using geostationary satellites links are (1) identification of requirements for RAL and mobile ATM functions; (2) study of the impact of satellite delay on traffic management and congestion control procedures; (3) development of requirements and specifications for bandwidth efficient operation of ATM speech over satellite links; (4) investigation of various WATM access scenarios; and (5) investigation of frequency spectrum availability issues.

One interesting SATATM application scenario has been proposed to provide ATM services to multiuser airborne platforms via satellite links for military and commercial applications. In this scenario, a satellite constellation is assumed to provide contiguous overlapping coverage regions along the flight path of the airborne platforms. A set of interworked ground stations form the mobile enhanced ATM network that provides connectivity between airborne platforms and the fixed terrestrial ATM network via bent-pipe satellite links. Key WATM requirements for this scenario have been proposed, which envisage among other things modifications to existing PNNI signalling and routing mechanisms to allow for mobility of (ATM) switches.

In related work, the Telecommunications Industry Association (TIA) TR34.1 WG has also proposed to develop technical specifications for SATATM networks. Three ATM network architectures are proposed for bent-pipe satellites and three others for satellites with onboard ATM switches [23]. Among the technical issues that are likely to be addressed by TR34.1 are protocol reference models and architecture specifications, RAL specifications for SATATM, and support for routing, rerouting, and handoff of active connections. A liaison has been established between the ATM Forum and TR34.1, which is likely to lead to the WATM WG working closely with TR34.1 to develop certain aspects of the TR34.1 SATATM specifications.

Ad Hoc Networks

The term ad hoc network is used to characterize wireless networks that do not have dedicated terminals to perform traditional BS and/or wireless resource control functions. Instead, any mobile terminal (or a subset of mobile terminals) can be configured to perform these functions at any time. Ad hoc networking topologies have been investigated by wireless LAN designers, and are part of the HIPERLAN and IEEE 802.11 wireless LAN specifications [13]. As far as its application to WATM is concerned, low-cost, plug-and-play, and flexibility of system architecture are essential requirements. Potential application service categories include rapidly deployable networks for government use (e.g., military tactical networks, rescue missions in times of natural disasters, law enforcement operations, etc.), ad hoc business conferencing devoid of any dedicated coordinating device, and ad hoc residential network for transfer of information between compatible home appliances.

There are some unique interworking features inherent in ad hoc networks. For example, there is a need for location management functions not only to identify the location of terminals but also to identify the current mode of operation of such terminals. Hence, the WATM WG is considering proposals to develop separate requirements for ad hoc networks independent of the underlying wireless access technology. It is likely that requirements will be developed for an ad hoc RAL, mobility management signalling functions, and location management functions for supporting interworking of ad hoc networks that provide access to ATM infrastructure networks.

The range of potential wireless access service features and wireless interworking scenarios presented in the five wireless access technologies discussed above is quite large. For example, unlicensed satellite systems could provide 32 kb/s voice, and perhaps up to 10 Mb/s for wireless data services. The main problem centers around the feasibility of developing specifications for IWFs to accommodate the range of applications and service features associated with the wireless access options shown in Fig. 34.3. The WATM WG might consider forming sub-working groups which would work in parallel to develop the network interface specifications for each (or a subset of each) of the above non-ATM wireless access options.

34.4   The PCS-to-ATM Interworking Scenario

This section presents a more detailed view of potential near-term and longer-term architectures and reference models for PCS-to-ATM interworking. A signalling link evolution strategy to support mobility is also described. Mobility and location management signalling starts with the current CCS network, which is based on the SS7 protocol and 56 kb/s signalling links, and eventually migrates to ATM signalling. The system level architectures, and the mobility and location management signalling issues addressed in this section serve to illustrate the range of technical issues that need to be addressed for the other WATM interworking options.

34.4.1   Architecture and Reference Model

The near-term approach for the PCS-to-ATM interworking scenario is shown in Fig. 34.4. The near-term approach targets existing PCS providers with network infrastructures in place, who wish to continue to use them while upgrading specific network elements (e.g., MSCs) to provide ATM transport for user data. The existing MSCs in the PCS network are upgraded to include fixed ATM interfaces. ATM is used for transport and switching of user data, while mobility/location management and call control signalling is carried by the SS7 network. No mobility support is required in the ATM network. Synchronization problems might develop because different traffic types relating to the same call may traverse the narrowband SS7 network and the broadband ATM network and arrive at the destination in an uncoordinated manner. Upgrading the SS7 network to broadband SS7 (e.g., T1 speeds or higher) should partially alleviate this potential problem.

A longer-term approach for PCS-to-ATM interworking has been proposed in some technical contributions to the WATM WG. This is illustrated in Fig. 34.5, together with the protocol stacks for both data and signalling. The ATM UNI is placed at the BSC, which acts as the PCS-to-ATM gateway. The ATM network carries both data and signalling. ATM cells are not transmitted over the PCS link. Communications between the BS and BSC are specific to the PCS access network, and could be a proprietary interface. Compared with existing/emerging BSCs, additional protocol layer functionality is required in the BSC to provide (1) transfer/translation and/or encapsulation of PCS **protocol data units (PDUs),** (2) ATM to wireless PDU conversion, and (3) a limited amount of ATM multiplexing/demultiplexing capabilities.

The BSC is connected to the ATM network through a ME-ATM access switch instead of a MSC. The ME-ATM switch provides switching and signalling protocol functions to support ATM connections together with mobility and location management. These functions could

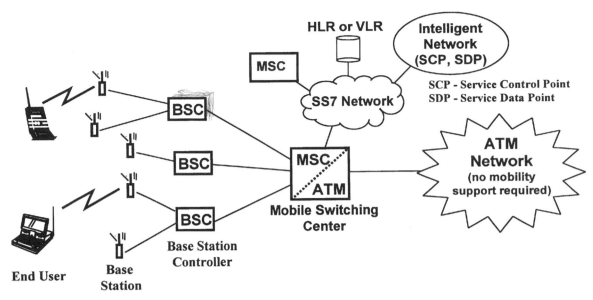

FIGURE 34.4 A near-term architecture for PCS-to-ATM interworking.

be implemented in servers that are physically separate from, but logically connected to, the ME-ATM switch. The WATM WG is expected to formulate requirements and specifications for these mobile-specific functions. On the other hand, an IWF can be introduced between the BSC and the ME-ATM switch shown in Fig. 34.5. In this case, the UNI is between the IWF and the ME-ATM switch, and another standard interface (not necessarily ATM) can be used to connect the BSC to the IWF. The BSC then requires no modification, but a new entity (i.e., the IWF) is required. The IWF will perform protocol conversion and it may serve multiple BSCs.

A unique feature of this architecture is that modifications to network entities to allow for interworking of PCS with the transit ATM network are only required at the edges of the ATM and wireless access networks, i.e., to the BSC and the ME-ATM access switch. In order to minimize the technical impact of mobility on existing/emerging transit ATM networks and PCS specifications, an initial interworking scenario is envisaged in which there are no (or only very limited) interactions between PCS and ATM signalling entities. PCS signalling would be maintained over the air interface, traditional ATM signalling would be carried in the control plane (C-Plane), and PCS signalling would be carried over the ATM network as user traffic in the user plane (U-Plane). In the long term, this architecture is expected to eventually evolve to an end-to-end ATM capability.

On the ATM network side, mobility and location management signalling is implemented in the user plane as the **mobile application part (MAP)** layer above the ATM adaptation layer (AAL), e.g., AAL5. The MAP can be based on the MAP defined in the existing PCS standards (e.g., IS41-MAP or GSM-MAP), or based on a new set of mobile ATM protocols. The U-plane is logically divided into two parts, one for handling mobility and location management signalling messages and the other for handling traditional user data. This obviates the need to make modifications to the ATM UNI and NNI signalling protocols, currently being standardized. The MAP layer also provides the necessary end-to-end reliability management for the signalling because it is implemented in the U-Plane, where reliable communication is not provided in the lower layers of the ATM protocol stack as is done in the signalling AAL (SAAL) layer. The MAP functionality can be distributed at

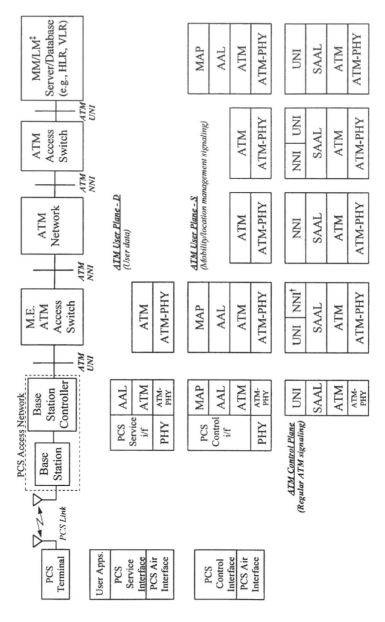

FIGURE 34.5   A longer-term architecture and reference model PCS-to-ATM interworking.

† NNI here is used in a generic sense, to refer to both public and private networks. In public networks, there will be an additional MTP3 layer between the NNI and SAAL layers.

‡ MM—Mobility Management, LM—Location Management.

individual BSCs or ME-ATM access switches or centralized in a separate server to further reduce modifications to existing implementations. The setup and release of ATM connections are still handled by the existing ATM signalling layer. This approach allows easy evolution to future wireless ATM architectures, which will eventually integrate the ATM and PCS network segments to form a homogeneous network to carry both mobility/location management and traditional ATM signalling in the C-Plane. Issues relating to mobility support for PCS interworking with ATM networks are addressed in more detail in Cheng [7].

34.4.2 Signalling Link Evolution

Signalling will play a crucial role in supporting end-to-end connections (without location restrictions) in an integrated WATM network infrastructure. The signalling and control message exchanges required to support mobility will occur more frequently than in wireline ATM networks. Today's CCS network, which is largely based on the SS7 protocol and 56 kb/s signalling links, will not be able to support the long-term stringent service and reliability requirements of WATM. Two broad deployment alternatives have been proposed for migrating the CCS/SS7 network (for wireline services) to using the broadband signalling platform [22].

- Migration to high-speed signalling links using the narrowband signalling platform supported by current digital transmission (e.g., DS1) facilities. The intent of this approach is to support migration to high-speed signalling links using the existing CCS infrastructure with modified signalling network elements. This would allow the introduction of high-speed (e.g., 1.5 Mb/s) links with possible minimal changes in the CCS network. One option calls for modifications of existing MTP2 procedures while maintaining the current protocol layer structure. Another option replaces MTP2 with some functionality of the ATM/SAAL link layer, while continuing to use the same transport infrastructure as DS1 and transport messages over the signalling links in variable length signal units delimited by flags.

- Alternative 2 supports migration of signalling links to a broadband/ATM signalling network architecture. Signalling links use both ATM cells and the ATM SAAL link layer, with signalling message transported over synchronous optical network (SONET) or existing DS1/DS3 facilities at rates of 1.5 Mb/s or higher. This alternative is intended primarily to upgrade the current CCS network elements to support an ATM-based interface, but could also allow for the inclusion of signalling transfer point (STP) functions in the ATM network elements to allow for internetworking between ATM and existing CCS networks.

The second alternative provides a good vehicle for the long-term goal of providing high-speed signalling links on a broadband signalling platform supported by the ATM technology. One signalling network configuration and protocol option is the extended Q.93B signalling protocol over ATM in associated mode [22]. The PSTN's CCS networks are currently quasi-associated signalling networks. Q.93B is primarily intended for point-to-point bearer control in user-to-network access, but can also be extended for (link-to-link) switch-to-switch and some network-to-network applications. Standards activities to define high-speed signalling link characteristics in the SS7 protocol have been largely finalized. Standards to support SS7 over ATM are at various stages of completion. These activities provide a good basis for further evolving the SS7 protocol to provide the mobility and location management features that will be required to support PCS (and other wireless systems) access to the ATM network.

The functionality required to support mobility in cellular networks is currently defined as part of the MAP. Both IS-41 and GSM MAPs are being evolved to support PCS services with SS7 as the signalling transport protocol [14]. Quite similar to the two alternatives described above, three architectural alternatives have been proposed for evolving today's IS-41 MAP on SS7 to a future modified (or new) IS-41 on ATM signalling transport platform [28]. They are illustrated in Fig. 34.6. In the first (or near-term) approach, user data is carried over the ATM network, while signalling is carried over existing SS7 links. The existing SS7 network can also be upgraded to broadband SS7 network (e.g., using T1 links) to alleviate some of the capacity and delay constraints in the narrowband SS7 network. This signalling approach can support the near-term PCS-to-ATM interworking described in the previous subsection. In the second (or midterm) approach, a hybrid-mode operation is envisaged, with the introduction of broadband SS7 network elements into the ATM network. This results in an SS7-over-ATM signalling transport platform. Taking advantage of the ATM's switching and routing capabilities, the MTP3 layer could also be modified to utilize these capabilities and eliminate the B-STP functionality from the network. In the third phase (or long-term approach), ATM replaces the SS7 network with a unified network for both signalling and user data. No SS7 functionality exists in this approach. Here, routing of signalling messages is completely determined by the ATM layer, and the MAP may be implemented in a format other than the transaction capability application part (TCAP), so that the unique features of ATM are best utilized. The longer-term PCS-to-ATM interworking approach is best supported for this signalling approach. The performance of several signalling protocols for PCS mobility support for this long-term architecture is presented in Cheng [6, 7]. The above discussion mainly focuses on public networks. In private networks, there are no existing standard signalling networks. Therefore, the third approach can be deployed immediately in private networks.

There are several ways to achieve the third signalling approach in ATM networks. The first approach is to overlay another "network layer" protocol (e.g., IP) over the ATM network, but this requires the management of an extra network. The second approach is to enhance the current ATM network-network interface (e.g., B-ICI for public networks and PNNI for private networks) to handle the new mobility/location management signalling messages and information elements, but this requires modifications in the existing ATM network. The third approach is to use dedicated channels (PVCs or SVCs) between the mobility control signalling points. This does not require any modifications in existing ATM specifications. However, signalling latency may be high in the case of SVCs, and a full mesh of PVCs between all mobility control signalling points is difficult to manage. The fourth approach is to use the generic functional protocols (e.g., connection oriented-bearer independent (CO-BI) or connectionless-BI (CL-BI) transport mechanism) defined in ITU-T's Q.2931.2 [12]. This cannot be done in existing ATM networks without modifications, but these functions are being included in the next version of PNNI (PNNI 2.0) to provide a generic support for supplementary services. There are also technical contributions to the ATM Forum proposing "connectionless ATM" [26], which attempt to route ATM cells in a connectionless manner by using the routing information obtained through the PNNI protocol. However, the "connectionless ATM" concept is still being debated.

34.5   QoS Support

One immediate impact of adding mobility to an otherwise fixed ATM infrastructure network is the need to manage the changing QoS levels that are inherent in a mobile environment due to the vagaries of the wireless link, the need for rerouting of traffic due to handoff,

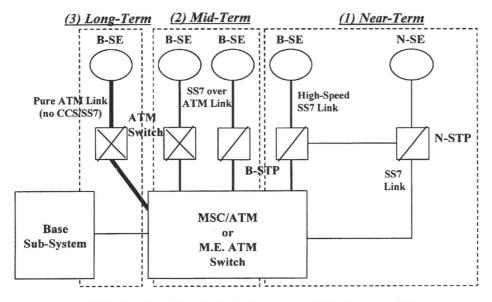

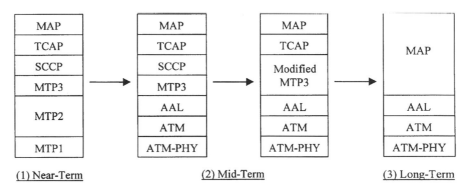

FIGURE 34.6 Signalling link evolution for mobility/location management over a public ATM network.

and available bandwidth, etc. Dynamic QoS negotiation and flow control will be required to flexibly support QoS guarantees for multimedia service applications that are likely to be encountered in this environment. QoS provisioning is based on the notion that the wireless channel is likely to demand more stringent measures than the fixed ATM network to support end-to-end QoS. QoS metrics include (1) throughput, (2) delay sensitivity, (3) loss sensitivity, and, (4) BER performance. Issues relating to end-to-end QoS provisioning in multimedia wireless networks are discussed in Naghshineh [15] and articles therein. Here, the focus is on BER maintenance in the context of PCS-to-ATM interworking using forward error correction (FEC) at the radio PHY layer. This is the first step towards developing a hybrid automatic repeat request (ARQ)/FEC protocol for error control of the wireless link, with FEC at the PHY supplemented by ARQ at the DLC layer. A comparison of commonly used FEC and ARQ techniques and their potential application to WATM is presented in Ayanoglu [3].

One adaptive FEC coding scheme that has been proposed for PCS-to-ATM interworking is based on the use of **rate-compatible punctured convolution (RCPC),** punctured Bose-Chaudhuri-Hocquenghem (BCH) or Reed-Solomon (RS) coding at the wireless PHY layer to provide unequal error protection of the wireless PDU [5]. These coding schemes can support a broad range of QoS levels consistent with the requirements of multimedia services, minimize the loss of information on the wireless access segment, and prevent misrouting of cells on the fixed ATM network. Code rate puncturing is a procedure used to periodically discard a set of predetermined coded bits from the sequence generated by an encoder for the purposes of constructing a higher rate code. With the rate-compatibility restriction, higher rate codes are embedded in the lower rate codes, allowing for continuous code rate variation within a data frame.

An example set of three wireless PDU formats that might be appropriate for a PCS system that provides access to the ATM network is shown in Table 34.1. It is desirable to establish a tight relationship between the wireless PDU and wireline ATM cell to minimize incompatibilities between them. This will reduce the complexity of the IWF at the PCS-to-ATM gateway by limiting the amount of processing required for protocol conversion. The wireless PDU can be tightly coupled to the wireline ATM cell in two ways.

1. Encapsulate the equivalent of a full 48-byte ATM information payload, along with wireless-specific overhead (and a full/compressed ATM header, if required by the mobile terminal), in the wireless PDU (e.g., PDU-3).

2. Transmit a submultiple of 48-byte ATM information payload, along with wireless-specific overhead (and a compressed ATM header, if required by the mobile terminal), in the wireless PDU (e.g., PDU-1 and PDU-2).

If the second option is used, then the network has to decide whether to send partially filled ATM cells over the ATM infrastructure network, or wait until enough wireless PDUs arrive to fill an ATM cell.

The performance of RCPC and punctured BCH (and RS) codes have been evaluated in terms of the decoded bit BER [5] on a Rayleigh fading channel. For RCPC coding, the Viterbi upper bound on the decoded BER is given by:

$$P_b \leq \frac{1}{P} \sum_{d=d_f}^{\infty} \beta_d P_d \tag{34.1}$$

where P is the puncturing period, β_d is the weight coefficient of paths having distance d, P_d is the probability of selecting the wrong path, and d_f is the free distance of the RCPC code.

TABLE 34.1 Examples of Wireless Protocol Data Unit (PDU) Formats

PDU type	PDU header (bits)	Information payload (bits)	PDU trailer (bits)	PDU size (bits)
PDU-1	24	128	—	152
PDU-2	40	256	8	304
PDU-3	56	384	16	456

Note: Information payloads are limited to submultiples of a 48-byte ATM cell information payload.

Closed-form expressions for P_d are presented in Hagenauer [9] for a flat Rayleigh fading channel, along with tables for determining β_d. Since the code weight structure of BCH codes is known only for a small subset of these codes, an upper BER performance bound is derived in the literature independent of the structure. Assume that for a t-error correcting BCH(n,k) code, a pattern of i channel errors $(i > t)$ will cause the decoded word to differ from the correct word in $i+t$ bit positions. For flat Rayleigh fading channel conditions and differential quadrature phase-shift keying (DQPSK) modulation, the decoded BER for the BCH(n,k) code is [18]:

$$P_b = \frac{2}{3} \sum_{i=t+1}^{n} \frac{i+t}{n} (1 - P_s)^{n-1} P_s^i \tag{34.2}$$

where P_s is the raw symbol error rate (SER) on the channel. An upper bound on the decoded BER for a t-error correcting RS(n,k) code with DQPSK signalling on a flat Rayleigh fading channel is similar in form to (34.2). However, P_s should be replaced with the RS-coded digit error rate P_e, and the unit of measure for the data changed from symbols do digits. A transmission system that uses the RS code [from GF(2^m)]^2 with M-ary signalling generates $r = m/\log_2 M$ symbols per m-bit digit. For statistically independent symbol errors, $P_e = 1 - (1 - P_s)^r$. For the binary BCH(n,k) code, $n = 2^m - 1$.

Table 34.2 shows performance results for RCPC and punctured BCH codes that are used to provide adaptive FEC on the PHY layer of a simulated TDMA-based PCS system that accesses the fixed ATM network. The FEC codes provide different levels of error protection for the header, information payload, and trailer of the wireless PDU. This is particularly useful when the wireless PDU header contains information required to route cells in the fixed ATM network, for example. Using a higher level of protection for the wireless PDU header increases the likelihood that the PDU will reach its destination, and not be misrouted in the ATM network.

The numerical results show achievable average code rates for the three example PDU formats in Table 34.1. The PCS system operates in a microcellular environment at 2 GHz, with a transmission bit rate of 384 kb/s, using DQPSK modulation. The channel is modeled as a time-correlated Rayleigh fading channel. The PCS transmission model assumes perfect symbol and frame synchronization, as well as perfect frequency tracking. Computed code rates are shown with and without the use of diversity combining. All overhead blocks in the wireless PDUs are assumed to require a target BER of 10^{-9}. On the other hand, the information payload has target BERs of 10^{-3} and 10^{-6}, which might be typical for voice

---

^{2}GF(2^m) denotes the Galois Field of real numbers from which the RS code is constructed. Multiplication and addition of elements in this field are based on modulo-2 arithmetic, and each RS code word consists of m bits.

TABLE 34.2   Average Code Rates for RCPC Coding[a] and Punctured BCH Coding[b] for a 2-GHz TDMA-Based PCS System with Access to an ATM Infrastructure Network

Coding scheme	Average code rate (10^{-3} BER for information payload)			Average code rate (10^{-6} BER for information payload)		
	PDU-1	PDU-2	PDU-3	PDU-1	PDU-2	PDU-3
RCPC: no diversity	0.76	0.77	0.78	0.61	0.62	0.63
RCPC: 2-branch diversity	0.91	0.93	0.93	0.83	0.84	0.85
Punctured BCH: no diversity	0.42	0.44	0.51	0.34	0.39	0.46
Punctured BCH: 2-branch Diversity	0.83	0.81	0.87	0.75	0.76	0.82

[a] with soft decision decoding and no channel state information at the receiver.

[b] with and without 2-branch frequency diversity.

and data, respectively. Associated with this target BERs is a design goal of 20 dB for the SNR. The mother code rate for the RCPC code is $R = 1/3$, the puncturing period $P = 8$, and the memory length $M = 6$. For BCH coding, the parameter $m \geq 8$.

The numerical results in Table 34.2 show the utility of using code rate puncturing to improve the QoS performance of the wireless access segment. The results for punctured RS coding are quite similar to those for punctured BCH coding. Adaptive PHY layer FEC coding can be further enhanced by implementing an ARQ scheme at the DLC sublayer, which is combined with FEC to form a hybrid ARQ/FEC protocol [9] to supplement FEC at the PHY layer. This approach allows adaptive FEC to be distributed between the wireless PHY and DLC layers.

34.6   Conclusions

The ATM Forum is developing specifications intended to facilitate the use of ATM technology for a broad range of wireless network access and interworking scenarios, both public and private. These specifications are intended to cover requirements for seamless extension of ATM to mobile devices and mobility control in ATM infrastructure networks to allow interworking of non-ATM wireless terminals with the fixed ATM network. A mobility-enhanced ATM network that is developed from specifications for WATM interworking may be used in near-term cellular/PCS/satellite/wireless LAN deployments, while providing a smooth migration path to the longer-term end-to-end WATM application scenario. It is likely to be cost-competitive with other approaches that adopt non-ATM overlay transport network topologies.

This chapter describes various WATM interworking scenarios where the ATM infrastructure might be a public (or private) transit ATM network, designed primarily to support broadband wireline services. A detailed description of a generic PCS-to-ATM architectural interworking scenario is presented, along with an evolution strategy to eventually provide end-to-end WATM capabilities in the long term. One approach is described for providing QoS support using code rate puncturing at the wireless PHY layer, along with numerical results. The network architectures, protocol reference models, signalling protocols, and QoS management strategies described for PCS-to-ATM interworking can be applied, in varying degrees, to the other WATM interworking scenarios described in this chapter.

Defining Terms

Broadband integrated services digital network (BISDN): A cell-relay-based information transfer technology upon which the next-generation telecommunications infrastructure is to be based.

BISDN intercarrier interface (B-ICI): A carrier-to-carrier public interface that supports multiplexing of different services such as SMDS, frame relay, circuit emulation, and cell relay services.

High-Performance Radio LAN (HIPERLAN): Family of standards being developed by the European Telecommunications Standards Institute (ETSI) for high-speed wireless LANs, to provide short-range and remote wireless access to ATM networks and for wireless ATM interconnection.

Interworking functions (IWFs): A set of network functional entities that provide interaction between dissimilar subnetworks, end systems, or parts thereof, to support end-to-end communications.

Location management: A set of registration and call delivery functions.

Mobile application part (MAP): Application layer protocols and processes that are defined to support mobility services such as intersystem roaming and hand-offs.

Mobility-enabled ATM (ME-ATM): An ATM switch with additional capabilities and features to support location and mobility management.

Mobility management: The handoff process associated with switching (rerouting) of the communication link from the old coverage area to the new coverage area when a wireless user moves during active communication.

Personal communications services (PCS): Emerging digital wireless systems which support mobility in microcellular and other environments, which have a set of capabilities that allow some combination of terminal mobility, personal mobility, and service profile management.

Protocol data units (PDUs): The physical layer message structure used to carry information across the communications link.

Private network–network interface (PNNI): The interface between two private networks.

User network interface (UNI): A standardized interface providing basic call control functions for subscriber access to the telecommunications network.

Radio access layer (RAL): A reference to the physical, medium access control, and data link control layers of the radio link.

Rate-compatible punctured convolution (RCPC): Periodic discarding of predetermined coded bits from the sequence generated by a convolutional encoder for the purposes of constructing a higher rate code. The rate-compatibility restriction insures that the higher rate codes are embedded in the lower rate codes, allowing for continuous code rate variation to change from low to high error protection within a data frame.

Satellite ATM (SATATM): A satellite network that provides ATM network access to fixed or mobile terminals, high-speed links to interconnect fixed or mobile ATM networks, or form an ATM network in the sky to provide user access and network interconnection services.

Wireless asynchronous transfer mode (WATM): An emerging wireless networking technology that extends ATM over the wireless access segment, and/or uses the ATM infrastructure as a transport network for a broad range of wireless network access scenarios, both public and private.

References

[1] ATM. B-ICI Specification, V 2.0. ATM Forum, 1995.

[2] ATM. Private Network-Network Interface (PNNI) Specification, Version 1.0. ATM Forum, 1996.

[3] Ayanoglu, E., et al., Wireless ATM: Limits, challenges, and protocols. *IEEE Personal Commun.*, 3(4), 18–34, Aug. 1996.

[4] Barton, M., Architecture for wireless ATM networks. *PIMRC'95.* 778–782, Sep. 1995.

[5] Barton, M., Unequal error protection for wireless ATM applications. *GLOBECOM'96*, 1911–1915, Nov. 1996.

[6] Cheng, M., Performance comparison of mobile assisted network controlled, and mobile-controlled hand-off signalling for TDMA PCS with an ATM backbone. *ICC'97.* Jun. 1977a.

[7] Cheng, M., Rajagopalan, S., Chang, L.F., Pollini, G.P., and Barton, M., PCS mobility support over fixed ATM networks. *IEEE Commun. Mag.*, 35(11), 82–92, Nov. 1997b.

[8] Cox, D.C., Wireless personal communications: what is it?. *IEEE Personal Commun. Mag.*, 2(2), 20–35, Apr. 1995.

[9] Hagenauer, J., Rate-compatible punctured convolution codes (RCPC codes) and their applications. *IEEE Trans. Commun.*, COM-36(4), 389–400, Apr. 1988.

[10] ITU. Message Transfer Part Level 3 Functions and Messages Using the Service of ITU-T Recommendations, Q.2140. International Telecommunications Union-Telecommunications Standardization Sector, Geneva, Switzerland. TD PL/11–97. 1995a.

[11] ITU. B-ISDN Adaptation Layer—Service Specific Coordination Function for Signaling at the Network Node Interface (SCCF) at NNI. ITU-T Q.2140. International Telecommunications Union, Telecommunications Standardization Sector, Geneva, Switzerland. 1995b.

[12] ITU. Digital Subscriber Signaling System No. 2—Generic Functional Protocol: Core Functions. ITU-T Recommendation Q.2931.2. International Telecommunications Union, Telecommunications Standardization Sector, Geneva, Switzerland. 1996.

[13] LaMaire, R.O., Krishna, A., Bhagwat, P., and Panian, J., Wireless LANs and mobile networking: standards and future directions. *IEEE Commun. Mag.*, 34(8), 86–94, Aug. 1996.

[14] Lin, Y.B. and Devries, S.K., PCS network signalling using SS7. *IEEE Commun. Mag.*, 2(3), 44–55, Jun. 1995.

[15] Naghshineh, M. and Willebeek-LeMair, M., End-to-end QoS provisioning in multimedia wireless/mobile networks using an adaptive framework. *IEEE Commun. Mag.*, 72–81, Nov. 1997.

[16] NSTC. *Strategic Planning Document—Information and Communications,* National Science and Technology Council. 10, Mar. 1995.

[17] Pahlavan, K., Zahedi, A., and Krishnamurthy, P., Wideband local wireless: wireless LAN and wireless ATM. *IEEE Commun. Mag.*, 34–40, Nov. 1997.

[18] Proakis, J.G., *Digital Communications.* McGraw-Hill, New York. 1989.

[19] Raychaudhuri, D. and Wilson, N.D., ATM-based transport architecture for multiservices wireless personal communication networks. *IEEE J. Select. Areas Commun.*, 12(8), 1401–1414, Oct. 1994.

[20] Raychaudhuri, D., Wireless ATM networks: architecture, system design, and prototyping. *IEEE Personal Commun.*, 3(4), 42–49, Aug. 1996.

[21] Schmidt, W.R., et al., Optimization of ATM and legacy LAN for high speed satellite communications. Transport Protocols for High-Speed Broadband Networks Workshop. *GLOBECOM'96.* Nov. 1996.

[22] SR. Alternatives for Signaling Link Evolution. Bellcore Special Report. *Bellcore SR-NWT-002897.* (1), Feb. 1994.

[23] TIA. TIA/EIA Telecommunications Systems Bulletin (TSB)—91. Satellite ATM Networks: Architectures and Guidelines. Telecommunications Industry Association. TIA/EIA/TSB-91. Apr. 1998.

[24] Truong, H.L. et al., LAN emulation on an ATM network. *IEEE Commun. Mag.,* 70–85, May 1995.

[25] Umehira, M. et al., An ATM wireless system for tetherless multimedia services. *ICUPC'95.* Nov. 1995.

[26] Veeraraghavan, M., Pancha, P., and Eng, K.Y., Connectionless Transport in ATM Networks. ATM Forum Contribution. ATMF/97-0141, 9–14, Feb. 1997.

[27] Wilkinson, T. et al., A report on HIPERLAN standardization. *Intl. J. Wireless Inform. Networks.* 2(2), 99–120, 1995.

[28] Wu, T.H. and Chang, L.F., Architecture for PCS mobility management on ATM transport networks. *ICUPC'95.* 763–768, Nov. 1995.

Further Information

Information supplementing the wireless ATM standards work may be found in the ATM Forum documents relating to the Wireless ATM Working Group's activities (web page `http://www.atmforum.com`). Special issues on wireless ATM have appeared in the August 1996 issue of *IEEE Personal Communications,* the January 1997 issue of the *IEEE Journal on Selected Areas in Communications,* and the November 1997 issue of *IEEE Communications Magazine.*

Reports on proposals for higher speed wireless LAN extensions in the 2.4 GHz and 5 GHz bands can be found at the IEEE 802.11 web site (`http://grouper.ieee.org/groups/802/11/Reports`). Additional information on HIPERLAN and related activities in ETSI BRAN can be obtained from their web site (`http://www.etsi.fr/bran`).

35

Wireless ATM: QoS and Mobility Management

Bala Rajagopalan
NEC C&C Research Laboratories

Daniel Reininger
NEC C&C Research Laboratories

35.1 Introduction

Wireless ATM (WATM) refers to the technology that enables ATM end-system mobility as well as tetherless access to ATM core networks. Wireless ATM has two distinct components: the radio access technology, and enhancements to existing ATM technology to support end-system mobility. The latter component is referred to as "MATM" (mobility-enhanced ATM) and it is independent of the radio access technology. The rationale for wireless ATM has been discussed at length elsewhere [1, 2]. In this chapter, we restrict our discussion to two challenging issues in wireless ATM: the provisioning of ATM quality of service (QoS) for connections that terminate on mobile end systems over a radio link, and the protocols for mobility management in the MATM infrastructure.

Figure 35.1 illustrates the WATM reference model considered in this chapter [3]. "W" UNI in this figure indicates the ATM user-network interface established over the wireless link. "M" NNI refers to the ATM network-node interface supporting mobility management protocols. The figure depicts the scenario where an MATM network has both end system mobility-supporting ATM switches (EMAS) and traditional ATM switches with no mobility support. Thus, one of the features of mobility management protocols in an MATM network is the ability to work transparently over switches that do not implement mobility support.

35.2 QoS in Wireless ATM

The type of QoS guarantees to be provided in wireless ATM systems is debatable [4]. On the one hand, the QoS model for traditional ATM networks is based on fixed terminals and

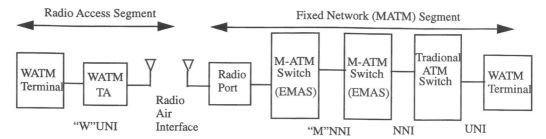

FIGURE 35.1   WATM reference model.

high quality links. Terminal mobility and error-prone wireless links introduce numerous problems [5]. On the other hand, maintaining the existing QoS model allows the transparent extension of fixed ATM applications into the domain of mobile networking. Existing prototype implementations have chosen the latter approach [6]–[8]. This is also the decision of the ATM Forum wireless ATM working group [9]. Our discussion, therefore, is oriented in the same direction, and to this end we first briefly summarize the existing ATM QoS model.

35.2.1   ATM QoS Model

Five service categories have been defined under ATM [10]. These categories are differentiated according to whether they support constant or variable rate traffic, and real-time or non-real-time constraints. The service parameters include a characterization of the traffic and a reservation specification in the form of QoS parameters. Also, traffic is policed to ensure that it conforms to the traffic characterization, and rules are specified for how to treat nonconforming traffic. ATM provides the ability to tag nonconforming cells and specify whether tagged cells are policed (and dropped) or provided with best-effort service.

Under UNI 4.0, the service categories are constant bit rate (CBR), real-time variable bit rate (rt-VBR), non-real-time variable bit rate (nrt-VBR), unspecified bit rate (UBR) and available bit rate (ABR). The definition of these services can be found in [10]. Table 35.1 summarizes the traffic descriptor parameters and QoS parameters relevant to each service category in ATM traffic management specifications version 4.0 [11]. Here, the traffic parameters are peak cell rate (PCR), cell delay variation tolerance (CDVT), sustainable cell rate (SCR), maximum burst size (MBS) and minimum cell rate (MCR). The QoS parameters are cell loss ratio (CLR), maximum cell transfer delay (max CTD) and cell delay variation (CDV). The explanation of these parameters can be found in [11].

Functions related to the implementation of QoS in ATM networks are usage parameter control (UPC) and connection admission control (CAC). In essence, the UPC function (implemented at the network edge) ensures that the traffic generated over a connection conforms to the declared traffic parameters. Excess traffic may be dropped or carried on a best-effort basis (i.e., QoS guarantees do not apply). The CAC function is implemented by each switch in an ATM network to determine whether the QoS requirements of a connection can be satisfied with the available resources. Finally, ATM connections can be either point-to-point or point-to-multipoint. In the former case, the connection is bidirectional, with separate traffic and QoS parameters for each direction, while in the latter case it is unidirectional. In this chapter, we consider only point-to-point connections for the sake of simplicity.

TABLE 35.1 ATM Traffic and QoS Parameters

	ATM Service Category				
Attribute	CBR	rt-VBR	nrt-VBR	UBR	ABR
Traffic Parameters					
PCR and CDVT	Yes	Yes	Yes	Yes	Yes
SCR and MBS	N/A	Yes	Yes	N/A	N/A
MCR	N/A	N/A	N/A	N/A	Yes
QoS Parameters					
CDV	Yes	Yes	No	No	No
Maximum CTD	Yes	Yes	No	No	No
CLR	Yes	Yes	Yes	No	No

35.2.2 QoS Approach in Wireless ATM

QoS in wireless ATM requires a combination of several mechanisms acting in concert. Figure 35.2 illustrates the various points in the system where QoS mechanisms are needed:

- *At the radio interface:* A QoS-capable medium access control (MAC) layer is required. The mechanisms here are resource reservation and allocation for ATM virtual circuits under various service categories, and scheduling to meet delay requirements. Furthermore, an error control function is needed to cope with radio link errors that can otherwise degrade the link quality. Finally, a CAC mechanism is required to limit access to the multiple access radio link in order to maintain QoS for existing connections.

- *In the network:* ATM QoS mechanisms are assumed in the network. In addition, a capability for QoS renegotiation will be useful. This allows the network or the mobile terminal (MT) to renegotiate the connection QoS when the existing connection QoS cannot be maintained during handover. Renegotiation may also be combined with *soft* QoS mechanisms, as described later. Finally, mobility management protocols must include mechanisms to maintain QoS of connections rerouted within the network during handover.

- *At the MT:* The MT implements the complementary functions related to QoS provisioning in the MAC and network layers. In addition, application layer functions may be implemented to deal with variations in the available QoS due to radio link degradation and/or terminal mobility. Similar functions may be implemented in fixed terminals communicating with MTs.

In the following, we consider the QoS mechanisms in some detail. We focus on the MAC layer and the new network layer functions such as QoS renegotiation and soft QoS. The implementation of existing ATM QoS mechanisms have been described in much detail by others (for example, see [12]).

35.2.3 MAC Layer Functions

The radio link in a wireless ATM system is typically a broadcast multiple access channel shared by a number of MTs. Different multiple access technologies are possible, for instance frequency, time, or code division multiple access (FDMA, TDMA, and CDMA, respectively). A combination of FDMA and *dynamic* TDMA is popular in wireless ATM implementations.

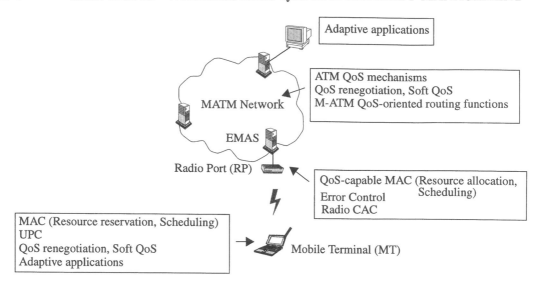

FIGURE 35.2   QoS mechanisms in wireless ATM.

That is, each radio port (RP) operates on a certain frequency band and this bandwidth is shared dynamically among ATM connections terminating on multiple MTs using a TDMA scheme [13]. ATM QoS is achieved under dynamic TDMA using a combination of a resource reservation/allocation scheme and a scheduling mechanism. This is further explained using the example of two wireless ATM implementations: NEC's WATMnet 2.0 prototype [13] and the European Union's Magic WAND (Wireless ATM Network Demonstrator) project [8].

Resource Reservation and Allocation Mechanisms (WATMnet 2.0)

WATMnet utilizes a TDMA/TDD (time division duplexing) scheme for medium access. The logical transmission frame structure under this scheme is shown in Fig. 35.3. As shown, this scheme allows the flexibility to partition the frame dynamically for downlink (from EMAS to MTs) and uplink (from MTs to EMAS) traffic, depending on the traffic load in each direction. Other notable features of this scheme are:

- A significant portion of each slot is used for forward error control (FEC)
- A separate contention region in the frame is used for MTs to communicate with the EMAS
- 8-byte control packets are used for bandwidth request and allocation announcements. An MT can tag request packets along with the WATM cells it sends or in the contention slots
- WATM cells are modified ATM cells with data link control (DLC) and cyclic redundancy check (CRC) information added

In the downlink direction, the WATM cells transported belong to various ATM connections terminating on different MTs. After such cells arrive at the EMAS from the fixed network, the allocation of TDMA slots for specific connections is done at the EMAS based on the connections' traffic and QoS parameters. This procedure is described in the next section. In the uplink direction, the allocation is based on requests from MTs. For bursty traffic, an MT makes a request only after each burst is generated. Once uplink slots are allocated to specific MTs, the transmission of cells from multiple active connections at an MT is again subject to the scheduling scheme. Both the request for uplink slot allocation

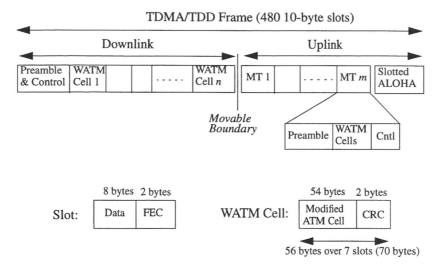

FIGURE 35.3   TDMA logical frame format.

from the MTs and the results from the EMAS are carried in control packets whose format is shown in Fig. 35.4. Here, the numbers indicate the bits allocated for various fields. The sequence number is used to recover from transmission losses. Request and allocation types indicate one of four types: CBR, VBR, ABR or UBR. The allocation packet has a start slot field which indicates where in the frame the MT should start transmission. The number of allocated slots is also indicated.

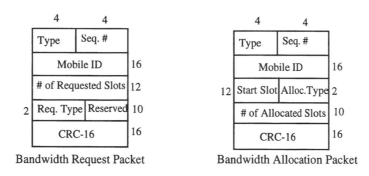

FIGURE 35.4   Bandwidth control packet formats.

The DLC layer implementation in WATMnet is used to reduce the impact of errors that cannot be corrected using the FEC information. The DLC layer is responsible for selective retransmission of cells with uncorrectable errors or lost cells. Furthermore, the DLC layer provides request/reply control interface to the ATM layer to manage the access to the wireless bandwidth, based on the instantaneous amount of traffic to be transmitted. The wireless ATM cell sent over the air interface is a modified version of the standard ATM cell with DLC and CRC information, as shown in Fig. 35.5. The same figure also shows the acknowledgment packet format for implementing selective retransmission. In this packet, the VCI field specifies the ATM connection for which the acknowledgment is being sent.

The sequence number field indicates the beginning sequence number from which the 16-bit acknowledgment bitmap indicates the cells correctly received (a "1" in the bit map indicates correct reception).

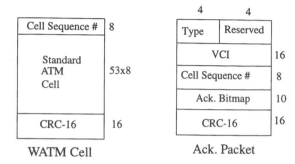

FIGURE 35.5   WATM formats at the DLC layer.

The TDMA/TDD scheme thus provides a mechanism for dynamic bandwidth allocation to multiple ATM connections. How active connections are serviced at the EMAS and the MTs to maintain their QoS needs is another matter. This is described next using Magic WAND as the example.

Scheduling (Magic WAND)

The Magic WAND system utilizes a TDMA scheme similar to that used by the WATM-net. This is shown in Fig. 35.6. Here, each MAC Protocol Data Unit (MPDU) consists of a header and a sequence of WATM cells from the same MT (or the EMAS) referred to as a *cell train*. In the Magic WAND system, the scheduling of both uplink and downlink transmissions is done at the EMAS. Furthermore, the scheduling is based on the fact that the frame length is variable. A simplified description of the scheduling scheme is presented below. More details can be found in [14, 15].

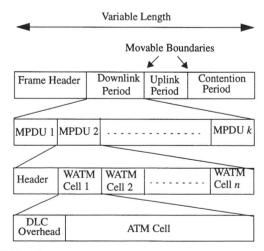

FIGURE 35.6   Magic WAND TDMA frame structure.

At the beginning of each TDMA frame, the scheduling function at the EMAS considers pending transmission requests, uplink and downlink, from active connections. The scheduler addresses two issues:

1. The determination of the number of cells to be transmitted from each connection in the frame, and
2. The transmission sequence of the selected cells within the frame

The objective of the scheduler is to regulate the traffic over the radio interface as per the declared ATM traffic parameters of various connections and to ensure that the delay constraints (if any) are met for these connections over this interface.

The selection of cells for transmission is done based on the service categories of active connections as well as their traffic characteristics. First, for each connection, a priority based on its service category is assigned. CBR connections are assigned the highest priority, followed by rt-VBR, nrt-VBR, ABR. In addition, for each active connection that is not of type UBR, a token pool is implemented. Tokens for a connection are generated at the declared SCR of the connection and tokens may be accumulated in the pool as long as their number does not exceed the declared MBS for the connection. The scheduler services active connections in two passes: in the first pass, connections are considered in priority order from CBR to ABR (UBR is omitted) and within each priority class only connections with a positive number of tokens in their pools are considered. Such connections are serviced in the decreasing order of the number of tokens in their pools. Whenever a cell belonging to a connection is selected for transmission, a token is removed from its pool. At the end of the first pass, either all the slots in the downlink portion of the frame are used up or there are still some slots available. In the latter case, the second pass is started. In this pass, the scheduler services remaining excess traffic in each of CBR, rt and nrt-VBR and ABR classes, and UBR traffic in the priority order. It is clear that in order to adequately service active connections, the mean bandwidth requirement of the connections cannot exceed the number of downlink slots available in each frame. The CAC function is used to block the setting up of new connections over a radio interface when there is a danger of overloading. Another factor that can result in overloading is the handover of connections. The CAC must have some knowledge of the expected load due to handovers so that it can limit new connection admissions. Preserving the QoS for handed over connections while not degrading existing connections at a radio interface requires good network engineering. In addition, mechanisms such as QoS renegotiation and soft QoS (Section 35.2.4) may be helpful.

Now, at the end of the selection phase, the scheduler has determined the number of cells to be transmitted from each active connection. Some of these cells are to be transmitted uplink while the others are to be transmitted downlink. The scheduler attempts to place a cell for transmission within the frame such that the cell falls in the appropriate portion of the frame (uplink or downlink, Fig. 35.6) and the delay constraint (CDT) of the corresponding connection is met. To do this, first the delay allowed over the radio segment is determined for each connection with a delay constraint (it is assumed that this value can be obtained during the connection routing phase by decomposing the path delay into delays for each hop). Then, for downlink cells, the arrival time for the cell from the fixed network is marked. For uplink cells, the arrival time is estimated from the time at which the request was received from the MT. The deadline for the transmission of a cell (uplink or downlink) is computed as the arrival time plus the delay allowed over the radio link.

The final placement of the cells in the frame is based on a three-step process, as illustrated by an example with six connections (Fig. 35.7). Here, Dn and Un indicate downlink and uplink cells with deadline = slot n, respectively, and D_j^i and U_j^i indicate downlink and uplink cells of the ith connection with deadline = slot j, respectively. In the first step, the cells are ordered based on their deadlines [Fig. 35.7(a)]. Several cells belonging to the same connection may have been selected for transmission in a frame. When assigning a slot for the first cell of such a "cell train" the scheduler positions the cell such that its transmission will be before and as close to its deadline as possible. Some cells may have to be shifted from their previously allocated slots to make room for the new allocation. This is done only if the action does not violate the deadline of any cell. When assigning a slot for another cell in the train, the scheduler attempts to place it in the slot next to the one allocated to the previous cell. This may require shifting of existing allocations as before. This is shown in Figs. 35.7(b)–35.7(d). Here, the transmission frame is assumed to begin at slot 5.

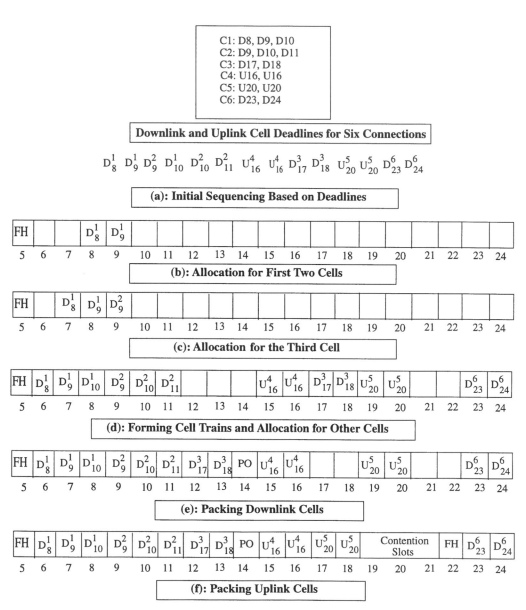

FIGURE 35.7   Scheduling example.

At the end of the first step, the cell sequencing may be such that uplink and downlink cells may be interleaved. The second step builds the downlink portion of the frame by first shifting all downlink cells occurring before the first uplink cell as close to the beginning of the frame as possible. In the space between the last such downlink cell and the first uplink cell, as many downlink cells as possible are packed. This is illustrated in Fig. 35.7(e). A slot for *period overhead* (PO) is added between the downlink and uplink portions. Finally, in the last step, the uplink cells are packed, by moving all uplink cells occurring before the next downlink cell as shown in Fig. 35.7(f). The contention slots are added after the last uplink cell and the remaining cells are left for the next frame.

Thus, scheduling can be a rather complicated function. The specific scheduling scheme used in the Magic WAND system relies on the fact that the frame length is variable. Scheduling schemes for other frame structures could be different.

35.2.4 Network and Application Layer Functions

Wireless broadband access is subject to sudden variations in bandwidth availability due to the dynamic nature of the service demand (e.g., terminals moving in and out of RPs coverage area, variable bit-rate interactive multimedia connections) and the natural constraints of the physical channel (e.g, fading and other propagation conditions). QoS control mechanisms should be able to handle efficiently both the mobility and the heterogeneous and dynamic bandwidth needs of multimedia applications. In addition, multimedia applications themselves should be able to adapt to terminal heterogeneity, computing limitations, and varying availability of network resources [16].

In this section, the network and application layer QoS control functions are examined in the context of NECs WATMnet system [6]. In this system, a concept called *soft-QoS* is used to effectively support terminal mobility, maintain acceptable application performance and high network capacity utilization. Soft-QoS relies on a QoS control framework that permits the allocation of network resources to dynamically match the varying demands of mobile multimedia applications. Network mechanisms under this framework for connection admission, QoS renegotiation, and handoff control are described. The soft-QoS concept and the realization of the soft-QoS controller based on this concept are described next. Finally, experimental results on the impact of network utilization and soft-QoS provisioning for video applications are discussed.

A Dynamic Framework for QoS Control

The bit-rates of multimedia applications vary significantly among sessions and within a session due to user interactivity and traffic characteristics. Contributing factors include the presence of heterogeneous media (e.g., video, audio, and images) compression schemes (e.g., MPEG, JPEG), presentation quality requirements (e.g., quantization, display size), and session interactivity (e.g., image scaling, VCR-like control). Consider, for example, a multimedia application using several media components or media objects (such as a multi-window multimedia user interface or future MPEG-4 encoded video) which allows users to vary the relative importance-of-presence (IoP) of a given media object to match the current viewing priorities. In this case there would be a strong dependency of user/application interaction on the bandwidth requirements of individual media components. Figure 35.8 shows the bit-rate when the user changes the video level-of-detail (LoD) during a session. A suitable network service for these applications should support bandwidth renegotiation to simultaneously achieve high network utilization and maintain acceptable performance. For this purpose, an efficient network service model should support traffic contract renegotiation during a session. It has been experimentally verified that bandwidth renegotiation is key for efficient QoS support of network-aware adaptive multimedia applications, and that the added implementation complexity is reasonable [17, 18].

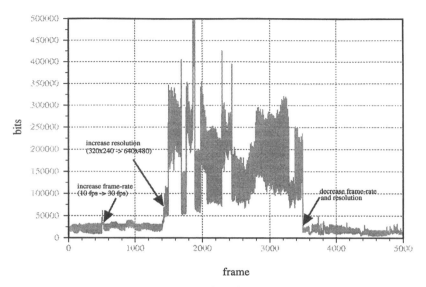

FIGURE 35.8   Video bit-rate changes on an interactive multimedia session.

In the mobile multimedia communication scenario, bandwidth renegotiation is particularly important. Conventional network services use static bandwidth allocation models that lack the flexibility needed to cope with multimedia interactivity and session mobility. These session properties enlarge the dynamic range of bandwidth requirements and make dynamic bandwidth management protocols a requirement for effective end-to-end QoS support. Renegotiation may be required during handover, as well as when resource allocation changes are warranted due to instantaneous application needs and sudden changes in network resource availability.

Figure 35.9 shows the system and API model for QoS control with bandwidth renegotiation. The application programming interface (API) between the adaptive application and the QoS control module is dynamic, i.e., its parameters can be modified during the session. For example, the Winsock 2 API under Microsoft Windows [19] allows the dynamic specification of QoS parameters suitable for the application. In addition, the API between the QoS controller and the network allows the traffic descriptor to be varied to track the bit-rate requirements of the bitstream. A new network service, called VBR$^+$ allows renegotiation of traffic descriptors between the network elements and the terminals [20]. VBR$^+$ allows multimedia applications to request "bandwidth-on-demand" suitable for their needs.

Soft-QoS Model

Although multimedia applications have a wide range of bandwidth requirements, most can gracefully adapt to sporadic network congestion while still providing acceptable performance. This graceful adaptation can be quantified by a softness profile [17]. Figure 35.10 shows the characteristics of a softness profile. The softness profile is a function defined on the scales of two parameters: satisfaction index and bandwidth ratio. The satisfaction index is based on the subjective mean-opinion-score (MOS), graded from 1 to 5; a minimum

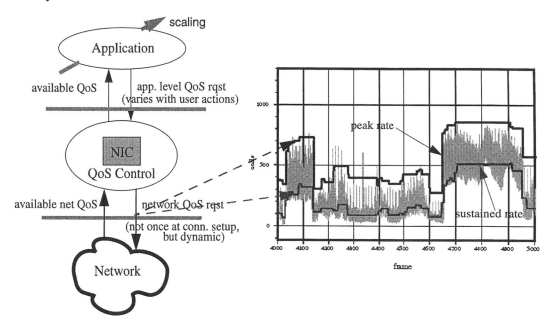

FIGURE 35.9   System and API model for QoS control with bandwidth renegotiation.

satisfaction divides the scale in two operational regions: the acceptable satisfaction region and the low satisfaction region. The bandwidth ratio is defined by dividing the current bandwidth allocated by the network to the bandwidth requested to maintain the desired application performance. Thus, the bandwidth ratio is graded from 0 to 1; a value of 1 means that the allocated bandwidth is sufficient to achieve the desired application performance. The point indicated as B is called the critical bandwidth ratio since it is the value that results in minimum acceptable satisfaction. As shown in Fig. 35.10, the softness profile is approximated by piecewise linear "S-shaped" function consisting of three linear segments. The slope of each linear segment represents the rate at which applications performance degrades (satisfaction index decreases) when the network allocates only a portion of the requested bandwidth: the steeper the slope is, the "harder" the corresponding profile is.

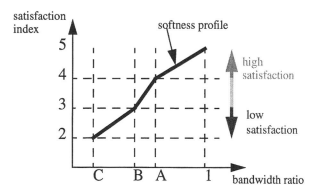

FIGURE 35.10   Example softness profile.

The softness profile allows efficient match of application requirements to network resource availability. With the knowledge of the softness profile, network elements can perform soft-QoS control—QoS-fair allocation of resources among contending applications when congestion arises. Applications can define a softness profile that best represents their needs. For example, the softness profile for digital compressed video is based on the nonlinear relationship between coding bit-rate and quality, and the satisfaction index is correlated to the user perception of quality [21, 22]. While video-on-demand (VoD) applications may, in general, tolerate bit-rate regulations within a small dynamic range, applications such as surveillance or teleconference may have a larger dynamic range for bit-rate control. Other multimedia applications may allow a larger range of bit-rate control by resolution scaling [18]. In these examples, VoD applications are matched to a "harder" profile than the other, more adaptive multimedia applications. Users on wireless mobile terminals may select a "softer" profile for an application in order to reduce the connection's cost, while a "harder" profile may be selected when the application is used on wired desktop terminal. Thus, adaptive multimedia applications able to scale their video quality could specify their soft-QoS requirements dynamically to control the session's cost.

Figure 35.11 conceptually illustrates the role of application QoS/bandwidth renegotiation, service contract, and session cost in the service model. The soft-QoS service model is suitable for adaptive multimedia applications capable of gracefully adjusting their performance to variable network conditions. The service definition is needed to match the requirements of the application with the capabilities of the network. The service definition consists of two parts: a usage profile that specifies the target regime of operation and the service contract that statistically quantifies the soft-QoS service to be provided by the network. The usage profile, for example, can describe the media type (e.g., MPEG video), interactivity model (e.g., multimedia browsing, video conference), mobility model (indoors, urban semi-mobile, metropolitan coverage area), traffic, and softness profiles. The service contract quantifies soft-QoS in terms of the probability that the satisfaction of a connection will fall outside the acceptable range (given in its softness profile), the expected duration of "satisfaction outage," and the new connection blocking probability.

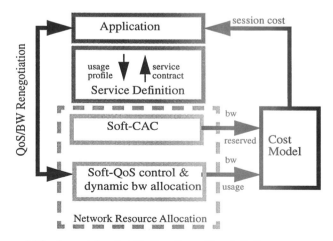

FIGURE 35.11   Service model for dynamic bandwidth allocation with soft-QoS.

Network resource allocation is done in two phases. First, a connection admission control procedure, called soft-CAC, checks the availability of resources on the terminals coverage area at connection set-up time. The necessary resources are estimated based on the service definition. The new connection is accepted if sufficient resources are estimated to be available for the connection to operate within the service contract without affecting the service of other ongoing connections. Otherwise the connection is blocked. Second, while the connection is in progress, dynamic bandwidth allocation is performed to match the requirements of interactive variable bit-rate traffic. When congestion occurs, the soft-QoS control mechanism (re)-allocates bandwidth among connections to maintain the service of all ongoing connections within their service contracts. The resulting allocation improves the satisfaction of undersatisfied connections while maintaining the overall satisfaction of other connections as high as possible [17]. Under this model, connections compete for bandwidth in a "socially responsible" manner based on their softness profiles. Clearly, if a cost model is not in place, users would request the maximum QoS possible. The cost model provides feedback on session cost to the applications; the user can adjust the long-term QoS requirements to maintain the session cost within budget.

Soft-QoS Control in the WATMnet System

In the WATMnet system, soft-QoS control allows effective support of mobile multimedia applications with high network capacity utilization. When congestion occurs, the soft-QoS controller at the EMASs allocates bandwidth to connections based on their relative robustness to congestion given by the applications softness profiles. This allocation improves the satisfaction of undersatisfied connections while maintaining the overall satisfaction of other connections as high as possible. Within each EMAS, connections compete for bandwidth in a "socially responsible" manner based on their softness profiles.

ATM UNI signalling extensions are used in the WATMnet system to support dynamic bandwidth management. These extensions follow ITU-T recommendations for ATM traffic parameter modification while the connection is active [23]. Although these procedures are not finalized at the time of this writing, an overview of the current state of the recommendation is given next with an emphasis on its use to support soft-QoS in the mobile WATM scenario.

ITU-T Q.2963 allows all three ATM traffic parameters, PCR, SCR, and MBS, to be modified during a call. All traffic parameters must be increased or decreased; it is not possible to increase a subset of the parameters while decreasing the others. The user who initiates the modification request expects to receive from the network a new set of traffic parameters that are greater than or equal to (or less than) the existing traffic parameters if the modification request is an increase (or decrease). Traffic parameter modification is applicable only to point-to-point connections and may be requested only by the terminal that initiated the connection while in the active state.

The following messages are added to the UNI:

- MODIFY REQUEST message is sent by the connection owner to request modification of the traffic descriptor; its information element (IE) is the ATM traffic descriptor.

- MODIFY ACKNOWLEDGE message is sent by the called user or network to indicate that the modify request is accepted. The broadband report type IE is included in the message when the called user requires confirmation of the success of modification.

- CONNECTION AVAILABLE is an optional message issued by the connection owner to confirm the connection modification performed in the addressed user to requesting user direction. The need for explicit confirmation of modification is

indicated by the "modification confirmation" field in the MODIFY ACKNOWL-
EDGE broadband report IE.

- MODIFY REJECT message is sent by the called user or network to indicate that
  the modify connection request is rejected. The cause of the rejection is informed
  through the cause IE.

Figures 35.12, 35.13, and 35.14 show the use of these messages for successful, addressed
user rejection and network rejection of modification request, respectively.

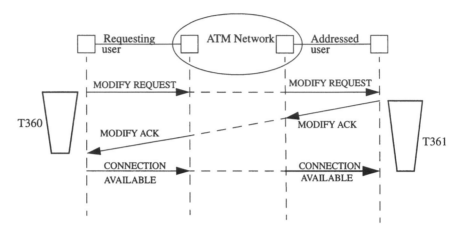

FIGURE 35.12   Successful Q2963 modification of ATM traffic parameters with (optional) confirmation.

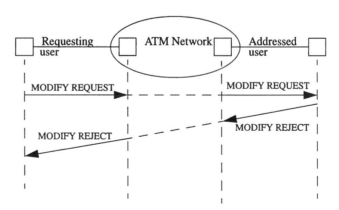

FIGURE 35.13   Addressed user rejection of modification.

Additionally, the soft-QoS control framework of the WATMnet system uses the following
modifications to the Q2963 signalling mechanisms:

- BANDWIDTH CHANGE INDICATION (BCI) message supports network-initiated
  and called user-initiated modification. The message is issued by the network or
  called user to initiate a modification procedure. The traffic descriptor to be used
  by the connection owner when issuing the corresponding MODIFY REQUEST
  message is specified in the BCIs ATM traffic descriptor IE. Figure 35.15 illus-

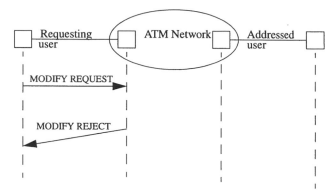

FIGURE 35.14 Network rejection of modification request.

trates the use of BCI for called user-initiated modification. Timer T362 is set when issuing the BCI message and cleared when the corresponding MODIFY REQUEST message is received; if T362 expires, the terminal and/or network element can modify the traffic policers to use the ATM traffic descriptor issued in the BCI message.

- Specification of softness profile and associated minimum acceptable satisfaction level (sat_{min}) in the MODIFY REQUEST message. The softness profile and sat_{min} are used for QoS-fair allocation within the soft-QoS control algorithm.

- Specification of available bandwidth fraction (ABF) for each ATM traffic descriptor parameter. ABF is defined as the ratio of the available to requested traffic descriptor parameter. This results in ABF-PCR, ABF-SCR, and ABF-MBS for the peak, sustained, and maximum burst size, respectively. These parameters are included in the MODIFY REJECT message. Using the ABF information, the connection owner may recompute the requested ATM traffic descriptor and reissue an appropriate MODIFY REQUEST message.

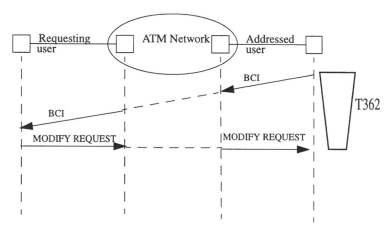

FIGURE 35.15 Use of BCI message for addressed user-initiated modification.

Two additional call states are defined to support modification. An entity enters the modify request state when it issues a MODIFY REQUEST of BCI message to the other side of the interface. An entity enters the modify received state when it receives a MODIFY REQUEST of BCI message from the other side of the interface.

Soft-QoS control is particularly useful during the handover procedure as a new MT moves into a cell and places demands on resources presently allocated to connections from other MTs. In the present WATM baseline handover specification [9], an MT sends a prioritized list of VCs to the network during handover, but there is no specification as to what the MT or the network should do if not all these VCs can be accommodated at the new cell. A flexible way of prioritizing the bandwidth allocation to various session VCs is through their softness profiles. If a mobile terminal faces significant drop in bandwidth availability as it moves from one cell to another, rather than dropping the handover connections, the EMAS might be able to reallocate bandwidth among selected active connections in the new cell.

Within the soft-QoS framework, the soft-QoS controller selects a set of connections, called donors, and changes their bandwidth reservation so as to ensure satisfactory service for all [17]. This process is called network-initiated renegotiation. Network-initiated renegotiation improves the session handover success probability since multiple connections within and among sessions can share the available resources at the new EMAS, maintaining the satisfaction of individual connections above the minimum required. This mechanism allows multimedia sessions to transparently migrate the relative priority of connections as the MT moves across cells without a need to further specify details of the media session's content.

Figure 35.16 shows an MT moving into the coverage area of a new RP under a new EMAS and issuing a MODIFY REQUEST message (MRE). As a result, the EMAS might have

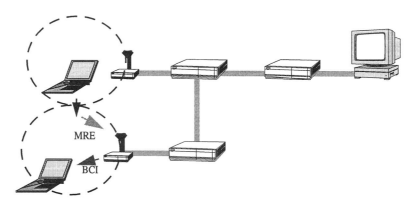

FIGURE 35.16   Handover procedure with soft-QoS in the WATMnet system.

to reallocate bandwidth of other connections under the RP to successfully complete the handover. This is accomplished by issuing BCI messages to a selected set of connections, called donors. At the time of receiving the first MRE message for a connection being handed over, no state exists for that connection within the new EMAS. This event differentiates the MRE messages from ongoing connections and connections being handed over. Different algorithmic provisions can be made to expedite bandwidth allocation to MRE messages of connections being handed over, reducing the probability of handover drop. For example, upon identifying a MRE from such connection, the soft QoS controller can use cached bandwidth reserves to maintain the satisfaction of the connection above the minimum. The size of the bandwidth cache is made adaptive to the ratio of handed-over to local bandwidth demand. The bandwidth cache for each RP can be replenished off-line using the network-initiated modification procedure. In this way, a handed-over connection need not wait for the network-initiated modification procedure to end before being able to use the bandwidth. The outcome of the reallocation enables most connections to sustain a better than minimum application performance while resources become available. Short-term congestion may occur due to statistical multiplexing. If long-term congestion arises due to the creation of a hot spot, dynamic channel allocation (DCA) may be used to provide additional resources. It is also possible that if connection rerouting is required inside the network for handover, the required resources to support the original QoS request may not be available within the network along the new path. Renegotiation is a useful feature in this case also.

An important performance metric for the soft-QoS service model is the low satisfaction rate (LSR). LSR measures the probability of failing to obtain link capacity necessary to maintain acceptable application performance. Figure 35.17 compares the LSR with and without network-initiated modification over a wide range of link utilization for an MPEG-based interactive video application. The softness profiles for MPEG video were derived from

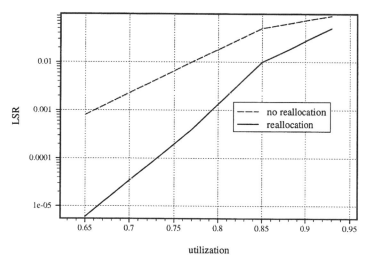

FIGURE 35.17 Effect of soft-QoS control with and without network-initiated modification.

empirical results reported in [21, 22]. The figure shows that network-initiated modification has an important contribution to soft-QoS control performance: robust operation (LSR $<$ 10^{-3}) is achievable while maintaining 70 to 80% utilization. In the WATM scenario, the handoff success probability with soft-QoS control is related to the LSR by Prob(handoff success) $=$ P(sat$^+$ $>$ sat$_{min}$) $>$ $1 -$ LSR $>>$ $1 -$ P$_b$, where sat$^+$ represents the satisfaction after handover to the new AP completes. Since it is better to block a new connection than to drop an existing connection for lack of capacity, the condition LSR $\ll$ P$_b$ is used, where P$_b$ is the connection blocking probability. The operating goal for the system is to maintain utilization $>$ 70%, LSR $\sim$ 10^{-5}, P$_b$ $\sim$ 10^{-3}.

Although the results presented are based on softness profiles for video, the definition of soft-QoS is appropriate for adaptive multimedia applications in general. Various representative softness profiles can be defined and refined as users' experience with distributed multimedia applications grows. New profiles can easily be incorporated within the framework as they become available.

35.3   Mobility Management in Wireless ATM

Allowing end system mobility in ATM networks gives rise to the problem of mobility management, i.e., maintaining service to end systems regardless of their location or movement. A fundamental design choice here is whether mobility management deals with user mobility or terminal mobility. When a network supports user mobility, it recognizes the user as the subscriber with an associated service profile. The user can then utilize any MT for access to the subscribed services. This results in flexibility in service provisioning and usage, but some extra complexity is introduced in the system implementation, as exemplified by the GSM system [24]. Support for user mobility implies support for terminal mobility. A network may support only terminal mobility and not recognize the user of the terminal, resulting in a simpler implementation. In either case, the mobility management tasks include:

- *Location Management:* Keeping track of the current location of an MT in order to permit correspondent systems to set up connections to it. A key requirement here is that the correspondent systems need not be aware of the mobility or the current location of the MT.

- *Connection Handover:* Maintaining active connections to an MT as it moves between different points of attachment in the network. The handover function requires protocols at both the radio layer and at the network layer. The issue of preserving QoS during handovers was described earlier and this introduces some complexity in handover implementations.

- *Security Management:* Authentication of mobile users (or terminals) and establishing cryptographic procedures for secure communications based on the user (or terminal) profile [25].

- *Service Management:* Maintaining service features as a user (or terminal) roams among networks managed by different administrative entities. Security and service management can be incorporated as part of location management procedures [24].

Early wireless ATM implementations have considered only terminal mobility [6, 7]. This has been to focus the initial effort on addressing the core technical problems of mobility management, i.e., location management and handover [26]. Flexible service management in wide-area settings, in the flavor of GSM, has not been an initial concern in these systems. The mobility management protocol standards being developed by the ATM Forum WATM working group may include support for user mobility. But the details are yet to be specified. In the following, therefore, we concentrate on the location management and handover functions required to support terminal mobility in wireless ATM. Our description follows along the lines of the ongoing ATM Forum specifications [9] with examples from wireless ATM implementations.

35.3.1   Location Management in Wireless ATM

Location management (LM) in WATM networks is based on the notions of permanent and temporary ATM addresses. A permanent ATM address is a location-invariant, unique address assigned to each MT. As the MT attaches to different points in a WATM network, it may be assigned different temporary ATM addresses. As all ATM end system addresses, both permanent and temporary addresses are derived from the addresses of switches in the network, in this case EMASs. This allows connection set-up messages to be routed towards the MT, as described later. The EMAS whose address is used to derive the permanent address of an MT is referred to as the home EMAS of that MT. The LM function in essence keeps track of the current temporary address corresponding to the permanent address of each MT. Using this function, it becomes possible for correspondent systems to establish connections to an MT using only its permanent address and without knowledge of its location.

35.3.2 Network Entities Involved in LM

The LM functions are distributed across four entities:

- *The Location Server (LS):* This is a logical entity maintaining the database of associations between the permanent and temporary addresses of mobile terminals. The LS responds to query and update requests from EMASs to retrieve and modify database entries. The LS may also keep track of service-specific information for each MT.

- *The AUthentication Server (AUS):* This is a logical entity maintaining a secure database of authentication and privacy related information for each MT. The authentication protocol may be implemented between EMASs and the AUS, or directly between MTs and the AUS.

- *The Mobile Terminal:* The MT is required to execute certain functions to initiate location updates and participate in authentication and privacy protocols.

- *The EMAS:* Certain EMASs are required to identify connection set-up messages destined to MTs and invoke location resolution functions. These can be home EMASs or certain intermediate EMASs in the connection path. All EMASs in direct contact with MTs (via their RPs) may be required to execute location update functions. Home EMASs require the ability to redirect a connection set-up message. In addition, all EMASs may be required to participate in the redirection of a connection set-up message to the current location of an MT.

There could be multiple LSs and AUSs in a WATM network. Specifically, an LS and an AUS may be incorporated with each home EMAS, containing information on all the MTs that the EMAS is home to. On the other hand, an LS or an AUS may be shared between several EMASs, by virtue of being separate entities. These choices are illustrated in Fig. 35.18, where the terms "integrated" and "modular" are used to indicate built-in and separated LS and AUS. In either case, protocols must be implemented for reliably querying and updating the LS, and mechanisms to maintain the integrity and security of the AUS. NEC's WATMnet [6] and BAHAMA [7] are examples of systems implementing integrated servers. The modular approach is illustrated by GSM [24] and next-generation wireless network proposals [27].

35.3.3 Location Management Functions and Control Flow

At a high level, the LM functions are registration, location update, connection routing to home or gateway EMAS, location query, and connection redirect.

Registration and Location Update

When an MT connects to a WATM network, a number of resources must be instantiated for that mobile. This instantiation is handled by two radio layer functions: association, which establishes a channel for the MT to communicate with the edge EMAS, and registration, which binds the permanent address of the MT to a temporary address. In addition,

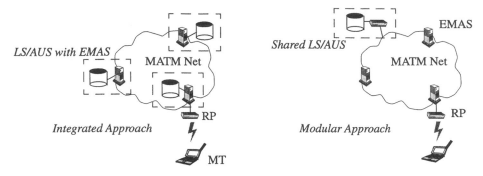

FIGURE 35.18   Server organizations.

the routing information pertaining to the mobile at one or more location servers must be updated whenever a new temporary address is assigned. This is done using location updates.

The authentication of a mobile terminal and the establishment of encryption parameters for further communication can be done during the location updating procedure. This is illustrated in Fig. 35.19 which shows one possible control flow when an MT changes location from one EMAS to another. Here, the Broadcast_ID indicates the identity of the network, the location area, and the current radio port. Based on this information, e.g., by comparing its access rights and the network ID, the MT can decide to access the network. After an association phase, which includes the setting up of the signalling channel to the EMAS, the MT sends a registration message to the switch. This message includes the MT's home address and authentication information. The location update is initiated by the visited EMAS and the further progression is as shown. The LS/AUS are shown logically separate from the home EMAS for generality. They can be integrated with the home EMAS. Details on the implementation of a similar location update scheme can be found in [28].

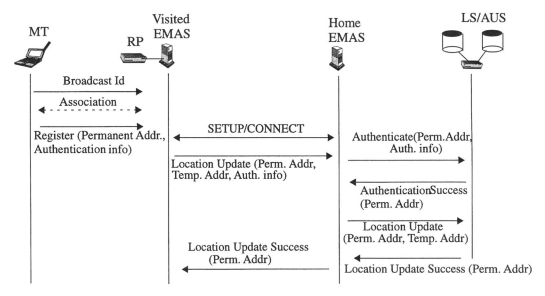

FIGURE 35.19   Location update control flow.

Now, there are other possible configurations of LSs that give rise to different control message flow. For example, a two-level hierarchical LS arrangement can be used. Under this organization, the LS in the visiting network is updated as long as the MT remains in this network, and the home LS is updated only when the MT moves to a different WATM network. The information kept in the home LS must, therefore, point to a gateway EMAS in the visited network, since precise location in the visited network will not be available at the home LS. GSM location management is an example of this scheme [24].

Connection Forwarding, Location Query, and Connection Redirect

After a location update, a location server handling the MT has the correct association between its permanent and temporary ATM addresses. When a new connection to the MT is established, the set-up message must be routed to some EMAS that can query the LS to determine the current address of the MT. This is the connection forwarding function. Depending on how MT addresses are assigned, the location query can occur very close to the origination of the connection or it must progress to the home EMAS of the MT. For instance, if MT addresses are assigned from a separately reserved ATM address space within a network, a gateway EMAS in the network can invoke location query when it processes a set-up message with a destination address known to be an MT address. To reach some EMAS that can interpret an MT address, it is sufficient to always forward connection set-up messages towards the home EMAS. This ensures that at least the home EMAS can invoke the query if no other EMAS enroute can do this. The location query is simply a reliable control message exchange between an EMAS and an LS. If the LS is integrated with the EMAS, this is a trivial operation. Otherwise, it requires a protocol to execute this transaction.

The control flow for connection establishment when the MT is visiting a foreign network is shown in Fig. 35.20. The addresses of various entities shown have been simplified for illustration purposes. Here, a fixed ATM terminal (A.1.1.0) issues a SETUP towards the MT whose permanent address is C.2.1.1. The SETUP message is routed towards the home EMAS whose address is C.2.1. It is assumed that no other EMAS in the path to the home EMAS can detect MT addresses. Thus, the message reaches the home EMAS which determines that the end system whose address is C.2.1.1 is an MT. It then invokes a location query to the LS which returns the temporary address for the MT (B.3.3). The home EMAS issues a redirected SETUP towards the temporary address. In this message, the MT is identified by its permanent address thereby enabling the visited EMAS to identify the MT and proceed with the connection SETUP signalling.

It should be noted that in the topology shown in Fig. 35.20, the redirection of the connection set-up does not result in a nonoptimal path. But, in general, this may not be the case. To improve the overall end-to-end path, redirection can be done with partial tear-down in which case a part of the established path is released and the connection set-up is redirected from an EMAS that occurs further upstream of the home EMAS. This is shown in Fig. 35.21. Here, the EMAS labelled COS (Cross Over Switch) occurs in the original connection path upstream of the home EMAS. To redirect the set-up to B.3.3, the connection already established to the home EMAS is torn down up to COS, and new segment is established from the COS to B.3.3. This requires additional signalling procedures.

Finally, in Fig. 35.22, the case of hierarchical location servers is illustrated. Here, as long as the MT is in the visited network, the address of the gateway EMAS (B.1.1) is registered in its home LS. The connection set-up is sent via the home EMAS to the gateway EMAS. The gateway then queries its local LS to obtain the exact location (B.3.3) of the MT. It is assumed that the gateway can distinguish the MT address (C.2.1.1) in a SETUP message

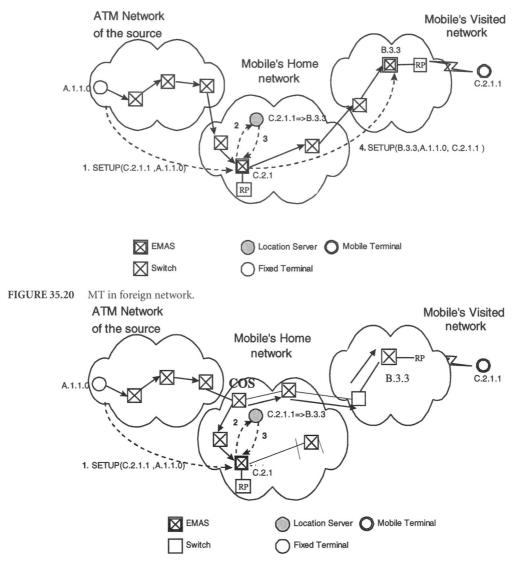

FIGURE 35.20   MT in foreign network.

FIGURE 35.21   Connection redirect.

from the fact that this address has a different network prefix (C) than the one used in the visited network (B).

Signalling and Control Messages for LM

The signalling and control messages required can be derived from the scenarios above. Specifically, interactions between the EMAS and the LS require control message exchange over a VC established for this purpose. This is described in [9]. The ATM signalling support needed for connection set-up and redirection is described in [28].

35.3.4   Connection Handover in Wireless ATM

Wireless ATM implementations, as well as the standards being developed by the ATM Forum, rely on mobile-initiated handovers whereby the MT is responsible for monitoring

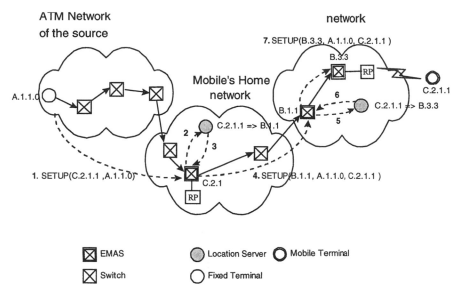

FIGURE 35.22 Hierarchical LS configuration.

the radio link quality and decides when to initiate a handover [9, 26]. A handover process typically involves the following steps:

1. *Link quality monitoring:* When there are active connections, the MT constantly monitors the strength of the signal it receives from each RP within range.

2. *Handover trigger:* At a given instance, all the connections from/to the MT are routed through the same RP, but deterioration in the quality of the link to this RP triggers the handover procedure.

3. *Handover initiation:* Once a handover is triggered, the MT initiates the procedure by sending a signal to the edge EMAS with which it is in direct contact. This signal indicates to the EMAS the list of candidate RPs to which active connections can be handed over.

4. *Target RP selection:* The edge EMAS selects one RP as the handover target from the list of candidates sent by the MT. This step may make use of network-specific criteria for spreading the traffic load among various RPs and interaction between the edge EMAS and other EMASs housing the candidate RPs.

5. *Connection rerouting:* Once the target RP is selected, the edge EMAS initiates the rerouting of all connections from/to the MT within the MATM network to the target RP. The complexity of this step depends on the specific procedures chosen for rerouting connections, as described next. Due to constraints on the network or radio resources, it is possible that not all connections are successfully rerouted at the end of this step.

6. *Handover completion:* The MT is notified of the completion of handover for one or more active connections. The MT may then associate with the new RP and begin sending/receiving data over the connections successfully handed over.

Specific implementations may differ in the precise sequence of events during handover. Furthermore, the handover complexity and capabilities may be different. For instance, some systems may implement lossless handover whereby cell loss and missequencing of cells are avoided during handover by buffering cells inside the network [29]. The handover control flow is described in detail below for two types of handovers:

- *Backward handover:* The MT initiates handover through the current RP it is connected to. This is the normal scenario.
- *Forward handover:* The MT loses connectivity to the current RP due to a sudden degeneration of the radio link. It then chooses a new RP and initiates the handover of active connections.

Our description is based on the handover model being considered for standardization by the ATM Forum [9]. This model presently allows only hard handovers, i.e., active connections are routed via exactly one RP at a given instance, as opposed to soft handovers in which the MT can receive data for active connections simultaneously from more than one RP during handover.

Backward Handover Control Flow

Figure 35.23 depicts the control sequence for backward handover when the handover involves two different EMASs. Here, "old" and "new" EMAS refer to the current EMAS and the target EMAS, respectively. The figure does not show handover steps (1) and (2), which are radio layer functions, but starts with step (3). The following actions take place:

1. The MT initiates handover by sending an *HO_REQUEST* message to the old EMAS. With this message, the MT identifies a set of candidate RPs. Upon receiving the message, the old EMAS identifies a set of candidate EMASs that house the indicated RPs. It then sends an *HO_REQUEST_QUERY* to each candidate EMAS, identifying the candidate RP as well as the set of connections (including the traffic and QoS parameters) to be handed over. The connection identifiers are assumed to be unique within the network [28].

2. After receiving the *HO_REQUEST_QUERY* message, a candidate EMAS checks the radio resources available on all the candidate RPs it houses and selects the one that can accommodate the most number of connections listed. It then sends an *HO_REQUEST_RESPONSE* message to the old EMAS identifying the target RP chosen and a set of connections that can be accommodated (this may be a subset of connections indicated in the *QUERY* message).

3. After receiving an *HO_REQUEST_RESPONSE* message from all candidate EMASs, the old EMAS selects one target RP, based on some local criteria (e.g., traffic load spreading). It then sends an *HO_RESPONSE* message to the MT, indicating the target RP. At the same time, it also sends an *HO_COMMAND* to the new EMAS. This message identifies the target RP and all the connections to be handed over along with their ATM traffic and QoS parameters. This message may also indicate the connection rerouting method. Rerouting involves first the selection of a cross-over switch (COS) which is an EMAS in the existing connection path. A new connection segment is created from the new EMAS to the COS and the existing segment from the COS to the old EMAS is deleted. Some COS selection options are:

   VC Extension: The old EMAS-E itself serves as the COS.

   Anchor-Based Rerouting: The COS is determined apriori (e.g., a designated EMAS in the network) or during connection set-up (e.g., the EMAS that first served the MT when the connection was set up). The selected COS is used for all handovers during the lifetime of the connection.

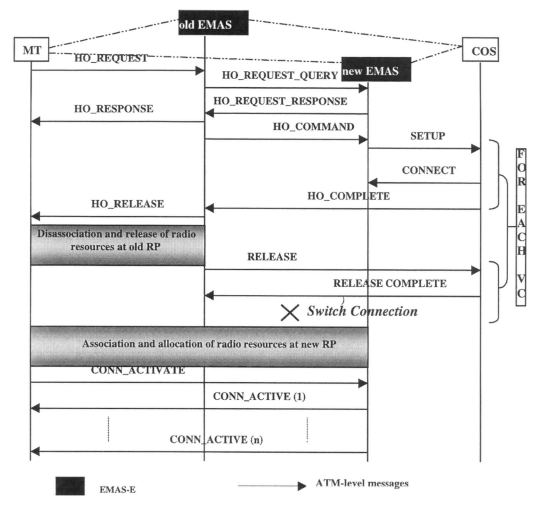

FIGURE 35.23 Backward handover control flow.

Dynamic COS Discovery: The COS is dynamically determined during each handover.

These procedures are illustrated in Fig. 35.24. While anchor-based rerouting and VC extension result in the same COS being used for all the connections being handed over, dynamic COS selection may result in different COSs for different connections. The *HO_COMMAND* message indicates which COS selection method is used, and if VC extension or anchor-based rerouting is used, it also includes the identity of the COS. If dynamic COS selection is used, the message includes the identity of the first EMAS in the connection path from the source (this information is collected during connection set-up [28]). For illustrative purposes, we assume that the dynamic COS selection procedure is used.

4. Upon receiving the *HO_COMMAND* message, the new EMAS allocates radio resources in the target RP for as many connections in the list as possible. It then sends a *SETUP* message towards the COS of each such connection. An EMAS in the existing connection path that first processes this message becomes the actual COS. This is illustrated in Fig. 35.25. This action, if successful, establishes a

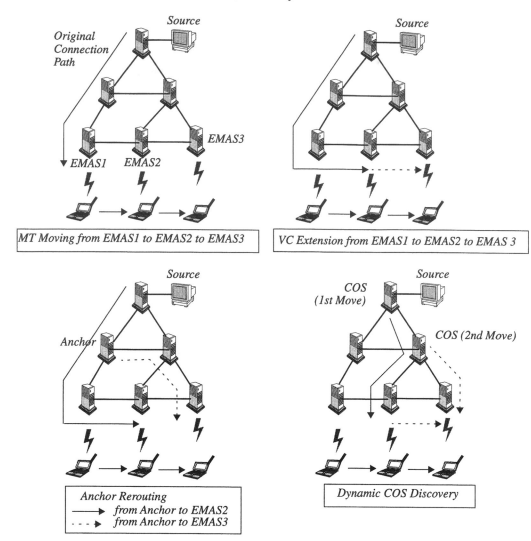

FIGURE 35.24   COS discovery.

new segment for each connection from the new EMAS to the COS.

5. If the set-up attempt is not successful, new EMAS sends an *HO_FAILURE* message to the old EMAS, after releasing all the local resources reserved in step (4). This message identifies the connection in question and it is forwarded by the old EMAS to the MT. What the MT does in response to an *HO_FAILURE* message is not part of the backward handover specification.

6. If the COS successfully receives the *SETUP* message, it sends a *CONNECT* message in reply, thereby completing the partial connection establishment procedure. It then sends an *HO_COMPLETE* message to old EMAS-E. The *HO_COMPLETE* message is necessary to deal with the situation when handover is simultaneously initiated by both ends of the connection when two MTs communicate (for the sake of simplicity, we omit further description of this situation, but the reader may refer to [30] for further details).

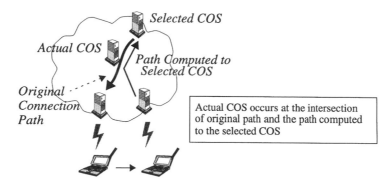

FIGURE 35.25 COS selection when MT moves.

7. The old EMAS-E waits to receive *HO_COMPLETE* messages for all the connections being handed over. However, the waiting period is limited by the expiry of a timer. Upon receiving the *HO_COMPLETE* message for the last connection, or if the timer expires, the old EMAS sends an *HO_RELEASE* message to MT. Waiting for the *HO_RELEASE* message allows the MT to utilize the existing connection segment as long as possible threreby minimizing data loss. However, if the radio link deteriorates rapidly, the MT can switch over to the new RP without receiving the *HO_RELEASE* message.

8. The old EMAS initiates the release of each connection for which an *HO_COMPLETE* was received by sending a *RELEASE* message to the corresponding COS.

9. Upon receiving the *RELEASE* message, the COS sends a *RELEASE COMPLETE* to the previous switch in the path (as per regular ATM signalling) and switches the data flow from the old to the new connection segment.

10. Meanwhile, after receiving the *HO_RELEASE* message from the old EMAS or after link deterioration, the MT dissociates from the old RP and associates with the new RP. This action triggers the assignment of radio resources for the signalling channel and user data connections for which resources were reserved in step (4).

11. Finally, the MT communicates to the new EMAS its readiness to send and receive data on all connections that have been handed over by sending a *CONN_ACTIVATE* message.

12. Upon receiving the *CONN_ACTIVATE* message from the MT, new EMAS responds with a *CONN_ACTIVE* message. This message contains the identity of the connections that have been handed over, including their new ATM VC identifiers. Multiple *CONN_ACTIVE* messages may be generated, if all the connections have not been handed over when the *CONN_ACTIVATE* message was received. However, handover of remaining connections and the subsequent generation of *CONN_ACTIVE* signals are timer-bound: if the MT does not receive information about a connection in a *CONN_ACTIVE* message before the corresponding timer expires, it assumes that the connection was not successfully handed over. The recovery in this case is left up to the MT.

The description above has left open some questions. Among them: what mechanisms are used to reliably exchange control messages between the various entities that take part in handover? What actions are taken when network or radio link failures occur during handover? How can lossless handover be included in the control flow? What effect do

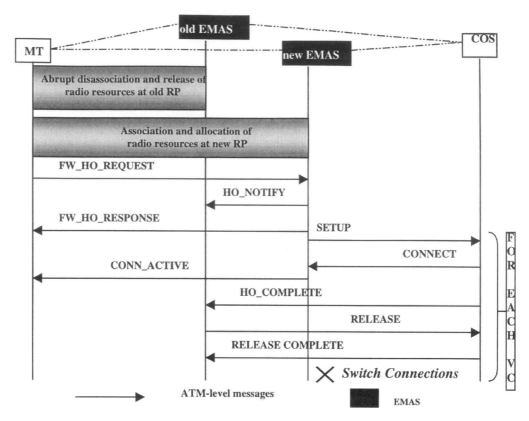

FIGURE 35.26   Forward handover control flow.

transient disruptions in service during handover have on application behavior?, and what are the performance impacts of signalling for handover? The short answers to these questions are: reliability can be incorporated by implementing a reliable transfer protocol for those control messages that do not already use such a transport (*SETUP, RELEASE,* etc., do, but *HO_REQUEST_QUERY* for example, requires attention). Actions taken during network failures require further analysis, but forward handover can be used to recover from radio link failures during handover. Lossless handover requires inband signalling within each connection and buffering in the network. Details on this can be found in [29]. The effect of transient disruptions on applications can be minimal, depending on how rerouting is implemented during handover. This is described in detail in [31]. Finally, some of the performance issues related to mobility management are investigated in [32].

Forward Handover

Forward handover is considered as the measure of last resort, to be invoked when the current radio link deteriorates suddenly. The forward handover procedure is simpler than backward handover, as illustrated in Fig. 35.26. Here,

1. The radio link with the old EMAS degrades abruptly. This results in the MT being dissociated with the old RP. The MT then chooses a new RP and associates

with it. In the example shown, the new RP is housed by a different EMAS (the new EMAS).

2. The MT initiates forward handover by sending a *FW_HO_REQUEST* message to the new EMAS. This message indicates the active connections by their network-wide unique identifiers, along with their ATM traffic and QoS parameters, the identity of the previous EMAS (the "old" EMAS), and the COS information (this information may be obtained when the connection is initially set up).

3. The new EMAS sends an *HO_NOTIFY* message to the old EMAS indicating the initiation of handover. This serves to keep the old EMAS from prematurely releasing the existing connections.

4. The new EMAS reserves radio resources for as many listed connections as possible on the radio port to which the MT is associated. It then sends a *FW_HO_RESPONSE* message to the MT identifying the connections that can be handed over.

5. For each such connection, the new EMAS generates a SETUP message towards the COS to establish the new connection segment. This message includes the identity of the new EMAS. An EMAS in the existing connection path that first processes this message becomes the COS (Fig. 35.25).

6. Upon receiving the *SETUP* message, the COS completes the establishment of the new connection segment by sending a *CONNECT* message to the new EMAS.

7. After receiving the *CONNECT* message, new EMAS sends a *CONN_ACTIVE* message to the MT, indicating the connection has become active. Reception of *CONN_ACTIVE* by the MT is subject to a timer expiry: if it does not receive information about a connection in any *CONN_ACTIVE* message before the corresponding timer expires, it may initiate any locally defined recovery procedure.

8. If the new connection segment cannot be setup, the new EMAS sends an *HO_FAILURE* message to the old EMAS and the MT, after releasing all the local resources reserved for the connection. Recovery in this case is left up to the MT.

9. If the COS did send a *CONNECT* in step 7, it switches the connection data to the new segment and sends an *HO_COMPLETE* message to old EMAS. As in the case of backward handover, the *HO_COMPLETE* message is necessary to resolve conflicts in COS selection when handover is simultaneously initiated by both ends of the connection when two MTs communicate.

10. Upon receiving the *HO_COMPLETE* message, the old EMAS releases the existing connection segment by sending a *RELEASE* message to the COS. In response, the COS sends a *RELEASE COMPLETE* to the previous switch.

35.4 Summary and Conclusions

In this chapter, the QoS and mobility management aspects of wireless ATM were described. WATM implementations, as well as the WATM standards being developed, allow the same ATM service categories in WATM networks as found in fixed ATM networks. The support for these classes of service in wireless ATM requires a variety of QoS control mechanisms acting in concert. The implementation of QoS in the wireless MAC layer, as well as the new QoS control mechanisms in the network and application layers, were described. The role of QoS renegotiation during handover was also described.

While mobility management in wide area involves various service-related aspects, the wireless ATM implementations and the standards efforts have so far focused on the core technical problems of location management and connection handover. These were described in some detail, along the lines of the specifications being developed by the ATM forum with examples from WATM implementations. In conclusion, we note that wireless ATM is still an evolving area and much work is needed to fully understand how to efficiently support QoS and mobility management.

References

[1] Raychaudhuri, D. and Wilson, N., ATM based transport architecture for multiservices wireless personal communication network, *IEEE J. Selected Areas in Commun.*, 1401–1414, Oct. 1994.

[2] Raychaudhuri, D., Wireless ATM: An enabling technology for personal multimedia communications, in *Proc. Mobile Multimedia Commun. Workshop*, Bristol, U.K., Apr. 1995.

[3] Raychaudhuri, D., Wireless ATM networks: Architecture, system design and prototyping, *IEEE Personal Commun.*, 42–49, Aug. 1996.

[4] Singh, S., Quality of service guarantees in mobile computing, *Computer Communications*, 19, 359–371, 1996.

[5] Naghshineh, M., Schwartz, M., and Acampora, A.S., Issues in wireless access broadband networks, in *Wireless Information Networks, Architecture, Resource Management, and Mobile Data*, Holtzman, J.M., Ed., Kluwer, 1996.

[6] Raychaudhuri, D., French, L.J., Siracusa, R.J., Biswas, S.K., Yuan, R., Narasimhan, P., and Johnston, C.A., WATMnet: A prototype wireless ATM system for multimedia personal communication, *IEEE J. Selected Areas in Commun.*, 83–95, Jan. 1997.

[7] Veeraraghavan, M., Karol, M.J., and Eng, K.Y., Mobility and connection management in a wireless ATM LAN, *IEEE J. Selected Areas in Commun.*, 50–68, Jan. 1997.

[8] Ala-Laurila, J. and Awater, G., The magic WAND: Wireless ATM network demonstrator, in *Proc. ACTS Mobile Summit 97*, Denmark, Oct. 1997.

[9] Rauhala, K., Ed., ATM Forum BTD-WATM-01.07, *Wireless ATM Baseline Text*, Apr. 1998.

[10] The ATM Forum Technical Committee, *ATM User-Network Signalling Specification*, Version 4.0, AF-95-1434R9, Jan. 1996.

[11] The ATM Forum Technical Committee, *Traffic Management Specification*, Version 4.0, AF-95-0013R11, Mar, 1996.

[12] Liu, K., Petr, D.W., Frost, V.S., Zhu, H., Braun, C., and Edwards, W.L., A bandwidth management framework for ATM-based broadband ISDN, *IEEE Communications Mag.*, 138–145, May 1997.

[13] Johnston, C.A., Narasimhan, P., and Kokudo, J., Architecture and implementation of radio access protocols in wireless ATM networks, in *Proc. IEEE ICC 98*, Atlanta, Jun. 1998.

[14] Passas, N., Paskalis, S., Vali., D., and Merakos, L., Quality-of-service-oriented medium access control for wireless ATM networks, *IEEE Communications Mag.*, 42–50, Nov. 1997.

[15] Passas, N., Merakos, L., and Skyrianoglou, D., Traffic scheduling in wireless ATM networks, in *Proc. IEEE ATM 97 Workshop*, Lisbon, May 1997.

[16] Raychaudhuri, D., Reininger, D., Ott, M., and Welling, G., Multimedia processing and transport for the wireless personal terminal scenario, *Proceedings SPIE Visual Communications and Image Processing Conference*, VCIP95, May 1995.

[17] Reininger, D. and Izmailov, R., Soft Quality-of-Service with VBR$^+$ video, *Proceedings of 8th International Workshop on Packet Video (AVSPN97)*, Aberdeen, Scotland, Sept. 1997.

[18] Ott, M., Michelitsch, G., Reininger, D., and Welling, G., An architecture for adaptive QoS and its application to multimedia systems design *Computers and Communications*, 1997.

[19] Microsoft Corporation., Windows Quality of Service Technology, White paper available on-line at `http://www.microsoft.com/ntserver/`

[20] Reininger, D., Raychaudhuri, D., and Hui, J., Dynamic bandwidth allocation for VBR video over ATM networks, *IEEE Journal on Selected Areas in Communications*, 14(6), 1076–1086, Aug. 1996.

[21] Lourens, J.G., Malleson, H.H., and Theron, C.C., Optimization of bit-rates, for digitally compressed television services as a function of acceptable picture quality and picture complexity, *Proceedings IEE Colloquium on Digitally Compressed TV by Satellite*, 1995.

[22] Nakasu, E., Aoi, K., Yajima, R., Kanatsugu, Y., and Kubota, K., A statistical analysis of MPEG-2 picture quality for television broadcasting, *SMPTE Journal*, 702–711, Nov. 1996.

[23] ITU-T, ATM Traffic Descriptor Modification by the connection owner, ITU-T Q.2963.2, Sept. 1997.

[24] Mouly, M. and Pautet, M-B., *The GSM System for Mobile Communications*, Cell & Sys, Palaiseau, France, 1992.

[25] Brown, D., Techniques for privacy and authentication in personal communication systems, *IEEE Personal Comm.*, Aug. 1985.

[26] Acharya, A., Li, J., Rajagopalan, B., and Raychaudhuri, D., Mobility management in wireless ATM networks, *IEEE Communications Mag.*, 100–109, Nov. 1997.

[27] Tabbane, S., Location management methods for third-generation mobile systems, *IEEE Communications Mag.*, Aug. 1997.

[28] Acharya, A., Li, J., Bakre, A., and Raychaudhuri, D., Design and prototyping of location management and handoff protocols for wireless ATM networks, in *Proc. ICUPC*, San Diego, Nov. 1997.

[29] Mitts, H., Hansen, H., Immonen, J., and Veikkolainen, Lossless handover in wireless ATM, *Mobile Networks and Applications*, 299–312, Dec. 1996.

[30] Rajagopalan, B., Mobility management in integrated wireless ATM networks, *Mobile Networks and Applications*, 273–286, Dec. 1996.

[31] Mishra, P. and Srivastava, M., Effect of connection rerouting on application performance in mobile networks, in *Proc. IEEE Conf. on Distributed Computing Syst.*, May 1997.

[32] Pollini, G.P., Meier-Hellstern, K.S., and Goodman, D.J., Signalling traffic volume generated by mobile and personal communications, *IEEE Communications Mag.*, Jun. 1995.

36

An Overview of cdma2000, WCDMA, and EDGE

Tero Ojanperä
Nokia Research Center

Steven D. Gray
Nokia Research Center

Abstract—In response to the International Telecommunications Union's (ITU) call for proposals, third generation cellular technologies are evolving at a rapid pace where different proposals are vying for the future market place in digital wireless multimedia communications. While the original intent for third generation was to have a convergence of cellular based technologies, this appears to be an unrealistic expectation. As such, three technologies key for the North American and European markets are the third generation extension of TIA/EIA-95B based Code Division Multiple Access (CDMA) called cdma2000, the European third generation CDMA called WCDMA, and the third generation Time Division Multiple Access (TDMA) system based on EDGE. For packet data, EDGE is one case where second generation technologies converged to a single third generation proposal with convergence of the US TDMA system called TIA/EIA-136 and the European system GSM. This chapter provides an overview of the air interfaces of these key technologies. Particular attention is given to the channel structure, modulation, and offered data rates of each technology. A comparison is also made between cdma2000 and WCDMA to help the reader understand the similarities and differences of these two CDMA approaches for third generation.

36.1   Introduction

The promise of third generation is a world where the subscriber can access the World Wide Web (WWW) or perform file transfers over packet data connections capable of providing 144 kbps for high mobility, 384 kbps with restricted mobility, and 2 Mbps in an indoor office environment [1]. With these guidelines on rate from the ITU, standards bodies started the task of developing an air interface for their third generation system. In North America, the Telecommunications Industry Association (TIA) evaluated proposals from TIA members pertaining to the evolution of TIA/EIA-95B and TIA/EIA-136. In Europe, the European Telecommunications Standards Institute (ETSI) evaluated proposals from ETSI members pertaining to the evolution of GSM.

While TIA and ETSI were still discussing various targets for third generation systems, Japan began to roll out their contributions for third generation technology and develop proof-of-concept prototypes. In the beginning of 1997, the Association for Radio Industry and Business (ARIB), a body responsible for standardization of the Japanese air interface, decided to proceed with the detailed standardization of a wideband CDMA system. The technology push from Japan accelerated standardization in Europe and the U.S. During 1997, joint parameters for Japanese and European wideband CDMA proposals were agreed. The air interface is commonly referred to as WCDMA. In January 1998, the strong support behind wideband CDMA led to the selection of WCDMA as the UMTS terrestrial air interface scheme for FDD (Frequency Division Duplex) frequency bands in ETSI. In the U.S., third generation CDMA came through a detailed proposal process from vendors interested in the evolution of TIA/EIA-95B. In February 1998, the TIA committee TR45.5 responsible for TIA/EIA-95B standardization adopted a framework that combined the different vendors' proposals and later became known as cdma2000.

For TDMA, the focus has been to offer IS-136 and GSM operators a competitive third generation evolution. WCDMA is targeted toward GSM evolution; however, Enhanced Data Rates for Global TDMA Evolution (EDGE) allows the operators to supply IMT-2000 data rates without the spectral allocation requirements of WCDMA. Thus, EDGE will be deployed by those operators who wish to maintain either IS-136 or GSM for voice services and augment these systems with a TDMA-based high rate packet service. TDMA convergence occurred late in 1997 when ETSI approved standardization of the EDGE concept and in February 1998 when TIA committee TR45.3 approved the UWC-136 EDGE-based proposal.

The push to third generation was initially focused on submission of an IMT-2000 radio transmission techniques (RTT) proposal. To date, the evaluation process has recently started in ITU [2] where Fig. 36.1 depicts the time schedule of the ITU RTT development. Since at the same time regional standards have started the standards writing process, it is not yet clear what is the relationship between the ITU and regional standards. Based upon actions in TIA and ETSI, it is reasonable to assume that standards will exist for cdma2000, WCDMA, and EDGE and all will be deployed based upon market demands.

The chapter is organized as follows: issues effecting third generation CDMA are discussed followed by a brief introduction of cdma2000, WCDMA, and EDGE. A table comparing cdma2000 and WCDMA is given at the end of the CDMA section. For TDMA, an overview of the IS-136-based evolution is given including the role played by EDGE.

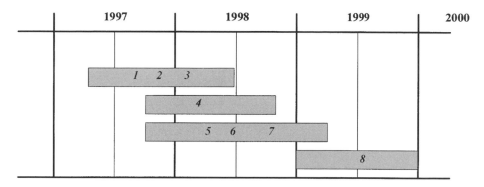

FIGURE 36.1 ITU timelines: 1, 2, 3—RTTs request, development, & submission; 4—RTT evaluation; 5—review outside evaluation; 6—assess compliance with performance parameters; 7—consideration of evaluation results and consensus on key characteristics; 8—development of detailed radio interface specifications.

36.2 CDMA-Based Schemes

Third generation CDMA system descriptions in TIA and ETSI have similarities and differences. Some of the similarities between cdma2000 and WCDMA are variable spreading, convolutional coding, and QPSK data modulation. The major differences between cdma2000 and WCDMA occur with the channel structure, including the structure of the pilot used on the forward link. To aid in comparison of the two CDMA techniques, a brief overview is given to some important third generation CDMA issues, the dedicated channel structure of cdma2000 and WCDMA, and a table comparing air interface characteristics.

36.3 CDMA System Design Issues

36.3.1 Bandwidth

An important design goal for all third generation proposals is to limit spectral emissions to a 5 MHz dual-sided passband. There are several reasons for choosing this bandwidth. First, data rates of 144 and 384 kbps, the main targets of third generation systems, are achievable within 5 MHz bandwidth with reasonable coverage. Second, lack of spectrum calls for limited spectrum allocation, especially if the system has to be deployed within the existing frequency bands already occupied by the second generation systems. Third, the 5 MHz bandwidth improves the receiver's ability to resolve multipath when compared to narrower bandwidths, increasing diversity and improving performance. Larger bandwidths of 10, 15, and 20 MHz have been proposed to support highest data rates more effectively.

36.3.2 Chip Rate

Given the bandwidth, the choice of chip rate depends on spectrum deployment scenarios, pulse shaping, desired maximum data rate and dual-mode terminal implementation. Figure 36.2 shows the relation between chip rate (CR), pulse shaping filter roll-off factor (α) and channel separation (Δf). If raised cosine filtering is used, spectrum is zero (in theory) after CR/2(1+α). In Fig. 36.2, channel separation is selected such that two adjacent channel spectra do not overlap. Channel separation should be selected this way, if there can be high power level differences between the adjacent carriers. For example, for WCDMA parameters minimum channel separation (Δf_{min}) for nonoverlapping carriers is

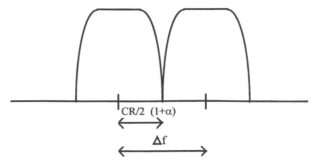

FIGURE 36.2   Relationship between chip rate (CR), roll-off factor (α), and channel separation (Δf).

$\Delta f_{\min} = 4.096(1+0.22) = 4.99712$ MHz. If channel separation is selected in such a way that the spectrum of two adjacent channel signals overlap, some power leaks from one carrier to another. Partly overlapping carrier spacing can be used, for example, in micro cells where the same antenna masts are used for both carriers.

A designer of dual-mode terminals needs to consider the relation between the different clock frequencies of different modes. Especially important are the transmitter and receiver sampling rates and the carrier raster. A proper selection of these frequencies for the standard would ease the dual mode terminal implementation. The different clock frequencies in a terminal are normally derived from a common reference oscillator by either direct division or synthesis by the use of a PLL. The use of a PLL will add some complexity. The WCDMA chip rate has been selected based on consideration of backward compatibility with GSM and PDC. cdma2000 chip rate is a direct derivation of the TIA/EIA-95B chip rate.

36.3.3   Multirate

Multirate design means multiplexing different connections with different quality of service requirements in a flexible and spectrum efficient way. The provision for flexible data rates with different quality of service requirements can be divided into three subproblems: how to map different bit rates into the allocated bandwidth, how to provide the desired quality of service, and how to inform the receiver about the characteristics of the received signal. The first problem concerns issues like multicode transmission and variable spreading. The second problem concerns coding schemes. The third problem concerns control channel multiplexing and coding.

Multiple services belonging to the same session can be either time- or code-multiplexed as depicted in Fig. 36.3. The time multiplexing avoids multicode transmissions thus reducing peak-to-average power of the transmission. A second alternative for service multiplexing is to treat parallel services completely separate with separate channel coding/interleaving. Services are then mapped to separate physical data channels in a multicode fashion as illustrated in the lower part of Fig. 36.3. With this alternative scheme, the power, and consequently the quality, of each service can be controlled independently.

36.3.4   Spreading and Modulation Solutions

A complex spreading circuit as shown in Fig. 36.4 helps to reduce the peak-to-average power and thus improves power efficiency.

The spreading modulation can be either balanced- or dual-channel QPSK. In the balanced QPSK spreading the same data signal is split into I and Q channels. In dual-channel QPSK spreading the symbol streams on the I and Q channels are independent of each other. In

Time multiplexing

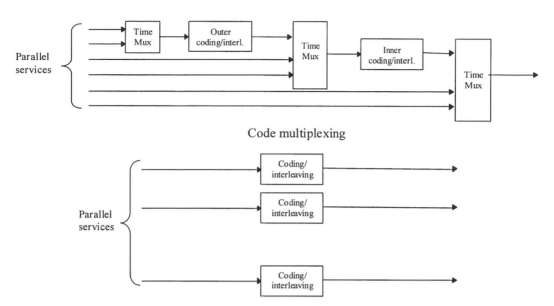

Code multiplexing

FIGURE 36.3 Time and code multiplexing principles.

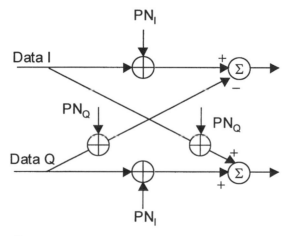

FIGURE 36.4 Complex spreading.

the forward link, QPSK data modulation is used in order to save code channels and allow the use of the same orthogonal sequence for I and Q channels. In the reverse link, each mobile station uses the same orthogonal codes; this allows for efficient use of BPSK data modulation and balanced QPSK spreading.

36.3.5 Coherent Detection in the Reverse Link

Coherent detection can improve the performance of the reverse link up to 3 dB compared to noncoherent reception used by the second generation CDMA system. To facilitate coherent detection a pilot signal is required. The actual performance improvement depends on the proportion of the pilot signal power to the data signal power and the fading environment.

36.3.6 Fast Power Control in Forward Link

To improve the forward link performance fast power control is used. The impact of the fast power control in the forward link is twofold. First, it improves the performance in a fading multipath channel. Second, it increases the multiuser interference variance within the cell since orthogonality between users is not perfect due to multipath channel. The net effect, however, is improved performance at low speeds.

36.3.7 Additional Pilot Channel in the Forward Link for Beamforming

An additional pilot channel on the forward link that can be assigned to a single mobile or to a group of mobiles enables deployment of adaptive antennas for beamforming since the pilot signal used for channel estimation needs to go through the same path as the data signal. Therefore, a pilot signal transmitted through an omnicell antenna cannot be used for the channel estimation of a data signal transmitted through an adaptive antenna.

36.3.8 Seamless Interfrequency Handover

For third generation systems hierarchical cell structures (HCS), constructed by overlaying macro cells on top of smaller micro or pico cells, have been proposed to achieve high capacity. The cells belonging to different cell layers will be in different frequencies, and thus an interfrequency handover is required. A key requirement for the support of seamless interfrequency handover is the ability of the mobile station to carry out cell search on a carrier frequency different from the current one, without affecting the ordinary data flow. Different methods have been proposed to obtain multiple carrier frequency measurements. For mobile stations with receiver diversity, there is a possibility for one of the receiver branches to be temporarily reallocated from diversity reception and instead carry out reception on a different carrier. For single-receiver mobile stations, slotted forward link transmission could allow interfrequency measurements. In the slotted mode, the information normally transmitted during a certain time, e.g., a 10 ms frame, is transmitted in less than that time, leaving an idle time that the mobile can use to measure on other frequencies.

36.3.9 Multiuser Detection

Multiuser Detection (MUD) has been the subject of extensive research since 1986 when Verdu formulated an optimum multiuser detector for AWGN channel, maximum likelihood sequence estimation (MLSE) [3]. In general, it is easier to apply MUD in a system with short spreading codes since cross-correlations do not change every symbol as with long spreading codes. However, it seems that the proposed CDMA schemes would all use long spreading codes. Therefore, the most feasible approach seems to be interference cancellation algorithms that carry out the interference cancellation at the chip level, thereby avoiding explicit calculation of the cross-correlation between spreading codes from different users [4]. Due to complexity, MUD is best suited for the reverse link. In addition, the mobile station is interested in detecting its own signal in contrast to the base station, which needs to demodulate the signals of all users. Therefore, a simpler interference suppression scheme could be applied in the mobile station. Furthermore, if short spreading codes are used, the receiver could exploit the cyclostationarity, i.e., the periodic properties of the signal, to suppress interference without knowing the interfering codes.

36.3.10 Transmit Diversity

The forward link performance can be improved in many cases by using transmit diversity. For direct spread CDMA schemes, this can be performed by splitting the data stream and spreading the two streams using orthogonal sequences or switching the entire data stream between two antennas. For multicarrier CDMA, the different carriers can be mapped into different antennas.

36.4 WCDMA

To aid in the comparison of cdma2000 and WCDMA, the dedicated frame structure of WCDMA is illustrated in Figs. 36.5 and 36.6. The approach follows a time multiplex philosophy where the Dedicated Physical Control Channel (DPCCH) provides the pilot, power control, and rate information and the Dedicated Physical Data Channel (DPDCH) is the portion used for data transport. The forward and reverse DPDCH channels have been convolutional encoded and interleaved prior to framing. The major difference between the forward and reverse links is that the reverse channel structure of the DPCCH is a separate code channel from the DPDCH.

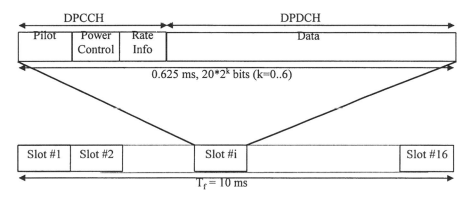

FIGURE 36.5 Forward link dedicated channel structure in WCDMA.

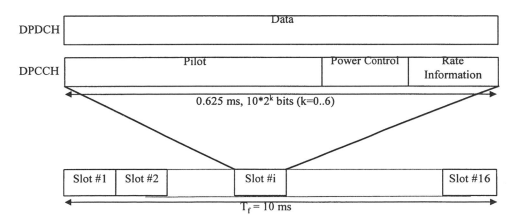

FIGURE 36.6 Reverse link dedicated channel structure in WCDMA.

After framing, the forward and reverse link channels are spread as shown in Figs. 36.7 and 36.8. On the forward link orthogonal, variable rate codes, c_{ch}, are used to separate channels and pseudo random scrambling sequences, c_{scramb}, are used to spread the signal evenly across the spectrum and separate different base stations. On the reverse link, the orthogonal channelization codes are used as in the forward link to separate CDMA channels. The scrambling codes, c'_{scramb} and c''_{scramb}, are used to identify mobile stations and to spread the signal evenly across the band. The optional scrambling code is used as a means to group mobiles under a common scrambling sequence.

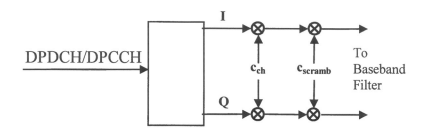

c_{ch}: channelization code
c_{scramb}: scrambling code

FIGURE 36.7   Forward link spreading of DPDCH and DPCCH.

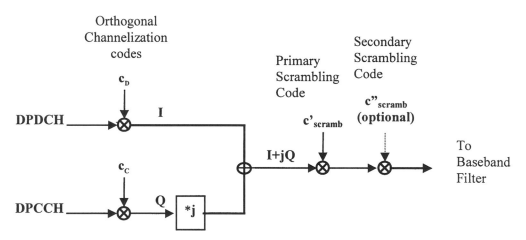

FIGURE 36.8   Reverse link spreading for the DPDCH and DPCCH.

36.4.1   Spreading Codes

WCDMA employs long spreading codes. Different spreading codes are used for cell separation in the forward link and user separation in the reverse link. In the forward link Gold codes of length 2^{18} are truncated to form cycles of 2^{16} times 10 ms frames. In order to minimize the cell search time, a special short code mask is used. The synchronization channel of WCDMA is masked with an orthogonal short Gold code of length 256 chips spanning

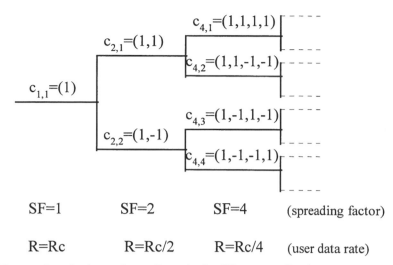

SF=1   SF=2   SF=4   (spreading factor)

R=Rc   R=Rc/2   R=Rc/4   (user data rate)

FIGURE 36.9   Construction of orthogonal spreading codes for different spreading factors.

one symbol. The mask symbols carry information about the BS long code group. Thus, the mobile station first acquires the short mask code and then searches the corresponding long code. A short VL-Kasami code has been proposed for the reverse link to ease the implementation of multiuser detection. In this case, code planning would also be negligible because the number of VL-Kasami sequences is more than one million. However, in certain cases, the use of short codes may lead to bad correlation properties, especially with very small spreading factors. If multiuser detection were not used, adaptive code allocation could be used to restore the cross correlation properties. The use of short codes to ease the implementation of advanced detection techniques is more beneficial in the forward link since the cyclostationarity of the signal could be utilized for adaptive implementation of the receiver.

Orthogonality between the different spreading factors can be achieved by tree-structured orthogonal codes whose construction is illustrated in Fig. 36.9 [5]. The tree-structured codes are generated recursively according to the following equation,

$$
c_{2n} = \begin{pmatrix} c_{2n,1} \\ c_{2n,2} \\ \vdots \\ c_{2n,2n} \end{pmatrix} = \begin{pmatrix} \begin{pmatrix} c_{n,1} & c_{n,1} \\ c_{n,1} & -c_{n,1} \end{pmatrix} \\ \vdots \\ \begin{pmatrix} c_{n,n} & c_{n,n} \\ c_{n,n} & -c_{n,n} \end{pmatrix} \end{pmatrix}
$$

where C_{2n} is the orthogonal code set of size $2n$. The generated codes within the same layer constitute a set of orthogonal functions and are thus orthogonal. Furthermore, any two codes of different layers are also orthogonal except for the case that one of the two codes is a mother code of the other. For example code $c_{4,4}$ is not orthogonal with codes $c_{1,1}$ and $c_{2,2}$.

36.4.2   Coherent Detection and Beamforming

In the forward link, time-multiplexed pilot symbols are used for coherent detection. Because the pilot symbols are user dedicated, they can be used for channel estimation with adaptive antennas as well. In the reverse link, WCDMA employs pilot symbols multiplexed with power control and rate information for coherent detection.

36.4.3   Multirate

WCDMA traffic channel structure is based on a single code transmission for small data rates and multicode for higher data rates. Multiple services belonging to the same connection are, in normal cases, time multiplexed as was depicted in the upper part of Fig. 36.3. After service multiplexing and channel coding, the multiservice data stream is mapped to one or more dedicated physical data channels. In the case of multicode transmission, every other data channel is mapped into Q and every other into I channel. The channel coding of WCDMA is based on convolutional and concatenated codes. For services with BER = 10^{-3}, a convolutional code with constraint length of 9 and different code rates (between 1/2–1/4) is used. For services with BER = 10^{-6}, a concatenated coding with an outer Reed-Solomon code has been proposed. Typically, block interleaving over one frame is used. WCDMA is also capable of interframe interleaving, which improves the performance for services allowing longer delay. Turbo codes for data services are under study. Rate matching is performed by puncturing or symbol repetition.

36.4.4   Packet Data

WCDMA has two different types of packet data transmission possibilities. Short data packets can be appended directly to a random access burst. The WCDMA random-access burst is 10 ms long, it is transmitted with fixed power, and the access principle is based on the slotted Aloha scheme. This method, called common channel packet transmission, is used for short infrequent packets, where the link maintenance needed for a dedicated channel would lead to an unacceptable overhead. Larger or more frequent packets are transmitted on a dedicated channel. A large single packet is transmitted using a single-packet scheme where the dedicated channel is released immediately after the packet has been transmitted. In a multipacket scheme the dedicated channel is maintained by transmitting power control and synchronization information between subsequent packets.

36.5   cdma2000

The dedicated channels used in cdma2000 system are the fundamental, supplemental, pilot,[1] and dedicated control channels. Shown for the forward link in Fig. 36.10 and for the reverse in Fig. 36.11, the fundamental channel provides for the communication of voice, low rate data, and signalling where power control information for the reverse channels is punctured on the forward fundamental channel. For high rate data services, the supplemental channel is used where one important difference between the supplemental and the fundamental channel is the addition of parallel-concatenated turbo codes. For different service options, multiple supplemental channels can be used. The code multiplex pilot channel allows for

---

[1]Dedicated for the reverse link and common for the forward link.

phase coherent detection. In addition, the pilot channel on the forward link is used for determining soft handoff and the pilot channel on the reverse is used for carrying power control information for the forward channels. Finally, the dedicated control channel, also shown in Fig. 36.10 for the forward link and in Fig. 36.11, for the reverse, is used primarily for exchange of high rate Media Access Control (MAC) layer signalling.

36.5.1 Multicarrier

In addition to direct spread, a multicarrier approach has been proposed for the cdma2000 forward link since it would maintain orthogonality between the cdma2000 and TIA/EIA-95B carriers [6]. The multicarrier variant is achieved by using three 1.25 MHz carriers for a 5 MHz bandwidth where all carriers have separate channel coding and are power controlled in unison.

36.5.2 Spreading Codes

On the forward link, the cell separation for cdma2000 is performed by two M-sequences of length 3×2^{15}, one for I and one for Q channel, which are phase shifted by PN-offset for different cells. Thus, during the cell search process only these sequences are searched. Because there are a limited number of PN-offsets, they need to be planned in order to avoid PN-confusion [7]. In the reverse link, user separation is performed by different phase shifts of M-sequence of length 2^{41}. The channel separation is performed using variable spreading factor Walsh sequences, which are orthogonal to each other.

36.5.3 Coherent Detection

In the forward link, cdma2000 has a common pilot channel, which is used as a reference signal for coherent detection when adaptive antennas are not employed. When adaptive antennas are used, an auxiliary pilot is used as a reference signal for coherent detection. Code multiplexed auxiliary pilots are generated by assigning a different orthogonal code to each auxiliary pilot. This approach reduces the number of orthogonal codes available for the traffic channels. This limitation is alleviated by expanding the size of the orthogonal code set used for the auxiliary pilots. Since a pilot signal is not modulated by data, the pilot orthogonal code length can be extended, thereby yielding an increased number of available codes, which can be used as additional pilots. In the reverse link, the pilot signal is time multiplexed with power control and erasure indicator bit (EIB).

36.5.4 Multirate Scheme

cdma2000 has two traffic channel types, the fundamental and the supplemental channel, which are code multiplexed. The fundamental channel is a variable rate channel which supports basic rates of 9.6 kbps and 14.4 kbps and their corresponding subrates, i.e., Rate Set 1 and Rate Set 2 of TIA/EIA-95B. It conveys voice, signalling, and low rate data. The supplemental channel provides high data rates. Services with different QoS requirements are code multiplexed into supplemental channels. The user data frame length of cdma2000 is 20 ms. For the transmission of control information, 5 and 20 ms frames can be used on the fundamental channel or dedicated control channel. On the fundamental channel a convolutional code with constraint length 9 is used. On supplemental channels convolutional coding is used up to 14.4 kbps. For higher rates Turbo codes with constraint length 4 and

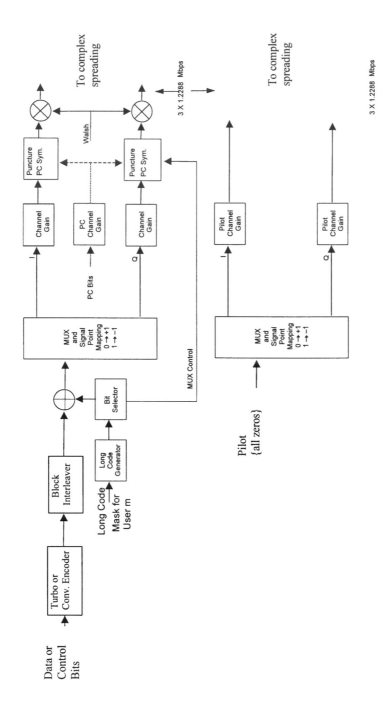

FIGURE 36.10 Forward link channel structure in cdma2000 for direct spread. (*Note*: dashed line indicates that it is only used for the fundamental channel).

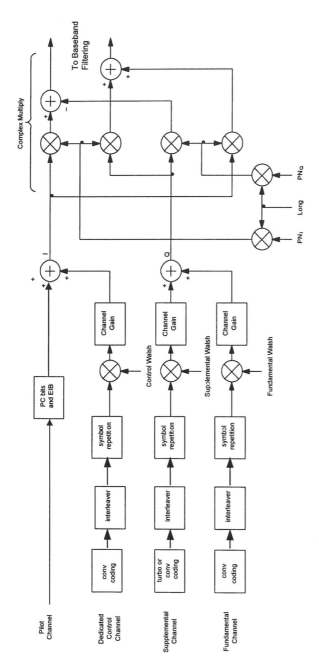

FIGURE 36.11 Reverse link channel structure in cdma2000.

rate 1/4 are preferred. Rate matching is performed by puncturing, symbol repetition, and sequence repetition.

36.5.5   Packet Data

cdma2000 also allows short data burst using the slotted Aloha principle. However, instead of fixed transmission power it increases the transmission power for the random access burst after an unsuccessful access attempt. When the mobile station has been allocated a traffic channel, it can transmit without scheduling up to a predefined bit rate. If the transmission rate exceeds the defined rate, a new access request has to be made. When the mobile station stops transmitting, it releases the traffic channel but not the dedicated control channel. After a while it also releases the dedicated control channel as well but maintains the link layer and network layer connections in order to shorten the channel set-up time when new data needs to be transmitted.

36.5.6   Parametric Comparison

For comparison, Table 36.1 lists the parameters of cdma2000 and WCDMA. cdma2000 uses a chip rate of 3.6864 Mcps for the 5 MHz band allocation with the direct spread forward link option and a 1.2288 Mcps chip rate with three carriers for the multicarrier option. WCDMA uses direct spread with a chip rate of 4.096 Mcps. The multicarrier approach is motivated by a spectrum overlay of cdma2000 carriers with existing TIA/EIA-95B carriers [6]. Similar to EIA/TIA-95B, the spreading codes of cdma2000 are generated using different phase shifts of the same M-sequence. This is possible due to the synchronous network operation. Since WCDMA has an asynchronous network, different long codes rather than different phase shifts of the same code are used for the cell and user separation. The code structure determines how code synchronization, cell acquisition, and handover synchronization are performed.

36.6   TDMA-Based Schemes

As discussed, TIA/EIA-136 and GSM evolution have similar paths in the form of EDGE. The UWC-136 IMT 2000 proposal contains, in addition to the TIA/EIA-136 30 kHz carriers, the high rate capability provided by the 200 kHz and 1.6 MHz carriers shown in Table 36.2. The targets for the IS-136 evolution were to meet IMT-2000 requirements and an initial deployment within 1 MHz spectrum allocation. UWC-136 meets these targets via modulation enhancement to the existing 30 kHz channel (136+) and by defining complementary wider band TDMA carriers with bandwidths of 200 kHz for vehicular/outdoor environments and 1.6 MHz for indoor environments. The 200 kHz carrier, 136 HS (vehicular/outdoor) with the same parameters as EDGE provides medium bit rates up to 384 kbps and the 1.6 MHz carrier, 136 HS (indoor), highest bit rates up to 2 Mbps. The parameters of the 136 HS proposal submitted to ITU are listed in Table 36.2 and the different carrier types of UWC-136 are shown in Fig. 36.12.

36.6.1   Carrier Spacing and Symbol Rate

The motivation for the 200 kHz carrier is twofold. First, the adoption of the same physical layer for 136 HS (Vehicular/Outdoor) and GSM data carriers provides economics of scale and therefore cheaper equipment and faster time to market. Second, the 200 kHz carrier

TABLE 36.1 Parameters of WCDMA and cdma2000

	WCDMA	cdma2000
Channel bandwidth	5, 10, 20 MHz	1.25, 5, 10, 15, 20 MHz
Forward link RF channel structure	Direct spread	Direct spread or multicarrier
Chip rate	4.096/8.192/16.384 Mcps	1.2288/3.6864/7.3728/11.0593/14.7456 Mcps for direct spread $n \times 1.2288$ Mcps (n=1,3,6,9,12) for multicarrier
Roll-off factor	0.22	Similar to TIA/EIA-95B
Frame length	10 ms / 20 ms (optional)	20 ms for data and control/5ms for control information on the fundamental and dedicated control channel
Spreading modulation	Balanced QPSK (forward link) Dual channel QPSK (reverse link) Complex spreading circuit	Balanced QPSK (forward link) Dual channel QPSK (reverse link) Complex spreading circuit
Data modulation	QPSK (forward link) BPSK (reverse link)	QPSK (forward link) BPSK (reverse link)
Coherent detection	User dedicated time multiplexed pilot (forward link and reverse link), common pilot in forward link	Pilot time multiplexed with PC and EIB (reverse link) Common continuous pilot channel and auxiliary pilot (forward link)
Channel multiplexing in reverse link	Control and pilot channel time multiplexed I&Q multiplexing for data and control channel	Control, pilot fundamental, and supplemental code multiplexed I&Q multiplexing for data and control channels
Multirate	Variable spreading and multicode	Variable spreading and multicode
Spreading factors	4-256 (4.096 Mcps)	4-256 (3.6864 Mcps)
Power Control	Open and fast closed loop (1.6 kHz)	Open loop and fast closed loop (800 Hz)
Spreading (forward link)	Variable length orthogonal sequences for channel separation. Gold sequences for cell and user separation	Variable length Walsh sequences for channel separation, M-sequence 3×2^{15} (same sequence with time shift utilized in different cells different sequence in I&Q channel)
Spreading (reverse link)	Variable length orthogonal sequences for channel separation. Gold sequence 2^{41} for user separation (different time shifts in I and Q channel, cycle 2^{16} 10 ms radio frames)	Variable length orthogonal sequences for channel separation, M-sequence 2^{15} (same for all users different sequences in I&Q channels), M-sequence 2^{41} for user separation (different time shifts for different users)
Handhover	Soft handover Interfrequency handover	Soft handover Interfrequency handover

with higher order modulation can provide bit rates of 144 and 384 kbps with reasonable range and capacity fulfilling IMT-2000 requirements for pedestrian and vehicular environments. The 136 HS (Indoor) carrier can provide 2 Mbit/s user data rate with a reasonably strong channel coding.

36.6.2 Modulation

First proposed modulation methods were Quaternary Offset QAM (Q-O-QAM) and Binary Offset QAM (B-O-QAM). Q-O-QAM could provide higher data rates and good spectral efficiency. For each symbol two bits are transmitted and consecutive symbols are shifted by $\pi/2$. An offset modulation was proposed, because it causes smaller amplitude variations than 16QAM, which can be beneficial when using amplifiers that are not completely linear. The second modulation B-O-QAM has been introduced, which has the same symbol rate

TABLE 36.2   Parameters of 136 HS

	136 HS (Vehicular/Outdoor)	136 HS (Indoor)
Duplex method	FDD	FDD and TDD
Carrier spacing	200 kHz	1.6 MHz
Modulation	Q-O-QAM	Q-O-QAM
	B-O-QAM	B-O-QAM
	8 PSK	
	GMSK	
Modulation bit rate	722.2 kbps (Q-O-QAM)	5200 kbps (Q-O-QAM)
	361.1 kbps (B-O-QAM)	2600 kbps (B-O-QAM)
	812.5 kbps (8 PSK)	
	270.8 kbps (GMSK)	
Payload	521.6 kbps (Q-O-QAM)	4750 kbps (Q-O-QAM)
	259.2 kbps (B-O-QAM)	2375 kbps (B-O-QAM)
	547.2 kbps (8 PSK)	
	182.4 kbps (GMSK)	
Frame length	4.615 ms	4.615 ms
Number of slots	8	64 (72 μs)
		16 (288 μs)
Coding	Convolutional	Convolutional
	1/2, 1/4, 1/3, 1/1	1/2, 1/4, 1/3, 1/1
	ARQ	Hybrid Type II ARQ
Frequency hopping	Optional	Optional
Dynamic channel allocation	Optional	Optional

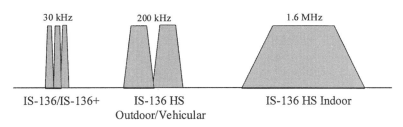

FIGURE 36.12   UWC-136 carrier types.

of 361.111 ksps, but where only the outer signal points of the Q-O-QAM modulation are used. For each symbol one bit is transmitted and consecutive symbols are shifted by $\pi/2$. A second modulation scheme with the characteristic of being a subset of the first modulation scheme and having the same symbol rate as the first modulation allows seamless switching between the two modulation types between bursts. Both modulation types can be used in the same burst. From a complexity point of view the addition of a modulation, which is subset of the first modulation, adds no new requirements for the transmitter or receiver.

In addition to the originally proposed modulation schemes, Quaternary Offset QAM (Q-O-QAM) and Binary Offset QAM (B-O-QAM), other modulation schemes, CPM (Continuous Phase Modulation) and 8-PSK, have been evaluated in order to select the modulation best suited for EDGE. The outcome of this evaluation is that 8 PSK was considered to have implementation advantages over Q-O-QAM. Parties working on EDGE are in the process of revising the proposals so that 8 PSK would replace the Q-O-QAM and GMSK can be used as the lower level modulation instead of B-O-QAM. The symbol rate of the 8 PSK will be the same as for GMSK and the detailed bit rates will be specified early in 1999.

36.6.3 Frame Structures

The 136 HS (Vehicular/Outdoor) data frame length is 4.615 ms and one frame consists of eight slots. The burst structure is suitable for transmission in a high delay spread environment. The frame and slot structures of the 136 HS (Indoor) carrier were selected for cell coverage for high bit rates. The HS-136 Indoor supports both FDD and TDD duplex methods. Figure 36.13 illustrates the frame and slot structure. The frame length is 4.615 ms and it can consist of

- 64 1/64 time slots of length 72 μs
- 16 1/16 time slots of length 288 μs

In the TDD mode, the same burst types as defined for the FDD mode are used. The 1/64 slot can be used for every service from low rate speech and data to high rate data services. The 1/16 slot is to be used for medium to high rate data services. Figure 36.13 also illustrates the dynamic allocation of resources between the reverse link and the forward link in the TDD mode.

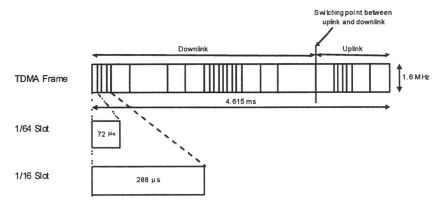

FIGURE 36.13 Wideband TDMA frame and slot structure.

The physical contents of the time slots are bursts of corresponding length. Three types of traffic bursts are defined. Each burst consists of a training sequence, two data blocks, and a guard period. The bursts differ in the length of the burst (72 μs and 288 μs) and in the length of the training sequence (27 symbols and 49 symbols) leading to different numbers of payload symbols and different multipath delay performances (Fig. 36.14). The number of required reference symbols in the training sequence depends on the length of the channel's impulse response, the required signal-to-noise ratio, the expected maximum Doppler frequency shift, and the number of modulation levels. The number of reference symbols should be matched to the channel characteristics, remain practically stable within the correlation window, and have good correlation properties. All 136 based schemes can use interference cancellation as a means to improve performance [8]. For 136 HS (Indoor), the longer sequence can handle about 7 μs of time dispersion and the shorter one 2.7 μs. It should be noted that if the time dispersion is larger, the drop in performance is slow and depends on the power delay profile.

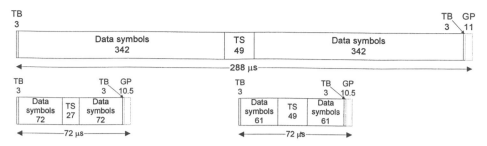

FIGURE 36.14   Burst structure.

36.6.4   Multirate Scheme

The UWC-136 multirate scheme is based on a variable slot, code, and modulation structure. Data rates up to 43.2 kbps can be offered using the 136+ 30 kHz carrier and multislot transmission. Depending on the user requirements and channel conditions a suitable combination of modulation, coding, and number of data slots is selected. 136 HS can offer packet switched, and both transparent and nontransparent circuit switched data services. Asymmetrical data rates are provided by allocating a different number of time slots in the reverse and forward links. For packet switched services the RLC/MAC protocol provides fast medium access via a reservation based medium access scheme, supplemented by selective ARQ for efficient retransmission.

Similar to 136 HS (Outdoor/Vehicular), the 136 HS (Indoor) uses two modulation schemes and different coding schemes to provide variable data rates. In addition, two different slot sizes can be used. For delay tolerant packet data services, error control is based on a Type II hybrid ARQ (automatic repeat request) scheme [5]. The basic idea is to first send all data blocks using a simple error control coding scheme. If decoding at the receiver fails, a retransmission is requested using a stronger code. After the second retransmission, diversity combining can be performed between the first and second transmissions prior to hard decisions. This kind of ARQ procedure can be used due to the ability of the RLC/MAC protocol to allocate resources fast and to send transmission requests reliably in the feedback channel [5].

36.6.5   Radio Resource Management

The radio resource management schemes of UWC-136 include link adaptation, frequency hopping, power control, and dynamic channel allocation. Link adaptation offers a mechanism for choosing the best modulation and coding alternative according to channel and interference conditions. Frequency hopping averages interference and improves link performance against fast fading. For 136 HS (Indoor) fast power control (frame-by-frame) could be used to improve the performance in cases where frequency hopping cannot be applied, for example, when only one carrier is available. Dynamic channel allocation can be used for channel assignments. However, when deployment with minimum spectrum is desired, reuse 1/3 and fractional loading with fixed channel allocation is used.

36.7   Time Division Duplex (TDD)

The main discussion about the IMT-2000 air interface has been concerned with technologies for FDD. However, there are several reasons why TDD would be desirable. First, there will likely be dedicated frequency bands for TDD within the identified UMTS frequency bands.

Furthermore, FDD requires exclusive paired bands and spectrum is, therefore, hard to find. With a proper design including powerful FEC, TDD can be used even in outdoor cells. The second reason for using TDD is flexibility in radio resource allocation, i.e., bandwidth can be allocated by changing the number of time slots for the reverse link and forward link. However, the asymmetric allocation of radio resources leads to two interference scenarios that will impact the overall spectrum efficiency of a TDD scheme:

- asymmetric usage of TDD slots will impact the radio resource in neighboring cells, and
- asymmetric usage of TDD slots will lead to blocking of slots in adjacent carriers within their own cells.

Figure 36.15 depicts the first scenario. MS2 is transmitting at full power at the cell border. Since MS1 has a different asymmetric slot allocation than MS2, its forward link slots received at the sensitivity limit are interfered by MS1, which causes blocking. On the other hand, since the BS1 can have much higher EIRP (effective isotropically radiated power) than MS2, it will interfere BS2's ability to receive MS2. Hence, the radio resource algorithm needs to avoid this situation.

In the second scenario, two mobiles would be connected into the same cell but using different frequencies. The base station receives MS1 on the frequency f1 using the same time slot it uses on the frequency f2 to transmit into MS2. As shown in Table 36.3, the transmission will block the reception due to the irreducible noise floor of the transmitter regardless of the frequency separation between f1 and f2.

TABLE 36.3   Adjacent Channel Interference Calculation

BTS transmission power for MS2 in forward link 1W	30 dBm
Received power for MS1	−100 dBm
Adjacent channel attenuation due to irreducible noise floor	50 to 70 dB
Signal to adjacent channel interference ratio	−60 to −80 dB

Both TDMA- and CDMA-based schemes have been proposed for TDD. Most of the TDD aspects are common to TDMA- and CDMA-based air interfaces. However, in CDMA-based TDD systems the slot duration on the forward and reverse links must be equal to enable the use of soft handoff and prevent the interference situation described in the first scenario. Because TDMA systems do not have soft handoff on a common frequency, slot imbalances from one BS to the next are easier to accommodate. Thus, TDMA-based solutions have higher flexibility. The frame structure for the wide band TDMA for the TDD system was briefly discussed in the previous section. WCDMA has been proposed for TDD in Japan and Europe. The frame structure is the same as for the FDD component, i.e., a 10 ms frame split into 16 slots of 0.625 ms each. Each slot can be used either for reverse link or forward link. For cdma2000, the TDD frame structure is based on a 20 ms frame split into 16 slots of 1.25 ms each.

36.8   Conclusions

Third generation cellular systems are a mechanism for evolving the telecommunications business based primarily on voice telephony to mobile wireless datacomm. In light of events in TIA, ETSI, and ARIB, cdma2000, WCDMA, and EDGE will be important technologies

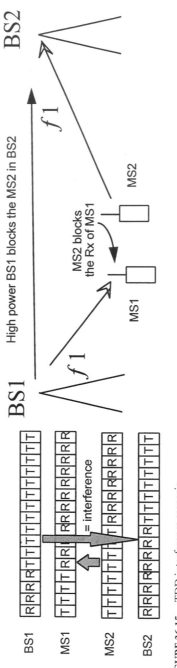

FIGURE 36.15   TDD interference scenario.

used to achieve the datacomm goal. Standardization related to radio access technologies discussed in this chapter were underway at the time of writing and will offer the European, U.S., and Japanese markets both CDMA and TDMA third-generation options. In comparing CDMA evolution, the European, U.S., and Japanese based systems have some similarities, but differ in the chip rate and channel structure. In the best circumstances, some harmonization will occur between cdma2000 and WCDMA making deployment of hardware capable of supporting both systems easier. In TDMA, the third generation paths of GSM and TIA/EIA-136 are through a common solution. This alignment will offer TDMA systems an advantage in possible global roaming for data services. In spite of the regional standards differences, third generation will be the mechanism for achieving wireless multimedia enabling services beyond the comprehension of second generation systems.

Acknowledgments

The authors would like to thank Harri Holma, Pertti Lukander, and Antti Toskala from Nokia Telecommunications, George Fry, Kari Kalliojarvi, Riku Pirhonen, Rauno Ruismaki, and Zhigang Rong from Nokia Research Center, Kari Pehkonen from Nokia Mobile Phones, and Kari Pulli from the University of Stanford for helpful comments. In addition, contributions related to spectrum and modulation aspects from Harri Lilja from Nokia Mobile Phones are acknowledged.

References

[1] ITU-R M.1225, Guidelines for Evaluation of Radio Transmission Technologies for IMT-2000, 1998.

[2] Special Issue on IMT-2000: Standards Efforts of the ITU, *IEEE Pers. Commun.*, 4(4), Aug. 1997.

[3] Verdu, S., Minimum probability of error for asynchronous gaussian multiple access, *IEEE Trans. on IT.*, IT-32(1), 85–96, Jan. 1986.

[4] Monk, A.M. et al., A noise-whitening approach to multiple access noise rejection—Pt I: Theory and background, *IEEE J. Select. Areas Commun.*, 12(5), 817–827, Jun. 1997.

[5] Nikula, E., Toskala, A., Dahlman, E., Girard, L., and Klein, A., FRAMES multiple access for UMTS and IMT-2000, *IEEE Pers. Commun.*, Apr. 1998.

[6] Tiedemann, E.G., Jr., Jou, Y-C., and Odenwalder, J.P., The evolution of IS-95 to a third generation system and to the IMT-2000 era, *Proc. ACTS Summit*, Aalborg, Denmark, 924–929, Oct. 1997.

[7] Chang, C.R., Van, J.Z., and Yee, M.F., PN offset planning strategies for nonuniform CDMA networks, *Proc. VTC'97*, 3, Phoenix, Arizona, 1543–1547, May 4–7, 1997.

[8] Ranta, P., Lappeteläinen, A., and Honkasalo, Z-C., Interference cancellation by joint detection in random frequency hopping TDMA networks, *Proc. ICUPC96*, 1, Cambridge, MA, 428–432, Sept./Oct. 1996.

Index